Normal curve areas
Standard normal probability in right-hand
tail (for negative values of z areas are found by symmetry)

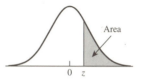

Area

| z | Second decimal place of z | | | | | | | | | |
---	.00	.01	.02	.03	.04	.05	.06	.07	.08	.09
0.0	.5000	.4960	.4920	.4880	.4840	.4801	.4761	.4721	.4681	.4641
0.1	.4602	.4562	.4522	.4483	.4443	.4404	.4364	.4325	.4286	.4247
0.2	.4207	.4168	.4129	.4090	.4052	.4013	.3974	.3936	.3897	.3859
0.3	.3821	.3783	.3745	.3707	.3669	.3632	.3594	.3557	.3520	.3483
0.4	.3446	.3409	.3372	.3336	.3300	.3264	.3228	.3192	.3156	.3121
0.5	.3085	.3050	.3015	.2981	.2946	.2912	.2877	.2843	.2810	.2776
0.6	.2743	.2709	.2676	.2643	.2611	.2578	.2546	.2514	.2483	.2451
0.7	.2420	.2389	.2358	.2327	.2296	.2266	.2236	.2206	.2177	.2148
0.8	.2119	.2090	.2061	.2033	.2005	.1977	.1949	.1922	.1894	.1867
0.9	.1841	.1814	.1788	.1762	.1736	.1711	.1685	.1660	.1635	.1611
1.0	.1587	.1562	.1539	.1515	.1492	.1469	.1446	.1423	.1401	.1379
1.1	.1357	.1335	.1314	.1292	.1271	.1251	.1230	.1210	.1190	.1170
1.2	.1151	.1131	.1112	.1093	.1075	.1056	.1038	.1020	.1003	.0985
1.3	.0968	.0951	.0934	.0918	.0901	.0885	.0869	.0853	.0838	.0823
1.4	.0808	.0793	.0778	.0764	.0749	.0735	.0722	.0708	.0694	.0681
1.5	.0668	.0655	.0643	.0630	.0618	.0606	.0594	.0582	.0571	.0559
1.6	.0548	.0537	.0526	.0516	.0505	.0495	.0485	.0475	.0465	.0455
1.7	.0446	.0436	.0427	.0418	.0409	.0401	.0392	.0384	.0375	.0367
1.8	.0359	.0352	.0344	.0336	.0329	.0322	.0314	.0307	.0301	.0294
1.9	.0287	.0281	.0274	.0268	.0262	.0256	.0250	.0244	.0239	.0233
2.0	.0228	.0222	.0217	.0212	.0207	.0202	.0197	.0192	.0188	.0183
2.1	.0179	.0174	.0170	.0166	.0162	.0158	.0154	.0150	.0146	.0143
2.2	.0139	.0136	.0132	.0129	.0125	.0122	.0119	.0116	.0113	.0110
2.3	.0107	.0104	.0102	.0099	.0096	.0094	.0091	.0089	.0087	.0084
2.4	.0082	.0080	.0078	.0075	.0073	.0071	.0069	.0068	.0066	.0064
2.5	.0062	.0060	.0059	.0057	.0055	.0054	.0052	.0051	.0049	.0048
2.6	.0047	.0045	.0044	.0043	.0041	.0040	.0039	.0038	.0037	.0036
2.7	.0035	.0034	.0033	.0032	.0031	.0030	.0029	.0028	.0027	.0026
2.8	.0026	.0025	.0024	.0023	.0023	.0022	.0021	.0021	.0020	.0019
2.9	.0019	.0018	.0017	.0017	.0016	.0016	.0015	.0015	.0014	.0014
3.0	.00135									
3.5	.000 233									
4.0	.000 031 7									
4.5	.000 003 40									
5.0	.000 000 287									

From R. E. Walpole, *Introduction to Statistics* (New York: Macmillan, 1968).

MATHEMATICAL STATISTICS WITH APPLICATIONS

SIXTH EDITION

Mathematical Statistics with Applications

Dennis D. Wackerly

University of Florida

William Mendenhall III

University of Florida, Emeritus

Richard L. Scheaffer

University of Florida

DUXBURY

TM

THOMSON LEARNING

Australia • Canada • Mexico • Singapore • Spain • United Kingdom • United States

DUXBURY

THOMSON LEARNING ™

Sponsoring Editor: *Carolyn Crockett*
Editorial Assistant: *Jennifer Jenkins*
Marketing: *Tom Ziokowski/Samantha Cabaluna*
Assistant Editor: *Ann Day*
Production Editor: *Tessa Avila*
Production Service: *Integre Technical Publishing
    Company, Inc.*

Permissions Editor: *Karyn Morrison*
Manuscript Editor: *Linda Thompson*
Print Buyer: *Jessica Reed*
Typesetting: *Integre Technical Publishing
    Company, Inc.*
Printing and Binding: *RR Donnelley*

For more information about this or any other Duxbury product, contact:
DUXBURY
511 Forest Lodge Road
Pacific Grove, CA 93950 USA
www.duxbury.com
1-800-423-0563 (Thomson Learning Academic Resource Center)

Printed in the United States of America

10   9   8   7

Library of Congress Cataloging-in-Publication Data

Wackerly, Dennis D.
    Mathematical statistics with applications / Dennis D. Wackerly, William Mendenhall III,
    Richard L. Scheaffer. — 6th ed.
        p.   cm.
    Includes bibliographical references and index.
    ISBN 0-534-37741-6
    1. Mathematical statistics. I. Mendenhall, William. II. Scheaffer, Richard. III. title.

QA276 . M426 2002
519.5—dc21

CONTENTS

Preface   xiii

Note to the Student   xvii

1   What Is Statistics?   1

    1.1   Introduction   1

    1.2   Characterizing a Set of Measurements: Graphical Methods   3

    1.3   Characterizing a Set of Measurements: Numerical Methods   8

    1.4   How Inferences Are Made   12

    1.5   Theory and Reality   13

    1.6   Summary   14

2   Probability   19

    2.1   Introduction   19

    2.2   Probability and Inference   20

    2.3   A Review of Set Notation   22

    2.4   A Probabilistic Model for an Experiment: The Discrete Case   25

    2.5   Calculating the Probability of an Event: The Sample-Point Method   34

    2.6   Tools for Counting Sample Points   39

    2.7   Conditional Probability and the Independence of Events   50

    2.8   Two Laws of Probability   55

2.9 Calculating the Probability of an Event: The Event-Composition Method 59

2.10 The Law of Total Probability and Bayes' Rule 67

2.11 Numerical Events and Random Variables 72

2.12 Random Sampling 74

2.13 Summary 76

3 Discrete Random Variables and Their Probability Distributions 83

3.1 Basic Definition 83

3.2 The Probability Distribution for a Discrete Random Variable 84

3.3 The Expected Value of a Random Variable or a Function of a Random Variable 88

3.4 The Binomial Probability Distribution 97

3.5 The Geometric Probability Distribution 110

3.6 The Negative Binomial Probability Distribution (Optional) 116

3.7 The Hypergeometric Probability Distribution 119

3.8 The Poisson Probability Distribution 124

3.9 Moments and Moment-Generating Functions 131

3.10 Probability-Generating Functions (Optional) 136

3.11 Tchebysheff's Theorem 139

3.12 Summary 142

4 Continuous Random Variables and Their Probability Distributions 150

4.1 Introduction 150

4.2 The Probability Distribution for a Continuous Random Variable 151

4.3 Expected Values for Continuous Random Variables 162

4.4 The Uniform Probability Distribution 166

4.5 The Normal Probability Distribution 170

4.6 The Gamma Probability Distribution 175

4.7 The Beta Probability Distribution 182

4.8 Some General Comments 187

4.9 Other Expected Values 188

4.10   Tchebysheff's Theorem   194

4.11   Expectations of Discontinuous Functions and Mixed Probability Distributions (Optional)   197

4.12   Summary   201

5   Multivariate Probability Distributions   209

5.1   Introduction   209

5.2   Bivariate and Multivariate Probability Distributions   210

5.3   Marginal and Conditional Probability Distributions   222

5.4   Independent Random Variables   233

5.5   The Expected Value of a Function of Random Variables   241

5.6   Special Theorems   244

5.7   The Covariance of Two Random Variables   249

5.8   The Expected Value and Variance of Linear Functions of Random Variables   255

5.9   The Multinomial Probability Distribution   263

5.10   The Bivariate Normal Distribution (Optional)   268

5.11   Conditional Expectations   270

5.12   Summary   275

6   Functions of Random Variables   279

6.1   Introduction   279

6.2   Finding the Probability Distribution of a Function of Random Variables   280

6.3   The Method of Distribution Functions   281

6.4   The Method of Transformations   294

6.5   The Method of Moment-Generating Functions   302

6.6   Multivariable Transformations Using Jacobians (Optional)   310

6.7   Order Statistics   317

6.8   Summary   325

7   Sampling Distributions and the Central Limit Theorem   330

7.1   Introduction   330

7.2   Sampling Distributions Related to the Normal Distribution   331

7.3   The Central Limit Theorem   346

7.4   A Proof of the Central Limit Theorem (Optional)   352

7.5   The Normal Approximation to the Binomial Distribution   354

7.6   Summary   360

8   Estimation   364

8.1   Introduction   364

8.2   The Bias and Mean Square Error of Point Estimators   366

8.3   Some Common Unbiased Point Estimators   370

8.4   Evaluating the Goodness of a Point Estimator   373

8.5   Confidence Intervals   380

8.6   Large-Sample Confidence Intervals   385

8.7   Selecting the Sample Size   395

8.8   Small-Sample Confidence Intervals for μ and $\mu_1 - \mu_2$ 399

8.9   Confidence Intervals for σ^2 407

8.10   Summary   410

9   Properties of Point Estimators and Methods of Estimation   416

9.1   Introduction   416

9.2   Relative Efficiency   417

9.3   Consistency   420

9.4   Sufficiency   429

9.5   The Rao–Blackwell Theorem and Minimum-Variance Unbiased
              Estimation   435

9.6   The Method of Moments   444

9.7   The Method of Maximum Likelihood   448

9.8   Some Large-Sample Properties of MLEs (Optional)   455

9.9   Summary   457

10   Hypothesis Testing   460

10.1   Introduction   460

10.2   Elements of a Statistical Test   461

10.3   Common Large-Sample Tests   467

10.4   Calculating Type II Error Probabilities and Finding the Sample
Size for the Z Test   477

10.5   Relationships Between Hypothesis-Testing Procedures and Confidence
Intervals   480

10.6   Another Way to Report the Results of a Statistical Test: Attained
Significance Levels or p-Values   482

10.7   Some Comments on the Theory of Hypothesis Testing   487

10.8   Small-Sample Hypothesis Testing for μ and $\mu_1 - \mu_2$ 489

10.9   Testing Hypotheses Concerning Variances   498

10.10   Power of Tests and the Neyman–Pearson Lemma   507

10.11   Likelihood Ratio Tests   517

10.12   Summary   524

11   Linear Models and Estimation by Least Squares   531

11.1   Introduction   531

11.2   Linear Statistical Models   534

11.3   The Method of Least Squares   537

11.4   Properties of the Least Squares Estimators: Simple Linear
Regression   544

11.5   Inferences Concerning the Parameters β_i 552

11.6   Inferences Concerning Linear Functions of the Model Parameters:
Simple Linear Regression   557

11.7   Predicting a Particular Value of Y Using Simple Linear Regression   562

11.8   Correlation   567

11.9   Some Practical Examples   571

11.10   Fitting the Linear Model by Using Matrices   578

11.11   Linear Functions of the Model Parameters: Multiple Linear
Regression   584

11.12   Inferences Concerning Linear Functions of the Model Parameters:
Multiple Linear Regression   585

11.13   Predicting a Particular Value of Y Using Multiple Regression   591

11.14   A Test for $H_0: \beta_{g+1} = \beta_{g+2} = \cdots = \beta_k = 0$ 593

11.15   Summary and Concluding Remarks   600

12   Considerations in Designing Experiments   607

12.1   The Elements Affecting the Information in a Sample   607

12.2   Designing Experiments to Increase Accuracy   608

12.3   The Matched Pairs Experiment   612

12.4   Some Elementary Experimental Designs   618

12.5   Summary   624

13   The Analysis of Variance   628

13.1   Introduction   628

13.2   The Analysis of Variance Procedure   629

13.3   Comparison of More than Two Means: Analysis of Variance for a One-Way Layout   635

13.4   An Analysis of Variance Table for a One-Way Layout   639

13.5   A Statistical Model for the One-Way Layout   645

13.6   Proof of Additivity of the Sums of Squares and E(MST) for a One-Way Layout (Optional)   647

13.7   Estimation in the One-Way Layout   649

13.8   A Statistical Model for the Randomized Block Design   653

13.9   The Analysis of Variance for a Randomized Block Design   655

13.10   Estimation in the Randomized Block Design   662

13.11   Selecting the Sample Size   663

13.12   Simultaneous Confidence Intervals for More than One Parameter   665

13.13   Analysis of Variance Using Linear Models   668

13.14   Summary   672

14   Analysis of Categorical Data   680

14.1   A Description of the Experiment   680

14.2   The Chi-Square Test   682

14.3   A Test of a Hypothesis Concerning Specified Cell Probabilities: A Goodness-of-Fit Test   684

14.4   Contingency Tables   688

14.5 $r \times c$ Tables with Fixed Row or Column Totals   697

14.6   Other Applications   701

14.7   Summary and Concluding Remarks   703

15   Nonparametric Statistics   708

15.1   Introduction   708

15.2   A General Two-Sample Shift Model   709

15.3   The Sign Test for a Matched Pairs Experiment   711

15.4   The Wilcoxon Signed-Rank Test for a Matched Pairs Experiment   716

15.5   The Use of Ranks for Comparing Two Population Distributions: Independent Random Samples   722

15.6   The Mann–Whitney U Test: Independent Random Samples   724

15.7   The Kruskal–Wallis Test for the One-Way Layout   732

15.8   The Friedman Test for Randomized Block Designs   738

15.9   The Runs Test: A Test for Randomness   743

15.10   Rank Correlation Coefficient   750

15.11   Some General Comments on Nonparametric Statistical Tests   756

Appendix 1   Matrices and Other Useful Mathematical Results   763

A1.1   Matrices and Matrix Algebra   763

A1.2   Addition of Matrices   764

A1.3   Multiplication of a Matrix by a Real Number   765

A1.4   Matrix Multiplication   766

A1.5   Identity Elements   768

A1.6   The Inverse of a Matrix   769

A1.7   The Transpose of a Matrix   770

A1.8   A Matrix Expression for a System of Simultaneous Linear Equations   771

A1.9   Inverting a Matrix   772

A1.10   Solving a System of Simultaneous Linear Equations   777

A1.11   Other Useful Mathematical Results   778

Appendix 2   Common Probability Distributions, Means, Variances, and Moment-Generating Functions   781

A2.1   Discrete Distributions   781

A2.2   Continuous Distributions   782

Appendix 3   Tables   783

Table 1   Binomial Probabilities   783

Table 2   Table of e^{-x} 786

Table 3   Poisson Probabilities   787

Table 4   Normal Curve Areas   792

Table 5   Percentage Points of the t Distributions   793

Table 6   Percentage Points of the χ^2 Distributions   794

Table 7   Percentage Points of the F Distributions   796

Table 8   Distribution Function of U 806

Table 9   Critical Values of T in the Wilcoxon Matched-Pairs, Signed-Ranks Test   812

Table 10   Distribution of the Total Number of Runs R in Samples of Size (n_1, n_2); $P(R \leq a)$ 814

Table 11   Critical Values of Spearman's Rank Correlation Coefficient   816

Table 12   Random Numbers   817

Answers to Exercises   821

Index   839

PREFACE

The Purpose and Prerequisites of this Book

Mathematical Statistics with Applications was written for use with an undergraduate one-year sequence of courses (9 quarter or 6 semester hours) on mathematical statistics. The intent of the text is to present a solid undergraduate foundation in statistical theory and, at the same time, to provide an indication of the relevance and importance of the theory in solving practical problems in the real world. We think a course of this type is suitable for most undergraduate disciplines, including mathematics, where contact with applications may provide a refreshing and motivating experience. The only mathematical prerequisite is a thorough knowledge of first-year college calculus—including sums of infinite series, differentiation, and single and double integration.

Our Approach

Talking with students taking or having completed a beginning course in mathematical statistics reveals a major flaw in many courses. Students can take the course and leave it without a clear understanding of the nature of statistics. Many see the theory as a collection of topics, weakly or strongly related, but fail to see that statistics is a theory of information with inference as its goal. Further, they may leave the course without an understanding of the important role played by statistics in scientific investigations.

These considerations led us to develop a text that differs from others in three ways:

- First, the presentation of probability is preceded by a clear statement of the objective of statistics—statistical inference—and its role in scientific research. As students proceed through the theory of probability (Chapters 2 through 7), they are reminded frequently of the role that major topics play in statistical inference. The cumulative effect is that statistical inference is the dominating theme of the course.

- The second feature of the text is **connectivity**. We explain not only how major topics play a role in statistical inference, but also how the topics are related to one another. These integrating discussions appear most frequently in chapter introductions and conclusions.

- Finally, the text is unique in its **practical emphasis**, both in exercises throughout the text and in the useful statistical methodological topics contained in the last five chapters, whose goal is to reinforce the elementary but sound theoretical foundation developed in the initial chapters.

The book can be used in a variety of ways and adapted to the tastes of students and instructors. The difficulty of the material can be increased or decreased by controlling the assignment of exercises, by eliminating some topics, and by varying the amount of time devoted to each topic. A stronger applied flavor can be added by the elimination of some topics—for example, some sections of Chapters 6 and 7—and by devoting more time to the applied chapters at the end.

Changes in the Sixth Edition

Chapter 2 contains many small changes intended to improve the exposition. We now include the concept of a "partition" of the sample space and use this concept in the law of total probability and in Bayes rule. Maximum likelihood estimates are introduced in Chapter 3 via examples for the estimates of the parameters of the binomial, geometric and negative binomial distributions based on specific observed numerical values of random variables that possess these distributions. Follow-up problems at the end of the respective sections expand on these examples. In Chapter 4, we maintain emphasis on the χ^2 distribution, including some theoretical results that are useful in the development of the t and F distributions. There are many small changes throughout Chapter 5. We have clarified that conditional densities are undefined for values of the conditioning variable where the marginal density is 0 and have paid careful attention to when this occurs. Some interesting new problems about covariance are included. We have also retained the discussion of the "conditional variance" and its use in finding the variance of a random variable. Hierarchical models are briefly discussed. Chapter 6 has undergone some fairly substantial changes. Notably, we have introduced the concept of the *support* of a density and emphasize that a transformation method can be used when the transformation is monotone on the region of support. A new section uses the Jacobian to implement a bivariate transformation. In Chapter 7 we establish the independence of the sample mean and sample variance for a sample of size 2 from a normal distribution. As before, the proof of this result for general n is contained in an optional exercise. Problems in Chapter 7 take students through step-by-step derivations of the mean and variance of the t and F distributions.

Throughout Chapter 8, we have stressed the assumptions associated with confidence intervals based on the t-distributions. We have also included a brief discussion of the robustness of the t procedures and the lack of such for the intervals based on the χ^2 and F distributions. The Rao-Blackwell theorem is stated, proved, and used in Chapter 9. We include an optional section dealing with large-sample confidence intervals based on maximum-likelihood estimates of functions of the parameter of interest. In Chapter 10 we make clear that the tests based on larger sample sizes typically have smaller probabilities of Type II errors if the level of the tests stays fixed. We also illustrate explicitly that the power of a uniformly most powerful test can be smaller (although the largest possible) than desired. There are fairly substantial changes in the section on generalized likelihood ratio tests.

In Chapter 11, the simple linear regression model is thoroughly discussed (including confidence intervals, prediction intervals, and correlation) before the matrix approach to multiple linear regression model is introduced. Chapter 12 includes a separate section on the matched-pairs experiment. Although many possible sets of "dummy variables" can be used to cast the analysis of variance into a regression context, in Chapter 13 we focus on the dummy variables typically used by SAS and other statistical analysis computing packages. The text still focuses primarily on the randomized block design with fixed (nonrandom) block effects. If an instructor wishes, a series of supplemental exercises dealing with the randomized block design with random block effect can be used to illustrate the similarities and differences of these two versions of the randomized block design. Many other less obvious changes have been made throughout the text.

The Exercises

This edition contains more than 150 new exercises. As in previous editions, some of the new exercises are theoretical, whereas others contain data from documented sources that deal with research in a variety of fields. We continue to believe that exercises based on real data or actual experimental scenarios permit students to see the practical uses of the various statistical and probabilistic methods presented in the text. As they work through these exercises, students gain insight into the real-life applications of the theoretical results developed in the text. This insight makes learning the necessary theory more enjoyable and produces a deeper understanding of the theoretical methods. As in previous editions, the more challenging exercises are marked with an asterisk (*). Answers to the odd-numbered exercises are provided in the back of the book. The solutions manuals that accompany the text have been substantially improved.

Tables and Appendices

We have maintained the use of the upper-tail normal tables because the users of the text find them to be more convenient. We have also maintained the format of the table of the F distributions that we introduced in previous editions. This table of the F distributions provides critical values corresponding to upper-tail areas of .100, .050, .025, .010, and .005 in a single table. Because tests based on statistics possessing the F distribution occur quite often, this table facilitates the computation of attained significance levels, or p-values, associated with observed values of these statistics.

We have also maintained our practice of providing easy access to often-used information. Because the normal and t tables are the most frequently used statistical tables in the text, copies of these tables are given in Appendix III and inside the front cover of the text. Users of previous editions have often remarked favorably about the utility of tables of the common probability distributions, means, variances, and moment-generating functions provided in Appendix II and inside the back cover of the text. In addition, we have included some frequently used mathematical results in a supplement to Appendix I. These results include the binomial expansion of $(x + y)^n$, the series expansion of e^x, sums of geometric series, definitions of the gamma and beta functions, and so on. As before, each chapter begins with a "local" table of contents containing the titles of the major sections in that chapter.

Acknowledgments

The authors wish to thank the many colleagues, friends, and students who have made helpful suggestions concerning the revisions of this text. In particular, we are indebted to P. V. Rao, J. G. Saw, Malay Ghosh, Andrew Rosalsky, and Brett Presnell for their technical comments.

We wish to thank E. S. Pearson, W. H. Beyer, I. Olkin, R. A. Wilcox, C. W. Dunnett, and A. Hald. We profited substantially from the suggestions of the reviewers of the Fifth edition of the text: Sitadri N. Bagchi, University of Nevada, Reno; Walter Freiberger, Brown University; James Hartman, College of Wooster; Melvin Lax, California State University, Long Beach; Steven E. Rigdon, Southern Illinois University; and Joe Schafer, Pennsylvania State University. Thanks are also due to the reviewer of the present edition: Mary Parker, University of Texas, Austin.

We also wish to acknowledge the contributions of Carolyn Crockett, our editor; Ann Day, assistant editor; Jennifer Jenkins, editorial assistant; and of those involved in the production of the text: Tessa Avila, Duxbury's project manager; Leslie Galen of Integre Technical Publishing; and Linda Thompson, copyeditor.

Finally, we appreciate the support of our families during the writing of the various editions of this text.

DENNIS D. WACKERLY
WILLIAM MENDENHALL III
RICHARD L. SCHEAFFER

NOTE TO THE STUDENT

As the title *Mathematical Statistics with Applications* implies, this text is concerned with statistics, in both theory and application, and only deals with mathematics as a necessary tool to give you a firm understanding of statistical techniques. The following suggestions for using the text will increase your learning and save your time.

The connectivity of the book is provided by the introductions and summaries in each chapter. These sections explain how each chapter fits into the overall picture of statistical inference and how each chapter relates to the preceding ones.

Within the chapters, important concepts are set off as definitions. These should be read and reread until they are clearly understood, because they form the framework on which everything else is built. The main theoretical results are set off as theorems. Although it is not necessary to understand the proof of each theorem, a clear understanding of the meaning and implications of the theorems is essential.

It is also essential that you work many of the exercises—for at least three reasons. First, you can be certain that you understand what you have read only by putting your knowledge to the test of working problems. Second, many of the exercises are of a practical nature and shed light on the applications of probability and statistics. Third, some of the exercises present new concepts and thus extend the material covered in the chapter.

D. D. W.
W. M.
R. L. S.

CHAPTER 1

What Is Statistics?

1.1   Introduction

1.2   Characterizing a Set of Measurements: Graphical Methods

1.3   Characterizing a Set of Measurements: Numerical Methods

1.4   How Inferences Are Made

1.5   Theory and Reality

1.6   Summary

     References and Further Readings

1.1   Introduction

Statistical techniques are employed in almost every phase of life. Surveys are designed to collect early returns on election day and forecast the outcome of an election. Consumers are sampled to provide information for predicting product preferences. Research physicians conduct experiments to determine the effect of various drugs and controlled environmental conditions on humans in order to infer the appropriate treatment for various illnesses. Engineers sample a product quality characteristic and various controllable process variables to identify key variables related to product quality. Newly manufactured fuses are sampled before shipping to decide whether to ship or hold individual lots. Economists observe various indices of economic health over a period of time and use the information to forecast the condition of the economy in the future. Statistical techniques play an important role in achieving the objective of each of these practical problems. The development of the theory underlying these techniques is the focus of this text.

A prerequisite to a discussion of the theory of statistics is a definition of *statistics* and a statement of its objectives. *Webster's New Collegiate Dictionary* defines statistics as "a branch of mathematics dealing with the collection, analysis, interpretation, and presentation of masses of numerical data." Stuart and Ord (1991) state: "Statistics is the branch of the scientific method which deals with the data obtained by counting or measuring the properties of populations." Rice (1995), commenting on experimentation and statistical applications, states that statistics is "essentially concerned with procedures for analyzing data, especially data that in some vague sense have a random character." Freund and Walpole (1987), among others, view statistics as encompassing "the science of basing inferences on observed data and the entire problem of making decisions in the face of uncertainty." And Mood, Graybill, and Boes (1974) define statistics as "the technology of the scientific method" and add that statistics is concerned with "(1) the design of experiments and investigations, (2) statistical inference." A superficial examination of these definitions suggests a substantial lack of agreement, but all possess common elements. Each description implies that data are collected, with inference as the objective. Each requires selecting a subset of a large collection of data, either existent or conceptual, in order to infer the characteristics of the complete set. All the authors imply that *statistics is a theory of information, with inference making as its objective.*

The large body of data that is the target of our interest is called a *population*, and the subset selected from it is a *sample*. The preferences of voters for a gubernatorial candidate, Jones, expressed in quantitative form (1 for "prefer" and 0 for "do not prefer") provide a real, finite, and existing population of great interest to Jones. In order to determine the true fraction who favor his election, Jones would need to interview *all* eligible voters—a task that is practically impossible. The voltage at a particular point in the guidance system for a spacecraft may be tested in the only three systems that have been built. The resulting data could be used to estimate the voltage characteristics for other systems that might be manufactured some time in the future. In this case the population is *conceptual*. We think of the sample of three as being representative of a large population of guidance systems that could be built using the same method. Presumably, this population would possess characteristics similar to the three systems in the sample. Analogously, measurements on patients in a medical experiment represent a sample from a conceptual population consisting of all patients similarly afflicted today, as well as those who will be afflicted in the near future. You will find it useful to clearly define the populations of interest for each of the statistical problems described earlier in this section and to clarify the inferential objective for each.

It is interesting to note that billions of dollars are spent each year by U.S. industry and government for data from experimentation, sample surveys, and other data collection procedures. This money is expended solely to obtain information about phenomena susceptible to measurement in areas of business, science, or the arts. The implication of this statement provides a key to the nature of the very valuable contribution that the discipline of statistics makes to research and development in all areas of society. Information useful in inferring some characteristic of a population (either existing or conceptual) is purchased in a specified quantity and results in an inference (estimation or decision) with an associated *degree of goodness*. For example, if Jones arranges for a sample of voters to be interviewed, the information in the sample can be used to estimate the true fraction of all voters who favor Jones's election. In addition to the estimate itself, Jones should also be concerned with the likelihood (chance) that the estimate provided is close to the true fraction of eligible voters who favor his election. Intuitively, the larger the number of eligible voters in the sample, the higher will be the likelihood of an accurate estimate. Similarly, if a decision is made regarding the relative merits of two manufacturing processes based on examination of samples of products from both processes, we should be interested in the decision regarding which is better *and* the likelihood that the decision is correct. In general, the study of statistics is concerned with the design of experiments or sample surveys to obtain a specified quantity of information at minimum cost and the optimum use of this information in making an inference about a population. *The objective of statistics is to make an inference about a population based on information contained in a sample from that population and to provide an associated measure of goodness for the inference.*

Exercises

1.1 For each of the following situations, discuss the nature of the population of interest, the inferential objective, and how you might go about collecting a sample.

  a A university researcher wants to estimate the proportion of U.S. citizens from "Generation X" who are interested in starting their own businesses.

  b For more than a century, normal body temperature for humans has been accepted to be 98.6°. Is it really? Researchers want to estimate the average temperature of healthy adults in the United States.

  c A city engineer wants to estimate the average weekly water consumption for single-family dwelling units in the city.

  d The National Highway Safety Council wants to estimate the proportion of automobile tires with unsafe tread among all tires manufactured by a specific company during the current production year.

  e A political scientist wants to determine whether a majority of adult residents of a state favor a unicameral legislature.

  f A medical scientist wants to estimate the average length of time until the recurrence of a certain disease.

  g An electrical engineer wants to determine whether the average length of life of transistors of a certain type is greater than 500 hours.

1.2 Characterizing a Set of Measurements: Graphical Methods

In the broadest sense, making an inference implies partially or completely describing a phenomenon or physical object. Little difficulty is encountered when appropriate and meaningful descriptive measures are available, but this is not always the case. For example, we might characterize a person by using height, weight, color of hair and eyes, and other descriptive measures of the person's physiognomy. Identifying a set of descriptive measures to characterize an oil painting would be a comparatively more difficult task. Characterizing a population that consists of a set of measurements is equally challenging. Consequently, a necessary prelude to a discussion of inference making is the acquisition of a method for characterizing a set of numbers. The characterizations must be meaningful so that knowledge of the descriptive measures enables us to clearly visualize the set of numbers. In addition, we require that the characterizations possess practical significance, so that knowledge of the descriptive measures for a population can be used to solve a practical, nonstatistical problem. We will develop our ideas on this subject by examining a process that generates a population.

Consider a study to determine important variables affecting profit in a business that manufactures custom-made machined devices. Some of these variables might be the dollar size of the contract, the type of industry with which the contract is negotiated, the degree of competition in acquiring contracts, the salesperson who estimates the contract, fixed dollar costs, and the supervisor who is assigned the task of organizing and conducting the manufacturing operation. The statistician will wish to measure the response or dependent variable, profit per contract, for several jobs (the sample). Along with recording the profit, the statistician will obtain measurements on the variables that might be related to profit—the independent variables. His or her objective is to use information in the sample to infer the approximate relationship of the independent variables just described to the dependent variable, profit, and to measure the strength of this relationship. The manufacturer's objective is to determine optimum conditions for maximizing profit.

The population of interest in the manufacturing problem is conceptual and consists of all measurements of profit (per unit of capital and labor invested) that might be made on contracts, now and in the future, for fixed values of the independent variables (size of the contract, measure of competition, etc.). The profit measurements will vary from contract to contract in a seemingly random manner as a result of variations in materials, time needed to complete individual segments of the work, and other uncontrollable variables affecting the job. Consequently, we view the population as being represented by a *distribution* of profit measurements, with the form of the distribution depending on specific values of the independent variables. Our wish to determine the relationship between the dependent variable, profit, and a set of independent variables is, therefore, translated into a desire to determine the effect of the independent variables on the conceptual distribution of population measurements.

An individual population (or any set of measurements) can be characterized by a *relative frequency distribution*, which can be represented by a *relative frequency histogram*. A graph is constructed by subdividing the axis of measurement into intervals of equal width. A rectangle is constructed over each interval, such that the height of the rectangle is proportional to the *fraction* of the total number of measurements falling in each cell. For example, to characterize the ten measurements 2.1, 2.4, 2.2, 2.3, 2.7, 2.5, 2.4, 2.6, 2.6, and 2.9, we could divide the axis of measurement into intervals of equal width (say, .2 unit), commencing with 2.05. The relative frequencies (fraction of total number of measurements), calculated for each interval, are shown in Figure 1.1. Notice that the figure gives a clear pictorial description of the entire set of ten measurements.

Observe that we have not given precise rules for selecting the number, widths, or locations of the intervals used in constructing a histogram. This is because the selection of these items is somewhat at the discretion of the person who is involved in the construction.

Although they are arbitrary, a few guidelines can be very helpful in selecting the intervals. *Points of subdivision of the axis of measurement should be chosen so that it is impossible for a measurement to fall on a point of division.* This eliminates a source of confusion and is easily accomplished, as indicated in Figure 1.1. The sec-

FIGURE **1.1**
Relative frequency
histogram

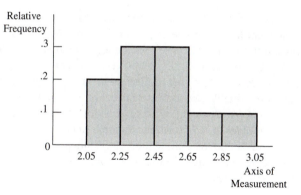

ond guideline involves the width of each interval and, consequently, the minimum number of intervals needed to describe the data. Generally speaking, we wish to obtain information on the form of the distribution of the data. Many times the form will be mound-shaped, as illustrated in Figure 1.2. (Others prefer to refer to distributions such as these as bell-shaped, or normal.) Using many intervals with a small amount of data results in little summarization and presents a picture very similar to the data in their original form. The larger the amount of data, the greater the number of included intervals can be while still presenting a satisfactory picture of the data. *We suggest spanning the range of the data with from 5 to 20 intervals and using the larger number of intervals for larger quantities of data.*

Some people feel that the description of data is an end in itself. Histograms are often used for this purpose, but there are many other graphical methods that provide meaningful summaries of the information contained in a set of data. Some excellent references for the general topic of graphical descriptive methods are given in the references at the end of this chapter. Keep in mind, however, that the usual objective of statistics is to make inferences. The relative frequency distribution associated with a data set and the accompanying histogram are sufficient for our objectives in developing the material in this text. This is primarily due to the probabilistic interpretation that can be derived from the frequency histogram, Figure 1.1. We have already stated that the area of a rectangle over a given interval is proportional to the fraction of the total number of measurements falling in that interval. Let's extend this idea one step further. If a measurement is selected at random from the original data set, the probability that it will fall in a given interval is proportional to the area under the histogram lying over that interval. (At this point we rely on the layperson's concept of probability. This term is discussed in greater detail in Chapter 2.) For example,

FIGURE **1.2**
Relative frequency
distribution

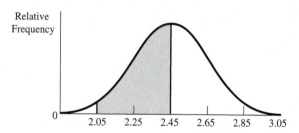

for the data used to construct Figure 1.1, the probability that a randomly selected measurement falls in the interval from 2.05 to 2.45 is .5, because half the measurements fall in this interval. Correspondingly, the *area under the histogram* in Figure 1.1 over the interval from 2.05 to 2.45 is .5. It is clear that this interpretation applies to the distribution of any set of measurements—a population or a sample. If Figure 1.2 gives the relative frequency distribution of profit (in millions of dollars) for a conceptual population of profit responses for contracts at given settings of the independent variables (size of contract, measure of competition, etc.), the probability that the next contract (at the same setting of the independent variables) yields a profit that falls in the interval from 2.05 to 2.45 million is proportional to the shaded area under the distribution curve.

Exercises

1.2 The following data represent the federal government's income from federal income taxes (on a per capita basis) for each of the 50 states in fiscal year 1991:

964	3168	1256	997	1477	951	1514	1713
1040	1080	2325	1159	1151	1102	1233	1120
1358	1013	1261	1317	1615	1185	1588	948
968	1007	1103	1312	565	1500	1347	1567
1165	1189	1056	1216	1037	1088	1251	1104
786	870	923	1051	1207	1090	1592	1292
1416	1385.						

Source: U.S. Department of Commerce, 1991.

Construct a relative frequency histogram for these data.

1.3 Of great importance to residents of central Florida is the amount of radioactive material present in the soil of reclaimed phosphate mining areas. Measurements of the amount of ^{238}U in 25 soil samples were as follows (measurements in picocuries per gram):

.74	6.47	1.90	2.69	.75
.32	9.99	1.77	2.41	1.96
1.66	.70	2.42	.54	3.36
3.59	.37	1.09	8.32	4.06
4.55	.76	2.03	5.70	12.48.

Construct a relative frequency histogram for these data.

1.4 Acquired immunodeficiency syndrome (AIDS) has become one of the most devastating diseases in modern society. The numbers of cases of AIDS (in thousands) reported in 25 major cities in the United States during 1992 are as follows:

38.3	6.2	3.7	2.6	2.1
14.6	5.6	3.7	2.3	2.0
11.9	5.5	3.4	2.2	2.0
6.6	4.6	3.1	2.2	1.9
6.3	4.5	2.7	2.1	1.8.

Source: Wright (1992), p. 209.

a   Construct a relative frequency histogram to describe these data.

b   What proportion of these cities reported more than 10,000 cases of AIDS in 1992?

c   If one of the cities is selected at random from the 25 for which the preceding data were taken, what is the probability that it will have reported fewer than 3000 cases of AIDS in 1992?

1.5   Given here is the relative frequency histogram associated with grade point averages (GPAs) of a sample of 30 students.

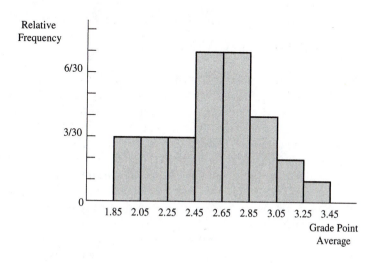

a   Which of the GPA categories identified on the horizontal axis are associated with the largest proportion of students?

b   What proportion of students had GPAs in each of the categories that you identified?

c   What proportion of the students had GPAs less than 2.65?

1.6   The relative frequency histogram given next was constructed from data obtained from a random sample of 25 families. Each was asked the number of quarts of milk that had been purchased the previous week.

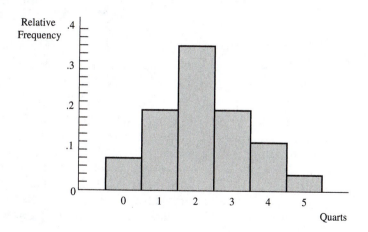

a Use this relative frequency histogram to determine the number of quarts of milk purchased by the largest proportion of the 25 families. The category associated with the largest relative frequency is called the *modal category*.

b What proportion of the 25 families purchased more than 2 quarts of milk?

c What proportion purchased more than 0 but fewer than 5 quarts?

1.3 Characterizing a Set of Measurements: Numerical Methods

The relative frequency histograms presented in Section 1.2 provide useful information regarding the distribution of sets of measurement, but histograms are usually not adequate for the purpose of making inferences. Indeed, many similar histograms could be formed from the same set of measurements. To make inferences about a population based on information contained in a sample and to measure the goodness of the inferences, we need rigorously defined quantities for summarizing the information contained in a sample. These sample quantities typically have mathematical properties that allow us to make probability statements regarding the goodness of our inferences. These mathematical properties are developed in the chapters that follow.

The quantities we define are *numerical descriptive measures* of a set of data. We seek some numbers that have meaningful interpretations and that can be used to describe the frequency distribution for any set of measurements. We will confine our attention to two types of descriptive numbers: *measures of central tendency* and *measures of dispersion or variation*.

Probably the most common measure of central tendency used in statistics is the arithmetic mean. (Because this is the only type of mean discussed in this text, we shall omit the word *arithmetic*.)

DEFINITION 1.1

The *mean* of a sample of n measured responses $y_1, y_2, \ldots, y_n$ is given by

$$\overline{y} = \frac{1}{n} \sum_{i=1}^{n} y_i.$$

The corresponding population mean is denoted μ.

The symbol $\overline{y}$, read "y bar," refers to a sample mean. We usually cannot measure the value of the population mean, μ; rather, μ is an unknown constant that we may want to estimate from sample information.

The mean of a set of measurements only locates the center of the distribution of data; by itself, it does not provide an adequate description of a set of measurements. Two sets of measurements could have widely different frequency distributions but equal means, as pictured in Figure 1.3. The difference between distributions I and II

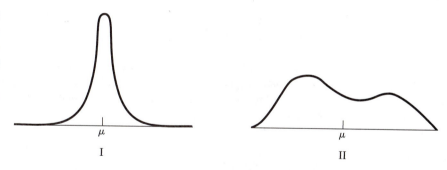

in the figure lies in the variation or dispersion of measurements on either side of the mean. To describe data adequately, we must define measures of data variability.

The most common measure of variability used in statistics is the *variance*, which is a function of the deviations (or distances) of the sample measurements from their mean.

DEFINITION 1.2

The *variance* of a sample of measurements $y_1, y_2, \ldots, y_n$ is the sum of the square of the differences between the measurements and their mean, divided by $n - 1$. Symbolically, the sample variance is

$$s^2 = \frac{1}{n-1} \sum_{i=1}^{n} (y_i - \bar{y})^2.$$

The corresponding population variance is denoted by the symbol σ^2.

Notice that we divided by $n-1$ instead of by n in our definition of s^2. The theoretical reason for this choice of divisor is provided in Chapter 8, where we will show that s^2 defined this way provides a "better" estimator for the true population variance, σ^2. Nevertheless, it is useful to think of s^2 as "almost" the average of the squared deviations of the observed values from their mean. The larger the variance of a set of measurements, the greater the amount of variation within the set. The variance is of value in comparing the relative variation of two sets of measurements, but it gives information about the variation in a single set only when interpreted in terms of the standard deviation.

DEFINITION 1.3

The *standard deviation* of a sample of measurements is the positive square root of the variance; that is,

$$s = \sqrt{s^2}.$$

The corresponding *population* standard deviation is denoted by $\sigma = \sqrt{\sigma^2}$.

Although it is closely related to the variance, the standard deviation can be used to give a fairly accurate picture of data variation for a single set of measurements. It can be interpreted using Tchebysheff's theorem (which is discussed in Exercise 1.30 and will be presented formally in Chapter 3) and by the empirical rule (which we now explain).

Many distributions of data in real life are mound-shaped. That is, they can be approximated by a bell-shaped frequency distribution known as a normal curve. Data possessing mound-shaped distributions have definite characteristics of variation, as expressed in the following statement.

Empirical Rule

For a distribution of measurements that is approximately normal (bell-shaped), it follows that the interval with endpoints

$\mu \pm \sigma$ contains approximately 68% of the measurements.

$\mu \pm 2\sigma$ contains approximately 95% of the measurements.

$\mu \pm 3\sigma$ contains almost all of the measurements.

As was mentioned in Section 1.2, once the frequency distribution of a set of measurements is known, probability statements regarding the measurements can be made. These probabilities were shown as areas under a frequency histogram. Thus the probabilities contained in the empirical rule are areas under the normal curve shown in Figure 1.4.

Use of the empirical rule is illustrated by the following example. Suppose that the scores on an achievement test given to all high school seniors in a certain state are known to have, approximately, a normal distribution with mean $\mu = 64$ and standard deviation $\sigma = 10$. It can then be deduced that approximately 68% of the scores are between 54 and 74, that approximately 95% of the scores are between 44 and 84, and that almost all of the scores are between 34 and 94. Thus, knowledge of the mean and the standard deviation gives us a fairly good picture of the frequency distribution of scores.

Suppose that a single high school student is randomly selected from those who took the test. What is the probability that his score will be between 54 and 74? Based on the empirical rule, we find that 0.68 is a reasonable answer to this probability question.

The utility and value of the empirical rule are due to the common occurrence of approximately normal distributions of data in nature—more so because the rule

FIGURE **1.4**
Normal curve

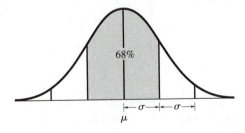

applies to distributions that are not exactly normal but just mound-shaped. You will find that approximately 95% of a set of measurements will be within 2σ of μ for a variety of distributions.

Exercises

1.7 Resting breathing rates for college-age students are approximately normally distributed with mean 12 and standard deviation 2.3 breaths per minute. What fraction of all college-age students have breathing rates in the following intervals?

  a   9.7 to 14.3 breaths per minute

  b   7.4 to 16.6 breaths per minute

  c   9.7 to 16.6 breaths per minute

  d   less than 5.1 or more than 18.9 breaths per minute

1.8 It has been projected that the average and standard deviation of the amount of time spent on-line using the Internet are, respectively, 14 and 17 hours per person per year.

  a   What value is exactly one standard deviation below the mean?

  b   If the amount of time spent on-line using the Internet is approximately normally distributed, what proportion of the users spend an amount of time on-line that is less than the value you found in part (a)?

  c   Is the amount of time spent on-line using the Internet approximately normally distributed? Why?

1.9 The following results on summations will help us in calculating the sample variance s^2. For any constant c,

  a $\displaystyle\sum_{i=1}^{n} c = nc.$

  b $\displaystyle\sum_{i=1}^{n} cy_i = c\sum_{i=1}^{n} y_i.$

  c $\displaystyle\sum_{i=1}^{n}(x_i + y_i) = \sum_{i=1}^{n} x_i + \sum_{i=1}^{n} y_i.$

Use (a), (b), and (c) to show that

$$s^2 = \frac{1}{n-1}\left[\sum_{i=1}^{n} y_i^2 - \frac{1}{n}\left(\sum_{i=1}^{n} y_i\right)^2\right].$$

1.10 Use the result of Exercise 1.9 to calculate s for the $n = 6$ sample measurements 1, 4, 2, 1, 3, and 3.

1.11 Refer to Exercise 1.2.

  a   Calculate $\bar{y}$ and s for the data given.

  b   Calculate the interval $\bar{y} \pm ks$ for $k = 1, 2,$ and 3. Count the number of measurements that fall within each interval, and compare this result with the number that you would expect according to the empirical rule.

1.12 Refer to Exercise 1.3, and repeat (a) and (b) of Exercise 1.11.

1.13 Refer to Exercise 1.4, and repeat (a) and (b) of Exercise 1.11.

1.14 In Exercise 1.4, there is one extremely large value (38.3). Eliminate this value and calculate $\bar{y}$ and s for the remaining 24 observations. Also, calculate the intervals $\bar{y} \pm ks$ for $k = 1, 2,$ and 3; count the number of measurements in each; and compare these results with those predicted by the empirical rule. Compare the answers here to those found in Exercise 1.13. Note the effect of a single large observation on $\bar{y}$ and s.

1.15 The *range* of a set of measurements is the difference between the largest and the smallest values. The empirical rule suggests that the standard deviation of a set of measurements may be roughly approximated by one-fourth of the range (i.e., range/4). Calculate this approximation to s for the data sets in Exercises 1.2, 1.3, and 1.4. Compare the result in each case to the actual, calculated value of s.

1.16 The College Board's verbal and mathematics scholastic aptitude tests are scored on a scale of 200 to 800. It seems reasonable to assume that the distribution of test scores are approximately normally distributed for both tests. Use the result from Exercise 1.15 to approximate the standard deviation for scores on the verbal test.

1.17 According to the EPA, chloroform, which in its gaseous form is suspected to be a cancer-causing agent, is present in small quantities in all the country's 240,000 public water sources. If the mean and standard deviation of the amounts of chloroform present in water sources are 34 and 53 μg/L (micrograms per liter), respectively, explain why chloroform amounts do not have a normal distribution.

1.18 Weekly maintenance costs for a factory, recorded over a long period of time and adjusted for inflation, tend to have an approximately normal distribution with an average of $420 and a standard deviation of $30. If $450 is budgeted for next week, what is an approximate probability that this budgeted figure will be exceeded?

1.19 The manufacturer of a new food additive for beef cattle claims that 80% of the animals fed a diet including this additive should have monthly weight gains in excess of 20 pounds. A large sample of measurements on weight gains for cattle fed this diet exhibits an approximately normal distribution with mean 22 pounds and standard deviation 2 pounds. Do you think the sample information contradicts the manufacturer's claim? (Calculate the probability of a weight gain exceeding 20 pounds.)

1.4 How Inferences Are Made

The mechanism instrumental in making inferences can be well illustrated by analyzing our own intuitive inference-making procedures.

Suppose that two candidates are running for a public office in our community and that we wish to determine whether our candidate, Jones, is favored to win. The population of interest is the set of responses from all eligible voters who will vote on election day and we wish to determine whether the fraction favoring Jones exceeds .5. For the sake of simplicity, suppose that all eligible voters will go to the polls and that we randomly select a sample of 20 from the courthouse rosters. All 20 are contacted and all favor Jones. What do you conclude about Jones's prospects for winning the election?

There is little doubt that most of us would immediately infer that Jones will win. This was an easy inference to make, but this inference itself is not our immediate goal. Rather, we wish to examine the mental processes that were employed in reaching this conclusion about the prospective behavior of a large voting population on the basis of a sample of only 20 people.

Winning means acquiring more than 50% of the votes. Did we conclude that Jones would win because we thought that the fraction favoring Jones in the sample was identical to the fraction favoring Jones in the population? We know that this is probably not true. A simple experiment will verify that the fraction in the sample favoring Jones need not be the same as the fraction of the population who favor him. If a balanced coin is tossed, it is intuitively obvious that the true proportion of times it will turn up heads is .5. Yet if we sample the outcomes for our coin by tossing it 20 times, the proportion of heads will vary from sample to sample. That is, on one occasion we might observe 12 heads out of 20 flips, for a sample proportion of $12/20 = .6$. On another occasion we might observe 8 heads out of 20 flips, for a sample proportion of $8/20 = .4$. In fact, the sample proportion of heads could be $0, .05, .10, \ldots, 1.0$.

Did we conclude that Jones would win because it would be impossible for 20 out of 20 sample voters to favor him if, in fact, less than 50% of the electorate intended to vote for him? The answer to this question is certainly no, but it provides the key to our hidden line of logic. It is not *impossible* to draw 20 out of 20 favoring Jones when less than 50% of the electorate favor him, but it is *highly improbable*. As a result, we concluded that he would win.

This example illustrates the potent role played by probability in making inferences. Probabilists assume that they know the structure of the population of interest and use the theory of probability to compute the probability of obtaining a particular sample. Assuming that they know the structure of a population generated by random drawings of five cards from a standard deck, probabilists compute the probability that the draw will yield three aces and two kings. Statisticians use probability to make the trip in reverse—from the sample to the population. Observing five aces in a sample of five cards, they immediately infer that the deck (which generates the population) is loaded and not standard. The probability of drawing five aces from a standard deck is zero! This is an exaggerated case, but it makes the point. Basic to inference making is the problem of calculating the probability of an observed sample. As a result, probability is the mechanism used in making statistical inferences.

One final comment is in order. If you did not think that the sample justified an inference that Jones would win, do not feel too chagrined. One can easily be misled when making intuitive evaluations of the probabilities of events. If you decided that the probability was very low that 20 voters out of 20 would favor Jones, assuming that Jones would lose, you were correct. However, it is not difficult to concoct an example in which an intuitive assessment of probability would be in error. Intuitive assessments of probabilities are unsatisfactory, and we need a rigorous theory of probability in order to develop methods of inference.

1.5   Theory and Reality

Theories are conjectures proposed to explain phenomena in the real world. As such, theories are approximations or models for reality. These models or explanations of reality are presented in verbal form in some less quantitative fields and as mathematical relationships in others. Whereas a theory of social change might be expressed verbally in sociology, a description of the motion of a vibrating string is presented in a precise mathematical manner in physics. When we choose a mathematical model for a physical process, we hope that the model reflects faithfully, in mathematical terms, the attributes of the physical process. If so, the mathematical model can be used to arrive at conclusions about the process itself. If we could develop an equation to predict the position of a vibrating string, the quality of the prediction would depend on how well the equation fit the motion of the string. The process of finding a good equation is not necessarily simple and usually requires several simplifying assumptions (uniform string mass, no air resistance, etc.). The final criterion for deciding whether a model is "good" is whether it yields good and useful information. The motivation for using mathematical models lies primarily in their utility.

This text is concerned with the theory of statistics and hence with models of reality. We will postulate theoretical frequency distributions for populations and will develop a theory of probability and inference in a precise mathematical manner. The net result will be a theoretical or mathematical model for acquiring and utilizing information in real life. The model will not be an exact representation of nature, but this should not disturb us. Its utility, like that of other theories, will be measured by its ability to assist us in understanding nature and in solving problems in the real world.

1.6 Summary

The objective of statistics is to make an inference about a population based on information contained in a sample taken from that population. The theory of statistics is a theory of information concerned with quantifying information, designing experiments or procedures for data collection, and analyzing data. Our goal is to minimize the cost of a specified quantity of information and to use this information to make inferences. Most importantly, we have viewed making an inference about the unknown population as a two-step procedure. First, we enlist a suitable inferential procedure for the given situation. Second, we seek a measure of the goodness of the resulting inference. For example, every estimate of a population characteristic based on information contained in the sample might have associated with it a probabilistic bound on the error of estimation.

A necessary prelude to making inferences about a population is the ability to describe a set of numbers. Frequency distributions provide a graphic and useful method for characterizing conceptual or real populations of numbers. Numerical descriptive measures are often more useful when we wish to make an inference and measure the goodness of that inference.

The mechanism for making inferences is the theory of probability. The probabilist reasons from a known population to the outcome of a single experiment, the sample. In contrast, the statistician utilizes the theory of probability to calculate the probability of an observed sample and to infer from this the characteristics of an unknown population. Thus probability is the foundation of the theory of statistics.

Finally, we have noted the difference between theory and reality. In this text we study the mathematical theory of statistics, which is an idealization of nature. It is rigorous, mathematical, and subject to study in a vacuum completely isolated from the real world. Or it can be tied very closely to reality and can be useful in making inferences from data in all fields of science. In this text we will be utilitarian. We will not regard statistics as a branch of mathematics but as an area of science concerned with developing a practical theory of information. We will consider statistics as a separate field, analogous to physics—not as a branch of mathematics but as a theory of information that utilizes mathematics heavily.

Subsequent chapters will expand on the topics that we have encountered in this introduction. We will begin with a study of the mechanism employed in making inferences, the theory of probability. This theory provides theoretical models for generating experimental data and thereby provides the basis for our study of statistical inference.

References and Further Readings

Cleveland, W. S. 1994. *The Elements of Graphing Data.* Murray Hill, N.J.: AT&T Bell Laboratories.

———. *Visualizing Data.* 1993. Summit, N.J.: Hobart Press.

Fraser, D. A. S. 1958. *Statistics, an Introduction.* New York: John Wiley.

Freund, J. E., and R. E. Walpole. 1987. *Mathematical Statistics,* 4th ed. Englewood Cliffs, N.J.: Prentice Hall.

Iman, R. L. 1994. *A Data-Based Approach to Statistics.* Belmont, Calif.: Duxbury Press.

Mendenhall, W., R. J. Beaver, and B. M. Beaver. 1999. *Introduction to Probability and Statistics,* 10th ed. Pacific Grove, Calif.: Duxbury Press.

Mood, A. M., F. A. Graybill, and D. Boes. 1974. *Introduction to the Theory of Statistics,* 3d ed. New York: McGraw-Hill.

Moore, D. S., and G. P. McCabe. 1999. *Introduction to the Practice of Statistics,* 3d ed. New York: W. H. Freeman.

Rice, J. A. *Mathematical Statistics and Data Analysis,* 2d ed. Belmont, Calif.: Duxbury Press, 1995.

Stuart, A., and J. K. Ord. 1991. *Kendall's Theory of Statistics,* 5th ed., vol. 1. London: Edward Arnold.

Wright, J. W. (ed.) 1992. *The Universal Almanac, 1993.* Kansas City, Mo.: Andrews & McMeel.

Supplementary Exercises

1.20 Prove that the sum of the deviations of a set of measurements about their mean is equal to zero; that is,

$$\sum_{i=1}^{n}(y_i - \overline{y}) = 0.$$

1.21 There were 1521 NCAA tournament basketball games played between 1939 and 1992. If the total number of points is computed for each game, the resultant totals are approximately normally distributed with mean 143 and standard deviation 26.

    a What percentage of the games ended with a total score in excess of 169 points?

    b What percentage of the games ended with a total score between 117 and 195?

    c Would you expect the total score to exceed 225? Why or why not?

1.22 Aqua running has been suggested as a method of cardiovascular conditioning for injured athletes and others who desire a low-impact aerobics program. In a study to investigate the relationship between exercise cadence and heart rate,[1] the heart rates of 20 healthy volunteers were measured at a cadence of 48 cycles per minute (a cycle consisted of two steps). The data are as follows:

87	109	79	80	96	95	90	92	96	98
101	91	78	112	94	98	94	107	81	96.

    a Use the range of the measurements to obtain an estimate of the standard deviation.

    b Construct a frequency histogram for the data. Use the histogram to obtain a visual approximation to $\bar{y}$ and s.

    c Calculate $\bar{y}$ and s. Compare these results with the calculation checks provided by (a) and (b).

    d Construct the intervals $\bar{y} \pm ks, k = 1, 2, 3$, and count the number of measurements falling in each interval. Compare the fractions falling in the intervals with the fractions that you would expect according to the empirical rule.

1.23 The following data give the lengths of time to failure for $n = 88$ radio transmitter-receivers:

16	224	16	80	96	536	400	80
392	576	128	56	656	224	40	32
358	384	256	246	328	464	448	716
304	16	72	8	80	72	56	608
108	194	136	224	80	16	424	264
156	216	168	184	552	72	184	240
438	120	308	32	272	152	328	480
60	208	340	104	72	168	40	152
360	232	40	112	112	288	168	352
56	72	64	40	184	264	96	224
168	168	114	280	152	208	160	176.

    a Use the range to approximate s for the $n = 88$ lengths of time to failure.

    b Construct a frequency histogram for the data. [Notice the tendency of the distribution to tail outward (skew) to the right.]

    c Use a calculator (or computer) to calculate $\bar{y}$ and s. (Hand calculation is much too tedious for this exercise.)

    d Calculate the intervals $\bar{y} \pm ks, k = 1, 2, 3$, and count the number of measurements falling in each interval. Compare your results with the empirical rule results. Note that the empirical rule provides a rather good description of these data, even though the distribution is highly skewed.

[1] *Source:* R. P. Wilder, D. Breenan, and D. E. Schotte, "A Standard Measure for Exercise Prescription for Aqua Running." *American Journal of Sports Medicine* 21(1) (1993): 45.

1.24 Compare the ratio of the range to s for the three sample sizes ($n = 6, 20$, and 88) for Exercises 1.10, 1.22, and 1.23. Note that the ratio tends to increase as the amount of data increases. The greater the amount of data, the greater will be their tendency to contain a few extreme values that will inflate the range and have relatively little effect on s. We ignored this phenomenon and suggested that you use 4 as the ratio for finding a guessed value of s in checking calculations.

1.25 A set of 340 examination scores exhibiting a bell-shaped relative frequency distribution has a mean of $\overline{y} = 72$ and a standard deviation of $s = 8$. Approximately how many of the scores would you expect to fall in the interval from 64 to 80? The interval from 56 to 88?

1.26 The discharge of suspended solids from a phosphate mine is normally distributed with mean daily discharge 27 mg/L and standard deviation 14 mg/L. In what proportion of the days will the daily discharge be less than 13 mg/L?

1.27 A machine produces bearings with mean diameter 3.00 inches and standard deviation 0.01 inches. Bearings with diameters in excess of 3.02 inches or less than 2.98 inches will fail to meet quality specifications. Approximately what fraction of this machine's production will fail to meet specifications? What assumptions did you make concerning the distribution of bearing diameters in order to answer this question?

1.28 Compared to their stay-at-home peers, employed women have higher levels of high-density lipoproteins (HDL), the "good" cholesterol associated with lower risk for heart attacks. A study of cholesterol levels in 2000 women, aged 25–64, living in Augsburg, Germany, was conducted by Ursula Haertel, Ulrigh Keil, and colleagues[2] at the GSF-Medis Institut in Munich. Of these 2000 women, the 48% who worked outside the home had HDL levels that were between 2.5 and 3.6 milligrams per deciliter (mg/dL) higher than the HDL levels of their stay-at-home counterparts. Suppose that the difference in HDL levels is normally distributed, with mean 0 (indicating no difference between the two groups of women) and standard deviation 1.2 mg/dL. If you were to select a working woman and a stay-at-home counterpart at random, what is the probability that the difference in their HDL levels would be between 1.2 and 2.4?

1.29 Over the past year, a fertilizer production process has shown an average daily yield of 60 tons with a variance in daily yields of 100. If the yield should fall to less than 40 tons tomorrow, should this result cause you to suspect an abnormality in the process? (Calculate the probability of obtaining less than 40 tons.) What assumptions did you make concerning the distribution of yields?

*1.30 Let $k \geq 1$. Show that, for any set of n measurements, the fraction included in the interval $\overline{y} - ks$ to $\overline{y} + ks$ is at least $(1 - 1/k^2)$. [*Hint:*

$$s^2 = \frac{1}{n-1} \left(\sum_{i=1}^{n} (y_i - \overline{y})^2 \right).$$

In this expression replace all deviations for which $|y_i - \overline{y}| \geq ks$ with ks. Simplify.] This result is known as *Tchebysheff's theorem.*[3]

1.31 A personnel manager for a certain industry has records of the number of employees absent per day. The average number absent is 5.5, and the standard deviation is 2.5. Because there are many days with zero, one, or two absent and only a few with more than ten absent, the frequency distribution is highly skewed. The manager wants to publish an interval in which as least 75% of these values lie. Use the result in Exercise 1.30 to find such an interval.

[2] *Source: Science News* 135 (June 1989): 389.
[3] Exercises preceded by an asterisk are optional.

1.32 For the data discussed in Exercise 1.31, give an upper bound to the fraction of days when there are more than 13 absentees.

1.33 Studies indicate that drinking water supplied by some old lead-lined city piping systems may contain harmful levels of lead. Based on data presented by Karalekas and colleagues (1983),[4] it appears that the distribution of lead content readings for individual water specimens has mean .033 mg/L and standard deviation .10 mg/L. Explain why it is obvious that the lead content readings are *not* normally distributed.

1.34 In Exercise 1.17 the mean and standard deviation of the amount of chloroform present in water sources were given to be 34 and 53, respectively. You argued that the amounts of chloroform could therefore not be normally distributed. Use Tchebysheff's Theorem (Exercise 1.30) to describe the distribution of chloroform amounts in water sources.

[4]*Source:* P. C. Karalekas, Jr., C. R. Ryan, and F. B. Taylor, "Control of Lead, Copper and Iron Pipe Corrosion in Boston." *American Water Works Journal* (February 1983): 92.

CHAPTER **2**

Probability

2.1   Introduction

2.2   Probability and Inference

2.3   A Review of Set Notation

2.4   A Probabilistic Model for an Experiment: The Discrete Case

2.5   Calculating the Probability of an Event: The Sample-Point Method

2.6   Tools for Counting Sample Points

2.7   Conditional Probability and the Independence of Events

2.8   Two Laws of Probability

2.9   Calculating the Probability of an Event: The Event-Composition Method

2.10   The Law of Total Probability and Bayes' Rule

2.11   Numerical Events and Random Variables

2.12   Random Sampling

2.13   Summary

References and Further Readings

2.1   Introduction

In everyday conversation, the term *probability* is a measure of one's belief in the occurrence of a future event. We accept this as a meaningful and practical interpretation of probability but seek a clearer understanding of its context, how it is measured, and how it assists in making inferences.

The concept of probability is necessary in work with physical, biological, or social mechanisms that generate observations that cannot be predicted with certainty. For example, the blood pressure of a person at a given point in time cannot be predicted with certainty, and we never know the exact load that a bridge will endure before collapsing into a river. Such random events cannot be predicted with certainty, but the relative frequency with which they occur in a long series of trials is often remarkably stable. Events possessing this property are called *random*, or *stochastic*, events. This stable long-term relative frequency provides an intuitively meaningful measure of our belief in the occurrence of a random event if a future observation is to be made. It is impossible, for example, to predict with certainty the occurrence of heads on a single toss of a balanced coin, but we would be willing to state with a fair measure of confidence that the fraction of heads in a long series of trials would be very near .5. That this relative frequency is commonly used as a measure of belief in the outcome for a single toss is evident when we consider chance from a gambler's perspective. He risks money on the single toss of a coin, not a long series of tosses. The relative frequency of a head in a long series of tosses, which a gambler calls the probability of a head, gives him a measure of the chance of winning on a single toss. If the coin were unbalanced and gave 90% heads in a long series of tosses, the gambler would say that the probability of a head is .9, and he would be fairly confident in the occurrence of a head on a single toss of the coin.

The preceding example possesses some realistic and practical analogies. In many respects all people are gamblers. The research physician gambles time and money on a research project, and she is concerned with her success on a single flip of this symbolic coin. Similarly, the investment of capital in a new manufacturing plant is a gamble that represents a single flip of a coin on which the entrepreneur has high hopes for success. The fraction of similar investments that are successful in a long series of trials is of interest to the entrepreneur only insofar as it provides a measure of belief in the successful outcome of a single individual investment.

The relative frequency concept of probability, although intuitively meaningful, does not provide a rigorous definition of probability. Many other concepts of probability have been proposed, including that of subjective probability, which allows the probability of an event to vary depending upon the person performing the evaluation. Nevertheless, for our purposes we accept an interpretation based on relative frequency as a meaningful measure of our belief in the occurrence of an event. Next, we will examine the link that probability provides between observation and inference.

2.2   Probability and Inference

The role that probability plays in making inferences will be discussed in detail after an adequate foundation has been laid for the theory of probability. At this point we will present an elementary treatment of this theory through an example and an appeal to your intuition.

The example selected is similar to that presented in Section 1.4 but simpler and less practical. It was chosen because of the ease with which we can visualize the population and sample and because it provides an observation-producing mechanism for which a probabilistic model will be constructed in Section 2.3.

Consider a gambler who wishes to make an inference concerning the balance of a die. The conceptual population of interest is the set of numbers that would be

FIGURE **2.1**
Frequency
distribution for the
population generated
by a balanced die

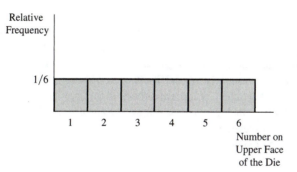

generated if the die were rolled over and over again, ad infinitum. If the die were perfectly balanced, one-sixth of the measurements in this population would be 1s, one-sixth, 2s, one-sixth, 3s, and so on. The corresponding frequency distribution is shown in Figure 2.1.

Using the scientific method, the gambler proposes the hypothesis *that the die is balanced*, and he seeks observations from nature to contradict the theory, if false. A sample of ten tosses is selected from the population by rolling the die ten times. All ten tosses result in 1s. The gambler looks upon this output of nature with a jaundiced eye and concludes that his hypothesis is not in agreement with nature and hence that the die is *not* balanced.

The reasoning employed by the gambler identifies the role that probability plays in making inferences. The gambler rejected his hypothesis (and concluded that the die is unbalanced) not because it is *impossible* to throw ten 1s in ten tosses of a balanced die but because it is highly *improbable*. His evaluation of the probability was most likely subjective. That is, the gambler may not have known how to calculate the probability of ten 1s in ten tosses, but he had an intuitive feeling that this event was highly unlikely if the die were balanced. The point to note is that his decision was based on the probability of the observed sample.

The need for a theory of probability that will provide a rigorous method for finding a number (a probability) that will agree with the actual relative frequency of occurrence of an event in a long series of trials is apparent if we imagine a different result for the gambler's sample. Suppose, for example, that instead of ten 1s, he observed five 1s along with two 2s, one 3, one 4, and one 6. Is this result so improbable that we should reject our hypothesis that the die is balanced and conclude that the die is loaded in favor of 1s? If we must rely solely on experience and intuition to make our evaluation, it is not so easy to decide whether the probability of five 1s in ten tosses is large or small. The probability of throwing four 1s in ten tosses would be even more difficult to guess. We will not deny that experimental results often are obviously inconsistent with a given hypothesis and lead to its rejection. However, many experimental outcomes fall in a gray area where we require a rigorous assessment of the probability of their occurrence. Indeed, it is not difficult to show that intuitive evaluations of probabilities often lead to answers that are substantially in error and result in incorrect inferences about the target population. For example, if there are 20 people in a room, most people would guess that it is very unlikely that there would be two or more persons with the same birthday. Yet, under certain reasonable assumptions, in Example 2.18 we will show that the probability of such an occurrence is larger than .4, a number that is surprisingly large to many.

We need a theory of probability that will permit us to calculate the probability (or a quantity proportional to the probability) of observing specified outcomes, assuming that our hypothesized model is correct. This topic will be developed in detail in subsequent chapters. Our immediate goal is to present an introduction to the theory of probability, which provides the foundation for modern statistical inference. We will begin by reviewing some set notation that will be used in constructing probabilistic models for experiments.

2.3 A Review of Set Notation

To proceed with an orderly development of probability theory, we need some basic concepts of set theory. We will use capital letters, $A, B, C, \ldots$, to denote sets of points. If the elements in the set A are a_1, a_2, and a_3, we will write

$$A = \{a_1,\ a_2,\ a_3\}.$$

Let S denote the set of all elements under consideration; that is, S is the *universal set*. For any two sets A and B we will say that A is a *subset* of B, or A is contained in B (denoted $A \subset B$), if every point in A is also in B. The *null, or empty, set*, denoted by $\emptyset$, is the set consisting of no points. Thus $\emptyset$ is a subset of every set.

Sets and relationships between sets can be conveniently portrayed by using *Venn diagrams*. The Venn diagram in Figure 2.2 shows two sets, A and B, in the universal set S. Set A is the set of all points inside the triangle; set B is the set of all points inside the circle. Note that in Figure 2.2, $A \subset B$.

Consider now two arbitrary sets of points. The *union* of A and B, denoted by $A \cup B$, is the set of all points in A or B or both. That is, the union of A and B contains all points that are in at least one of the sets. The Venn diagram in Figure 2.3 shows two sets A and B, where A is the set of points in the left-hand circle and B is the set of points in the right-hand circle. The set $A \cup B$ is the shaded region consisting of all points inside either circle (or both). The key word for expressing the union of two sets is *or* (meaning A or B or both).

FIGURE **2.2**
Venn diagram for
$A \subset B$

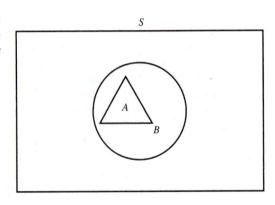

FIGURE **2.3**
Venn diagram for
$A \cup B$

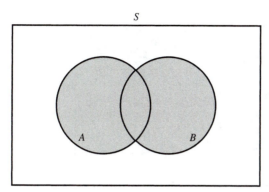

FIGURE **2.4**
Venn diagram for AB

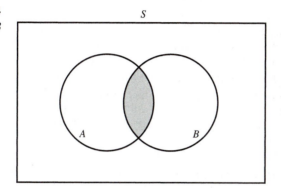

The *intersection* of A and B, denoted by $A \cap B$ or by AB, is the set of all points in both A and B. The Venn diagram of Figure 2.4 shows two sets A and B, with $A \cap B$ consisting of the points in the shaded region where the two sets overlap. The key word for expressing intersections is *and* (meaning A and B *simultaneously*).

If A is a subset of S, then the *complement* of A, denoted by $\overline{A}$, is the set of points that are in S but not in A. Figure 2.5 is a Venn diagram illustrating that the shaded area in S but not in A is $\overline{A}$. Note that $A \cup \overline{A} = S$.

FIGURE **2.5**
Venn diagram for $\overline{A}$

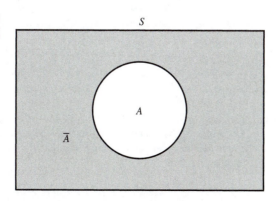

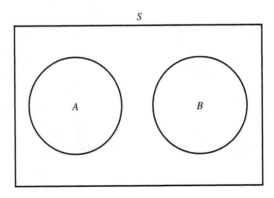

FIGURE **2.6**
Venn diagram for
mutually exclusive
sets *A* and *B*

Two sets, *A* and *B*, are said to be *disjoint*, or *mutually exclusive*, if $A \cap B = \emptyset$. That is, mutually exclusive sets have no points in common. The Venn diagram in Figure 2.6 illustrates two sets *A* and *B* that are mutually exclusive. Referring to Figure 2.5, it is easy to see that, for any set *A*, *A* and $\overline{A}$ are mutually exclusive.

Consider the die-tossing problem of Section 2.2 and let *S* denote the set of all possible numerical observations for a single toss of a die. That is, $S = \{1, 2, 3, 4, 5, 6\}$. Let $A = \{1, 2\}$, $B = \{1, 3\}$, and $C = \{2, 4, 6\}$. Then $A \cup B = \{1, 2, 3\}$, $A \cap B = \{1\}$, and $\overline{A} = \{3, 4, 5, 6\}$. Also, note that *B* and *C* are mutually exclusive, whereas *A* and *C* are not.

We will not attempt a thorough review of set algebra, but we mention four equalities of considerable importance. These are the *distributive laws*, given by

$$A \cap (B \cup C) = (A \cap B) \cup (A \cap C)$$

$$A \cup (B \cap C) = (A \cup B) \cap (A \cup C)$$

and *DeMorgan's laws*:

$$\overline{(A \cap B)} = \overline{A} \cup \overline{B}$$

$$\overline{(A \cup B)} = \overline{A} \cap \overline{B}.$$

In the next section we will proceed with an elementary discussion of probability theory.

Exercises

2.1 Suppose a family contains two children of different ages, and we are interested in the gender of these children. Let *F* denote that a child is female and *M* that the child is male, and let a pair such as *FM* denote that the older child is female and the younger is male. There are four points in the set *S* of possible observations:

$$S = \{FF, \ FM, \ MF, \ MM\}.$$

Let A denote the subset of possibilities containing no males, B, the subset containing two males, and C, the subset containing at least one male. List the elements of $A, B, C, A \cap B$, $A \cup B, A \cap C, A \cup C, B \cap C, B \cup C$, and $C \cap \overline{B}$.

2.2 Suppose that A and B are two events. Write expressions involving unions, intersections, and complements that describe the following:

a Both events occur.

b At least one occurs.

c Neither occurs.

d Exactly one occurs.

2.3 Draw Venn diagrams to verify DeMorgan's Laws. That is, for any two sets A and B, $\overline{(A \cup B)} = \overline{A} \cap \overline{B}$ and $\overline{(A \cap B)} = \overline{A} \cup \overline{B}$.

2.4 Suppose two dice are tossed and the numbers on the upper faces are observed. Let S denote the set of all possible pairs that can be observed. [These pairs can be listed, for example, by letting (2, 3) denote that a 2 was observed on the first die and a 3 on the second.] Define the following subsets of S:

   A: The number on the second die is even.

   B: The sum of the two numbers is even.

   C: At least one number in the pair is odd.

List the points in $A, \overline{C}, A \cap B, A \cap \overline{B}, \overline{A} \cup B$, and $\overline{A} \cap C$.

2.5 A group of five applicants for a pair of identical jobs consists of three men and two women. The employer is to select two of the five applicants for the jobs. Let S denote the set of all possible outcomes for the employer's selection. Let A denote the subset of outcomes corresponding to the selection of two men and B the subset corresponding to the selection of at least one woman. List the outcomes in $A, \overline{B}, A \cup B, A \cap B$, and $A \cap \overline{B}$. (Denote the different men and women by M_1, M_2, M_3 and W_1, W_2, respectively.)

2.6 From a survey of 60 students attending a university, it was found that 9 were living off campus, 36 were undergraduates, and 3 were undergraduates living off campus.

a Find the number of these students who were undergraduates, were living off campus, or both.

b Find the number of these students who were undergraduates living on campus.

c Find the number of these students who were graduate students living on campus.

2.4 A Probabilistic Model for an Experiment: The Discrete Case

In Section 2.2 we referred to the die-tossing *experiment* when we observed the number appearing on the upper face. We will use the term *experiment* to include observations obtained from completely uncontrollable situations (such as observations on the daily price of a particular stock) as well as those made under controlled laboratory conditions. We have the following definition:

DEFINITION **2.1**   An *experiment* is the process by which an observation is made.

Examples of experiments include coin and die tossing, measuring the IQ score of an individual, or determining the number of bacteria per cubic centimeter in a portion of processed food.

When an experiment is performed, it can result in one or more outcomes, which are called *events*. In our discussions, events will be denoted by capital letters. If the experiment consists of counting the number of bacteria in a portion of food, some events of interest could be:

A: Exactly 110 bacteria are present.

B: More than 200 bacteria are present.

C: The number of bacteria present is between 100 and 300.

Some events associated with a single toss of a balanced die are these:

A: Observe an odd number.

B: Observe a number less than 5.

C: Observe a 2 or a 3.

E_1: Observe a 1.

E_2: Observe a 2.

E_3: Observe a 3.

E_4: Observe a 4.

E_5: Observe a 5.

E_6: Observe a 6.

You can see that there is a distinct difference among some of the events associated with the die-tossing experiment. For example, if you observe event A (an odd number), at the same time you will have observed E_1, E_3, or E_5. Thus event A, which can be decomposed into three other events, is called a *compound event*. In contrast, the events E_1, E_2, E_3, E_4, E_5, and E_6 cannot be decomposed and are called *simple events*. A simple event can happen only in one way, whereas a compound event can happen in more than one distinct way.

Certain concepts from set theory are useful for expressing the relationships between various events associated with an experiment. Because sets are collections of points, we associate a distinct point, called a *sample point*, with each and every simple event associated with an experiment.

DEFINITION 2.2   A *simple event* is an event that cannot be decomposed. Each simple event corresponds to one and only one *sample point*. The letter E with a subscript will be used to denote a simple event or the corresponding sample point.

Thus we can think of a simple event as a set consisting of a single point—namely, the single sample point associated with the event.

DEFINITION **2.3**

The *sample space* associated with an experiment is the set consisting of all possible sample points. A sample space will be denoted by S.

We can easily see that the sample space S associated with the die-tossing experiment consists of six sample points corresponding to the six simple events $E_1, E_2, E_3, E_4, E_5,$ and E_6. That is, $S = \{E_1, E_2, E_3, E_4, E_5, E_6\}$. A Venn diagram exhibiting the sample space for the die-tossing experiment is given in Figure 2.7.

For the microbiology example of counting bacteria in a food specimen, let E_0 correspond to observing 0 bacteria, E_1 correspond to observing 1 bacterium, and so on. Then the sample space is

$$S = \{E_0, E_1, E_2, \ldots\}$$

because no integer number of bacteria can be ruled out as a possible outcome.

Both sample spaces we examined have the property that they consist of either a finite or a countable number of sample points. In the die-tossing example there are six (a finite number) sample points. The number of sample points associated with the bacteria-counting experiment is infinite, but the number of distinct sample points can be put into a one-to-one correspondence with the integers (that is, the number of sample points is countable). Such sample spaces are said to be discrete.

DEFINITION **2.4**

A *discrete sample space* is one that contains either a finite or a countable number of distinct sample points.

When an experiment is conducted a single time, you will observe one and only one simple event. For example, if you toss a die and observe a 1, you cannot at the same time observe a 2. Thus the single sample point E_1 associated with observing a 1 and the single sample point E_2 associated with observing a 2 are distinct, and the sets $\{E_1\}$ and $\{E_2\}$ are mutually exclusive sets. Thus events E_1 and E_2 are mutually exclusive events. Similarly, all distinct simple events correspond to mutually exclusive sets of simple events and are thus mutually exclusive events.

FIGURE **2.7**
Venn diagram for the sample space associated with the die-tossing experiment

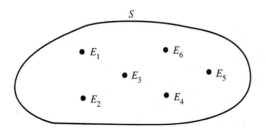

For experiments with discrete sample spaces, *compound events* can be viewed as collections (sets) of sample points or, equivalently, as unions of the sets of single sample points corresponding to the appropriate simple events. For example, the die tossing event A (observe an odd number) will occur if and only if one of the simple events E_1, E_3, or E_5 occurs. Thus

$$A = \{E_1, E_3, E_5\} \quad \text{or} \quad A = E_1 \cup E_3 \cup E_5.$$

Similarly, B (observe a number less than 5) can be written as

$$B = \{E_1, E_2, E_3, E_4\} \quad \text{or} \quad B = E_1 \cup E_2 \cup E_3 \cup E_4.$$

The rule for determining which simple events to include in a compound event is very precise. *A simple event E_i is included in event A if and only if A occurs whenever E_i occurs.*

DEFINITION 2.5

An *event* in a discrete sample space S is a collection of sample points—that is, any subset of S.

Figure 2.8 gives a Venn diagram representing the sample space and events A (observe an odd number) and B (observe a number less than 5) for the die-tossing experiment. Notice that it is easy to visualize the relationship between events by using a Venn diagram.

By Definition 2.5, any event in a discrete sample space S is a subset of S. In the example concerning counting bacteria in a portion of food, the event B (the number of bacteria is more than 200) can be expressed as

$$B = \{E_{201}, E_{202}, E_{203}, \ldots\}$$

where E_i denotes the simple event that there are i bacteria present in the food sample and $i = 0, 1, 2, \ldots$.

A probabilistic model for an experiment with a discrete sample space can be constructed by assigning a numerical probability to each simple event in the sample space S. We will select this number, a measure of our belief in the event's occurrence on a single repetition of the experiment, in such a way that it will be consistent

FIGURE 2.8
Venn diagram for the
die-tossing
experiment

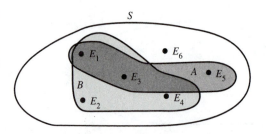

with the relative frequency concept of probability. Although relative frequency does not provide a rigorous definition of probability, any definition applicable to the real world should agree with our intuitive notion of the relative frequencies of events.

On analyzing the relative frequency concept of probability, we see that three conditions must hold.

1. The relative frequency of occurrence of any event must be greater than or equal to zero. A negative relative frequency does not make sense.

2. The relative frequency of the whole sample space S must be unity. Because every possible outcome of the experiment is a point in S, it follows that S must occur every time the experiment is performed.

3. If two events are mutually exclusive, the relative frequency of their union is the sum of their respective relative frequencies. (For example, if the experiment of tossing a balanced die yields a 1 on $1/6$ of the tosses, it should yield a 1 or a 2 on $1/6 + 1/6 = 1/3$ of the tosses.)

These three conditions form the basis of the following definition of probability.

DEFINITION **2.6**

Suppose S is a sample space associated with an experiment. To every event A in S (A is a subset of S) we assign a number, $P(A)$, called the *probability* of A, so that the following axioms hold:

Axiom 1: $P(A) \geq 0$.

Axiom 2: $P(S) = 1$.

Axiom 3: If A_1, A_2, A_3, ... form a sequence of pairwise mutually exclusive events in S (that is, $A_i \cap A_j = \emptyset$ if $i \neq j$), then

$$P(A_1 \cup A_2 \cup A_3 \cup \cdots) = \sum_{i=1}^{\infty} P(A_i).$$

We can easily show that Axiom 3, which is stated in terms of an infinite sequence of events, implies a similar property for a finite sequence. Specifically, if $A_1, A_2, \ldots, A_n$ are pairwise mutually exclusive events, then

$$P(A_1 \cup A_2 \cup A_3 \cup \cdots \cup A_n) = \sum_{i=1}^{n} P(A_i).$$

Notice that the definition states only the conditions an assignment of probabilities must satisfy; it does not tell us how to assign specific probabilities to events. For example, suppose that a coin has yielded 800 heads in 1000 previous tosses. Consider the experiment of one more toss of the same coin. There are two possible outcomes, head or tail, and hence two simple events. The definition of probability allows us to assign to these simple events any two nonnegative numbers that add to 1. For example, each simple event could have the probability $1/2$. In light of the past history of this coin, however, it might be more reasonable to assign a probability nearer .8 to the outcome involving a head. Specific assignments of probabilities must be consistent with reality if the probabilistic model is to serve a useful purpose.

For discrete sample spaces it suffices to assign probabilities to each simple event. If a balanced die is used for the die-tossing example, it seems reasonable to assume that all simple events would have the same relative frequency in the long run. We will assign a probability of $1/6$ to each simple event: $P(E_i) = 1/6$, for $i = 1, 2, \ldots, 6$. This assignment of probabilities agrees with Axiom 1. To see that Axiom 2 is satisfied, write

$$P(S) = P(E_1 \cup E_2 \cup \cdots \cup E_6) = P(E_1) + P(E_2) + \cdots + P(E_6) = 1.$$

The second equality follows because Axiom 3 must hold. Axiom 3 also tells us that we can calculate the probability of any event by summing the probabilities of the simple events contained in that event (recall that distinct simple events are mutually exclusive). Event A was defined to be "observe an odd number." Hence

$$P(A) = P(E_1 \cup E_3 \cup E_5) = P(E_1) + P(E_3) + P(E_5) = 1/2.$$

EXAMPLE **2.1**   A manufacturer has five seemingly identical computer terminals available for shipping. Unknown to her, two of the five are defective. A particular order calls for two of the terminals and is filled by randomly selecting two of the five that are available.

a   List the sample space for this experiment.

b   Let A denote the event that the order is filled with two nondefective terminals. List the sample points in A.

c   Construct a Venn diagram for the experiment that illustrates event A.

d   Assign probabilities to the simple events in such a way that the information about the experiment is used and the axioms in Definition 2.6 are met.

e   Find the probability of event A.

Solution   a   Let the two defective terminals be labeled D_1 and D_2, and let the three good terminals be labeled G_1, G_2, and G_3. Any single sample point will consist of a list of the two terminals selected for shipment. The simple events may be denoted by

$$
\begin{array}{llll}
E_1 = \{D_1,\ D_2\} & E_5 = \{D_2,\ G_1\} & E_8 = \{G_1,\ G_2\} & E_{10} = \{G_2,\ G_3\} \\
E_2 = \{D_1,\ G_1\} & E_6 = \{D_2,\ G_2\} & E_9 = \{G_1,\ G_3\} & \\
E_3 = \{D_1,\ G_2\} & E_7 = \{D_2,\ G_3\} & & \\
E_4 = \{D_1,\ G_3\} & & &
\end{array}
$$

Thus there are ten sample points in S, and $S = \{E_1, E_2, \ldots, E_{10}\}$

b   Event $A = \{E_8, E_9, E_{10}\}$.

c

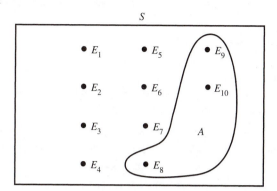

d Because the terminals are selected at random, any pair of terminals is as likely to be selected as any other pair. Thus, $P(E_i) = 1/10$, for $i = 1, 2, \ldots, 10$, is a reasonable assignment of probabilities.

e Because $A = E_8 \cup E_9 \cup E_{10}$, Axiom 3 implies that

$$P(A) = P(E_8) + P(E_9) + P(E_{10}) = 3/10.$$

The next section contains an axiomatic description of the method for calculating $P(A)$ that we just used.

Before we proceed, let us note that there are experiments for which the sample space is not countable and hence is not discrete. Suppose, for example, that the experiment consists of measuring the blood glucose level of a diabetic patient. The sample space for this experiment would contain an interval of real numbers, and any such interval contains an uncountable number of values. Thus the sample space is not discrete. Situations like the latter will be discussed in Chapter 4. The remainder of this chapter is devoted to developing methods for calculating the probabilities of events defined on discrete sample spaces.

Exercises

2.7 Every person's blood type is A, B, AB or O. In addition, each individual either has the Rhesus (Rh) factor (+) or does not (−). A medical technician records a person's blood type and Rh factor. List the sample space for this experiment.

2.8 The proportions of blood phenotypes, A, B, AB, and O, in the population of all Caucasians in the United States are approximately .41, .10, .04, and .45, respectively. A single Caucasian is chosen at random from the population.

a List the sample space for this experiment.

b Make use of the information given above to assign probabilities to each of the simple events.

c What is the probability that the person chosen at random has either type A or type AB blood?

2.9 A sample space consists of five simple events, E_1, E_2, E_3, E_4, and E_5.

a If $P(E_1) = P(E_2) = 0.15$, $P(E_3) = 0.4$, and $P(E_4) = 2P(E_5)$, find the probabilities of E_4 and E_5.

b If $P(E_1) = 3P(E_2) = 0.3$, find the probabilities of the remaining simple events if you know that the remaining events are equally probable.

2.10 A vehicle arriving at an intersection can turn right, turn left, or continue straight ahead. The experiment consists of observing the movement of a single vehicle through the intersection.

a List the sample space for this experiment.

b Assuming that all sample points are equally likely, find the probability that the vehicle turns.

2.11 Americans can be quite suspicious, especially when it comes to government conspiracies. On the question of whether the U.S. Air Force has withheld proof of the existence of intelligent life on other planets, the proportions of Americans with varying opinions are given in the table.

Opinion	Proportion
Very likely	.24
Somewhat likely	.24
Unlikely	.40
Other	.12

Suppose that one American is selected and his or her opinion is recorded.

a What are the simple events for this experiment?

b Are the simple events that you gave in part (a) all equally likely? If not, what are the probabilities that should be assigned to each?

c What is the probability that the person selected finds it at least somewhat likely that the Air Force is withholding information about intelligent life on other planets?

2.12 A survey classified a large number of adults according to whether they were diagnosed as needing eyeglasses to correct their reading vision and whether they use eyeglasses when reading. The proportions falling into the four resulting categories are given in the table.

Needs Glasses	Uses Eyeglasses for Reading	
	Yes	No
Yes	.44	.14
No	.02	.40

If a single adult is selected from the large group, find the probabilities of the events defined below.

a The adult needs glasses.

b The adult needs glasses but does not use them.

c The adult uses glasses whether the glasses are needed or not.

2.13 An oil prospecting firm hits oil or gas on 10% of its drillings. If the firm drills two wells, the four possible simple events and three of their associated probabilities are as given in the accompanying table.

Simple Event	Outcome of First Drilling	Outcome of Second Drilling	Probability
E_1	Hit (oil or gas)	Hit (oil or gas)	.01
E_2	Hit	Miss	?
E_3	Miss	Hit	.09
E_4	Miss	Miss	.81

a Find the probability that the company will hit oil or gas on the first drilling and miss on the second.

b Find the probability that the company will hit oil or gas on at least one of the two drillings.

2.14 Of the volunteers coming into a blood center, 1 in 3 have O^+ blood, 1 in 15 have O^-, 1 in 3 have A^+, and 1 in 16 have A^-. The name of one person who previously has donated blood is selected from the records of the center. What is the probability that the person selected has

a O^+ blood?

b type O blood?

c type A blood?

d neither type A nor type O blood?

2.15 Hydraulic landing assemblies coming from an aircraft rework facility are each inspected for defects. Historical records indicate that 8% have defects in shafts only, 6% have defects in bushings only, and 2% have defects in both shafts and bushings. One of the hydraulic assemblies is selected randomly. What is the probability that the assembly has

a a bushing defect?

b a shaft or bushing defect?

c exactly one of the two types of defects?

d neither type of defect?

2.16 Suppose two balanced coins are tossed and the upper faces are observed.

a List the sample points for this experiment.

b Assign a reasonable probability to each sample point. (Are the sample points equally likely?)

c Let A denote the event that *exactly* one head is observed and B the event that *at least* one head is observed. List the sample points in both A and B.

d From your answer to (c) find $P(A)$, $P(B)$, $P(A \cap B)$, $P(A \cup B)$, and $P(\overline{A} \cup B)$.

2.17 A business office orders paper supplies from one of three vendors, V_1, V_2, or V_3. Orders are to be placed on two successive days, one order per day. Thus (V_2, V_3) might denote that vendor V_2 gets the order on the first day and vendor V_3 gets the order on the second day.

a List the sample points in this experiment of ordering paper on two successive days.

b Assume the vendors are selected at random each day and assign a probability to each sample point.

c Let A denote the event that the same vendor gets both orders and B the event that V_2 gets at least one order. Find $P(A)$, $P(B)$, $P(A \cup B)$, and $P(A \cap B)$ by summing the probabilities of the sample points in these events.

2.5 Calculating the Probability of an Event: The Sample-Point Method

Finding the probability of an event defined on a sample space that contains a finite or denumerable (countably infinite) set of sample points can be approached in two ways, the *sample-point* and the *event-composition* methods. Both methods use the sample space model, but they differ in the *sequence* of steps necessary to obtain a solution and in the tools that are used. Separation of the two procedures may not be palatable to the unity-seeking theorist, but it can be extremely useful to a beginner attempting to find the probability of an event. In this section we consider the sample-point method. The event-composition method requires additional results and will be presented in Section 2.9.

> The sample-point method is outlined in Section 2.4. The following steps are used to find the probability of an event:
>
> 1. Define the experiment and clearly determine how to describe one simple event.
> 2. List the simple events associated with the experiment and test each to make certain that it cannot be decomposed. This defines the sample space S.
> 3. Assign reasonable probabilities to the sample points in S, making certain that $P(E_i) \geq 0$ and $\sum P(E_i) = 1$.
> 4. Define the event of interest, A, as a specific collection of sample points. (A sample point is in A if A occurs when the sample point occurs. Test *all* sample points in S to identify those in A.)
> 5. Find $P(A)$ by summing the probabilities of the sample points in A.

We will illustrate these steps with three examples.

EXAMPLE 2.2   Consider the problem of selecting two applicants for a job out of a group of five and imagine that the applicants vary in competence, 1 being the best, 2 second best, and so on, for 3, 4, and 5. These ratings are, of course, unknown to the employer. Define two events A and B as

A: The employer selects the best and one of the two poorest applicants (applicants 1 and 4 or 1 and 5).

B: The employer selects at least one of the two best.

Find the probabilities of these events.

Solution   The steps are as follows:

1. The experiment involves randomly selecting two applicants out of five. Denote the selection of applicants 3 and 5 by {3, 5}.

2. The ten simple events, with {i, j} denoting the selection of applicants i and j, are

E_1: {1, 2} E_5: {2, 3} E_8: {3, 4} E_{10}: {4, 5}.
E_2: {1, 3} E_6: {2, 4} E_9: {3, 5}
E_3: {1, 4} E_7: {2, 5}
E_4: {1, 5}

3. A random selection of two out of five gives each pair an equal chance for selection. Hence we will assign each sample point the probability 1/10. That is,

$$P(E_i) = 1/10 = .1, \quad i = 1, 2, \ldots, 10.$$

4. Checking the sample points, we see that B occurs whenever E_1, E_2, E_3, E_4, E_5, E_6, or E_7 occurs. Hence these sample points are included in B.

5. Finally, $P(B)$ is equal to the sum of the probabilities of the sample points in B, or

$$P(B) = \sum_{i=1}^{7} P(E_i) = \sum_{i=1}^{7} .1 = .7.$$

Similarly, we see that event $A = E_3 \cup E_4$ and that $P(A) = .1 + .1 = .2$.

The solution of this and similar problems would be of importance to a company personnel director.

EXAMPLE 2.3   A balanced coin is tossed three times. Calculate the probability that exactly two of the three tosses result in heads.

Solution   The five steps of the sample-point method are as follows:

1. The experiment consists of observing the outcomes (heads or tails) for each of three tosses of a coin. A simple event for this experiment can be symbolized by a three-letter sequence of H's and T's, representing heads and tails, respectively. The first letter in the sequence represents the observation on the first coin. The second letter represents the observation on the second coin, and so on.

2. The eight simple events in S are

E_1: HHH E_3: HTH E_5: HTT E_7: TTH
E_2: HHT E_4: THH E_6: THT E_8: TTT.

3. Because the coin is balanced, we would expect the simple events to be equally likely; that is,

$$P(E_i) = 1/8, \quad i = 1, 2, \ldots, 8.$$

4. The event of interest, A, is the event that exactly two of the tosses result in heads. An examination of the sample points will verify that

$$A = \{E_2, \ E_3, \ E_4\}.$$

5. Finally,

$$P(A) = P(E_2) + P(E_3) + P(E_4) = 1/8 + 1/8 + 1/8 = 3/8.$$

Although the sample points in the sample spaces associated with Examples 2.2 and 2.3 are equally likely, it is important to realize that sample points need not be equally likely. An example to illustrate this point follows.

EXAMPLE 2.4   The odds are two to one that, when A and B play tennis, A wins. Suppose that A and B play two matches. What is the probability that A wins at least one match?

Solution   1. The experiment consists of observing the winner (A or B) for each of two matches. Let AB denotes the event that player A wins the first match and player B wins the second.

2. The sample space for the experiment consists of four sample points:

$$E_1: \ AA \quad E_2: \ AB \quad E_3: \ BA \quad E_4: \ BB$$

3. Because A has a better chance of winning any match, it does not seem appropriate to assign equal probabilities to these sample points. As you will see in Section 2.9, under certain conditions it is reasonable to make the following assignment of probabilities:

$$P(E_1) = 4/9, \quad P(E_2) = 2/9, \quad P(E_3) = 2/9, \quad P(E_4) = 1/9.$$

Notice that, even though the probabilities assigned to the simple events are not all equal, $P(E_i) \geq 0$ for $i = 1, 2, 3, 4$ and $\sum_S P(E_i) = 1$.

4. The event of interest is that A wins at least one game. Thus, if we denote the event of interest as C, it is easily seen that

$$C = E_1 \cup E_2 \cup E_3.$$

5. Finally,

$$P(C) = P(E_1) + P(E_2) + P(E_3) = 4/9 + 2/9 + 2/9 = 8/9.$$

The sample-point method for solving a probability problem is direct and powerful and in some respects is a bulldozer approach. It can be applied to find the probability of any event defined over a sample space containing a finite or countable set of sample points, but it is not resistant to human error. Common errors include incorrectly diagnosing the nature of a simple event and failing to list all the sample points in S. A second complication occurs because many sample spaces contain a very large number of sample points, and a complete itemization is tedious and time consuming and might be practically impossible.

Fortunately, many sample spaces generated by experimental data contain subsets of sample points that are equiprobable. (The sample spaces for Examples 2.1, 2.2, and 2.3 possess this property.) When this occurs, we need not list the points but may simply count the number in each subset. If such counting methods are inapplicable, an orderly method should be use to list the sample points (notice the listing schemes for Examples 2.1, 2.2, and 2.3). The listing of large numbers of sample points can be accomplished by using a computer.

Tools that reduce the effort and error associated with the sample-point approach for finding the probability of an event include orderliness, a computer, and the mathematical theory of counting, called *combinatorial analysis*. Computer programming and applications form a topic for separate study. The mathematical theory of combinatorial analysis is also a broad subject, but some quite useful results can be given succinctly. Hence our next topic concerns some elementary results in combinatorial analysis and their application to the sample-point approach for the solution of probability problems.

Exercises

2.18 According to *Webster's New Collegiate Dictionary*, a divining rod is "a forked rod believed to indicate [divine] the presence of water or minerals by dipping downward when held over a vein." To test the claims of a divining rod expert, skeptics bury four cans in the ground, two empty and two filled with water. The expert is led to the four cans and told that two contain water. He uses the divining rod to test each of the four cans and decide which two contain water.

    a List the sample space for this experiment.

    b If the divining rod is completely useless for locating water, what is the probability that the expert will correctly identify (by guessing) both of the cans containing water?

2.19 In Exercise 2.10 we considered a situation where cars entering an intersection each could turn right, turn left, or go straight. An experiment consists of observing two vehicles moving through the intersection.

    a How many sample points are there in the sample space? List them.

    b Assuming that all sample points are equally likely, what is the probability that at least one car turns left?

    c Again assuming equally likely sample points, what is the probability that at most one vehicle turns?

2.20   Four equally qualified people apply for two identical positions in a company. One and only one applicant is a member of a minority group. The positions are filled by choosing two of the applicants at random.

a   List the possible outcomes for this experiment.

b   Assign reasonable probabilities to the sample points.

c   Find the probability that the applicant from the minority group is selected for a position.

2.21   Two additional jurors are needed to complete a jury for a criminal trial. There are six prospective jurors, two women and four men. Two jurors are randomly selected from the six available.

a   Define the experiment and describe one sample point. Assume that you need describe only the two jurors chosen and not the order in which they were selected.

b   List the sample space associated with this experiment.

c   What is the probability that both of the jurors selected are women?

2.22   Three imported wines are to be ranked from lowest to highest by a purported wine expert. That is, one wine will be identified as best, another as second best, and the remaining wine as worst.

a   Describe one sample point for this experiment.

b   List the sample space.

c   Assume that the "expert" really knows nothing about wine and randomly assigns ranks to the three wines. One of the wines is of much better quality than the others. What is the probability that the expert ranks the best wine no worse than second best?

2.23   A boxcar contains six complex electronic systems. Two of the six are to be randomly selected for thorough testing and then classified as defective or not defective.

a   If two of the six systems are actually defective, find the probability that at least one of the two systems tested will be defective. Find the probability that both are defective.

b   If four of the six systems are actually defective, find the probabilities indicated in (a).

2.24   A retailer sells only two styles of stereo consoles, and experience shows that these are in equal demand. Four customers in succession come into the store to order stereos. The retailer is interested in their preferences.

a   List the possibilities for preference arrangements among the four customers (that is, list the sample space).

b   Assign probabilities to the sample points.

c   Let A denote the event that all four customers prefer the same style. Find $P(A)$.

2.25   The Bureau of the Census reports that the median family income for all families in the United States during the year 1991 was $35,353. That is, half of all American families had incomes exceeding this amount and half had incomes equal to or below this amount (Wright 1992, 242). Suppose that four families are surveyed and that each one reveals whether its income exceeded $35,353 in 1991.

a   List the points in the sample space.

b   Identify the simple events in each of the following events:

  A: At least two had incomes exceeding $35,353.

  B: Exactly two had incomes exceeding $35,353.

  C: Exactly one had income less than or equal to $35,353.

c   Make use of the given interpretation for the median to assign probabilities to the simple events and find $P(A)$, $P(B)$, and $P(C)$.

2.26   Patients arriving at a hospital outpatient clinic can select one of three stations for service. Suppose that physicians are assigned randomly to the stations and that the patients therefore have no station preference. Three patients arrive at the clinic and their selection of stations is observed.

a   List the sample points for the experiment.

b   Let A be the event that each station receives a patient. List the sample points in A.

c   Make a reasonable assignment of probabilities to the sample points and find $P(A)$.

2.6   Tools for Counting Sample Points

This section presents some useful results from the theory of combinatorial analysis and illustrates their application to the sample-point method for finding the probability of an event. In many cases, these results enable you to count the total number of sample points in the sample space S and in an event of interest, thereby providing a confirmation of your listing of simple events. When the number of simple events in a sample space is very large and manual enumeration of every sample point is tedious or even impossible, counting the number of points in the sample space and in the event of interest may be the only efficient way to calculate the probability of an event. Indeed, if a sample space contains N equiprobable sample points and an event A contains exactly n_a sample points, it is easily seen that $P(A) = n_a/N$.

The first result from combinatorial analysis that we present, often called the *mn rule*, is stated as follows:

THEOREM **2.1**   With m elements a_1, a_2, $\ldots$, a_m and n elements b_1, b_2, $\ldots$, b_n it is possible to form $mn = m \times n$ pairs containing one element from each group.

Proof   Verification of the theorem can be seen by observing the rectangular table in Figure 2.9. There is one square in the table for each a_i, b_j pair and hence a total of $m \times n$ squares.   ∎

FIGURE **2.9**
Table indicating the number of pairs (a_i, b_j)

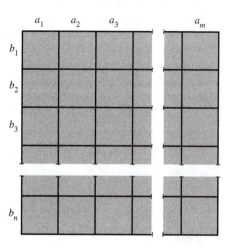

The *mn* rule can be extended to any number of sets. Given three sets of elements, $a_1, a_2, \ldots, a_m, b_1, b_2, \ldots, b_n$, and $c_1, c_2, \ldots, c_p$, the number of distinct triplets containing one element from each set is equal to *mnp*. The proof of the theorem for three sets involves two applications of Theorem 2.1. We think of the first set as an (a_i, b_j) pair and unite each of these pairs with elements of the third set, $c_1, c_2, \ldots, c_p$. Theorem 2.1 implies that there are *mn* pairs (a_i, b_j). Because there are *p* elements $c_1, c_2, \ldots, c_p$, another application of Theorem 2.1 implies that there are $(mn)(p) = mnp$ triplets $a_i b_j c_k$.

EXAMPLE 2.5   An experiment involves tossing a pair of dice and observing the numbers on the upper faces. Find the number of sample points in *S*, the sample space for the experiment.

Solution   A sample point for this experiment can be represented symbolically as an ordered pair of numbers representing the outcomes on the first and second die, respectively. Thus (4, 5) denotes the event that the uppermost face on the first die was a 4 and on the second die, a 5. The sample space *S* consists of the set of all possible pairs (x, y), where *x* and *y* are both integers between 1 and 6.

The first die can result in one of six numbers. These represent $a_1, a_2, \ldots, a_6$. Likewise, the second die can fall in one of six ways, and these correspond to $b_1, b_2, \ldots, b_6$. Then $m = n = 6$ and the total number of sample points in *S* is $mn = (6)(6) = 36$.

EXAMPLE 2.6   Refer to the coin-tossing experiment in Example 2.3. We found for this example that the total number of sample points was eight. Use the extension of the *mn* rule to confirm this result.

Solution   Each sample point in *S* was identified by a sequence of three letters, where each position in the sequence contained one of two letters, an *H* or a *T*. The problem therefore involves the formation of triples, with an element (an *H* or a *T*) from each of three sets. For this example the sets are identical and all contain two elements (*H* and *T*). Thus the number of elements in each set is $m = n = p = 2$, and the total number of triples that can be formed is $mnp = (2)^3 = 8$.

EXAMPLE 2.7   Consider an experiment that consists of recording the birthday for each of 20 randomly selected persons. Ignoring leap years and assuming that there are only 365 possible distinct birthdays, find the number of points in the sample space *S* for this experiment. If we assume that each of the possible sets of birthdays is equiprobable, what is the probability that each person in the 20 has a different birthday?

Solution   Number the days of the year 1, 2, ..., 365. A sample point for this experiment can be represented by an ordered sequence of 20 numbers, where the first number denotes the number of the day that is the first person's birthday, the second number denotes the number of the day that is the second person's birthday, and so on. We are concerned with the number of twenty-tuples that can be formed, selecting a number representing one of the 365 days in the year from each of 20 sets. The sets are all identical, and each contains 365 elements. Repeated applications of the *mn* rule tell us there are $(365)^{20}$ such twenty-tuples. Thus the sample space S contains $N = (365)^{20}$ sample points. Although we could not feasibly list all the sample points, if we assume them to be equiprobable, $P(E_i) = 1/(365)^{20}$ for each simple event.

If we denote the event that each person has a different birthday by A, the probability of A can be calculated if we can determine n_a, the number of sample points in A. A sample point is in A if the corresponding 20-tuple is such that no two positions contain the same number. Thus the set of numbers from which the first element in a 20-tuple in A can be selected contains 365 numbers, the set from which the second element can be selected contains 364 numbers (all but the one selected for the first element), the set from which the third can be selected contains 363 (all but the two selected for the first two elements), ..., and the set from which the twentieth element can be selected contains 346 elements (all but those selected for the first 19 elements). An extension of the *mn* rule yields

$$n_a = (365) \times (364) \times \cdots \times (346).$$

Finally, we may determine that

$$P(A) = \frac{n_a}{N} = \frac{365 \times 364 \times \cdots \times 346}{(365)^{20}} = .5886.$$

Notice that for Examples 2.5 and 2.6 the numbers of sample points in the respective sample spaces are both relatively small and that listings for these sample spaces could easily be written down. For instances like these, the *mn* rule provides a simple method to verify that the sample spaces contain the correct number of points. In contrast, it is not feasible to list the sample space in Example 2.7. However, the *mn* rule can be used to count the number of sample points in S and in the event of interest, permitting calculation of the probability of the event.

We have seen that the sample points associated with an experiment often can be represented symbolically as a sequence of numbers or symbols. In some instances it will be clear that the total number of sample points equals the number of distinct ways that the respective symbols can be arranged in sequence. The following theorem can be used to determine the number of ordered arrangements that can be formed.

DEFINITION **2.7**

An ordered arrangement of r distinct objects is called a *permutation*. The number of ways of ordering n distinct objects taken r at a time will be designated by the symbol P_r^n.

THEOREM **2.2**

$$P_r^n = n(n - 1)(n - 2) \cdots (n - r + 1) = \frac{n!}{(n - r)!}.$$

Proof   We are concerned with the number of ways of filling r positions with n distinct objects. Applying the extension of the mn rule, we see that the first object can be chosen in one of n ways. After the first is chosen, the second can be chosen in $(n-1)$ ways, the third in $(n-2)$, and the rth in $(n-r+1)$ ways. Hence the total number of distinct arrangements is

$$P_r^n = n(n-1)(n-2)\cdots(n-r+1).$$

Expressed in terms of factorials,

$$P_r^n = n(n-1)(n-2)\cdots(n-r+1)\frac{(n-r)!}{(n-r)!} = \frac{n!}{(n-r)!}$$

where $n! = n(n-1)\cdots(2)(1)$ and $0! = 1$. ∎

EXAMPLE 2.8   The names of 3 employees are to be randomly drawn, without replacement, from a bowl containing the names of 30 employees of a small company. The person whose name is drawn first receives \$100, and the individuals whose names are drawn second and third receive \$50 and \$25, respectively. How many sample points are associated with this experiment?

Solution   Because the prizes awarded are different, the number of sample points is the number of ordered arrangements of $r = 3$ out of the possible $n = 30$ names. Thus, the number of sample points in S is

$$P_3^{30} = \frac{30!}{27!} = (30)(29)(28) = 24{,}360.$$

EXAMPLE 2.9   Suppose that an assembly operation in a manufacturing plant involves four steps, which can be performed in any sequence. If the manufacturer wishes to compare the assembly time for each of the sequences, how many different sequences will be involved in the experiment?

Solution   The total number of sequences equals the number of ways of arranging the $n = 4$ steps taken $r = 4$ at a time, or

$$P_4^4 = \frac{4!}{(4-4)!} = \frac{4!}{0!} = 24.$$

The next result from combinatorial analysis can be used to determine the number of subsets of various sizes that can be formed by partitioning a set of n distinct objects into k nonoverlapping groups.

THEOREM 2.3

The number of ways of partitioning n distinct objects into k distinct groups containing $n_1, n_2, \ldots, n_k$ objects, respectively, where each object appears in exactly one group and $\sum_{i=1}^{k} n_i = n$, is

$$N = \binom{n}{n_1\, n_2\, \cdots\, n_k} = \frac{n!}{n_1!\, n_2!\, \cdots\, n_k!}.$$

Proof

N is the number of distinct arrangements of n objects in a row for a case in which rearrangement of the objects within a group does not count. For example, the letters a to l are arranged in three groups, where $n_1 = 3, n_2 = 4$, and $n_3 = 5$:

$$abc|defg|hijkl$$

is one such arrangement.

The number of distinct arrangements of the n objects, assuming all objects are distinct, is $P_n^n = n!$ (from Theorem 2.2). Then P_n^n equals the number of ways of partitioning the n objects into k groups (ignoring order within groups) multiplied by the number of ways of ordering the $n_1, n_2, \ldots, n_k$ elements within each group. This application of the extended mn rule gives

$$P_n^n = (N) \cdot (n_1!\, n_2!\, n_3! \cdots n_k!)$$

where $n_i!$ is the number of distinct arrangements of the n_i objects in group i.

Solving for N, we have

$$N = \frac{n!}{n_1!\, n_2! \cdots n_k!} \equiv \binom{n}{n_1\, n_2\, \cdots\, n_k}.$$

The terms $\binom{n}{n_1\, n_2\, \cdots\, n_k}$ are often called *multinomial coefficients* because they occur in the expansion of the *multinomial term* $y_1 + y_2 + \cdots + y_k$ raised to the nth power:

$$(y_1 + y_2 + \cdots + y_k)^n = \sum \binom{n}{n_1\, n_2\, \cdots\, n_k} y_1^{n_1} y_2^{n_2} \cdots y_k^{n_k}$$

where this sum is taken over all $n_i = 0, 1, \ldots, n$ such that $n_1 + n_2 + \cdots + n_k = n$.

EXAMPLE 2.10

A labor dispute has arisen concerning the distribution of 20 laborers to four different construction jobs. The first job (considered to be very undesirable) required 6 laborers; the second, third, and fourth utilized 4, 5, and 5 laborers, respectively. The dispute arose over an alleged random distribution of the laborers to the jobs which placed all four members of a particular ethnic group on job 1. In considering whether the assignment represented injustice, a mediation panel desired the probability of the observed event. Determine the number of sample points in the sample space S for this experiment. That is, determine the number of ways the 20 laborers can be divided into groups of the appropriate sizes to fill all of the jobs. Find the probability of the observed event if it is assumed that the laborers are randomly assigned to jobs.

Solution

The number of ways of assigning the 20 laborers to the four jobs is equal to the number of ways of partitioning the 20 into four groups of sizes $n_1 = 6, n_2 = 4, n_3 = n_4 = 5$. Then

$$N = \binom{20}{6\ 4\ 5\ 5} = \frac{20!}{6!\,4!\,5!\,5!}.$$

By a *random assignment* of laborers to the jobs we mean that each of the N sample points has probability equal to $1/N$. If A denotes the event of interest and n_a the number of sample points in A, the sum of the probabilities of the sample points in A is $P(A) = n_a(1/N) = n_a/N$. The number of sample points in A, n_a, is the number of ways of assigning laborers to the four jobs with the four members of the ethnic group all going to job 1. The remaining sixteen laborers need to be assigned to the remaining jobs. Because there remain two openings for job 1, this can be done in

$$n_a = \binom{16}{2\ 4\ 5\ 5} = \frac{16!}{2!\,4!\,5!\,5!}$$

ways. It follows that

$$P(A) = \frac{n_a}{N} = 0.0031.$$

Thus if laborers are randomly assigned to jobs, the probability that the four members of the ethnic group all go to the undesirable job is very small. There is reason to doubt that the jobs were randomly assigned.

In many situations the sample points are identified by an array of symbols in which the arrangement of symbols is *unimportant*. The sample points for the selection of applicants, Example 2.2, imply a selection of two applicants out of five. Each sample point is identified as a pair of symbols, and the order of the symbols used to identify the sample points is irrelevant.

DEFINITION **2.8**

The number of *combinations* of n objects taken r at a time is the number of subsets, each of size r, that can be formed from the n objects. This number will be denoted by C_r^n or $\binom{n}{r}$.

THEOREM **2.4**

The number of unordered subsets of size r chosen (without replacement) from n available objects is

$$\binom{n}{r} = C_r^n = \frac{P_r^n}{r!} = \frac{n!}{r!(n-r)!}.$$

Proof

The selection of r objects from a total of n is equivalent to partitioning the n objects into $k = 2$ groups, the r selected and the $(n - r)$ remaining. This is a special case of the general partitioning problem dealt with in Theorem 2.3. In the present case, $k = 2$, $n_1 = r$, and $n_2 = (n - r)$ and therefore,

$$\binom{n}{r} = C_r^n = \binom{n}{r \quad n - r} = \frac{n!}{r!(n-r)!}.$$

∎

The terms $\binom{n}{r}$ are generally referred to as *binomial coefficients* because they occur in the *binomial expansion*

$$(x + y)^n = \binom{n}{0}x^n y^0 + \binom{n}{1}x^{n-1}y^1 + \binom{n}{2}x^{n-2}y^2 + \cdots + \binom{n}{n}x^0 y^n$$

$$= \sum_{i=0}^{n} \binom{n}{i}x^{n-i}y^i.$$

EXAMPLE **2.11**   Find the number of ways of selecting two applicants out of five and hence the total number of sample points in S for Example 2.2.

Solution

$$\binom{5}{2} = \frac{5!}{2!3!} = 10.$$

(Notice that this agrees with the number of sample points listed in Example 2.2.)

EXAMPLE **2.12**   Let A denote the event that exactly one of the two best applicants appears in a selection of two out of five. Find the number of sample points in A and $P(A)$.

Solution   Let n_a denote the number of sample points in A. Then n_a equals the number of ways of selecting one of the two best (call this number m) times the number of ways of selecting one of the three low-ranking applicants (call this number n). Then $m = \binom{2}{1}$, $n = \binom{3}{1}$, and applying the mn rule,

$$n_a = \binom{2}{1} \cdot \binom{3}{1} = \frac{2!}{1!1!} \cdot \frac{3!}{1!2!} = 6.$$

(This number can be verified by counting the sample points in A from the listing in Example 2.2.)

In Example 2.11 we found the total number of sample points in S to be $N = 10$. If each selection is equiprobable $P(E_i) = 1/10 = .1, i = 1, 2, \ldots, 10$, and

$$P(A) = \sum_{E_i \subset A} P(E_i) = \sum_{E_i \subset A} (.1) = n_a(.1) = 6(.1) = .6.$$

EXAMPLE **2.13** A company orders supplies from M distributors and wishes to place n orders ($n <$ M). Assume that the company places the orders in a manner that allows every distributor an equal chance of obtaining any one order and there is no restriction on the number of orders that can be placed with any distributor. Find the probability that a particular distributor, say distributor I, gets exactly k orders ($k \leq n$).

Solution Because any of the M distributors can be selected to receive any one of the orders, there are M ways that each order can be placed, and the number of different ways that the n orders can be placed is $M \cdot M \cdot M \cdots M = (M)^n$. Consequently, there are $(M)^n$ sample points in S. All these points are equally likely; hence $P(E_i) = 1/(M)^n$.

Let A denote the event that distributor I receives exactly k orders from among the n. The k orders assigned to distributor I can be chosen from the n in $\binom{n}{k}$ ways. It remains to determine the number of ways the remaining $(n - k)$ orders can be assigned to the other $M - 1$ distributors. Because each of these $(n - k)$ orders can go to any of the $(M - 1)$ distributors, this assignment can be made in $(M - 1)^{n-k}$ ways. Thus A contains

$$n_a = \binom{n}{k}(M - 1)^{n-k}$$

sample points, and because the sample points are equally likely,

$$P(A) = \sum_{E_i \subset A} P(E_i) = \sum_{E_i \subset A} \left(\frac{1}{M^n}\right) = n_a \left(\frac{1}{M^n}\right) = \frac{\binom{n}{k}(M - 1)^{n-k}}{M^n}.$$

Theorems 2.1 through 2.4 provide a few of the many useful counting rules found in the theory of combinatorial analysis. A few additional theorems appear in the exercises at the end of the chapter. If you are interested in extending your knowledge of combinatorial analysis, refer to one of the numerous texts on this subject.

We will next direct our attention to the concept of conditional probability. Conditional probability plays an important role in the event-composition approach for finding the probability of an event and is sometimes useful in finding the probabilities of sample points (for sample spaces with sample points that are not equally likely).

Exercises

2.27 An airline has six flights from New York to California and seven flights from California to Hawaii per day. If the flights are to be made on separate days, how many different flight arrangements can the airline offer from New York to Hawaii?

2.28 An assembly operation in a manufacturing plant requires three steps that can be performed in any sequence. How many different ways can the assembly be performed?

2.29 A businesswoman in Philadelphia is preparing an itinerary for a visit to six major cities. The distance traveled, and hence the cost of the trip, will depend on the order in which she plans her route.

  a How many different itineraries (and trip costs) are possible?

  *b If the businesswoman randomly selects one of the possible itineraries and Denver and San Francisco are two of the cities that she plans to visit, what is the probability that she will visit Denver before San Francisco?[1]

2.30 An upscale restaurant offers a special *fixe prix* menu in which, for a fixed dinner cost, a diner can select from four appetizers, three salads, four entrees and five desserts. How many different dinners are available if a dinner consists of one appetizer, one salad, one entree and one dessert?

2.31 An experiment consists of tossing a pair of dice.

  a Use the combinatorial theorems to determine the number of sample points in the sample space S.

  b Find the probability that the sum of the numbers appearing on the dice is equal to 7.

2.32 A brand of automobile comes in five different styles, with four types of engines, with two types of transmissions, and in eight colors.

  a How many autos would a dealer have to stock if he included one for each style-engine-transmission combination?

  b How many would a distribution center have to carry if all colors of cars were stocked for each combination in (a)?

2.33 How many different seven-digit telephone numbers can be formed if the first digit cannot be zero?

2.34 A personnel director for a corporation has hired ten new engineers. If three (distinctly different) positions are open at a Cleveland plant, in how many ways can she fill the positions?

2.35 A fleet of nine taxis is to be dispatched to three airports in such a way that three go to airport A, five go to airport B, and one goes to airport C. In how many distinct ways can this be accomplished?

2.36 Refer to Exercise 2.35. Assume that taxis are allocated to airports at random.

  a If exactly one of the taxis is in need of repair, what is the probability that it is dispatched to airport C?

  b If exactly three of the taxis are in need of repair, what is the probability that every airport receives one of the taxis requiring repairs?

2.37 Suppose that we wish to expand $(x + y + z)^{17}$. What is the coefficient of $x^2 y^5 z^{10}$?

2.38 If we wish to expand $(x + y)^8$, what is the coefficient of $x^5 y^3$? What is the coefficient of $x^3 y^5$?

2.39 Students attending the University of Florida can select from 130 major areas of study. A student's major is identified in the registrar's records with a two- or three-letter code (for example, statistics majors are identified by STA, math majors by MS). Some students opt for a double major and complete the requirements for both of the major areas before graduation. The registrar was asked to consider assigning these double majors a distinct two- or three-letter code so that they could be identified through the student records' system.

---

[1]Exercises preceded by an asterisk are optional.

    a What is the maximum number of possible double majors available to University of Florida students?

    b If any two- or three-letter code is available to identify majors or double majors, how many major codes are available?

    c How many major codes are required to identify students who have either a single major or a double major?

    d Are there enough major codes available to identify all single and double majors at the University of Florida?

2.40 Probability played a role in the rigging of the April 24, 1980, Pennsylvania state lottery (*Los Angeles Times*, September 8, 1980). To determine each digit of the three-digit winning number, each of the numbers 0, 1, 2, ..., 9 is placed on a ping-pong ball, the ten balls are blown into a compartment, and the number selected for the digit is the one on the ball that floats to the top of the machine. To alter the odds, the conspirators injected a liquid into all balls used in the game except those numbered 4 and 6, making it almost certain that the lighter balls would be selected and determine the digits in the winning number. Then they bought lottery tickets bearing the potential winning numbers. How many potential winning numbers were there (666 was the eventual winner)?

2.41 A local fraternity is conducting a raffle where 50 tickets are to be sold—one per customer. There are three prizes to be awarded. If the four organizers of the raffle each buy one ticket, what is the probability that the four organizers

    a win all of the prizes?

    b win exactly two of the prizes?

    c win exactly one of the prizes?

    d win none of the prizes?

2.42 An experimenter wishes to investigate the effect of three variables, pressure, temperature, and the type of catalyst, on the yield in a refining process. If the experimenter intends to use three settings each for temperature and pressure and two types of catalysts, how many experimental runs will have to be conducted if he wishes to run all possible combinations of pressure, temperature, and types of catalysts?

2.43 Five firms, $F_1, F_2, \ldots, F_5$, each offer bids on three separate contracts, C_1, C_2, and C_3. Any one firm will be awarded at most one contract. The contracts are quite different, so an assignment of C_1 to F_1, say, is to be distinguished from an assignment of C_2 to F_1.

    a How many sample points are there altogether in this experiment involving assignment of contracts to the firms? (No need to list them all.)

    b Under the assumption of equally likely sample points, find the probability that F_3 is awarded a contract.

2.44 A group of three undergraduate and five graduate students are available to fill certain student government posts. If four students are to be randomly selected from this group, find the probability that exactly two undergraduates will be among the four chosen.

2.45 A study is to be conducted in a hospital to determine the attitudes of nurses toward various administrative procedures. A sample of 10 nurses is to be selected from a total of the 90 nurses employed by the hospital.

    a How many different samples of 10 nurses can be selected?

    b Twenty of the 90 nurses are male. If 10 nurses are randomly selected from those employed by the hospital, what is the probability that the sample of ten will include exactly 4 male (and 6 female) nurses?

2.46   A student prepares for an exam by studying a list of ten problems. She can solve six of them. For the exam, the instructor selects five problems at random from the ten on the list given to the students. What is the probability that the student can solve all five problems on the exam?

2.47   Two cards are drawn from a standard 52-card playing deck. What is the probability that the draw will yield an ace and a face card?

2.48   A manufacturer has nine distinct motors in stock, two of which came from a particular supplier. The motors must be divided among three production lines, with three motors going to each line. If the assignment of motors to lines is random, find the probability that both motors from the particular supplier are assigned to the first line.

2.49   The eight-member Human Relations Advisory Board of Gainesville, Florida, considered the complaint of a woman who claimed discrimination, based on sex, on the part of a local company. The board, composed of five women and three men, voted 5–3 in favor of the plaintiff, the five women voting in favor of the plaintiff, the three men against. The attorney representing the company appealed the board's decision by claiming sex bias on the part of the board members. If there was no sex bias among the board members, it might be reasonable to conjecture that any group of five board members would be as likely to vote for the complainant as any other group of five. If this were the case, what is the probability that the vote would split along sex lines (five women for, three men against)?

2.50   A balanced die is tossed six times and the number on the uppermost face is recorded each time. What is the probability that the numbers recorded are 1, 2, 3, 4, 5, and 6 in any order?

2.51   Refer to Exercise 2.50. Suppose that the die has been altered so that the faces are 1, 2, 3, 4, 5, and 5. If the die is tossed five times, what is the probability that the numbers recorded are 1, 2, 3, 4, and 5 in any order?

2.52   Refer to Example 2.10.

    a   What is the probability that an ethnic group member is assigned to each type of job?

    b   What is the probability that no ethnic group member is assigned to a type 4 job?

2.53   Refer to Example 2.13. Suppose that the number of distributors is $M = 10$ and that there are $n = 7$ orders to be placed.

    a   What is the probability that all of the orders go to different distributors?

    *b   What is the probability that distributor I gets exactly two orders *and* distributor II gets exactly three orders?

    *c   What is the probability that distributors I, II and III get exactly two, three, and one order(s), respectively?

2.54   Show that, for any integer $n \geq 1$,

    a $\binom{n}{n} = 1$. Interpret this result.

    b $\binom{n}{0} = 1$. Interpret this result.

    c $\binom{n}{r} = \binom{n}{n-r}$. Interpret this result.

    d $\sum_{i=0}^{n} \binom{n}{i} = 2^n$. [*Hint:* Consider the binomial expansion of $(x + y)^n$ with $x = y = 1$.]

2.55   Prove that $\binom{n+1}{k} = \binom{n}{k} + \binom{n}{k-1}$.

*2.56   Consider the situation where n items are to be partitioned into $k < n$ distinct subsets. The multinomial coefficients $\binom{n}{n_1 \ n_2 \ \cdots \ n_k}$ provide the number of distinct partitions where n_1

items are in group 1, n_2 are in group 2, ..., n_k are in group k. Prove that the total number of distinct partitions equals k^n. [*Hint:* Recall Exercise 2.54(d).]

2.7 Conditional Probability and the Independence of Events

The probability of an event will sometimes depend upon whether we know that other events have occurred. For example, Florida sport fishermen are vitally interested in the probability of rain. The probability of rain on a given day, ignoring the daily atmospheric conditions or any other events, is the fraction of days in which rain occurs over a long period of time. This is the *unconditional probability* of the event "rain on a given day." Now suppose that we wish to consider the probability of rain tomorrow. It has rained almost continuously for two days in succession and a tropical storm is heading up the coast. We have extra information related to whether or not it rains tomorrow, and are interested in the *conditional* probability that it will rain *given* this information. A Floridian would tell you that the conditional probability of rain (given that it has rained two preceding days and that a tropical storm is predicted) is much larger than the unconditional probability of rain.

The unconditional probability of a 1 in the toss of one balanced die is 1/6. If we know that an odd number has fallen, the number on the die must be 1, 3, or 5 and the relative frequency of occurrence of a 1 is 1/3. The conditional probability of an event is the probability (relative frequency of occurrence) of the event given the fact that one or more events have already occurred. A careful perusal of this example will indicate the agreement of the following definition with the relative frequency concept of probability.

DEFINITION 2.9

The *conditional probability of an event A*, given that an event B has occurred, is equal to

$$P(A|B) = \frac{P(A \cap B)}{P(B)}$$

provided $P(B) > 0$. [The symbol $P(A|B)$ is read "probability of A given B."]

Further confirmation of the consistency of Definition 2.9 with the relative frequency concept of probability can be obtained from the following construction. Suppose that an experiment is repeated a large number, N, of times, resulting in both A and B, $A \cap B$, n_{11} times; A and not B, $A \cap \overline{B}$, n_{21} times; B and not A, $\overline{A} \cap B$, n_{12} times; and neither A nor B, $\overline{A} \cap \overline{B}$, n_{22} times. These results are contained in Table 2.1.

Note that $n_{11} + n_{12} + n_{21} + n_{22} = N$. Then it follows that

$$P(A) \approx \frac{n_{11} + n_{21}}{N}, \quad P(B) \approx \frac{n_{11} + n_{12}}{N} \quad P(A|B), \approx \frac{n_{11}}{n_{11} + n_{12}},$$

Table 2.1. Table for events A and B

	A	$\overline{A}$	
B	n_{11}	n_{12}	$n_{11} + n_{12}$
$\overline{B}$	n_{21}	n_{22}	$n_{21} + n_{22}$
	$n_{11} + n_{21}$	$n_{12} + n_{22}$	N

$$P(B|A) \approx \frac{n_{11}}{n_{11} + n_{21}}, \quad \text{and } P(A \cap B) \approx \frac{n_{11}}{N},$$

where $\approx$ is read *approximately equal to*.

With these probabilities it is easy to see that

$$P(B|A) \approx \frac{P(A \cap B)}{P(A)} \quad \text{and} \quad P(A|B) \approx \frac{P(A \cap B)}{P(B)}.$$

Hence Definition 2.9 is consistent with the relative frequency concept of probability.

EXAMPLE 2.14   Suppose that a balanced die is tossed once. Use Definition 2.9 to find the probability of a 1, given that an odd number was obtained.

Solution   Define these events:

A: Observe a 1.

B: Observe an odd number.

We seek the probability of A given that the event B has occurred. The event $A \cap B$ requires the observance of both a 1 and an odd number. In this instance, $A \subset B$ so $A \cap B = A$ and $P(A \cap B) = P(A) = 1/6$. Also, $P(B) = 1/2$ and, using Definition 2.9,

$$P(A|B) = \frac{P(A \cap B)}{P(B)} = \frac{1/6}{1/2} = \frac{1}{3}.$$

Notice that this result is in complete agreement with our earlier intuitive evaluation of this probability.

Suppose that probability of the occurrence of an event A is unaffected by the occurrence or nonoccurrence of event B. When this happens, we would be inclined to say that events A and B are independent. This event relationship is expressed by the following definition.

DEFINITION **2.10**

Two events A and B are said to be *independent* if any one of the following holds:

$$P(A|B) = P(A)$$

$$P(B|A) = P(B)$$

$$P(A \cap B) = P(A)P(B).$$

Otherwise, the events are said to be *dependent*.

The notion of independence as a probabilistic concept is in agreement with our everyday usage of the word if we carefully consider the events in question. Most would agree that "smoking" and "contracting lung cancer" are not independent events and would intuitively feel that the probability of contracting lung cancer, given that a person smokes, is greater than the (unconditional) probability of contracting lung cancer. In contrast, the events "rain today" and "rain a month from today" may well be independent.

EXAMPLE **2.15**

Consider the following events in the toss of a single die:

A: Observe an odd number.
B: Observe an even number.
C: Observe a 1 or 2.

a Are A and B independent events?
b Are A and C independent events?

Solution
a To decide whether A and B are independent, we must see whether they satisfy the conditions of Definition 2.10. In this example, $P(A) = 1/2$, $P(B) = 1/2$ and $P(C) = 1/3$. Because $A \cap B = \emptyset$, $P(A|B) = 0$ and it is clear that $P(A|B) \neq P(A)$. Events A and B are dependent events.

b Are A and C independent? Note that $P(A|C) = 1/2$ and, as before, $P(A) = 1/2$. Therefore, $P(A|C) = P(A)$, and A and C are independent.

EXAMPLE **2.16**

Three brands of coffee, X, Y, and Z, are to be ranked according to taste by a judge. Define the following events:

A: Brand X is preferred to Y.
B: Brand X is ranked best.
C: Brand X is ranked second best.
D: Brand X is ranked third best.

If the judge actually has no taste preference and randomly assigns ranks to the brands, is event A independent of events B, C, and D?

Solution   The six equally likely sample points for this experiment are given by

$$E_1: XYZ \quad E_3: YXZ \quad E_5: ZXY$$
$$E_2: XZY \quad E_4: YZX \quad E_6: ZYX$$

where XYZ denotes that X is ranked best, Y is second best, and Z is last.

Then $A = \{E_1, E_2, E_5\}$, $B = \{E_1, E_2\}$, $C = \{E_3, E_5\}$, $D = \{E_4, E_6\}$ and it follows that

$$P(A) = 1/2, \quad P(A|B) = \frac{P(A \cap B)}{P(B)} = 1, \quad P(A|C) = 1/2, \quad \text{and} \quad P(A|D) = 0.$$

Thus events A and C are independent but events A and B are dependent. Events A and D are also dependent.

Exercises

2.57   If two events, A and B, are such that $P(A) = .5$, $P(B) = .3$, and $P(A \cap B) = .1$, find the following:

a $P(A|B)$
b $P(B|A)$
c $P(A|A \cup B)$
d $P(A|A \cap B)$
e $P(A \cap B|A \cup B)$

2.58   For a certain population of employees the percentage passing or failing a job competency exam, listed according to sex, were as shown in the accompanying table. That is, of all the people taking the exam, 24% were in the male-pass category, 16% were in the male-fail category, and so forth. An employee is to be selected randomly from this population. Let A be the event that the employee scores a passing grade on the exam, and let M be the event that a male is selected. Are the events A and M independent? Are the events $\bar{A}$ and F independent?

| | Sex | | |
Outcome	Male (M)	Female (F)	Total
Pass (A)	24	36	60
Fail ($\bar{A}$)	16	24	40
Total	40	60	100

2.59   Gregor Mendel was a monk who, in 1865, suggested a theory of inheritance based on the science of genetics. He identified heterozygous individuals for flower color that had two alleles (one r = recessive white color allele and one R = dominant red color allele). When these individuals were mated, 3/4 of the offspring were observed to have red flowers and 1/4 had white flowers. The table summarizes this mating; each parent gives one of its alleles to form the gene of the offspring.

Parent 1	Parent 2	
	r	R
r	rr	rR
R	Rr	RR

We assume that each parent is equally likely to give either of the two alleles and that, if either one or two of the alleles in a pair is dominant (R), the offspring will have red flowers.

a   What is the probability that an offspring in this mating has at least one dominant allele?

b   What is the probability that an offspring has at least one recessive allele?

c   What is the probability that an offspring has one recessive allele, given that the offspring has red flowers?

2.60   A survey of consumers in a particular community showed that 10% were dissatisfied with plumbing jobs done in their homes. Half the complaints dealt with plumber A, who does 40% of the plumbing jobs in the town.

a   Find the probability that a consumer will obtain an unsatisfactory plumbing job, given that the plumber was A.

b   Find the probability that a consumer will obtain a satisfactory plumbing job, given that the plumber was A.

2.61   A study of the posttreatment behavior of a large number of drug abusers suggests that the likelihood of conviction within a 2-year period after treatment may depend upon the offenders education. The proportions of the total number of cases falling in four education-conviction categories in the following table.

Education	Status Within Two Years After Treatment		Totals
	Convicted	Not Convicted	
10 years or more	.10	.30	.40
9 years or less	.27	.33	.60
Totals	.37	.63	1.00

Suppose that a single offender is selected from the treatment program. Define the events:

　　A: The offender has 10 or more years of education.

　　B: The offender is convicted within 2 years after completion of treatment.

Find :

a $P(A)$

b $P(B)$

c $P(AB)$

d $P(A \cup B)$

e $P(\overline{A})$

f $P(\overline{A \cup B})$

g $P(\overline{AB})$

h $P(A|B)$

i $P(B|A)$

2.62   In the definition of the independence of two events, you were given three equalities to check: $P(A|B) = P(A)$ or $P(B|A) = P(B)$ or $P(A \cap B) = P(A)P(B)$. If any one of these equalities holds, A and B are independent. Show that if any of these equalities hold, the other two also hold.

2.63   Suppose A and B are mutually exclusive events, with $P(A) > 0$ and $P(B) < 1$. Are A and B independent? Give a proof for your answer.

2.64 Suppose that $A \subset B$ and that $P(A) > 0$ and $P(B) > 0$. Are A and B independent? Prove your answer.

2.65 If $P(A) > 0$, $P(B) > 0$, and $P(A) < P(A|B)$, show that $P(B) < P(B|A)$.

2.8 Two Laws of Probability

The following two laws give the probabilities of unions and intersections of events. As such, they play an important role in the event-composition approach to the solution of probability problems.

THEOREM 2.5

(The Multiplicative Law of Probability) The probability of the intersection of two events A and B is

$$P(A \cap B) = P(A)P(B|A)$$
$$= P(B)P(A|B).$$

If A and B are independent, then

$$P(A \cap B) = P(A)P(B).$$

Proof

The multiplicative law follows directly from Definition 2.9, the definition of conditional probability. ∎

Notice that the multiplicative law can be extended to find the probability of the intersection of any number of events. Thus, twice applying Theorem 2.5, we obtain

$$P(A \cap B \cap C) = P[(A \cap B) \cap C] = P(A \cap B)P(C|A \cap B)$$
$$= P(A)P(B|A)P(C|A \cap B).$$

The probability of the intersection of any number of, say, k events, can be obtained in the same manner:

$$P(A_1 \cap A_2 \cap A_3 \cap \cdots \cap A_k) = P(A_1)P(A_2|A_1)P(A_3|A_1 \cap A_2)$$
$$\cdots P(A_k|A_1 \cap A_2 \cap \cdots \cap A_{k-1}).$$

The additive law of probability gives the probability of the union of two events.

THEOREM 2.6

(The Additive Law of Probability) The probability of the union of two events A and B is

$$P(A \cup B) = P(A) + P(B) - P(A \cap B).$$

FIGURE 2.10
Venn diagram for the
union of A and B

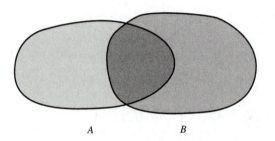

A B

If A and B are mutually exclusive events, $P(A \cap B) = 0$ and

$$P(A \cup B) = P(A) + P(B).$$

Proof   The proof of the additive law can be followed by inspecting the Venn diagram in Figure 2.10.

Notice that $A \cup B = A \cup (\overline{A} \cap B)$, where A and $(\overline{A} \cap B)$ are mutually exclusive events. Further, $B = (\overline{A} \cap B) \cup (A \cap B)$, where $(A \cap \overline{B})$ and $(A \cap B)$ are mutually exclusive events. Then, by Axiom 3,

$$P(A \cup B) = P(A) + P(\overline{A} \cap B) \quad \text{and} \quad P(B) = P(\overline{A} \cap B) + P(A \cap B).$$

The equality given on the right implies that $P(\overline{A} \cap B) = P(B) - P(A \cap B)$. Substituting this expression for $P(\overline{A} \cap B)$ into the expression for $P(A \cup B)$ given in the left-hand equation of the preceding pair, we obtain the desired result:

$$P(A \cup B) = P(A) + P(B) - P(A \cap B). \qquad \blacksquare$$

The probability of the union of three events can be obtained by making use of Theorem 2.6. Observe that

$$P(A \cup B \cup C) = P[A \cup (B \cup C)]$$
$$= P(A) + P(B \cup C) - P[A \cap (B \cup C)]$$
$$= P(A) + P(B) + P(C) - P(B \cap C) - P[(A \cap B) \cup (A \cap C)]$$
$$= P(A) + P(B) + P(C) - P(B \cap C) - P(A \cap B) - P(A \cap C)$$
$$+ P(A \cap B \cap C)$$

because $(A \cap B) \cap (A \cap C) = A \cap B \cap C$.

Another useful result expressing the relationship between the probability of an event and its complement is immediately available from the axioms of probability.

THEOREM 2.7   If A is an event, then

$$P(A) = 1 - P(\overline{A}).$$

Proof   Observe that $S = A \cup \overline{A}$. Because A and $\overline{A}$ are mutually exclusive events, it follows that $P(S) = P(A) + P(\overline{A})$. Therefore, $P(A) + P(\overline{A}) = 1$ and the result follows.   ∎

As we will see in Section 2.9, it is sometimes easier to calculate $P(\overline{A})$ than to calculate $P(A)$. In such cases, it is easier to find $P(A)$ by the relationship $P(A) = 1 - P(\overline{A})$ than to find $P(A)$ directly.

Exercises

2.66   If A_1, A_2, and A_3 are three events and $P(A_1 \cap A_2) = P(A_1 \cap A_3) \neq 0$ but $P(A_2 \cap A_3) = 0$, show that

$$P(\text{at least one } A_i) = P(A_1) + P(A_2) + P(A_3) - 2P(A_1 \cap A_2).$$

2.67   If A and B are independent events, show that A and $\overline{B}$ are also independent. Are $\overline{A}$ and $\overline{B}$ independent?

2.68   A policy requiring all hospital employees to take lie detector tests may reduce losses due to theft, but some employees regard such tests as a violation of their rights. Reporting on a particular hospital that uses this procedure, the *Orlando Sentinel Star* (August 3, 1981)[2] notes that lie detectors have accuracy rates that vary from 92% to 99%. To gain some insight into the risks that employees face when taking a lie detector test, suppose that the probability is .05 that a particular lie detector concludes that a person is lying who, in fact, is telling the truth, and suppose that any pair of tests are independent.

    a   What is the probability that a machine will conclude that each of three employees is lying when all are telling the truth?

    b   What is the probability that the machine will conclude that at least one of the three employees is lying when all are telling the truth?

2.69   In a game, a participant is given three attempts to hit a ball. On each try, she either scores a hit, H, or a miss, M. The game requires that the player must alternate which hand she uses in successive attempts. That is, if she makes her first attempt with her right hand, she must use her left hand for the second attempt and her right hand for the third. Her chance of scoring a hit with her right hand is .7 and with her left hand is .4. Assume that the results of successive attempts are independent and that she wins the game if she scores at least two hits in a row. If she makes her first attempt with her right hand what is the probability that she wins the game?

2.70   A smoke detector system uses two devices, A and B. If smoke is present, the probability that it will be detected by device A is .95; by device B, .90; and by both devices, .88.

    a   If smoke is present, find the probability that the smoke will be detected by either device A or B or both devices.

    b   Find the probability that the smoke will be undetected.

2.71   Two events A and B are such that $P(A) = .2$, $P(B) = .3$, and $P(A \cup B) = .4$. Find the following:

---

[2]*Source:* Reprinted by permission. ©1980 Sentinel Communications Co. All rights reserved.

a $P(A \cap B)$

b $P(\overline{A} \cup \overline{B})$

c $P(\overline{A} \cap \overline{B})$

d $P(\overline{A}|B)$

2.72 If A and B are independent events with $P(A) = .5$ and $P(B) = .2$, find the following:

a $P(A \cup B)$

b $P(\overline{A} \cap \overline{B})$

c $P(\overline{A} \cup \overline{B})$

2.73 Consider the following portion of an electric circuit with three relays. Current will flow from point a to point b if there is at least one closed path when the relays are activated. The relays may malfunction and not close when activated. Suppose that the relays act independently of one another and close properly when activated, with a probability of .9.

a What is the probability that current will flow when the relays are activated?

b Given that current flowed when the relays were activated, what is the probability that relay 1 functioned?

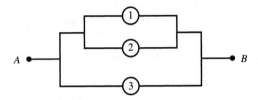

2.74 With relays operating as in Exercise 2.73, compare the probability of current flowing from a to b in the series system shown

with the probability of flow in the parallel system shown.

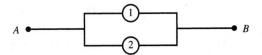

2.75 Suppose that A and B are independent events such that the probability that neither occurs is a and the probability of B is b. Show that $P(A) = \dfrac{1 - b - a}{1 - b}$.

*2.76 Show that Theorem 2.6, the additive law of probability, holds for *conditional* probabilities. That is, if A, B, and C are events such that $P(C) > 0$, prove that $P(A \cup B|C) = P(A|C) + P(B|C) - P(A \cap B|C)$. [*Hint:* Make use of the distributive law $(A \cup B) \cap C = (A \cap C) \cup (B \cap C)$.]

2.77 Articles coming through an inspection line are visually inspected by two successive inspectors. When a defective article comes through the inspection line, the probability that it gets by the first inspector is .1. The second inspector will "miss" five out of ten of the defective items that get past the first inspector. What is the probability that a defective item gets by both inspectors?

2.78 Diseases I and II are prevalent among people in a certain population. It is assumed that 10% of the population will contract disease I sometime during their lifetime, 15% will contract disease II eventually, and 3% will contract both diseases.

    a   Find the probability that a randomly chosen person from this population will contract at least one disease.

    b   Find the conditional probability that a randomly chosen person from this population will contract both diseases, given that he or she has contracted at least one disease.

2.79   Refer to Exercise 2.40. Hours after the rigging of the Pennsylvania state lottery was announced, Connecticut state lottery officials were stunned to learn that their winning number for the day was 666 (*Los Angeles Times*, September 21, 1980).

    a   All evidence indicates that the Connecticut selection of 666 was due to pure chance. What is the probability that a 666 would be drawn in Connecticut, given that a 666 had been selected in the April 24, 1980, Pennsylvania lottery?

    b   What is the probability of drawing a 666 in the April 24, 1980, Pennsylvania lottery (remember, this drawing was rigged) and a 666 in the September 19, 1980, Connecticut lottery?

2.80   If A and B are two events, prove that $P(A \cap B) \geq 1 - P(\overline{A}) - P(\overline{B})$. [*Note*: This is a simplified version of the *Bonferroni inequality*.]

2.81   If the probability of injury on each individual parachute jump is .05, use the result in Exercise 2.80 to provide a lower bound for the probability of landing safely on both of two jumps.

2.82   If A and B are equally likely events and we require that the probability of their intersection be at least .98, what is $P(A)$?

2.83   Let A, B, and C be events such that $P(A) > P(B)$ and $P(C) > 0$. Construct an example to demonstrate that it is possible that $P(A|C) < P(B|C)$.

2.84   If A, B and C are three events, use two applications of the result in Exercise 2.80 to prove that $P(A \cap B \cap C) \geq 1 - P(\overline{A}) - P(\overline{B}) - P(\overline{C})$.

2.85   If A, B, and C are three equally likely events and we require that $P(A \cap B \cap C) \geq .95$, what is $P(A)$?

2.9 Calculating the Probability of an Event: The Event-Composition Method

We learned in Section 2.4 that sets (events) can often be expressed as unions, intersections, or complements of other sets. The event-composition method for calculating the probability of an event, A, expresses A, as a composition involving unions and/or intersections of other events. The laws of probability are then applied to find $P(A)$. We will illustrate this method with an example.

EXAMPLE 2.17   Of the voters in a city, 40% are Republicans and 60% are Democrats. Among the Republicans 70% are in favor of a bond issue, whereas 80% of the Democrats favor the issue. If a voter is selected at random in the city, what is the probability that he or she will favor the bond issue?

Solution   Let F denote the event "favor the bond issue," R the event "a Republican is selected," and D the event "a Democrat is selected." Then $P(R) = .4$, $P(D) = .6$, $P(F|R) =$

FIGURE **2.11**
Venn diagram
for events of
Example 2.17

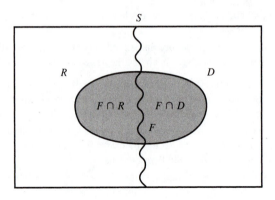

.7, and $P(F|D) = .8$. Notice that

$$P(F) = P[(F \cap R) \cup (F \cap D)] = P(F \cap R) + P(F \cap D)$$

because $(F \cap R)$ and $(F \cap D)$ are mutually exclusive events. Figure 2.11 will help you visualize the result that $F = (F \cap R) \cup (F \cap D)$. Now

$$P(F \cap R) = P(F|R)P(R) = (.7)(.4) = .28 \quad \text{and}$$

$$P(F \cap D) = P(F|D)P(D) = (.8)(.6) = .48.$$

It follows that

$$P(F) = .28 + .48 = .76.$$

EXAMPLE **2.18**   In Example 2.7 we considered an experiment wherein the birthdays of 20 randomly selected persons were recorded. Under certain conditions we found that $P(A) = .5886$, where A denotes the event that each person has a different birthday. Let B denote the event that at least one pair of individuals share a birthday. Find $P(B)$.

Solution   The event B is the set of all sample points in S that are not in A, that is, $B = \overline{A}$. Therefore,

$$P(B) = 1 - P(A) = 1 - .5886 = .4114.$$

(Most would agree that this probability is surprisingly high!)

Let us refer to Example 2.4, which involves the two tennis players, and let D_1 and D_2 denote the events that player A wins the first and second games, respectively. The information given in the example implies that $P(D_1) = P(D_2) = 2/3$. Further, if we make the assumption that D_1 and D_2 are independent, it follows that $P(D_1 \cap D_2) = 2/3 \times 2/3 = 4/9$. In that example we identified the simple event E_1, which we denoted AA, as meaning that player A won both games. With the present notation,

$$E_1 = D_1 \cap D_2,$$

and thus $P(E_1) = 4/9$. The probabilities assigned to the other simple events in Example 2.4 can be verified in a similar manner.

The event-composition approach will not be successful unless the probabilities of the events that appear in $P(A)$ (after the additive and multiplicative laws have been applied) are known. If one or more of these probabilities is unknown, the method fails. Often it is desirable to form compositions of mutually exclusive or independent events. Mutually exclusive events simplify the use of the additive law and the multiplicative law of probability is easier to apply to independent events.

A summary of the steps used in the event-composition method follows:

1. Define the experiment.
2. Visualize the nature of the sample points. Identify a few to clarify your thinking.
3. Write an equation expressing the event of interest, say, A, as a composition of two or more events, using unions, intersections, and/or complements. (Notice that this equates point sets.) Make certain that event A and the event implied by the composition represent the same set of sample points.
4. Apply the additive and multiplicative laws of probability to the compositions obtained in step 3 to find $P(A)$.

Step 3 is the most difficult because we can form many compositions that will be equivalent to event A. The trick is to form a composition in which all the probabilities appearing in step 4 are known.

The event-composition approach does not require listing the sample points in S, but it does require a clear understanding of the nature of a typical sample point. The major error students tend to make in applying the event-composition approach occurs in writing the composition. That is, the point-set equation that expresses A as union and/or intersection of other events is frequently incorrect. Always test your equality to make certain that the composition implies an event that contains the same set of sample points as those in A.

A comparison of the sample-point and event-composition methods for calculating the probability of an event can be obtained by applying both methods to the same problem. We will apply the event-composition approach to the problem of selecting applicants that was solved by the sample-point method in Examples 2.11 and 2.12.

EXAMPLE **2.19**   Two applicants are randomly selected from among five who have applied for a job. Find the probability that exactly one of the two best applicants is selected, event A.

Solution   Define the following two events:

B: Draw the best and one of the three poorest applicants.
C: Draw the second best and one of the three poorest applicants.

Events B and C are mutually exclusive and $A = B \cup C$. Also, let $D_1 = B_1 \cap B_2$, where

$B_1 = $ Draw the best on the first draw

$B_2 = $ Draw one of the three poorest applicants on the second draw

and $D_2 = B_3 \cap B_4$, where

$B_3 = $ Draw one of the three poorest applicants on the first draw

$B_4 = $ Draw the best on the second draw.

Note that $B = D_1 \cup D_2$.

Similarly, let $G_1 = C_1 \cap C_2$ and $G_2 = C_3 \cap C_4$, where C_1, C_2, C_3, and C_4 are defined like B_1, B_2, B_3, and B_4, with the words *second best* replacing *best*. Notice that D_1 and D_2 and G_1 and G_2 are pairs of mutually exclusive events and that

$$A = B \cup C = (D_1 \cup D_2) \cup (G_1 \cup G_2)$$
$$A = (B_1 \cap B_2) \cup (B_3 \cap B_4) \cup (C_1 \cap C_2) \cup (C_3 \cap C_4).$$

Applying the additive law of probability to these four mutually exclusive events, we have

$$P(A) = P(B_1 \cap B_2) + P(B_3 \cap B_4) + P(C_1 \cap C_2) + P(C_3 \cap C_4).$$

Applying the multiplicative law, we have

$$P(B_1 \cap B_2) = P(B_1)P(B_2|B_1).$$

The probability of drawing the best on the first draw is

$$P(B_1) = 1/5.$$

Similarly, the probability of drawing one of the three poorest on the second draw, given that the best was drawn on the first selection, is

$$P(B_2|B_1) = 3/4.$$

Then

$$P(B_1 \cap B_2) = P(B_1)P(B_2|B_1) = (1/5)(3/4) = 3/20.$$

The probabilities of all other intersections in $P(A)$, $P(B_3 \cap B_4)$, $P(C_1 \cap C_2)$, and $P(C_3 \cap C_4)$ are obtained in exactly the same manner, and all equal 3/20. Then

$$P(A) = P(B_1 \cap B_2) + P(B_3 \cap B_4) + P(C_1 \cap C_2) + P(C_3 \cap C_4)$$
$$= (3/20) + (3/20) + (3/20) + (3/20) = 3/5.$$

This answer is identical to that obtained in Example 2.12, where $P(A)$ was calculated by using the sample-point approach.

EXAMPLE **2.20**   It is known that a patient with a disease will respond to treatment with probability equal to .9. If three patients with the disease are treated and respond independently, find the probability that at least one will respond.

Solution   Define the following events:

> A: At least one of the three patients will respond.
> B_1: The first patient will not respond.
> B_2: The second patient will not respond.
> B_3: The third patient will not respond.

Then observe that $\overline{A} = B_1 \cap B_2 \cap B_3$. Theorem 2.7 implies that

$$P(A) = 1 - P(\overline{A})$$
$$= 1 - P(B_1 \cap B_2 \cap B_3).$$

Applying the multiplicative law, we have

$$P(B_1 \cap B_2 \cap B_3) = P(B_1)P(B_2|B_1)P(B_3|B_1 \cap B_2)$$

where, because the events are independent,

$$P(B_2|B_1) = P(B_2) = 0.1 \quad \text{and} \quad P(B_3|B_1 \cap B_2) = P(B_3) = 0.1.$$

Substituting $P(B_i) = .1, i = 1, 2, 3$, we obtain

$$P(A) = 1 - (.1)^3 = .999.$$

Notice that we have demonstrated the utility of complementary events. This result is important because frequently it is easier to find the probability of the complement, $P(\overline{A})$, than to find $P(A)$ directly.

EXAMPLE **2.21**   Observation of a waiting line at a medical clinic indicates the probability that a new arrival will be an emergency case is $p = 1/6$. Find the probability that the rth patient is the first emergency case. (Assume that conditions of arriving patients represent independent events.)

Solution   The experiment consists of watching patient arrivals until the first emergency case appears. Then the sample points for the experiment are

> E_i: The ith patient is the first emergency case, for $i = 1, 2, \ldots$.

Because only one sample point falls in the event of interest,

$$P(r\text{th patient is the first emergency case }) = P(E_r).$$

Now define A_i to denote the event that the ith arrival is not an emergency case. Then we can represent E_r as the intersection

$$E_r = A_1 \cap A_2 \cap A_3 \cap \cdots \cap A_{r-1} \cap \overline{A}_r.$$

Applying the multiplicative law, we have

$$P(E_r) = P(A_1)P(A_2|A_1)P(A_3|A_1 \cap A_2) \cdots P(\overline{A}_r|A_1 \cap \cdots \cap A_{r-1})$$

and because the events $A_1, A_2, \ldots, A_{r-1}$, and $\overline{A}_r$ are independent, it follows that

$$P(E_r) = P(A_1)P(A_2) \cdots P(A_{r-1})P(\overline{A}_r) = (1-p)^{r-1}p$$
$$= (5/6)^{r-1}(1/6), \quad r = 1, 2, 3, \ldots .$$

Notice that

$$P(S) = P(E_1) + P(E_2) + P(E_3) + \cdots + P(E_i) + \cdots$$
$$= (1/6) + (5/6)(1/6) + (5/6)^2(1/6) + \cdots + (5/6)^{i-1}(1/6) + \cdots$$
$$= \frac{1}{6} \sum_{i=0}^{\infty} (5/6)^i = \frac{1/6}{1-(5/6)} = 1.$$

This result follows from the formula for the sum of a *geometric series* given in Appendix A1.11. This formula, which states that if $|r| < 1$, $\sum_{i=0}^{\infty} r^i = \frac{1}{1-r}$, is useful in many simple probability problems.

EXAMPLE 2.22 A monkey is to demonstrate that she recognizes colors by tossing one red, one black, and one white ball into boxes of the same respective colors, one ball to a box. If the monkey has not learned the colors and merely tosses one ball into each box at random, find the probabilities of the following results:

a There are no color matches.

b There is exactly one color match.

Solution This problem can be solved by listing sample points because only three balls are involved, but a more general method will be illustrated. Define the following events:

A_1: A color match occurs in the red box.
A_2: A color match occurs in the black box.
A_3: A color match occurs in the white box.

There are $3! = 6$ equally likely ways of randomly tossing the balls into the boxes with one ball in each box. Also, there are only $2! = 2$ ways of tossing the balls into the boxes if one particular box is required to have a color match. Hence

$$P(A_1) = P(A_2) = P(A_3) = 2/6 = 1/3.$$

Similarly, it follows that

$$P(A_1 \cap A_2) = P(A_1 \cap A_3) = P(A_2 \cap A_3) = P(A_1 \cap A_2 \cap A_3) = 1/6.$$

We can now answer (a) and (b) by using the event-composition method.

a   Notice that

$$P(\text{no color matches}) = 1 - P(\text{at least one color match})$$
$$= 1 - P(A_1 \cup A_2 \cup A_3)$$
$$= 1 - [P(A_1) + P(A_2) + P(A_3) - P(A_1 \cap A_2)$$
$$- P(A_1 \cap A_3) - P(A_2 \cap A_3) + P(A_1 \cap A_2 \cap A_3)]$$
$$= 1 - [3(1/3) - 3(1/6) + (1/6)] = 2/6 = 1/3.$$

b   We leave it to you to show that

$$P(\text{exactly one match}) = P(A_1) + P(A_2) + P(A_3)$$
$$- 2[P(A_1 \cap A_2) + P(A_1 \cap A_3) + P(A_2 \cap A_3)]$$
$$+ 3[P(A_1 \cap A_2 \cap A_3)]$$
$$= (3)(1/3) - (2)(3)(1/6) + (3)(1/6) = 1/2.$$

The best way to learn how to solve probability problems is to learn by doing. To assist you in developing your skills, many exercises are provided at the end of this section, at the end of the chapter, and in the references.

Exercises

2.86   Of the items produced daily by a factory, 40% come from line I and 60%, from line II. Line I has a defect rate of 8%, whereas line II has a defect rate of 10%. If an item is chosen at random from the day's production, find the probability that it will not be defective.

2.87   An advertising agency notices that approximately 1 in 50 potential buyers of a product sees a given magazine ad, and 1 in 5 sees a corresponding ad on television. One in 100 sees both. One in 3 actually purchases the product after seeing the ad, 1 in 10 without seeing it. What is the probability that a randomly selected potential customer will purchase the product?

2.88   Three radar sets, operating independently, are set to detect any aircraft flying through a certain area. Each set has a probability of .02 of failing to detect a plane in its area.

a   If an aircraft enters the area, what is the probability that it goes undetected?

b   If an aircraft enters the area, what is the probability that it is detected by all three radar sets?

2.89   Consider one of the radar sets of Exercise 2.88. What is the probability that it will correctly detect exactly three aircraft before it fails to detect one, if aircraft arrivals are independent single events occurring at different times?

2.90 A lie detector will show a positive reading (indicate a lie) 10% of the time when a person is telling the truth and 95% of the time when the person is lying. Suppose two people are suspects in a one-person crime and (for certain) one is guilty and will lie. Assume further that the lie detector operates independently for the truthful person and the liar.

    a What is the probability that the detector shows a positive reading for both suspects?

    b What is the probability that the detector shows a positive reading for the guilty suspect and a negative reading for the innocent suspect?

    c What is the probability that the detector is completely wrong—that is, that it gives a positive reading for the innocent suspect and a negative reading for the guilty?

    d What is the probability that it gives a positive reading for either or both of the two suspects?

2.91 A football team has a probability of .75 of winning when playing any of the other four teams in its conference. If the games are independent, what is the probability the team wins all its conference games?

2.92 A communications network has a built-in safeguard system against failures. In this system if line I fails, it is bypassed and line II is used. If line II also fails, it is bypassed and line III is used. The probability of failure of any one of these three lines is .01, and the failures of these lines are independent events. What is the probability that this system of three lines does not completely fail?

2.93 A state auto-inspection station has two inspection teams. Team 1 is lenient and passes all automobiles of a recent vintage; team 2 rejects all autos on a first inspection because their "headlights are not properly adjusted." Four unsuspecting drivers take their autos to the station for inspection on four different days and randomly select one of the two teams.

    a If all four cars are new and in excellent condition, what is the probability that three of the four will be rejected?

    b What is the probability that all four will pass?

2.94 An accident victim will die unless in the next 10 minutes he receives some type A, Rh-positive blood, which can be supplied by a single donor. The hospital requires 2 minutes to type a prospective donor's blood and 2 minutes to complete the transfer of blood. Many untyped donors are available, and 40% of them have type A, Rh positive blood. What is the probability that the accident victim will be saved if only one blood-typing kit is available? Assume that the typing kit is reusable but can process only one donor at a time.

*2.95 Suppose that two balanced dice are tossed repeatedly and the sum of the two uppermost faces is determined on each toss.

    a What is the probability that we obtain a sum of 3 before we obtain a sum of 7?

    b What is the probability that we obtain a sum of 4 before we obtain a sum of 7?

2.96 Suppose that two defective refrigerators have been included in a shipment of six refrigerators. The buyer begins to test the six refrigerators one at a time.

    a What is the probability that the last defective refrigerator is found on the fourth test?

    b What is the probability that no more than four refrigerators need to be tested to locate both of the defective refrigerators?

    c When given that exactly one of the two defective refrigerators has been located in the first two tests, what is the probability that the remaining defective refrigerator is found in the third or fourth test?

2.97 A new secretary has been given n computer passwords, only one of which will permit access to a computer file. Because the secretary has no idea which password is correct, he chooses one of the passwords at random and tries it. If the password is incorrect, he discards it and randomly selects another password from among those remaining, proceeding in this manner until he finds the correct password.

a   What is the probability that he obtains the correct password on the first try?

b   What is the probability that he obtains the correct password on the second try? The third try?

c   A security system has been set up so that if three incorrect passwords are tried before the correct one, the computer file is locked and access to it denied. If $n = 7$, what is the probability that the secretary will gain access to the file?

2.10   The Law of Total Probability and Bayes' Rule

The event-composition approach to solving probability problems is sometimes facilitated by viewing the sample space, S, as a union of mutually exclusive subsets and using the following *law of total probability*. The results of this section are based on the following construction.

DEFINITION 2.11

For some positive integer k, let the sets $B_1, B_2, \ldots, B_k$ be such that

1. $S = B_1 \cup B_2 \cup \cdots \cup B_k.$
2. $B_i \cap B_j = \emptyset$ for $i \neq j.$

Then the collection of sets $\{B_1, B_2, \ldots, B_k\}$ is said to be a *partition* of S.

If A is any subset of S and $\{B_1, B_2, \ldots, B_k\}$ is a partition of S, A can be *decomposed* as follows:

$$A = (A \cap B_1) \cup (A \cap B_2) \cup \cdots \cup (A \cap B_k).$$

Figure 2.12 illustrates this decomposition for $k = 3$.

FIGURE 2.12
Decomposition of
event A.

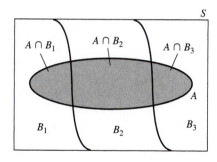

THEOREM 2.8   Assume that $\{B_1, B_2, \ldots, B_k\}$ is a partition of S (see Definition 2.11) such that $P(B_i) > 0$, for $i = 1, 2, \ldots, k$. Then for any event A

$$P(A) = \sum_{i=1}^{k} P(A|B_i)P(B_i).$$

Proof   Any subset A of S can be written as

$$A = A \cap S = A \cap (B_1 \cup B_2 \cup \cdots \cup B_k)$$
$$= (A \cap B_1) \cup (A \cap B_2) \cup \cdots \cup (A \cap B_k).$$

Notice that, because $\{B_1, B_2, \cdots, B_k\}$ is a partition of S, if $i \neq j$,

$$(A \cap B_i) \cap (A \cap B_j) = A \cap (B_i \cap B_j) = A \cap \emptyset = \emptyset$$

and that $(A \cap B_i)$ and $(A \cap B_j)$ are mutually exclusive events. Thus,

$$P(A) = P(A \cap B_1) + P(A \cap B_2) + \cdots + P(A \cap B_k)$$
$$= P(A|B_1)P(B_1) + P(A|B_2)P(B_2) + \cdots + P(A|B_k)P(B_k)$$
$$= \sum_{i=1}^{k} P(A|B_i)P(B_i). \qquad \blacksquare$$

In the examples and exercises that follow, you will see that it is sometimes much easier to calculate the conditional probabilities $P(A|B_i)$ for suitably chosen B_i than it is to compute $P(A)$ directly. In such cases, the law of total probability can be applied to determine $P(A)$. Using the result of Theorem 2.8, it is a simple matter to derive the result known as *Bayes' rule*.

THEOREM 2.9   **Bayes' Rule**   Assume that $\{B_1, B_2, \ldots, B_k\}$ is a partition of S (see Definition 2.11) such that $P(B_i) > 0$, for $i = 1, 2, \ldots, k$. Then

$$P(B_j|A) = \frac{P(A|B_j)P(B_j)}{\sum_{i=1}^{k} P(A|B_i)P(B_i)}.$$

Proof   The proof follows directly from the definition of conditional probability and the law of total probability. Note that

$$P(B_j|A) = \frac{P(A \cap B_j)}{P(A)} = \frac{P(A|B_j)P(B_j)}{\sum_{i=1}^{k} P(A|B_i)P(B_i)}. \qquad \blacksquare$$

EXAMPLE **2.23**   An electronic fuse is produced by five production lines in a manufacturing operation. The fuses are costly, are quite reliable, and are shipped to suppliers in 100-unit lots. Because testing is destructive, most buyers of the fuses test only a small number of fuses before deciding to accept or reject lots of incoming fuses.

All five production lines produce fuses at the same rate and normally produce only 2% defective fuses, which are dispersed randomly in the output. Unfortunately, production line 1 suffered mechanical difficulty and produced 5% defectives during the month of March. This situation became known to the manufacturer after the fuses had been shipped. A customer received a lot produced in March and tested three fuses. One failed. What is the probability that the lot was produced on line 1? What is the probability that the lot came from one of the four other lines?

Solution   Let $P(L_i)$ denote the probability that a fuse was drawn from line i and let D denote the event that a fuse was defective. Then it follows directly that

$$P(L_1) = 0.2 \quad \text{and} \quad P(D|L_1) = 3(.05)(.95)^2 = .135375.$$

Similarly,

$$P(\overline{L_1}) = 0.8 \quad \text{and} \quad P(D|\overline{L_1}) = 3(.02)(.98)^2 = .057624.$$

Note that these conditional probabilities were very easy to calculate. Using the law of total probability,

$$P(D) = P(D|L_1)P(L_1) + P(D|\overline{L_1})P(\overline{L_1})$$

$$= (.135375)(.2) + (.057624)(.8) = .0731742.$$

Finally,

$$P(L_1|D) = \frac{P(L_1 \cap D)}{P(D)} = \frac{P(D|L_1)P(L_1)}{P(D)} = \frac{(.135375)(.2)}{.0731742} = .37$$

and

$$P(\overline{L_1}|D) = 1 - P(L_1|D) = 1 - .37 = .63.$$

Exercises

2.98   A population of voters contains 40% Republicans and 60% Democrats. It is reported that 30% of the Republicans and 70% of the Democrats favor an election issue. A person chosen at random from this population is found to favor the issue in question. Find the conditional probability that this person is a Democrat.

2.99   A diagnostic test for a disease is said to be 90% accurate in that if a person has the disease, the test will detect it with probability .9. Also, if a person does not have the disease, the test will report that he or she does not have it with probability .9. Only 1% of the population has the disease in question. If a person is chosen at random from the population and the diagnostic test indicates that she has the disease, what is the conditional probability that she does, in fact, have the disease? Are you surprised by the answer? Would you call this diagnostic test reliable?

2.100 We know the following about a colormetric method used to test lake water for nitrates. If water specimens contain nitrates, a solution dropped into the water will cause the specimen to turn red 95% of the time. When used on water specimens without nitrates, the solution causes the water to turn red 10% of the time (because chemicals other than nitrates are sometimes present and they also react to the agent). Past experience in a lab indicates that nitrates are contained in 30% of the water specimens that are sent to the lab for testing.

a If a water specimen is randomly selected from among those sent to the lab, what is the probability that it will turn red when tested?

b If a water specimen is randomly selected and turns red when tested, what is the probability that it actually contain nitrates?

2.101 Medical case histories indicate that different illnesses may produce identical symptoms. Suppose that a particular set of symptoms, denoted H, occurs only when any one of three illnesses, I_1, I_2, or I_3, occurs. Assume that the simultaneous occurrence of more that one of these illnesses is impossible and that

$$P(I_1) = .01, \quad P(I_2) = .005, \quad \text{and} \quad P(I_3) = .02.$$

The probabilities of developing the set of symptoms H, given each of these illnesses, are known to be

$$P(H|I_1) = .90, \quad P(H|I_2) = .95, \quad \text{and} \quad P(H|I_3) = .75.$$

Assuming that an ill person exhibits the symptoms, H, what is the probability that the person has illness I_1?

2.102 Use Theorem 2.8, the law of total probability, to prove the following.

a If $P(A|B) = P(A|\overline{B})$, then A and B are independent.

b If $P(A|C) > P(B|C)$ and $P(A|\overline{C}) > P(B|\overline{C})$, then $P(A) > P(B)$.

2.103 Males and females are observed to react differently to a given set of circumstances. It has been observed that 70% of the females react positively to these circumstances, whereas only 40% of males react positively. A group of 20 people, 15 female and 5 male, was subjected to these circumstances, and the subjects were asked to describe their reactions on a written questionnaire. A response picked at random from the 20 was negative. What is the probability that it was that of a male?

2.104 A study of Georgia residents suggests that those who worked in shipyards during World War II were subjected to a significantly higher risk of lung cancer (*Wall Street Journal*, September 21, 1978).[3] It was found that approximately 22% of those persons who had lung cancer worked at some prior time in a shipyard. In contrast, only 14% of those who had no lung cancer worked at some prior time in a shipyard. Suppose that the proportion of all Georgians living during World War II who have or will have contracted lung cancer is .04%. Find the percentage of Georgians living during the same period who will contract (or have contracted) lung cancer, given that they have at some prior time worked in a shipyard.

---

[3] *Source:* Reprinted by permission of the *Wall Street Journal*, © Dow Jones & Company, Inc. 1981. All rights reserved worldwide.

2.105   A study of the residents of region showed that 20% were smokers. The probability of death due to lung cancer, given that a person smoked, was ten times the probability of death due to lung cancer, given that the person did not smoke. If the probability of death due to lung cancer in the region is .006, what is the probability of death due to lung cancer given that the person is a smoker?

2.106   A bowl contains w white balls and b black balls. One ball is selected at random from the bowl, its color is noted, and it is returned to the bowl along with n additional balls of the same color. Another single ball is randomly selected from the bowl (now containing $w + b + n$ balls) and it is observed that the ball is black. Show that the (conditional) probability that the first ball selected was white is $\dfrac{w}{w + b + n}$.

2.107   The *symmetric difference* between two events A and B is the set of all sample points that are in *exactly one* of the sets and is often denoted $A \triangle B$. Note that $A \triangle B = (A \cap \overline{B}) \cup (\overline{A} \cap B)$. Prove that $P(A \triangle B) = P(A) + P(B) - 2P(A \cap B)$.

2.108   A plane is missing and is presumed to have equal probability of going down in any of three regions. If a plane is actually down in region i, let $1 - \alpha_i$ denote the probability that the plane will be found upon a search of the ith region, $i = 1, 2, 3$.

    a   What is the conditional probability that the plane is in region 1 given that the search of region 1 was unsuccessful?

    b   What is the conditional probability that the plane is in region 2 given that the search of region 1 was unsuccessful?

    c   What is the conditional probability that the plane is in region 3 given that the search of region 1 was unsuccessful?

2.109   As items come to the end of a production line, an inspector chooses which items are to go through a complete inspection. Ten percent of all items produced are defective. Sixty percent of all defective items go through a complete inspection, and 20% of all good items go through a complete inspection. Given that an item is completely inspected, what is the probability it is defective?

2.110   Many public schools are implementing a "no-pass, no-play" rule for athletes. Under this system, a student who fails a course is disqualified from participating in extracurricular activities during the next grading period. Suppose that the probability is .15 that an athlete who has not previously been disqualified will be disqualified next term. For athletes who have been previously disqualified, the probability of disqualification next term is .5. If 30% of the athletes have been disqualified in previous terms, what is the probability that a randomly selected athlete will be disqualified during the next grading period?

2.111   A student answers a multiple-choice examination question that offers four possible answers. Suppose that the probability that the student knows the answer to the question is .8 and the probability that the student will guess is .2. Assume that if the student guesses, the probability of selecting the correct answer is .25. If the student correctly answers a question, what is the probability that the student really knew the correct answer?

2.112   Two methods, A and B, are available for teaching a certain industrial skill. The failure rate is 20% for A and 10% for B. However, B is more expensive and hence is used only 30% of the time. (A is used the other 70%.) A worker was taught the skill by one of the methods but failed to learn it correctly. What is the probability that she was taught by method A?

2.113   Of the travelers arriving at a small airport, 60% fly on major airlines, 30% fly on privately owned planes, and the remainder fly on commercially owned planes not belonging to a major airline. Of those traveling on major airlines, 50% are traveling for business reasons, whereas

60% of those arriving on private planes and 90% of those arriving on other commercially owned planes are traveling for business reasons. Suppose that we randomly select one person arriving at this airport. What is the probability that the person

a is traveling on business?

b is traveling for business on a privately owned plane?

c arrived on a privately owned plane, given that the person is traveling for business reasons?

d is traveling on business, given that the person is flying on a commercially owned plane?

2.114 A personnel director has two lists of applicants for jobs. List 1 contains the names of five women and two men, whereas list 2 contains the names of two women and six men. A name is randomly selected from list 1 and added to list 2. A name is then randomly selected from the augmented list 2. Given that the name selected is that of a man, what is the probability that a woman's name was originally selected from list 1?

2.115 Five identical bowls are labeled 1, 2, 3, 4, and 5. Bowl i contains i white and $5 - i$ black balls, with $i = 1, 2, \ldots, 5$. A bowl is randomly selected and two balls are randomly selected (without replacement) from the contents of the bowl.

a What is the probability that both balls selected are white?

b Given that both balls selected are white, what is the probability that bowl 3 was selected?

*2.116 Following is a description of the game of *craps*. A player rolls two dice and computes the total of the spots showing. If the player's first toss is a 7 or an 11, the player wins the game. If the first toss is a 2, 3, or 12, the player loses the game. If the player rolls anything else (4, 5, 6, 8, 9 or 10) on the first toss, that value becomes the player's *point*. If the player does not win or lose on the first toss, he tosses the dice repeatedly until he obtains either his point or a 7. He wins if he tosses his point before tossing a 7 and loses if he tosses a 7 before his point. What is the probability that the player wins a game of craps? (*Hint:* Recall Exercise 2.95.)

2.11 Numerical Events and Random Variables

Events of major interest to the scientist, engineer, or businessperson are those identified by numbers, called *numerical events*. The research physician is interested in the event that ten of ten treated patients survive an illness; the businessperson is interested in the event that sales next year will reach $5 million. Let Y denote a variable to be measured in an experiment. Because the value of Y will vary depending on the outcome of the experiment, it is called a *random variable*.

To each point in the sample space we will assign a real number denoting the value of the variable Y. The value assigned to Y will vary from one sample point to another, but some points may be assigned the same numerical value. Thus we have defined a variable that is a function of the sample points in S, and {all sample points where $Y = a$} is the numerical event assigned the number a. Indeed, the sample space S can be partitioned into subsets so that points within a subset are all assigned the same value of Y. These subsets are mutually exclusive since no point is assigned two different numerical values. The partitioning of S is symbolically indicated in Figure 2.13 for a random variable that can assume values 0, 1, 2, 3, 4.

FIGURE **2.13**
Partitioning S into
subsets that define
the events
$Y = 0, 1, 2, 3$, and 4

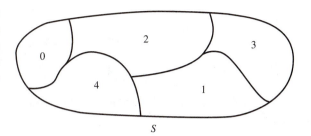

DEFINITION **2.12**

A *random variable* is a real-valued function for which the domain is a sample space.

EXAMPLE **2.24**   Define an experiment as tossing two coins and observing the results. Let Y equal the number of heads obtained. Identify the sample points in S, assign a value of Y to each sample point, and identify the sample points associated with each value of the random variable Y.

Solution   Let H and T represent head and tail, respectively; and let an ordered pair of symbols identify the outcome for the first and second coins. (Thus HT implies a head on the first coin and a tail on the second.) Then the four sample points in S are E_1: HH, E_2: HT, E_3: TH and E_4: TT. The values of Y assigned to the sample points depend on the number of heads associated with each point. For E_1: HH, two heads were observed, and E_1 is assigned the value $Y = 2$. Similarly, we assign the values $Y = 1$ to E_2 and E_3 and $Y = 0$ to E_4. Summarizing, the random variable Y can take three values, $Y = 0, 1, 2$, which are events defined by specific collections of sample points:

$$\{Y = 0\} = \{E_4\}, \qquad \{Y = 1\} = \{E_2, E_3\}, \quad \text{and} \quad \{Y = 2\} = \{E_1\}.$$

Let y denote an observed value of the random variable Y. Then $P(Y = y)$ is the sum of the probabilities of the sample points that are assigned the value y.

EXAMPLE **2.25**   Compute the probabilities for each value of Y in Example 2.24.

Solution   The event $\{Y = 0\}$ results only from sample point E_4. If the coins are balanced, the sample points are equally likely and hence

$$P(Y = 0) = P(E_4) = 1/4.$$

Similarly,

$$P(Y = 1) = P(E_2) + P(E_3) = 1/2 \quad \text{and} \quad P(Y = 2) = P(E_1) = 1/4.$$

A more detailed examination of random variables will be undertaken in the next two chapters.

Exercises

2.117 Refer to Exercise 2.88. Let the random variable Y represent the number of radar sets that detect a particular aircraft. Compute the probabilities associated with each value of Y.

2.118 Refer to Exercise 2.96. Let the random variable Y represent the number of defective refrigerators found after three refrigerators have been tested. Compute the probabilities for each value of Y.

2.119 Refer again to Exercise 2.96. Let the random variable Y represent the number of the test in which the last defective refrigerator is identified. Compute the probabilities for each value of Y.

2.120 A spinner can land in any of four positions, A, B, C, and D, with equal probability. The spinner is used twice and the position is noted each time. Let the random variable Y denote the number of positions on which the spinner did *not* land. Compute the probabilities for each value of Y.

2.12 Random Sampling

As our final topic in this chapter, we move from theory to application and examine the nature of experiments conducted in statistics. A statistical experiment involves the observation of a sample selected from a larger body of data, existing or conceptual, called a *population*. The measurements in the sample, viewed as observations of the values of one or more random variables, are then employed to make an inference about the characteristics of the target population.

How are these inferences made? An exact answer to this question is deferred until later, but a general observation follows from our discussion in Section 2.2. There we learned that the probability of the observed sample plays a major role in making an inference and evaluating the credibility of the inference.

Without belaboring the point, it is clear that the method of sampling will affect the probability of a particular sample outcome. For example, suppose that a fictitious population contains only $N = 5$ elements, from which we plan to take a sample of size $n = 2$. You could mix the elements thoroughly and select two in such a way that all pairs of elements possess an equal probability of selection. A second sampling procedure might require selecting a single element, replacing it in the population, and then drawing a single element again. The two methods of sample selection are called *sampling without* and *with replacement*, respectively.

If all the $N = 5$ population elements are distinctly different, the probability of drawing a specific pair, when sampling without replacement, is $1/10$. The probability of drawing the same specific pair, when sampling with replacement, is $2/25$. You can easily verify these results.

The point that we make is that the method of sampling, known as the *design of an experiment*, affects both the quantity of information in a sample and the probability of observing a specific sample result. Hence every sampling procedure must be clearly described if we wish to make valid inferences from sample to population.

The study of the design of experiments, the various types of designs along with their properties, is a course in itself. Hence at this early stage of study we introduce only the simplest sampling procedure, *simple random sampling*. The notion of simple random sampling will be needed in subsequent discussions of the probabilities associated with random variables, and it will inject some realism into our discussion of statistics. This is because simple random sampling is often employed in practice. Now let us define the term *random sample*.

DEFINITION **2.13**

> Let N and n represent the numbers of elements in the population and sample, respectively. If the sampling is conducted in such a way that each of the $\binom{N}{n}$ samples has an equal probability of being selected, the sampling is said to be random and the result is said to be a *random sample*.

Perfect random sampling is difficult to achieve in practice. If the population is not too large, we might write each of the N numbers on a poker chip, mix all the chips, and select a sample of n chips. The numbers on the poker chips would specify the measurements to appear in the sample.

Tables of random numbers have been formed by computer to expedite the selection of random samples. An example of such a table is Table 12, Appendix III. A random number table is a set of integers $(0, 1, \ldots, 9)$ generated so that, in the long run, the table will contain all ten integers in approximately equal proportions, with no trends in the patterns in which the digits were generated. Thus if one digit is selected from a random point on the table, it is equally likely to be any of the digits 0 through 9.

Choosing numbers from the table is analogous to drawing numbered poker chips from the mixed pile, as mentioned earlier. Suppose we want a random sample of three persons to be selected from a population of seven persons. We could number the people from 1 to 7, put the numbers on chips, thoroughly mix the chips, and then draw three out. Analogously, we could drop a pencil point on a random starting point in Table 12, Appendix III. Suppose the point falls on the fifteenth line of column 9 and we decide to use the rightmost digit of the group of five, which is a 5 in this case. This process is like drawing the chip numbered 5. We may now proceed in any direction to obtain the remaining numbers in the sample. If we decide to proceed down the page, the next number (immediately below the 5) is a 2. So our second sampled person would be number 2. Proceeding, we next come to an 8, but there are only seven elements in the population. Thus the 8 is ignored and we continue down the column. Two more 5s then appear, but they must both be ignored because person 5 has already been selected. (The chip numbered 5 has been removed from the pile.)

Finally, we come to a 1, and our sample of three is completed with persons numbered 5, 2, and 1.

Any starting point can be used in a random number table, and we may proceed in any direction from the starting point. However, if more than one sample is to be used in any problem, each should have a unique starting point.

In many situations the population is conceptual, as in an observation made during a laboratory experiment. Here the population is envisioned to be the infinitely many measurements that would be obtained if the experiment were to be repeated over and over again. If we wish a sample of $n = 10$ measurements from this population, we repeat the experiment ten times and hope that the results represent, to a reasonable degree of approximation, a random sample.

Although the primary purpose of this discussion was to clarify the meaning of a random sample, we would like to mention that some sampling techniques are only partially random. For instance, if we wish to determine the voting preference of the nation in a presidential election, we would not likely choose a random sample from the population of voters. By pure chance, all the voters appearing in the sample might be drawn from a single city, say, San Francisco, which might not be at all representative of the population of all voters in the United States. We would prefer a random selection of voters from smaller political districts, perhaps states, allotting a specified number to each state. The information from the randomly selected subsamples drawn from the respective states would be combined to form a prediction concerning the entire population of voters in the country. In general, we want to select a sample so as to obtain a specified quantity of information at minimum cost.

2.13 Summary

This chapter has been concerned with providing a model for the repetition of an experiment and, consequently, a model for the population frequency distributions of Chapter 1. The acquisition of a probability distribution is the first step in forming a theory to model reality and to develop the machinery for making inferences.

An experiment was defined as the process of making an observation. The concepts of an event, a simple event, the sample space, and the probability axioms have provided a probabilistic model for calculating the probability of an event. Numerical events and the definition of a random variable were introduced in Section 2.11.

Inherent in the model is the sample-point approach for calculating the probability of an event (Section 2.5). Counting rules useful in applying the sample-point method were discussed in Section 2.6. The concept of conditional probability, the operations of set algebra, and the laws of probability set the stage for the event-composition method for calculating the probability of an event (Section 2.9).

Of what value is the theory of probability? It provides the theory and the tools for calculating the probabilities of numerical events and hence the probability distributions for the random variables that will be discussed in Chapter 3. The numerical events of interest to us appear in a sample, and we will wish to calculate the probability of an observed sample to make an inference about the target population. Probability provides both the foundation and the tools for statistical inference, the objective of statistics.

References and Further Readings

Cramer, H. 1973. *The Elements of Probability Theory and Some of Its Applications*, 2d ed. Huntington, NY: Krieger.

Feller, W. 1968. *An Introduction to Probability Theory and Its Applications*, 3d. ed., vol. 1. New York: John Wiley.

———. 1971. *An Introduction to Probability Theory and Its Applications*, 2d. ed., vol. 2. New York: John Wiley.

Meyer, P. L. 1970. *Introductory Probability and Statistical Applications*, 2d ed. Reading, Mass.: Addison-Wesley.

Parzen, E. 1960. *Modern Probability Theory and Its Applications*. New York: John Wiley.

Riordan, J. 1980. *An Introduction to Combinatorial Analysis*. Princeton, N.J.: Princeton University Press.

Wright, J. W., ed. 1992. *The Universal Almanac, 1993*. Kansas City, Mo., and New York: Andrews & McMeel.

Supplementary Exercises

2.121 Show that Theorem 2.7 holds for *conditional* probabilities. That is, if $P(B) > 0$, then $P(A|B) = 1 - P(\overline{A}|B)$.

2.122 Let S contain four sample points, E_1, E_2, E_3, and E_4.

a List all possible events in S (include the null event).

b In Exercise 2.54(d), you showed that $\sum_{i=1}^{n} \binom{n}{i} = 2^n$. Use this result to give the total number of events in S.

c Let A and B be the events $\{E_1, E_2, E_3\}$ and $\{E_2, E_4\}$, respectively. Give the sample points in the following events: $A \cup B$, $A \cap B$, $\overline{A} \cap \overline{B}$, and $\overline{A} \cup B$.

2.123 A patient receiving a yearly physical examination must have 18 checks or tests performed. The sequence in which the tests are conducted is important because the time lost between tests will vary depending on the sequence. If an efficiency expert were to study the sequences to find the one that required the minimum length of time, how many sequences would be included in her study if all possible sequences were admissible?

2.124 Five cards are drawn from a standard 52-card playing deck. What is the probability that all 5 cards will be of the same suit?

2.125 Refer to Exercise 2.124. What is the probability of a full house'? (Recall that a full house is a hand of five cards containing one pair and three of another kind, e.g., two kings and three tens.)

2.126 A bin contains three components from supplier A, four from supplier B, and five from supplier C. If four of the components are randomly selected for testing, what is the probability that each supplier would have at least one component tested?

2.127 A large group of people is to be checked for two common symptoms of a certain disease. It is thought that 20% of the people possess symptom A alone, 30% possess symptom B alone,

10% possess both symptoms, and the remainder have neither symptom. For one person chosen at random from this group, find these probabilities:

a   that the person has neither symptom.

b   that the person has at least one symptom.

c   that the person has both symptoms, given that he has symptom B.

2.128   Refer to Exercise 2.127. Let the random variable Y represent the number of symptoms possessed by a person chosen at random from the group. Compute the probabilities associated with each value of Y.

*2.129   A Model for the World Series : Two teams A and B play a series of games until one team wins four games. We assume that the games are played independently and that the probability that A wins any game is p. What is the probability that the series last exactly five games?

2.130   **a**   A drawer contains $n = 5$ different and distinguishable pairs of socks (a total of 10 socks). If a person (perhaps in the dark) randomly selects 4 socks, what is the probability that there is no matching pair in the sample?

*b   A drawer contains n different and distinguishable pairs of socks (a total of $2n$ socks). A person randomly selects $2r$ of the socks, where $2r < n$. In terms of n and r, what is the probability that there is no matching pair in the sample?

2.131   A group of men possesses the three characteristics of being married (A), having a college degree (B), and being a citizen of a specified state (C), according to the fractions given in the accompanying Venn diagram. That is, 5% of the men possess all three characteristics, whereas 20% have a college education but are not married and are not citizens of the specified state. One man is chosen at random from this group.

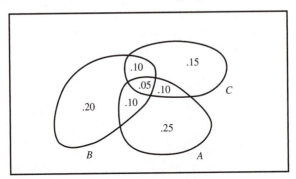

a   Find the probability that he is married.

b   Find the probability that he has a college degree and is married.

c   Find the probability that he is not from the specified state but is married and has a college degree.

d   Find the probability that he is not married or does not have a college degree, given that he is from the specified state.

2.132   The accompanying table lists accidental deaths by age and certain specific types for the United States in 1976.

a   A randomly selected person from the United States was known to have an accidental death in 1976.

i.   Find the probability that he was over the age of 15.

ii.   Find the probability that the cause of death was a motor vehicle accident.

| | Type of Accident | | |
Age	All Types	Motor Vehicle	Falls	Drowning
Under 5	4,692	1,532	201	720
5 to 14	6,308	3,175	121	1,050
15 to 24	24,316	16,650	463	2,090
25 to 34	13,868	7,888	426	1,060
35 to 44	8,531	4,224	534	520
45 to 54	9,434	4,118	931	500
55 to 64	9,566	3,652	1,340	420
65 to 74	8,823	3,082	1,997	270
75 and over	15,223	2,717	8,123	197
Total	100,761	47,038	14,136	6,827

Source: The World Almanac and Book of Facts, 1979 ed., copyright ©Newspaper Enterprise Association Inc., 1978, New York, NY 10166.

    iii. Find the probability that the cause of death was a motor vehicle accident, given that the person was between 15 and 24 years old.

    iv. Find the probability that the cause of death was a drowning accident, given that it was not a motor vehicle accident and the person was 34 or under.

  b From these figures can you determine the probability that a person selected at random from the U.S. population had a fatal motor vehicle accident in 1976?

2.133 The accompanying table is derived from 1970 U.S. census population figures.

| | 1970 U.S. Population | | |
Area	Total	White	Black and Other
Urban	149,325	128,773	20,552
Inside urbanized areas	118,447	100,952	17,495
Central cities	63,922	49,547	14,375
Urban fringe	54,525	51,405	3,120
Outside urbanized areas	30,878	27,821	3,057
Rural	53,887	48,976	4,911
Total	203,212	177,749	25,463

Source: The World Almanac and Book of Facts, 1979 ed., copyright ©Newspaper Enterprise Association Inc., 1978, New York, NY 10166.

(The figures are in thousands.) If a person is selected at random from the 1970 U.S. population, find the probabilities for the following:

a The person is white.

b The person lives in a central city area.

c The person lives in an urban fringe, given that she is white.

d The person is white, given that she lives in an urban fringe.

e The person lives outside an urbanized area, given that she is not white.

f The person either is not white living in a central city or is white and lives outside an urbanized area.

2.134 A machine for producing a new experimental electronic component generates defectives from time to time in a random manner. The supervising engineer for a particular machine has noticed that defectives seem to be grouping (hence appearing in a nonrandom manner), thereby suggesting a malfunction in some part of the machine. One test for nonrandomness is based on the number of *runs* of defectives and nondefectives (a run is an unbroken sequence of either defectives or nondefectives). The smaller the number of runs, the greater will be the amount of evidence indicating nonrandomness. Of 12 components drawn from the machine, the first 10 were not defective and the last 2 were defective ($NNNNNNNNNNDD$). Assume randomness.

a   What is the probability of observing this arrangement (resulting in two runs) given that 10 of the 12 components are not defective?

b   What is the probability of observing two runs?

2.135   Refer to Exercise 2.134. What is the probability that the number of runs, R, is less than or equal to 3?

2.136   Assume that there are nine parking spaces next to one another in a parking lot. Nine cars need to be parked by an attendant. Three of the cars are expensive sports cars, three are large domestic cars, and three are imported compacts. Assuming that the attendant parks the cars at random, what is the probability that the three expensive sports cars are parked adjacent to one another?

2.137   Relays used in the construction of electric circuits function properly with probability 0.9. Assuming that the circuits operate independently, which of the following circuit designs yields the higher probability that current will flow when the relays are activated?

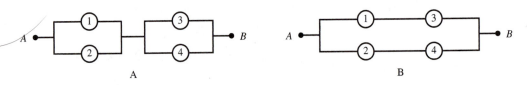

2.138   Eight tires of different brands are ranked from 1 to 8 (best to worst) according to mileage performance. If four of these tires are chosen at random by a customer, find the probability that the best tire among those selected by the customer is actually ranked third among the original eight.

2.139   Refer to Exercise 2.138. Let Y denote the actual quality rank of the best tire selected by the customer. In Exercise 2.138, you computed $P(Y = 3)$. Give the possible values of Y and the probabilities associated with all of these values.

2.140   As in Exercises 2.138 and 2.139, eight tires of different brands are ranked from 1 to 8 (best to worst) according to mileage performance. If four of these tires are chosen at random by a customer:

a   What is the probability that the best tire selected is ranked 3 and the worst is ranked 7?

b   In part (a) you computed the probability that the best tire selected is ranked 3 and the worst is ranked 7. If that is the case, the *range* of the ranks, R = largest rank–smallest rank = $7 - 3 = 4$. What is $P(R = 4)$?

c   Give all possible values for R and the probabilities associated with all of these values.

*2.141   Three beer drinkers (say I, II, and III) are to rank four different brands of beer (say A, B, C, and D) in a blindfold test. Each drinker ranks the four beers as 1 (for the beer he or she liked best), 2 (for the next best), 3, or 4.

a. Carefully describe a sample space for this experiment (note that we need to specify the ranking of all four beers for all three drinkers). How many sample points are in this sample space?

b. Assume that the drinkers cannot discriminate between the beers, so that each assignment of ranks to the beers is equally likely. After all the beers are ranked by all three drinkers, the ranks of each brand of beer are summed. What is the probability that some beer will receive a total rank of 4 or less?

2.142 Three names are to be selected from a list of seven names for a public opinion survey. Find the probability that the first name on the list is selected for the survey.

2.143 An AP news service story, printed in the *Gainesville Sun* of May 20, 1979, states the following with regard to debris from Skylab striking someone on the ground: "The odds are 1 in 150 that a piece of Skylab will hit someone. But 4 billion people ... live in the zone in which pieces could fall. So any one person's chances of being struck are one in 150 times 4 billion—or one in 600 billion." Do you see any inaccuracies in this reasoning?

2.144 Let A and B be any two events. Which of the following statements, in general, are false?

$$P(A|B) + P(\overline{A}|\overline{B}) = 1, \qquad P(A|B) + P(A|\overline{B}) = 1, \qquad P(A|B) + P(\overline{A}|B) = 1.$$

2.145 Three events, A, B, and C, are said to be mutually independent if

$$P(AB) = P(A)P(B), \qquad P(BC) = P(B)P(C),$$
$$P(AC) = P(A)P(C), \qquad P(ABC) = P(A)P(B)P(C).$$

Suppose that a balanced coin is independently tossed two times. Define the following events:

A: Head appears on the first toss.

B: Head appears on the second toss.

C: Both tosses yield the same outcome.

Are A, B, and C mutually independent?

*2.146 Suppose that the probability of exposure to the flu during an epidemic is .6. Experience has shown that a serum is 80% successful in preventing an inoculated person from acquiring the flu, if exposed to it. A person not inoculated faces a probability of .90 of acquiring the flu if exposed to it. Two persons, one inoculated and one not, perform a highly specialized task in a business. Assume that they are not at the same location, are not in contact with the same people, and cannot expose each other to the flu. What is the probability that at least one will get the flu?

*2.147 Two gamblers bet $1 each on the successive tosses of a coin. Each has a bank of $6.

a What is the probability that they break even after six tosses of the coin?

b What is the probability that one player, say, Jones, wins all the money on the tenth toss of the coin?

*2.148 Suppose that the streets of a city are laid out in a grid with streets running north-south and east-west. Consider the following scheme for patrolling an area of 16 blocks by 16 blocks. An officer commences walking at the intersection in the center of the area. At the corner of each block the officer randomly elects to go north, south, east, or west.

a What is the probability that the officer will reach the boundary of the patrol area after walking the first 8 blocks?

b   What is the probability that the officer will return to the starting point after walking exactly 4 blocks?

*2.149   Suppose that n indistinguishable balls are to be arranged in N distinguishable boxes so that each distinguishable arrangement is equally likely. If $n \geq N$, show that the probability no box will be empty is given by

$$\frac{\binom{n-1}{N-1}}{\binom{N+n-1}{N-1}}.$$

CHAPTER **3**

Discrete Random Variables and Their Probability Distributions

3.1 Basic Definition

3.2 The Probability Distribution for a Discrete Random Variable

3.3 The Expected Value of a Random Variable or a Function of a Random Variable

3.4 The Binomial Probability Distribution

3.5 The Geometric Probability Distribution

3.6 The Negative Binomial Probability Distribution (Optional)

3.7 The Hypergeometric Probability Distribution

3.8 The Poisson Probability Distribution

3.9 Moments and Moment-Generating Functions

3.10 Probability-Generating Functions (Optional)

3.11 Tchebysheff's Theorem

3.12 Summary

References and Further Readings

3.1 Basic Definition

As stated in Section 2.12, a random variable is a real-valued function defined over a sample space. Consequently, a random variable can be used to identify numerical events that are of interest in an experiment. For example, the event of interest in an opinion poll regarding voter preferences is not usually the particular people sampled or the order in which preferences were obtained but $Y =$ the *number* of voters favoring a certain candidate or issue. The observed value of this random variable must be zero or an integer between 1 and the sample size. Thus, this random variable can take on only a finite number of values with nonzero probability. A random variable of this type is said to be discrete.

DEFINITION **3.1**

A random variable Y is said to be *discrete* if it can assume only a finite or countably infinite[1] number of distinct values.

A less formidable characterization of discrete random variables can be obtained by considering some practical examples. The number of bacteria per unit area in the study of drug control on bacterial growth is a discrete random variable, as is the number of defective television sets in a shipment of 100 sets. Indeed, discrete random variables often represent counts associated with real phenomena.

Let us now consider the relation of the material in Chapter 2 to this chapter. Why study the theory of probability? The answer is that the probability of an observed event is needed to make inferences about a population. The events of interest are often numerical events that correspond to values of discrete random variables. Hence it is imperative that we know the probabilities of these numerical events. Because certain types of random variables occur so frequently in practice, it is useful to have at hand the probability for each value of a random variable. This collection of probabilities is called the *probability distribution* of the discrete random variable. We will find that many experiments exhibit similar characteristics and generate random variables with the same type of probability distribution. Consequently, knowledge of the probability distributions for random variables associated with common types of experiments will eliminate the need for solving the same probability problems over and over again.

3.2 The Probability Distribution for a Discrete Random Variable

Notationally, we will use *uppercase letters*, such as Y, to denote *random variables* and *lowercase letters*, such as y, to denote *particular values* that a random variable may assume. For example, let Y denote any one of the six possible values that could be observed on the upper face when a die is tossed. After the die is tossed, the number actually observed will be denoted by the symbol y. Note that Y is a random variable, but the specific observed value, y, is not random.

The expression $(Y = y)$ can be read, *the set of all points in S assigned the value y by the random variable Y*.

It is now meaningful to talk about the probability that Y takes on the value y, denoted by $P(Y = y)$. As in Section 2.11 this probability is defined as the sum of the probabilities of appropriate sample points in S.

---

[1] Recall that a set of elements is countably infinite if the elements in the set can be put into one-to-one correspondence with the positive integers.

DEFINITION 3.2

The probability that Y takes on the value y, $P(Y = y)$, is defined as the *sum of the probabilities of all sample points in S* that are assigned the value y. We will sometimes denote $P(Y = y)$ by $p(y)$.

Notice that $p(y)$ is nothing more than a function that assigns probabilities to each value y; hence it is sometimes called the *probability function* for Y.

DEFINITION 3.3

The *probability distribution* for a discrete variable Y can be represented by a formula, a table, or a graph, which provides $p(y) = P(Y = y)$ for all y.

Notice that $p(y) \geq 0$ for all y, but the probability distribution for a discrete random variable assigns nonzero probabilities to only a countable number of distinct y values. Any value y not explicitly assigned a positive probability is understood to be such that $p(y) = 0$. We illustrate these ideas with an example.

EXAMPLE 3.1

A supervisor in a manufacturing plant has three men and three women working for him. He wants to choose two workers for a special job. Not wishing to show any biases in his selection, he decides to select the two workers at random. Let Y denote the number of women in his selection. Find the probability distribution for Y.

Solution

The supervisor can select two workers from six in $\binom{6}{2} = 15$ ways. Hence S contains 15 sample points, which we assume to be equally likely because random sampling was employed. Thus $P(E_i) = 1/15$, for $i = 1, 2, \ldots, 15$. The values for Y that have nonzero probability are 0, 1, and 2. The number of ways of selecting $Y = 0$ women is $\binom{3}{0}\binom{3}{2}$ because the supervisor must select zero workers from the three women and two from the three men. Thus there are $\binom{3}{0}\binom{3}{2} = 1 \cdot 3 = 3$ sample points in the event $Y = 0$, and

$$p(0) = P(Y = 0) = \frac{\binom{3}{0}\binom{3}{2}}{15} = \frac{3}{15} = \frac{1}{5}.$$

Similarly,

$$p(1) = P(Y = 1) = \frac{\binom{3}{1}\binom{3}{1}}{15} = \frac{9}{15} = \frac{3}{5}$$

$$p(2) = P(Y = 2) = \frac{\binom{3}{2}\binom{3}{0}}{15} = \frac{3}{15} = \frac{1}{5}.$$

Notice that $(Y = 1)$ is by far the most likely outcome, which should seem reasonable since the number of women equals the number of men in the original group.

Table 3.1.
Probability distribution
for Example 3.1

y	$p(y)$
0	1/5
1	3/5
2	1/5

The table for the probability distribution of the random variable Y considered in Example 3.1 is summarized in Table 3.1. The same distribution is given in graphical form in Figure 3.1. If we regard the width at each bar in Figure 3.1 as one unit, then the area in a bar is equal to the probability that Y takes on the value over which the bar is centered. This concept of areas representing probabilities was introduced in Section 1.2.

The most concise method of representing discrete probability distributions is by means of a formula. For Example 3.1 we see that the formula for $p(y)$ can be written as

$$p(y) = \frac{\binom{3}{y}\binom{3}{2-y}}{\binom{6}{2}}, \qquad y = 0, 1, 2.$$

Notice that the probabilities associated with all distinct values of a discrete random variable must sum to 1. In summary, the following properties must hold for any discrete probability distribution:

THEOREM 3.1

For any discrete probability distribution, the following must be true:

1. $0 \le p(y) \le 1$ for all y.
2. $\sum_y p(y) = 1$, where the summation is over all values of y with nonzero probability.

FIGURE 3.1
Probability histogram
for Table 3.1

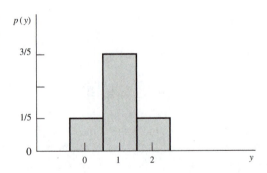

As mentioned in Section 1.5, the probability distributions we derive are *models*, not exact representations, for the frequency distributions of populations of real data that occur (or would be generated) in nature. Thus they are models for real distributions of data similar to the distributions discussed in Chapter 1. For example, if we were to randomly select two workers from among the six described in Example 3.1, we would observe a single y value. In this instance the observed y value would be 0, 1, or 2. If the experiment were repeated many times, many y values would be generated. A relative frequency histogram for the resulting data, constructed in the manner described in Chapter 1, would be very similar to the probability histogram of Figure 3.1. Such simulation studies are very useful. By repeating some experiments over and over again, we can generate measurements of discrete random variables that possess frequency distributions very similar to the probability distributions derived in this chapter, reinforcing the conviction that our models are quite accurate.

Exercises

3.1 When the health department tested private wells in a county for two impurities commonly found in drinking water, it found that 20% of the wells had neither impurity, 40% had impurity A, and 50% had impurity B. (Obviously, some had both impurities.) If a well is randomly chosen from those in the county, find the probability distribution for Y, the number of impurities found in the well.

3.2 You and a friend play a game where you each toss a balanced coin. If the upper faces on the coins are both tails, you win \$1; if the faces are both heads, you win \$2; if the coins do not match (one shows a head, the other a tail), you lose \$1 (win $(-\$1)$). Give the probability distribution for your winnings, Y, on a single play of this game.

3.3 A group of four components is known to contain two defectives. An inspector tests the components one at a time until the two defectives are located. Once she locates the two defectives, she stops testing, but the second defective is tested to ensure accuracy. Let Y denote the number of the test on which the second defective is found. Find the probability distribution for Y.

3.4 Consider a system of water flowing through valves from A to B. (See the accompanying diagram.) Valves 1, 2, and 3 operate independently, and each correctly opens on signal with probability .8. Find the probability distribution for Y, the number of open paths from A to B after the signal is given. (Note that Y can take on the values 0, 1, and 2.)

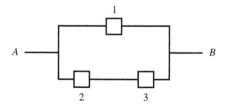

3.5 A problem in a test given to small children asks them to match each of three pictures of animals to the word identifying that animal. If a child assigns the three words at random to the three pictures, find the probability distribution for Y, the number of correct matches.

3.6 Five balls, numbered 1, 2, 3, 4, and 5, are placed in an urn. Two balls are randomly selected from the five, and their numbers noted. Find the probability distribution for the following:

a The *largest* of the two sampled numbers.

b The *sum* of the two sampled numbers.

3.7 In order to verify the accuracy of their financial accounts, companies use auditors on a regular basis to verify accounting entries. The company's employees make erroneous entries 5% of the time. Suppose that an auditor randomly checks three entries.

a Find the probability distribution for Y, the number of errors detected by the auditor.

b Construct a probability histogram for $p(y)$.

c Find the probability that the auditor will detect more than one error.

3.8 A rental agency, which leases heavy equipment by the day, has found that one expensive piece of equipment is leased, on the average, only one day in five. If rental on one day is independent of rental on any other day, find the probability distribution of Y, the number of days between a pair of rentals.

3.9 Persons entering a blood bank are such that 1 in 3 have type O^+ blood and 1 in 15 have type O^- blood. Consider three randomly selected donors for the blood bank. Let X denote the number of donors with type O^+ blood and Y denote the number with type O^- blood. Find the probability distributions for X and Y. Also find the probability distribution for $X + Y$, the number of donors who have type O blood.

3.3 The Expected Value of a Random Variable or a Function of a Random Variable

We have observed that the probability distribution for a random variable is a theoretical model for the empirical distribution of data associated with a real population. If the model is an accurate representation of nature, the theoretical and empirical distributions are equivalent. Consequently, as in Chapter 1, we attempt to find the mean and the variance for a random variable and thereby to acquire numerical descriptive measures, *parameters*, for the probability distribution $p(y)$ that are consistent with those discussed in Chapter 1.

DEFINITION 3.4

Let Y be a discrete random variable with the probability function $p(y)$. Then the *expected value* of Y, $E(Y)$, is defined to be[2]

$$E(Y) = \sum_y y p(y).$$

---

[2]To be precise, the expected value of a discrete random variable is said to exist if the sum, as given earlier, is absolutely convergent—that is, if

$$\sum_y |y| p(y) < \infty.$$

This absolute convergence will hold for all examples in this text and will not be mentioned each time an expected value is defined.

Table 3.2. Probability
distribution for Y

y	$p(y)$
0	1/4
1	1/2
2	1/4

If $p(y)$ is an accurate characterization of the population frequency distribution, then $E(Y) = \mu$, the population mean.

Definition 3.4 is completely consistent with the definition of the mean of a set of measurements that was given in Definition 1.1. For example, consider a discrete random variable Y that can assume values 0, 1, 2 with probability distribution $p(y)$ as shown in Table 3.2, and the probability histogram shown in Figure 3.2. A visual inspection will reveal the mean of the distribution to be located at $y = 1$.

To show that $E(Y) = \sum_y yp(y)$ is the mean of the probability distribution $p(y)$, suppose that the experiment were conducted 4 million times, yielding 4 million observed values for Y. Noting $p(y)$, in Figure 3.2, we would expect *approximately* 1 million of the 4 million repetitions to result in the outcome $Y = 0$, 2 million in $Y = 1$, and 1 million in $Y = 2$. To find the mean value of Y, we average these 4 million measurements and obtain

$$\mu \approx \frac{\sum_{i=1}^{n} y_i}{n} = \frac{(1,000,000)(0) + (2,000,000)(1) + (1,000,000)(2)}{4,000,000}$$

$$= (0)(1/4) + (1)(1/2) + (2)(1/4)$$

$$= \sum_{y=0}^{2} yp(y) = 1.$$

Thus $E(Y)$ is an average and Definition 3.4 is consistent with the definition of a mean given in Definition 1.1.

FIGURE **3.2**
Probability
distribution for Y

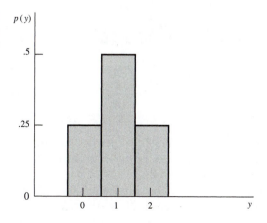

Similarly, we frequently are interested in the mean or expected value of a function of a random variable Y. For example, molecules in space move at varying velocities, where Y, the velocity of a given molecule, is a random variable. The energy imparted upon impact by a moving body is proportional to the square of the velocity. Consequently, to find the mean amount of energy transmitted by a molecule upon impact, we must find the mean value of Y^2. More importantly, we note in Definition 1.2 that the variance of a set of measurements is the mean of the square of the differences between each value in the set of measurements and their mean, or the mean value of $(Y - \mu)^2$.

THEOREM 3.2 Let Y be a discrete random variable with probability function $p(y)$ and $g(Y)$ be a real-valued function of Y. Then the expected value of $g(Y)$ is given by

$$E[g(Y)] = \sum_y g(y)p(y).$$

Proof We prove the result in the case where the random variable Y takes on the finite number of values $y_1, y_2, \ldots, y_n$. Because the function $g(y)$ may not be one to one, suppose that $g(Y)$ takes on values $g_1, g_2, \ldots, g_m$ (where $m \le n$). It follows that $g(Y)$ is a random variable such that for $i = 1, 2, \ldots, m$,

$$P[g(Y) = g_i] = \sum_{\substack{\text{all } y_j \text{ such that} \\ g(y_j)=g_i}} p(y_j) = p^*(g_i).$$

Thus by Definition 3.4,

$$E[g(Y)] = \sum_{i=1}^{m} g_i\, p^*(g_i)$$

$$= \sum_{i=1}^{m} g_i \left\{ \sum_{\substack{\text{all } y_j \text{ such that} \\ g(y_j)=g_i}} p(y_j) \right\}$$

$$= \sum_{i=1}^{m} \sum_{\substack{\text{all } y_j \text{ such that} \\ g(y_j)=g_i}} g_i\, p(y_j)$$

$$= \sum_{j=1}^{n} g(y_j)p(y_j). \qquad \blacksquare$$

Now let us return to our immediate objective, finding numerical descriptive measures (or *parameters*) to characterize $p(y)$. As previously discussed, $E(Y)$ provides the mean of the population with distribution given by $p(y)$. We next seek the variance and standard deviation of this population. You will recall from Chapter 1 that the variance of a set of measurements is the average of the square of the differences between the values in a set of measurements and their mean. Thus we wish to find the mean value of the function $g(Y) = (Y - \mu)^2$.

DEFINITION **3.5**

The variance of a random variable Y is defined to be the expected value of $(Y - \mu)^2$. That is,

$$V(Y) = E[(Y - \mu)^2].$$

The *standard deviation* of Y is the positive square root of $V(Y)$.

If $p(y)$ is an accurate characterization of the population frequency distribution (and to simplify notation, we will assume this to be true), then $E(Y) = \mu$, $V(Y) = \sigma^2$, the population variance, and σ is the population standard deviation.

EXAMPLE **3.2**   The probability distribution for a random variable Y is given in Table 3.3. Find the mean, variance, and standard deviation of Y.

Solution   By Definitions 3.4 and 3.5,

$$\mu = E(Y) = \sum_{y=0}^{3} yp(y) = (0)(1/8) + (1)(1/4) + (2)(3/8) + (3)(1/4) = 1.75$$

$$\sigma^2 = E[(Y - \mu)^2] = \sum_{y=0}^{3} (y - \mu)^2 p(y)$$

$$= (0 - 1.75)^2(1/8) + (1 - 1.75)^2(1/4) + (2 - 1.75)^2(3/8) + (3 - 1.75)^2(1/4)$$

$$= .9375$$

$$\sigma = +\sqrt{\sigma^2} = \sqrt{.9375} = .97.$$

The probability histogram is shown in Figure 3.3. Locate μ on the axis of measurement, and observe that it does locate the "center" of the nonsymmetrical probability distribution of Y. Also notice that the interval $(\mu \pm \sigma)$ contains the discrete points $Y = 1$ and $Y = 2$, which account for 5/8 of the probability. Thus the empirical rule (Chapter 1) provides a reasonable approximation to the probability of a measure-

Table 3.3. Probability distribution for Y

y	p(y)
0	1/8
1	1/4
2	3/8
3	1/4

FIGURE 3.3
Probability histogram
for Example 3.2

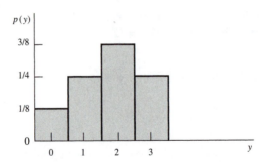

ment falling in this interval. (Keep in mind that the probabilities are concentrated at the points $Y = 0, 1, 2, 3$ because Y cannot take intermediate values.)

It will be helpful to acquire a few additional tools and definitions before attempting to find the expected values and variances of more complicated discrete random variables, such as the binomial or Poisson. Hence we present three useful expectation theorems that follow directly from the theory of summation. (Other useful techniques are presented in Sections 3.4 and 3.9.) For each theorem we assume that Y is a discrete random variable with probability function $p(y)$.

The first theorem states the rather obvious result that the mean or expected value of a nonrandom quantity c is equal to c.

THEOREM 3.3

Let Y be a discrete random variable with probability function $p(y)$ and c be a constant. Then $E(c) = c$.

Proof

Consider the function $g(Y) \equiv c$. By Theorem 3.2,

$$E(c) = \sum_y cp(y) = c \sum_y p(y).$$

But $\sum_y p(y) = 1$ (Theorem 3.1) and hence $E(c) = c(1) = c$. ∎

The second theorem states that the expected value of the product of a constant c times a function of a random variable is equal to the constant times the expected value of the function of the variable.

THEOREM 3.4

Let Y be a discrete random variable with probability function $p(y)$, $g(Y)$ be a function of Y, and let c be a constant. Then

$$E[cg(Y)] = cE[g(Y)].$$

Proof   By Theorem 3.2,

$$E[cg(Y)] = \sum_y cg(y)p(y) = c\sum_y g(y)p(y) = cE[g(Y)].$$ ∎

The third theorem states that the mean or expected value of a sum of functions of a random variable Y is equal to the sum of their respective expected values.

THEOREM 3.5   Let Y be a discrete random variable with probability function $p(y)$ and $g_1(Y), g_2(Y), \ldots, g_k(Y)$ be k functions of Y. Then

$$E[g_1(Y)+g_2(Y)+\cdots+g_k(Y)]=E[g_1(Y)]+E[g_2(Y)]+\cdots+E[g_k(Y)].$$

Proof   We will demonstrate the proof only for the case $k = 2$, but analogous steps will hold for any finite k. By Theorem 3.2,

$$E[g_1(Y) + g_2(Y)] = \sum_y [g_1(y) + g_2(y)]p(y)$$

$$= \sum_y g_1(y)p(y) + \sum_y g_2(y)p(y)$$

$$= E[g_1(Y)] + E[g_2(Y)].$$ ∎

Theorems 3.3, 3.4, and 3.5 can be used immediately to develop a theorem useful in finding the variance of a discrete random variable.

THEOREM 3.6   Let Y be a discrete random variable with probability function $p(y)$; then

$$V(Y) = \sigma^2 = E[(Y - \mu)^2] = E(Y^2) - \mu^2.$$

Proof

$$\sigma^2 = E[(Y - \mu)^2] = E(Y^2 - 2\mu Y + \mu^2)$$

$$= E(Y^2) - E(2\mu Y) + E(\mu^2) \quad \text{(by Theorem 3.5)}.$$

Noting that μ is a constant and applying Theorems 3.4 and 3.3 to the second and third terms, respectively, we have

$$\sigma^2 = E(Y^2) - 2\mu E(Y) + \mu^2.$$

But $\mu = E(Y)$ and, therefore,

$$\sigma^2 = E(Y^2) - 2\mu^2 + \mu^2 = E(Y^2) - \mu^2.$$ ∎

Theorem 3.6 often greatly reduces the labor in finding the variance of a discrete random variable. We will demonstrate the usefulness of this result by recomputing the variance of the random variable considered in Example 3.2.

EXAMPLE **3.3**   Use Theorem 3.6 to find the variance of the random variable Y in Example 3.2.

Solution   The mean $\mu = 1.75$ was found in Example 3.3. Because

$$E(Y^2) = \sum_y y^2 p(y) = (0)^2(1/8) + (1)^2(1/4) + (2)^2(3/8) + (3)^2(1/4) = 4,$$

Theorem 3.6 yields that

$$\sigma^2 = E(Y^2) - \mu^2 = 4 - (1.75)^2 = .9375.$$

EXAMPLE **3.4**   The manager of an industrial plant is planning to buy a new machine of either type A or type B. If t denotes the number of hours of daily operation, the number of daily repairs Y_1 required to maintain a machine of type A is a random variable with mean and variance both equal to .10t. The number of daily repairs Y_2 for a machine of type B is a random variable with mean and variance both equal to .12t. The daily cost of operating A is $C_A(t) = 10t + 30Y_1^2$; for B it is $C_B(t) = 8t + 30Y_2^2$. Assume that the repairs take negligible time and that each night the machines are tuned so that they operate essentially like new machines at the start of the next day. Which machine minimizes the expected daily cost if a workday consists of (a) 10 hours and (b) 20 hours?

Solution   The expected daily cost for A is

$$E[C_A(t)] = E[10t + 30Y_1^2] = 10t + 30E(Y_1^2)$$
$$= 10t + 30\{V(Y_1) + [E(Y_1)]^2\} = 10t + 30[.10t + (.10t)^2]$$
$$= 13t + .3t^2.$$

In this calculation, we used the known values for $V(Y_1)$ and $E(Y_1)$ and the fact that $V(Y_1) = E(Y_1^2) - [E(Y_1)]^2$ to obtain that $E(Y_1^2) = V(Y_1) + [E(Y_1)]^2 = .10t + (.10t)^2$. Similarly,

$$E[C_B(t)] = E[8t + 30Y_2^2] = 8t + 30E(Y_2^2)$$
$$= 8t + 30\{V(Y_2) + [E(Y_2)]^2\} = 8t + 30[.12t + (.12t)^2]$$
$$= 11.6t + .432t^2.$$

Thus, for scenario (a) where $t = 10$,

$$E[C_A(10)] = 160 \quad \text{and} \quad E[C_B(10)] = 159.2,$$

which results in the choice of machine B.

For scenario (b), $t = 20$ and

$$E[C_A(20)] = 380 \quad \text{and} \quad E[C_B(20)] = 404.8,$$

resulting in the choice of machine A.

In conclusion, machines of type B are more economical for short time periods because of their smaller hourly operating cost. For long time periods, however, machines of type A are more economical because they tend to be repaired less frequently.

The purpose of this section was to introduce the concept of an expected value and to develop some useful theorems for finding means and variances of random variables or functions of random variables. In the following sections, we present some specific types of discrete random variables and provide formulas for their probability distributions and their means and variances. As you will see, actually deriving some of these expected values requires skill in the summation of algebraic series and knowledge of a few tricks. We will illustrate some of these tricks in some of the derivations in the upcoming sections.

Exercises

3.10 Let Y be a random variable with $p(y)$ given in the accompanying table. Find $E(Y)$, $E(1/Y)$, $E(Y^2 - 1)$, and $V(Y)$.

y	1	2	3	4
$p(y)$	.4	.3	.2	.1

3.11 Refer to the coin-tossing game in Exercise 3.2. Calculate the mean and variance of Y, your winnings on a single play of the game. Note that $E(Y) > 0$. How much should you pay to play this game if your *net winnings*, the difference between the payoff and cost of playing, are to have mean 0?

3.12 The secretary in Exercise 2.97 was given n computer passwords and tries the passwords at random. Exactly one password will permit access to a computer file. Find the mean and the variance of Y, the number of trials required to open the file, if unsuccessful passwords are eliminated (as in Exercise 2.97).

3.13 An insurance company issues a 1-year $1000 policy insuring against an occurrence A that historically happens to 2 out of every 100 owners of the policy. Administrative fees are $15 per policy and are not part of the company's "profit". How much should the company charge for the policy if it requires that the expected profit per policy be $50? (*Hint:* If C is the premium for the policy, the company's "profit" is $C - 15$ if A does not occur and $C - 15 - 1000$ if A does occur.)

3.14 A manufacturing company ships its product in two different sizes of truck trailers. Each shipment is made in a trailer with dimensions 8ft × 10ft × 30ft or 8ft × 10ft × 40ft. If 30% of its shipments are made by using 30-foot trailers and 70% by using 40-foot trailers, find the mean volume shipped per trailer load. (Assume that the trailers are always full.)

3.15 The number N of residential homes that a fire company can serve depends on the distance r (in city blocks) that a fire engine can cover in a specified (fixed) period of time. If we assume that N is proportional to the area of a circle R blocks from the firehouse, then $N = C\pi R^2$, where C is a constant, $\pi = 3.1416\ldots$, and R, a random variable, is the number of blocks that a fire engine can move in the specified time interval. For a particular fire company, $C = 8$, the probability distribution for R is as shown in the accompanying table, and $p(r) = 0$ for $r \leq 20$ and $r \geq 27$.

r	21	22	23	24	25	26
$p(r)$	.05	.20	.30	.25	.15	.05

Find the expected value of N, the number of homes that the fire department can serve.

3.16 A single fair die is tossed once. Let Y be the number facing up. Find the expected value and variance of Y.

3.17 In a gambling game a person draws a single card from an ordinary 52-card playing deck. A person is paid $15 for drawing a jack or a queen and $5 for drawing a king or an ace. A person who draws any other card pays $4. If a person plays this game, what is the expected gain?

3.18 Approximately 10% of the glass bottles coming off a production line have serious flaws in the glass. If two bottles are randomly selected, find the mean and variance of the number of bottles that have serious flaws.

3.19 Two construction contracts are to be randomly assigned to one or more of three firms: I, II, and III. Any firm may receive both contracts. If each contract will yield a profit of $90,000 for the firm, find the expected profit for firm I. If firms I and II are actually owned by the same individual, what is the owner's expected total profit?

*3.20 A heavy-equipment salesperson can contact either one or two customers per day with probability 1/3 and 2/3, respectively. Each contact will result in either no sale or a $50,000 sale, with the probabilities .9 and .1, respectively. Give the probability distribution for daily sales. Find the mean and standard deviation of the daily sales.[3]

3.21 A potential customer for an $85,000 fire insurance policy possesses a home in an area that, according to experience, may sustain a total loss in a given year with probability of .001 and a 50% loss with probability .01. Ignoring all other partial losses, what premium should the insurance company charge for a yearly policy in order to break even on all $85,000 policies in this area?

3.22 Refer to Exercise 3.3. If the cost of testing a component is $2 and the cost of repairing a defective is $4, find the expected total cost for testing and repairing the lot.

3.23 Let Y be a discrete random variable with mean μ and variance σ^2. If a and b are constants, use Theorems 3.3 through 3.6 to prove that

a $E(aY + b) = aE(Y) + b = a\mu + b$
b $V(aY + b) = a^2 V(Y) = a^2\sigma^2$.

3.24 The manager of a stockroom in a factory has constructed the following probability distribution for the daily demand (number of times used) for a particular tool.

y	0	1	2
$p(y)$	.1	.5	.4

[3]Exercises preceded by an asterisk are optional.

It costs the factory $10 each time the tool is used. Find the mean and variance of the daily cost for use of the tool.

3.4   The Binomial Probability Distribution

Some experiments consist of the observation of a sequence of identical and independent trials, each of which can result in one of two outcomes. Each item leaving a manufacturing production line is either defective or nondefective. Each shot in a sequence of firings at a target can result in a hit or a miss, and each of n persons questioned prior to a local election either favors candidate Jones or does not. In this section we are concerned with experiments, known as *binomial experiments*, that exhibit the following characteristics.

DEFINITION **3.6**

A *binomial experiment* possesses the following properties:

1. The experiment consists of a fixed number, n, of identical trials.
2. Each trial results in one of two outcomes. We will call one outcome success, S, and the other failure, F.
3. The probability of success on a single trial is equal to some value p and remains the same from trial to trial. The probability of a failure is equal to $q = (1 - p)$.
4. The trials are independent.
5. The random variable of interest is Y, the number of successes observed during the n trials.

Determining whether a particular experiment is a binomial experiment requires examining the experiment for each of the characteristics just listed. Notice that the random variable of interest is the number of successes observed in the n trials. It is important to realize that a success is not necessarily "good" in the everyday sense of the word. In our discussions, success is merely a name for one of the two possible outcomes on a single trial of an experiment.

EXAMPLE **3.5**   An early-warning detection system for aircraft consists of four identical radar units operating independently of one another. Suppose that each has a probability of .95 of detecting an intruding aircraft. When an intruding aircraft enters the scene, the random variable of interest is Y, the number of radar units that *do not detect* the plane. Is this a binomial experiment?

Solution   To decide whether this is a binomial experiment, we must determine whether each of the five requirements in Definition 3.6 is met. Notice that the random variable of interest is Y, the number of radar units that do not detect an aircraft. The random variable of interest in a binomial experiment is always the number of successes; consequently, the present experiment can be binomial only if we call the event *do not detect* a success. We now examine the experiment for the five characteristics of the binomial experiment.

1. The experiment involves four identical trials. Each trial consists of determining whether (or not) a particular radar unit detects the aircraft.
2. Each trial results in one of two outcomes. Because the random variable of interest is the number of successes, S denotes that the aircraft was not detected, and F denotes that it was detected.
3. Because all the radar units detect aircraft with equal probability, the probability of an S on each trial is the same, and $p = P(S) = P(\text{do not detect}) = .05$.
4. The trials are independent because the units operate independently.
5. The random variable of interest is Y, the number of successes in four trials.

Thus the experiment is a binomial experiment, with $n = 4$, $p = .05$, and $q = 1 - .05 = .95$.

EXAMPLE 3.6   Suppose that 40% of a large population of registered voters favor candidate Jones. A random sample of $n = 10$ voters will be selected, and Y, the number favoring Jones, is to be observed. Does this experiment meet the requirements of a binomial experiment?

Solution   If each of the ten people is selected at random from the population, then we have ten nearly identical trials, with each trial resulting in a person either favoring Jones (S) or not favoring Jones (F). The random variable of interest is then the number of successes in the ten trials. For the first person selected, the probability of favoring Jones (S) is .4. But what can be said about the *unconditional* probability that the second person will favor Jones? In Exercise 3.25 you will show that *unconditionally* the probability that the second person favors Jones is also .4. Thus, the probability of a success S stays the same from trial to trial. However, the *conditional* probability of a success on later trials depends on the number of successes in the previous trials. If the population of voters is large, removal of one person will not substantially change the fraction of voters favoring Jones, and the *conditional* probability that the second person favors Jones will be very close to .4. In general, if the population is large and the sample size is relatively small, the *conditional* probability of success on a later trial given the number of successes on the previous trials will stay approximately the same regardless of the outcomes on previous trials. Thus, the trials will be approximately independent and so sampling problems of this type are approximately binomial.

If the sample size in Example 3.6 was large relative to the population size (say, 10% of the population), the *conditional* probability of selecting a supporter of Jones

on a later selection would be significantly altered by the preferences of persons selected earlier in the experiment, and the experiment would not be binomial. The
hypergeometric probability distribution, the topic of Section 3.7, is the appropriate
probability model to be used when the sample size is large relative to the population
size.

You may wish to refine your ability to identify binomial experiments by reex-
amining the exercises at the end of Chapter 2. Several of the experiments in those
exercises are binomial or approximately binomial experiments.

The binomial probability distribution $p(y)$ can be derived by applying the sample-
point approach to find the probability that the experiment yields y successes. Each
sample point in the sample space can be characterized by an n-tuple involving the
letters S and F, corresponding to success and failure. A typical sample point would
thus appear as

$$\underbrace{SSFSFFFSFS \ldots FS}_{n \text{ positions}}$$

where the letter in the ith position (proceeding from left to right) indicates the outcome of the ith trial.

Now let us consider a particular sample point corresponding to y successes and
hence contained in the numerical event $Y = y$. This sample point,

$$\underbrace{SSSSS \ldots SSS}_{y}\underbrace{FFF \ldots FF}_{n-y},$$

represents the intersection of n *independent* events (the outcomes of the n trials),
in which there were y successes followed by $(n - y)$ failures. Because the trials
were independent and the probability of S, p, stays the same from trial to trial, the
probability of this sample point is

$$\underbrace{ppppp \cdots ppp}_{y \text{ terms}}\underbrace{qqq \cdots qq}_{n-y \text{ terms}} = p^y q^{n-y}.$$

Every other sample point in the event $Y = y$ can be represented as an n-tuple containing y S's and $(n - y)$ F's in some order. Any such sample point also has probability
$p^y q^{n-y}$. Because the number of distinct n-tuples that contain y S's and $(n - y)$ F's
is (from Theorem 2.3)

$$\binom{n}{y} = \frac{n!}{y!(n - y)!}$$

it follows that the event $(Y = y)$ is made up of $\binom{n}{y}$ sample points, each with probability $p^y q^{n-y}$, and that $p(y) = \binom{n}{y}p^y q^{n-y}$, $y = 0, 1, 2, \ldots, n$. The result that we
have just derived is the formula for the *binomial probability distribution*.

DEFINITION **3.7**   A random variable Y is said to have a *binomial distribution* based on n trials with success probability p if and only if

$$p(y) = \binom{n}{y} p^y q^{n-y}, \qquad y = 0, 1, 2, \ldots, n \quad \text{and} \quad 0 \leq p \leq 1.$$

Figure 3.4 portrays $p(y)$ graphically as probability histograms, the first for $n = 10$, $p = .1$; the second for $n = 10$, $p = .5$; and the third for $n = 20$, $p = .5$.

FIGURE 3.4
Binomial probability
histograms

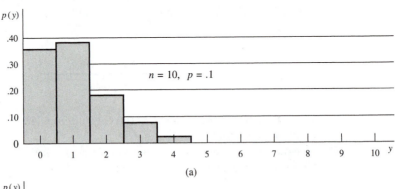

(a)

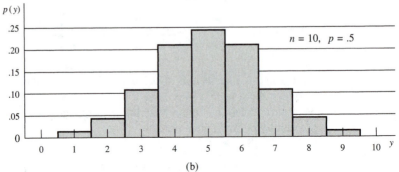

(b)

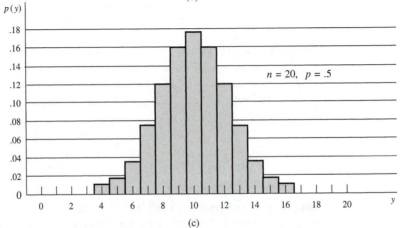

(c)

Before we proceed, let us reconsider the representation for the sample points in this experiment. We have seen that a sample point can be represented by a sequence of n letters, each of which is either S or F. If the sample point contains exactly one S, the probability associated with that sample point is pq^{n-1}. If another sample point contains 2 S's—and $(n-2)F$'s—the probability of this sample point is p^2q^{n-2}. Notice that the sample points for a binomial experiment are not equiprobable unless $p = .5$.

The term *binomial experiment* derives from the fact each trial results in *one of two* possible outcomes and that the probabilities $p(y)$, $y = 0, 1, 2, \ldots, n$, are terms of the binomial expansion

$$(q+p)^n = \binom{n}{0}q^n + \binom{n}{1}p^1q^{n-1} + \binom{n}{2}p^2q^{n-2} + \cdots + \binom{n}{n}p^n.$$

You will observe that $\binom{n}{0}q^n = p(0)$, $\binom{n}{1}p^1q^{n-1} = p(1)$, and, in general, $p(y) = \binom{n}{y}p^yq^{n-y}$. It also follows that $p(y)$ satisfies the necessary properties for a probability function, because $p(y)$ is positive for $y = 0, 1, \ldots, n$ and [because $(q+p) = 1$]

$$\sum_y p(y) = \sum_{y=0}^n \binom{n}{y}p^yq^{n-y} = (q+p)^n = 1^n = 1.$$

The binomial probability distribution has many applications because the binomial experiment occurs in sampling for defectives in industrial quality control, in the sampling of consumer preference or voting populations, and in many other physical situations. We will illustrate with a few examples. Other practical examples will appear in the exercises at the end of this section and at the end of the chapter.

EXAMPLE **3.7**   Suppose that a lot of 5000 electrical fuses contains 5% defectives. If a sample of five fuses is tested, find the probability of observing at least one defective.

Solution   It is reasonable to assume that Y, the number of defectives observed, has an approximate binomial distribution because the lot is large. Removing a few fuses does not change the composition of those remaining enough to cause us concern. Thus,

$$P(\text{at least one defective}) = 1 - p(0) = 1 - \binom{5}{0}p^0q^5$$

$$= 1 - (.95)^5 = 1 - .774 = .226.$$

Notice that there is a fairly large chance of seeing at least one defective, even though the sample is quite small.

EXAMPLE **3.8**   Experience has shown that 30% of all persons afflicted by a certain illness recover. A drug company has developed a new medication. Ten people with the illness were selected at random and received the medication; nine recovered shortly thereafter.

Suppose that the medication was absolutely worthless. What is the probability that at least nine of ten receiving the medication will recover?

Solution   Let Y denote the number of people who recover. If the medication is worthless, the probability that a single ill person will recover is $p = .3$. Then the number of trials is $n = 10$ and the probability of *exactly* nine recoveries is

$$P(Y = 9) = p(9) = \binom{10}{9}(.3)^9(.7) = .000138.$$

Similarly, the probability of exactly ten recoveries is

$$P(Y = 10) = p(10) = \binom{10}{10}(.3)^{10}(.7)^0 = .000006$$

and

$$P(Y \geq 9) = p(9) + p(10) = .000138 + .000006 = .000144.$$

If the medication is ineffective, the probability of observing at least nine recoveries is extremely small. If we administered the medication to 10 individuals and observed at least 9 recoveries, then either (1) the medication is worthless and we have observed a rare event or (2) the medication is indeed useful in curing the illness. We adhere to conclusion 2.

A tabulation of binomial probabilities in the form $\sum_{y=0}^a p(y)$, presented in Table 1, Appendix III, will greatly reduce the computations for some of the exercises. The references at the end of the chapter list several more extensive tabulations of binomial probabilities. We illustrate the use of Table 1 in the following example.

EXAMPLE **3.9**   The large lot of electrical fuses of Example 3.7 is supposed to contain only 5% defectives. If $n = 20$ fuses are randomly sampled from this lot, find the probability that at least four defectives will be observed.

Solution   Letting Y denote the number of defectives in the sample, we assume the binomial model for Y, with $p = .05$. Thus,

$$P(Y \geq 4) = 1 - P(Y \leq 3)$$

and using Table 1, Appendix III, we have

$$P(Y \leq 3) = \sum_{y=0}^3 p(y) = .984.$$

The value .984 is found in the table labeled $n = 20$ in Table 1, Appendix III. Specifically, it appears in the column labeled $P = .05$ and in the row labeled $a = 3$. It follows that

$$P(Y \geq 4) = 1 - .984 = .016.$$

This probability is quite small. If we did indeed observe more than three defectives out of 20 fuses, we might suspect that the reported 5% defect rate is erroneous.

The mean and variance associated with a binomial random variable are derived in the following theorem. As you will see in the proof of the theorem, it is necessary to evaluate the sum of some arithmetic series. In the course of the proof, we illustrate some of the techniques that are available for summing such series. In particular, we use the fact that $\sum_y p(y) = 1$ for any discrete random variable.

THEOREM 3.7   Let Y be a binomial random variable based on n trials and success probability p. Then

$$\mu = E(Y) = np \quad \text{and} \quad \sigma^2 = V(Y) = npq.$$

Proof   By Definitions 3.4 and 3.7,

$$E(Y) = \sum_y y p(y) = \sum_{y=0}^{n} y \binom{n}{y} p^y q^{n-y}.$$

Notice that the first term in the sum is 0 and hence that

$$E(Y) = \sum_{y=1}^{n} y \frac{n!}{(n-y)! y!} p^y q^{n-y}$$

$$= \sum_{y=1}^{n} \frac{n!}{(n-y)!(y-1)!} p^y q^{n-y}.$$

The summands in this last expression bear a striking resemblance to binomial probabilities. In fact, if we factor np out of each term in the sum and let $z = y - 1$,

$$E(Y) = np \sum_{y=1}^{n} \frac{(n-1)!}{(n-y)!(y-1)!} p^{y-1} q^{n-y}$$

$$= np \sum_{z=0}^{n-1} \frac{(n-1)!}{(n-1-z)! z!} p^z q^{n-1-z}$$

$$= np \sum_{z=0}^{n-1} \binom{n-1}{z} p^z q^{n-1-z}.$$

Notice that $p(z) = \binom{n-1}{z} p^z q^{n-1-z}$ is the binomial probability function based on $(n-1)$ trials. Thus $\sum_z p(z) = 1$, and it follows that

$$\mu = E(Y) = np.$$

From Theorem 3.6 we know that $\sigma^2 = V(Y) = E(Y^2) - \mu^2$. Thus, σ^2 can be calculated if we find $E(Y^2)$. Finding $E(Y^2)$ directly is difficult because

$$E(Y^2) = \sum_{y=0}^{n} y^2 p(y) = \sum_{y=0}^{n} y^2 \binom{n}{y} p^y q^{n-y} = \sum_{y=0}^{n} y^2 \frac{n!}{y!(n-y)!} p^y q^{n-y}$$

and the quantity y^2 does not appear as a factor of $y!$. Where do we go from here? Notice that

$$E[Y(Y-1)] = E(Y^2 - Y) = E(Y^2) - E(Y)$$

and, therefore,

$$E(Y^2) = E[Y(Y-1)] + E(Y) = E[Y(Y-1)] + \mu.$$

In this case

$$E[Y(Y-1)] = \sum_{y=0}^{n} y(y-1) \frac{n!}{y!(n-y)!} p^y q^{n-y}.$$

The first and second terms of this sum equal zero (when $y = 0$ and $y = 1$). Then

$$E[Y(Y-1)] = \sum_{y=2}^{n} \frac{n!}{(y-2)!(n-y)!} p^y q^{n-y}.$$

(Notice the cancellation that led to this last result. The anticipation of this cancellation is what actually motivated the consideration of $E[Y(Y-1)]$.) Again, the summands in the last expression look very much like binomial probabilities. Factor $n(n-1)p^2$ out of each term in the sum and let $z = y - 2$ to obtain

$$E[Y(Y-1)] = n(n-1)p^2 \sum_{y=2}^{n} \frac{(n-2)!}{(y-2)!(n-y)!} p^{y-2} q^{n-y}$$

$$= n(n-1)p^2 \sum_{z=0}^{n-2} \frac{(n-2)!}{z!(n-2-z)!} p^z q^{n-2-z}$$

$$= n(n-1)p^2 \sum_{z=0}^{n-2} \binom{n-2}{z} p^z q^{n-2-z}.$$

Again note that $p(z) = \binom{n-2}{z}p^z q^{n-2-z}$ is the binomial probability function based on $(n-2)$ trials. Then $\sum_{z=0}^{n-2} p(z) = 1$ (again using the device illustrated in the derivation of the mean) and

$$E[Y(Y-1)] = n(n-1)p^2.$$

Thus,

$$E(Y^2) = E[Y(Y-1)] + \mu = n(n-1)p^2 + np$$

and

$$\sigma^2 = E(Y^2) - \mu^2 = n(n-1)p^2 + np - n^2 p^2$$
$$= np[(n-1)p + 1 - np] = np(1-p) = npq.$$ ∎

In addition to providing formulas for the mean and variance of a binomial random variable, the derivation of Theorem 3.7 illustrates the use of two fairly common tricks, namely, to use the fact that $\sum p(y) = 1$ if $p(y)$ is a valid probability function and to find $E(Y^2)$ by finding $E[Y(Y-1)]$. These techniques also will be useful in the next sections, where we consider other discrete probability distributions and the associated means and variances.

A frequent source of error in applying the binomial probability distribution to practical problems is the failure to define which of the two possible results of a trial is the success. As a consequence, q may be used erroneously in place of p. Carefully define a success, and make certain that p equals the probability of a success for each application.

Thus far in this section we have assumed that the number of trials, n, and the probability of success, p, were known, and we used the formula $p(y) = \binom{n}{y}p^y q^{n-y}$ to compute probabilities associated with binomial random variables. In Example 3.8 we obtained a value for $P(Y \geq 9)$ and used this probability to reach a conclusion about the effectiveness of the medication. The next example exhibits another *statistical*, rather than probabilistic, use of the binomial distribution.

EXAMPLE **3.10**   Suppose that we survey 20 individuals working for a large company and ask each whether they favor implementation of a new policy regarding retirement funding. If, in our sample, 6 favored the new policy, find an estimate for p, the true but unknown proportion of employees that favor the new policy.

Solution   If Y denotes the number among the 20 who favor the new policy, it is reasonable to conclude that Y has a binomial distribution with $n = 20$ for some value of p. Whatever the true value for p, we conclude that the probability of observing 6 out of 20 in favor of the policy is

$$P(Y = 6) = \binom{20}{6}p^6(1-p)^{14}.$$

We will use as our estimate for p the value that maximizes the probability of observing the value that we *actually observed* (6 in favor in 20 trials). How do we find the value of p that maximizes $P(Y = 6)$?

Because $\binom{20}{6}$ is a constant (relative to p) and $\ln(w)$ is an increasing function of w, the value of p that maximizes $P(Y = 6) = \binom{20}{6} p^6 (1 - p)^{14}$ is the same as the value of p that maximizes $\ln(p^6(1 - p)^{14}) = \{6 \ln(p) + 14 \ln(1 - p)\}$.

If we take the derivative of $\{6 \ln(p) + 14 \ln(1 - p)\}$ with respect to p, we obtain

$$\frac{d[6 \ln(p) + 14 \ln(1 - p)]}{dp} = \left(\frac{6}{p}\right) - \left(\frac{14}{1 - p}\right).$$

The value of p that maximizes (or minimizes) $\{6 \ln(p) + 14 \ln(1 - p)\}$ [and, more importantly, $P(Y = 6)$] is the solution to the equation

$$\frac{6}{p} - \frac{14}{1 - p} = 0.$$

Solving, we obtain $p = 6/20$.

Because the second derivative of $\{6 \ln(p) + 14 \ln(1 - p)\}$ is negative when $p = 6/20$, it follows that $\{6 \ln(p) + 14 \ln(1 - p)\}$ [and $P(Y = 6)$] is *maximized* when $p = 6/20$. Our estimate for p, based on 6 "successes" in 20 trials is therefore 6/20.

The ultimate answer that we obtained should look very reasonable to you. Because p is the probability of a "success" on any given trial, a reasonable estimate is, indeed, the proportion of "successes" in our sample, in this case 6/20. In the next section, we will apply this same technique to obtain an estimate that is not initially so intuitive. As we will see in Chapter 9, the estimate that we just obtained is the *maximum likelihood* estimate for p and the procedure used above is an example the application of the *method of maximum likelihood*.

Exercises

3.25 Consider the population of voters described in Example 3.6. Suppose that there are $N = 5000$ voters in the population, 40% of whom favor Jones. Identify the event *favors Jones* as a success S. It is evident that the probability of S on trial 1 is .40. Consider the event B that S occurs on the second trial. Then B can occur two ways: the first two trials are both successes *or* the first trial is a failure and the second is a success. Show that $P(B) = .4$. What is $P(B|$ the first trial is S)? Does this *conditional* probability differ markedly from $P(B)$?

3.26 The manufacturer of a low-calorie dairy drink wishes to compare the taste appeal of a new formula (formula B) with that of the standard formula (formula A). Each of four judges is given three glasses in random order, two containing formula A and the other containing formula B. Each judge is asked to state which glass he or she most enjoyed. Suppose that the two formulas are equally attractive. Let Y be the number of judges stating a preference for the new formula.

a Find the probability function for Y.

b What is the probability that at least three of the four judges state a preference for the new formula?

    c Find the expected value of Y.

    d Find the variance of Y.

3.27 A complex electronic system is built with a certain number of backup components in its sub-systems. One subsystem has four identical components, each with a probability of .2 of failing in less than 1000 hours. The subsystem will operate if any two of the four components are operating. Assume that the components operate independently.

    a Find the probability that exactly two of the four components last longer than 1000 hours.

    b Find the probability that the subsystem operates longer than 1000 hours.

3.28 The probability that a patient recovers from a stomach disease is .8. Suppose 20 people are known to have contracted this disease.

    a What is the probability that exactly 14 recover?

    b What is the probability that at least 10 recover?

    c What is the probability that at least 14 but not more than 18 recover?

    d What is the probability that at most 16 recover?

3.29 A multiple-choice examination has 15 questions, each with five possible answers, only one of which is correct. Suppose that one of the students who takes the examination answers each of the questions with an independent random guess. What is the probability that he answers at least ten questions correctly?

3.30 Many employers are finding that some of the people they hire are not who and what they claim to be. Detecting job applicants who falsify their application information has spawned some new businesses: credential checking services. *U.S. News and World Report* (July 13, 1981) reported on this problem and noted that one service in a 2-month period found that 35% of all credentials examined were falsified. Suppose that you hired five new employees last week and that the probability that any one employee falsified the information on his or her application form is .35. What is the probability that at least one of the five application forms has been falsified? Two or more?

3.31 Many utility companies promote energy conservation by offering discount rates to consumers who keep their energy usage below certain established subsidy standards. A recent EPA report notes that 70% of the island residents of Puerto Rico have reduced their electricity usage sufficiently to qualify for discounted rates. If five residential subscribers are randomly selected from San Juan, Puerto Rico, find the probability of each of the following events:

    a All five qualify for the favorable rates.

    b At least four qualify for the favorable rates.

3.32 A new surgical procedure is successful with a probability of p. Assume the operation is performed five times and the results are independent of one another.

    a What is the probability that all five operations are successful if $p = .8$?

    b What is the probability that exactly four are successful if $p = .6$?

    c What is the probability that less than two are successful if $p = .3$?

3.33 A fire-detection device utilizes three temperature-sensitive cells acting independently of each other in such a manner that any one or more may activate the alarm. Each cell possesses a probability of $p = .8$ of activating the alarm when the temperature reaches $100°$ Celsius or more. Let Y equal the number of cells activating the alarm when the temperature reaches $100°$. Find the probability distribution for Y. Find the probability that the alarm will function when the temperature reaches $100°$.

3.34   Construct probability histograms for the binomial probability distributions for $n = 5$, $p = .1$, $.5$, and $.9$. (Table 1, Appendix III, will reduce the amount of calculation.) Notice the symmetry for $p = .5$ and the direction of skewness for $p = .1$ and $.9$.

3.35   Use Table 1, Appendix III, to construct a probability histogram for the binomial probability distribution for $n = 20$ and $p = .5$. Notice that almost all the probability falls in the interval $5 \le y \le 15$.

3.36   A missile protection system consists of n radar sets operating independently, each with a probability of .9 of detecting a missile entering a zone that is covered by all of the units.

    a   If $n = 5$ and a missile enters the zone, what is the probability that exactly four sets detect the missile? At least one set?

    b   How large must n be if we require that the probability of detecting a missile that enters the zone be .999?

3.37   A manufacturer of floor wax has developed two new brands, A and B, which she wishes to subject to homeowners' evaluation to determine which of the two is superior. Both waxes, A and B, are applied to floor surfaces in each of 15 homes. Assume that there is actually no difference in the quality of the brands.

    a   What is the probability that 10 or more homeowners would state a preference for brand A?

    b   What is the probability that 10 or more homeowners would state a preference for either brand A or brand B?

3.38   In Exercise 2.129, you considered a model for the World Series. Two teams A and B play a series of games until one team wins four games. We assume that the games are played independently and that the probability that A wins any game is p. Compute the probability that the series last exactly five games. Use what you know about the binomial random variable, Y, the number of games A wins among the first four games.

3.39   Suppose that Y is a binomial random variable with $n > 2$ trials and success probability p. Use the technique presented in Theorem 3.7 and the fact that $E\{Y(Y - 1)(Y - 2)\} = E(Y^3) - 3E(Y^2) + 2E(Y)$ to derive $E(Y^3)$.

3.40   An oil exploration firm is formed with enough capital to finance ten explorations. The probability of a particular exploration being successful is .1. Assume the explorations are independent. Find the mean and variance of the number of successful explorations.

3.41   Refer to Exercise 3.40. Suppose the firm has a fixed cost of $20,000 in preparing equipment prior to doing its first exploration. If each successful exploration costs $30,000 and each unsuccessful exploration costs $15,000, find the expected total cost to the firm for its ten explorations.

3.42   A particular sale involves four items randomly selected from a large lot that is known to contain 10% defectives. Let Y denote the number of defectives among the four sold. The purchaser of the items will return the defectives for repair, and the repair cost is given by $C = 3Y^2 + Y + 2$. Find the expected repair cost. [*Hint:* The result of Theorem 3.6 implies that, for any random variable Y, $E(Y^2) = \sigma^2 + \mu^2$.]

3.43   Ten motors are packaged for sale in a certain warehouse. The motors sell for $100 each, but a double-your-money-back guarantee is in effect for any defectives the purchaser may receive. Find the expected net gain for the seller if the probability of any one motor being defective is .08. (Assume the quality of any one motor is independent of that of the others.)

3.44   A particular concentration of a chemical found in polluted water has been found to be lethal to 20% of the fish that are exposed to the concentration for 24 hours. Twenty fish are placed in a tank containing this concentration of chemical in water.

   a   Find the probability that exactly 14 survive.

   b   Find the probability that at least 10 survive.

   c   Find the probability that at most 16 survive.

   d   Find the mean and variance of the number that survive.

3.45   Of the volunteers donating blood in a clinic, 80% have the Rhesus factor present in their blood.

   a   If five of the volunteers are randomly selected, what is the probability that at least one does not have the Rhesus factor?

   b   If five are randomly selected, what is the probability that at most four have the Rhesus factor?

   c   What is the smallest number of volunteers who must be selected if we want to be at least 90% certain that we obtain at least five donors with the Rhesus factor?

3.46   Goranson and Hall (1980) explain that the probability of detecting a crack in an airplane wing is the product of p_1, the probability of inspecting a plane with a wing crack; p_2, the probability of inspecting the detail in which the crack is located; and p_3, the probability of detecting the damage.

   a   What assumptions justify the multiplication of these probabilities?

   b   Suppose $p_1 = .9$, $p_2 = .8$, and $p_3 = .5$ for a certain fleet of planes. If three planes are inspected from this fleet, find the probability that a wing crack will be detected on at least one of them.

*3.47   Consider the binomial distribution with n trials and $P(S) = p$.

   a   Show that $\dfrac{p(y)}{p(y-1)} = \dfrac{(n-y+1)p}{yq}$ for $y = 1, 2, \ldots, n$. Equivalently, for $y = 1, 2, \ldots, n$, the equation $p(y) = \dfrac{(n-y+1)p}{yq} p(y-1)$ gives a recursive relationship between the probabilities associated with successive values of Y.

   b   Show that $\dfrac{p(y)}{p(y-1)} = \dfrac{(n-y+1)p}{yq} > 1$ if $y < (n+1)p$, that $\dfrac{p(y)}{p(y-1)} < 1$ if $y > (n+1)p$, and that $\dfrac{p(y)}{p(y-1)} = 1$ if $(n+1)p$ is an integer and $y = (n+1)p$. This establishes that $p(y) > p(y-1)$ if y is small $(y < (n+1)p)$ and $p(y) < p(y-1)$ if y is large $(y > (n+1)p)$. Thus, successive binomial probabilities increase for a while and decrease from then on.

   c   Show that the value of y assigned the largest probability is equal to the greatest integer less than or equal to $(n+1)p$. If $(n+1)p = m$ for some integer m, then $p(m) = p(m-1)$.

*3.48   Consider an extension of the situation discussed in Example 3.10. If there are n trials in a binomial experiment and we observe y_0 "successes," show that $P(Y = y_0)$ is maximized when $p = y_0/n$. Again, we are determining (in general this time) the value of p that maximizes the probability of the value of Y that we actually observed.

*3.49   Refer to Exercise 3.48. The *maximum likelihood estimator* for p is Y/n (note that Y is the binomial random variable, not a particular value of it).

   a   Derive $E(Y/n)$. In Chapter 9, we will see that this result implies that Y/n is an *unbiased* estimator for p.

   b   Derive $V(Y/n)$. What happends to $V(Y/n)$ as n gets large?

3.5 The Geometric Probability Distribution

The random variable with the geometric probability distribution is associated with an experiment that shares some of the characteristics of a binomial experiment. This experiment also involves identical and independent trials, each of which can result in one of two outcomes: success or failure. The probability of success is equal to p and is constant from trial to trial. However, instead of the number of successes that occur in n trials, the geometric random variable Y is the number of the trial on which the first success occurs. Thus the experiment consists of a series of trials that concludes with the first success. Consequently, the experiment could end with the first trial if a success is observed on the very first trial, or the experiment could go on indefinitely.

The sample space S for the experiment contains the countably infinite set of sample points:

$$E_1: \quad S \qquad \text{(success on first trial)}$$
$$E_2: \quad FS \qquad \text{(failure on first, success on second)}$$
$$E_3: \quad FFS \qquad \text{(first success on the third trial)}$$
$$E_4: \quad FFFS \qquad \text{(first success on the fourth trial)}$$
$$\vdots$$
$$E_k: \quad \underbrace{FFFF\ldots F}_{k-1}\, S \qquad \text{(first success on the } k^{\text{th}} \text{ trial)}$$
$$\vdots$$

Because the random variable Y is the number of trials up to and including the first success, the events $(Y = 1)$, $(Y = 2)$, and $(Y = 3)$ contain only the sample points E_1, E_2, and E_3, respectively. More generally, the numerical event $(Y = y)$ contains only E_y. Because the trials are independent, for any $y = 1, 2, 3, \ldots$,

$$p(y) = P(Y = y) = P(E_y) = P(\underbrace{FFFF\ldots F}_{y-1}\, S) = \underbrace{qqq\cdots q}_{y-1}\, p = q^{y-1}p.$$

DEFINITION 3.8

A random variable Y is said to have a *geometric probability distribution* if and only if

$$p(y) = q^{y-1}p, \qquad y = 1, 2, 3, \ldots, \qquad 0 \le p \le 1.$$

A probability histogram for $p(y)$, $p = .5$, is shown in Figure 3.5. Areas over intervals correspond to probabilities, as they did for the frequency distributions of data in Chapter 1, except that Y can assume only discrete values, $y = 1, 2, \ldots, \infty$. That $p(y) \ge 0$ is obvious by inspection of the respective values. In Exercise 3.50 you will show that these probabilities add up to 1, as is required for any valid discrete probability distribution.

FIGURE 3.5
The geometric
probability
distribution, $p = .5$

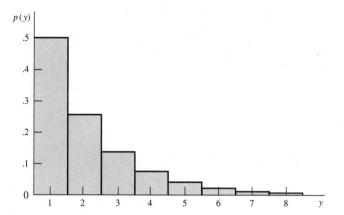

The geometric probability distribution is often used to model distributions of lengths of waiting times. For example, suppose that a commercial aircraft engine is serviced periodically so that its various parts are replaced at different points in time and hence are of varying ages. Then the probability of engine malfunction, p, during any randomly observed 1-hour interval of operation might be the same as for any other 1-hour interval. The length of time prior to engine malfunction is the number of 1-hour intervals, Y, until the first malfunction. (For this application, engine malfunction in a given 1-hour period is defined as a success. Notice that, as in the case of the binomial experiment, either of the two outcomes of a trial can be defined as a success. Again, a "success" is not necessarily what would be considered to be "good" in everyday conversation.)

EXAMPLE 3.11   Suppose that the probability of engine malfunction during any 1-hour period is $p = .02$. Find the probability that a given engine will survive 2 hours.

Solution   Letting Y denote the number of 1-hour intervals until the first malfunction, we have

$$P(\text{survive 2 hours}) = P(Y \geq 3) = \sum_{y=3}^{\infty} p(y).$$

Because $\sum_{y=1}^{\infty} p(y) = 1$,

$$P(\text{survive 2 hours}) = 1 - \sum_{y=1}^{2} p(y)$$

$$= 1 - p - qp = 1 - .02 - (.98)(.02) = .9604.$$

If you examine the formula for the geometric distribution given in Definition 3.8, you will see that larger values of p (and hence smaller values of q) lead to higher probabilities for the smaller values of Y and hence lower probabilities for the larger values of Y. Thus the mean value of Y appears to be inversely proportional to p. As we show in the next theorem, the mean of a random variable with a geometric distribution is actually equal to $1/p$.

THEOREM 3.8   If Y is a random variable with a geometric distribution,

$$\mu = E(Y) = \frac{1}{p} \quad \text{and} \quad \sigma^2 = V(Y) = \frac{1-p}{p^2}.$$

Proof

$$E(Y) = \sum_{y=1}^{\infty} yq^{y-1}p = p\sum_{y=1}^{\infty} yq^{y-1}.$$

This series might seem to be difficult to sum directly. Actually, it can be summed easily if we take into account that, for $y \geq 1$,

$$\frac{d}{dq}(q^y) = yq^{y-1}$$

and hence

$$\frac{d}{dq}\left(\sum_{y=1}^{\infty} q^y\right) = \sum_{y=1}^{\infty} yq^{y-1}.$$

(The interchanging of derivative and sum here can be justified.) Substituting, we obtain

$$E(Y) = p\sum_{y=1}^{\infty} yq^{y-1} = p\frac{d}{dq}\left(\sum_{y=1}^{\infty} q^y\right).$$

The latter sum is the geometric series, $q + q^2 + q^3 + \cdots$, which is equal to $q/(1-q)$ (see Appendix A1.11). Therefore,

$$E(Y) = p\frac{d}{dq}\left(\frac{q}{1-q}\right) = p\left(\frac{1}{(1-q)^2}\right) = \frac{p}{p^2} = \frac{1}{p}.$$

To summarize, our approach is to express a series that cannot be summed directly as the derivative of a series for which the sum can be readily obtained. Once we evaluate the more easily handled series, we differentiate to complete the process.

The derivation of the variance is left as an exercise. See Exercise 3.67. ∎

EXAMPLE **3.12**   If the probability of engine malfunction during any 1-hour period is $p = .02$ and Y denotes the number of 1-hour intervals until the first malfunction, find the mean and standard deviation of Y.

Solution   As in Example 3.11, it follows that Y has a geometric distribution with $p = .02$. Thus $E(Y) = 1/p = 1/(.02) = 50$, and we expect to wait quite a few hours before encountering a malfunction. Further, $V(Y) = .98/.0004 = 2450$, and it follows that the standard deviation of Y is $\sigma = \sqrt{2450} = 49.497$.

The next example, similar to Example 3.10, illustrates how knowledge of the geometric probability distribution can be used to estimate an unknown value of p, the probability of a success.

EXAMPLE **3.13**   Suppose that we interview successive individuals working for the large company discussed in Example 3.10 and stop interviewing when we find the first person who likes the policy. If the fifth person interviewed is the first one who favors the new policy, find an estimate for p, the true but unknown proportion of employees who favor the new policy.

Solution   If Y denotes the number of individuals interviewed until we find the first person who likes the new retirement plan, it is reasonable to conclude that Y has a geometric distribution for some value of p. Whatever the true value for p, we conclude that the probability of observing the first person in favor of the policy on the fifth trial is

$$P(Y = 5) = (1 - p)^4 p.$$

We will use as our estimate for p the value that maximizes the probability of observing the value that we *actually observed* (the first success on trial 5).

To find the value of p that maximizes $P(Y = 5)$, we again observe that the value of p that maximizes $P(Y = 5) = (1 - p)^4 p$ is the same as the value of p that maximizes $\ln((1 - p)^4 p) = \{4\ln(1 - p) + \ln(p)\}$.

If we take the derivative of $\{4\ln(1 - p) + \ln(p)\}$ with respect to p, we obtain

$$\frac{d[4\ln(1 - p) + \ln(p)]}{dp} = \left(\frac{-4}{1 - p}\right) + \left(\frac{1}{p}\right).$$

Setting this derivative equal to 0 and solving, we obtain $p = 1/5$.

Because the second derivative of $\{4\ln(1 - p) + \ln(p)\}$ is negative when $p = 1/5$, it follows that $\{4\ln(1 - p) + \ln(p)\}$ [and $P(Y = 5)$] is *maximized* when $p = 1/5$. Our estimate for p, based on observing the first success on the fifth trial is $1/5$.

Perhaps this result is a little more surprising than the answer we obtained in Example 3.10 where we estimated p on the basis of observing 6 in favor of the new plan in a sample of size 20. Again, this is an example of the use of the *method of maximum likelihood* that will be studied in more detail in Chapter 9.

Exercises

3.50 Suppose that Y is a random variable with a geometric distribution.

  a Show that $\sum_y p(y) = \sum_{y=1}^{\infty} q^{y-1} p = 1$.

  b Show that $\dfrac{p(y)}{p(y-1)} = q$, for $y = 2, 3, \ldots$. This ratio is less than 1, implying that the geometric probabilities are monotonically decreasing as a function of y. If Y has a geometric distribution, what value of Y is the most likely (has the highest probability)?

3.51 Suppose that 30% of the applicants for a certain industrial job possess advanced training in computer programming. Applicants are interviewed sequentially and are selected at random from the pool. Find the probability that the first applicant with advanced training in programming is found on the fifth interview.

3.52 Refer to Exercise 3.51. What is the expected number of applicants who need to be interviewed in order to find the first one with advanced training?

3.53 Shortly before the November 1992 presidential election, a Gallup poll indicated that a near record level of 73% of adults expressed dissatisfaction with the way things were going in the United States (*Gallup Monthly Poll*, September 1992). Suppose that you conducted your own telephone survey at the same time. You randomly called people and asked them whether they were dissatisfied with the state of the nation. Find the probability distribution for Y, the number of calls until the first person is found who *is* satisfied with the state of the nation.

3.54 An oil prospector will drill a succession of holes in a given area to find a productive well. The probability that he is successful on a given trial is .2.

  a What is the probability that the third hole drilled is the first to yield a productive well?

  b If the prospector can afford to drill at most ten wells, what is the probability that he will fail to find a productive well?

3.55 Let Y denote a geometric random variable with probability of success p.

  a Show that for a positive integer a,

$$P(Y > a) = q^a.$$

  b Show that for positive integers a and b,

$$P(Y > a + b \mid Y > a) = q^b = P(Y > b).$$

  This result implies that, for example, $P(Y > 7 \mid Y > 2) = P(Y > 5)$. Why do you think this property is called the *memoryless* property of the geometric distribution?

3.56 Given that we have already tossed a balanced coin ten times and obtained zero heads, what is the probability that we must toss it at least two more times to obtain the first head?

3.57 A certified public accountant (CPA) has found that nine of ten company audits contain substantial errors. If the CPA audits a series of company accounts, what is the probability that

  a the first account containing substantial errors is the third one to be audited?

  b the first account containing substantial errors will occur on or after the third audited account?

3.58 Refer to Exercise 3.57. What are the mean and standard deviation of the number of accounts that must be examined to find the first one with substantial errors?

3.59   The probability of a customer arrival at a grocery service counter in any 1 second is equal to .1. Assume that customers arrive in a random stream and hence that an arrival in any 1 second is independent of all others.

  a Find the probability that the first arrival will occur during the third 1-second interval.

  b Find the probability that the first arrival will not occur until at least the third 1-second interval.

3.60   Of a population of consumers, 60% are reputed to prefer a particular brand, A, of toothpaste. If a group of randomly selected consumers is interviewed, what is the probability that exactly five people have to be interviewed to encounter the first consumer who prefers brand A? At least five people?

3.61   In responding to a survey question on a sensitive topic (such as, "Have you ever tried marijuana?"), many people prefer not to respond in the affirmative. Derive the probability distribution of Y, the number of people you would need to question in order to obtain a single affirmative response, if 80% of the population will truthfully answer no to your question and if, of the 20% who should truthfully answer yes, 70% will lie.

3.62   Two people took turns tossing a fair die until one of them tossed a 6. Person A tossed first, B second, A third, and so on. Given that person B threw the first 6, what is the probability that B obtained the first 6 on her second toss (that is, on the fourth toss overall)?

3.63   How many times would you expect to toss a balanced coin in order to obtain the first head?

3.64   Refer to Exercise 3.54. The prospector drills holes until he finds a productive well. How many holes would the prospector expect to drill? Interpret your answer intuitively.

3.65   The secretary in Exercises 2.97 and 3.12 was given n computer passwords and tries the passwords at random. Exactly one of the passwords permits access to a computer file. Suppose now that the secretary selects a password, tries it, and—if it does not work—puts it back in with the other passwords before randomly selecting the next password to try (not a very clever secretary!). What is the probability that the correct password is found on the sixth try?

3.66   Refer to Exercise 3.65. Find the mean and the variance of Y, the number of the trial on which the correct password is first identified.

*3.67   Find $E[Y(Y-1)]$ for a geometric random variable Y by finding $d^2/dq^2\left(\sum_{y=1}^{\infty} q^y\right)$. Use this result to find the variance of Y.

*3.68   Consider an extension of the situation discussed in Example 3.13. If we observe y_0 as the value for a geometric random variable Y, show that $P(Y = y_0)$ is maximized when $p = 1/y_0$. Again, we are determining (in general this time) the value of p that maximizes the probability of the value of Y that we actually observed.

*3.69   Refer to Exercise 3.68. The *maximum likelihood estimator* for p is $1/Y$ (note that Y is the geometric random variable, not a particular value of it). Derive $E(1/Y)$. [*Hint:* if $|r| < 1$, $\sum_{i=1}^{\infty} r^i/i = -\ln(1-r)$.]

*3.70   If Y is a geometric random variable, define $Y^* = Y - 1$. If Y is interpreted as the number of the trial on which the first success occurs, then Y^* can be interpreted as the number of failures *before* the first success. If $Y^* = Y - 1$, $P(Y^* = y) = P(Y - 1 = y) = P(Y = y + 1)$ for $y = 0, 1, 2, \ldots$. Show that

$$P(Y^* = y) = q^y p, \qquad y = 0, 1, 2, \ldots .$$

The probability distribution of Y^* is sometimes used by actuaries as a model for the distribution of the number of insurance claims made in a specific time period.

*3.71 Refer to Exercise 3.70. Derive the mean and variance of the random variable Y^*

    a by using the result in Exercise 3.23 and the relationship $Y^* = Y - 1$, where Y is geometric.

  *b directly, using the probability distribution for Y^* given in Exercise 3.70.

3.6 The Negative Binomial Probability Distribution (Optional)

A random variable with a negative binomial distribution originates from a context much like the one that yields the geometric distribution. Again, we focus on independent and identical trials, each of which results in one of two outcomes: success or failure. The probability p of success stays the same from trial to trial. The geometric distribution handles the case where we are interested in the number of the trial on which the first success occurs. What if we are interested in knowing the number of the trial on which the second, third, or fourth success occurs? The distribution that applies to the random variable Y equal to the number of the trial on which the rth success occurs $(r = 2, 3, 4, \text{etc.})$ is the negative binomial distribution.

    The following steps closely resemble those in the previous section. Let us select fixed values for r and y and consider events A and B, where

$$A = \{\text{the first } (y-1) \text{ trials contain } (r-1) \text{ successes}\}$$

and

$$B = \{\text{trial } y \text{ results in a success}\}.$$

Because we assume that the trials are independent, it follows that A and B are independent events, and previous assumptions imply that $P(B) = p$. Therefore,

$$p(y) = p(Y = y) = P(A \cap B) = P(A) \times P(B).$$

Notice that $P(A)$ is 0 if $(y-1) < (r-1)$ or, equivalently, if $y < r$. If $y \geq r$, our previous work with the binomial distribution implies that

$$P(A) = \binom{y-1}{r-1} p^{r-1} q^{y-r}.$$

Finally,

$$p(y) = \binom{y-1}{r-1} p^r q^{y-r}, \qquad y = r, \, r+1, \, r+2, \ldots.$$

DEFINITION 3.9   A random variable Y is said to have a *negative binomial probability distribution* if and only if

$$p(y) = \binom{y-1}{r-1} p^r q^{y-r}, \qquad y = r,\, r+1,\, r+2, \ldots, \qquad 0 \le p \le 1.$$

EXAMPLE 3.14   A geological study indicates that an exploratory oil well drilled in a particular region should strike oil with probability .2. Find the probability that the third oil strike comes on the fifth well drilled.

Solution   Assuming independent drillings and probability .2 of striking oil with any one well, let Y denote the number of the trial on which the third oil strike occurs. Then it is reasonable to assume that Y has a negative binomial distribution with $p = .2$. Because we are interested in $r = 3$ and $y = 5$,

$$P(Y = 5) = p(5) = \binom{4}{2}(.2)^3(.8)^2$$

$$= 6(.008)(.64) = .0307.$$

The mean and variance of a random variable with a negative binomial distribution can be derived directly from Definitions 3.4 and 3.5 by using techniques like those previously illustrated. However, summing the resulting infinite series is somewhat tedious. These derivations will be much easier after we have developed some of the techniques of Chapter 5. For now, we state the following theorem without proof.

THEOREM 3.9   If Y is a random variable with a negative binomial distribution,

$$\mu = E(Y) = \frac{r}{p} \quad \text{and} \quad \sigma^2 = V(Y) = \frac{r(1-p)}{p^2}.$$

EXAMPLE 3.15   A large stockpile of used pumps contains 20% that are in need of repair. A maintenance worker is sent to the stockpile with three repair kits. She selects pumps at random and tests them one at a time. If the pump works, she sets it aside for future use. However, if the pump does not work, she uses one of her repair kits on it. Suppose that it takes 10 minutes to test a pump that is in working condition and 30 minutes to test and repair a pump that does not work. Find the mean and variance of the total time it takes the maintenance worker to use her three repair kits.

Solution   Let Y denote the number of the trial on which the third nonfunctioning pump is found. It follows that Y has a negative binomial distribution with $p = .2$. Thus,

$E(Y) = 3/(.2) = 15$ and $V(Y) = 3(.8)/(.2)^2 = 60$. Because it takes an additional 20 min to repair each defective pump, the total time necessary to use the three kits is

$$T = 10Y + 3(20).$$

Using the result derived in Exercise 3.23, we see that

$$E(T) = 10E(Y) + 60 = 10(15) + 60 = 210$$

and

$$V(T) = 10^2 V(Y) = 100(60) = 6000.$$

Thus, the total time necessary to use all three kits has mean 210 and standard deviation $\sqrt{6000} = 77.46$.

Exercises

3.72 The employees of a firm that manufactures insulation are being tested for indications of asbestos in their lungs. The firm is requested to send three employees who have positive indications of asbestos on to a medical center for further testing. If 40% of the employees have positive indications of asbestos in their lungs, find the probability that ten employees must be tested in order to find three positives.

3.73 Refer to Exercise 3.72. If each test costs $20, find the expected value and variance of the total cost of conducting the tests necessary to locate the three positives.

3.74 Ten percent of the engines manufactured on an assembly line are defective. If engines are randomly selected one at a time and tested, what is the probability that the first nondefective engine will be found on the second trial?

3.75 Refer to Exercise 3.74. What is the probability that the third nondefective engine will be found
  a on the fifth trial?
  b on or before the fifth trial?

3.76 Refer to Exercise 3.74. Find the mean and variance of the number of the trial on which
  a the first nondefective engine is found.
  b the third nondefective engine is found.

3.77 Refer to Exercise 3.74. Given that the first two engines tested were defective, what is the probability that at least two more engines must be tested before the first nondefective is found?

3.78 The telephone lines serving an airline reservation office are all busy about 60% of the time.
  a If you are calling this office, what is the probability that you will complete your call on the first try? The second try? The third try?
  b If you and a friend must both complete calls to this office, what is the probability that a total of four tries will be necessary for both of you to get through?

3.79 A geological study indicates that an exploratory oil well should strike oil with probability .2. What is the probability that

a   the first strike comes on the third well drilled?

b   the third strike comes on the seventh well drilled?

c   What assumptions did you make to obtain the answers to (a) and (b)?

d   Find the mean and variance of the number of wells that must be drilled if the company wants to set up three producing wells.

*3.80   Consider the negative binomial distribution given in Definition 3.9.

   a   Show that if $y \geq r + 1$, $\dfrac{p(y)}{p(y - 1)} = \left(\dfrac{y - 1}{y - r}\right) q$. This establishes a recursive relationship between successive negative binomial probabilities, because $p(y) = p(y - 1) \times \left(\dfrac{y - 1}{y - r}\right) q$.

   b   Show that $\dfrac{p(y)}{p(y - 1)} = \left(\dfrac{y - 1}{y - r}\right) q > 1$ if $y < \dfrac{r - q}{1 - q}$. Similarly, $\dfrac{p(y)}{p(y - 1)} < 1$ if $y > \dfrac{r - q}{1 - q}$.

   c   Apply the result in (b) for the case $r = 7$, $p = .5$ to determine the values of y for which $p(y) > p(y - 1)$.

*3.81   In a sequence of independent identical trials with two possible outcomes on each trial, S and F, and with $P(S) = p$, what is the probability that y trials will occur *before* the rth success?

*3.82   If Y is a negative binomial random variable, define $Y^* = Y - r$. If Y is interpreted as the number of the trial on which the rth success occurs, then Y^* can be interpreted as the number of failures *before* the rth success.

   a   If $Y^* = Y - r$, $P(Y^* = y) = P(Y - r = y) = P(Y = y + r)$ for $y = 0, 1, 2, \ldots$, show that $P(Y^* = y) = \left(\dbinom{y + r - 1}{r - 1}\right) p^r q^y$, $y = 0, 1, 2, \ldots$.

   b   Derive the mean and variance of the random variable Y^* by using the relationship $Y^* = Y - r$, where Y is negative binomial and the result in Exercise 3.23.

*3.83   **a**   We observe a sequence of independent identical trials with two possible outcomes on each trial, S and F, and with $P(S) = p$. The number of the trial on which we observe the fifth success, Y, has a negative binomial distribution with parameters $r = 5$ and p. Suppose that we observe the fifth success on the eleventh trial. Find the value of p that *maximizes* $P(Y = 11)$.

   b   Generalize the result from part (a) to find the value of p that *maximizes* $P(Y = y_0)$ when Y has a negative binomial distribution with parameters r (known) and p.

3.7   The Hypergeometric Probability Distribution

In Example 3.6 we considered a population of voters, 40% of whom favored candidate Jones. A sample of voters was selected, and Y (the number favoring Jones) was to be observed. We concluded that if the sample size n was small relative to the population size N, the distribution of Y could be approximated by a binomial distribution.

We also determined that if n was large relative to N, the *conditional* probability of selecting a supporter of Jones on a later draw would be significantly affected by the observed preferences of persons selected on earlier draws. Thus, the trials were not independent, and the probability distribution for Y could not be approximated adequately by a binomial probability distribution. Consequently, we need to develop the probability distribution for Y when n is large relative to N.

Suppose that a population contains a finite number N of elements that possess one of two characteristics. Thus r of the elements might be red and $b = N - r$, black. A sample of n elements is randomly selected from the population and the random variable of interest is Y, the number of red elements in the sample. This random variable has what is known as the *hypergeometric probability distribution*. For example, the number of workers who are women, Y, in Example 3.1 has the hypergeometric distribution.

The hypergeometric probability distribution can be derived by using the combinatorial theorems given in Section 2.6 and the sample-point approach. A sample point in the sample space S will correspond to a unique selection of n elements, some red and the remainder black. As in the binomial experiment, each sample point can be characterized by an n-tuple whose elements correspond to a selection of n elements from the total of N. If each element in the population were numbered from 1 to N, the sample point indicating the selection of items 5, 7, 8, 64, 17, ..., 87 would appear as the n-tuple

$$\underbrace{(5, 7, 8, 64, 17, \ldots, 87)}_{n \text{ positions}}.$$

The total number of sample points in S, therefore, will equal the number of ways of selecting a subset of n elements from a population of N, or $\binom{N}{n}$. Because random selection implies that all sample points are equiprobable, the probability of a sample point in S is

$$P(E_i) = \frac{1}{\binom{N}{n}}, \qquad \text{all } E_i \in S.$$

The total number of sample points in the numerical event $Y = y$ is the number of sample points in S that contain y red and $(n - y)$ black elements. This number can be obtained by applying the *mn* rule (Section 2.6). The number of ways of selecting y red elements to fill y positions in the n-tuple representing a sample point is the number of ways of selecting y from a total of r, or $\binom{r}{y}$. [We use the convention $\binom{a}{b} = 0$ if $b > a$.] The total number of ways of selecting $(n - y)$ black elements to fill the remaining $(n - y)$ positions in the n-tuple is the number of ways of selecting $(n - y)$ black elements from a possible $(N - r)$, or $\binom{N-r}{n-y}$. Then the number of sample points in the numerical event $Y = y$ is the number of ways of combining a set of y red and $(n - y)$ black elements. By the *mn* rule, this is the product $\binom{r}{y} \cdot \binom{N-r}{n-y}$. Summing the probabilities of the sample points in the numerical event $Y = y$ (multiplying the number of sample points by the common probability per sample point), we obtain the hypergeometric probability function.

DEFINITION **3.10**

A random variable Y is said to have a *hypergeometric probability distribution* if and only if

$$p(y) = \frac{\binom{r}{y}\binom{N-r}{n-y}}{\binom{N}{n}}$$

where y is an integer $0, 1, 2, \ldots, n$, subject to the restrictions $y \leq r$ and $n - y \leq N - r$.

With the convention $\binom{a}{b} = 0$ if $b > a$, it is clear that $p(y) \geq 0$ for the hypergeometric probabilities. The fact that the hypergeometric probabilities sum to 1 follows from the fact that

$$\sum_{i=0}^{n} \binom{r}{i}\binom{N-r}{n-i} = \binom{N}{n}.$$

A sketch of the proof of this result is outlined in Exercise 3.176.

EXAMPLE **3.16**

An important problem encountered by personnel directors and others faced with the selection of the best in a finite set of elements is exemplified by the following scenario. From a group of 20 Ph.D. engineers, 10 are randomly selected for employment. What is the probability that the 10 selected include all the 5 best engineers in the group of 20?

Solution

For this example $N = 20$, $n = 10$, and $r = 5$. That is, there are only 5 in the set of 5 best engineers, and we seek the probability that $Y = 5$, where Y denotes the number of best engineers among the ten selected. Then

$$p(5) = \frac{\binom{5}{5}\binom{15}{5}}{\binom{20}{10}} = \left(\frac{15!}{5!10!}\right)\left(\frac{10!10!}{20!}\right) = \frac{21}{1292} = .0162.$$

The mean and variance of a random variable with a hypergeometric distribution can be derived directly from Definitions 3.4 and 3.5. However, deriving closed form expressions for the resulting summations is somewhat tedious. In Chapter 5 we will develop methods that permit a much simpler derivation of the results presented in the following theorem.

THEOREM **3.10**

If Y is a random variable with a hypergeometric distribution,

$$\mu = E(Y) = \frac{nr}{N} \quad \text{and} \quad \sigma^2 = V(Y) = n\left(\frac{r}{N}\right)\left(\frac{N-r}{N}\right)\left(\frac{N-n}{N-1}\right).$$

Although the mean and the variance of the hypergeometric random variable seem to be rather complicated, they bear a striking resemblance to the mean and variance of a binomial random variable. Indeed, if we define $p = \frac{r}{N}$ and $q = 1 - p = \frac{N-r}{N}$, we can reexpress the mean and variance of the hypergeometric as $\mu = np$ and

$$\sigma^2 = npq \left(\frac{N-n}{N-1} \right).$$

You can view the factor

$$\frac{N-n}{N-1}$$

in $V(Y)$ as an adjustment that is appropriate when n is large relative to N. For fixed n, as $N \to \infty$,

$$\frac{N-n}{N-1} \to 1.$$

EXAMPLE **3.17**   A particular industrial product is shipped in lots of 20. Testing to determine whether an item is defective is costly, and hence the manufacturer samples his production rather than using a 100% inspection plan. A sampling plan constructed to minimize the number of defectives shipped to customers calls for sampling 5 items from each lot and rejecting the lot if more than 1 defective is observed. (If the lot is rejected, each item in it is later tested.) If a lot contains 4 defectives, what is the probability that it will be rejected? What is the expected number of defectives in the sample of size 5? What is the variance of the number of defectives in the sample of size 5?

Solution   Let Y equal the number of defectives in the sample. Then $N = 20$, $r = 4$, and $n = 5$. The lot will be rejected if $Y = 2, 3$, or 4. Then

$$P(\text{rejecting the lot}) = P(Y \geq 2) = p(2) + p(3) + p(4)$$

$$= 1 - p(0) - p(1)$$

$$= 1 - \frac{\binom{4}{0}\binom{16}{5}}{\binom{20}{5}} - \frac{\binom{4}{1}\binom{16}{4}}{\binom{20}{5}}$$

$$= 1 - .2817 - .4696 = .2487.$$

The mean and variance of the number of defectives in the sample of size 5 are

$$\mu = \frac{(5)(4)}{20} = 1 \quad \text{and} \quad \sigma^2 = 5 \left(\frac{4}{20} \right) \left(\frac{20-4}{20} \right) \left(\frac{20-5}{20-1} \right) = .632.$$

Example 3.17 involves sampling a lot of N industrial products, of which r are defective. The random variable of interest is Y, the number of defectives in a sample of size n. As noted in the beginning of this section, Y possesses an approximately binomial distribution when N is large and n is relatively small. Consequently, we would expect the probabilities assigned to values of Y by the hypergeometric distribution to approach those assigned by the binomial distribution as N becomes large and r/N, the fraction defective in the population, is held constant and equal to p. You can verify this expectation by using limit theorems encountered in your calculus courses to show that

$$\lim_{N \to \infty} \frac{\binom{r}{y}\binom{N-r}{n-y}}{\binom{N}{n}} = \binom{n}{y} p^y (1-p)^{n-y}$$

where

$$\frac{r}{N} = p.$$

(The proof of this result is omitted.) Hence, for a fixed fraction defective $p = r/N$, the hypergeometric probability function converges to the binomial probability function as N becomes large.

Exercises

3.84 An urn contains ten marbles, of which five are green, two are blue, and three are red. Three marbles are to be drawn from the urn, one at a time without replacement. What is the probability that all three marbles drawn will be green?

3.85 A warehouse contains ten printing machines, four of which are defective. A company selects five of the machines at random, thinking all are in working condition. What is the probability that all five of the machines are nondefective?

3.86 Refer to Exercise 3.85. The company repairs the defective ones at a cost of $50 each. Find the mean and variance of the total repair cost.

3.87 A group of six software packages available to solve a linear programming problem has been ranked from 1 to 6 (best to worst). An engineering firm, unaware of the rankings, randomly selected and then purchased two of the packages. Let Y denote the number of packages purchased by the firm that are ranked 3, 4, 5, or 6. Give the probability distribution for Y.

3.88 A corporation is sampling without replacement for $n = 3$ firms to determine the one from which to purchase certain supplies. The sample is to be selected from a pool of six firms, of which four are local and two are not local. Let Y denote the number of nonlocal firms among the three selected.

   a Find $P(Y = 1)$.
   b Find $P(Y \geq 1)$.
   c Find $P(Y \leq 1)$.

3.89 Specifications call for a thermistor to test out at between 9000 and 10,000 ohms at 25°C. Ten thermistors are available, and three of these are to be selected for use. Let Y denote the number among the three that do not conform to specifications. Find the probability distributions for Y (in tabular form) under the following conditions.

   a There are two thermistors not conforming to specifications among the ten that are available.
   b There are four thermistors not conforming to specifications among the ten that are available.

3.90 Used photocopy machines are returned to the supplier, cleaned, and then sent back out on lease agreements. Major repairs are not made, however, and as a result, some customers receive malfunctioning machines. Among eight used photocopiers available today, three are malfunctioning. A customer wants to lease four machines immediately. To meet the customer's deadline, four of the eight machines are randomly selected and, without further checking, shipped to the customer.

a What is the probability that the customer receives no malfunctioning machines?

b What is the probability that the customer receives at least one malfunctioning machine?

3.91 A jury of 6 persons was selected from a group of 20 potential jurors, of whom 8 were African American and 12 were white. The jury was supposedly randomly selected, but it contained only 1 African American member. Do you have any reason to doubt the randomness of the selection?

3.92 Refer to Exercise 3.91. If the selection process were really random, what would be the mean and variance of the number of African American members selected for the jury?

3.93 Suppose that a radio contains six transistors, two of which are defective. Three transistors are selected at random, removed from the radio, and inspected. Let Y equal the number of defectives observed, where $Y = 0, 1$, or 2. Find the probability distribution for Y. Express your results graphically as a probability histogram.

3.94 Simulate the experiment described in Exercise 3.93 by marking six marbles or coins so that two represent defectives and four represent nondefectives. Place the marbles in a hat, mix, draw three, and record Y, the number of defectives observed. Replace the marbles and repeat the process until $n = 100$ observations of Y have been recorded. Construct a relative frequency histogram for this sample and compare it with the population probability distribution (Exercise 3.93).

3.95 In an assembly-line production of industrial robots, gearbox assemblies can be installed in 1 minute each if holes have been properly drilled in the boxes and in 10 minutes if the holes must be redrilled. Twenty gearboxes are in stock, 2 with improperly drilled holes. Five gearboxes must be selected from the 20 that are available for installation in the next five robots.

a Find the probability that all 5 gearboxes will fit properly.

b Find the mean, variance, and standard deviation of the time it takes to install these 5 gearboxes.

*3.96 The sizes of animal populations are often estimated by using a capture–tag–recapture method. In this method k animals are captured, tagged, and then released into the population. Some time later n animals are captured, and Y, the number of tagged animals among the n, is noted. The probabilities associated with Y are a function of N, the number of animals in the population, so the observed value of Y contains information on this unknown N. Suppose that $k = 4$ animals are tagged and then released. A sample of $n = 3$ animals is then selected at random from the same population. Find $P(Y = 1)$ as a function of N. What value of N will *maximize* $P(Y = 1)$?

3.8 The Poisson Probability Distribution

Suppose that we want to find the probability distribution of the number of automobile accidents at a particular intersection during a time period of one week. At first glance this random variable, the number of accidents, may not seem even remotely related to a binomial random variable, but we will see that an interesting relationship exists.

Think of the time period, one week in this example, as being split up into n subintervals, *each of which is so small that at most one accident could occur in it with probability different from zero*. Denoting the probability of an accident in any subinterval by p, we have, for all practical purposes,

$$P(\text{no accidents occur in a subinterval}) = 1 - p$$

$$P(\text{one accident occurs in a subinterval}) = p$$

and

$$P(\text{more than one accident occurs in a subinterval}) = 0.$$

Then the total number of accidents in the week is just the total number of subintervals that contain one accident. If the occurrence of accidents can be regarded as independent from interval to interval, the total number of accidents has a binomial distribution.

Although there is no unique way to choose the subintervals, and we therefore know neither n nor p, it seems reasonable that as we divide the week into a greater number n of subintervals, the probability p of one accident in one of these shorter subintervals will decrease. Letting $\lambda = np$ and taking the limit of the binomial probability $p(y) = \binom{n}{y} p^y (1-p)^{n-y}$ as $n \to \infty$, we have

$$\lim_{n \to \infty} \binom{n}{y} p^y (1-p)^{n-y} = \lim_{n \to \infty} \frac{n(n-1)\cdots(n-y+1)}{y!} \left(\frac{\lambda}{n}\right)^y \left(1 - \frac{\lambda}{n}\right)^{n-y}$$

$$= \lim_{n \to \infty} \frac{\lambda^y}{y!} \left(1 - \frac{\lambda}{n}\right)^n \frac{n(n-1)\cdots(n-y+1)}{n^y} \left(1 - \frac{\lambda}{n}\right)^{-y}$$

$$= \frac{\lambda^y}{y!} \lim_{n \to \infty} \left(1 - \frac{\lambda}{n}\right)^n \left(1 - \frac{\lambda}{n}\right)^{-y} \left(1 - \frac{1}{n}\right) \times$$

$$\left(1 - \frac{2}{n}\right) \times \cdots \times \left(1 - \frac{y-1}{n}\right).$$

Noting that

$$\lim_{n \to \infty} \left(1 - \frac{\lambda}{n}\right)^n = e^{-\lambda}$$

and all other terms to the right of the limit have a limit of 1, we obtain

$$p(y) = \frac{\lambda^y}{y!} e^{-\lambda}.$$

(*Note*: $e = 2.718\ldots$) Random variables possessing this distribution are said to have a Poisson distribution. Hence Y, the number of accidents per week, has the Poisson distribution just derived.

Because the binomial probability function converges to the Poisson, the Poisson probabilities can be used to approximate their binomial counterparts for large n, small p, and $\lambda = np$ less than, roughly, 7. Exercise 3.106 requires you to calculate corresponding binomial and Poisson probabilities and will demonstrate the adequacy of the approximation.

The Poisson probability distribution often provides a good model for the probability distribution of the number Y of rare events that occur in space, time, volume, or any other dimension, where λ is the average value of Y. As we have noted, it provides a good model for the probability distribution of the number Y of automobile accidents, industrial accidents, or other types of accidents in a given unit of time. Other examples of random variables with approximate Poisson distributions are the number of telephone calls handled by a switchboard in a time interval, the number of radioactive particles that decay in a particular time period, the number of errors a typist makes in typing a page, and the number of automobiles using a freeway access ramp in a 10-minute interval.

DEFINITION 3.11

A random variable Y is said to have a *Poisson probability distribution* if and only if

$$p(y) = \frac{\lambda^y}{y!} e^{-\lambda}, \qquad y = 0, 1, 2, \ldots, \qquad \lambda > 0.$$

As we shall see in Theorem 3.11, the parameter λ that appears in the formula for the Poisson distribution is actually the mean of the distribution.

EXAMPLE 3.18   Show that the probabilities assigned by the Poisson probability distribution satisfy the requirements that $0 \leq p(y) \leq 1$ for all y and $\sum_y p(y) = 1$.

Solution   Because $\lambda > 0$, it is obvious that $p(y) > 0$ for $y = 0, 1, 2, \ldots$, and that $p(y) = 0$ otherwise. Further,

$$\sum_{y=0}^{\infty} p(y) = \sum_{y=0}^{\infty} \frac{\lambda^y}{y!} e^{-\lambda} = e^{-\lambda} \sum_{y=0}^{\infty} \frac{\lambda^y}{y!} = e^{-\lambda} e^{\lambda} = 1$$

because the infinite sum $\sum_{y=0}^{\infty} \lambda^y / y!$ is a series expansion of e^{λ}. Sums of special series are given in Appendix *A*1.11.

EXAMPLE 3.19   Suppose that a random system of police patrol is devised so that a patrol officer may visit a given beat location $Y = 0, 1, 2, 3, \ldots$ times per half-hour period, with each location being visited an average of once per time period. Assume that Y possesses, approximately, a Poisson probability distribution. Calculate the probability that the patrol officer will miss a given location during a half-hour period. What is the probability that it will be visited once? Twice? At least once?

Solution   For this example the time period is a half-hour and the mean number of visits per half-hour interval is $\lambda = 1$. Then

$$p(y) = \frac{(1)^y e^{-1}}{y!} = \frac{e^{-1}}{y!}, \qquad y = 0, 1, 2, \ldots.$$

The event that a given location is missed in a half-hour period corresponds to $(Y = 0)$ and

$$P(Y = 0) = p(0) = \frac{e^{-1}}{0!} = e^{-1} = .368.$$

Similarly,

$$p(1) = \frac{e^{-1}}{1!} = e^{-1} = .368$$

and

$$p(2) = \frac{e^{-1}}{2!} = \frac{e^{-1}}{2} = .184.$$

The probability that the location is visited *at least* once is the event $(Y \geq 1)$. Then

$$P(Y \geq 1) = \sum_{y=1}^{\infty} p(y) = 1 - p(0) = 1 - e^{-1} = .632.$$

EXAMPLE 3.20   A certain type of tree has seedlings randomly dispersed in a large area, with the mean density of seedlings being approximately five per square yard. If a forester randomly locates ten 1-square-yard sampling regions in the area, find the probability that none of the regions will contain seedlings.

Solution   If the seedlings really are randomly dispersed, the number of seedlings per region, Y, can be modeled as a Poisson random variable with $\lambda = 5$. (The average density is five per square yard.) Thus,

$$P(Y = 0) = p(0) = \frac{\lambda^0 e^{-\lambda}}{0!} = e^{-5} = .006738.$$

The probability that $Y = 0$ on ten independently selected regions is $(e^{-5})^{10}$, because the probability of the intersection of independent events is equal to the product of the respective probabilities. The resulting probability is extremely small. Thus, if this event actually occurred, we would seriously question the assumption of randomness, the stated average density of seedlings, or both.

For your convenience, we provide in Table 3, Appendix III, the partial sums $\sum_{y=0}^{a} p(y)$ for the Poisson probability distribution for many values of λ between .02 and 25. This table is laid out similarly to the table of partial sums for the binomial distribution, Table 1, Appendix III. The following example illustrates the use of

Table 3 and demonstrates that the Poisson probability distribution can approximate the binomial probability distribution.

EXAMPLE **3.21**   Suppose that Y possesses a binomial distribution with $n = 20$ and $p = .1$. Find the exact value of $P(Y \leq 3)$ using the table of binomial probabilities, Table 1, Appendix III. Use Table 3, Appendix III, to approximate this probability, using a corresponding probability given by the Poisson distribution. Compare the exact and approximate values for $P(Y \leq 3)$.

Solution   According to Table 1, Appendix III, the exact (accurate to three decimal places) value of $P(Y \leq 3) = .867$. If W is a Poisson-distributed random variable with $\lambda = np = 20(.1) = 2$, previous discussions indicate that $P(Y \leq 3)$ is approximately equal to $P(W \leq 3)$. Table 3, Appendix III, gives $P(W \leq 3) = .857$. Thus you can see that the Poisson approximation is quite good, yielding a value that differs from the exact value by only .01.

In our derivation of the mean and variance of a random variable with the Poisson distribution, we again use the fundamental property that $\sum_y p(y) = 1$ for any discrete probability distribution.

THEOREM **3.11**   If Y is a random variable possessing a Poisson distribution with parameter λ, then

$$\mu = E(Y) = \lambda \quad \text{and} \quad \sigma^2 = V(Y) = \lambda.$$

Proof   By definition,

$$E(Y) = \sum_y y p(y) = \sum_{y=0}^{\infty} y \, \frac{\lambda^y e^{-\lambda}}{y!}.$$

Notice that the first term in this sum is equal to 0 (when $y = 0$) and hence

$$E(Y) = \sum_{y=1}^{\infty} y \, \frac{\lambda^y e^{-\lambda}}{y!} = \sum_{y=1}^{\infty} \frac{\lambda^y e^{-\lambda}}{(y-1)!}.$$

As it stands, this quantity is not equal to the sum of the values of a probability function $p(y)$ over all values of y, but we can change it to the proper form by factoring λ out of the expression and letting $z = y - 1$. Then the limits of summation become $z = 0$ (when $y = 1$) and $z = \infty$ (when $y = \infty$), and

$$E(Y) = \lambda \sum_{y=1}^{\infty} \frac{\lambda^{y-1} e^{-\lambda}}{(y-1)!} = \lambda \sum_{z=0}^{\infty} \frac{\lambda^z e^{-\lambda}}{z!}.$$

Notice that $p(z) = \lambda^z e^{-\lambda}/z!$ is the probability function for a Poisson random variable, and $\sum_{z=0}^{\infty} p(z) = 1$. Therefore, $E(Y) = \lambda$. Thus the mean of a Poisson random variable is the single parameter λ that appears in the expression for the Poisson probability function.

We leave the derivation of the variance as an exercise (Exercise 3.110).   ∎

EXAMPLE 3.22   The number of industrial accidents at a particular manufacturing plant is found to average three per month. During the last month, six accidents occurred. Does this number seem highly improbable if μ is still equal to 3? Does it indicate an increase in the mean μ?

Solution   The number of accidents Y is likely to follow a Poisson probability distribution with $\lambda = 3$. The probability that Y is as large as 6 is

$$P(Y \geq 6) = \sum_{y=6}^{\infty} \frac{3^y e^{-3}}{y!}.$$

The tedious calculation required to find $P(Y \geq 6)$ can be avoided by using Table 3, Appendix III, or the empirical rule. From Theorem 3.11,

$$\mu = \lambda = 3, \qquad \sigma^2 = \lambda = 3, \qquad \sigma = \sqrt{3} = 1.73.$$

The empirical rule tells us that we should expect Y to take values in the interval $\mu \pm 2\sigma$ with a high probability.

Notice that $\mu + 2\sigma = 3 + (2)(1.73) = 6.5$. The observed number of accidents, $Y = 6$, does not lie more than 2σ from μ, but it is close to the boundary. Thus the observed result is not highly improbable, but it may be sufficiently improbable to warrant an investigation. See Exercise 3.170 for the exact probability $P(|Y - \lambda| \leq 2\sigma)$.

Exercises

3.97   Let Y denote a random variable that has a Poisson distribution with mean $\lambda = 2$. Find

   a $P(Y = 4)$.

   b $P(Y \geq 4)$.

   c $P(Y < 4)$.

   d $P(Y \geq 4 | Y \geq 2)$.

3.98   Customers arrive at a checkout counter in a department store according to a Poisson distribution at an average of seven per hour. During a given hour, what are the probabilities that

    a   no more than three customers arrive?

    b   at least two customers arrive?

    c   exactly five customers arrive?

3.99   Refer to Exercise 3.98. If it takes approximately 10 minutes to serve each customer, find the mean and variance of the total service time for customers arriving during a 1-hour period. (Assume that a sufficient number of servers are available so that no customer must wait for service.) Is it likely that the total service time will exceed 2.5 hours?

3.100   Refer to Exercise 3.98. Find the probability that exactly two customers arrive in the 2-hour period of time:

    a   between 2:00 P.M. and 4:00 P.M. (one continuous 2-hour period).

    b   between 1:00 P.M. and 2:00 P.M. or between 3:00 P.M. and 4:00 P.M. (two separate 1-hour periods that total 2 hours).

3.101   The number of typing errors made by a typist has a Poisson distribution with an average of four errors per page. If more than four errors appear on a given page, the typist must retype the whole page. What is the probability that a certain page does not have to be retyped?

3.102   A parking lot has two entrances. Cars arrive at entrance I according to a Poisson distribution at an average of three per hour and at entrance II according to a Poisson distribution at an average of four per hour. What is the probability that a total of three cars will arrive at the parking lot in a given hour? (Assume that the numbers of cars arriving at the two entrances are independent.)

3.103   The number of knots in a particular type of wood has a Poisson distribution with an average of 1.5 knots in 10 cubic feet of the wood. Find the probability that a 10-cubic-foot block of the wood has at most one knot.

3.104   The mean number of automobiles entering a mountain tunnel per 2-minute period is one. An excessive number of cars entering the tunnel during a brief period of time produces a hazardous situation. Find the probability that the number of autos entering the tunnel during a 2-minute period exceeds three. Does the Poisson model seem reasonable for this problem?

3.105   Assume that the tunnel in Exercise 3.104 is observed during ten 2-minute intervals, thus giving ten independent observations $Y_1, Y_2, \ldots, Y_{10}$, on the Poisson random variable. Find the probability that $Y > 3$ during at least one of the ten 2-minute intervals.

3.106   Consider a binomial experiment for $n = 20$, $p = .05$. Use Table 1, Appendix III, to calculate the binomial probabilities for $Y = 0, 1, 2, 3, 4$. Calculate the same probabilities by using the Poisson approximation with $\lambda = np$. Compare.

3.107   A salesperson has found that the probability of a sale on a single contact is approximately .03. If the salesperson contacts 100 prospects, what is the approximate probability of making at least one sale?

3.108   According to the U.S. Fire Administration, 185 people died in 12,438 hotel and motel fires in 1979, roughly 1.5 deaths per 100 fires (*U.S. News and World Report*, August 3, 1981).

    a   If 200 hotel and motel fires occurred in a given region, what is the probability that the number of deaths would exceed 8? [*Hint:* The mean number of deaths per 200 fires equals $2(1.5) = 3$.]

    b   If 200 hotel and motel fires occurred in a given region and the number of deaths exceeded 8, would you suspect that the region's mean death rate would exceed the national average? Explain.

3.109   The probability that a mouse inoculated with a serum will contract a certain disease is .2. Using the Poisson approximation, find the probability that at most 3 of 30 inoculated mice will contract the disease.

3.110   Let Y have a Poisson distribution with mean λ. Find $E[Y(Y - 1)]$ and then use this to show that $V(Y) = \lambda$.

3.111   In the daily production of a certain kind of rope, the number of defects per foot Y is assumed to have a Poisson distribution with mean $\lambda = 2$. The profit per foot when the rope is sold is given by X, where $X = 50 - 2Y - Y^2$. Find the expected profit per foot.

*3.112   A store owner has overstocked a certain item and decides to use the following promotion to decrease the supply. The item has a marked price of $100. For each customer purchasing the item during a particular day, the owner will reduce the price by a factor of one-half. Thus the first customer will pay $50 for the item, the second will pay $25, and so on. Suppose that the number of customers who purchase the item during the day has a Poisson distribution with mean 2. Find the expected cost of the item at the end of the day. [*Hint:* The cost at the end of the day is $100(1/2)^Y$, where Y is the number of customers who have purchased the item.]

3.113   A food manufacturer uses an extruder (a machine that produces bite-size cookies and snack food) that yields revenue for the firm at a rate of $200 per hour when in operation. However, the extruder breaks down an average of two times every day it operates. If Y denotes the number of breakdowns per day, the daily revenue generated by the machine is $R = 1600 - 50Y^2$. Find the expected daily revenue for the extruder.

*3.114   Let $p(y)$ denote the probability function associated with a Poisson random variable with mean λ.

   a   Show that the ratio of successive probabilities satisfies $\dfrac{p(y)}{p(y-1)} = \dfrac{\lambda}{y}$, for $y = 1, 2, \ldots$.

   b   For which values of y is $p(y) > p(y-1)$?

   c   Notice that the result in (a) implies that Poisson probabilities increase for awhile as y increases and decrease thereafter. For what value of y is $p(y)$ maximized?

3.9   Moments and Moment-Generating Functions

The parameters μ and σ are meaningful numerical descriptive measures that locate the center and describe the spread associated with the values of a random variable Y. They do not, however, provide a unique characterization of the distribution of Y. Many different distributions possess the same means and standard deviations. We now consider a set of numerical descriptive measures that (at least under certain conditions) uniquely determine $p(y)$.

DEFINITION 3.12

> The *k*th *moment of a random variable Y taken about the origin* is defined to be $E(Y^k)$ and is denoted by μ'_k.

Notice in particular that the first moment about the origin, is $E(Y) = \mu'_1 = \mu$ and that $\mu'_2 = E(Y^2)$ is employed in Theorem 3.6 for finding σ^2.

Another useful moment of a random variable is one taken about its mean.

DEFINITION 3.13

The *k*th *moment of a random variable Y taken about its mean*, or the *k*th *central moment of Y*, is defined to be $E[(Y - \mu)^k]$ and is denoted by μ_k.

In particular, $\sigma^2 = \mu_2$.

Let us concentrate on moments μ_k' about the origin where $k = 1, 2, 3, \ldots$. Suppose that two random variables Y and Z possess finite moments with $\mu_{1Y}' = \mu_{1Z}'$, $\mu_{2Y}' = \mu_{2Z}'$, $\ldots$, $\mu_{jY}' = \mu_{jZ}'$, where j can assume any integer value. That is, the two random variables possess identical corresponding moments about the origin. Under some fairly general conditions, it can be shown that Y and Z have identical probability distributions. Thus a major use of moments is to approximate the probability distribution of a random variable (usually an estimator or a decision maker). Consequently, the moments μ_k', where $k = 1, 2, 3, \ldots$, are primarily of theoretical value for $k > 3$.

Yet another interesting expectation is the moment-generating function for a random variable, which, figuratively speaking, packages all the moments for a random variable into one simple expression. We will first define the moment-generating function and then explain how it works.

DEFINITION 3.14

The *moment-generating function m(t) for a random variable Y* is defined to be $m(t) = E(e^{tY})$. We say that a moment-generating function for Y exists if there exists a positive constant b such that $m(t)$ is finite for $|t| \leq b$.

Why is $E(e^{tY})$ called the *moment-generating function for Y*? From a series expansion for e^{ty}, we have

$$e^{ty} = 1 + ty + \frac{(ty)^2}{2!} + \frac{(ty)^3}{3!} + \frac{(ty)^4}{4!} + \cdots.$$

Then, assuming that μ_k' is finite for $k = 1, 2, 3, \ldots$, we have

$$E(e^{tY}) = \sum_y e^{ty} p(y) = \sum_y \left[1 + ty + \frac{(ty)^2}{2!} + \frac{(ty)^3}{3!} + \cdots \right] p(y)$$

$$= \sum_y p(y) + t \sum_y y p(y) + \frac{t^2}{2!} \sum_y y^2 p(y) + \frac{t^3}{3!} \sum_y y^3 p(y) + \cdots$$

$$= 1 + t\mu_1' + \frac{t^2}{2!}\mu_2' + \frac{t^3}{3!}\mu_3' + \cdots.$$

This argument involves an interchange of summations, which is justifiable if $m(t)$ exists.

Thus $E(e^{tY})$ is a function of all the moments μ'_k about the origin, for $k = 1, 2, 3, \ldots$. In particular, μ'_k is the coefficient of $t^k/k!$ in the series expansion of $m(t)$.

The moment-generating function possesses two important applications. First, if we can find $E(e^{tY})$, we can find any of the moments for Y.

THEOREM 3.12

If $m(t)$ exists, then for any positive integer k,

$$\frac{d^k m(t)}{dt^k}\Bigg]_{t=0} = m^{(k)}(0) = \mu'_k.$$

In other words, if you find the kth derivative of $m(t)$ with respect to t and then set $t = 0$, the result will be μ'_k.

Proof $d^k m(t)/dt^k$, or $m^{(k)}(t)$, is the kth derivative of $m(t)$ with respect to t. Because

$$m(t) = E(e^{tY}) = 1 + t\mu'_1 + \frac{t^2}{2!}\mu'_2 + \frac{t^3}{3!}\mu'_3 + \cdots$$

it follows that

$$m^{(1)}(t) = \mu'_1 + \frac{2t}{2!}\mu'_2 + \frac{3t^2}{3!}\mu'_3 + \cdots,$$

$$m^{(2)}(t) = \mu'_2 + \frac{2t}{2!}\mu'_3 + \frac{3t^2}{3!}\mu'_4 + \cdots,$$

and, in general,

$$m^{(k)}(t) = \mu'_k + \frac{2t}{2!}\mu'_{k+1} + \frac{3t^2}{3!}\mu'_{k+2} + \cdots.$$

Setting $t = 0$ in each of the above derivatives, we obtain

$$m^{(1)}(0) = \mu'_1, \qquad m^{(2)}(0) = \mu'_2$$

and, in general,

$$m^{(k)}(0) = \mu'_k.$$

These operations involve interchanging derivatives and infinite sums, which can be justified if $m(t)$ exists.   ∎

EXAMPLE 3.23   Find the moment-generating function $m(t)$ for a Poisson distributed random variable with mean λ.

Solution

$$m(t) = E(e^{tY}) = \sum_{y=0}^{\infty} e^{ty} p(y) = \sum_{y=0}^{\infty} e^{ty} \frac{\lambda^y e^{-\lambda}}{y!}$$

$$= \sum_{y=0}^{\infty} \frac{(\lambda e^t)^y e^{-\lambda}}{y!} = e^{-\lambda} \sum_{y=0}^{\infty} \frac{(\lambda e^t)^y}{y!}.$$

To complete the summation, consult Appendix $A1.11$ to find the Taylor series expansion

$$\sum_{y=0}^{\infty} \frac{(\lambda e^t)^y}{y!} = e^{\lambda e^t}$$

or employ the method of Theorem 3.11. Thus, multiply and divide by $e^{\lambda e^t}$. Then

$$m(t) = e^{-\lambda} e^{\lambda e^t} \sum_{y=0}^{\infty} \frac{(\lambda e^t)^y e^{-\lambda e^t}}{y!}.$$

The quantity to the right of the summation sign is the probability function for a Poisson random variable with mean λe^t. Hence

$$\sum_{y} p(y) = 1 \quad \text{and} \quad m(t) = e^{-\lambda} e^{\lambda e^t} (1) = e^{\lambda(e^t - 1)}.$$

The calculations in Example 3.23 are no more difficult than those in Theorem 3.11, where only the expected value for a Poisson random variable Y was calculated. Direct evaluation of the variance of Y through the use of Theorem 3.6 required that $E(Y^2)$ be found by summing another series (actually, we obtained $E(Y^2)$ from $E(Y(Y - 1))$ in Exercise 3.110). Example 3.24 illustrates the use of the moment-generating function of the Poisson random variable to calculate its mean and variance.

EXAMPLE 3.24   Use the moment-generating function of Example 3.23 and Theorem 3.12 to find the mean, μ, and variance, σ^2, for the Poisson random variable.

Solution   According to Theorem 3.12, $\mu = \mu_1' = m^{(1)}(0)$ and $\mu_2' = m^{(2)}(0)$. Taking the first and second derivatives of $m(t)$, we obtain

$$m^{(1)}(t) = \frac{d}{dt}(e^{\lambda(e^t - 1)}) = e^{\lambda(e^t - 1)} \cdot \lambda e^t$$

$$m^{(2)}(t) = \frac{d^2}{dt^2}(e^{\lambda(e^t - 1)}) = \frac{d}{dt}(e^{\lambda(e^t - 1)} \cdot \lambda e^t)$$

$$= e^{\lambda(e^t - 1)} \cdot (\lambda e^t)^2 + e^{\lambda(e^t - 1)} \cdot \lambda e^t.$$

Then, because

$$\mu = m^{(1)}(0) = \left\{ e^{\lambda(e^t-1)} \cdot \lambda e^t \right\}\Big]_{t=0} = \lambda$$

$$\mu_2' = m^{(2)}(0) = \left\{ e^{\lambda(e^t-1)} \cdot (\lambda e^t)^2 + e^{\lambda(e^t-1)} \cdot \lambda e^t \right\}\Big]_{t=0} = \lambda^2 + \lambda,$$

Theorem 3.6 tells us that $\sigma^2 = E(Y^2) - \mu^2 = \mu_2' - \mu^2 = \lambda^2 + \lambda - (\lambda)^2 = \lambda$. Notice how easily we obtained μ_2' from $m(t)$.

The second (but primary) application of a moment-generating function is to prove that a random variable possesses a particular probability distribution $p(y)$. If $m(t)$ exists for a probability distribution $p(y)$, it is unique. Also, if the moment-generating functions for two random variables Y and Z are equal (for all $|t| < b$ for some $b > 0$), then Y and Z must have the same probability distribution. It follows that, if we can recognize the moment-generating function of a random variable Y to be one associated with a specific distribution, then Y must have that distribution.

In summary, a moment-generating function is a mathematical expression that sometimes (but not always) provides an easy way to find moments associated with random variables. More importantly, it can be used to establish the equivalence of two probability distributions.

EXAMPLE 3.25 Suppose that Y is a random variable with moment-generating function $m_Y(t) = e^{3.2(e^t-1)}$. What is the distribution of Y?

Solution In Example 3.23, we showed that the moment-generating function of a Poisson distributed random variable with mean λ is $m(t) = e^{\lambda(e^t-1)}$. Note that the moment-generating function of Y is exactly equal to the moment-generating function of a Poisson distributed random variable with $\lambda = 3.2$. Because moment-generating functions are unique, Y *must* have a Poisson distribution with mean 3.2.

Exercises

3.115 If Y has a binomial distribution with n trials and probability of success p, show that the moment-generating function for Y is

$$m(t) = (pe^t + q)^n, \qquad \text{where } q = 1 - p.$$

3.116 Differentiate the moment-generating function in Exercise 3.115 to find $E(Y)$ and $E(Y^2)$. Then find $V(Y)$.

3.117 If Y has a geometric distribution with probability of success p, show that the moment-generating function for Y is

$$m(t) = \frac{pe^t}{1 - qe^t} \qquad \text{where } q = 1 - p.$$

3.118 Differentiate the moment-generating function in Exercise 3.117 to find $E(Y)$ and $E(Y^2)$. Then find $V(Y)$.

3.119 Refer to Exercise 3.115. Use the uniqueness of moment-generating functions to give the distribution of a random variable with moment-generating function $m(t) = (.6e^t + .4)^3$.

3.120 Refer to Exercise 3.117. Use the uniqueness of moment-generating functions to give the distribution of a random variable with moment-generating function $m(t) = \dfrac{.3e^t}{1 - .7e^t}$.

3.121 Find the distributions of the random variables that have each of the following moment-generating functions:

   a $m(t) = [(1/3)e^t + (2/3)]^5$.

   b $m(t) = \dfrac{e^t}{2 - e^t}$.

   c $m(t) = e^{2(e^t - 1)}$.

3.122 Refer to Exercise 3.121. By inspection, give the mean and variance of the random variables associated with the moment-generating functions given in (a), (b), and (c).

3.123 Let $m(t) = (1/6)e^t + (2/6)e^{2t} + (3/6)e^{3t}$. Find the following:

   a $E(Y)$.

   b $V(Y)$.

   c the distribution of Y.

3.124 If Y is a random variable with moment-generating function $m(t)$, and if W is given by $W = aY + b$, show that the moment-generating function of W is $e^{tb}m(at)$.

3.125 Use the result in Exercise 3.124 to prove that, if $W = aY + b$, then $E(W) = aE(Y) + b$ and $V(W) = a^2V(Y)$.

*3.126 Let $r(t) = \ln[m(t)]$ and $r^{(k)}(0)$ denote the kth derivative of $r(t)$ evaluated for $t = 0$. Show that $r^{(1)}(0) = \mu_1' = \mu$ and $r^{(2)}(0) = \mu_2' - (\mu_1')^2 = \sigma^2$ [*Hint:* $m(0) = 1$.]

*3.127 Use the results of Exercise 3.125 to find the mean and variance of a Poisson random variable with $m(t) = e^{5(e^t - 1)}$. Notice that $r(t)$ is easier to differentiate than $m(t)$ in this case.

3.10 Probability-Generating Functions (Optional)

An important class of discrete random variables is one in which Y represents a count and consequently takes integer values: $Y = 0, 1, 2, 3, \ldots$. The binomial, geometric, hypergeometric, and Poisson random variables all fall in this class. The following examples give practical situations that result in integer-valued random variables. One, involving the theory of queues (waiting lines), is concerned with the number of persons (or objects) awaiting service at a particular point in time. Knowledge of the behavior of this random variable is important in designing manufacturing plants where production consists of a sequence of operations, each taking a different length of time to complete. An insufficient number of service stations for a particular production operation can result in a bottleneck, the formation of a queue of products waiting to be serviced, and a resulting slowdown in the manufacturing operation.

Queuing theory is also important in determining the number of checkout counters needed for a supermarket and in designing hospitals and clinics.

Integer-valued random variables are also important in studies of population growth. For example, epidemiologists are interested in the growth of bacterial populations and the growth of the number of persons afflicted by a particular disease. The numbers of elements in each of these populations are integer-valued random variables.

A mathematical device useful in finding the probability distributions and other properties of integer-valued random variables is the probability-generating function.

DEFINITION   3.15

Let Y be an integer-valued random variable for which $P(Y = i) = p_i$, where $i = 0, 1, 2, \ldots$. The *probability-generating function* $P(t)$ for Y is defined to be

$$P(t) = E(t^Y) = p_0 + p_1 t + p_2 t^2 + \cdots = \sum_{i=0}^{\infty} p_i t^i$$

for all values of t such that $P(t)$ is finite.

The reason for calling $P(t)$ a probability-generating function is clear when we compare $P(t)$ with the moment-generating function $m(t)$. In particular, the coefficient of t^i in $P(t)$ is the probability p_i. Correspondingly, the coefficient of t^i for $m(t)$ is a constant times the ith moment μ_i'. If we know $P(t)$ and can expand it into a series, we can determine $p(y)$ as the coefficient of t^y.

Repeated differentiation of $P(t)$ yields *factorial moments* for the random variable Y.

DEFINITION   3.16

The kth *factorial moment* for a random variable Y is defined to be

$$\mu_{[k]} = E[Y(Y - 1)(Y - 2) \cdots (Y - k + 1)]$$

where k is a positive integer.

Notice that $\mu_{[1]} = E(Y) = \mu$. The second factorial moment, $\mu_{[2]} = E[Y(Y - 1)]$, was useful in finding the variance for binomial, geometric, and Poisson random variables in Theorem 3.7, Exercise 3.67, and Exercise 3.110, respectively.

THEOREM   3.13

If $P(t)$ is the probability-generating function for an integer-valued random variable, Y, then the kth factorial moment of Y is given by

$$\frac{d^k P(t)}{dt^k} \bigg]_{t=1} = P^{(k)}(1) = \mu_{[k]}.$$

Proof   Because

$$P(t) = p_0 + p_1 t + p_2 t^2 + p_3 t^3 + p_4 t^4 + \cdots$$

it follows that

$$P^{(1)}(t) = \frac{dP(t)}{dt} = p_1 + 2p_2 t + 3p_3 t^2 + 4p_4 t^3 + \cdots,$$

$$P^{(2)}(t) = \frac{d^2 P(t)}{dt^2} = (2)(1)p_2 + (3)(2)p_3 t + (4)(3)p_4 t^2 + \cdots,$$

and, in general,

$$P^{(k)}(t) = \frac{d^k P(t)}{dt^k} = \sum_{y=k}^{\infty} y(y-1)(y-2)\cdots(y-k+1)p(y)t^{y-k}.$$

Setting $t = 1$ in each of these derivatives, we obtain

$$P^{(1)}(1) = p_1 + 2p_2 + 3p_3 + 4p_4 + \cdots = \mu_{[1]} = E(Y),$$

$$P^{(2)}(1) = (2)(1)p_2 + (3)(2)p_3 + (4)(3)p_4 + \cdots = \mu_{[2]} = E[Y(Y-1)],$$

and, in general,

$$P^{(k)}(1) = \sum_{y=k}^{\infty} y(y-1)(y-2)\cdots(y-k+1)p(y)$$

$$= E[Y(Y-1)(Y-2)\cdots(Y-k+1)] = \mu_{[k]}. \qquad \blacksquare$$

EXAMPLE 3.26   Find the probability-generating function for a geometric random variable.

Solution   Notice that $p_0 = 0$, because Y cannot assume this value. Then

$$P(t) = E(t^Y) = \sum_{y=1}^{\infty} t^y q^{y-1} p = \sum_{y=1}^{\infty} \frac{p}{q}(qt)^y$$

$$= \frac{p}{q}[qt + (qt)^2 + (qt)^3 + \cdots].$$

The terms in the series are those of an infinite geometric progression. If $qt < 1$, then

$$P(t) = \frac{p}{q}\left(\frac{qt}{1-qt}\right) = \frac{pt}{1-qt} \qquad \text{if } t < 1/q.$$

(For summation of the series, consult Appendix A1.11.)

EXAMPLE **3.27**   Use $P(t)$, Example 3.26, to find the mean of a geometric random variable.

Solution   From Theorem 3.13, $\mu_{[1]} = \mu = P^{(1)}(1)$. Using the result in Example 3.26,

$$P^{(1)}(t) = \frac{d}{dt}\left(\frac{pt}{1-qt}\right) = \frac{(1-qt)p - (pt)(-q)}{(1-qt)^2}.$$

Setting $t = 1$, we obtain

$$P^{(1)}(1) = \frac{p^2 + pq}{p^2} = \frac{p(p+q)}{p^2} = \frac{1}{p}.$$

Because we already have the moment-generating function to assist in finding the moments of a random variable, of what value is $P(t)$? The answer is that it may be difficult to find $m(t)$ but much easier to find $P(t)$. Thus $P(t)$ provides an additional tool for finding the moments of a random variable. It may or may not be useful in a given situation.

Finding the moments of a random variable is not the major use of the probability-generating function. Its primary application is in deriving the probability function (and hence the probability distribution) for other related integer-valued random variables. For these applications, see Feller (1968) and Parzen (1964).

Exercises

*3.128   Let Y denote a binomial random variable with n trials and probability of success p. Find the probability-generating function for Y, and use it to find $E(Y)$.

*3.129   Let Y denote a Poisson random variable with mean λ. Find the probability-generating function for Y, and use it to find $E(Y)$ and $V(Y)$.

*3.130   Refer to Exercise 3.129. Use the probability-generating function found there to find $E(Y^3)$.

3.11 Tchebysheff's Theorem

We have seen in Section 1.3 and Example 3.22 that if the probability or population histogram is roughly bell-shaped and the mean and variance are known, the empirical rule is of great help in approximating the probabilities of certain intervals. However, in many instances, the shapes of probability histograms differ markedly from a mound shape, and the empirical rule may not yield useful approximations to the probabilities of interest. The following result, known as Tchebysheff's theorem, can be used to determine a lower bound for the probability that the random variable Y of interest falls in an interval $\mu \pm k\sigma$.

THEOREM 3.14

Tchebysheff's Theorem Let Y be a random variable with mean μ and finite variance σ^2. Then, for any constant $k > 0$,

$$P(|Y - \mu| < k\sigma) \geq 1 - \frac{1}{k^2} \quad \text{or} \quad P(|Y - \mu| \geq k\sigma) \leq \frac{1}{k^2}.$$

Two important aspects of this result should be pointed out. First, the result applies for *any* probability distribution, whether the probability histogram is bell-shaped or not. Second, the results of the theorem are very conservative in the sense that the actual probability that Y is in the interval $\mu \pm k\sigma$ usually exceeds the lower bound for the probability, $1 - 1/k^2$, by a considerable amount. However, as discussed in Exercise 3.133, for any $k > 1$, it is possible to construct a probability distribution so that, for that k, the bound provided by Tchebysheff's theorem is actually attained. (You should verify that the results of the empirical rule do not contradict those given by Theorem 3.14.) The proof of this theorem will be deferred to Section 4.10. The usefulness of this theorem is illustrated in the following example.

EXAMPLE 3.28

The number of customers per day at a sales counter, Y, has been observed for a long period of time and found to have mean 20 and standard deviation 2. The probability distribution of Y is not known. What can be said about the probability that, tomorrow, Y will be greater than 16 but less than 24?

Solution

We want to find $P(16 < Y < 24)$. From Theorem 3.14 we know that, for any $k \geq 0$, $P(|Y - \mu| < k\sigma) \geq 1 - 1/k^2$, or

$$P[(\mu - k\sigma) < Y < (\mu + k\sigma)] \geq 1 - \frac{1}{k^2}.$$

Because $\mu = 20$ and $\sigma = 2$, it follows that $\mu - k\sigma = 16$ and $\mu + k\sigma = 24$ if $k = 2$. Thus

$$P(16 < Y < 24) = P(\mu - 2\sigma < Y < \mu + 2\sigma) \geq 1 - \frac{1}{(2)^2} = \frac{3}{4}.$$

In other words, tomorrow's customer total will be between 16 and 24 with a fairly high probability (at least 3/4).

Notice that if σ were 1, k would be 4 and

$$P(16 < Y < 24) = P(\mu - 4\sigma < Y < \mu + 4\sigma) \geq 1 - \frac{1}{(4)^2} = \frac{15}{16}.$$

Thus the value of σ has considerable effect on probabilities associated with intervals.

Exercises

3.131 Let Y be a random variable with mean 11 and variance 9. Using Tchebysheff's theorem, find

    a   a lower bound for $P(6 < Y < 16)$,

    b   the value of C such that $P(|Y - 11| \geq C) \leq .09$.

3.132 Suppose, as noted in Exercise 3.30, that approximately 35% of all applicants for jobs falsify the information on their application forms. Consider a company with 2300 employees.

    a   What is the expected value of the number Y of application forms that have been falsified?

    b   Find the standard deviation of Y.

    c   Calculate the interval $\mu \pm 2\sigma$.

    d   Suppose that the company had a credentials-checking firm verify the information on 2300 application forms, and 249 application forms contained falsified information. Is the company's application falsification rate consistent with the contention that 35% of all job applicants falsify information on their applications? Explain.

3.133 This exercise demonstrates that, in general, the results provided by Tchebysheff's theorem cannot be improved upon. Let Y be a random variable such that

$$p(-1) = \frac{1}{18}, \qquad p(0) = \frac{16}{18}, \quad \text{and} \quad p(1) = \frac{1}{18}.$$

    a   Show that $E(Y) = 0$ and $V(Y) = \frac{1}{9}$.

    b   Use the probability distribution of Y to calculate $P(|Y - \mu| \geq 3\sigma)$. Compare this exact probability with the upper bound provided by Tchebysheff's theorem to see that the bound provided by Tchebysheff's theorem is actually attained when $k = 3$.

    *c   In (b) we guaranteed $E(Y) = 0$ by placing all probability mass on the values -1, 0, and 1, with $p(-1) = p(1)$. The variance was controlled by the probabilities assigned to $p(-1)$ and $p(1)$. Using this same basic idea, construct a probability distribution for a random variable X that will yield $P(|X - \mu_X| \geq 2\sigma_X) = 1/4$.

    *d   If any $k > 1$ is specified, how can a random variable W be constructed so that $P(|W - \mu_W| \geq k\sigma_W) = 1/k^2$?

3.134 The U.S. mint produces dimes with an average diameter of .5 inches and standard deviation .01. Using Tchebysheff's theorem, find a lower bound for the number of coins in a lot of 400 coins that are expected to have a diameter between .48 and .52.

3.135 For a certain type of soil the number of wireworms per cubic foot has a mean of 100. Assuming a Poisson distribution of wireworms, give an interval that will include at least 5/9 of the sample values of wireworm counts obtained from a large number of 1-cubic-foot samples.

3.136 Refer to Exercise 3.93. Using the probability histogram, find the fraction of values in the population that fall within two standard deviations of the mean. Compare your result with that of Tchebysheff's theorem.

3.137 A balanced coin is tossed three times. Let Y equal the number of heads observed.

    a   Use the formula for the binomial probability distribution to calculate the probabilities associated with $Y = 0, 1, 2,$ and 3.

    b   Construct a probability distribution similar to the one in Table 3.1.

    c   Find the expected value and standard deviation of Y, using the formulas $E(Y) = np$ and $V(Y) = npq$.

d Using the probability distribution from part (b), find the fraction of the population measurements lying within one standard deviation of the mean. Repeat for two standard deviations. How do your results compare with the results of Tchebysheff's theorem and the empirical rule?

3.138 Suppose that a coin was definitely unbalanced and that the probability of a head was equal to $p = .1$. Follow instructions (a), (b), (c), and (d) as stated in Exercise 3.137. Notice that the probability distribution loses its symmetry and becomes skewed when p is not equal to 1/2.

3.139 The *Wall Street Journal* (July 29, 1981)[4] notes that hang gliding can be dangerous. Of the estimated 36,000 hang-gliding enthusiasts in the United States, 29 were killed last year in hang-gliding accidents and, it is estimated, thousands more were injured. Suppose that a hang-glider pilot, randomly selected from among all hang-glider pilots, has probability .0006 of being killed in any one year. If there are 40,000 pilots next year,

  a what is the expected number of fatalities?

  b what is the standard deviation of the number Y of fatalities?

  c is it likely that the number of fatalities would exceed 40?

3.140 A national poll of 1502 adults by pollster Louis Harris (*Environmental News*, EPA, September 1977) indicated that 71% "would rather live in an environment that is clean than in an area with a lot of jobs." If people really had no preference for environment over jobs, or vice versa, comment on the probability of observing this survey result (that is, observing 71% or more of a sample of 1502 in favor of environment over jobs). What assumption must be made about the sampling procedure in order to calculate this probability? (*Hint:* Recall Tchebysheff's theorem and the empirical rule.)

3.141 For a certain section of a pine forest, the number of diseased trees per acre, Y, has a Poisson distribution with mean $\lambda = 10$. The diseased trees are sprayed with an insecticide at a cost of $3.00 per tree, plus a fixed overhead cost for equipment rental of $50.00. Letting C denote the total spraying cost for a randomly selected acre, find the expected value and standard deviation for C. Within what interval would you expect C to lie with probability at least 0.75?

3.142 It is known that 10% of a brand of television tubes will burn out before their guarantee has expired. If 1000 tubes are sold, find the expected value and variance of Y, the number of original tubes that must be replaced. Within what limits would Y be expected to fall?

3.143 Refer to Exercise 3.73. In this exercise, we determined that the mean and variance of the costs necessary to find three employees with positive indications of asbestos poisoning were 150 and 4500, respectively. Do you think it is highly unlikely that the cost of completing the tests will exceed $350?

3.12 Summary

This chapter has explored discrete random variables, their probability distributions and their expected values. Calculating the probability distribution for a discrete random variable requires the use of the probabilistic methods of Chapter 2 to evaluate

---

[4] *Source:* Reprinted by permission of the *Wall Street Journal*, Dow Jones & Company, Inc. 1981. All rights reserved.

the probabilities of numerical events. Probability distribution functions were derived for the binomial, geometric, negative binomial, hypergeometric, and Poisson random variables.

The expected values of random variables and functions of random variables provided a method for finding the mean and variance of Y and consequently measures of centrality and variation for $p(y)$. Much of the remaining material in the chapter was devoted to the techniques for acquiring expectations, which sometimes involved summing apparently intractable series. The techniques for obtaining closed-form expressions for some of the resulting expected values included (1) use of the fact that $\sum_y p(y) = 1$ for any discrete random variable and (2) $E(Y^2) = E[Y(Y-1)] + E(Y)$. The means and variances of several of the more common discrete distributions are summarized in Table 3.4. These results and more are also found in Table A2.1 in Appendix II and inside the back cover of this book.

We then discussed the moment-generating function associated with a random variable. Although sometimes useful in finding μ and σ, the moment-generating function is of primary value to the theoretical statistician for deriving the probability distribution of a random variable. The moment-generating functions for most of the common random variables are found in Appendix II and inside the back cover of this book.

The probability-generating function is a useful device for deriving moments and probability distributions of integer-valued random variables.

Finally, we gave Tchebysheff's theorem, a very useful result that permits approximating certain probabilities when only the mean and variance are known.

To conclude this summary, we recall the primary objective of statistics: to make an inference about a population based on information contained in a sample. Drawing the sample from the population is the experiment. The sample is often a set of measurements of one or more random variables, and it is the observed event resulting from a single repetition of the experiment. Finally, making the inference about the population requires knowledge of the probability of occurrence of the observed sample, which in turn requires knowledge of the probability distributions of the random variables that generated the sample.

Table 3.4. Means and variances for some common random variables

Distribution	$E(Y)$	$V(Y)$
Binomial	np	$np(1-p) = npq$
Geometric	$\dfrac{1}{p}$	$\dfrac{1-p}{p^2} = \dfrac{q}{p^2}$
Hypergeometric	$n\left(\dfrac{r}{N}\right)$	$n\left(\dfrac{r}{N}\right)\left(\dfrac{N-r}{N}\right)\left(\dfrac{N-n}{N-1}\right)$
Poisson	λ	λ
Negative binomial	$\dfrac{r}{p}$	$\dfrac{r(1-p)}{p^2} = \dfrac{rq}{p^2}$

References and Further Readings

Feller, W. 1968. *An Introduction to Probability Theory and Its Applications*, 3d. ed., vol. 1. New York: John Wiley.

Goranson, U. G., and Hall, J. 1980. "Airworthiness of Long-Life Jet Transport Structures," *Aeronautical Journal* 84 (838): 279–80.

Johnson, N. L. 1992. *Univariate Discrete Distributions*, 2d ed. New York: John Wiley.

Mosteller, F.; R. E. K. Rourke, and G. B. Thomas. 1970. *Probability with Statistical Applications*, 2d ed. Reading, MA: Addison-Wesley.

Parzen, E. 1960. *Modern Probability Theory and Its Applications*. New York: John Wiley.

———. 1964. *Stochastic Processes*. San Francisco: Holden-Day.

Standard Mathematical Tables, 28th ed. 1987. Boca Raton, Fla.: CRC Press.

Supplementary Exercises

3.144   Four possibly winning numbers for a lottery—AB-4536, NH-7812, SQ-7855, and ZY-3221—arrive in the mail. You will win a prize if one of your numbers matches one of the winning numbers contained on a list held by those conducting the lottery. One first prize of $100,000, two second prizes of $50,000 each, and 10 third prizes of $1000 each will be awarded. To be eligible to win, you need to mail the coupon back to the company at a cost of 33¢ for postage. No purchase is required. From the structure of the numbers that you received, it is obvious the numbers sent out consist of two letters followed by four digits. Assuming that the numbers you received were generated at random, what are your expected winnings from the lottery? Is it worth 33¢ to enter this lottery?

3.145   Sampling for defectives from large lots of manufactured product yields a number of defectives, Y, that follows a binomial probability distribution. A sampling plan consists of specifying the number of items n to be included in a sample and an acceptance number a. The lot is accepted if $Y \leq a$ and rejected if $Y > a$. Let p denote the proportion of defectives in the lot. For $n = 5$ and $a = 0$, calculate the probability of lot acceptance if (a) $p = 0$, (b) $p = .1$, (c) $p = .3$, (d) $p = .5$, (e) $p = 1.0$. A graph showing the probability of lot acceptance as a function of lot fraction defective is called the *operating characteristic curve* for the sample plan. Construct the operating characteristic curve for the plan $n = 5, a = 0$. Notice that a sampling plan is an example of statistical inference. Accepting or rejecting a lot based on information contained in the sample is equivalent to concluding that the lot is either good or bad. "Good" implies that a low fraction is defective and that the lot is, therefore, suitable for shipment.

3.146   Refer to Exercise 3.145. Use Table 1, Appendix III, to construct the operating characteristic curves for the following sampling plans:

a $n = 10, a = 0$.

b $n = 10, a = 1$.

c $n = 10, a = 2$.

For each sampling plan, calculate P(lot acceptance) for $p = 0, .05, .1, .3, .5$, and 1.0. Our intuition suggests that sampling plan (a) would be much less likely to accept bad lots than plans (b) and (c). A visual comparison of the operating characteristic curves will confirm this intuitive conjecture.

3.147 A quality control engineer wishes to study alternative sampling plans: $n = 5$, $a = 1$ and $n = 25$, $a = 5$. On a sheet of graph paper, construct the operating characteristic curves for both plans, making use of acceptance probabilities at $p = .05$, $p = .10$, $p = .20$, $p = .30$, and $p = .40$ in each case.

a If you were a seller producing lots with fraction defective ranging from $p = 0$ to $p = .10$, which of the two sampling plans would you prefer?

b If you were a buyer wishing to be protected against accepting lots with fraction defective exceeding $p = .30$, which of the two sampling plans would you prefer?

3.148 Refer to Exercises 3.51 and 3.52. Let Y denote the number of the trial on which the first applicant with computer training was found. If each interview costs $30, find the expected value and variance of the total cost incurred interviewing candidates until an applicant with advanced computer training is found. Within what limits would you expect the interview costs to fall?

3.149 Consider the following game: a player throws a fair die repeatedly until he rolls a 2, 3, 4, 5, or 6. In other words, the player continues to throw the die as long as he rolls 1s. When he rolls a "non-1," he stops.

a What is the probability that the player tosses the die exactly three times?

b What is the expected number of rolls needed to obtain the first non-1?

c If he rolls a non-1 on the first throw, the player is paid $1. Otherwise, the payoff is doubled for each 1 that the player rolls before rolling a non-1. Thus the player is paid $2 if he rolls a 1 followed by a non-1; $4 if he rolls two 1s followed by a non-1; $8 if he rolls three 1s followed by a non-1; etc. In general, if we let Y be the number of throws needed to obtain the first non-1, then the player rolls $(Y - 1)$ 1s before rolling his first non-1, and he is paid 2^{Y-1} dollars. What is the expected amount paid to the player?

3.150 If Y is a binomial random variable based on n trials and success probability p, show

$$P(Y > 1|Y \geq 1) = \frac{1 - (1 - p)^n - np(1 - p)^{n-1}}{1 - (1 - p)^n}.$$

3.151 A starter motor used in a space vehicle has a high rate of reliability and was reputed to start on any given occasion with probability .99999. What is the probability of at least one failure in the next 10,000 starts?

3.152 Refer to Exercise 3.93. Find μ, the expected value of Y, for the theoretical population by using the probability distribution obtained in Exercise 3.93. Find the sample mean $\bar{y}$ for the $n = 100$ measurements generated in Exercise 3.94. Does $\bar{y}$ provide a good estimate of μ?

3.153 Find the population variance σ^2 for Exercise 3.93 and the sample variance s^2 for Exercise 3.94. Compare.

3.154 Toss a balanced die and let Y be the number of dots observed on the upper face. Find the mean and variance of Y. Construct a probability histogram, and locate the interval $\mu \pm 2\sigma$. Verify that Tchebysheff's theorem holds.

3.155 Two assembly lines I and II have the same rate of defectives in their production of voltage regulators. Five regulators are sampled from each line and tested. Among the total of ten tested regulators, four are defective. Find the probability that exactly two of the defective regulators came from line I.

3.156   One concern of a gambler is that she will go broke before achieving her first win. Suppose that ahe plays a game in which the probability of winning is .1 (and is unknown to her). It costs her $10 to play and she receives $80 for a win. If she commences with $30, what is the probability that she wins exactly once before she loses her initial capital?

3.157   The number of imperfections in the weave of a certain textile has a Poisson distribution with a mean of four per square yard.

    a   Find the probability that a 1-square-yard sample will contain at least one imperfection.

    b   Find the probability that a 3-square-yard sample will contain at least one imperfection.

3.158   Refer to Exercise 3.157. The cost of repairing the imperfections in the weave is $10 per imperfection. Find the mean and standard deviation of the repair cost for an 8-square-yard bolt of the textile.

3.159   The number of bacteria colonies of a certain type in samples of polluted water has a Poisson distribution with a mean of 2 per cubic centimeter.

    a   If four 1-cubic-centimeter samples are independently selected from this water, find the probability that at least one sample will contain one or more bacteria colonies.

    b   How many 1-cubic-centimeter samples should be selected in order to have a probability of approximately 0.95 of seeing at least one bacteria colony?

3.160   Use the fact that

$$e^z = 1 + z + \frac{z^2}{2!} + \frac{z^3}{3!} + \frac{z^4}{4!} + \cdots$$

to expand the moment-generating function for the binomial distribution

$$m(t) = (q + pe^t)^n$$

into a power series in t. (Acquire only the low-order terms in t.) Identify μ_i' as the coefficient of $t^i / i!$ appearing in the series. Specifically, find μ_1' and μ_2' and compare them with the results of Exercise 3.116.

3.161   Refer to Exercises 3.85 and 3.86. In what interval would you expect the repair costs on these five machines to lie? (Use Tchebysheff's theorem.)

*3.162   The number of cars driving past a parking area in a 1-minute time interval has a Poisson distribution with mean λ. The probability that any individual driver actually wants to park his or her car is p. Assume that individuals decide whether to park independently of one another.

    a   If one parking place is available and it will take you 1 minute to reach the parking area, what is the probability that a space will still be available when you reach the lot? (Assume that no one leaves the lot during the 1-minute interval.)

    b   Let W denote the number of drivers who wish to park during a 1-minute interval. Derive the probability distribution of W.

3.163   A type of bacteria cell divides at a constant rate λ over time. (That is, the probability that a cell divides in a small interval of time t is approximately λt.) Given that a population starts out at time zero with k cells of this bacteria and that cell divisions are independent of one another, the size of the population at time t, $Y(t)$, has the probability distribution

$$P[Y(t) = n] = \binom{n-1}{k-1} e^{-\lambda k t} \left(1 - e^{-\lambda t}\right)^{n-k} \qquad n = k, \, k+1, \, \ldots .$$

      a Find the expected value and variance of $Y(t)$ in terms of λ and t.

      b If, for a type of bacteria cell, $\lambda = .1$ per second and the population starts out with two cells at time zero, find the expected value and variance of the population after 5 seconds.

3.164 The probability that any single driver will turn left at an intersection is .2. The left turn lane at this intersection has room for three vehicles. If the left turn lane is empty when the light turns red and five vehicles arrive at this intersection while the light is red, find the probability that the left turn lane will hold the vehicles of all of the drivers who want to turn left.

3.165 An experiment consists of tossing a fair die until a 6 occurs four times. What is the probability that the process ends after exactly ten tosses with a 6 occurring on the ninth and tenth tosses?

3.166 Accident records collected by an automobile insurance company give the following information. The probability that an insured driver has an automobile accident is .15. If an accident has occurred, the damage to the vehicle amounts to 20% of its market value with a probability of .80, to 60% of its market value with a probability of .12, and to a total loss with a probability of .08. What premium should the company charge on a \$12,000 car so that the expected gain by the company is zero?

3.167 The number of people entering the intensive care unit at a hospital on any single day possesses a Poisson distribution with a mean equal to five persons per day.

      a What is the probability that the number of people entering the intensive care unit on a particular day is equal to 2? Is less than or equal to 2?

      b Is it likely that Y will exceed 10? Explain.

3.168 A recent survey suggests that Americans anticipate a reduction in living standards and that a steadily increasing level of consumption no longer may be as important as it was in the past. Suppose that a poll of 2000 people indicated 1373 in favor of forcing a reduction in the size of American automobiles by legislative means. Would you expect to observe as many as 1373 in favor of this proposition if, in fact, the general public was split 50–50 on the issue? Why?

3.169 A supplier of heavy construction equipment has found that new customers are normally obtained through customer requests for a sales call and that the probability of a sale of a particular piece of equipment is .3. If the supplier has three pieces of the equipment available for sale, what is the probability that it will take fewer than five customer contacts to clear the inventory?

3.170 Calculate $P(|Y - \lambda| \le 2\sigma)$ for the Poisson probability distribution of Example 3.22. Does this agree with the empirical rule?

*3.171 A merchant stocks a certain perishable item. She knows that on any given day she will have a demand for either two, three, or four of these items with probabilities .1, .4, and .5, respectively. She buys the items for \$1.00 each and sells them for \$1.20 each. If any are left at the end of the day, they represent a total loss. How many items should the merchant stock in order to maximize her expected daily profit?

*3.172 Show that the hypergeometric probability function approaches the binomial in the limit as $N \to \infty$ and $p = r/N$ remains constant. That is, show that

$$\lim_{N \to \infty} \frac{\binom{r}{y}\binom{N-r}{n-y}}{\binom{N}{n}} = \binom{n}{y} p^y q^{n-y}$$

for $p = r/N$ constant.

3.173 A lot of $N = 100$ industrial products contains 40 defectives. Let Y be the number of defectives in a random sample of size 20. Find $p(10)$ by using (a) the hypergeometric probability distribution and (b) the binomial probability distribution. Is N large enough that the the value for $p(10)$ obtained from the binomial distribution is a good approximation to that obtained using the hypergeometric distribution?

*3.174   For simplicity, let us assume that there are two kinds of drivers. The safe drivers, who are 70% of the population, have probability .1 of causing an accident in a year. The rest of the population are accident makers, who have probability .5 of causing an accident in a year. The insurance premium is $400 times one's probability of causing an accident in the following year. A new subscriber has an accident during the first year. What should be his insurance premium for the next year?

*3.175   It is known that 5% of the members of a population have disease A, which can be discovered by a blood test. Suppose that N (a large number) people are to be tested. This can be done in two ways: (1) Each person is tested separately; or (2) the blood samples of k people are pooled together and analyzed. (Assume that $N = nk$, with n an integer.) If the test is negative, all of them are healthy (that is, just this one test is needed). If the test is positive, each of the k persons must be tested separately (that is, a total of $k + 1$ tests are needed).

a   For fixed k what is the expected number of tests needed in (2)?

b   Find the k that will minimize the expected number of tests in (2).

c   If k is selected as in (b), on the average how many tests does (2) save in comparison with (1)?

*3.176   Let Y have a hypergeometric distribution

$$p(y) = \frac{\binom{r}{y}\binom{N-r}{n-y}}{\binom{N}{n}}, \qquad y = 0, 1, 2, \ldots, n.$$

a   Show that

$$P(Y = n) = p(n) = \left(\frac{r}{N}\right)\left(\frac{r-1}{N-1}\right)\left(\frac{r-2}{N-2}\right)\cdots\left(\frac{r-n+1}{N-n+1}\right).$$

b   Write $p(y)$ as $p(y|r)$. Show that if $r_1 < r_2$, then

$$\frac{p(y|r_1)}{p(y|r_2)} > \frac{p(y+1|r_1)}{p(y+1|r_2)}.$$

c   Apply the binomial expansion to each factor in the following equation:

$$(1+a)^{N_1}(1+a)^{N_2} = (1+a)^{N_1+N_2}.$$

Now compare the coefficients of a^n on both sides to prove that

$$\binom{N_1}{0}\binom{N_2}{n} + \binom{N_1}{1}\binom{N_2}{n-1} + \cdots + \binom{N_1}{n}\binom{N_2}{0} = \binom{N_1+N_2}{n}.$$

d   Using the result of (c), conclude that

$$\sum_{y=0}^{n} p(y) = 1.$$

*3.177   Use the result derived in Exercise 3.176(c) and Definition 3.4 to derive directly the mean of a hypergeometric random variable.

*3.178   Use the results of Exercises 3.176(c) and 3.177 to show that, for a hypergeometric random variable,

$$E[Y(Y - 1)] = \frac{r(r - 1)n(n - 1)}{N(N - 1)}.$$

CHAPTER 4

Continuous Random Variables and Their Probability Distributions

4.1 Introduction

4.2 The Probability Distribution for a Continuous Random Variable

4.3 Expected Values for Continuous Random Variables

4.4 The Uniform Probability Distribution

4.5 The Normal Probability Distribution

4.6 The Gamma Probability Distribution

4.7 The Beta Probability Distribution

4.8 Some General Comments

4.9 Other Expected Values

4.10 Tchebysheff's Theorem

4.11 Discontinuous Functions and Mixed Probability Distributions

4.12 Summary

References and Further Readings

4.1 Introduction

A moment of reflection on statistical problems encountered in the real world should convince you that not all random variables of interest are discrete random variables. The number of defectives in a lot of n manufactured items is a discrete random variable because the number of defectives must take one of the $n + 1$ values 0, 1, 2, . . . , or n. Now consider the daily rainfall at a specified geographical point. Theoretically, with measuring equipment of perfect accuracy, the amount of rainfall could take on any value between 0 and 5 inches. As a result, each of the uncountably infinite number of points in the interval (0, 5) represents a distinct possible value of the amount of rainfall in a day. The type of random variable that takes on any value in an interval is called *continuous*, and the purpose of this chapter is to study probability distributions for continuous random variables. The yield of an antibiotic in a fermentation process is a continuous random variable, as is the length of life, in years, of a washing machine. The line segments over which these two random variables are defined are contained in the positive half of the real line. This does not mean that, if we observed enough washing machines, we would eventually observe an outcome corresponding to every value in the interval (3, 7); rather it means that no value between 3 and 7 can be ruled out as as a possible value for the number of years that a washing machine remains in service.

The probability distribution for a discrete random variable can always be given by assigning a positive probability to each of the possible values the variable may assume. In every case, of course, the sum of the probabilities we assign must be equal to 1. Unfortunately, the probability distribution for a continuous random variable cannot be specified in the same way. It is mathematically impossible to assign nonzero probabilities to all the points on a line interval and at the same time satisfy the requirement that the probabilities of the distinct possible values sum to 1. As a result, we must develop a different method to describe the probability distribution for a continuous random variable.

4.2 The Probability Distribution for a Continuous Random Variable

Before we can state a formal definition for a continuous random variable, we must define a distribution function (or *cumulative* distribution function) associated with any random variable.

DEFINITION 4.1   Let Y denote any random variable. The *distribution function* of Y, denoted by $F(y)$, is given by $F(y) = P(Y \leq y)$ for $-\infty < y < \infty$.

The nature of the distribution function associated with a random variable determines whether the variable is continuous or discrete. Consequently, we will commence our discussion by examining the distribution function for a discrete random variable and noting the characteristics of this function.

EXAMPLE 4.1   Suppose that Y has a binomial distribution with $n = 2$ and $p = 1/2$. Find $F(y)$.

Solution   The probability function for Y is given by

$$p(y) = \binom{2}{y}\left(\frac{1}{2}\right)^y \left(\frac{1}{2}\right)^{2-y}, \qquad y = 0, 1, 2$$

FIGURE 4.1
Binomial distribution
function,
$n = 2, p = 1/2$

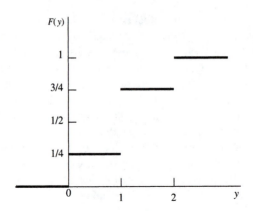

which yields

$$p(0) = 1/4, \qquad p(1) = 1/2, \quad \text{and} \quad p(2) = 1/4.$$

What is $F(-2) = P(Y \leq -2)$? Because the only values of Y that are assigned positive probabilities are 0, 1, and 2 and none of these values are less than or equal to -2, $F(-2) = 0$. Using similar logic, $F(y) = 0$ for all $y < 0$. What is $F(1.5)$? The only values of Y that are less than or equal to 1.5 and have nonzero probabilities are the values 0 and 1. Therefore,

$$F(1.5) = P(Y \leq 1.5) = P(Y = 0) + P(Y = 1)$$
$$= (1/4) + (1/2) = 3/4.$$

In general,

$$F(y) = P(Y \leq y) = \begin{cases} 0, & \text{for} \quad y < 0 \\ 1/4, & \text{for} \quad 0 \leq y < 1 \\ 3/4, & \text{for} \quad 1 \leq y < 2 \\ 1, & \text{for} \quad y \geq 2. \end{cases}$$

A graph of $F(y)$ is given in Figure 4.1.

In Example 4.1 the points between 0 and 1 or between 1 and 2 contributed nothing to the cumulative probability depicted by the distribution function. As a result, the cumulative distribution function stayed flat between the possible values of Y and increased in jumps or steps at each of the possible values of Y. Functions that behave in such a manner are called *step functions*. *Distribution functions for discrete random variables are always step functions because the cumulative distribution function increases only at a countable number of points.*

Because the distribution function associated with any random variable is such that $F(y) = P(Y \leq y)$, from a practical point of view it is clear that $F(-\infty) = \lim_{y \to -\infty} P(Y \leq y)$ must equal zero. If we consider any two values $y_1 < y_2$, then $P(Y \leq y_1) \leq P(Y \leq y_2)$—that is, $F(y_1) \leq F(y_2)$. So, a distribution function, $F(y)$, is always a monotonic, nondecreasing function. Further, it is clear that $F(\infty) = \lim_{y\to\infty} P(Y \leq y) = 1$. These three characteristics define the properties of any distribution function and are summarized in the following theorem.

THEOREM **4.1**

Properties of a Distribution Function.[1] If $F(y)$ is a distribution function, then

1. $F(-\infty) \equiv \lim_{y\to-\infty} F(y) = 0$.
2. $F(\infty) \equiv \lim_{y\to\infty} F(y) = 1$.
3. $F(y)$ is a nondecreasing function of y. (If y_1 and y_2 are *any* values such that $y_1 < y_2$, then $F(y_1) \leq F(y_2)$.)

You should check that the distribution function developed in Example 4.1 has each of these properties.

Let us now examine the distribution function for a continuous random variable. Suppose that, for all practical purposes, the amount of daily rainfall, Y, must be less than 12 inches. For every $0 \leq y_1 < y_2 \leq 12$, the interval (y_1, y_2) has a positive probability of including Y, no matter how close y_1 gets to y_2. It follows that $F(y)$ in this case should be a smooth, increasing function over some interval of real numbers, as graphed in Figure 4.2.

We are thus led to the definition of a continuous random variable.

FIGURE **4.2**
Distribution function
for a continuous
random variable

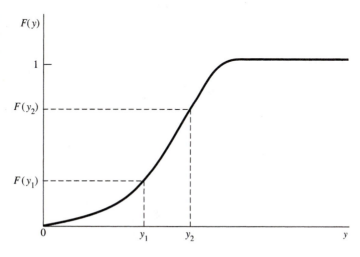

DEFINITION 4.2   Let Y denote a random variable with distribution function $F(y)$. Y is said to be *continuous* if the distribution function $F(y)$ is continuous, for $-\infty < y < \infty$.[2]

For a continuous random variable Y, we must have, for any real number y,

$$P(Y = y) = 0.$$

If this were not true, and $P(Y = y_0) = p_0 > 0$, then $F(y)$ would have a discontinuity (jump) of size p_0 at the point y_0. Practically speaking, the fact that continuous random variables have zero probability at discrete points should not bother us. Consider the example of measuring daily rainfall. What is the probability that we will see a daily rainfall measurement of exactly 2.193 inches? It is quite likely that we would never observe that exact value even if we took rainfall measurements for a lifetime, although we might see many days with measurements between 2 and 3 inches.

The derivative of $F(y)$ is another function of prime importance in probability theory and statistics.

DEFINITION 4.3   Let $F(y)$ be the distribution function for a continuous random variable Y. Then $f(y)$, given by

$$f(y) = \frac{dF(y)}{dy} = F'(y)$$

wherever the derivative exists, is called the *probability density function* for the random variable Y.

It follows from Definitions 4.2 and 4.3 that $F(y)$ can be written as

$$F(y) = \int_{-\infty}^{y} f(t)\, dt$$

where $f(\cdot)$ is the probability density function and t is used as the variable of integration. The relationship between the distribution and density functions is shown graphically in Figure 4.3.

FIGURE 4.3
The distribution
function

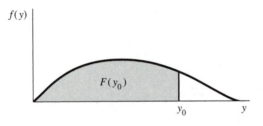

[2]To be mathematically precise, we also need the first derivative of $F(y)$ to exist and be continuous except for, at most, a finite number of points in any finite interval. The distribution functions for the continuous random variables discussed in this text satisfy this requirement.

The probability density function is a *theoretical model* for the frequency distribution of a population of measurements. For example, observations of the lengths of life of a particular brand of washing machines will generate measurements that can be characterized by a relative frequency histogram, as discussed in Chapter 1. Conceptually, the experiment could be repeated ad infinitum, thereby generating a relative frequency distribution (a smooth curve) that would characterize the population of interest to the manufacturer. This theoretical relative frequency distribution corresponds to the probability density function for the length of life of a single machine, Y.

Because the distribution function $F(y)$ for any random variable always has the properties given in Theorem 4.1, it follows that density functions also have some corresponding properties. Because $F(y)$ is a nondecreasing function, the derivative $f(y)$ is never negative. Further, we know that $F(\infty) = 1$ and, therefore, that $\int_{-\infty}^{\infty} f(t)\, dt = 1$. In summary, the properties of a probability density function are as given in the following theorem.

THEOREM **4.2**

Properties of a Density Function.   If $f(y)$ is a density function, then

1. $f(y) \geq 0$ for any value of y.
2. $\int_{-\infty}^{\infty} f(y)\, dy = 1$.

An example involving identification of a distribution function and a density function for a continuous random variable follows.

EXAMPLE **4.2**   Suppose that

$$
F(y) = \begin{cases} 0, & \text{for} \quad y < 0 \\ y, & \text{for} \quad 0 \leq y \leq 1 \\ 1, & \text{for} \quad y > 1. \end{cases}
$$

Find the probability density function for Y, and graph it.

Solution   Because the density function $f(y)$ is the derivative of the distribution function $F(y)$, when the derivative exists,

$$
f(y) = \frac{dF(y)}{dy} = \begin{cases} \frac{d(0)}{dy} = 0, & \text{for} \quad y < 0 \\ \frac{d(y)}{dy} = 1, & \text{for} \quad 0 < y < 1 \\ \frac{d(1)}{dy} = 0, & \text{for} \quad y > 1 \end{cases}
$$

and $f(y)$ is undefined at $y = 0$ and $y = 1$. A graph of $F(y)$ is shown in Figure 4.4.

FIGURE **4.4**
Distribution function
$F(y)$ for Example 4.2

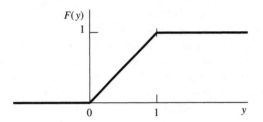

FIGURE **4.5**
Density function
$f(y)$ for Example 4.2

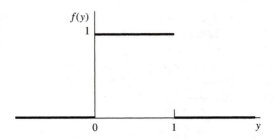

The graph of $f(y)$ for Example 4.2 is shown in Figure 4.5. Notice that the distribution and density functions given in Example 4.2 have all the properties required of distribution and density functions, respectively. Moreover, $F(y)$ is a continuous function of y, but $f(y)$ is discontinuous at the points $y = 0$, 1. In general, the distribution function for a continuous random variable must be continuous, but the density function need not be everywhere continuous.

EXAMPLE **4.3**   Let Y be a continuous random variable with probability density function given by

$$f(y) = \begin{cases} 3y^2, & 0 \le y \le 1 \\ 0, & \text{elsewhere.} \end{cases}$$

Find $F(y)$. Graph both $f(y)$ and $F(y)$.

Solution   The graph of $f(y)$ appears in Figure 4.6. Because

$$F(y) = \int_{-\infty}^{y} f(t)\,dt$$

we have, for this example,

$$F(y) = \begin{cases} \int_{-\infty}^{y} 0\,dt = 0, & \text{for } y < 0 \\ \int_{-\infty}^{0} 0\,dt + \int_{0}^{y} 3t^2\,dt = 0 + t^3\big]_{0}^{y} = y^3, & \text{for } 0 \le y \le 1 \\ \int_{-\infty}^{0} 0\,dt + \int_{0}^{1} 3t^2\,dt + \int_{1}^{y} 0\,dt = 0 + t^3\big]_{0}^{1} + 0 = 1, & \text{for } 1 < y. \end{cases}$$

Notice that some of the integrals that we evaluated yield a value of 0. These are included for completeness in this initial example. In future calculations, we will not

FIGURE **4.6**
Density function
for Example 4.3

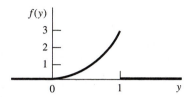

FIGURE **4.7**
Distribution function
for Example 4.3

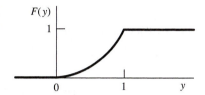

explicitly display any integral that has value 0. The graph of $F(y)$ is given in Figure 4.7.

$F(y_0)$ gives the probability that $Y \leq y_0$. The next step is to find the probability that Y falls in a specific interval; that is, $P(a \leq Y \leq b)$. From Chapter 1 we know that this probability corresponds to the area under the frequency distribution over the interval $a \leq y \leq b$. Because $f(y)$ is the theoretical counterpart of the frequency distribution, we would expect $P(a \leq Y \leq b)$ to equal a corresponding area under the density function $f(y)$. This indeed is true, because

$$P(a < Y \leq b) = P(Y \leq b) - P(Y \leq a) = F(b) - F(a) = \int_a^b f(y)\,dy.$$

Because $P(Y = a) = 0$, we have the following result.

THEOREM **4.3**

If the random variable Y has density function $f(y)$ and $a \leq b$, then the probability that Y falls in the interval $[a, b]$ is

$$P(a \leq Y \leq b) = \int_a^b f(y)\,dy.$$

This probability is the shaded area in Figure 4.8.

If Y is a continuous random variable and a and b are constants such that $a < b$, then $P(Y = a) = 0$ and $P(Y = b) = 0$ and Theorem 4.3 implies that

$$P(a < Y < b) = P(a \leq Y < b) = P(a < Y \leq b)$$

$$= P(a \leq Y \leq b) = \int_a^b f(y)\,dy.$$

FIGURE **4.8**
$P(a \leq Y \leq b)$

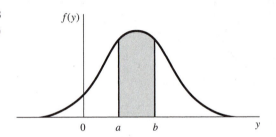

The fact that the above string of equalities is *not*, in general, true for discrete random variables is illustrated in Exercise 4.3.

EXAMPLE **4.4**   Given $f(y) = cy^2$, $0 \leq y \leq 2$, and $f(y) = 0$ elsewhere, find the value of c for which $f(y)$ is a valid density function.

Solution   We require a value for c such that

$$F(\infty) = \int_{-\infty}^{\infty} f(y)\,dy = 1$$

$$= \int_{0}^{2} cy^2\,dy = \left. \frac{cy^3}{3} \right]_{0}^{2} = \left(\frac{8}{3} \right) c.$$

Thus $(8/3)c = 1$, and we find that $c = 3/8$.

EXAMPLE **4.5**   Find $P(1 \leq Y \leq 2)$ for Example 4.4. Also find $P(1 < Y < 2)$

Solution
$$P(1 \leq Y \leq 2) = \int_{1}^{2} f(y)\,dy = \frac{3}{8} \int_{1}^{2} y^2\,dy = \left(\frac{3}{8} \right) \left. \frac{y^3}{3} \right]_{1}^{2} = \frac{7}{8}.$$

Because Y has a continuous distribution, it follows that $P(Y = 1) = P(Y = 2) = 0$ and, therefore, that

$$P(1 < Y < 2) = P(1 \leq Y \leq 2) = \frac{3}{8} \int_{1}^{2} y^2\,dy = \frac{7}{8}.$$

Probability statements regarding a continuous random variable Y are meaningful only if, first, the integral defining the probability exists and, second, the resulting probabilities agree with the axioms of Chapter 2. These two conditions will always be satisfied if we consider only probabilities associated with a finite or countable collection of intervals. Because we almost always are interested in probabilities that continuous variables fall in intervals, this consideration will cause us no practical difficulty. Some density functions that provide good models for population frequency distributions encountered in nature are presented in subsequent sections.

Exercises

4.1 Suppose that Y is a random variable that takes on only integer values 1, 2, ... and has distribution function $F(y)$. Show that the probability function $p(y) = P(Y = y)$ is given by

$$p(y) = \begin{cases} F(1), & y = 1 \\ F(y) - F(y-1), & y = 2, 3, \dots. \end{cases}$$

4.2 Consider a random variable with a geometric distribution (Section 3.5), that is,

$$p(y) = q^{y-1}p, \qquad y = 1, 2, 3, \dots, \qquad 0 < p < 1.$$

  a Show that Y has distribution function $F(y)$ such that $F(i) = 1 - q^i, i = 0, 1, 2, \dots$, and that, in general,

$$F(y) = \begin{cases} 0, & y < 0 \\ 1 - q^i, & i \leq y < i+1 \quad \text{for} \quad i = 0, 1, 2, \dots. \end{cases}$$

  b Show that the preceding cumulative distribution function has the properties given in Theorem 4.1.

4.3 Let Y be a binomial random variable with $n = 10$ and $p = .2$.

  a Use Table 1, Appendix III, to obtain $P(2 < Y < 5)$ and $P(2 \leq Y < 5)$. Are the probabilities that Y falls in the intevals (2, 5) and [2, 5) equal? Why or why not?

  b Use Table 1, Appendix III, to obtain $P(2 < Y \leq 5)$ and $P(2 \leq Y \leq 5)$. Are these two probabilities equal? Why or why not?

  c Earlier in this section, we argued that if Y is continuous and $a < b$ then $P(a < Y < b) = P(a \leq Y < b)$. Does the result in part (a) contradict this claim? Why?

4.4 Suppose that Y has density function

$$f(y) = \begin{cases} ky(1-y), & 0 \leq y \leq 1 \\ 0, & \text{elsewhere.} \end{cases}$$

  a Find the value of k that makes $f(y)$ a probability density function.

  b Find $P(.4 \leq Y \leq 1)$.

  c Find $P(.4 \leq Y < 1)$.

  d Find $P(Y \leq .4 | Y \leq .8)$.

  e Find $P(Y < .4 | Y < .8)$.

4.5 Suppose that Y possesses the density function

$$f(y) = \begin{cases} cy, & 0 \leq y \leq 2 \\ 0, & \text{elsewhere.} \end{cases}$$

  a Find the value of c that makes $f(y)$ a probability density function.

  b Find $F(y)$.

  c Graph $f(y)$ and $F(y)$.

  d Use $F(y)$ to find $P(1 \leq Y \leq 2)$.

  e Use $f(y)$ and geometry to find $P(1 \leq Y \leq 2)$.

4.6 The length of time to failure (in hundreds of hours) for a transistor is a random variable Y with distribution function given by

$$F(y) = \begin{cases} 0, & y < 0 \\ 1 - e^{-y^2}, & y \geq 0. \end{cases}$$

  a Show that $F(y)$ has the properties of a distribution function.
  b Find $f(y)$.
  c Find the probability that the transistor operates for at least 200 hours.
  d Find $P(Y > 100 | Y \leq 200)$.

4.7 A supplier of kerosene has a 150-gallon tank that is filled at the beginning of each week. His weekly demand shows a relative frequency behavior that increases steadily up to 100 gallons and then levels off between 100 and 150 gallons. If Y denotes weekly demand in hundreds of gallons, the relative frequency of demand can be modeled by

$$f(y) = \begin{cases} y, & 0 \leq y \leq 1 \\ 1, & 1 < y \leq 1.5 \\ 0, & \text{elsewhere.} \end{cases}$$

  a Find $F(y)$.
  b Find $P(0 \leq Y \leq .5)$.
  c Find $P(.5 \leq Y \leq 1.2)$.

4.8 A gas station operates two pumps, each of which can pump up to 10,000 gallons of gas in a month. The total amount of gas pumped at the station in a month is a random variable Y (measured in 10,000 gallons) with a probability density function given by

$$f(y) = \begin{cases} y, & 0 < y < 1 \\ 2 - y, & 1 \leq y < 2 \\ 0, & \text{elsewhere.} \end{cases}$$

  a Graph $f(y)$.
  b Find $F(y)$ and graph it.
  c Find the probability that the station will pump between 8000 and 12,000 gallons in a particular month.
  d Given that the station pumped more than 10,000 gallons in a particular month, find the probability that the station pumped more than 15,000 gallons during the month.

4.9 As a measure of intelligence, mice are timed when going through a maze to reach a reward of food. The time (in seconds) required for any mouse is a random variable Y with a density function given by

$$f(y) = \begin{cases} \dfrac{b}{y^2}, & y \geq b \\ 0, & \text{elsewhere,} \end{cases}$$

  where b is the minimum possible time needed to traverse the maze.
  a Show that $f(y)$ has the properties of a density function.
  b Find $F(y)$.

    c Find $P(Y > b + c)$ for a positive constant c.

    d If c and d are both positive constants such that $d > c$, find $P(Y > b + d | Y > b + c)$.

4.10 Let Y possess a density function

$$f(y) = \begin{cases} c(2 - y), & 0 \le y \le 2 \\ 0, & \text{elsewhere.} \end{cases}$$

    a Find c.

    b Find $F(y)$.

    c Graph $f(y)$ and $F(y)$.

    d Use $F(y)$ in (b) to find $P(1 \le Y \le 2)$.

    e Use the geometric figure for $f(y)$ to calculate $P(1 \le Y \le 2)$.

4.11 The length of time required by students to complete a 1-hour exam is a random variable with a density function given by

$$f(y) = \begin{cases} cy^2 + y, & 0 \le y \le 1 \\ 0, & \text{elsewhere.} \end{cases}$$

    a Find c.

    b Find $F(y)$.

    c Graph $f(y)$ and $F(y)$.

    d Use $F(y)$ in (b) to find $F(-1)$, $F(0)$, and $F(1)$.

    e Find the probability that a randomly selected student will finish in less than half an hour.

    f Given that a particular student needs at least 15 minutes to complete the exam, find the probability that she will require at least 30 minutes to finish.

4.12 Let Y have the density function given by

$$f(y) = \begin{cases} .2, & -1 < y \le 0 \\ .2 + cy, & 0 < y \le 1 \\ 0, & \text{elsewhere.} \end{cases}$$

    a Find c.

    b Find $F(y)$.

    c Graph $f(y)$ and $F(y)$.

    d Use $F(y)$ in (b) to find $F(-1)$, $F(0)$, and $F(1)$.

    e Find $P(0 \le Y \le .5)$.

    f Find $P(Y > .5 | Y > .1)$.

4.13 Let the distribution function of a random variable Y be

$$F(y) = \begin{cases} 0, & y \le 0 \\ \dfrac{y}{8}, & 0 < y < 2 \\ \dfrac{y^2}{16}, & 2 \le y < 4 \\ 1, & y \ge 4. \end{cases}$$

    a Find the density function of Y.

    b Find $P(1 \leq Y \leq 3)$.

    c Find $P(Y \geq 1.5)$.

    d Find $P(Y \geq 1 | Y \leq 3)$.

4.3 Expected Values for Continuous Random Variables

The next step in the study of continuous random variables is to find their means, variances, and standard deviations, thereby acquiring numerical descriptive measures associated with their distributions. Many times it is difficult to find the probability distribution for a random variable Y or a function of a random variable, $g(Y)$. Even if the density function for a random variable is known, it can be difficult to evaluate appropriate integrals (we will see this to be the case when a random variable has a gamma distribution, Section 4.6). When we encounter these situations, the approximate behavior of variables of interest can be established by using their moments and the empirical rule or Tchebysheff's theorem (Chapters 1 and 3).

DEFINITION 4.4

The expected value of a continuous random variable Y is

$$E(Y) = \int_{-\infty}^{\infty} y f(y) \, dy$$

provided that the integral exists.[3]

If the definition of the expected value for a discrete random variable Y, $E(Y) = \sum_y y p(y)$, is meaningful, then Definition 4.4 also should agree with our intuitive notion of a mean. The quantity $f(y) \, dy$ corresponds to $p(y)$ for the discrete case, and integration evolves from and is analogous to summation. Hence, $E(Y)$ in Definition 4.4 agrees with our notion of an average, or mean.

As in the discrete case, we are sometimes interested in the expected value of a function of a random variable. A result that permits us to evaluate such an expected value is given in the following theorem.

[3] Technically, $E(Y)$ is said to exist if

$$\int_{-\infty}^{\infty} |y| f(y) \, dy < \infty$$

This will be the case in all expectations we discuss, and we will not mention this additional condition each time we define an expected value.

THEOREM **4.4**

Let $g(Y)$ be a function of Y; then the expected value of $g(Y)$ is given by

$$E[g(Y)] = \int_{-\infty}^{\infty} g(y)f(y)\,dy$$

provided that the integral exists.

The proof of Theorem 4.4 is similar to that of Theorem 3.2 and is omitted. The expected values of three important functions of a continuous random variable Y evolve as a consequence of well-known theorems of integration. As expected, these results lead to conclusions analogous to those contained in Theorems 3.3, 3.4, and 3.5. As a consequence, the proof of Theorem 4.5 will be left as an exercise.

THEOREM **4.5**

Let c be a constant, and let $g(Y)$, $g_1(Y)$, $g_2(Y)$, $\ldots$, $g_k(Y)$ be functions of a continuous random variable Y. Then the following results hold:

1. $E(c) = c$.
2. $E[cg(Y)] = cE[g(Y)]$.
3. $E[g_1(Y)+g_2(Y)+\cdots+g_k(Y)] = E[g_1(Y)]+E[g_2(Y)]+\cdots+E[g_k(Y)]$.

As in the case of discrete random variables, we often seek the expected value of the function $g(Y) = (Y - \mu)^2$. As before, the expected value of this function is the variance of the random variable Y. That is, as in Definition 3.5, $V(Y) = E(Y - \mu)^2$. It is a simple exercise to show that Theorem 4.5 implies that $V(Y) = E(Y^2) - \mu^2$.

EXAMPLE **4.6**

In Example 4.4 we determined that $f(y) = (3/8)y^2$ for $0 \le y \le 2$, $f(y) = 0$ elsewhere, is a valid density function. If the random variable Y has this density function, find $\mu = E(Y)$ and $\sigma^2 = V(Y)$.

Solution

According to Definition 4.4,

$$E(Y) = \int_{-\infty}^{\infty} yf(y)\,dy$$

$$= \int_{0}^{2} y\left(\frac{3}{8}\right)y^2\,dy$$

$$= \left(\frac{3}{8}\right)\left(\frac{1}{4}\right)y^4\Big]_0^2 = 1.5.$$

The variance of Y can be found once we determine $E(Y^2)$. In this case,

$$E(Y^2) = \int_{-\infty}^{\infty} y^2 f(y)\,dy$$

$$= \int_0^2 y^2 \left(\frac{3}{8}\right) y^2 \, dy$$

$$= \left(\frac{3}{8}\right)\left(\frac{1}{5}\right) y^5 \Bigg]_0^2 = 2.4.$$

Thus, $\sigma^2 = V(Y) = E(Y^2) - [E(Y)]^2 = 2.4 - (1.5)^2 = 0.15$.

Exercises

4.14 If, as in Exercise 4.10, Y has density function

$$f(y) = \begin{cases} (1/2)(2-y), & 0 \le y \le 2 \\ 0, & \text{elsewhere,} \end{cases}$$

find the mean and variance of Y.

4.15 If, as in Exercise 4.11, Y has density function

$$f(y) = \begin{cases} (3/2)y^2 + y, & 0 \le y \le 1 \\ 0, & \text{elsewhere,} \end{cases}$$

find the mean and variance of Y.

4.16 If, as in Exercise 4.12, Y has density function

$$f(y) = \begin{cases} .2, & -1 < y \le 0 \\ .2 + (1.2)y, & 0 < y \le 1 \\ 0, & \text{elsewhere,} \end{cases}$$

find the mean and variance of Y.

4.17 Prove Theorem 4.5.

4.18 If Y is a continuous random variable with density function $f(y)$, use Theorem 4.5 to prove that $\sigma^2 = V(Y) = E(Y^2) - [E(Y)]^2$.

4.19 If, as in Exercise 4.13, Y has distribution function

$$F(y) = \begin{cases} 0, & y \le 0 \\ \dfrac{y}{8}, & 0 < y < 2 \\ \dfrac{y^2}{16}, & 2 \le y < 4 \\ 1, & y \ge 4, \end{cases}$$

find the mean and variance of Y.

4.20 If Y is a continuous random variable with mean μ and variance σ^2 and a and b are constants, use Theorem 4.5 to prove the following:

a $E(aY + b) = aE(Y) + b = a\mu + b$.

b $V(aY + b) = a^2 V(Y) = a^2\sigma^2$.

4.21   For certain ore samples, the proportion Y of impurities per sample is a random variable with density function given in Exercise 4.15. The dollar value of each sample is $W = 5 - .5Y$. Find the mean and variance of W.

4.22   The proportion of time per day that all checkout counters in a supermarket are busy is a random variable Y with density function

$$f(y) = \begin{cases} cy^2(1-y)^4, & 0 \le y \le 1 \\ 0, & \text{elsewhere.} \end{cases}$$

a   Find the value of c that makes $f(y)$ a probability density function.
b   Find $E(Y)$.

4.23   The temperature Y at which a thermostatically controlled switch turns on has probability density function given by

$$f(y) = \begin{cases} 1/2, & 59 \le y \le 61 \\ 0, & \text{elsewhere.} \end{cases}$$

Find $E(Y)$ and $V(Y)$.

4.24   The proportion of time Y that an industrial robot is in operation during a 40-hour week is a random variable with probability density function

$$f(y) = \begin{cases} 2y, & 0 \le y \le 1 \\ 0, & \text{elsewhere.} \end{cases}$$

a   Find $E(Y)$ and $V(Y)$.
b   For the robot under study, the profit X for a week is given by $X = 200Y - 60$. Find $E(X)$ and $V(X)$.
c   Find an interval in which the profit should lie for at least 75% of the weeks that the robot is in use.

4.25   Daily total solar radiation for a specified location in Florida in October has probability density function given by

$$f(y) = \begin{cases} (3/32)(y-2)(6-y), & 2 \le y \le 6 \\ 0, & \text{elsewhere} \end{cases}$$

with measurements in hundreds of calories. Find the expected daily solar radiation for October.

4.26   Weekly CPU time used by an accounting firm has probability density function (measured in hours) given by

$$f(y) = \begin{cases} (3/64)y^2(4-y), & 0 \le y \le 4 \\ 0, & \text{elsewhere.} \end{cases}$$

a   Find the expected value and variance of weekly CPU time.
b   The CPU time costs the firm $200 per hour. Find the expected value and variance of the weekly cost for CPU time.
c   Would you expect the weekly cost to exceed $600 very often? Why?

4.27   The pH of water samples from a specific lake is a random variable Y with probability density function given by

$$f(y) = \begin{cases} (3/8)(7-y)^2, & 5 \leq y \leq 7 \\ 0, & \text{elsewhere.} \end{cases}$$

a   Find $E(Y)$ and $V(Y)$.

b   Find an interval shorter than $(5, 7)$ in which at least three-fourths of the pH measurements must lie.

c   Would you expect to see a pH measurement below 5.5 very often? Why?

4.4   The Uniform Probability Distribution

Suppose that a bus always arrives at a particular stop between 8:00 and 8:10 A.M. and that the probability that the bus will arrive in any given subinterval of time is proportional only to the length of the subinterval. That is, the bus is as likely to arrive between 8:00 and 8:02 as it is to arrive between 8:06 and 8:08. Let Y denote the length of time a person must wait for the bus if that person arrived at the bus stop at exactly 8:00. If we carefully measured in minutes how long after 8:00 the bus arrived for several mornings, we could develop a relative frequency histogram for the data. From the description just given, it should be clear that the relative frequency with which we observed a value of Y between 0 and 2 would be approximately the same as the relative frequency with which we observed a value of Y between 6 and 8. A reasonable model for the density function of Y is given in Figure 4.9. Because areas under curves represent probabilities for continuous random variables and $A_1 = A_2$ (by inspection), it follows that $P(0 \leq Y \leq 2) = P(6 \leq Y \leq 8)$, as desired.

The random variable Y just discussed is an example of a random variable that has a uniform distribution. The general form for the density function of a random variable with a uniform distribution is as follows.

DEFINITION 4.5   If $\theta_1 < \theta_2$, a random variable Y is said to have a continuous *uniform probability distribution* on the interval (θ_1, θ_2) if and only if the density function of Y is

FIGURE 4.9
Density function
for Y

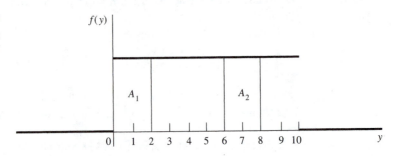

$$f(y) = \begin{cases} \dfrac{1}{\theta_2 - \theta_1}, & \theta_1 \leq y \leq \theta_2 \\ 0, & \text{elsewhere.} \end{cases}$$

In the bus problem we can take $\theta_1 = 0$ and $\theta_2 = 10$, because we are interested only in a particular 10-minute interval. The density function discussed in Example 4.2 is a uniform distribution with $\theta_1 = 0$ and $\theta_2 = 1$. Graphs of the distribution function and density function for the random variable in Example 4.2 are given in Figures 4.4 and 4.5, respectively.

DEFINITION **4.6**   The constants that determine the specific form of a density function are called *parameters* of the density function.

The quantities θ_1 and θ_2 are parameters of the uniform density function and are clearly meaningful numerical values associated with the theoretical density function. Both the range and the probability that Y will fall in any given interval depend on the values of θ_1 and θ_2.

Some continuous random variables in the physical, management, and biological sciences have approximately uniform probability distributions. For example, suppose that the number of events, such as calls coming into a switchboard, that occur in the time interval $(0, t)$ has a Poisson distribution. If it is known that exactly one such event has occurred in the interval $(0, t)$, then the actual time of occurrence is distributed uniformly over this interval.

EXAMPLE **4.7**   Arrivals of customers at a checkout counter follow a Poisson distribution. It is known that, during a given 30-minute period, one customer arrived at the counter. Find the probability that the customer arrived during the last 5 min of the 30-minute period.

Solution   As just mentioned, the actual time of arrival follows a uniform distribution over the interval of $(0, 30)$. If Y denotes the arrival time, then

$$P(25 \leq Y \leq 30) = \int_{25}^{30} \frac{1}{30} \, dy = \frac{30 - 25}{30} = \frac{5}{30} = \frac{1}{6}.$$

The probability of the arrival occurring in any other 5-minute interval is also $1/6$.

As we will see, the uniform distribution is very important for theoretical reasons. Simulation studies are valuable techniques for validating models in statistics. If we desire a set of observations on a random variable Y with distribution function $F(y)$, we often can obtain the desired results by transforming a set of observations on a uniform random variable. For this reason most computer systems contain a random number generator, which generates observed values for a random variable that has a continuous uniform distribution.

THEOREM 4.6

If $\theta_1 < \theta_2$ and Y is a random variable uniformly distributed on the interval (θ_1, θ_2), then

$$\mu = E(Y) = \frac{\theta_1 + \theta_2}{2} \quad \text{and} \quad \sigma^2 = V(Y) = \frac{(\theta_2 - \theta_1)^2}{12}.$$

Proof

By Definition 4.4,

$$E(Y) = \int_{-\infty}^{\infty} y f(y) \, dy$$

$$= \int_{\theta_1}^{\theta_2} y \left(\frac{1}{\theta_2 - \theta_1} \right) dy$$

$$= \left(\frac{1}{\theta_2 - \theta_1} \right) \frac{y^2}{2} \Bigg]_{\theta_1}^{\theta_2} = \frac{\theta_2^2 - \theta_1^2}{2(\theta_2 - \theta_1)}$$

$$= \frac{\theta_2 + \theta_1}{2}.$$

Note that the mean of a uniform random variable is simply the value midway between the two parameter values, θ_1 and θ_2. The derivation of the variance is left as an exercise.

Exercises

4.28 Suppose that Y has a uniform distribution over the interval $(0, 1)$.

    a Find $F(y)$.

    b Show that $P(a \leq Y \leq a + b)$ for $a \geq 0, b \geq 0$, and $a + b \leq 1$ depends only upon the value of b.

4.29 If a parachutist lands at a random point on a line between markers A and B, find the probability that she is closer to A than to B. Find the probability that her distance to A is more than three times her distance to B.

4.30 Suppose that three parachutists operate independently as described in Exercise 4.29. What is the probability that exactly one of the three lands past the midpoint between A and B?

4.31 A random variable Y has a uniform distribution over the interval (θ_1, θ_2). Derive the variance of Y.

4.32 The change in depth of a river from one day to the next, measured (in feet) at a specific location, is a random variable Y with the following density function:

$$f(y) = \begin{cases} k, & -2 \leq y \leq 2 \\ 0, & \text{elsewhere.} \end{cases}$$

    a Determine the value of k.

    b Obtain the distribution function for Y.

4.33   Upon studying low bids for shipping contracts, a microcomputer manufacturing company finds that intrastate contracts have low bids that are uniformly distributed between 20 and 25, in units of thousands of dollars. Find the probability that the low bid on the next intrastate shipping contract

a   is below $22,000.

b   is in excess of $24,000.

4.34   Refer to Exercise 4.33. Find the expected value of low bids on contracts of the type described there.

4.35   The failure of a circuit board interrupts work that utilizes a computing system until a new board is delivered. The delivery time, Y, is uniformly distributed on the interval 1 to 5 days. The cost of a board failure and interruption includes the fixed cost c_0 of a new board and a cost that increases proportionally to Y^2. If C is the cost incurred, $C = c_0 + c_1 Y^2$.

a   Find the probability that the delivery time exceeds 2 days.

b   In terms of c_0 and c_1, find the expected cost associated with a single failed circuit board.

4.36   If a point is *randomly* located in an interval (a, b), and if Y denotes the location of the point, then Y is assumed to have a uniform distribution over (a, b). A plant efficiency expert randomly selects a location along a 500-foot assembly line from which to observe the work habits of the workers on the line. What is the probability that the point she selects

a   is within 25 feet of the end of the line?

b   is within 25 feet of the beginning of the line?

c   is closer to the beginning of the line than to the end of the line?

4.37   A telephone call arrived at a switchboard at random within a 1-minute interval. The switchboard was fully busy for 15 seconds into this 1-minute period. What is the probability that the call arrived when the switchboard was not fully busy?

4.38   Beginning at 12:00 midnight, a computer center is up for 1 hour and then down for 2 hours on a regular cycle. A person who is unaware of this schedule dials the center at a random time between 12:00 midnight and 5:00 A.M. What is the probability that the center is up when the person's call comes in?

4.39   The cycle time for trucks hauling concrete to a highway construction site is uniformly distributed over the interval 50 to 70 minutes. What is the probability that the cycle time exceeds 65 minutes if it is known that the cycle time exceeds 55 minutes?

4.40   Refer to Exercise 4.39. Find the mean and variance of the cycle times for the trucks.

4.41   The number of defective circuit boards coming off a soldering machine follows a Poisson distribution. During a specific 8-hour day, one defective circuit board was found.

a   Find the probability that it was produced during the first hour of operation during that day.

b   Find the probability that it was produced during the last hour of operation during that day.

c   Given that no defective circuit boards were produced during the first 4 hours of operation, find the probability that the defective board was manufactured during the fifth hour.

4.42   In using the triangulation method to determine the range of an acoustic source, the test equipment must accurately measure the time at which the spherical wave front arrives at a receiving sensor. According to Perruzzi and Hilliard (1984), measurement errors in these times can be modeled as possessing a uniform distribution from -0.05 to $+0.05$ μs (microseconds).

a   What is the probability that a particular arrival-time measurement will be accurate to within 0.01 μs?

b   Find the mean and variance of the measurement errors.

4.43 Refer to Exercise 4.42. Suppose that measurement errors are uniformly distributed between -0.02 to $+0.05$ μs.

a What is the probability that a particular arrival-time measurement will be accurate to within 0.01 μs?

b Find the mean and variance of the measurement errors.

4.44 Refer to Example 4.7. Find the conditional probability that a customer arrives during the last 5 minutes of the 30-minute period if it is known that no one arrives during the first 10 minutes of the period.

4.45 According to Y. Zimmels (1983), the sizes of particles used in sedimentation experiments often have a uniform distribution. In sedimentation involving mixtures of particles of various sizes, the larger particles hinder the movements of the smaller ones. Thus, it is important to study both the mean and the variance of particle sizes. Suppose that spherical particles have diameters that are uniformly distributed between .01 and .05 cm. Find the mean and variance of the *volumes* of these particles. (Recall that the volume of a sphere is $(4/3)\pi r^3$.)

4.5 The Normal Probability Distribution

The most widely used continuous probability distribution is the normal distribution, a distribution with the familiar "bell" shape that was discussed in connection with the empirical rule. The examples and exercises in this section illustrate some of the many random variables that have distributions that are closely approximated by a normal probability distribution. In Chapter 7 we will present an argument that, at least partially, explains the common occurrence of normal distributions of data in nature. The normal density function is as follows:

DEFINITION 4.7

A random variable Y is said to have a *normal probability distribution* if and only if, for $\sigma > 0$ and $-\infty < \mu < \infty$, the density function of Y is

$$f(y) = \frac{1}{\sigma\sqrt{2\pi}} e^{-(y-\mu)^2/(2\sigma^2)}, \qquad -\infty < y < \infty.$$

Notice that the normal density function contains two parameters, μ and σ.

THEOREM 4.7

If Y is a normally distributed random variable with parameters μ and σ, then

$$E(Y) = \mu \quad \text{and} \quad V(Y) = \sigma^2.$$

The proof of this theorem will be deferred to Section 4.9, where we derive the moment-generating function of a normally distributed random variable. The results contained in Theorem 4.7 imply that the parameter μ locates the center of the distribution and that σ measures its spread. A graph of a normal density function is shown in Figure 4.10.

FIGURE **4.10**
The normal
probability
density function

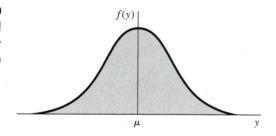

$f(y)$

μ

y

Areas under the normal density function corresponding to $P(a \leq Y \leq b)$ require evaluation of the integral

$$\int_a^b \frac{1}{\sigma\sqrt{2\pi}} e^{-(y-\mu)^2/(2\sigma^2)} dy.$$

Unfortunately, a closed-form expression for this integral does not exist; hence its evaluation requires the use of numerical integration techniques. We will make heavy use of Table 4, Appendix III, when computing areas under normal densities.

The normal density function is symmetric around the value μ, so areas need be tabulated on only one side of the mean. The tabulated areas are to the right of points z, where z is the distance from the mean, measured in standard deviations. This area is shaded in Figure 4.11.

EXAMPLE **4.8**   Let Z denote a normal random variable with mean 0 and standard deviation 1.

    a Find $P(Z > 2)$.
    b Find $P(-2 \leq Z \leq 2)$.
    c Find $P(0 \leq Z \leq 1.73)$.

Solution   **a** Since $\mu = 0$ and $\sigma = 1$, the value 2 is actually $z = 2$ standard deviations above the mean. Proceed down the first (z) column in Table 4, and read the area opposite $z = 2.0$. This area, denoted by the symbol $A(z)$, is $A(2.0) = .0228$. Thus, $P(Z > 2) = .0228$.

    b Refer to Figure 4.12, where we have shaded the area of interest. In (a) we determined that $A_1 = A(2.0) = .0228$. Because the density function is symmetric about the mean $\mu = 0$, it follows that $A_2 = A_1 = .0228$ and hence

FIGURE **4.11**
Tabulated area
for the normal
density function

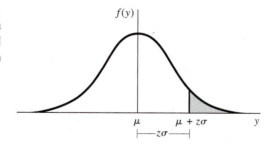

$f(y)$

μ $\mu + z\sigma$ y

$\vdash\!\!-z\sigma-\!\!\dashv$

FIGURE **4.12**
Desired area for
Example 4.8(b)

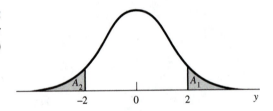

that

$$P(-2 \le Z \le 2) = 1 - A_1 - A_2 = 1 - 2(.0228) = .9544.$$

c Because $P(Z > 0) = A(0) = .5$, we obtain that $P(0 \le Z \le 1.73) = .5 - A(1.73)$, where $A(1.73)$ is obtained by proceeding down the z column in Table 4 to the entry 1.7 and then across the top of the table to the column labeled .03 to read $A(1.73) = .0418$. Thus,

$$P(0 \le Z \le 1.73) = .5 - .0418 = .4582.$$

EXAMPLE **4.9** The achievement scores for a college entrance examination are normally distributed with mean 75 and standard deviation 10. What fraction of the scores lies between 80 and 90?

Solution Recall that z is the distance from the mean of a normal distribution expressed in units of standard deviation. Thus,

$$z = \frac{y - \mu}{\sigma}.$$

Then the desired fraction of the population is given by the area between

$$z_1 = \frac{80 - 75}{10} = .5 \quad \text{and} \quad z_2 = \frac{90 - 75}{10} = 1.5.$$

This area is shaded in Figure 4.13.
    You can see from Figure 4.13 that $A = A(.5) - A(1.5) = .3085 - .0668 = .2417.$

FIGURE **4.13**
Required area for
Example 4.9

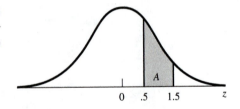

We can always transform a normal random variable Y to a *standard normal* random variable Z by using the relationship

$$Z = \frac{Y - \mu}{\sigma}.$$

Table 4 can then be used to compute probabilities, as shown here. Z locates a point measured from the mean of a normal random variable, with the distance *expressed in units of the standard deviation* of the original normal random variable. Thus, the mean value of Z must be 0, and its standard deviation must equal 1. The proof that the *standard normal random variable*, Z, is normally distributed with mean 0 and standard deviation 1 is given in Chapter 6.

Exercises

4.46   Use Table 4, Appendix III, to find the following probabilities for a standard normal random variable Z:

a $P(0 \le Z \le 1.2)$.

b $P(-.9 \le Z \le 0)$.

c $P(.3 \le Z \le 1.56)$.

d $P(-.2 \le Z \le .2)$.

e $P(-1.56 \le Z \le -.2)$.

4.47   a   Find the value z_0 such that $P(Z > z_0) = .5$.

b   Find the value z_0 such that $P(Z < z_0) = .8643$.

c   Find the value z_0 such that $P(-z_0 < Z < z_0) = .90$.

d   Find the value z_0 such that $P(-z_0 < Z < z_0) = .99$.

4.48   A normally distributed random variable has density function

$$f(y) = \frac{1}{\sigma\sqrt{2\pi}}\, e^{-(y-\mu)^2/(2\sigma^2)}, \qquad -\infty < y < \infty.$$

Using the fundamental properties associated with any density function, argue that the parameter σ must be such that $\sigma > 0$.

4.49   A company that manufactures and bottles apple juice uses a machine that automatically fills 16-ounce bottles. There is some variation, however, in the amounts of liquid dispensed into the bottles that are filled. The amount dispensed has been observed to be approximately normally distributed with mean 16 ounces and standard deviation 1 ounce. What proportion of bottles will have more than 17 ounces dispensed into them?

4.50   The weekly amount of money spent on maintenance and repairs by a company was observed, over a long period of time, to be approximately normally distributed with mean $400 and standard deviation $20. If $450 is budgeted for next week, what is the probability that the actual costs will exceed the budgeted amount?

4.51   In Exercise 4.50, how much should be budgeted for weekly repairs and maintenance to provide that the probability the budgeted amount will be exceeded in a given week is only .1?

4.52 A machining operation produces bearings with diameters that are normally distributed with mean 3.0005 inches and standard deviation .0010 inch. Specifications require the bearing diameters to lie in the interval $3.000 \pm .0020$ inches. Those outside the interval are considered scrap and must be remachined. With the existing machine setting, what fraction of total production will be scrap?

4.53 In Exercise 4.52, what should the mean diameter be in order that the fraction of bearings scrapped be minimized?

4.54 The grade point averages of a large population of college students are approximately normally distributed with mean 2.4 and standard deviation .8. What fraction of the students will possess a grade point average in excess of 3.0?

4.55 Refer to Exercise 4.54. If students possessing a grade point average less than 1.9 are dropped from college, what percentage of the students will be dropped?

4.56 Refer to Exercise 4.54. Suppose that three students are randomly selected from the student body. What is the probability that all three will possess a grade point average in excess of 3.0?

4.57 Wires manufactured for use in a computer system are specified to have resistances of between .12 and .14 ohms. The actual measured resistances of the wires produced by company A have a normal probability distribution with mean .13 ohm and standard deviation .005 ohm.

    a What is the probability that a randomly selected wire from company A's production will meet the specifications?

    b If four of these wires are used in each computer system and all are selected from company A, what is the probability that all four in a randomly selected system will meet the specifications?

4.58 One method of arriving at economic forecasts is to use a consensus approach. A forecast is obtained from each of a large number of analysts; the average of these individual forecasts is the consensus forecast. Suppose that the individual 1996 January prime-interest-rate forecasts of all economic analysts are approximately normally distributed with mean 7% and standard deviation 2.6%. If a single analyst is randomly selected from among this group, what is the probability that the analyst's forecast of the prime interest rate will

    a exceed 11%?

    b be less than 9%?

4.59 The width of bolts of fabric is normally distributed with mean 950 mm (millimeters) and standard deviation 10 mm.

    a What is the probability that a randomly chosen bolt has a width of between 947 and 958 mm?

    b What is the appropriate value for C such that a randomly chosen bolt has a width less than C with probability .8531?

4.60 Scores on an examination are assumed to be normally distributed with mean 78 and variance 36.

    a What is the probability that a person taking the examination scores higher than 72?

    b Suppose that students scoring in the top 10% of this distribution are to receive an A grade. What is the minimum score a student must achieve to earn an A grade?

    c What must be the cutoff point for passing the examination if the examiner wants only the top 28.1% of all scores to be passing?

    d Approximately what proportion of students have scores 5 or more points above the score that cuts off the lowest 25%?

e   If it is known that a student's score exceeds 72, what is the probability that his or her score exceeds 84?

4.61   A soft-drink machine can be regulated so that it discharges an average of μ ounces per cup. If the ounces of fill are normally distributed with standard deviation 0.3 ounce, give the setting for μ so that 8-ounce cups will overflow only 1% of the time.

4.62   The machine described in Exercise 4.61 has standard deviation σ that can be fixed at certain levels by carefully adjusting the machine. What is the largest value of σ that will allow the actual amount dispensed to fall within 1 ounce of the mean with probability at least .95?

4.63   The SAT and ACT college entrance exams are taken by thousands of students each year. The mathematics portions of each of these exams produce scores that are approximately normally distributed. In recent years, SAT mathematics exam scores have averaged 480 with standard deviation 100. The average and standard deviation for ACT mathematics scores are 18 and 6, respectively.

a   An engineering school sets 550 as the minimum SAT math score for new students. What percentage of students will score below 550 in a typical year?

b   What score should the engineering school set as a comparable standard on the ACT math test?

4.64   Show that the maximum value of the normal density with parameters μ and σ is $1/(\sigma\sqrt{2\pi})$ and occurs when $y = \mu$.

4.65   Show that the normal density with parameters μ and σ has inflection points at the values $\mu - \sigma$ and $\mu + \sigma$. (Recall that an inflection point is a point where the curve changes direction from concave up to concave down, or vice versa, and occurs when the second derivative changes sign. Such a change in sign may occur when the second derivative equals zero.)

4.66   Assume that Y is normally distributed with mean μ and standard deviation σ. After observing a value of Y, a mathematician constructs a rectangle with length $L = |Y|$ and width $W = 3|Y|$. Let A denote the area of the resulting rectangle. What is $E(A)$?

4.6   The Gamma Probability Distribution

Some random variables are always nonnegative and for various reasons yield distributions of data that are skewed (nonsymmetric) to the right. That is, most of the area under the density function is located near the origin, and the density function drops gradually as y increases. A skewed probability distribution is shown in Figure 4.14.

The lengths of time between malfunctions for aircraft engines possess a skewed frequency distribution, as do the lengths of time between arrivals at a supermarket

FIGURE **4.14**
A skewed probability
distribution

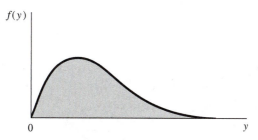

$f(y)$

0 y

checkout queue (that is, the line at the checkout counter). Similarly, the lengths of time to complete a maintenance checkup for an automobile or aircraft engine possess a skewed frequency distribution. The populations associated with these random variables frequently possess distributions that are adequately modeled by a gamma density function.

DEFINITION 4.8

A random variable Y is said to have a *gamma distribution with parameters* $\alpha > 0$ and $\beta > 0$ if and only if the density function of Y is

$$f(y) = \begin{cases} \dfrac{y^{\alpha-1}e^{-y/\beta}}{\beta^{\alpha}\Gamma(\alpha)}, & 0 \le y < \infty \\ 0, & \text{elsewhere,} \end{cases}$$

where

$$\Gamma(\alpha) = \int_0^{\infty} y^{\alpha-1}e^{-y}\,dy.$$

The quantity $\Gamma(\alpha)$ is known as the *gamma function*. Direct integration will verify that $\Gamma(1) = 1$. Integration by parts will verify that $\Gamma(\alpha) = (\alpha - 1)\Gamma(\alpha - 1)$ for any $\alpha > 1$ and that $\Gamma(n) = (n - 1)!$, provided that n is an integer.

Graphs of gamma density functions for $\alpha = 1$, 2, and 4 and $\beta = 1$ are given in Figure 4.15. Notice in Figure 4.15 that the shape of the gamma density differs for the different values of α. For this reason, α is sometimes called the *shape parameter* associated with a gamma distribution. The parameter β is generally called the *scale parameter,* because multiplying a gamma-distributed random variable by a positive constant (and thereby changing the scale on which the measurement is made) produces a random variable that also has a gamma distribution with the same value of α (shape parameter) but with an altered value of β.

In the special case when α is an integer, the distribution function of a gamma-distributed random variable can be expressed as a sum of certain Poisson probabilities. You will find this representation in Exercise 4.79. If α is not an integer and

FIGURE 4.15
Gamma density
functions, $\beta = 1$

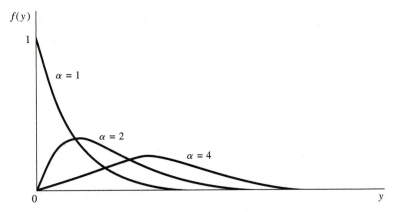

$0 < c < d < \infty$, it is impossible to give a closed-form expression for

$$\int_c^d \frac{y^{\alpha-1}e^{-y/\beta}}{\beta^\alpha\Gamma(\alpha)}\,dy.$$

As a result, it is impossible to obtain areas under the gamma density function by direct integration. Tabulated values for integrals like the above are given in *Tables of the Incomplete Gamma Function* (Pearson 1965).

THEOREM 4.8   If Y has a gamma distribution with parameters α and β, then

$$\mu = E(Y) = \alpha\beta \quad \text{and} \quad \sigma^2 = V(Y) = \alpha\beta^2.$$

Proof

$$E(Y) = \int_{-\infty}^{\infty} yf(y)\,dy = \int_0^{\infty} y\left(\frac{y^{\alpha-1}e^{-y/\beta}}{\beta^\alpha\Gamma(\alpha)}\right)dy.$$

By definition, the gamma density function is such that

$$\int_0^{\infty} \frac{y^{\alpha-1}e^{-y/\beta}}{\beta^\alpha\Gamma(\alpha)}\,dy = 1.$$

Hence,

$$\int_0^{\infty} y^{\alpha-1}e^{-y/\beta}\,dy = \beta^\alpha\Gamma(\alpha)$$

and

$$E(Y) = \int_0^{\infty} \frac{y^\alpha e^{-y/\beta}\,dy}{\beta^\alpha\Gamma(\alpha)} = \frac{1}{\beta^\alpha\Gamma(\alpha)}\int_0^{\infty} y^\alpha e^{-y/\beta}\,dy$$

$$= \frac{1}{\beta^\alpha\Gamma(\alpha)}[\beta^{\alpha+1}\Gamma(\alpha+1)] = \frac{\beta\alpha\Gamma(\alpha)}{\Gamma(\alpha)} = \alpha\beta.$$

From Exercise 4.18, $V(Y) = E[Y^2] - [E(Y)]^2$. Further,

$$E(Y^2) = \int_0^{\infty} y^2\left(\frac{y^{\alpha-1}e^{-y/\beta}}{\beta^\alpha\Gamma(\alpha)}\right)dy = \frac{1}{\beta^\alpha\Gamma(\alpha)}\int_0^{\infty} y^{\alpha+1}e^{-y/\beta}\,dy$$

$$= \frac{1}{\beta^\alpha\Gamma(\alpha)}[\beta^{\alpha+2}\Gamma(\alpha+2)] = \frac{\beta^2(\alpha+1)\alpha\Gamma(\alpha)}{\Gamma(\alpha)} = \alpha(\alpha+1)\beta^2.$$

Then $V(Y) = E[Y^2] - [E(Y)]^2$ where, from the earlier part of the derivation, $E(Y) = \alpha\beta$. Substituting $E[Y^2]$ and $E(Y)$ into the formula for $V(Y)$, we obtain

$$V(Y) = \alpha(\alpha+1)\beta^2 - (\alpha\beta)^2 = \alpha^2\beta^2 + \alpha\beta^2 - \alpha^2\beta^2 = \alpha\beta^2. \qquad \blacksquare$$

Two special cases of gamma-distributed random variables merit particular consideration.

DEFINITION **4.9**

Let v be a positive integer. A random variable Y is said to have a *chi-square distribution with v degrees of freedom* if and only if Y is a gamma-distributed random variable with parameters $\alpha = v/2$ and $\beta = 2$.

A random variable with a chi-square distribution is called a *chi-square (χ^2) random variable*. Such random variables occur often in statistical theory. The motivation behind calling the parameter v the degrees of freedom of the chi-square distribution rests on one of the major ways for generating a random variable with this distribution and is given in Theorem 6.4. The mean and variance of a χ^2 random variable follow directly from Theorem 4.8.

THEOREM **4.9**

If Y is a chi-square random variable with v degrees of freedom, then

$$\mu = E(Y) = v \quad \text{and} \quad \sigma^2 = V(Y) = 2v.$$

Proof

Apply Theorem 4.8 with $\alpha = v/2$ and $\beta = 2$.

Tables that give probabilities associated with χ^2 distributions are readily available in most statistics texts. Table 6, Appendix III, gives percentage points associated with χ^2 distributions for many choices of v. Tables of the general gamma distribution are not so readily available. However, we will show in Exercise 6.40 that if Y has a gamma distribution with $\alpha = n/2$ for some integer n, then $2Y/\beta$ has a χ^2 distribution with n degrees of freedom. Hence, for example, if Y has a gamma distribution with $\alpha = 1.5 = 3/2$ and $\beta = 4$, then $2Y/\beta = 2Y/4 = Y/2$ has a χ^2 distribution with 3 degrees of freedom. Thus, $P(Y < 3.5) = P([Y/2] < 1.75)$ can be found by using readily available tables of the χ^2 distribution.

The gamma density function in which $\alpha = 1$ is called the *exponential density function*.

DEFINITION **4.10**

A random variable Y is said to have an *exponential distribution with parameter* $\beta > 0$ if and only if the density function of Y is

$$f(y) = \begin{cases} \dfrac{1}{\beta} e^{-y/\beta}, & 0 \le y < \infty \\ 0, & \text{elsewhere.} \end{cases}$$

The exponential density function is often useful for modeling the length of life of electronic components. Suppose that the length of time a component already has operated does not affect its chance of operating for at least b additional time units. That is, the probability that the component will operate for more than $a + b$ time units, given that it has already operated for at least a time units, is the same as the probability that a new component will operate for at least b time units if the new component is put into service at time 0. A fuse is an example of a component for which this assumption often is reasonable. We will see in the next example that the exponential distribution provides a model for the distribution of the lifetime of such a component.

THEOREM 4.10   If Y is an exponential random variable with parameter β, then

$$\mu = E(Y) = \beta \quad \text{and} \quad \sigma^2 = V(Y) = \beta^2.$$

Proof   The proof follows directly from Theorem 4.8 with $\alpha = 1$.   ∎

EXAMPLE 4.10   Suppose that Y has an exponential probability density function. Show that, if $a > 0$ and $b > 0$,

$$P(Y > a + b | Y > a) = P(Y > b).$$

Solution   From the definition of conditional probability, we have that

$$P(Y > a + b | Y > a) = \frac{P(Y > a + b)}{P(Y > a)}$$

because the intersection of the events $(Y > a + b)$ and $(Y > a)$ is the event $(Y > a + b)$. Now

$$P(Y > a + b) = \int_{a+b}^{\infty} \frac{1}{\beta} e^{-y/\beta} \, dy = -e^{-y/\beta} \Big]_{a+b}^{\infty} = e^{-(a+b)/\beta}.$$

Similarly,

$$P(Y > a) = \int_{a}^{\infty} \frac{1}{\beta} e^{-y/\beta} \, dy = e^{-a/\beta}$$

and

$$P(Y > a + b | Y > a) = \frac{e^{-(a+b)/\beta}}{e^{-a/\beta}} = e^{-b/\beta} = P(Y > b).$$

This property of the exponential distribution is often called the *memoryless property* of the distribution.

You will recall from Chapter 3 that the geometric distribution, a discrete distribution, also had this *memoryless* property. An interesting relationship between the exponential and geometric distributions is given in Exercise 4.75.

Exercises

4.67 If $\alpha > 0$, $\Gamma(\alpha)$ is defined by $\Gamma(\alpha) = \int_0^\infty y^{\alpha-1} e^{-y}\, dy$.

    a Show that $\Gamma(1) = 1$.

    *b If $\alpha > 1$, integrate by parts to prove that $\Gamma(\alpha) = (\alpha - 1)\Gamma(\alpha - 1)$.

4.68 Use the results obtained in Exercise 4.67 to prove that if n is a positive integer, then $\Gamma(n) = (n - 1)!$. What are the numerical values of $\Gamma(2)$, $\Gamma(4)$, and $\Gamma(7)$?

4.69 The magnitude of earthquakes recorded in a region of North America can be modeled as having an exponential distribution with mean 2.4, as measured on the Richter scale. Find the probability that an earthquake striking this region will

    a exceed 3.0 on the Richter scale.

    b fall between 2.0 and 3.0 on the Richter scale.

4.70 Refer to Exercise 4.69. Of the next ten earthquakes to strike this region, what is the probability that at least one will exceed 5.0 on the Richter scale?

4.71 The operator of a pumping station has observed that demand for water during early afternoon hours has an approximately exponential distribution with mean 100 cfs (cubic feet per second).

    a Find the probability that the demand will exceed 200 cfs during the early afternoon on a randomly selected day.

    b What water pumping capacity should the station maintain during early afternoons so that the probability that demand will exceed capacity on a randomly selected day is only .01?

4.72 The length of time Y necessary to complete a key operation in the construction of houses has an exponential distribution with mean 10 hours. The formula $C = 100 + 40Y + 3Y^2$ relates the cost C of completing this operation to the square of the time to completion. Find the mean and variance of C.

4.73 Times between accidents for all fatal accidents on scheduled American domestic passenger flights during the years 1948 through 1961 were found to have an approximately exponential distribution with mean 44 days (Pyke 1965, 426).

    a If one of the accidents occurred on July 1 of a randomly selected year in the study period, what is the probability that another accident occurred that same month?

    b What is the variance of the times between accidents for the years just indicated?

4.74 One-hour carbon monoxide concentrations in air samples from a large city have an approximately exponential distribution with mean 3.6 parts per million (Zamurs 1984, 637).

    a Find the probability that the carbon monoxide concentration exceeds 9 parts per million during a randomly selected 1-hour period.

    b A traffic-control strategy reduced the mean to 2.5 parts per million. Now find the probability that the concentration exceeds 9 parts per million.

4.75 Let Y be an exponentially distributed random variable with mean β. Define a random variable X in the following way: $X = k$ if $k - 1 \le Y < k$ for $k = 1, 2, \ldots$.

    a Find $P(X = k)$ for each $k = 1, 2, \ldots$.

    b Show that your answer to (a) can be written as

$$P(X = k) = \left(e^{-1/\beta}\right)^{k-1}\left(1 - e^{-1/\beta}\right), \qquad k = 1, 2, \ldots$$

    and that X has a geometric distribution with $p = \left(1 - e^{-1/\beta}\right)$.

4.76 Suppose that a random variable Y has a probability density function given by

$$f(y) = \begin{cases} ky^3 e^{-y/2}, & y > 0 \\ 0, & \text{elsewhere.} \end{cases}$$

Find the value of k that makes $f(y)$ a density function.

4.77 A manufacturing plant uses a specific bulk product. The amount of product used in one day can be modeled by an exponential distribution with $\beta = 4$ (measurements in tons). Find the probability that the plant will use more than 4 tons on a given day.

4.78 Consider the plant of Exercise 4.77. How much of the bulk product should be stocked so that the plant's chance of running out of the product is only .05?

4.79 If $\lambda > 0$ and α is a positive integer, the relationship between incomplete gamma integrals and sums of Poisson probabilities is given by

$$\frac{1}{\Gamma(\alpha)} \int_\lambda^\infty y^{\alpha-1} e^{-y}\, dy = \sum_{x=0}^{\alpha-1} \frac{\lambda^x e^{-\lambda}}{x!}.$$

If Y has a gamma distribution with $\alpha = 2$ and $\beta = 1$, find $P(Y > 1)$ by using the preceding equality and Table 3 of Appendix III.

*4.80 Let Y be a gamma-distributed random variable where α is a positive integer and $\beta = 1$. The result given in Exercise 4.79 implies that that if $y > 0$

$$\sum_{x=0}^{\alpha-1} \frac{y^x e^{-y}}{x!} = P(Y > y).$$

Suppose that X_1 is Poisson distributed with mean λ_1 and X_2 is Poisson distributed with mean λ_2 where $\lambda_2 > \lambda_1$.

a Show that $P(X_1 = 0) > P(X_2 = 0)$.

b Let k be any fixed positive integer. Show that $P(X_1 \le k) = P(Y > \lambda_1)$ and $P(X_2 \le k) = P(Y > \lambda_2)$ where Y is has a gamma distribution with $\alpha = k + 1$ and $\beta = 1$.

c Let k be any fixed positive integer. Use the result derived in part (b) and the fact that $\lambda_2 > \lambda_1$ to show that $P(X_1 \le k) > P(X_2 \le k)$.

d Because the result in part (c) is valid for any $k = 1, 2, 3, \ldots$, and part (a) is also valid, we have established that $P(X_1 \le k) > P(X_2 \le k)$ for all $k = 0, 1, 2, \ldots$. Interpret this result.

4.81 Explosive devices used in mining operations produce nearly circular craters when detonated. The radii of these craters are exponentially distributed with mean 10 feet. Find the mean and variance of the areas produced by these explosive devices.

4.82 The lifetime (in hours) Y of an electronic component is a random variable with density function given by

$$f(y) = \begin{cases} \dfrac{1}{100} e^{-y/100}, & y > 0 \\ 0, & \text{elsewhere.} \end{cases}$$

Three of these components operate independently in a piece of equipment. The equipment fails if at least two of the components fail. Find the probability that the equipment will operate for at least 200 hours without failure.

4.83   Four-week summer rainfall totals in a section of the Midwest United States have approximately a gamma distribution with $\alpha = 1.6$ and $\beta = 2.0$. Find the mean and variance of the 4-week rainfall totals.

4.84   The response times on an online computer terminal have approximately a gamma distribution with mean 4 seconds and variance 8 seconds^2. Write the probability density function for the response times.

4.85   Refer to Exercise 4.84. Use Tchebysheff's theorem to give an interval that contains at least 75% of the response times.

4.86   Annual incomes for heads of household in a section of a city have approximately a gamma distribution with $\alpha = 1000$ and $\beta = 20$. Find the mean and variance of these incomes. Would you expect to find many incomes in excess of \$40,000 in this section of the city?

4.87   The weekly amount of down time Y (in hours) for an industrial machine has approximately a gamma distribution with $\alpha = 3$ and $\beta = 2$. The loss L (in dollars) to the industrial operation as a result of this down time is given by $L = 30Y + 2Y^2$. Find the expected value and variance of L.

4.88   If Y has a probability density function given by

$$f(y) = \begin{cases} 4y^2 e^{-2y}, & y > 0 \\ 0, & \text{elsewhere}, \end{cases}$$

obtain $E(Y)$ and $V(Y)$ by inspection.

4.89   Suppose that Y has a gamma distribution with parameters α and β.
   a   If a is any positive or negative value such that $\alpha + a > 0$, show tha

$$E(Y^a) = \frac{\beta^a \Gamma(\alpha + a)}{\Gamma(\alpha)}.$$

   b   Why did your answer in (a) require that $\alpha + a > 0$?
   c   Show that, with $a = 1$, the result in (a) gives $E(Y) = \alpha\beta$.
   d   Use the result in (a) to give an expression for $E(\sqrt{Y})$. What do you need to assume about α?
   e   Use the result in (a) to give an expression for $E(1/Y)$, $E(1/\sqrt{Y})$, and $E(1/Y^2)$. What do you need to assume about α in each case?

4.90   Suppose that Y has a χ^2 distribution with ν degrees of freedom. Use the results in Exercise 4.89 in your answers to the following. These results will be useful when we study the t and F distributions in Chapter 7.
   a   Give an expression for $E(Y^a)$ if $\nu > -2a$.
   b   Why did your answer in (a) require that $\nu > -2a$?
   c   Use the result in (a) to give an expression for $E(\sqrt{Y})$. What do you need to assume about ν?
   d   Use the result in (a) to give an expression for $E(1/Y)$, $E(1/\sqrt{Y})$, and $E(1/Y^2)$. What do you need to assume about ν in each case?

4.7   The Beta Probability Distribution

The beta density function is a two-parameter density function defined over the closed interval $0 \le y \le 1$. It is often used as a model for proportions, such as the proportion of impurities in a chemical product or the proportion of time that a machine is under repair.

DEFINITION **4.11**

A random variable Y is said to have a *beta probability distribution with parameters* $\alpha > 0$ *and* $\beta > 0$ if and only if the density function of Y is

$$f(y) = \begin{cases} \dfrac{y^{\alpha-1}(1-y)^{\beta-1}}{B(\alpha, \beta)}, & 0 \le y \le 1 \\ 0, & \text{elsewhere,} \end{cases}$$

where

$$B(\alpha, \beta) = \int_0^1 y^{\alpha-1}(1-y)^{\beta-1}\, dy = \frac{\Gamma(\alpha)\Gamma(\beta)}{\Gamma(\alpha+\beta)}.$$

The graphs of beta density functions assume widely differing shapes for various values of the two parameters α and β. Some of these are shown in Figure 4.16.

Notice that defining y over the interval $0 \le y \le 1$ does not restrict the use of the beta distribution. If $c \le y \le d$, then $y^* = (y - c)/(d - c)$ defines a new variable such that $0 \le y^* \le 1$. Thus the beta density function can be applied to a random variable defined on the interval $c \le y \le d$ by translation and a change of scale.

The cumulative distribution function for the beta random variable is commonly called the *incomplete beta function* and is denoted by

$$F(y) = \int_0^y \frac{t^{\alpha-1}(1-t)^{\beta-1}}{B(\alpha, \beta)}\, dt = I_y(\alpha, \beta).$$

A tabulation of $I_y(\alpha, \beta)$ is given in *Tables of the Incomplete Beta Function* (Pearson 1968). When α and β are both positive integers, $I_y(\alpha, \beta)$ is related to the binomial

FIGURE **4.16**
Beta density
functions

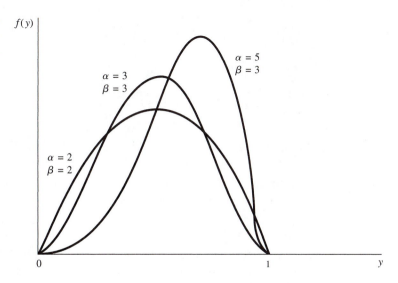

$f(y)$

$\alpha = 5$
$\beta = 3$

$\alpha = 3$
$\beta = 3$

$\alpha = 2$
$\beta = 2$

0 1 y

probability function. Integration by parts can be used to show that for $0 < y < 1$, and α and β both integers,

$$F(y) = \int_0^y \frac{t^{\alpha-1}(1-t)^{\beta-1}}{B(\alpha, \beta)}\, dt = \sum_{i=\alpha}^n \binom{n}{i} y^i (1-y)^{n-i},$$

where $n = \alpha + \beta - 1$. Notice that the sum on the right-hand side of this expression is just the sum of probabilities associated with a binomial random variable with $n = \alpha + \beta - 1$ and $p = y$. The binomial cumulative distribution function is presented in Table 1, Appendix III, for $n = 5, 10, 15, 20, 25$ and $p = .01, .05, .10, .20, .30, .40, .50, .60, .70, .80, .90, .95,$ and $.99$. A more extensive tabulation of the binomial probability function is given in *Tables of the Binomial Probability Distribution* (1950).

THEOREM 4.11   If Y is a beta distributed random variable with parameters $\alpha > 0$ and $\beta > 0$, then

$$\mu = E(Y) = \frac{\alpha}{\alpha + \beta} \quad \text{and} \quad \sigma^2 = V(Y) = \frac{\alpha\beta}{(\alpha + \beta)^2(\alpha + \beta + 1)}.$$

Proof   By definition,

$$E(Y) = \int_{-\infty}^{\infty} y f(y)\, dy$$

$$= \int_0^1 y \left(\frac{y^{\alpha-1}(1-y)^{\beta-1}}{B(\alpha, \beta)} \right) dy$$

$$= \frac{1}{B(\alpha, \beta)} \int_0^1 y^{\alpha}(1-y)^{\beta-1}\, dy$$

$$= \frac{B(\alpha + 1, \beta)}{B(\alpha, \beta)} \qquad (\text{because } \alpha > 0 \text{ implies that } \alpha + 1 > 0)$$

$$= \frac{\Gamma(\alpha + \beta)}{\Gamma(\alpha)\Gamma(\beta)} \times \frac{\Gamma(\alpha + 1)\Gamma(\beta)}{\Gamma(\alpha + \beta + 1)}$$

$$= \frac{\Gamma(\alpha + \beta)}{\Gamma(\alpha)\Gamma(\beta)} \times \frac{\alpha\Gamma(\alpha)\Gamma(\beta)}{(\alpha + \beta)\Gamma(\alpha + \beta)} = \frac{\alpha}{(\alpha + \beta)}.$$

The derivation of the variance is left to the reader (see Exercise 4.98).

We will see in the next example that the beta density function can be integrated directly in the case when α and β are both integers.

EXAMPLE 4.11   A gasoline wholesale distributor has bulk storage tanks that hold fixed supplies and are filled every Monday. Of interest to the wholesaler is the proportion of this supply that is sold during the week. Over many weeks of observation, the distributor found that this proportion could be modeled by a beta distribution with $\alpha = 4$ and $\beta = 2$. Find the probability that the wholesaler will sell at least 90% of her stock in a given week.

Solution   If Y denotes the proportion sold during the week, then

$$f(y) = \begin{cases} \dfrac{\Gamma(4+2)}{\Gamma(4)\Gamma(2)} y^3 (1-y), & 0 \le y \le 1 \\ 0, & \text{elsewhere} \end{cases}$$

and

$$P(Y > .9) = \int_{.9}^{\infty} f(y)\, dy = \int_{.9}^{1} 20(y^3 - y^4)\, dy$$

$$= 20 \left\{ \left. \frac{y^4}{4} \right]_{.9}^{1} - \left. \frac{y^5}{5} \right]_{.9}^{1} \right\} = 20(.004) = .08.$$

It is *not* very likely that 90% of the stock will be sold in a given week.

Exercises

4.91   The relative humidity Y, when measured at a location, has a probability density function given by

$$f(y) = \begin{cases} ky^3(1-y)^2, & 0 \le y \le 1 \\ 0, & \text{elsewhere.} \end{cases}$$

Find the value of k that makes $f(y)$ a density function.

4.92   The percentage of impurities per batch in a chemical product is a random variable Y with density function

$$f(y) = \begin{cases} 12y^2(1-y), & 0 \le y \le 1 \\ 0, & \text{elsewhere.} \end{cases}$$

A batch with more than 40% impurities cannot be sold. What is the probability that a randomly selected batch cannot be sold because of excessive impurities?

4.93   Refer to Exercise 4.92. Find the mean and variance of the percentage of impurities in a randomly selected batch of chemical.

4.94   Suppose that a random variable Y has a probability density function given by

$$f(y) = \begin{cases} 6y(1-y), & 0 \le y \le 1 \\ 0, & \text{elsewhere.} \end{cases}$$

a   Find $F(y)$.
b   Graph $F(y)$ and $f(y)$.
c   Find $P(.5 \le Y \le .8)$.

4.95 Verify that if Y has a beta distribution with $\alpha = \beta = 1$, then Y has a uniform distribution over $(0, 1)$. That is, the uniform distribution over the interval $(0, 1)$ is a special case of a beta distribution.

4.96 The weekly repair cost Y for a machine has a probability density function given by

$$f(y) = \begin{cases} 3(1-y)^2, & 0 < y < 1 \\ 0, & \text{elsewhere} \end{cases}$$

with measurements in hundreds of dollars. How much money should be budgeted each week for repair costs so that the actual cost will exceed the budgeted amount only 10% of the time?

4.97 During an 8-hour shift, the proportion of time Y that a sheet-metal stamping machine is down for maintenance or repairs has a beta distribution with $\alpha = 1$ and $\beta = 2$. That is,

$$f(y) = \begin{cases} 2(1-y), & 0 \le y \le 1, \\ 0, & \text{elsewhere.} \end{cases}$$

The cost (in hundreds of dollars) of this downtime, due to lost production and cost of maintenance and repair, is given by $C = 10 + 20Y + 4Y^2$. Find the mean and variance of C.

4.98 Prove that the variance of a beta distributed random variable with parameters α and β is

$$\sigma^2 = \frac{\alpha\beta}{(\alpha + \beta)^2(\alpha + \beta + 1)}.$$

4.99 Errors in measuring the time of arrival of a wave front from an acoustic source sometimes have an approximate beta distribution (Perruzzi and Hilliard 1984, 197). Suppose that these errors, measured in microseconds, have approximately a beta distribution with $\alpha = 1$ and $\beta = 2$.

a What is the probability that the measurement error in a randomly selected instance is less than .5 μs?

b Give the mean and standard deviation of the measurement errors.

4.100 Proper blending of fine and coarse powders prior to copper sintering is essential for uniformity in the finished product. One way to check the homogeneity of the blend is to select many small samples of the blended powders and measure the proportion of the total weight contributed by the fine powders in each. These measurements should be relatively stable if a homogeneous blend has been obtained.

a Suppose that the proportion of total weight contributed by the fine powders has a beta distribution with $\alpha = \beta = 3$. Find the mean and variance of the proportion of weight contributed by the fine powders.

b Repeat (a) if $\alpha = \beta = 2$.

c Repeat (a) if $\alpha = \beta = 1$.

d Which of the cases, (a), (b), or (c), yields the most homogeneous blending?

4.101 The proportion of time per day that all checkout counters in a supermarket are busy is a random variable Y with a density function given by

$$f(y) = \begin{cases} cy^2(1-y)^4, & 0 \le y \le 1 \\ 0, & \text{elsewhere.} \end{cases}$$

a Find the value of c that makes $f(y)$ a probability density function.

b Find $E(Y)$.

(Use what you have learned about the beta-type distribution. Compare your answers to those obtained in Exercise 4.22.)

4.102   In the text of this section, we noted the relationship between the distribution function of a beta distributed random variable and sums of binomial probabilities. Specifically, if Y has a beta distribution with positive integer values for α and β and $0 < y < 1$,

$$F(y) = \int_0^y \frac{t^{\alpha-1}(1-t)^{\beta-1}}{B(\alpha, \beta)}\, dt = \sum_{i=\alpha}^{n} \binom{n}{i} y^i (1-y)^{n-i},$$

where $n = \alpha + \beta - 1$.

a   If Y has a beta distribution with $\alpha = 4$ and $\beta = 7$, use the appropriate binomial tables to find $P(Y \le .7) = F(.7)$.

b   If Y has a beta distribution with $\alpha = 12$ and $\beta = 14$, use the appropriate binomial tables to find $P(Y \le .6) = F(.6)$.

*4.103   Suppose that Y_1 and Y_2 are binomial random variables with parameters (n, p_1) and (n, p_2), respectively, where $p_1 < p_2$. (Note that the parameter n is the same for the two variables.)

a.   Use the binomial formula to deduce that $P(Y_1 = 0) > P(Y_2 = 0)$.

b.   Use the relationship between the beta distribution function and sums of binomial probabilities given in Exercise 4.102 to deduce that, if k is an integer between 1 and $n - 1$

$$P(Y_1 \le k) = \sum_{i=0}^{k} \binom{n}{i} (p_1)^i (1-p_1)^{n-i} = \int_{p_1}^{1} \frac{t^k (1-t)^{n-k-1}}{B(k+1, n-k)}\, dt.$$

c.   If k is an integer between 1 and $n - 1$, the same argument used in part (b), yields that

$$P(Y_2 \le k) = \sum_{i=0}^{k} \binom{n}{i} (p_2)^i (1-p_2)^{n-i} = \int_{p_2}^{1} \frac{t^k (1-t)^{n-k-1}}{B(k+1, n-k)}\, dt.$$

Show that, if k is any integer between 1 and $n - 1$, $P(Y_1 \le k) > P(Y_2 \le k)$. Interpret this result.

4.8   Some General Comments

Keep in mind that density functions are theoretical models for populations of real data that occur in nature. How do we know which model to use? How much does it matter if we use the wrong density as our model for reality?

To answer the latter question first, we are unlikely ever to select a density function that provides a perfect representation of nature; but goodness of fit is not the criterion for assessing the adequacy of our model. The purpose of a probabilistic model is to provide the mechanism for making inferences about a population based on information contained in a sample. The probability of the observed sample (or a quantity proportional to it) is instrumental in making an inference about the population. It follows that a density function that provides a poor fit to the population frequency distribution could (but does not necessarily) yield incorrect probability statements and lead to erroneous inferences about the population. A good model is one that yields good inferences about the population of interest.

Selecting a reasonable model is sometimes a matter of acting on theoretical considerations. Often, for example, a situation in which the discrete Poisson random variable is appropriate is indicated by the random behavior of events in time. Knowing this, we can show that the length of time between any adjacent pair of events follows an exponential distribution. Similarly, if a and b are integers, $a < b$, then the length of time between the occurrences of the ath and bth events possesses a gamma distribution with $\alpha = b - a$. We will later encounter a theorem (called the *central limit theorem*) that outlines some conditions that imply that a normal distribution would be a suitable approximation for the distribution of data.

A second way to select a model is to form a frequency histogram (Chapter 1) for data drawn from the population and to choose a density function that would visually appear to give a similar frequency curve. For example, if a set of $n = 100$ sample measurements yielded a bell-shaped frequency distribution, we might conclude that the normal density function would adequately model the population frequency distribution.

Not all model selection is completely subjective. Statistical procedures are available to test a hypothesis that a population frequency distribution is of a particular type. We can also calculate a measure of goodness of fit for several distributions and select the best. Studies of many common inferential methods have been made to determine the magnitude of the errors of inference introduced by incorrect population models. It is comforting to know that many statistical methods of inference are insensitive to assumptions about the form of the underlying population frequency distribution.

The uniform, normal, gamma, and beta distributions offer an assortment of density functions that fit many population frequency distributions. Another, the Weibull distribution, appears in the exercises at the end of the chapter.

4.9 Other Expected Values

Moments for continuous random variables have definitions analogous to those given for the discrete case.

DEFINITION 4.12

If Y is a continuous random variable, then the kth *moment about the origin* is given by

$$\mu'_k = E(Y^k), \qquad k = 1, 2, \ldots.$$

The kth *moment about the mean*, or the kth *central moment*, is given by

$$\mu_k = E[(Y - \mu)^k], \qquad k = 1, 2, \ldots.$$

Notice that for $k = 1$, $\mu'_1 = \mu$, and for $k = 2$, $\mu_2 = V(Y) = \sigma^2$.

EXAMPLE **4.12**   Find μ_k' for the uniform random variable with $\theta_1 = 0$ and $\theta_2 = \theta$.

Solution   By definition,

$$\mu_k' = E(Y^k) = \int_{-\infty}^{\infty} y^k f(y)\, dy = \int_0^{\theta} y^k \left(\frac{1}{\theta}\right) dy = \frac{y^{k+1}}{\theta(k+1)}\Bigg]_0^{\theta} = \frac{\theta^k}{k+1}.$$

Thus,

$$\mu_1' = \mu = \frac{\theta}{2}, \qquad \mu_2' = \frac{\theta^2}{3}, \qquad \mu_3' = \frac{\theta^3}{4},$$

and so on.

DEFINITION **4.13**

If Y is a continuous random variable, then the *moment-generating function of Y* is given by

$$m(t) = E(e^{tY}).$$

The moment-generating function is said to exist if there exists a constant $b > 0$ such that $m(t)$ is finite for $|t| \le b$.

This is simply the continuous analogue of Definition 3.14. That $m(t)$ generates moments is established in exactly the same manner as in Section 3.9. If $m(t)$ exists, then

$$E(e^{tY}) = \int_{-\infty}^{\infty} e^{ty} f(y)\, dy = \int_{-\infty}^{\infty} \left(1 + ty + \frac{t^2 y^2}{2!} + \frac{t^3 y^3}{3!} + \cdots\right) f(y)\, dy$$

$$= \int_{-\infty}^{\infty} f(y)\, dy + t \int_{-\infty}^{\infty} yf(y)\, dy + \frac{t^2}{2!} \int_{-\infty}^{\infty} y^2 f(y)\, dy + \cdots$$

$$= 1 + t\mu_1' + \frac{t^2}{2!}\mu_2' + \frac{t^3}{3!}\mu_3' + \cdots.$$

Notice that the moment-generating function,

$$m(t) = 1 + t\mu_1' + \frac{t^2}{2!}\mu_2' + \cdots,$$

takes the same form for both discrete and continuous random variables. Hence Theorem 3.12 holds for continuous random variables and

$$\frac{d^k m(t)}{dt^k}\Bigg]_{t=0} = \mu_k'.$$

EXAMPLE **4.13**   Find the moment-generating function for a gamma distributed random variable.

Solution

$$m(t) = E(e^{tY}) = \int_0^\infty e^{ty} \left(\frac{y^{\alpha-1} e^{-y/\beta}}{\beta^\alpha \Gamma(\alpha)} \right) dy$$

$$= \frac{1}{\beta^\alpha \Gamma(\alpha)} \int_0^\infty y^{\alpha-1} \exp \left[-y \left(\frac{1}{\beta} - t \right) \right] dy$$

$$= \frac{1}{\beta^\alpha \Gamma(\alpha)} \int_0^\infty y^{\alpha-1} \exp \left[\frac{-y}{\beta/(1 - \beta t)} \right] dy.$$

[The term exp($\cdot$) is simply a more convenient way to write $e^{(\cdot)}$ when the term in the exponent is long or complex.]

To complete the integration, notice that the integral of the variable factor of any density function must equal the reciprocal of the constant factor. That is, if $f(y) = cg(y)$, where c is a constant, then

$$\int_{-\infty}^\infty f(y)\, dy = \int_{-\infty}^\infty cg(y)\, dy = 1 \quad \text{and so} \quad \int_{-\infty}^\infty g(y)\, dy = \frac{1}{c}.$$

Applying this result to the integral in $m(t)$ and noting that if $[\beta/(1 - \beta t)] > 0$ (or, equivalently, if $t < 1/\beta$),

$$g(y) = y^{\alpha-1} \times \exp\{-y/[\beta/(1 - \beta t)]\}$$

is the variable factor of a gamma density function with parameters $\alpha > 0$ and $[\beta/(1 - \beta t)] > 0$, we obtain

$$m(t) = \frac{1}{\beta^\alpha \Gamma(\alpha)} \left[\left(\frac{\beta}{1 - \beta t} \right)^\alpha \Gamma(\alpha) \right] = \frac{1}{(1 - \beta t)^\alpha} \quad \text{for} \quad t < \frac{1}{\beta}.$$

The moments μ_k' can be extracted from the moment-generating function by differentiating with respect to t (in accordance with Theorem 3.12) or by expanding the function into a power series in t. We will demonstrate the latter approach.

EXAMPLE **4.14**   Expand the moment-generating function of Example 4.13 into a power series in t and thereby obtain μ_k'.

Solution   From Example 4.13, $m(t) = 1/(1 - \beta t)^\alpha = (1 - \beta t)^{-\alpha}$. Using the expansion for a binomial term of the form $(x + y)^{-c}$, we have

$$m(t) = (1 - \beta t)^{-\alpha} = 1 + (-\alpha)(1)^{-\alpha-1}(-\beta t)$$

$$+ \frac{(-\alpha)(-\alpha - 1)(1)^{-\alpha-2}(-\beta t)^2}{2!} + \cdots$$

$$= 1 + t(\alpha\beta) + \frac{t^2[\alpha(\alpha + 1)\beta^2]}{2!} + \frac{t^3[\alpha(\alpha + 1)(\alpha + 2)\beta^3]}{3!} + \cdots .$$

Because μ'_k is the coefficient of $t^k/k!$, we find, by inspection,

$$\mu'_1 = \mu = \alpha\beta$$

$$\mu'_2 = \alpha(\alpha+1)\beta^2$$

$$\mu'_3 = \alpha(\alpha+1)(\alpha+2)\beta^3$$

and, in general, $\mu'_k = \alpha(\alpha+1)(\alpha+2)\cdots(\alpha+k-1)\beta^k$. Notice that μ'_1 and μ'_2 agree with the results of Theorem 4.8. Moreover, these results agree with the result of Exercise 4.89(a).

We have already explained the importance of the expected values of Y^k, $(Y-\mu)^k$, and e^{tY}, all of which provide important information about the distribution of Y. Sometimes, however, we are interested in the expected value of a function of a random variable as an end in itself. (We also may be interested in the probability distribution of functions of random variables, but we defer discussion of this topic until Chapter 6.)

EXAMPLE **4.15**   The kinetic energy k associated with a mass m moving at velocity v is given by the expression

$$k = \frac{mv^2}{2}.$$

Consider a device that fires a serrated nail into concrete at a mean velocity of 2000 feet per second, where the random velocity V possesses a density function given by

$$f(v) = \frac{v^3 e^{-v/500}}{(500)^4\Gamma(4)}, \qquad v \geq 0.$$

Find the expected kinetic energy associated with a nail of mass m.

Solution   Let K denote the random kinetic energy associated with the nail. Then

$$E(K) = E\left(\frac{mV^2}{2}\right) = \frac{m}{2}E(V^2),$$

by Theorem 4.5, part 2. The random variable V has a gamma distribution with $\alpha = 4$ and $\beta = 500$. Therefore, $E(V^2) = \mu'_2$ for the random variable V. Referring to Example 4.14, we have $\mu'_2 = \alpha(\alpha+1)\beta^2 = 4(5)(500)^2 = 5{,}000{,}000$. Therefore,

$$E(K) = \frac{m}{2}E(V^2) = \frac{m}{2}(5{,}000{,}000) = 2{,}500{,}000\ m.$$

Finding the moments of a function of a random variable is frequently facilitated by using its moment-generating function.

THEOREM 4.12

Let Y be a random variable with density function $f(y)$ and $g(Y)$ be a function of Y. Then the moment-generating function for $g(Y)$ is

$$E[e^{tg(Y)}] = \int_{-\infty}^{\infty} e^{tg(y)} f(y)\, dy.$$

This theorem follows directly from Definition 4.13 and Theorem 4.4.

EXAMPLE 4.16

Let $g(Y) = Y - \mu$, where Y is a normally distributed random variable with mean μ and variance σ^2. Find the moment-generating function for $g(Y)$.

Solution

The moment-generating function of $g(Y)$ is given by

$$m(t) = E[e^{tg(Y)}] = E[e^{t(Y-\mu)}] = \int_{-\infty}^{\infty} e^{t(y-\mu)} \left(\frac{\exp[-(y-\mu)^2/2\sigma^2]}{\sigma\sqrt{2\pi}} \right) dy.$$

To integrate, let $u = y - \mu$. Then $du = dy$ and

$$m(t) = \frac{1}{\sigma\sqrt{2\pi}} \int_{-\infty}^{\infty} e^{tu} e^{-u^2/(2\sigma^2)}\, du$$

$$= \frac{1}{\sigma\sqrt{2\pi}} \int_{-\infty}^{\infty} \exp\left[-\left(\frac{1}{2\sigma^2}\right)(u^2 - 2\sigma^2 tu) \right] du.$$

Complete the square in the exponent of e by multiplying and dividing by $e^{t^2\sigma^2/2}$. Then

$$m(t) = e^{t^2\sigma^2/2} \int_{-\infty}^{\infty} \frac{\exp[-(1/2\sigma^2)(u^2 - 2\sigma^2 tu + \sigma^4 t^2)]}{\sigma\sqrt{2\pi}}\, du$$

$$= e^{t^2\sigma^2/2} \int_{-\infty}^{\infty} \frac{\exp[-(u - \sigma^2 t)^2/2\sigma^2]}{\sigma\sqrt{2\pi}}\, du.$$

The function inside the integral is a normal density function with mean $\sigma^2 t$ and variance σ^2. (See the equation for the normal density function in Section 4.5.) Hence the integral is equal to 1 and

$$m(t) = e^{(t^2/2)\sigma^2}.$$

The moments of $U = Y - \mu$ can be obtained from $m(t)$ by differentiating $m(t)$ in accordance with Theorem 3.12 or by expanding $m(t)$ into a series.

The purpose of the preceding discussion of moments is twofold. First, moments can be used as numerical descriptive measures to describe the data that we obtain in an experiment. Second, they can be used in a theoretical sense to prove that a random variable possesses a particular probability distribution. It can be shown that if two random variables Y and Z possess identical moment-generating functions, then Y and Z possess identical probability distributions. This latter application of moments was mentioned in the discussion of moment-generating functions for discrete random variables in Section 3.9; it applies to continuous random variables as well.

For your convenience, the probability and density functions, means, variances, and moment-generating functions for some common random variables are given in Appendix II and inside the back cover of this text.

Exercises

4.104   Suppose that the waiting time for the first customer to enter a retail shop after 9:00 A.M. is a random variable Y with an exponential density function given by

$$f(y) = \begin{cases} \left(\dfrac{1}{\theta}\right)e^{-y/\theta}, & y > 0 \\ 0, & \text{elsewhere.} \end{cases}$$

   a   Find the moment-generating function for Y.
   b   Use the answer from (a) to find $E(Y)$ and $V(Y)$.

4.105   Show that the result given in Exercise 3.124 also holds for continuous random variables. That is, show that, if Y is a random variable with moment-generating function $m(t)$ and U is given by $U = aY + b$, the moment-generating function of U is $e^{tb}m(at)$. If Y has mean μ and variance σ^2, use the moment-generating function of U to derive the mean and variance of U.

4.106   Example 4.16 derives the moment-generating function for $Y - \mu$, where Y is normally distributed with mean μ and variance σ^2.

   a   Use the results in Example 4.16 and Exercise 4.105 to find the moment-generating function for Y.
   b   Differentiate the moment-generating function found in (a) to show that $E(Y) = \mu$ and $V(Y) = \sigma^2$.

4.107   The moment generating function of a normally distributed random variable , Y, with mean μ and variance σ^2 was shown in Exercise 4.106 to be $m(t) = e^{\mu t + \frac{1}{2}t^2\sigma^2}$. Use the result in Exercise 4.105 to derive the moment generating function of $X = -3Y + 4$. What is the distribution of X? Why?

4.108   Identify the distributions of the random variables with the following moment-generating functions.
   a $m(t) = (1 - 4t)^{-2}$.
   b $m(t) = 1/(1 - 3.2t)$.
   c $m(t) = e^{-5t+6t^2}$.

4.109   If $\theta_1 < \theta_2$, derive the moment-generating function of a random variable that has a uniform distribution on the interval (θ_1, θ_2).

4.110   Refer to Exercises 109 and 105. Suppose that Y is uniformly distributed on the interval $(0, 1)$ and that $a > 0$ is a constant.

    a Give the moment-generating function for Y.

    b Derive the moment-generating function of $W = aY$. What is the distribution of W? Why?

    c Derive the moment-generating function of $X = -aY$. What is the distribution of X? Why?

    d If b is a fixed constant, derive the moment-generating function of $V = aY + b$. What is the distribution of V? Why?

4.111 The moment-generating function for the gamma random variable is derived in Example 4.13. Differentiate this moment-generating function to find the mean and variance of the gamma distribution.

4.112 Consider a random variable Y with density function given by

$$f(y) = ke^{-y^2/2}, \qquad -\infty < y < \infty.$$

    a Find k.

    b Find the moment-generating function of Y.

    c Find $E(Y)$ and $V(Y)$.

4.113 A random variable Y has the density function

$$f(y) = \begin{cases} e^y, & y < 0 \\ 0, & \text{elsewhere.} \end{cases}$$

    a Find $E(e^{3Y/2})$.

    b Find the moment-generating function for Y.

    c Find $V(Y)$.

4.10 Tchebysheff's Theorem

As was the case for discrete random variables, an interpretation of μ and σ for continuous random variables is provided by the empirical rule and Tchebysheff's theorem. Even if the exact distributions are unknown for random variables of interest, knowledge of the associated means and standard deviations permits us to deduce meaningful bounds for the probabilities of events that are often of interest.

We stated and utilized Tchebysheff's theorem in Section 3.11. We now restate this theorem and give a proof applicable to a continuous random variable.

THEOREM 4.13   **Tchebysheff's Theorem.** Let Y be a random variable with finite mean μ and variance σ^2. Then, for any $k > 0$,

$$P(|Y - \mu| < k\sigma) \geq 1 - \frac{1}{k^2} \quad \text{or} \quad P(|Y - \mu| \geq k\sigma) \leq \frac{1}{k^2}.$$

Proof   We will give the proof for a continuous random variable. The proof for the discrete case proceeds similarly. Let $f(y)$ denote the density function of Y. Then

$$V(Y) = \sigma^2 = \int_{-\infty}^{\infty} (y - \mu)^2 f(y)\, dy$$

$$= \int_{-\infty}^{\mu - k\sigma} (y - \mu)^2 f(y)\, dy + \int_{\mu - k\sigma}^{\mu + k\sigma} (y - \mu)^2 f(y)\, dy$$

$$+ \int_{\mu + k\sigma}^{\infty} (y - \mu)^2 f(y)\, dy.$$

The second integral is always greater than or equal to zero and $(y - \mu)^2 \geq k^2 \sigma^2$ for all values of y between the limits of integration for the first and third integrals; that is, the regions of integration are in the tails of the density function and cover only values of y for which $(y - \mu)^2 \geq k^2 \sigma^2$. Replace the second integral by zero, and substitute $k^2 \sigma^2$ for $(y - \mu)^2$ in the first and third integrals to obtain the inequality

$$V(Y) = \sigma^2 \geq \int_{-\infty}^{\mu - k\sigma} k^2 \sigma^2 f(y)\, dy + \int_{\mu + k\sigma}^{\infty} k^2 \sigma^2 f(y)\, dy.$$

Then

$$\sigma^2 \geq k^2 \sigma^2 \left[\int_{-\infty}^{\mu - k\sigma} f(y)\, dy + \int_{\mu + k\sigma}^{+\infty} f(y)\, dy \right]$$

or

$$\sigma^2 \geq k^2 \sigma^2 \{ P(Y \leq \mu - k\sigma) + P(Y \geq \mu + k\sigma) \} = k^2 \sigma^2 P(|Y - \mu| \geq k\sigma).$$

Dividing by $k^2 \sigma^2$, we obtain

$$P(|Y - \mu| \geq k\sigma) \leq \frac{1}{k^2},$$

or, equivalently,

$$P(|Y - \mu| < k\sigma) \geq 1 - \frac{1}{k^2}.$$ ∎

One real value of Tchebysheff's theorem is that it enables us to find bounds for probabilities that ordinarily would have to be obtained by tedious mathematical manipulations (integration or summation). Further, we often can obtain means and variances of random variables (see Example 4.15) without specifying the *distribution* of the variable. In situations like these, Tchebysheff's theorem still provides meaningful bounds for probabilities of interest.

EXAMPLE **4.17**   Suppose that experience has shown that the length of time Y (in minutes) required to conduct a periodic maintenance check on a dictating machine follows a gamma distribution with $\alpha = 3.1$ and $\beta = 2$. A new maintenance worker takes 22.5 min to check the machine. Does this length of time to perform a maintenance check disagree with prior experience?

Solution The mean and variance for the length of maintenance check times (based on prior experience) are (from Theorem 4.8),

$$\mu = \alpha\beta = (3.1)(2) = 6.2 \quad \text{and} \quad \sigma^2 = \alpha\beta^2 = (3.1)(2^2) = 12.4.$$

It follows that $\sigma = \sqrt{12.4} = 3.52$. Notice that $y = 22.5$ minutes exceeds the mean $\mu = 6.2$ minutes by 16.3 minutes, or $k = 16.3/3.52 = 4.63$ standard deviations. Then from Tchebysheff's theorem

$$P(|Y - 6.2| \geq 16.3) = P(|Y - \mu| \geq 4.63\sigma) \leq \frac{1}{(4.63)^2} = .0466.$$

This probability is based on the assumption that the distribution of maintenance times has not changed from prior experience. Then, observing that $P(Y \geq 22.5)$ is small, we must conclude either that our new maintenance worker has generated by chance a lengthy maintenance time that occurs with low probability or that the new worker is somewhat slower than preceding ones. Considering the low probability for $P(Y \geq 22.5)$, we favor the latter view.

The exact probability, $P(Y \geq 22.5)$, for Example 4.17 would require evaluation of the integral

$$P(Y \geq 22.5) = \int_{22.5}^{\infty} \frac{y^{2.1} e^{-y/2}}{2^{3.1} \Gamma(3.1)} \, dy.$$

Although we could utilize tables given by Pearson (1965) to evaluate this integral, we cannot evaluate it directly. Similar integrals are difficult to evaluate for the beta density and for many other density functions. Tchebysheff's theorem often provides quick bounds for probabilities while circumventing laborious integration or searches for appropriate tables.

Exercises

4.114 A manufacturer of tires wants to advertise a mileage interval that excludes no more than 10% of the mileage on tires he sells. All he knows is that, for a large number of tires tested, the mean mileage was 25,000 miles, and the standard deviation was 4000 miles. What interval would you suggest?

4.115 A machine used to fill cereal boxes dispenses, on the average, μ ounces per box. The manufacturer wants the actual ounces dispensed Y to be within 1 ounce of μ at least 75% of the time. What is the largest value of σ, the standard deviation of Y, that can be tolerated if the manufacturer's objectives are to be met?

4.116 Find $P(|Y - \mu| \leq 2\sigma)$ for Exercise 4.10. Compare with the corresponding probabilistic statements given by Tchebysheff's theorem and the empirical rule.

4.117   Find $P(|Y - \mu| \leq 2\sigma)$ for the uniform random variable. Compare with the corresponding probabilistic statements given by Tchebysheff's theorem and the empirical rule.

4.118   Find $P(|Y - \mu| \leq 2\sigma)$ for the exponential random variable. Compare with the corresponding probabilistic statements given by Tchebysheff's theorem and the empirical rule.

4.119   Refer to Exercise 4.72. Would you expect C to exceed 2000 very often?

4.120   Refer to Exercise 4.87. Find an interval that will contain L for at least 89% of the weeks that the machine is in use.

4.121   Refer to Exercise 4.97. Find an interval for which the probability that C will lie within it is at least .75.

4.122   Suppose that Y is a χ^2 distributed random variable with $\nu = 7$ degrees of freedom.
   a   What are the mean and variance of Y?
   b   Is it likely that Y will take on a value of 23 or more?

4.11   Expectations of Discontinuous Functions and Mixed Probability Distributions (Optional)

Problems in probability and statistics sometimes involve functions that are partly continuous and partly discrete, in one of two ways. First, we may be interested in the properties, perhaps the expectation, of a random variable $g(Y)$ that is a discontinuous function of a discrete or continuous random variable Y. Second, the random variable of interest itself may have a distribution function that is continuous over some intervals and such that some isolated points have positive probabilities.

We illustrate these ideas with the following examples.

EXAMPLE 4.18   A retailer for a petroleum product sells a random amount Y each day. Suppose that Y, measured in thousands of gallons, has the probability density function

$$f(y) = \begin{cases} (3/8)y^2, & 0 \leq y \leq 2 \\ 0, & \text{elsewhere.} \end{cases}$$

The retailer's profit turns out to be $100 for each 1000 gallons sold (10¢ per gallon) if $Y \leq 1$ and $40 *extra* per 1000 gallons (an extra 4¢ per gallon) if $Y > 1$. Find the retailer's expected profit for any given day.

Solution   Let $g(Y)$ denote the retailer's daily profit. Then

$$g(Y) = \begin{cases} 100Y, & 0 \leq Y \leq 1 \\ 140Y, & 1 < Y \leq 2. \end{cases}$$

We want to find expected profit; and by Theorem 4.4, the expectation is

$$E[g(Y)] = \int_{-\infty}^{\infty} g(y)f(y)\,dy$$

$$= \int_0^1 100y\left[\left(\frac{3}{8}\right)y^2\right]dy + \int_1^2 140y\left[\left(\frac{3}{8}\right)y^2\right]dy$$

$$= \frac{300}{(8)(4)}\,y^4\Big]_0^1 + \frac{420}{(8)(4)}\,y^4\Big]_1^2$$

$$= \frac{300}{32}(1) + \frac{420}{32}(15) = 206.25.$$

Thus the retailer can expect a profit of $206.25 on the daily sale of this particular product.

Suppose that Y denotes the amount paid out per policy in one year by an insurance company that provides automobile insurance. For many policies, $Y = 0$ because the insured individuals are not involved in accidents. For insured individuals who *do* have accidents, the amount paid by the company might be modeled with one of the density functions that we have previously studied. A random variable Y that has some of its probability at discrete points (0 in this example) and the remainder spread over intervals is said to have a *mixed distribution*. Let $F(y)$ denote a distribution function of a random variable Y that has a mixed distribution. For all practical purposes, any mixed distribution function $F(y)$ can be written uniquely as

$$F(y) = c_1 F_1(y) + c_2 F_2(y)$$

where $F_1(y)$ is a step distribution function, $F_2(y)$ is a continuous distribution function, c_1 is the accumulated probability of all discrete points, and $c_2 = 1 - c_1$ is the accumulated probability of all continuous portions.

The following example gives an illustration of a mixed distribution.

EXAMPLE 4.19   Let Y denote the length of life (in hundreds of hours) of electronic components. These components frequently fail immediately upon insertion into a system. It has been observed that the probability of immediate failure is $1/4$. If a component does not fail immediately, the distribution for its length of life has the exponential density function

$$f(y) = \begin{cases} e^{-y}, & y > 0 \\ 0, & \text{elsewhere.} \end{cases}$$

Find the distribution function for Y, and evaluate $P(Y > 10)$.

Solution   There is only one discrete point, $y = 0$, and this point has probability $1/4$. Hence $c_1 = 1/4$ and $c_2 = 3/4$. It follows that Y is a mixture of two random variables, X_1 and X_2, where X_1 has probability 1 at point 0 and X_2 has the given exponential

FIGURE **4.17**
Distribution
function $F(y)$ for
Example 4.19

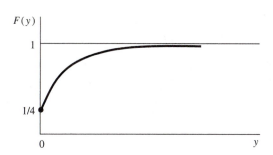

density. That is,

$$F_1(y) = \begin{cases} 0, & y < 0 \\ 1, & y \geq 0 \end{cases}$$

and

$$F_2(y) = \begin{cases} 0, & y < 0 \\ \int_0^y e^{-x}dx = 1 - e^{-y}, & y \geq 0. \end{cases}$$

Now

$$F(y) = (1/4)F_1(y) + (3/4)F_2(y)$$

and hence

$$P(Y > 10) = 1 - P(Y \leq 10) = 1 - F(10)$$
$$= 1 - [(1/4) + (3/4)(1 - e^{-10})]$$
$$= (3/4)[1 - (1 - e^{-10})] = (3/4)e^{-10}.$$

A graph of $F(y)$ is given in Figure 4.17.

An easy method for finding expectations of random variables with mixed distributions is given in Definition 4.14.

DEFINITION **4.14**

Let Y have the mixed distribution function

$$F(y) = c_1 F_1(y) + c_2 F_2(y)$$

and suppose that X_1 is a discrete random variable with distribution function $F_1(y)$ and that X_2 is a continuous random variable with distribution function $F_2(y)$. Let $g(Y)$ denote a function of Y. Then

$$E[g(Y)] = c_1 E[g(X_1)] + c_2 E[g(X_2)].$$

EXAMPLE **4.20** Find the mean and variance of the random variable defined in Example 4.19.

Solution With all definitions as in Example 4.19, it follows that

$$E(X_1) = 0 \quad \text{and} \quad E(X_2) = \int_0^\infty y e^{-y} \, dy = 1.$$

Therefore,

$$\mu = E(Y) = (1/4)E(X_1) + (3/4)E(X_2) = 3/4.$$

Also,

$$E(X_1^2) = 0 \quad \text{and} \quad E(X_2^2) = \int_0^\infty y^2 e^{-y} \, dy = 2.$$

Therefore,

$$E(Y^2) = (1/4)E(X_1^2) + (3/4)E(X_2^2) = (1/4)(0) + (3/4)(2) = 3/2.$$

Then

$$V(Y) = E(Y^2) - \mu^2 = (3/2) - (3/4)^2 = 15/16.$$

Exercises

***4.123** A builder of houses needs to order some supplies that have a waiting time Y for delivery, with a continuous uniform distribution over the interval from 1 to 4 days. Because she can get by without them for 2 days, the cost of the delay is fixed at $100 for any waiting time up to 2 days. After 2 days, however, the cost of the delay is $100 plus $20 per day (prorated) for each additional day. That is, if the waiting time is 3.5 days, the cost of the delay is $100 + $20(1.5) = $130. Find the expected value of the builder's cost due to waiting for supplies.[4]

***4.124** The duration Y of long-distance telephone calls (in minutes) monitored by a station is a random variable with the properties that

$$P(Y = 3) = .2 \quad \text{and} \quad P(Y = 6) = .1.$$

Otherwise, Y has a continuous density function given by

$$f(y) = \begin{cases} (1/4)y e^{-y/2}, & y > 0 \\ 0, & \text{elsewhere.} \end{cases}$$

The discrete points at 3 and 6 are due to the fact that the length of call is announced to the caller in 3-minute intervals and the caller must pay for 3 minutes even if he or she talks less than 3 minutes. Find the expected duration of a randomly selected long-distance call.

[4]Exercises preceded by an asterisk are optional.

*4.125 The life length Y of a component used in a complex electronic system is known to have an exponential density with a mean of 100 hours. The component is replaced at failure or at age 200 hours, whichever comes first.

a Find the distribution function for X, the length of time the component is in use.

b Find $E(X)$.

*4.126 Consider the nail-firing device of Example 4.15. When the device works, the nail is fired with velocity, V, with density

$$f(v) = \frac{v^3 e^{-v/500}}{(500)^4 \Gamma(4)}.$$

The device misfires 2% of the time it is used, resulting a velocity of 0. Find the expected kinetic energy associated with a nail of mass m. Recall that the kinetic energy, k, of a mass m moving at velocity v is $k = (mv^2)/2$.

*4.127 A random variable Y has distribution function

$$F(y) = \begin{cases} 0, & \text{if } y < 0 \\ y^2 + 0.1, & \text{if } 0 \le y < 0.5 \\ y, & \text{if } 0.5 \le y < 1 \\ 1, & \text{if } y \ge 1. \end{cases}$$

a. Give $F_1(y)$ and $F_2(y)$, the discrete and continuous components of $F(y)$.

b. Write $F(y)$ as $c_1 F_1(y) + c_2 F_2(y)$.

c. Find the expected value and variance of Y.

4.12 Summary

This chapter presented probabilistic models for continuous random variables. The density function, which provides a model for a population-frequency distribution associated with a continuous random variable, subsequently will yield a mechanism for inferring characteristics of the population based on measurements contained in a sample taken from that population. As a consequence, the density function provides a model for a real distribution of data that exist or could be generated by repeated experimentation. Similar distributions for small sets of data (samples from populations) were discussed in Chapter 1.

Four specific types of density functions—uniform, normal, gamma (with the χ^2 and exponential as special cases), and beta—were presented, providing a wide assortment of models for population frequency distributions. Many other density functions could be employed to fit real situations, but the four described suit many situations adequately. A few other density functions are presented in the exercises at the end of the chapter.

The adequacy of a density function in modeling the frequency distribution for a random variable depends upon the inference-making technique to be employed. If modest disagreement between the model and the real population frequency distribution does not affect the goodness of the inferential procedure, the model is adequate.

The latter part of the chapter concerned expectations, particularly moments and moment-generating functions. It is important to focus attention on the reason for presenting these quantities and to avoid excessive concentration on the mathematical aspects of the material. Moments, particularly the mean and variance, are numerical descriptive measures for random variables. Particularly, we will subsequently see that it is sometimes difficult to find the probability distribution for a random variable Y or a function $g(Y)$, and we already have observed that integration over intervals for many density functions (the normal and gamma, for example) is very difficult. When this occurs, we can approximately describe the behavior of the random variable by using its moments along with Tchebysheff's theorem and the empirical rule (Chapter 1).

References and Further Readings

Hogg, R. V., and A. T. Craig. 1995. *Introduction to Mathematical Statistics*, 5th ed. Upper Saddle River, N.J.: Prentice Hall.

Johnson, N.L., S. Kotz, and N. Balakrishnan. 1994. *Continuous Univariate Distributions*, 2d ed. New York, John Wiley.

Parzen, E. 1960. *Modern Probability Theory and Its Applications*. New York: John Wiley.

Pearson, K., ed. 1965. *Tables of the Incomplete Gamma Function*. London: Cambridge University Press.

———. 1968. *Tables of the Incomplete Beta Function*. London: Cambridge University Press.

Perruzzi, J. J., and E. J. Hilliard. 1984. "Modeling Time-Delay Measurement Errors Using a Generalized Beta Density Function." *Journal of the Acoustical Society of America* 75(1): 197–201.

Pyke, R. 1965. "Spacings." *Journal of the Royal Statistical Society* B, 27(3): 426.

Standard Mathematical Tables. 1987. 28th ed. Boca Raton, Fla.: CRC Press.

Tables of the Binomial Probability Distribution. 1950. Department of Commerce, National Bureau of Standards, Applied Mathematics Series 6.

Zamurs, J. 1984. "Assessing the Effect of Transportation Control Strategies on Urban Carbon Monoxide Concentrations." *Air Pollution Control Association Journal* 34(6): 637.

Zimmels, Y. 1983. "Theory of Kindered Sedimentation of Polydisperse Mixtures." *American Institute of Chemical Engineers Journal* 29(4): 669–76.

Supplementary Exercises

4.128 Let the density function of a random variable Y be given by

$$f(y) = \begin{cases} \dfrac{2}{\pi(1+y^2)}, & -1 \le y \le 1 \\ 0, & \text{elsewhere.} \end{cases}$$

a Find the distribution function.

b Find $E(Y)$.

4.129 The length of time required to complete a college achievement test is found to be normally distributed with mean 70 minutes and standard deviation 12 minutes. When should the test be terminated if we wish to allow sufficient time for 90% of the students to complete the test?

4.130 A manufacturing plant utilizes 3000 electric lightbulbs whose length of life is normally distributed with mean 500 hours and standard deviation 50 hours. To minimize the number of bulbs that burn out during operating hours, all the bulbs are replaced after a given period of operation. How often should the bulbs be replaced if we want not more than 1% of the bulbs to burn out between replacement periods?

4.131 Refer to Exercise 4.52. Suppose that five bearings are randomly drawn from production. What is the probability that at least one is defective?

4.132 The length of life of oil-drilling bits depends upon the types of rock and soil that the drill encounters, but it is estimated that the mean length of life is 75 hours. An oil exploration company purchases drill bits whose length of life is approximately normally distributed with mean 75 hours and standard deviation 12 hours.

a What proportion of the company's drill bits will fail before 60 hours of use?

b What proportion will last at least 60 hours?

c What proportion will have to be replaced after more than 90 hours of use?

4.133 Let Y have density function

$$f(y) = \begin{cases} cye^{-2y}, & 0 \le y \le \infty \\ 0, & \text{elsewhere.} \end{cases}$$

a Find the value of c that makes $f(y)$ a density function.

b Give the mean and variance for Y.

c Give the moment-generating function for Y.

4.134 Use the fact that

$$e^z = 1 + z + \frac{z^2}{2!} + \frac{z^3}{3!} + \frac{z^4}{4!} + \cdots$$

to expand the moment-generating function of Example 4.16 into a series to find μ_1, μ_2, μ_3, and μ_4 for the normal random variable.

4.135 Find an expression for $\mu'_k = E(Y^k)$, where the random variable Y has a beta distribution.

4.136 The number of arrivals N at a supermarket checkout counter in the time interval from 0 to t follows a Poisson distribution with mean λt. Let T denote the length of time until the first arrival. Find the density function for T. [*Note:* $P(T > t_0) = P(N = 0 \text{ at } t = t_0)$.]

4.137 An argument similar to that of Exercise 4.136 can be used to show that if events are occurring in time according to a Poisson distribution with mean λt, then the interarrival times between events have an exponential distributions with mean $1/\lambda$. If calls come into a police emergency center at the rate of 10 per hour, what is the probability that more than 15 minutes will elapse between the next two calls?

*4.138 Refer to Exercise 4.136.

a If U is the time until the *second* arrival, show that U has a gamma density function with $\alpha = 2$ and $\beta = 1/\lambda$.

b Show that the time until the kth arrival has a gamma density with $\alpha = k$ and $\beta = 1/\lambda$.

4.139 Suppose that customers arrive at a checkout counter at a rate of 2 per minute.

    a What are the mean and variance of the of the waiting times between successive customer arrivals?

    b If a clerk takes 3 minutes to serve the first customer arriving at the counter, what is the probability that at least one more customer will be waiting when the service to the first customer is completed?

4.140 Calls for dial-in connections to a computer center arrive at an average rate of four per minute. The calls follow a Poisson distribution. If a call arrives at the beginning of a 1-minute interval, what is the probability that a second call will not arrive in the next 20 seconds?

4.141 Suppose that plants of a particular species are randomly dispersed over an area, so that the number of plants in a given area follows a Poisson distribution with a mean density of λ plants per unit area. If a plant is randomly selected in this area, find the probability density function of the distance to the *nearest* neighboring plant. [*Hint:* If R denotes the distance to the nearest neighbor, then $P(R > r)$ is the same as the probability of seeing no plants in a circle of radius r.]

4.142 The time (in hours) a manager takes to interview a job applicant has an exponential distribution with $\beta = 1/2$. The applicants are scheduled at quarter-hour intervals, beginning at 8:00 A.M., and the applicants arrive exactly on time. When the applicant with an 8:15 A.M. appointment arrives at the manager's office, what is the probability that he will have to wait before seeing the manager?

4.143 The median value y of a continuous random variable is that value such that $F(y) = .5$. Find the median value of the random variable in Exercise 4.5.

4.144 Graph the beta probability density function for $\alpha = 3$ and $\beta = 2$. If Y has this beta density function, find $P(.1 \le Y \le .2)$ by using binomial probabilities to evaluate $F(y)$. (See Section 4.7.)

*4.145 A retail grocer has a daily demand Y for a certain food sold by the pound, where Y (measured in hundreds of pounds) has a probability density function given by

$$f(y) = \begin{cases} 3y^2, & 0 \le y \le 1 \\ 0, & \text{elsewhere.} \end{cases}$$

(She cannot stock over 100 lb.) The grocer wants to order $100k$ pounds of food. She buys the food at 6¢ per pound and sells it at 10¢ per pound. What value of k will maximize her expected daily profit?

4.146 Suppose that Y has a gamma distribution with $\alpha = 3$ and $\beta = 1$. Use Poisson probabilities to evaluate $P(Y \le 4)$. (See Exercise 4.79.)

4.147 Suppose that Y is a normally distributed random variable with mean μ and variance σ^2. Use the results of Example 4.16 to find the moment-generating function, mean, and variance of

$$Z = \frac{Y - \mu}{\sigma}.$$

What is the distribution of Z? Why?

*4.148 A random variable Y is said to have a log-normal distribution if $X = \ln(Y)$ has a normal distribution. (The symbol ln denotes natural logarithm.) In this case Y must be nonnegative. The shape of the log-normal probability density function is similar to that of the gamma distribution, with a long tail to the right. The equation of the log-normal density function is given by

$$f(y) = \begin{cases} \dfrac{1}{\sigma y \sqrt{2\pi}} e^{-(\ln(y)-\mu)^2/(2\sigma^2)}, & y > 0 \\ 0, & \text{elsewhere.} \end{cases}$$

Because $\ln(y)$ is a monotonic function of y,

$$P(Y \le y) = P[\ln(Y) \le \ln(y)] = P[X \le \ln(y)]$$

where X has a normal distribution with mean μ and variance σ^2. Thus probabilities for random variables with a log-normal distribution can be found by transforming them into probabilities that can be computed using the ordinary normal distribution. If Y has a log-normal distribution with $\mu = 4$ and $\sigma^2 = 1$, find

a $P(Y \le 4)$.

b $P(Y > 8)$.

4.149 If Y has a log-normal distribution with parameters μ and σ^2, it can be shown that

$$E(Y) = e^{\mu + \sigma^2/2}$$

and

$$V(Y) = e^{2\mu + \sigma^2}(e^{\sigma^2} - 1).$$

The grains composing polycrystalline metals tend to have weights that follow a log-normal distribution. For a type of aluminum, gram weights have a log-normal distribution with $\mu = 3$ and $\sigma = 4$ (in units of 10^{-2} g).

a Find the mean and variance of the grain weights.

b Find an interval in which at least 75% of the grain weights should lie. (Use Tchebysheff's theorem.)

c Find the probability that a randomly chosen grain weighs less than the mean grain weight.

4.150 Let Y denote a random variable with probability density function given by

$$f(y) = (1/2)e^{-|y|}, \qquad -\infty < y < \infty.$$

Find the moment-generating function of Y, and use it to find $E(Y)$.

*4.151 Let $f_1(y)$ and $f_2(y)$ be density functions, and let a be a constant such that $0 \le a \le 1$. Consider the function $f(y) = af_1(y) + (1 - a)f_2(y)$.

a Show that $f(y)$ is a density function. Such a density function is often referred to as a mixture of two density functions.

b Suppose that Y_1 is a random variable with density function $f_1(y)$, and that $E(Y_1) = \mu_1$ and $\text{Var}(Y_1) = \sigma_1^2$; and similarly suppose that Y_2 is a random variable with density function $f_2(y)$, and that $E(Y_2) = \mu_2$ and $\text{Var}(Y_2) = \sigma_2^2$. Assume that Y is a random variable whose density is a mixture of the densities corresponding to Y_1 and Y_2.

   i. Show that $E(Y) = a\mu_1 + (1 - a)\mu_2$.

   ii. Show that $\text{Var}(Y) = a\sigma_1^2 + (1 - a)\sigma_2^2 + a(1 - a)[\mu_1 - \mu_2]^2$.

   [Hint: $E(Y_i^2) = \mu_i^2 + \sigma_i^2$, $i = 1, 2$.]

*4.152 The random variable Y, with a density function given by

$$f(y) = \frac{my^{m-1}}{\alpha} e^{-y^m/\alpha}, \qquad 0 \le y < \infty; \quad \alpha, m > 0$$

is said to have a *Weibull* distribution. The Weibull density function provides a good model for the distribution of length of life for many mechanical devices and biological plants and animals. Find the mean and variance for a Weibull distributed random variable with $m = 2$.

*4.153 Refer to Exercise 4.152. Resistors used in the construction of an aircraft guidance system have life lengths that follow a Weibull distribution with $m = 2$ and $\alpha = 10$ (with measurements in thousands of hours).

  a Find the probability that the life length of a randomly selected resistor of this type exceeds 5000 hours.

  b If three resistors of this type are operating independently, find the probability that exactly one of the three will burn out prior to 5000 hours of use.

*4.154 Refer to Exercise 4.152.

  a What is the usual name of the distribution of a random variable that has a Weibull distribution with $m = 1$?

  b Derive, in terms of the parameters α and m, the mean and variance of a Weibull distributed random variable.

*4.155 If $n > 2$ is an integer, the distribution with density given by

$$f(y) = \begin{cases} \dfrac{1}{B(1/2, [n-2]/2)}(1-y^2)^{(n-4)/2}, & -1 \le y \le 1 \\ 0, & \text{elsewhere.} \end{cases}$$

is called the r distribution. Derive the mean and variance of a random variable with the r distribution.

*4.156 A function sometimes associated with continuous nonnegative random variables is the failure rate (or hazard rate) function, which is defined by

$$r(t) = \frac{f(t)}{1 - F(t)}$$

for a density function $f(t)$ with corresponding distribution function $F(t)$. If we think of the random variable in question as being the length of life of a component, $r(t)$ is proportional to the probability of failure in a small interval after t, given that the component has survived up to time t.

  a Show that for an exponential density function, $r(t)$ is constant.

  b Show that for a Weibull density function with $m > 1$, $r(t)$ is an increasing function of t. (See Exercise 4.152.)

*4.157 Suppose that Y is a continuous random variable with distribution function given by $F(y)$ and probability density function $f(y)$. We often are interested in conditional probabilities of the form $P(Y \le y | Y \ge c)$ for a constant c.

  a Show that, for $y \ge c$,

$$P(Y \le y | Y \ge c) = \frac{F(y) - F(c)}{1 - F(c)}.$$

  b Show that the function in (a) has all the properties of a distribution function.

c If the length of life Y for a battery has a Weibull distribution with $m = 2$ and $\alpha = 3$ (with measurements in years), find the probability that the battery will last less than 4 years, given that it is now 2 years old.

*4.158 The velocities of gas particles can be modeled by the Maxwell distribution, whose probability density function is given by

$$f(v) = 4\pi \left(\frac{m}{2\pi KT} \right)^{3/2} v^2 e^{-v^2 (m/[2KT])}, \qquad v > 0,$$

where m is the mass of the particle, K is Boltzmann's constant, and T is the absolute temperature.

a Find the mean velocity of these particles.

b The kinetic energy of a particle is given by $(1/2)mV^2$. Find the mean kinetic energy for a particle.

*4.159 Because

$$P(Y \le y | Y \ge c) = \frac{F(y) - F(c)}{1 - F(c)}$$

has the properties of a distribution function, its derivative will have the properties of a probability density function. This derivative is given by

$$\frac{f(y)}{1 - F(c)}, \qquad y \ge c.$$

We can thus find the expected value of Y, given that Y is greater than c, by using

$$E(Y|Y \ge c) = \frac{1}{1 - F(c)} \int_c^\infty y f(y) dy.$$

If Y, the length of life of an electronic component, has an exponential distribution with mean 100 hours, find the expected value of Y, given that this component already has been in use for 50 hours.

*4.160 We can show that the normal density function integrates to unity by showing that, if $u > 0$,

$$\frac{1}{\sqrt{2\pi}} \int_{-\infty}^\infty e^{-(1/2)uy^2} \, dy = \frac{1}{\sqrt{u}}.$$

This, in turn, can be shown by considering the product of two such integrals:

$$\frac{1}{2\pi} \left(\int_{-\infty}^\infty e^{-(1/2)uy^2} \, dy \right) \left(\int_{-\infty}^\infty e^{-(1/2)ux^2} \, dx \right) = \frac{1}{2\pi} \int_{-\infty}^\infty \int_{-\infty}^\infty e^{-(1/2)u(x^2+y^2)} \, dx \, dy.$$

By transforming to polar coordinates, show that the preceding double integral is equal to $1/u$.

*4.161 Let Z be a standard normal random variable and $W = (Z^2 + 3Z)^2$.

a Use the moments of Z to derive the mean of W.

b Use the result given in Exercise 4.164 to find a value of w that is such that $P(W \le w) \ge .90$.

*4.162 Show that $\Gamma(1/2) = \sqrt{\pi}$ by writing

$$\Gamma(1/2) = \int_0^\infty y^{-1/2} e^{-y} \, dy$$

by making the transformation $y = (1/2)x^2$ and by employing the result of Exercise 4.160.

*4.163 The function $B(\alpha, \beta)$ is defined by

$$B(\alpha, \beta) = \int_0^1 y^{\alpha-1}(1 - y)^{\beta-1}\, dy.$$

a Letting $y = \sin^2\theta$, show that

$$B(\alpha, \beta) = 2\int_0^{\pi/2} \sin^{2\alpha-1}\theta \cos^{2\beta-1}\theta\, d\theta.$$

b Write $\Gamma(\alpha)\Gamma(\beta)$ as a double integral, transform to polar coordinates, and conclude that

$$B(\alpha, \beta) = \frac{\Gamma(\alpha)\Gamma(\beta)}{\Gamma(\alpha + \beta)}.$$

*4.164 Let $g(Y)$ be a function of the random variable Y, with $E(|g(Y)|) < \infty$. Show that, for every positive constant k,

$$P(|g(Y)| \le k) \ge 1 - \frac{E(|g(Y)|)}{k}.$$

CHAPTER 5

Multivariate Probability Distributions

5.1 Introduction

5.2 Bivariate and Multivariate Probability Distributions

5.3 Marginal and Conditional Probability Distributions

5.4 Independent Random Variables

5.5 The Expected Value of a Function of Random Variables

5.6 Special Theorems

5.7 The Covariance of Two Random Variables

5.8 The Expected Value and Variance of Linear Functions of Random Variables

5.9 The Multinomial Probability Distribution

5.10 The Bivariate Normal Distribution (Optional)

5.11 Conditional Expectations

5.12 Summary

References and Further Readings

5.1 Introduction

The intersection of two or more events is frequently of interest to an experimenter. For example, a gambler playing blackjack is interested in the event of drawing both an ace and a face card from a 52-card deck. A biologist, observing the number of animals surviving in a litter, is concerned about the intersection of these events:

A: The litter contains n animals, and B: y animals survive.

Similarly, observing both the height and the weight of an individual represents the intersection of a specific pair of events associated with height–weight measurements.

Most important to statisticians are intersections that occur in the course of sampling. Suppose that $Y_1, Y_2, \ldots, Y_n$ denote the outcomes of n successive trials of an experiment. For example, this sequence could represent the weights of n people or the measurements of n physical characteristics of a single person. A specific set of outcomes, or sample measurements, may be expressed in terms of the intersection of the n events $(Y_1 = y_1)$, $(Y_2 = y_2)$, $\ldots$, $(Y_n = y_n)$, which we will denote as $(Y_1 = y_1, Y_2 = y_2, \ldots, Y_n = y_n)$, or, more compactly, as $(y_1, y_2, \ldots, y_n)$. Calculation of the probability of this intersection is essential in making inferences about the population from which the sample was drawn and is a major reason for studying multivariate probability distributions.

5.2 Bivariate and Multivariate Probability Distributions

Many random variables can be defined over the same sample space. For example, consider the experiment of tossing a pair of dice. The sample space contains 36 sample points, corresponding to the $mn = (6)(6) = 36$ ways in which numbers may appear on the faces of the dice. Any one of the following random variables could be defined over the sample space and might be of interest to the experimenter:

Y_1:   The number of dots appearing on die 1.

Y_2:   The number of dots appearing on die 2.

Y_3:   The sum of the number of dots on the dice.

Y_4:   The product of the number of dots appearing on the dice.

The 36 sample points associated with the experiment are equiprobable and correspond to the 36 numerical events $(y_1, \ y_2)$. Thus, throwing a pair of 1s is the simple event $(1, 1)$. Throwing a 2 on die 1 and a 3 on die 2 is the simple event $(2, 3)$. Because all pairs $(y_1, \ y_2)$ occur with the same relative frequency, we assign probability $1/36$ to each sample point. For this simple example, the intersection $(y_1, \ y_2)$ contains at most one sample point. Hence the bivariate probability function is

$$p(y_1, y_2) = P(Y_1 = y_1, Y_2 = y_2) = 1/36, \qquad y_1 = 1, 2, \ldots, 6, \quad y_2 = 1, 2, \ldots, 6.$$

A graph of the bivariate probability distribution for the die-tossing experiment is shown in Figure 5.1. Notice that a nonzero probability is assigned to a point $(y_1, \ y_2)$ in the plane if and only if $y_1 = 1, 2, \ldots, 6$ and $y_2 = 1, 2, \ldots, 6$. Thus, exactly 36 points in the plane are assigned nonzero probabilities. Further, the probabilities are assigned in such a way that the sum of the nonzero probabilities is equal to 1. In Figure 5.1 the points assigned nonzero probabilities are represented in the $(y_1, \ y_2)$ plane, whereas the probabilities associated with these points are given by the lengths of the lines above them. Figure 5.1 may be viewed as a theoretical, three-dimensional relative frequency histogram for the pairs of observations $(y_1, \ y_2)$. As in the single-variable discrete case, the theoretical histogram provides a model for the sample histogram that would be obtained if the die-tossing experiment were repeated a large number of times.

FIGURE **5.1**
Bivariate probability
distribution; $y_1 =$
number of dots on
die 1, y_2 = number
of dots on die 2

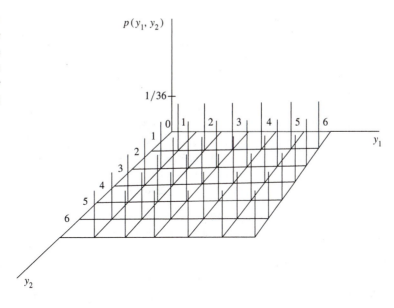

DEFINITION **5.1**

Let Y_1 and Y_2 be discrete random variables. The *joint* (or bivariate) *probability distribution* for Y_1 and Y_2 is given by

$$p(y_1, y_2) = P(Y_1 = y_1, \; Y_2 = y_2), \qquad -\infty < y_1 < \infty, \; -\infty < y_2 < \infty.$$

The function $p(y_1, y_2)$ will be referred to as the *joint probability function*.

In the single-variable case discussed in Chapter 3, we saw that the probability function for a discrete random variable Y assigns nonzero probabilities to a countable number of distinct values of Y in such a way that the sum of the probabilities is equal to 1. Similarly, in the bivariate case the joint probability function $p(y_1, y_2)$ assigns nonzero probabilities to only a countable number of pairs of values (y_1, y_2). Further, the nonzero probabilities must sum to 1.

THEOREM **5.1**

If Y_1 and Y_2 are discrete random variables with joint probability function $p(y_1, y_2)$, then

1. $p(y_1, y_2) \geq 0$ for all y_1, y_2.
2. $\sum_{y_1, y_2} p(y_1, y_2) = 1$, where the sum is over all values (y_1, y_2) that are assigned nonzero probabilities.

Once the joint probability function has been determined for discrete random variables Y_1 and Y_2, calculating joint probabilities involving Y_1 and Y_2 is straightforward.

For the die-tossing experiment, $P(2 \leq Y_1 \leq 3, \; 1 \leq Y_2 \leq 2)$ is

$$P(2 \leq Y_1 \leq 3, \; 1 \leq Y_2 \leq 2) = p(2, 1) + p(2, 2) + p(3, 1) + p(3, 2)$$
$$= 4/36 = 1/9.$$

EXAMPLE **5.1**   A local supermarket has three checkout counters. Two customers arrive at the counters at different times when the counters are serving no other customers. Each customer chooses a counter at random, independently of the other. Let Y_1 denote the number of customers who choose counter 1 and Y_2, the number who select counter 2. Find the joint distribution of Y_1 and Y_2.

Solution   We might proceed with the derivation in many ways. The most direct is to consider the sample space associated with the experiment. Let the pair $\{i, j\}$ denote the simple event that the first customer chose counter i and the second customer chose counter j, where $i, j = 1, 2, 3$. Using the *mn* rule, the sample space consists of $3 \times 3 = 9$ sample points. Under the assumptions given earlier, each sample point is equally likely and has probability $1/9$. The sample space associated with the experiment is

$$S = [\{1, 1\}, \{1, 2\}, \{1, 3\}, \{2, 1\}, \{2, 2\}, \{2, 3\}, \{3, 1\}, \{3, 2\}, \{3, 3\}].$$

Notice that sample point $\{1, 1\}$ is the only sample point corresponding to ($Y_1 = 2, Y_2 = 0$) and hence $P(Y_1 = 2, Y_2 = 0) = 1/9$. Similarly, $P(Y_1 = 1, Y_2 = 1) = P(\{1, 2\}$ or $\{2, 1\}) = 2/9$. Table 5.1 contains the probabilities associated with each possible pair of values for Y_1 and Y_2—that is, the joint probability distribution for Y_1 and Y_2. As always, the results of Theorem 5.1 hold for this example.

As in the case of univariate random variables, the distinction between jointly discrete and jointly continuous random variables may be characterized in terms of their (joint) distribution functions.

DEFINITION **5.2**   For any random variables Y_1 and Y_2, the joint (bivariate) distribution function $F(y_1, y_2)$ is given by

$$F(y_1, y_2) = P(Y_1 \leq y_1, Y_2 \leq y_2), \qquad -\infty < y_1 < \infty, \quad -\infty < y_2 < \infty.$$

Table 5.1. Probability distribution for Y_1 and Y_2, Example 5.1

	y_1		
y_2	0	1	2
0	1/9	2/9	1/9
1	2/9	2/9	0
2	1/9	0	0

For two discrete variables Y_1 and Y_2, $F(y_1, y_2)$ has the form

$$F(y_1, y_2) = \sum_{t_1=-\infty}^{y_1} \sum_{t_2=-\infty}^{y_2} p(t_1, t_2).$$

For the die-tossing experiment,

$$F(2, 3) = P(Y_1 \le 2, Y_2 \le 3)$$
$$= p(1, 1) + p(1, 2) + p(1, 3) + p(2, 1) + p(2, 2) + p(2, 3).$$

Because $p(y_1, y_2) = 1/36$ for all pairs of values of y_1 and y_2 under consideration, $F(2, 3) = 6/36 = 1/6$.

EXAMPLE **5.2**   Consider the random variables Y_1 and Y_2 of Example 5.1. Find $F(-1, 2)$, $F(1.5, 2)$, and $F(5, 7)$.

Solution   Using the results in Table 5.1, we see that

$$F(-1, 2) = P(Y_1 \le -1, Y_2 \le 2) = P(\emptyset) = 0.$$

Further,

$$F(1.5, 2) = P(Y_1 \le 1.5, Y_2 \le 2)$$
$$= p(0, 0) + p(0, 1) + p(0, 2) + p(1, 0) + p(1, 1) + p(1, 2) = 8/9.$$

Similarly,

$$F(5, 7) = P(Y_1 \le 5, Y_2 \le 7) = 1.$$

Notice that $F(y_1, y_2) = 1$ for all y_1, y_2 such that $\min\{y_1, y_2\} \ge 2$. Also, $F(y_1, y_2) = 0$ if $\min\{y_1, y_2\} < 0$.

Two random variables are said to be jointly continuous if their joint distribution function $F(y_1, y_2)$ is continuous in both arguments.

DEFINITION **5.3**   Let Y_1 and Y_2 be continuous random variables with joint distribution function $F(y_1, y_2)$. If there exists a nonnegative function $f(y_1, y_2)$ such that

$$F(y_1, y_2) = \int_{-\infty}^{y_1} \int_{-\infty}^{y_2} f(t_1, t_2)\, dt_2\, dt_1$$

for all $-\infty < y_1 < \infty, -\infty < y_2 < \infty$,

then Y_1 and Y_2 are said to be *jointly continuous random variables*. The function $f(y_1, y_2)$ is called the *joint probability density function*.

Bivariate cumulative distribution functions satisfy a set of properties similar to those specified for univariate cumulative distribution functions.

THEOREM 5.2

If Y_1 and Y_2 are random variables with joint distribution function $F(y_1, y_2)$, then

1. $F(-\infty, -\infty) = F(-\infty, y_2) = F(y_1, -\infty) = 0$.
2. $F(\infty, \infty) = 1$.
3. if $y_1^* \geq y_1$ and $y_2^* \geq y_2$, then

$$F(y_1^*, y_2^*) - F(y_1^*, y_2) - F(y_1, y_2^*) + F(y_1, y_2) \geq 0.$$

Part 3 follows because

$$F(y_1^*, y_2^*) - F(y_1^*, y_2) - F(y_1, y_2^*) + F(y_1, y_2)$$
$$= P(y_1 < Y_1 \leq y_1^*, \; y_2 < Y_2 \leq y_2^*) \geq 0.$$

Notice that $F(\infty, \infty) \equiv \lim_{y_1 \to \infty} \lim_{y_2 \to \infty} F(y_1, y_2) = 1$ implies that the joint density function $f(y_1, y_2)$ must be such that the integral of $f(y_1, y_2)$ over all values of (y_1, y_2) is 1.

THEOREM 5.3

If Y_1 and Y_2 are jointly continuous random variables with a joint density function given by $f(y_1, y_2)$, then

1. $f(y_1, y_2) \geq 0$ for all y_1, y_2.
2. $\displaystyle\int_{-\infty}^{\infty} \int_{-\infty}^{\infty} f(y_1, y_2) \, dy_1 \, dy_2 = 1.$

As in the univariate continuous case discussed in Chapter 4, the joint density function may be intuitively interpreted as a model for the joint relative frequency histogram for Y_1 and Y_2.

For the univariate continuous case, areas under the probability density over an interval correspond to probabilities. Similarly, the bivariate probability density function $f(y_1, y_2)$ traces a probability density surface over the (y_1, y_2) plane (Figure 5.2). Volumes under this surface correspond to probabilities. Thus, $P(a_1 \leq Y_1 \leq a_2, \; b_1 \leq Y_2 \leq b_2)$ is the shaded volume shown in Figure 5.2 and is equal to

$$\int_{b_1}^{b_2} \int_{a_1}^{a_2} f(y_1, y_2) \, dy_1 \, dy_2.$$

FIGURE **5.2**
A bivariate density
function $f(y_1, y_2)$

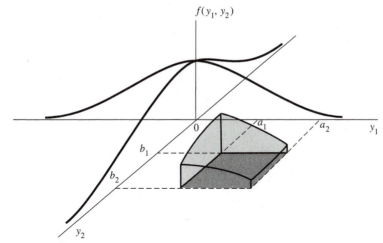

EXAMPLE **5.3**   Suppose that a radioactive particle is randomly located in a square with sides of unit
length. That is, if two regions of equal area are considered, the particle is equally
likely to be in either region. Let Y_1 and Y_2 denote the coordinates of the particle's
location. A reasonable model for the relative frequency histogram for Y_1 and Y_2 is
the bivariate analogue of the univariate uniform distribution:

$$f(y_1, y_2) = \begin{cases} 1, & 0 \le y_1 \le 1, \quad 0 \le y_2 \le 1 \\ 0, & \text{elsewhere.} \end{cases}$$

a Sketch the probability density surface.

b Find $F(.2, .4)$.

c Find $P(.1 \le Y_1 \le .3, \ 0 \le Y_2 \le .5)$.

Solution   **a** The sketch is shown in Figure 5.3.

b
$$F(.2, .4) = \int_{-\infty}^{.4} \int_{-\infty}^{.2} f(y_1, y_2)\, dy_1\, dy_2$$

$$= \int_{0}^{.4} \int_{0}^{.2} (1)\, dy_1\, dy_2$$

$$= \int_{0}^{.4} \left(y_1 \Big]_{0}^{.2} \right) dy_2 = \int_{0}^{.4} .2\, dy_2 = .08.$$

The probability $F(.2, .4)$ corresponds to the volume under $f(y_1, y_2) = 1$,
which is shaded in Figure 5.3. As geometric considerations indicate, the desired probability (volume) is equal to .08, which we formally obtained at the
beginning of this part.

FIGURE **5.3**
Geometric
representation
of $f(y_1, y_2)$,
Example 5.3

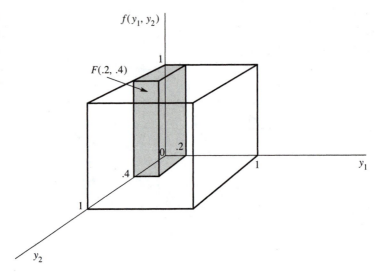

$$\mathbf{c} \qquad P(.1 \le Y_1 \le .3,\, 0 \le Y_2 \le .5) = \int_0^{.5}\!\!\int_{.1}^{.3} f(y_1,\, y_2)\,dy_1\,dy_2$$

$$= \int_0^{.5}\!\!\int_{.1}^{.3} 1\,dy_1\,dy_2 = .10.$$

This probability corresponds to the volume under the density function $f(y_1,\, y_2) = 1$ that is above the region $.1 \le y_1 \le .3$, $0 \le y_2 \le .5$. Like the solution in (b), the current solution can be obtained by using elementary geometric concepts. The density or height of the surface is equal to 1, and hence the desired probability (volume) is

$$P(.1 \le Y_1 \le .3,\, 0 \le Y_2 \le .5) = (.2)(.5)(1) = .10.$$

A slightly more complicated bivariate model is illustrated in the following example.

EXAMPLE **5.4**   Gasoline is to be stocked in a bulk tank once at the beginning of each week and then sold to individual customers. Let Y_1 denote the proportion of the capacity of the bulk tank that is available after the tank is stocked at the beginning of the week. Because of the limited supplies, Y_1 varies from week to week. Let Y_2 denote the proportion of the capacity of the bulk tank that is sold during the week. Because Y_1 and Y_2 are both proportions, both variables take on values between 0 and 1. Further, the amount sold, y_2, cannot exceed the amount available, y_1. Suppose that a model for the relative frequency histogram for Y_1 and Y_2 is given by

FIGURE **5.4**
The joint density
function for
Example 5.4

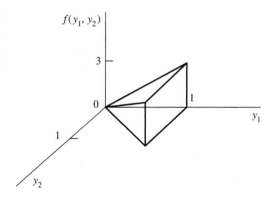

$$f(y_1, y_2) = \begin{cases} 3y_1, & 0 \le y_2 \le y_1 \le 1 \\ 0, & \text{elsewhere.} \end{cases}$$

A sketch of this function is given in Figure 5.4.

Find the probability that less than one-half of the tank will be stocked but more than one-quarter of the tank will be sold.

Solution   We want to find $P(0 \le Y_1 \le .5, Y_2 > .25)$. For any continuous random variable, the probability of observing a value in a region is the volume under the density function above the region of interest. The density function $f(y_1, y_2)$ is positive only in the large triangular portion of the (y_1, y_2) plane shown in Figure 5.5. We are interested only in values of y_1 and y_2 such that $0 \le y_1 \le .5$ and $y_2 > .25$. The intersection of this region and the region where the density function is positive is given by the small (shaded) triangle in Figure 5.5. Consequently, the probability we desire is the volume under the density function of Figure 5.4 above the shaded region in the (y_1, y_2) plane shown in Figure 5.5.

Thus, we have

$$P(0 \le Y_1 \le .5, .25 \le Y_2) = \int_{1/4}^{1/2} \int_{1/4}^{y_1} 3y_1 \, dy_2 \, dy_1$$

$$= \int_{1/4}^{1/2} 3y_1 \left(y_2 \Big]_{1/4}^{y_1} \right) dy_1$$

$$= \int_{1/4}^{1/2} 3y_1 (y_1 - 1/4) \, dy_1$$

$$= [y_1^3 - (3/8)y_1^2] \Big]_{1/4}^{1/2}$$

$$= [(1/8) - (3/8)(1/4)] - [(1/64) - (3/8)(1/16)]$$

$$= 5/128.$$

FIGURE **5.5**
Region of integration
for Example 5.4

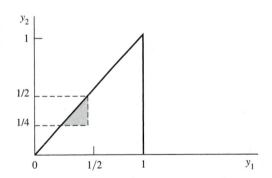

Calculating the probability specified in Example 5.4 involved integrating the joint density function for Y_1 and Y_2 over the appropriate region. The specification of the limits of integration was made easier by sketching the region of integration in Figure 5.5. This approach, sketching the appropriate region of integration, often facilitates setting up the appropriate integral.

The methods discussed in this section can be used to calculate the probability of the intersection of two events $(Y_1 = y_1, \ Y_2 = y_2)$. In a like manner, we can define a probability function (or probability density function) for the intersection of n events $(Y_1 = y_1, \ Y_2 = y_2, \ldots, Y_n = y_n)$. The probability function corresponding to the discrete case is given by

$$p(y_1, \ y_2, \ldots, \ y_n) = P(Y_1 = y_1, \ Y_2 = y_2, \ldots, \ Y_n = y_n).$$

The joint density function of $Y_1, Y_2, \ldots, Y_n$ is given by $f(y_1, \ y_2, \ldots, \ y_n)$. As in the bivariate case, these functions provide models for the joint relative frequency distributions of the populations of joint observations $(y_1, y_2, \ldots, y_n)$ for the discrete case and the continuous case, respectively. In the continuous case,

$$P(Y_1 \le y_1, \ Y_2 \le y_2, \ldots, \ Y_n \le y_n) = F(y_1, \ldots, y_n)$$

$$= \int_{-\infty}^{y_1} \int_{-\infty}^{y_2} \cdots \int_{-\infty}^{y_n} f(t_1, \ t_2, \ldots, \ t_n) \, dt_n \ldots dt_1$$

for every set of real numbers $(y_1, y_2, \ldots, y_n)$. Multivariate distribution functions defined by this equality satisfy properties similar to those specified for the bivariate case.

Exercises

5.1 Contracts for two construction jobs are randomly assigned to one or more of three firms, A, B, and C. Let Y_1 denote the number of contracts assigned to firm A, and Y_2 the number of contracts assigned to firm B. Recall that each firm can receive 0, 1, or 2 contracts.

a Find the joint probability function for Y_1 and Y_2.

b Find $F(1, 0)$.

5.2 Three balanced coins are tossed independently. One of the variables of interest is Y_1, the number of heads. Let Y_2 denote the amount of money won on a side bet in the following manner. If the first head occurs on the first toss, you win $1. If the first head occurs on toss 2 or on toss 3 you win $2 or $3, respectively. If no heads appear, you lose $1 (that is, win −$1).

    a   Find the joint probability function for Y_1 and Y_2.

    b   What is the probability that fewer than three heads will occur and you will win $1 or less? [That is, find $F(2, 1)$.]

5.3 Of nine executives in a certain business firm, four are married, three have never married, and two are divorced. Three of the executives are to be selected for promotion. Let Y_1 denote the number of married executives and Y_2, the number of never-married executives among the three selected for promotion. Assuming that the three are randomly selected from the nine available, find the joint probability distribution of Y_1 and Y_2.

5.4 Given here is the joint probability function associated with data obtained in a study of automobile accidents in which a child (under age 5) was in the car and at least one fatality occurred. Specifically, the study focused on whether or not the child survived and what type of seatbelt (if any) he or she used. Define

$$Y_1 = \begin{cases} 0, & \text{if the child survived} \\ 1, & \text{if not} \end{cases} \quad \text{and} \quad Y_2 = \begin{cases} 0, & \text{if no belt used} \\ 1, & \text{if adult belt used} \\ 2, & \text{if car-seat belt used.} \end{cases}$$

Notice that Y_1 is the number of fatalities per child and, since children's car seats usually utilize two belts, Y_2 is the number of seatbelts in use at the time of the accident.

y_2	y_1 0	1	
0	.38	.17	.55
1	.14	.02	.16
2	.24	.05	.29
	.76	.24	1.00

    a   Verify that the preceding probability function satisfies Theorem 5.1.

    b   Find $F(1, 2)$. What is the interpretation of this value?

5.5 Refer to Example 5.4. The joint density of Y_1, the proportion of the capacity of the tank that is stocked at the beginning of the week, and Y_2, the proportion of the capacity sold during the week, is given by

$$f(y_1, y_2) = \begin{cases} 3y_1, & 0 \le y_2 \le y_1 \le 1 \\ 0, & \text{elsewhere.} \end{cases}$$

    a   Find $F(1/2, 1/3) = P(Y_1 \le 1/2, Y_2 \le 1/3)$.

    b   Find $P(Y_2 \le Y_1/2)$, the probability that the amount sold is less than half the amount purchased.

5.6 Let Y_1 and Y_2 have the joint probability density function given by

$$f(y_1, y_2) = \begin{cases} ky_1y_2, & 0 \le y_1 \le 1, \quad 0 \le y_2 \le 1 \\ 0, & \text{elsewhere.} \end{cases}$$

    a Find the value of k that makes this a probability density function.

    b Find the joint distribution function for Y_1 and Y_2.

    c Find $P(Y_1 \leq 1/2, Y_2 \leq 3/4)$.

5.7   Let Y_1 and Y_2 have the joint probability density function given by

$$f(y_1, y_2) = \begin{cases} k(1 - y_2), & 0 \leq y_1 \leq y_2 \leq 1 \\ 0, & \text{elsewhere.} \end{cases}$$

    a Find the value of k that makes this a probability density function.

    b Find $P(Y_1 \leq 3/4, Y_2 \geq 1/2)$.

5.8   An environmental engineer measures the amount (by weight) of particulate pollution in air samples of a certain volume collected over two smokestacks at a coal-operated power plant. One of the stacks is equipped with a cleaning device. Let Y_1 denote the amount of pollutant per sample collected above the stack that has no cleaning device, and let Y_2 denote the amount of pollutant per sample collected above the stack that is equipped with the cleaning device. Suppose that the relative frequency behavior of Y_1 and Y_2 can be modeled by

$$f(y_1, y_2) = \begin{cases} k, & 0 \leq y_1 \leq 2, \quad 0 \leq y_2 \leq 1, \quad 2y_2 \leq y_1 \\ 0, & \text{elsewhere.} \end{cases}$$

That is, Y_1 and Y_2 are uniformly distributed over the region inside the triangle bounded by $y_1 = 2$, $y_2 = 0$, and $2y_2 = y_1$.

    a Find the value of k that makes this function a probability density function.

    b Find $P(Y_1 \geq 3Y_2)$. That is, find the probability that the cleaning device reduces the amount of pollutant by one-third or more.

5.9   Suppose that Y_1 and Y_2 are uniformly distributed over the triangle shaded in the accompanying diagram.

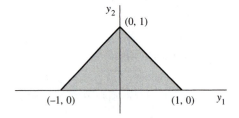

    a Find $P(Y_1 \leq 3/4, Y_2 \leq 3/4)$.

    b Find $P(Y_1 - Y_2 \geq 0)$.

5.10 Let Y_1 and Y_2 denote the proportions of two different types of components in a sample from a mixture of chemicals used as an insecticide. Suppose that Y_1 and Y_2 have the joint density function given by

$$f(y_1, y_2) = \begin{cases} 2, & 0 \leq y_1 \leq 1, \quad 0 \leq y_2 \leq 1, \quad 0 \leq y_1 + y_2 \leq 1 \\ 0, & \text{elsewhere.} \end{cases}$$

(Notice that $Y_1 + Y_2 \leq 1$ because the random variables denote proportions within the same sample.)

   a Find $P(Y_1 \leq 3/4, Y_2 \leq 3/4)$.

   b Find $P(Y_1 \leq 1/2, Y_2 \leq 1/2)$.

5.11 The joint density function of Y_1 and Y_2 is given by

$$f(y_1, y_2) = \begin{cases} 30y_1 y_2^2, & y_1 - 1 \leq y_2 \leq 1 - y_1, \quad 0 \leq y_1 \leq 1 \\ 0, & \text{elsewhere.} \end{cases}$$

   a Find $F(1/2, 1/2)$.

   b Find $F(1/2, 2)$.

   c Find $P(Y_1 > Y_2)$.

5.12 Suppose that the random variables Y_1 and Y_2 have joint probability density function $f(y_1, y_2)$ given by

$$f(y_1, y_2) = \begin{cases} 6y_1^2 y_2, & 0 \leq y_1 \leq y_2, \quad y_1 + y_2 \leq 2 \\ 0, & \text{elsewhere.} \end{cases}$$

   a Verify that this is a valid joint density function.

   b What is the probability that $Y_1 + Y_2$ is less than 1?

5.13 The management at a fast-food outlet is interested in the joint behavior of the random variables Y_1, defined as the total time between a customer's arrival at the store and departure from the service window, and Y_2, the time a customer waits in line before reaching the service window. Because Y_1 includes the time a customer waits in line, we must have $Y_1 \geq Y_2$. The relative frequency distribution of observed values of Y_1 and Y_2 can be modeled by the probability density function

$$f(y_1, y_2) = \begin{cases} e^{-y_1}, & 0 \leq y_2 \leq y_1 < \infty \\ 0, & \text{elsewhere} \end{cases}$$

with time measured in minutes.

   a Find $P(Y_1 < 2, Y_2 > 1)$.

   b Find $P(Y_1 \geq 2Y_2)$.

   c Find $P(Y_1 - Y_2 \geq 1)$. (Notice that $Y_1 - Y_2$ denotes the time spent at the service window.)

5.14 Let Y_1 and Y_2 denote the proportions of time (out of one workday) during which employees I and II, respectively, perform their assigned tasks. The joint relative frequency behavior of Y_1 and Y_2 is modeled by the density function

$$f(y_1, y_2) = \begin{cases} y_1 + y_2, & 0 \leq y_1 \leq 1, \quad 0 \leq y_2 \leq 1 \\ 0, & \text{elsewhere.} \end{cases}$$

   a Find $P(Y_1 < 1/2, Y_2 > 1/4)$.

   b Find $P(Y_1 + Y_2 \leq 1)$.

5.15 Let (Y_1, Y_2) denote the coordinates of a point chosen at random inside a unit circle whose center is at the origin. That is, Y_1 and Y_2 have a joint density function given by

$$f(y_1, y_2) = \begin{cases} \dfrac{1}{\pi}, & y_1^2 + y_2^2 \leq 1 \\ 0, & \text{elsewhere.} \end{cases}$$

Find $P(Y_1 \leq Y_2)$.

5.16 An electronic system has one each of two different types of components in joint operation. Let Y_1 and Y_2 denote the random lengths of life of the components of type I and type II, respectively. The joint density function is given by

$$f(y_1, y_2) = \begin{cases} (1/8)y_1 e^{-(y_1+y_2)/2}, & y_1 > 0, \quad y_2 > 0 \\ 0, & \text{elsewhere.} \end{cases}$$

(Measurements are in hundreds of hours.) Find $P(Y_1 > 1, \ Y_2 > 1)$.

5.3 Marginal and Conditional Probability Distributions

Recall that the distinct values assumed by a discrete random variable represent mutually exclusive events. Similarly, for all distinct pairs of values y_1, y_2, the bivariate events $(Y_1 = y_1, Y_2 = y_2)$, represented by (y_1, y_2), are mutually exclusive events. It follows that the univariate event $(Y_1 = y_1)$ is the union of bivariate events of the type $(Y_1 = y_1, Y_2 = y_2)$, with the union being taken over all possible values for y_2.

For example, reconsider the die-tossing experiment of Section 5.2, where

$$Y_1 = \text{number of dots on the upper face of die 1}$$

$$Y_2 = \text{number of dots on the upper face of die 2.}$$

Then

$$P(Y_1 = 1) = p(1, 1) + p(1, 2) + p(1, 3) + \cdots + p(1, 6)$$

$$= 1/36 + 1/36 + 1/36 + \cdots + 1/36 = 6/36 = 1/6$$

$$P(Y_1 = 2) = p(2, 1) + p(2, 2) + p(2, 3) + \cdots + p(2, 6) = 1/6$$

$$\vdots$$

$$P(Y_1 = 6) = p(6, 1) + p(6, 2) + p(6, 3) + \cdots + p(6, 6) = 1/6.$$

Expressed in summation notation, probabilities about the variable Y_1 alone are

$$P(Y_1 = y_1) = p_1(y_1) = \sum_{y_2=1}^{6} p(y_1, y_2).$$

Similarly, probabilities corresponding to values of the variable Y_2 alone are given by

$$p_2(y_2) = P(Y_2 = y_2) = \sum_{y_1=1}^{6} p(y_1, y_2).$$

Summation in the discrete case corresponds to integration in the continuous case, which leads us to the following definition.

DEFINITION 5.4

a Let Y_1 and Y_2 be jointly discrete random variables with probability function $p(y_1, y_2)$. Then the *marginal probability functions* of Y_1 and Y_2, respectively, are given by

$$p_1(y_1) = \sum_{y_2} p(y_1, y_2) \quad \text{and} \quad p_2(y_2) = \sum_{y_1} p(y_1, y_2).$$

b Let Y_1 and Y_2 be jointly continuous random variables with joint density function $f(y_1, y_2)$. Then the *marginal density functions* of Y_1 and Y_2, respectively, are given by

$$f_1(y_1) = \int_{-\infty}^{\infty} f(y_1, y_2)\, dy_2 \quad \text{and} \quad f_2(y_2) = \int_{-\infty}^{\infty} f(y_1, y_2)\, dy_1.$$

The term *marginal*, as applied to the univariate probability distributions of Y_1 and Y_2, has intuitive meaning. To find $p_1(y_1)$, we sum $p(y_1, y_2)$ over all values of y_2 and hence accumulate the probabilities on the y_1 axis (or margin). The discrete and continuous cases are illustrated in the following two examples.

EXAMPLE 5.5   From a group of three Republicans, two Democrats, and one independent, a committee of two people is to be randomly selected. Let Y_1 denote the number of Republicans and Y_2, the number of Democrats on the committee. Find the joint probability distribution of Y_1 and Y_2, and then find the marginal distribution of Y_1.

Solution   The probabilities sought here are similar to the hypergeometric probabilities of Chapter 3. For example,

$$P(Y_1 = 1, Y_2 = 1) = p(1, 1) = \frac{\binom{3}{1}\binom{2}{1}\binom{1}{0}}{\binom{6}{2}} = \frac{3(2)}{15} = \frac{6}{15}$$

because there are 15 equally likely sample points; for the event in question we must select one Republican from the three, one Democrat from the two, and zero independents. Similar calculations lead to the other probabilities shown in Table 5.2.

To find $p_1(y_1)$, we must sum over the values of Y_2, as Definition 5.4 indicates. Hence these probabilities are given by the column totals in Table 5.2. That is,

$$p_1(0) = p(0, 0) + p(0, 1) + p(0, 2) = 0 + 2/15 + 1/15 = 3/15.$$

Table 5.2. Joint probability function for Y_1 and Y_2, Example 5.5

		y_1		
y_2	0	1	2	Total
0	0	3/15	3/15	6/15
1	2/15	6/15	0	8/15
2	1/15	0	0	1/15
Total	3/15	9/15	3/15	1

Similarly,

$$p_1(1) = 9/15 \quad \text{and} \quad p_1(2) = 3/15.$$

Notice that the marginal distribution of Y_2 is given by the row totals.

EXAMPLE **5.6**   Let

$$f(y_1,\ y_2) = \begin{cases} 2y_1, & 0 \le y_1 \le 1, \quad 0 \le y_2 \le 1 \\ 0, & \text{elsewhere.} \end{cases}$$

Sketch $f(y_1,\ y_2)$ and find the marginal density functions for Y_1 and Y_2.

Solution   Viewed geometrically, $f(y_1,\ y_2)$ traces a wedge-shaped surface, as sketched in Figure 5.6.

Before applying Definition 5.4 to find $f_1(y_1)$ and $f_2(y_2)$, we will use Figure 5.6 to visualize the result. If the probability represented by the wedge were accumulated on the y_1 axis (accumulating probability along lines parallel to the y_2 axis), the result would be a triangular probability distribution that would look like the side of the wedge in Figure 5.6. If the probability were accumulated along the y_2 axis (ac-

FIGURE **5.6**
Geometric
representation
of $f(y_1, y_2)$,
Example 5.6

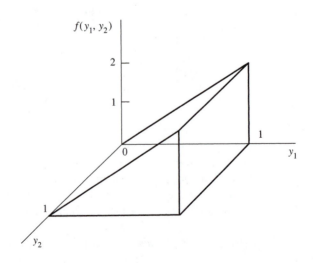

cumulating along lines parallel to the y_1 axis), the resulting distribution would be uniform. We will confirm these visual solutions by applying Definition 5.4. Then, if $0 \leq y_1 \leq 1$,

$$f_1(y_1) = \int_{-\infty}^{\infty} f(y_1, y_2) \, dy_2 = \int_0^1 2y_1 \, dy_2 = 2y_1 \left(y_2 \Big]_0^1 \right)$$

and if $y_1 < 0$ or $y_1 > 1$,

$$f_1(y_1) = \int_{-\infty}^{\infty} f(y_1, y_2) \, dy_2 = \int_0^1 0 \, dy_2 = 0.$$

Thus,

$$f_1(y_1) = \begin{cases} 2y_1, & 0 \leq y_1 \leq 1 \\ 0, & \text{elsewhere.} \end{cases}$$

Similarly, if $0 \leq y_2 \leq 1$,

$$f_2(y_2) = \int_{-\infty}^{\infty} f(y_1, y_2) \, dy_1 = \int_0^1 2y_1 \, dy_1 = y_1^2 \Big]_0^1$$

and if $y_2 < 0$ or $y_2 > 1$,

$$f_2(y_2) = \int_{-\infty}^{\infty} f(y_1, y_2) \, dy_1 = \int_0^1 0 \, dy_1 = 0.$$

Summarizing,

$$f_2(y_2) = \begin{cases} 1, & 0 \leq y_2 \leq 1 \\ 0, & \text{elsewhere.} \end{cases}$$

Graphs of $f_1(y_1)$ and $f_2(y_2)$ trace triangular and uniform probability distributions, respectively, as expected.

We now turn our attention to conditional distributions, looking first at the discrete case.

The multiplicative law (Section 2.8) gives the probability of the intersection $A \cap B$ as

$$P(A \cap B) = P(A)P(B|A)$$

where $P(A)$ is the unconditional probability of A and $P(B|A)$ is the probability of B given that A has occurred. Now consider the intersection of the two numerical events, $(Y_1 = y_1)$ and $(Y_2 = y_2)$, represented by the bivariate event (y_1, y_2). It follows directly from the multiplicative law of probability that the bivariate probability for the intersection $(y_1,\ y_2)$ is

$$p(y_1,\ y_2) = p_1(y_1)p(y_2|y_1)$$
$$= p_2(y_2)p(y_1|y_2).$$

The probabilities $p_1(y_1)$ and $p_2(y_2)$ are associated with the univariate probability distributions for Y_1 and Y_2 individually (recall Chapter 3). Using the interpretation of conditional probability discussed in Chapter 2, $p(y_1|y_2)$ is the probability that the random variable Y_1 equals y_1, given that Y_2 takes on the value y_2.

DEFINITION **5.5**   If Y_1 and Y_2 are jointly discrete random variables with joint probability function $p(y_1,\ y_2)$ and marginal probability functions $p_1(y_1)$ and $p_2(y_2)$, respectively, then the *conditional discrete probability function* of Y_1 given Y_2 is

$$p(y_1|y_2) = P(Y_1 = y_1|Y_2 = y_2) = \frac{P(Y_1 = y_1, Y_2 = y_2)}{P(Y_2 = y_2)} = \frac{p(y_1,\ y_2)}{p_2(y_2)}$$

provided that $p_2(y_2) > 0$.

Thus, $P(Y_1 = 2|Y_2 = 3)$ is the conditional probability that $Y_1 = 2$ given that $Y_2 = 3$. A similar interpretation can be attached to the conditional probability $p(y_2|y_1)$. Note that $p(y_1|y_2)$ is *undefined* if $p_2(y_2) = 0$.

EXAMPLE **5.7**   Refer to Example 5.5 and find the conditional distribution of Y_1 given that $Y_2 = 1$. That is, given that one of the two people on the committee is a Democrat, find the conditional distribution for the number of Republicans selected for the committee.

Solution   The joint probabilities are given in Table 5.2. To find $p(y_1|Y_2 = 1)$, we concentrate on the row corresponding to $Y_2 = 1$. Then

$$P(Y_1 = 0|Y_2 = 1) = \frac{p(0,\ 1)}{p_2(1)} = \frac{2/15}{8/15} = \frac{1}{4}$$

$$P(Y_1 = 1|Y_2 = 1) = \frac{p(1,\ 1)}{p_2(1)} = \frac{6/15}{8/15} = \frac{3}{4}$$

and

$$P(Y_1 \geq 2|Y_2 = 1) = \frac{p(2,\ 1)}{p_2(1)} = 0.$$

In the randomly selected committee, if one person is a Democrat (equivalently, if $Y_2 = 1$), there is a high probability that the other will be a Republican (equivalently, $Y_1 = 1$).

In the continuous case, we can obtain an appropriate analogue of the conditional probability function $p(y_1|y_2)$, but it is not obtained in such a straightforward manner. If Y_1 and Y_2 are continuous, $P(Y_1 = y_1|Y_2 = y_2)$ cannot be defined as in the discrete case, because both $(Y_1 = y_1)$ and $(Y_2 = y_2)$ are events with zero probability. The following considerations, however, do lead to a useful and consistent definition for a conditional density function.

Assuming that Y_1 and Y_2 are jointly continuous with density function $f(y_1, y_2)$, we might be interested in a probability of the form

$$P(Y_1 \leq y_1|Y_2 = y_2) = F(y_1|y_2)$$

which, as a function of y_1 for a fixed y_2, is called the *conditional distribution function* of Y_1 given $Y_2 = y_2$.

DEFINITION   5.6

If Y_1 and Y_2 are jointly continuous random variables with joint density function $f(y_1, y_2)$, then the *conditional distribution function* of Y_1 given $Y_2 = y_2$ is

$$F(y_1|y_2) = P(Y_1 \leq y_1|Y_2 = y_2)$$

Notice that $F(y_1|y_2)$ is a function of y_1 for a fixed value of y_2.

If we could take $F(y_1|y_2)$, multiply by $P(Y_2 = y_2)$ for each possible value of Y_2, and sum all the resulting probabilities, we would obtain $F(y_1)$. This is not possible, because the number of values for y_2 is uncountable and all probabilities $P(Y_2 = y_2)$ are zero. But we can do something analogous by multiplying by $f_2(y_2)$ and then integrating to obtain

$$F(y_1) = \int_{-\infty}^{\infty} F(y_1|y_2) f_2(y_2) \, dy_2.$$

The quantity $f_2(y_2) \, dy_2$ can be thought of as the approximate probability that Y_2 takes on a value in a small interval about y_2, and the integral is a generalized sum.

Now from previous considerations, we know that

$$F(y_1) = \int_{-\infty}^{y_1} f_1(t_1) \, dt_1 = \int_{-\infty}^{y_1} \left[\int_{-\infty}^{\infty} f(t_1, y_2) \, dy_2 \right] dt_1$$

$$= \int_{-\infty}^{\infty} \int_{-\infty}^{y_1} f(t_1, y_2) \, dt_1 \, dy_2.$$

From these two expressions for $F(y_1)$, we must have

$$F(y_1|y_2) f_2(y_2) = \int_{-\infty}^{y_1} f(t_1, y_2) \, dt_1$$

or

$$F(y_1|y_2) = \int_{-\infty}^{y_1} \frac{f(t_1, y_2)}{f_2(y_2)} \, dt_1.$$

We will call the integrand of this expression the *conditional density function* of Y_1 given $Y_2 = y_2$, and we will denote it by $f(y_1|y_2)$.

DEFINITION 5.7

Let Y_1 and Y_2 be jointly continuous random variables with joint density $f(y_1, y_2)$ and marginal densities $f_1(y_1)$ and $f_2(y_2)$, respectively. For any y_2 such that $f_2(y_2) > 0$, the conditional density of Y_1 given $Y_2 = y_2$ is given by

$$f(y_1|y_2) = \frac{f(y_1, y_2)}{f_2(y_2)}$$

and, for any y_1 such that $f_1(y_1) > 0$, the conditional density of Y_2 given $Y_1 = y_1$ is given by

$$f(y_2|y_1) = \frac{f(y_1, y_2)}{f_1(y_1)}.$$

Note that the conditional density $f(y_1|y_2)$ is undefined for all y_2 such that $f_2(y_2) = 0$. Similarly, $f(y_2|y_1)$ is undefined if y_1 is such that $f_1(y_1) = 0$.

EXAMPLE 5.8

A soft-drink machine has a random amount Y_2 in supply at the beginning of a given day and dispenses a random amount Y_1 during the day (with measurements in gallons). It is not resupplied during the day, and hence $Y_1 \leq Y_2$. It has been observed that Y_1 and Y_2 have a joint density given by

$$f(y_1, y_2) = \begin{cases} 1/2, & 0 \leq y_1 \leq y_2 \leq 2 \\ 0 & \text{elsewhere.} \end{cases}$$

That is, the points (y_1, y_2) are uniformly distributed over the triangle with the given boundaries. Find the conditional density of Y_1 given $Y_2 = y_2$. Evaluate the probability that less than $1/2$ gallons will be sold, given that the machine contains 1.5 gallons at the start of the day.

Solution

The marginal density of Y_2 is given by

$$f_2(y_2) = \int_{-\infty}^{\infty} f(y_1, y_2)\, dy_1.$$

Thus,

$$f_2(y_2) = \begin{cases} \int_0^{y_2} (1/2)\, dy_1 = (1/2)y_2, & 0 \leq y_2 \leq 2 \\ \int_{-\infty}^{\infty} 0\, dy_1 = 0, & \text{elsewhere.} \end{cases}$$

Note that $f_2(y_2) > 0$ if and only if $0 < y_2 \leq 2$. Thus, for any $0 < y_2 \leq 2$, using Definition 5.7,

$$f(y_1|y_2) = \frac{f(y_1, y_2)}{f_2(y_2)} = \frac{1/2}{(1/2)(y_2)} = \frac{1}{y_2}, \qquad 0 \le y_1 \le y_2.$$

Also, $f(y_1|y_2)$ is undefined if $y_2 \le 0$ or $y_2 > 2$. The probability of interest is

$$P(Y_1 \le 1/2 | Y_2 = 1.5) = \int_{-\infty}^{1/2} f(y_1|y_2 = 1.5)\, dy_1 = \int_0^{1/2} \frac{1}{1.5}\, dy_1 = \frac{1/2}{1.5} = \frac{1}{3}.$$

If the machine contains 2 gallons at the start of the day, then

$$P(Y_1 \le 1/2 | Y_2 = 2) = \int_0^{1/2} \frac{1}{2}\, dy_1 = \frac{1}{4}.$$

Thus, the conditional probability that $Y_1 \le 1/2$ given $Y_2 = y_2$ changes appreciably depending on the particular choice of y_2.

Exercises

5.17   In Exercise 5.1 we determined that the joint distribution of Y_1, the number of contracts awarded to firm A, and Y_2, the number of contracts awarded to firm B, is given by the entries in the following table.

	y_1		
y_2	0	1	2
0	1/9	2/9	1/9
1	2/9	2/9	0
2	1/9	0	0

a   Find the marginal probability distribution of Y_1.

b   According to results in Chapter 4, Y_1 has a binomial distribution with $n = 2$ and $p = 1/3$. Is there any conflict between this result and the answer you provided in (a)?

5.18   Refer to Exercise 5.2.

a   Derive the marginal probability distribution for your winnings on the side bet.

b   What is the probability that you obtained three heads, given that you won \$1 on the side bet?

5.19   In Exercise 5.3, we determined that the joint probability distribution of Y_1, the number of married executives, and Y_2, the number of never-married executives, is given by

$$p(y_1, y_2) = \frac{\binom{4}{y_1}\binom{3}{y_2}\binom{2}{3 - y_1 - y_2}}{\binom{9}{3}}$$

where y_1 and y_2 are integers, $0 \le y_1 \le 3$, $0 \le y_2 \le 3$, and $1 \le y_1 + y_2 \le 3$.

a Find the marginal probability distribution of Y_1, the number of married executives among the three selected for promotion.

b Find $P(Y_1 = 1|Y_2 = 2)$.

c If we let Y_3 denote the number of divorced executives among the three selected for promotion, then $Y_3 = 3 - Y_1 - Y_2$. Find $P(Y_3 = 1|Y_2 = 1)$.

d Compare the marginal distribution derived in (a) with the hypergeometric distributions with $N = 9, n = 3,$ and $r = 4$ encountered in Section 3.7.

5.20 In Exercise 5.4 you were given the following joint probability function for

$$Y_1 = \begin{cases} 0, & \text{if child survived} \\ 1, & \text{if not} \end{cases} \quad \text{and} \quad Y_2 = \begin{cases} 0, & \text{if no belt used} \\ 1, & \text{if adult belt used} \\ 2, & \text{if car-seat belt used.} \end{cases}$$

	y_1		
y_2	0	1	
0	.38	.17	.55
1	.14	.02	.16
2	.24	.05	.29
	.76	.24	1.00

a Give the marginal probability functions for Y_1 and Y_2.

b Give the conditional probability function for Y_2 given $Y_1 = 0$.

c What is the probability that a child survived given that he or she was in a car-seat belt?

5.21 In Example 5.4 and Exercise 5.5 we considered the joint density of Y_1, the proportion of the capacity of the tank that is stocked at the beginning of the week, and Y_2, the proportion of the capacity sold during the week, given by

$$f(y_1, y_2) = \begin{cases} 3y_1, & 0 \le y_2 \le y_1 \le 1 \\ 0, & \text{elsewhere.} \end{cases}$$

a Find the marginal density function for Y_2.

b For what values of y_2 is the conditional density $f(y_1|y_2)$ defined?

c What is the probability that more than half a tank is sold given that three-fourths of a tank is stocked?

5.22 In Exercise 5.6, we derived the fact that

$$f(y_1, y_2) = \begin{cases} 4y_1y_2, & 0 \le y_1 \le 1, \quad 0 \le y_2 \le 1 \\ 0, & \text{elsewhere} \end{cases}$$

is a valid joint probability density function.

a Find the marginal density functions for Y_1 and Y_2.

b Find $P(Y_1 \le 1/2|Y_2 \ge 3/4)$.

c Find the conditional density function of Y_1 given $Y_2 = y_2$.

d Find the conditional density function of Y_2 given $Y_1 = y_1$.

e Find $P(Y_1 \le 3/4|Y_2 = 1/2)$.

5.23   In Exercise 5.7, we determined that

$$f(y_1, y_2) = \begin{cases} 6(1 - y_2), & 0 \le y_1 \le y_2 \le 1 \\ 0, & \text{elsewhere} \end{cases}$$

is a valid joint probability density function.

a   Find the marginal density functions for Y_1 and Y_2.

b   Find $P(Y_2 \le 1/2 | Y_1 \le 3/4)$.

c   Find the conditional density function of Y_1 given $Y_2 = y_2$.

d   Find the conditional density function of Y_2 given $Y_1 = y_1$.

e   Find $P(Y_2 \ge 3/4 | Y_1 = 1/2)$.

5.24   In Exercise 5.8, we proved that

$$f(y_1, y_2) = \begin{cases} 1, & 0 \le y_1 \le 2, \quad 0 \le y_2 \le 1, \quad 2y_2 \le y_1 \\ 0, & \text{elsewhere} \end{cases}$$

is a valid joint probability density function for Y_1, the amount of pollutant per sample collected above the stack without the cleaning device, and for Y_2, the amount collected above the stack with the cleaner.

a   If we consider the stack with the cleaner installed, find the probability that the amount of pollutant in a given sample will exceed 0.5.

b   Given that the amount of pollutant in a sample taken above the stack with the cleaner is observed to be 0.5, find the probability that the amount of pollutant exceeds 1.5 above the other stack (without the cleaner).

5.25   Refer to Exercise 5.9.

a   Find the marginal density functions for Y_1 and Y_2.

b   Find $P(Y_2 > 1/2 | Y_1 = 1/4)$.

5.26   In Exercise 5.10, we were given the following joint probability density function for the random variables Y_1 and Y_2, which were the proportions of two components in a sample from a mixture of insecticide:

$$f(y_1, y_2) = \begin{cases} 2, & 0 \le y_1 \le 1, \quad 0 \le y_2 \le 1, \quad 0 \le y_1 + y_2 \le 1 \\ 0, & \text{elsewhere.} \end{cases}$$

a   Find $P(Y_1 \ge 1/2 | Y_2 \le 1/4)$.

b   Find $P(Y_1 \ge 1/2 | Y_2 = 1/4)$.

5.27   In Exercise 5.11, the joint density function of Y_1 and Y_2 is given by

$$f(y_1, y_2) = \begin{cases} 30y_1 y_2^2, & y_1 - 1 \le y_2 \le 1 - y_1, 0 \le y_1 \le 1, \\ 0, & \text{elsewhere.} \end{cases}$$

a   Show that the marginal density of Y_1 is a beta density with $\alpha = 2$ and $\beta = 4$.

b   Derive the marginal density of Y_2.

c   Derive the conditional density of Y_2 given $Y_1 = y_1$.

d   Find $P(Y_2 > 0 | Y_1 = .75)$.

5.28   Suppose that the random variables Y_1 and Y_2 have joint probability density function, $f(y_1, y_2)$, given by (see Exercise 5.12)

$$f(y_1, y_2) = \begin{cases} 6y_1^2 y_2, & 0 \le y_1 \le y_2, \quad y_1 + y_2 \le 2 \\ 0, & \text{elsewhere.} \end{cases}$$

a Show that the marginal density of Y_1 is a beta density with $\alpha = 3$ and $\beta = 2$.
b Derive the marginal density of Y_2.
c Derive the conditional density of Y_2 given $Y_1 = y_1$.
d Find $P(Y_2 < 1.1 | Y_1 = .60)$.

5.29 If Y_1 is the total time between a customer's arrival in the store and departure from the service window, and if Y_2 is the time spent in line before reaching the window, the joint density of these variables was given in Exercise 5.13 as

$$f(y_1, y_2) = \begin{cases} e^{-y_1}, & 0 \le y_2 \le y_1 \le \infty \\ 0, & \text{elsewhere.} \end{cases}$$

If 2 minutes elapse between a customer's arrival at the store and his departure from the service window, find the probability that he waited in line less than 1 minute to reach the window.

5.30 In Exercise 5.14, Y_1 and Y_2 denoted the proportions of time during which employees I and II actually performed their assigned tasks during a workday. The joint density of Y_1 and Y_2 is given by

$$f(y_1, y_2) = \begin{cases} y_1 + y_2, & 0 \le y_1 \le 1, \quad 0 \le y_2 \le 1 \\ 0, & \text{elsewhere.} \end{cases}$$

a Find the marginal density functions for Y_1 and Y_2.
b Find $P(Y_1 \ge 1/2 | Y_2 \ge 1/2)$.
c If employee II spends exactly 50% of the day working on assigned duties, find the probability that employee I spends more than 75% of the day working on similar duties.

5.31 In Exercise 5.16, Y_1 and Y_2 denoted the lengths of life, in hours, for components of types I and II, respectively, in an electronic system. The joint density of Y_1 and Y_2 is given by

$$f(y_1, y_2) = \begin{cases} (1/8)y_1 e^{-(y_1+y_2)/2}, & y_1 > 0, \; y_2 > 0 \\ 0, & \text{elsewhere.} \end{cases}$$

Find the probability that a component of type II will have a life length in excess of 200 hours.

5.32 Let Y_1 denote the weight (in tons) of a bulk item stocked by a supplier at the beginning of a week, and suppose that Y_1 has a uniform distribution over the interval $0 \le y_1 \le 1$. Let Y_2 denote the amount (by weight) of this item sold by the supplier during the week, and suppose that Y_2 has a uniform distribution over the interval $0 \le y_2 \le y_1$, where y_1 is a specific value of Y_1.

a Find the joint density function for Y_1 and Y_2.
b If the supplier stocks a half-ton of the item, what is the probability that she sells more than a quarter-ton?
c If it is known that the supplier sold a quarter-ton of the item, what is the probability that she had stocked more than a half-ton?

*5.33 Suppose that Y_1 and Y_2 are independent Poisson distributed random variables with means λ_1 and λ_2, respectively. Let $W = Y_1 + Y_2$. In Chapter 6 you will show that W has a Poisson distribution with mean $\lambda_1 + \lambda_2$. Use this result to show that the conditional distribution of Y_1, given that $W = w$, is a binomial distribution with $n = w$ and $p = \lambda_1/(\lambda_1 + \lambda_2)$.[1]

*5.34   Suppose that Y_1 and Y_2 are independent binomial distributed random variables based on samples of sizes n_1 and n_2, respectively. Suppose that $p_1 = p_2 = p$. That is, the probability of "success" is the same for the two random variables. Let $W = Y_1 + Y_2$. In Chapter 6 you will prove that W has a binomial distribution with success probability p and sample size $n_1 + n_2$. Use this result to show that the conditional distribution of Y_1, given that $W = w$, is a hypergeometric distribution with $N = n_1 + n_2, n = w$, and $r = n_1$.

*5.35   A quality control plan calls for randomly selecting three items from the daily production (assumed large) of a certain machine and observing the number of defectives. However, the proportion p of defectives produced by the machine varies from day to day and is assumed to have a uniform distribution on the interval $(0, 1)$. For a randomly chosen day, find the unconditional probability that exactly two defectives are observed in the sample.

*5.36   The number of defects per yard Y for a certain fabric is known to have a Poisson distribution with parameter λ. However, λ itself is a random variable with probability density function given by

$$f(\lambda) = \begin{cases} e^{-\lambda}, & \lambda \geq 0 \\ 0, & \text{elsewhere.} \end{cases}$$

Find the unconditional probability function for Y.

5.4   Independent Random Variables

In Example 5.8 we saw two dependent random variables, for which probabilities associated with Y_1 depended on the observed value of Y_2. We now present a formal definition of independence of random variables.

Two events A and B are independent if $P(A \cap B) = P(A)P(B)$. When discussing random variables, we are often concerned with events of the type $(a \leq Y_1 \leq b) \cap (c \leq Y_2 \leq d)$. For consistency with the earlier definition of independent events, if Y_1 and Y_2 are independent, we would like to have

$$P(a \leq Y_1 \leq b, \ c \leq Y_2 \leq d) = P(a \leq Y_1 \leq b)P(c \leq Y_2 \leq d)$$

for any choice of real numbers $a, b, c,$ and d. That is, if Y_1 and Y_2 are independent, the joint probability can be written as the product of the marginal probabilities. This property will be satisfied if Y_1 and Y_2 are independent in the sense detailed in the following definition.

DEFINITION 5.8   Let Y_1 have distribution function $F_1(y_1)$, Y_2 have distribution function $F_2(y_2)$, and Y_1 and Y_2 have joint distribution function $F(y_1, y_2)$. Then Y_1 and Y_2 are said to be *independent* if and only if

---

[1]Exercises preceded by an asterisk are optional.

$$F(y_1, y_2) = F_1(y_1)F_2(y_2)$$

for every pair of real numbers (y_1, y_2).

If Y_1 and Y_2 are not independent, they are said to be *dependent*.

It usually is convenient to establish independence, or the lack of it, by using the result contained in the following theorem. The proof is omitted; see the references at the end of the chapter.

THEOREM **5.4**   If Y_1 and Y_2 are discrete random variables with joint probability function $p(y_1, y_2)$ and marginal probability functions $p_1(y_1)$ and $p_2(y_2)$, respectively, then Y_1 and Y_2 are the independent if and only if

$$p(y_1, y_2) = p_1(y_1)p_2(y_2)$$

for all pairs of real numbers (y_1, y_2).

If Y_1 and Y_2 are continuous random variables with joint density function $f(y_1, y_2)$ and marginal density functions $f_1(y_1)$ and $f_2(y_2)$, respectively, then Y_1 and Y_2 are independent if and only if

$$f(y_1, y_2) = f_1(y_1)f_2(y_2)$$

for all pairs of real numbers (y_1, y_2).

We now illustrate the concept of independence with some examples.

EXAMPLE **5.9**   For the die-tossing problem of Section 5.2, show that Y_1 and Y_2 are independent.

Solution   In this problem each of the 36 sample points was given probability 1/36. Consider, for example, the point $(1, 2)$. We know that $p(1, 2) = 1/36$. Also, $p_1(1) = P(Y_1 = 1) = 1/6$ and $p_2(2) = P(Y_2 = 2) = 1/6$. Hence,

$$p(1, 2) = p_1(1)p_2(2).$$

The same is true for *all other values* for y_1 and y_2, and it follows that Y_1 and Y_2 are independent.

EXAMPLE **5.10**   Refer to Example 5.5. Is the number of Republicans in the sample independent of the number of Democrats? (Is Y_1 independent of Y_2?)

Solution   Independence of discrete random variables requires that $p(y_1, y_2) = p_1(y_1)p_2(y_2)$ for every choice (y_1, y_2). Thus, if this equality is violated for any pair of values,

(y_1, y_2), the random variables are dependent. Looking in the upper left-hand corner of Table 5.2, we see

$$p(0, 0) = 0.$$

But $p_1(0) = 3/15$ and $p_2(0) = 6/15$. Hence,

$$p(0, 0) \neq p_1(0)p_2(0)$$

so Y_1 and Y_2 are dependent.

EXAMPLE **5.11**   Let

$$f(y_1, y_2) = \begin{cases} 6y_1y_2^2, & 0 \leq y_1 \leq 1; \quad 0 \leq y_2 \leq 1 \\ 0, & \text{elsewhere.} \end{cases}$$

Show that Y_1 and Y_2 are independent.

Solution   We have

$$f_1(y_1) = \begin{cases} \displaystyle\int_{-\infty}^{\infty} f(y_1, y_2)\, dy_2 = \int_0^1 6y_1y_2^2 \, dy_2 = 6y_1 \left(\dfrac{y_2^3}{3} \right]_0^1 \\ \qquad\qquad\qquad\qquad\qquad = 2y_1, & 0 \leq y_1 \leq 1 \\ \displaystyle\int_{-\infty}^{\infty} f(y_1, y_2)\, dy_2 = \int_{-\infty}^{\infty} 0 \, dy_1 = 0, & \text{elsewhere.} \end{cases}$$

Similarly,

$$f_2(y_2) = \begin{cases} \displaystyle\int_{-\infty}^{\infty} f(y_1, y_2)\, dy_1 = \int_0^1 6y_1y_2^2 \, dy_1 = 3y_2^2, & 0 \leq y_2 \leq 1 \\ \displaystyle\int_{-\infty}^{\infty} f(y_1, y_2)\, dy_1 = \int_{-\infty}^{\infty} 0 \, dy_1 = 0, & \text{elsewhere.} \end{cases}$$

Hence,

$$f(y_1, y_2) = f_1(y_1)f_2(y_2)$$

for all real numbers (y_1, y_2), and therefore Y_1 and Y_2 are independent.

EXAMPLE **5.12**   Let

$$f(y_1, y_2) = \begin{cases} 2, & 0 \leq y_2 \leq y_1 \leq 1 \\ 0, & \text{elsewhere.} \end{cases}$$

Show that Y_1 and Y_2 are dependent.

FIGURE 5.7
Region over which
$f(y_1, y_2)$ is positive,
Example 5.12

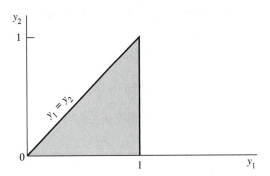

Solution We see that $f(y_1, y_2) = 2$ over the shaded region shown in Figure 5.7. Therefore,

$$f_1(y_1) = \begin{cases} \left. \int_0^{y_1} 2\,dy_2 = 2y_2 \right]_0^{y_1} = 2y_1, & 0 \le y_1 \le 1 \\ 0, & \text{elsewhere.} \end{cases}$$

Similarly,

$$f_2(y_2) = \begin{cases} \left. \int_{y_2}^1 2\,dy_1 = 2y_1 \right]_{y_2}^1 = 2(1 - y_2), & 0 \le y_2 \le 1 \\ 0, & \text{elsewhere.} \end{cases}$$

Hence,

$$f(y_1, y_2) \ne f_1(y_1)f_2(y_2)$$

for some pair of real numbers (y_1, y_2), and therefore Y_1 and Y_2 are dependent.

You will observe a distinct difference in the limits of integration employed in finding the marginal density functions obtained in Examples 5.11 and 5.12. The limits of integration for y_2 involved in finding the marginal density of Y_1 in Example 5.12 depended on y_1. In contrast, the limits of integration were constants when we found the marginal density functions in Example 5.11. If the limits of integration are constants, the following theorem provides an easy way to show independence of two random variables.

THEOREM 5.5
Let Y_1 and Y_2 have a joint density $f(y_1, y_2)$ that is positive if and only if $a \le y_1 \le b$ and $c \le y_2 \le d$, for constants $a, b, c,$ and d; and $f(y_1, y_2) = 0$ otherwise. Then Y_1 and Y_2 are independent random variables if and only if

$$f(y_1, y_2) = g(y_1)h(y_2)$$

where $g(y_1)$ is a nonnegative function of y_1 alone and $h(y_2)$ is a nonnegative function of y_2 alone.

The proof of this theorem is omitted. (See the references at the end of the chapter.) The key benefit of the result given in Theorem 5.5 is that we do not actually need to derive the marginal densities. Indeed, the functions $g(y_1)$ and $h(y_2)$ need not, themselves, be density functions (although they will be constant multiples of the marginal densities, should we go to the bother of determining the marginal densities).

EXAMPLE 5.13   Let Y_1 and Y_2 have a joint density given by

$$f(y_1, y_2) = \begin{cases} 2y_1, & 0 \le y_1 \le 1; \quad 0 \le y_2 \le 1 \\ 0, & \text{elsewhere.} \end{cases}$$

Are Y_1 and Y_2 independent variables?

Solution   Notice that $f(y_1, y_2)$ is positive if and only if $0 \le y_1 \le 1$ and $0 \le y_2 \le 1$. Further, $f(y_1, y_2) = g(y_1)h(y_2)$, where

$$g(y_1) = \begin{cases} y_1, & 0 \le y_1 \le 1 \\ 0, & \text{elsewhere} \end{cases} \quad \text{and} \quad h(y_2) = \begin{cases} 2, & 0 \le y_2 \le 1 \\ 0, & \text{elsewhere.} \end{cases}$$

Therefore, Y_1 and Y_2 are independent random variables. Notice that $g(y_1)$ and $h(y_2)$, as defined here, *are not* density functions, although $2g(y_1)$ and $h(y_2)/2$ are densities.

EXAMPLE 5.14   Refer to Example 5.4. Is Y_1, the amount in stock, independent of Y_2, the amount sold?

Solution   Because the density function is positive if and only if $0 \le y_2 \le y_1 \le 1$, there do not exist *constants* $a, b, c,$ and d such that the density is positive over the region $a \le y_1 \le b, c \le y_2 \le d$. Thus Theorem 5.5 cannot be applied. However, Y_1 and Y_2 can be shown to be dependent random variables because the joint density is not the product of the marginal densities.

Definition 5.8 easily can be generalized to n dimensions. Suppose that we have n random variables, $Y_1, \ldots, Y_n$, where Y_i has distribution function $F_i(y_i)$, for $i = 1, 2, \ldots, n$; and where $Y_1, Y_2, \ldots, Y_n$ have joint distribution function $F(y_1, y_2, \ldots, y_n)$. Then $Y_1, Y_2, \ldots, Y_n$ are independent if and only if

$$F(y_1, y_2, \ldots, y_n) = F_1(y_1) \cdots F_n(y_n)$$

for all real numbers $y_1, y_2, \ldots, y_n$, with the obvious equivalent forms for the discrete and continuous cases.

Exercises

5.37   Let Y_1 and Y_2 have joint density function $f(y_1, y_2)$ and marginal densities $f_1(y_1)$ and $f_2(y_2)$, respectively. Show that Y_1 and Y_2 are independent if and only if $f(y_1|y_2) = f_1(y_1)$ for all values of y_1 and for all y_2 such that $f_2(y_2) > 0$. A completely analogous argument establishes that Y_1 and Y_2 are independent if and only if $f(y_2|y_1) = f_2(y_2)$ for all values of y_2 and for all y_1 such that $f_1(y_1) > 0$.

5.38   Prove that the results in Exercise 5.37 also hold for discrete random variables.

5.39   In Exercise 5.1, we determined that the joint distribution of Y_1, the number of contracts awarded to firm A, and Y_2, the number of contracts awarded to firm B, is given by the entries in the following table.

		y_1	
y_2	0	1	2
0	1/9	2/9	1/9
1	2/9	2/9	0
2	1/9	0	0

The marginal probability function of Y_1 was derived in Exercise 5.17 to be binomial with $n = 2$ and $p = 1/3$. Are Y_1 and Y_2 independent? Why?

5.40   Refer to Exercise 5.2. The number of heads in three coin tosses is binomially distributed with $n = 3$, $p = 1/2$. Are the total number of heads and your winnings on the side bet independent? [Examine your answer to Exercise 5.18(b).]

5.41   In Exercise 5.3, we determined that the joint probability distribution of Y_1, the number of married executives, and Y_2, the number of never-married executives, is given by

$$p(y_1, y_2) = \frac{\binom{4}{y_1}\binom{3}{y_2}\binom{2}{3 - y_1 - y_2}}{\binom{9}{3}}$$

where y_1 and y_2 are integers, $0 \le y_1 \le 3$, $0 \le y_2 \le 3$, and $1 \le y_1 + y_2 \le 3$. Are Y_1 and Y_2 independent? (Recall your answer to Exercise 5.19.)

5.42   In Exercise 5.4 you were given the following joint probability function for

$$Y_1 = \begin{cases} 0, & \text{if child survived} \\ 1, & \text{if not} \end{cases} \quad \text{and} \quad Y_2 = \begin{cases} 0, & \text{if no belt used} \\ 1, & \text{if adult belt used} \\ 2, & \text{if car-seat belt used.} \end{cases}$$

	y_1		
y_2	0	1	
0	.38	.17	.55
1	.14	.02	.16
2	.24	.05	.29
	.76	.24	1.00

Are Y_1 and Y_2 independent? Why or why not?

5.43 In Example 5.4 and Exercise 5.5 we considered the joint density of Y_1, the proportion of the capacity of the tank that is stocked at the beginning of the week and Y_2, the proportion of the capacity sold during the week, given by

$$f(y_1, y_2) = \begin{cases} 3y_1, & 0 \le y_2 \le y_1 \le 1 \\ 0, & \text{elsewhere.} \end{cases}$$

Show that Y_1 and Y_2 are dependent.

5.44 In Exercise 5.6, we derived the fact that

$$f(y_1, y_2) = \begin{cases} 4y_1y_2, & 0 \le y_1 \le 1, \quad 0 \le y_2 \le 1 \\ 0, & \text{elsewhere} \end{cases}$$

is a valid joint probability density function. Are Y_1 and Y_2 independent?

5.45 In Exercise 5.7, we determined that

$$f(y_1, y_2) = \begin{cases} 6(1 - y_2), & 0 \le y_1 \le y_2 \le 1 \\ 0, & \text{elsewhere} \end{cases}$$

is a valid joint probability density function. Are Y_1 and Y_2 independent?

5.46 In Exercise 5.8, we proved that

$$f(y_1, y_2) = \begin{cases} 1, & 0 \le y_1 \le 2, \quad 0 \le y_2 \le 1, \quad 2y_2 \le y_1 \\ 0, & \text{elsewhere} \end{cases}$$

is a valid joint probability density function for Y_1, the amount of pollutant per sample collected above the stack without the cleaning device, and Y_2, the amount collected above the stack with the cleaner. Are the amounts of pollutants per sample collected with and without the cleaning device independent?

5.47 Refer to Exercise 5.9. Are Y_1 and Y_2 independent?

5.48 In Exercise 5.10, we were given the following joint probability density function for the random variables Y_1 and Y_2, which were the proportions of two components in a sample from a mixture of insecticide:

$$f(y_1, y_2) = \begin{cases} 2, & 0 \le y_1 \le 1, \quad 0 \le y_2 \le 1, \quad 0 \le y_1 + y_2 \le 1 \\ 0, & \text{elsewhere.} \end{cases}$$

Are Y_1 and Y_2 independent?

5.49 In Exercises 5.11 and 5.27, the joint density function of Y_1 and Y_2 was given by

$$f(y_1, y_2) = \begin{cases} 30y_1y_2^2, & y_1 - 1 \le y_2 \le 1 - y_1, \quad 0 \le y_1 \le 1 \\ 0, & \text{elsewhere.} \end{cases}$$

Are the random variables Y_1 and Y_2 independent?

5.50 Suppose that the random variables Y_1 and Y_2 have joint probability density function, $f(y_1, y_2)$, given by (see Exercises 5.12 and 5.28)

$$f(y_1, y_2) = \begin{cases} 6y_1^2 y_2, & 0 \le y_1 \le y_2, \quad y_1 + y_2 \le 2 \\ 0, & \text{elsewhere.} \end{cases}$$

Show that Y_1 and Y_2 are dependent random variables.

5.51 If Y_1 is the total time between a customer's arrival in the store and leaving the service window and if Y_2 is the time spent in line before reaching the window, the joint density of these variables, according to Exercise 5.13, is

$$f(y_1, y_2) = \begin{cases} e^{-y_1}, & 0 \le y_2 \le y_1 \le \infty \\ 0, & \text{elsewhere.} \end{cases}$$

Are Y_1 and Y_2 independent?

5.52 In Exercise 5.14, Y_1 and Y_2 denoted the proportions of time that employees I and II actually spent working on their assigned tasks during a workday. The joint density of Y_1 and Y_2 is given by

$$f(y_1, y_2) = \begin{cases} y_1 + y_2, & 0 \le y_1 \le 1, \quad 0 \le y_2 \le 1 \\ 0, & \text{elsewhere.} \end{cases}$$

Are Y_1 and Y_2 independent?

5.53 In Exercise 5.16, Y_1 and Y_2 denoted the lengths of life, in hours, for components of types I and II, respectively, in an electronic system. The joint density of Y_1 and Y_2 is

$$f(y_1, y_2) = \begin{cases} (1/8)y_1 e^{-(y_1+y_2)/2}, & y_1 > 0, \quad y_2 > 0 \\ 0, & \text{elsewhere.} \end{cases}$$

Are Y_1 and Y_2 independent?

5.54 Suppose that the probability that head appears when a coin is tossed is p and the probability that a tail occurs is $q = 1 - p$. Person A tosses the coin until the first head appears and stops. Person B does likewise. The results obtained by persons A and B are assumed to be independent. What is the probability that A and B stop on exactly the same number toss?

5.55 Let Y_1 and Y_2 be independent exponentially distributed random variables, each with mean 1. Find $P(Y_1 > Y_2 \mid Y_1 < 2Y_2)$.

5.56 Let Y_1 and Y_2 be independent random variables that are both uniformly distributed on the interval $(0, 1)$. Find $P(Y_1 < 2Y_2 \mid Y_1 < 3Y_2)$.

5.57 Suppose that, for $-1 \le \alpha \le 1$, the probability density function of (Y_1, Y_2) is given by

$$f(y_1, y_2) = \begin{cases} [1 - \alpha\{(1 - 2e^{-y_1})(1 - 2e^{-y_2})\}]e^{-y_1-y_2} & 0 \le y_1, \quad 0 \le y_2 \\ 0, & \text{elsewhere.} \end{cases}$$

a Show that the marginal distribution of Y_1 is exponential with mean 1.

b What is the marginal distribution of Y_2?

c Show that Y_1 and Y_2 are independent if and only if $\alpha = 0$.

5.58 A supermarket has two customers waiting to pay for their purchases at counter I and one customer waiting to pay at counter II. Let Y_1 and Y_2 denote the numbers of customers who spend more than \$50 on groceries at the respective counters. Suppose that Y_1 and Y_2 are independent binomial random variables, with the probability that a customer at counter I will spend more than \$50 equal to .2 and the probability that a customer at counter II will spend more than \$50 equal to .3.

a Find the joint probability distribution for Y_1 and Y_2.

b Find the probability that not more than one of the three customers will spend more than \$50.

5.59   The length of life Y for fuses of a certain type is modeled by the exponential distribution, with

$$f(y) = \begin{cases} (1/3)e^{-y/3}, & y > 0 \\ 0, & \text{elsewhere.} \end{cases}$$

(The measurements are in hundreds of hours.)

a   If two such fuses have independent lengths of life Y_1 and Y_2, find the joint probability density function for Y_1 and Y_2.

b   One fuse in (a) is in a primary system, and the other is in a backup system that comes into use only if the primary system fails. The total effective length of life of the two fuses is then $Y_1 + Y_2$. Find $P(Y_1 + Y_2 \le 1)$.

5.60   A bus arrives at a bus stop at a uniformly distributed time over the interval 0 to 1 hour. A passenger also arrives at the bus stop at a uniformly distributed time over the interval 0 to 1 hour. Assume that the arrival times of the bus and passenger are independent of one another and that the passenger will wait for up to $1/4$ hour for the bus to arrive. What is the probability that the passenger will catch the bus? [*Hint:* Let Y_1 denote the bus arrival time and Y_2 the passenger arrival time; determine the joint density of Y_1 and Y_2, and find $P(Y_2 \le Y_1 \le Y_2 + 1/4)$.]

5.61   Two telephone calls come into a switchboard at random times in a fixed 1-hour period. Assume that the calls are made independently of one another.

a   What is the probability that both calls are made in the first half hour?

b   What is the probability that the calls are made within 5 minutes of each other?

5.5   The Expected Value of a Function of Random Variables

You need only construct the multivariate analogue to the univariate situation to justify the following definition.

DEFINITION 5.9

Let $g(Y_1, Y_2, \ldots, Y_k)$ be a function of the discrete random variables, $Y_1, Y_2, \ldots, Y_k$, which have probability function $p(y_1, y_2, \ldots, y_k)$. Then the *expected value* of $g(Y_1, Y_2, \ldots, Y_k)$ is

$$E[g(Y_1, Y_2, \ldots, Y_k)] = \sum_{y_k} \cdots \sum_{y_2} \sum_{y_1} g(y_1, y_2, \ldots, y_k) p(y_1, y_2, \ldots, y_k).$$

If $Y_1, Y_2, \ldots, Y_k$ are continuous random variables with joint density function $f(y_1, y_2, \ldots, y_k)$, then[2]

$$E[g(Y_1, Y_2, \ldots, Y_k)] = \int_{y_k} \cdots \int_{y_2} \int_{y_1} g(y_1, y_2, \ldots, y_k)$$
$$\times f(y_1, y_2, \ldots, y_k) \, dy_1 \, dy_2 \ldots dy_k.$$

[2] Again we say the expectations exist if $\sum \cdots \sum |g(y_1, y_2, \ldots, y_n)| p(y_1, y_2, \ldots, y_k)$ or $\int \cdots \int |g(y_1, y_2, \ldots, y_n)| f(y_1, y_2, \ldots, y_k) dy_1 \ldots dy_k$ is finite.

EXAMPLE **5.15** Let Y_1 and Y_2 have joint density given by

$$f(y_1, y_2) = \begin{cases} 2y_1, & 0 \le y_1 \le 1, \quad 0 \le y_2 \le 1 \\ 0, & \text{elsewhere.} \end{cases}$$

Find $E(Y_1 Y_2)$.

Solution From Definition 5.9 we obtain

$$E(Y_1 Y_2) = \int_{-\infty}^{\infty} \int_{-\infty}^{\infty} y_1 y_2 f(y_1, y_2) \, dy_1 \, dy_2 = \int_0^1 \int_0^1 y_1 y_2 (2y_1) \, dy_1 \, dy_2$$

$$= \int_0^1 y_2 \left(\frac{2y_1^3}{3} \Big]_0^1 \right) dy_2 = \int_0^1 \left(\frac{2}{3} \right) y_2 \, dy_2 = \frac{2}{3} \frac{y_2^2}{2} \Big]_0^1 = \frac{1}{3}.$$

We will show that Definition 5.9 is consistent with Definition 4.4, in which we defined the expected value of a univariate random variable. Consider two random variables Y_1 and Y_2 with density function $f(y_1, y_2)$. We wish to find the expected value of $g(Y_1, Y_2) = Y_1$.

Then from Definition 5.9 we have

$$E(Y_1) = \int_{-\infty}^{\infty} \int_{-\infty}^{\infty} y_1 f(y_1, y_2) \, dy_2 \, dy_1$$

$$= \int_{-\infty}^{\infty} y_1 \left[\int_{-\infty}^{\infty} f(y_1, y_2) \, dy_2 \right] dy_1.$$

The quantity within the brackets, by definition, is the marginal density function for Y_1. Therefore, we obtain

$$E(Y_1) = \int_{-\infty}^{\infty} y_1 f_1(y_1) \, dy_1$$

which agrees with Definition 4.4.

EXAMPLE **5.16** Let Y_1 and Y_2 have a joint density given by

$$f(y_1, y_2) = \begin{cases} 2y_1, & 0 \le y_1 \le 1, \quad 0 \le y_2 \le 1 \\ 0, & \text{elsewhere.} \end{cases}$$

Find the expected value of Y_1.

Solution

$$E(Y_1) = \int_0^1 \int_0^1 y_1 (2y_1) \, dy_1 \, dy_2$$

$$= \int_0^1 \left(\frac{2y_1^3}{3} \Big]_0^1 \right) dy_2 = \int_0^1 \frac{2}{3} \, dy_2 = \frac{2}{3} y_2 \Big]_0^1 = \frac{2}{3}.$$

Refer to Figure 5.6 and estimate the expected value of Y_1. The value $E(Y_1) = 2/3$ appears to be quite reasonable.

EXAMPLE 5.17   In Figure 5.6 the mean value of Y_2 appears to be equal to .5. Let us confirm this visual estimate. Find $E(Y_2)$.

Solution
$$E(Y_2) = \int_0^1 \int_0^1 y_2(2y_1)\, dy_1\, dy_2 = \int_0^1 y_2 \left(\frac{2y_1^2}{2} \Big]_0^1 \right) dy_2$$

$$= \int_0^1 y_2\, dy_2 = \frac{y_2^2}{2} \Big]_0^1 = \frac{1}{2}.$$

EXAMPLE 5.18   Let Y_1 and Y_2 be random variables with density function

$$f(y_1, y_2) = \begin{cases} 2y_1, & 0 \le y_1 \le 1, \quad 0 \le y_2 \le 1 \\ 0, & \text{elsewhere.} \end{cases}$$

Find $V(Y_1)$.

Solution   The marginal density for Y_1 obtained in Example 5.6 is

$$f_1(y_1) = \begin{cases} 2y_1, & 0 \le y_1 \le 1 \\ 0, & \text{elsewhere.} \end{cases}$$

Then $V(Y_1) = E(Y_1^2) - [E(Y_1)]^2$, and

$$E(Y_1^k) = \int_{-\infty}^{\infty} y_1^k f_1(y_1)\, dy_1 = \int_0^1 y_1^k (2y_1)\, dy_1 = \frac{2y_1^{k+2}}{k+2} \Big]_0^1 = \frac{2}{k+2}.$$

If we let $k = 1$ and $k = 2$, it follows that $E(Y_1)$ and $E(Y_1^2)$ are 2/3 and 1/2, respectively. Then $V(Y_1) = E(Y_1^2) - [E(Y_1)]^2 = 1/2 - (2/3)^2 = 1/18$.

EXAMPLE 5.19   A process for producing an industrial chemical yields a product containing two types of impurities. For a specified sample from this process, let Y_1 denote the proportion of impurities in the sample and let Y_2 denote the proportion of type I impurities among all impurities found. Suppose that the joint distribution of Y_1 and Y_2 can be modeled by the following probability density function:

$$f(y_1, y_2) = \begin{cases} 2(1 - y_1), & 0 \le y_1 \le 1, \quad 0 \le y_2 \le 1 \\ 0, & \text{elsewhere.} \end{cases}$$

Find the expected value of the proportion of type I impurities in the sample.

Solution   Because Y_1 is the proportion of impurities in the sample and Y_2 is the proportion of type I impurities among the sample impurities, it follows that $Y_1 Y_2$ is the proportion of type I impurities in the entire sample. Thus, we want to find $E(Y_1 Y_2)$:

$$E(Y_1 Y_2) = \int_0^1 \int_0^1 2 y_1 y_2 (1 - y_1)\, dy_2\, dy_1 = 2 \int_0^1 y_1 (1 - y_1) \left(\frac{1}{2}\right) dy_1$$

$$= \int_0^1 (y_1 - y_1^2)\, dy_1 = \left(\frac{y_1^2}{2} - \frac{y_1^3}{3}\right) \Bigg]_0^1 = \frac{1}{2} - \frac{1}{3} = \frac{1}{6}.$$

Therefore, we would expect $1/6$ of the sample to be made up of type I impurities.

5.6 Special Theorems

Theorems concerning the expected value of a constant, the expected value of a constant times a function of random variables, and the expected value of the sum of functions of random variables are similar to those for the univariate case.

THEOREM 5.6   Let c be a constant. Then

$$E(c) = c.$$

THEOREM 5.7   Let $g(Y_1, Y_2)$ be a function of the random variables Y_1, Y_2, and let c be a constant. Then

$$E[cg(Y_1, Y_2)] = cE[g(Y_1, Y_2)].$$

THEOREM 5.8   Let Y_1 and Y_2 be random variables and $g_1(Y_1, Y_2), g_2(Y_1, Y_2), \ldots, g_k(Y_1, Y_2)$ be functions of Y_1 and Y_2. Then

$$E[g_1(Y_1, Y_2) + g_2(Y_1, Y_2) + \cdots + g_k(Y_1, Y_2)]$$
$$= E[g_1(Y_1, Y_2)] + E[g_2(Y_1, Y_2)] + \cdots + E[g_k(Y_1, Y_2)].$$

The proofs of these three theorems are analogous to the univariate cases discussed in Chapters 3 and 4.

EXAMPLE 5.20   Refer to Example 5.4. The random variable $Y_1 - Y_2$ denotes the proportional amount of gasoline remaining at the end of the week. Find $E(Y_1 - Y_2)$.

Solution   Employing Theorem 5.8 with $g_1(Y_1, Y_2) = Y_1$ and $g(Y_1, Y_2) = -Y_2$, we see that

$$E(Y_1 - Y_2) = E(Y_1) + E(-Y_2).$$

Theorem 5.7 applies, yielding $E(-Y_2) = -E(Y_2)$; therefore,

$$E(Y_1 - Y_2) = E(Y_1) - E(Y_2).$$

Also,

$$E(Y_1) = \int_0^1 \int_0^{y_1} y_1(3y_1)\, dy_2\, dy_1 = \int_0^1 3y_1^3\, dy_1 = \frac{3}{4}y_1^4 \Big]_0^1 = \frac{3}{4}$$

$$E(Y_2) = \int_0^1 \int_0^{y_1} y_2(3y_1)\, dy_2\, dy_1 = \int_0^1 3y_1 \left(\frac{y_2^2}{2} \Big]_0^{y_1} \right) dy_1 = \int_0^1 \frac{3}{2}y_1^3\, dy_1$$

$$= \frac{3}{8}y_1^4 \Big]_0^1 = \frac{3}{8}.$$

Thus,

$$E(Y_1 - Y_2) = (3/4) - (3/8) = 3/8$$

so we would expect 3/8 of the tank to be filled at the end of the week's sales.

If the random variables under study are independent, we sometimes can simplify the work involved in finding expectations. The following theorem is quite useful in this regard.

THEOREM 5.9   Let Y_1 and Y_2 be independent random variables and $g(Y_1)$ and $h(Y_2)$ be functions of only Y_1 and Y_2, respectively. Then

$$E[g(Y_1)h(Y_2)] = E[g(Y_1)]\, E[h(Y_2)]$$

provided that the expectations exist.

Proof   We will give the proof of the result for the continuous case. Let $f(y_1, y_2)$ denote the joint density of Y_1 and Y_2. The product $g(Y_1)h(Y_2)$ is a function of Y_1 and Y_2. Hence, by Definition 5.9,

$$E[g(Y_1)h(Y_2)] = \int_{-\infty}^{\infty} \int_{-\infty}^{\infty} g(y_1)h(y_2)f(y_1, y_2)\, dy_2\, dy_1$$

$$= \int_{-\infty}^{\infty} \int_{-\infty}^{\infty} g(y_1)h(y_2)f_1(y_1)f_2(y_2)\, dy_2\, dy_1$$

$$= \int_{-\infty}^{\infty} g(y_1) f_1(y_1) \left[\int_{-\infty}^{\infty} h(y_2) f_2(y_2) \, dy_2 \right] dy_1$$

$$= \int_{-\infty}^{\infty} g(y_1) f_1(y_1) E\left[h(Y_2)\right] dy_1$$

$$= E\left[h(Y_2)\right] \int_{-\infty}^{\infty} g(y_1) f_1(y_1) \, dy_1 = E\left[g(Y_1)\right] E\left[h(Y_2)\right].$$

The second equation in the preceding set follows from the fact that y_1 and y_2 are independent. The proof for the discrete case follows in an analogous manner.

EXAMPLE 5.21   Refer to Example 5.19. In that example we found $E(Y_1 Y_2)$ directly. By investigating the form of the joint density function given there, we can see that Y_1 and Y_2 are independent. Find $E(Y_1 Y_2)$ by using the result that $E(Y_1 Y_2) = E(Y_1) E(Y_2)$ if Y_1 and Y_2 are independent.

Solution   The joint density function is given by

$$f(y_1, y_2) = \begin{cases} 2(1 - y_1), & 0 \le y_1 \le 1, \quad 0 \le y_2 \le 1 \\ 0, & \text{elsewhere.} \end{cases}$$

Hence,

$$f_1(y_1) = \begin{cases} \int_0^1 2(1 - y_1) \, dy_2 = 2(1 - y_1), & 0 \le y_1 \le 1 \\ 0, & \text{elsewhere} \end{cases}$$

and

$$f_2(y_2) = \begin{cases} \int_0^1 2(1 - y_1) \, dy_1 = -(1 - y_1)^2 \Big]_0^1 = 1, & 0 \le y_2 \le 1 \\ 0, & \text{elsewhere.} \end{cases}$$

We then have

$$E(Y_1) = \int_0^1 y_1 \left[2(1 - y_1)\right] dy_1 = 2 \left(\frac{y_1^2}{2} - \frac{y_1^3}{3} \right) \Big]_0^1 = \frac{1}{3}$$

$$E(Y_2) = 1/2$$

because Y_2 is uniformly distributed over $(0, 1)$.

It follows that

$$E(Y_1 Y_2) = E(Y_1) E(Y_2) = (1/3)(1/2) = 1/6$$

which agrees with the answer in Example 5.19.

Exercises

5.62 In Exercise 5.1 we determined that the joint distribution of Y_1, the number of contracts awarded to firm A, and Y_2, the number of contracts awarded to firm B, is given by the entries in the following table.

	y_1		
y_2	0	1	2
0	1/9	2/9	1/9
1	2/9	2/9	0
2	1/9	0	0

The marginal probability function of Y_1 was derived in Exercise 5.17 to be binomial with $n = 2$ and $p = 1/3$.

 a Find $E(Y_1)$.
 b Find $V(Y_1)$.
 c Find $E(Y_1 - Y_2)$.

5.63 In Exercise 5.3, we determined that the joint probability distribution of Y_1, the number of married executives, and Y_2, the number of never-married executives, is given by

$$p(y_1,\ y_2) = \frac{\binom{4}{y_1}\binom{3}{y_2}\binom{2}{3-y_1-y_2}}{\binom{9}{3}},$$

where y_1 and y_2 are integers, $0 \le y_1 \le 3, 0 \le y_2 \le 3$, and $1 \le y_1+y_2 \le 3$. Find the expected number of married executives among the three selected for promotion. (See Exercise 5.19.)

5.64 In Exercise 5.6, we derived the fact that

$$f(y_1,\ y_2) = \begin{cases} 4y_1y_2, & 0 \le y_1 \le 1, \quad 0 \le y_2 \le 1 \\ 0, & \text{elsewhere.} \end{cases}$$

 a Find $E(Y_1)$.
 b Find $V(Y_1)$.
 c Find $E(Y_1 - Y_2)$.

5.65 In Exercise 5.7, we determined that

$$f(y_1,\ y_2) = \begin{cases} 6(1-y_2), & 0 \le y_1 \le y_2 \le 1 \\ 0, & \text{elsewhere} \end{cases}$$

is a valid joint probability density function.

 a Find $E(Y_1)$ and $E(Y_2)$.
 b Find $V(Y_1)$ and $V(Y_2)$.
 c Find $E(Y_1 - 3Y_2)$.

5.66 In Exercise 5.8, we proved that

$$f(y_1,\ y_2) = \begin{cases} 1, & 0 \le y_1 \le 2, \quad 0 \le y_2 \le 1, \quad 2y_2 \le y_1 \\ 0, & \text{elsewhere} \end{cases}$$

is a valid joint probability density function for Y_1, the amount of pollutant per sample collected above the stack without the cleaning device, and Y_2, the amount collected above the stack with the cleaner.

a   Find $E(Y_1)$ and $E(Y_2)$.

b   Find $V(Y_1)$ and $V(Y_2)$.

c   The random variable $Y_1 - Y_2$ represents the amount by which the weight of pollutant can be reduced by using the cleaning device. Find $E(Y_1 - Y_2)$.

d   Find $V(Y_1 - Y_2)$. Within what limits would you expect $Y_1 - Y_2$ to fall?

5.67   Refer to Exercise 5.9. Find $E(Y_1 Y_2)$.

5.68   In Exercise 5.14, Y_1 and Y_2 denoted the proportions of time that employees I and II actually spent working on their assigned tasks during a workday. The joint density of Y_1 and Y_2 is given by

$$f(y_1, y_2) = \begin{cases} y_1 + y_2, & 0 \le y_1 \le 1, 0 \le y_2 \le 1 \\ 0, & \text{elsewhere.} \end{cases}$$

Employee I has a higher productivity rating than employee II and a measure of the total productivity of the pair of employees is $30Y_1 + 25Y_2$. Find the expected value of this measure of productivity.

5.69   In Exercise 5.16, Y_1 and Y_2 denoted the lengths of life, in hours, for components of types I and II, respectively, in an electronic system. The joint density of Y_1 and Y_2 is

$$f(y_1, y_2) = \begin{cases} (1/8)y_1 e^{-(y_1+y_2)/2}, & y_1 > 0, y_2 > 0 \\ 0, & \text{elsewhere.} \end{cases}$$

One way to measure the relative efficiency of the two components is to compute the ratio Y_2/Y_1. Find $E(Y_2/Y_1)$. (*Hint:* In Exercise 5.53, we proved that Y_1 and Y_2 are independent.)

5.70   In Exercise 5.32 we determined that the joint density function for Y_1, the weight in tons of a bulk item stocked by a supplier, and Y_2, the weight of the item sold by the supplier, has joint density

$$f(y_1, y_2) = \begin{cases} 1/y_1, & 0 \le y_2 \le y_1 \le 1 \\ 0, & \text{elsewhere.} \end{cases}$$

In this case, the random variable $Y_1 - Y_2$ measures the amount of stock remaining at the end of the week, a quantity of great importance to the supplier. Find $E(Y_1 - Y_2)$.

5.71   In Exercise 5.36, we determined that the unconditional probability distribution for Y, the number of defects per yard in a certain fabric, is

$$p(y) = (1/2)^{y+1}, \qquad y = 0, 1, 2, \dots.$$

Find the expected number of defects per yard.

5.72   In Exercise 5.54 we considered two individuals who each tossed a coin until the first head appears. Let Y_1 and Y_2 denote the number of times that persons A and B toss the coin, respectively. If heads occurs with probability p and tails occurs with probability $q = 1 - p$, it is reasonable to conclude that Y_1 and Y_2 are independent and that each has a geometric distribution with parameter p. Consider $Y_1 - Y_2$, the difference in the number of tosses require by the two individuals.

a Find $E(Y_1)$, $E(Y_2)$, and $E(Y_1 - Y_2)$.

b Find $E(Y_1^2)$, $E(Y_2^2)$, and $E(Y_1 Y_2)$ (recall that Y_1 and Y_2 are independent).

c Find $E((Y_1 - Y_2)^2)$ and $V(Y_1 - Y_2)$.

d Give an interval that will contain $Y_1 - Y_2$ with probability at least 8/9.

5.73 In Exercise 5.57, we considered random variables Y_1 and Y_2 that, for $-1 \leq \alpha \leq 1$, have joint density function given by

$$f(y_1, y_2) = \begin{cases} [1 - \alpha\{(1 - 2e^{-y_1})(1 - 2e^{-y_2})\}]e^{-y_1-y_2} & 0 \leq y_1, \quad 0 \leq y_2 \\ 0, & \text{elsewhere} \end{cases}$$

and established that the marginal distributions of Y_1 and Y_2 are both exponential with mean 1.

a Find $E(Y_1)$ and $E(Y_2)$.

b Find $V(Y_1)$ and $V(Y_2)$.

c Find $E(Y_1 - Y_2)$.

d Find $E(Y_1 Y_2)$.

e Find $V(Y_1 - Y_2)$. Within what limits would you expect $Y_1 - Y_2$ to fall?

*5.74 Suppose that Z is a standard normal random variable and that Y_1 and Y_2 are χ^2 distributed random variables with v_1 and v_2 degrees of freedom, respectively. Further, assume that Z, Y_1, and Y_2 are independent.

a Define $W = \frac{Z}{\sqrt{Y_1}}$. Find $E(W)$ and $V(W)$. What assumptions do you need about the value of v_1? [*Hint:* $W = Z\left(\frac{1}{\sqrt{Y_1}}\right) = g(Z)h(Y_1)$. Use Theorem 5.9. The results of Exercise 4.90(d) will also be useful.]

b Define $U = \frac{Y_1}{Y_2}$. Find $E(U)$ and $V(U)$. What assumptions about v_1 and v_2 do you need? Use the hint from (a).

5.7 The Covariance of Two Random Variables

Intuitively, we think of the dependence of two random variables Y_1 and Y_2 as implying that one variable, say, Y_1, either increases or decreases as Y_2 changes. We will confine our attention to two measures of dependence: the covariance between two random variables and their correlation coefficient. In Figure 5.8(a) and (b) we give plots of the observed values of two variables, Y_1 and Y_2, for samples of $n = 10$ experimental units drawn from each of two populations. If all the points fall along

FIGURE 5.8
Dependent and
independent
observations
for (y_1, y_2)

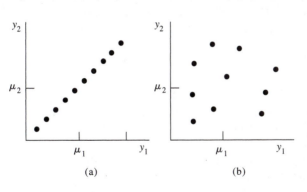

(a)   (b)

a straight line, as indicated in Figure 5.8(a), Y_1 and Y_2 are obviously dependent. In contrast, Figure 5.8(b) indicates little or no dependence between Y_1 and Y_2.

Suppose that we knew the values of $E(Y_1) = \mu_1$ and $E(Y_2) = \mu_2$ and located this point on the graph in Figure 5.8. Now locate a plotted point, (y_1, y_2), on Figure 5.8(a) and measure the deviations $(y_1 - \mu_1)$ and $(y_2 - \mu_2)$. Both deviations assume the same algebraic sign for any point, (y_1, y_2), and their product $(y_1 - \mu_1)(y_2 - \mu_2)$ is positive. Points to the right of μ_1 yield pairs of positive deviations; points to the left produce pairs of negative deviations; and the average of the product of the deviations $(y_1 - \mu_1)(y_2 - \mu_2)$ is large and positive. If the linear relation indicated in Figure 5.8(a) had sloped downward to the right, all corresponding pairs of deviations would have been of the opposite sign, and the average value of $(y_1 - \mu_1)(y_2 - \mu_2)$ would have been a large negative number.

The situation just described will not occur for Figure 5.8(b), where little dependence exists between Y_1 and Y_2. Their corresponding deviations $(y_1 - \mu_1)$ and $(y_2 - \mu_2)$ will assume the same algebraic sign for some points and opposite signs for others. Thus, the product $(y_1 - \mu_1)(y_2 - \mu_2)$ will be positive for some points, negative for others, and will average to some value near zero.

Clearly the average value of $(Y_1 - \mu_1)(Y_2 - \mu_2)$ provides a measure of the linear dependence between Y_1 and Y_2. This quantity, $E[(Y_1 - \mu_1)(Y_2 - \mu_2)]$, is called the *covariance* of Y_1 and Y_2.

DEFINITION **5.10**

If Y_1 and Y_2 are random variables with means μ_1 and μ_2, respectively, the *covariance* of Y_1 and Y_2 is given by

$$\text{Cov}(Y_1, Y_2) = E[(Y_1 - \mu_1)(Y_2 - \mu_2)].$$

The larger the absolute value of the covariance of Y_1 and Y_2, the greater the linear dependence between Y_1 and Y_2. Positive values indicate that Y_1 increases as Y_2 increases; negative values indicate that Y_1 decreases as Y_2 increases. A zero value of the covariance indicates no linear dependence between Y_1 and Y_2.

Unfortunately, it is difficult to employ the covariance as an absolute measure of dependence because its value depends upon the scale of measurement. As a result, it is difficult to determine at first glance whether a particular covariance is large or small. This problem can be eliminated by standardizing its value and using the *correlation coefficient*, ρ, a quantity related to the covariance and defined as

$$\rho = \frac{\text{Cov}(Y_1, Y_2)}{\sigma_1 \sigma_2}$$

where σ_1 and σ_2 are the standard deviations of Y_1 and Y_2, respectively. Supplemental discussions of the correlation coefficient may be found in Hogg and Craig (1995) and Myers (1990).

A proof that the correlation coefficient ρ satisfies the inequality $-1 \leq \rho \leq 1$ is outlined in Exercise 5.137.

The sign of the correlation coefficient is the same as the sign of the covariance. Thus $\rho > 0$ indicates that Y_2 increases as Y_1 increases and $\rho = +1$ implies perfect correlation, with all points falling on a straight line with positive slope. A value of $\rho = 0$ implies zero covariance and no correlation. A negative coefficient of correlation implies a decrease in Y_2 as Y_1 increases and $\rho = -1$ implies perfect correlation, with all points falling on a straight line with negative slope. A convenient computational formula for the covariance is contained in the next theorem.

THEOREM 5.10 Let Y_1 and Y_2 be random variables with means μ_1 and μ_2, respectively, then

$$\text{Cov}(Y_1, Y_2) = E\left[(Y_1 - \mu_1)(Y_2 - \mu_2)\right] = E(Y_1 Y_2) - E(Y_1)E(Y_2).$$

Proof
$$\text{Cov}(Y_1, Y_2) = E\left[(Y_1 - \mu_1)(Y_2 - \mu_2)\right]$$
$$= E(Y_1 Y_2 - \mu_1 Y_2 - \mu_2 Y_1 + \mu_1 \mu_2).$$

From Theorem 5.8, the expected value of a sum is equal to the sum of the expected values; and from Theorem 5.7, the expected value of a constant times a function of random variables is the constant times the expected value. Thus

$$\text{Cov}(Y_1, Y_2) = E(Y_1 Y_2) - \mu_1 E(Y_2) - \mu_2 E(Y_1) + \mu_1 \mu_2.$$

Because $E(Y_1) = \mu_1$ and $E(Y_2) = \mu_2$, it follows that

$$\text{Cov}(Y_1, Y_2) = E(Y_1 Y_2) - E(Y_1)E(Y_2)$$

or

$$\text{Cov}(Y_1, Y_2) = E(Y_1 Y_2) - \mu_1 \mu_2. \qquad \blacksquare$$

EXAMPLE 5.22 Refer to Example 5.4. Find the covariance between the amount in stock Y_1 and amount of sales Y_2.

Solution Recall that Y_1 and Y_2 have joint density function given by

$$f(y_1, y_2) = \begin{cases} 3y_1, & 0 \le y_2 \le y_1 \le 1 \\ 0, & \text{elsewhere.} \end{cases}$$

Thus,

$$E(Y_1 Y_2) = \int_0^1 \int_0^{y_1} y_1 y_2 (3y_1)\, dy_2\, dy_1 = \int_0^1 3y_1^2 \left(\frac{y_2^2}{2}\right]_0^{y_1}\right) dy_1$$

$$= \int_0^1 \frac{3}{2} y_1^4\, dy_1 = \frac{3}{2}\left(\frac{y_1^5}{5}\right]_0^1\right) = \frac{3}{10}.$$

From Example 5.20, we know that $E(Y_1) = 3/4$ and $E(Y_2) = 3/8$. Thus, using Theorem 5.10, we obtain

$$\text{Cov}(Y_1, \, Y_2) = E(Y_1 Y_2) - E(Y_1)E(Y_2) = (3/10) - (3/4)(3/8) = .30 - .28 = .02.$$

In this example, large values of Y_2 can occur only with large values of Y_1 and the density, $f(y_1, y_2)$, is larger for larger values of Y_1 (see Figure 5.4). Thus, it is intuitive that the covariance between Y_1 and Y_2 should be positive.

EXAMPLE **5.23** Let Y_1 and Y_2 have joint density given by

$$f(y_1, \, y_2) = \begin{cases} 2y_1, & 0 \le y_1 \le 1, \quad 0 \le y_2 \le 1 \\ 0, & \text{elsewhere.} \end{cases}$$

Find the covariance of Y_1 and Y_2.

Solution From Example 5.15, $E(Y_1 Y_2) = 1/3$. Also, from Examples 5.16 and 5.17, $\mu_1 = E(Y_1) = 2/3$ and $\mu_2 = E(Y_2) = 1/2$, so

$$\text{Cov}(Y_1, \, Y_2) = E(Y_1 Y_2) - \mu_1 \mu_2 = (1/3) - (2/3)(1/2) = 0.$$

Example 5.23 furnishes a specific example of the general result given in Theorem 5.11.

THEOREM **5.11** If Y_1 and Y_2 are independent random variables, then

$$\text{Cov}(Y_1, \, Y_2) = 0.$$

Proof Theorem 5.10 establishes that

$$\text{Cov}(Y_1, \, Y_2) = E(Y_1 Y_2) - \mu_1 \mu_2.$$

Because Y_1 and Y_2 are independent, Theorem 5.9 implies that

$$E(Y_1 Y_2) = E(Y_1)E(Y_2) = \mu_1 \mu_2$$

and the desired result follows immediately. ∎

Notice that the random variables Y_1 and Y_2 of Example 5.23 are independent; hence, by Theorem 5.11, their covariance must be zero. The converse of Theorem 5.11 is not true, as will be illustrated in the following example.

EXAMPLE **5.24** Let Y_1 and Y_2 be discrete random variables with joint probability distribution as shown in Table 5.3. Show that Y_1 and Y_2 are dependent but have zero covariance.

Table 5.3. Joint probability distribution, Example 5.24

		y_1	
y_2	-1	0	$+1$
-1	1/16	3/16	1/16
0	3/16	0	3/16
$+1$	1/16	3/16	1/16

Solution   Calculation of marginal probabilities yields $p_1(-1) = p_1(1) = 5/16 = p_2(-1) = p_2(1)$, and $p_1(0) = 6/16 = p_2(0)$. Looking at the upper left-hand cell, we have $p(-1, -1) = 1/16$. Obviously,

$$p(-1, -1) \neq p_1(-1)p_2(-1)$$

and this is sufficient to show that Y_1 and Y_2 are dependent.

Again looking at the marginal probabilities, we see that $E(Y_1) = E(Y_2) = 0$. Also,

$$
\begin{aligned}
E(Y_1 Y_2) &= \sum_{y_1}\sum_{y_2} y_1 y_2 p(y_1, \ y_2) \\
&= (-1)(-1)(1/16) + (-1)(0)(3/16) + (-1)(1)(1/16) \\
&\quad + (0)(-1)(3/16) + (0)(0)(0) + (0)(1)(3/16) \\
&\quad + (1)(-1)(1/16) + (1)(0)(3/16) + (1)(1)(1/16) \\
&= (1/16) - (1/16) - (1/16) + (1/16) = 0.
\end{aligned}
$$

Thus,

$$\text{Cov}(Y_1, \ Y_2) = E(Y_1 Y_2) - E(Y_1)E(Y_2) = 0 - 0(0) = 0.$$

This example shows that the converse of Theorem 5.11 is not true. If the covariance of two random variables is zero, the variables *need not* be independent.

Exercises

5.75   In Exercise 5.1 we determined that the joint distribution of Y_1, the number of contracts awarded to firm A, and Y_2, the number of contracts awarded to firm B, is given by the entries in the following table.

		y_1	
y_2	0	1	2
0	1/9	2/9	1/9
1	2/9	2/9	0
2	1/9	0	0

Find $\text{Cov}(Y_1, Y_2)$. Does it surprise you that $\text{Cov}(Y_1, Y_2)$ is negative? Why?

5.76   In Exercise 5.3, we determined that the joint probability distribution of Y_1, the number of married executives, and Y_2, the number of never married executives, is given by

$$p(y_1, y_2) = \frac{\binom{4}{y_1}\binom{3}{y_2}\binom{2}{3 - y_1 - y_2}}{\binom{9}{3}}$$

where y_1 and y_2 are integers, $0 \le y_1 \le 3, 0 \le y_2 \le 3$, and $1 \le y_1 + y_2 \le 3$. Find $\text{Cov}(Y_1, Y_2)$.

5.77   In Exercise 5.6, we derived the fact that

$$f(y_1, y_2) = \begin{cases} 4y_1 y_2, & 0 \le y_1 \le 1, 0 \le y_2 \le 1 \\ 0, & \text{elsewhere.} \end{cases}$$

Show that $\text{Cov}(Y_1, Y_2) = 0$. Does it surprise you that $\text{Cov}(Y_1, Y_2)$ is zero? Why?

5.78   In Exercise 5.7, we determined that

$$f(y_1, y_2) = \begin{cases} 6(1 - y_2), & 0 \le y_1 \le y_2 \le 1 \\ 0, & \text{elsewhere} \end{cases}$$

is a valid joint probability density function. Find $\text{Cov}(Y_1, Y_2)$. Are Y_1 and Y_2 independent?

5.79   Refer to Exercises 5.9 and 5.67. Find the covariance of Y_1 and Y_2. Find the coefficient of correlation for Y_1 and Y_2

5.80   Let Y_1 and Y_2 be uncorrelated random variables. Find the covariance and correlation between $U_1 = Y_1 + Y_2$ and $U_2 = Y_1 - Y_2$ in terms of the variances of Y_1 and Y_2.

5.81   Let the discrete random variables Y_1 and Y_2 have the joint probability function

$$p(y_1, y_2) = 1/3 \quad \text{for} \quad (y_1, y_2) = (-1, 0), (0, 1), (1, 0).$$

Find $\text{Cov}(Y_1, Y_2)$. Notice that Y_1 and Y_2 are dependent. (Why?) This is another example of uncorrelated random variables that are not independent.

5.82   Suppose that the random variables Y_1 and Y_2 have means μ_1 and μ_2 and variances σ_1^2 and σ_2^2, respectively. Use the basic definition of the covariance of two random variables to establish that:

a $\text{Cov}(Y_1, Y_2) = \text{Cov}(Y_2, Y_1)$.

b $\text{Cov}(Y_1, Y_1) = V(Y_1) = \sigma_1^2$. That is, the covariance of a random variable and itself is just the variance of the random variable.

5.83   The random variables Y_1 and Y_2 are such that $E(Y_1) = 4$, $E(Y_2) = -1$, $V(Y_1) = 2$ and $V(Y_2) = 8$.

a   What is $\text{Cov}(Y_1, Y_1)$?

b   Assuming that the means and variances are correct, as given, is it possible that $\text{Cov}(Y_1, Y_2) = 7$? [*Hint:* If $\text{Cov}(Y_1, Y_2) = 7$, what is the value of ρ, the coefficient of correlation?]

c   Assuming that the means and variances are correct, what is the largest possible value for $\text{Cov}(Y_1, Y_2)$? If $\text{Cov}(Y_1, Y_2)$ achieves this largest value, what does that imply about the relationship between Y_1 and Y_2?

d   Assuming that the means and variances are correct, what is the smallest possible value for $\text{Cov}(Y_1, Y_2)$? If $\text{Cov}(Y_1, Y_2)$ achieves this smallest value, what does that imply about the relationship between Y_1 and Y_2?

5.84   Let Z be a standard normal random variable and let $Y_1 = Z$ and $Y_2 = Z^2$.

    a   What are $E(Y_1)$ and $E(Y_2)$?

    b   What is $E(Y_1 Y_2)$? [*Hint:* $E(Y_1 Y_2) = E(Z^3)$.]

    c   What is $\text{Cov}(Y_1, Y_2)$?

    d   Notice that $P(Y_2 > 1 | Y_1 > 1) = 1$. Are Y_1 and Y_2 independent?

5.85   In Exercise 5.57, we considered random variables Y_1 and Y_2 that, for $-1 \le \alpha \le 1$, have joint density function given by

$$f(y_1, y_2) = \begin{cases} [1 - \alpha\{(1 - 2e^{-y_1})(1 - 2e^{-y_2})\}]e^{-y_1-y_2} & 0 \le y_1, \quad 0 \le y_2 \\ 0 & \text{elsewhere.} \end{cases}$$

We established that the marginal distributions of Y_1 and Y_2 are both exponential with mean 1 and showed that Y_1 and Y_2 are independent if and only if $\alpha = 0$. In Exercise 5.73, we derived $E(Y_1 Y_2)$.

    a   Derive $\text{Cov}(Y_1, Y_2)$.

    b   Show that $\text{Cov}(Y_1, Y_2) = 0$ if and only if $\alpha = 0$.

    c   Argue that Y_1 and Y_2 are independent if and only if $\rho = 0$.

5.8 The Expected Value and Variance of Linear Functions of Random Variables

In later chapters in this text, especially Chapters 9 and 11, we will frequently encounter parameter estimators that are linear functions of the measurements in a sample, $Y_1, Y_2, \ldots, Y_n$. If $a_1, a_2, \ldots, a_n$ are constants, we will need to find the expected value and variance of a linear function of the random variables $Y_1, Y_2, \ldots, Y_n$,

$$U_1 = a_1 Y_1 + a_2 Y_2 + a_3 Y_3 + \cdots + a_n Y_n = \sum_{i=1}^{n} a_i Y_i.$$

We also may be interested in the covariance between two such linear combinations. Results that simplify the calculation of these quantities are summarized in the following theorem.

THEOREM 5.12   Let $Y_1, Y_2, \ldots, Y_n$ and $X_1, X_2, \ldots, X_m$ be random variables with $E(Y_i) = \mu_i$ and $E(X_j) = \xi_j$. Define

$$U_1 = \sum_{i=1}^{n} a_i Y_i \quad \text{and} \quad U_2 = \sum_{j=1}^{m} b_j X_j$$

for constants $a_1, a_2, \ldots, a_n$ and $b_1, b_2, \ldots, b_m$. Then the following hold:

    a $E(U_1) = \sum_{i=1}^{n} a_i \mu_i$.

> **b** $V(U_1) = \sum_{i=1}^{n} a_i^2 V(Y_i) + 2 \sum \sum_{i<j} a_i a_j \text{Cov}(Y_i, Y_j)$, where the double
> sum is over all pairs (i, j) with $i < j$.
>
> **c** $\text{Cov}(U_1, U_2) = \sum_{i=1}^{n} \sum_{j=1}^{m} a_i b_j \text{Cov}(Y_i, X_j)$. ∎

Before proceeding with the proof of Theorem 5.12, we illustrate the use of the theorem with an example.

EXAMPLE 5.25   Let Y_1, Y_2, and Y_3 be random variables, where $E(Y_1) = 1$, $E(Y_2) = 2$, $E(Y_3) = -1$, $V(Y_1) = 1$, $V(Y_2) = 3$, $V(Y_3) = 5$, $\text{Cov}(Y_1, Y_2) = -0.4$, $\text{Cov}(Y_1, Y_3) = 1/2$, and $\text{Cov}(Y_2, Y_3) = 2$. Find the expected value and variance of $U = Y_1 - 2Y_2 + Y_3$. If $W = 3Y_1 + Y_2$, find $\text{Cov}(U, W)$.

Solution $U = a_1 Y_1 + a_2 Y_2 + a_3 Y_3$, where $a_1 = 1$, $a_2 = -2$, and $a_3 = 1$. Then by Theorem 5.12,

$$E(U) = a_1 E(Y_1) + a_2 E(Y_2) + a_3 E(Y_3)$$
$$= (1)(1) + (-2)(2) + (1)(-1) = -4.$$

Similarly,

$$V(U) = a_1^2 V(Y_1) + a_2^2 V(Y_2) + a_3^2 V(Y_3) + 2a_1 a_2 \text{Cov}(Y_1, Y_2)$$
$$+ 2a_1 a_3 \text{Cov}(Y_1, Y_3) + 2a_2 a_3 \text{Cov}(Y_2, Y_3)$$
$$= (1)^2(1) + (-2)^2(3) + (1)^2(5) + (2)(1)(-2)(-0.4)$$
$$+ (2)(1)(1)(1/2) + (2)(-2)(1)(2)$$
$$= 12.6.$$

Notice that $W = b_1 Y_1 + b_2 Y_2$, where $b_1 = 3$ and $b_2 = 1$. Thus,

$$\text{Cov}(U, W) = a_1 b_1 \text{Cov}(Y_1, Y_1) + a_1 b_2 \text{Cov}(Y_1, Y_2) + a_2 b_1 \text{Cov}(Y_2, Y_1)$$
$$+ a_2 b_2 \text{Cov}(Y_2, Y_2) + a_3 b_1 \text{Cov}(Y_3, Y_1) + a_3 b_2 \text{Cov}(Y_3, Y_2).$$

Notice that, as established in Exercise 5.82, $\text{Cov}(Y_i, Y_j) = \text{Cov}(Y_j, Y_i)$ and $\text{Cov}(Y_i, Y_i) = V(Y_i)$. Therefore,

$$\text{Cov}(U, W) = (1)(3)(1) + (1)(1)(-0.4) + (-2)(3)(-0.4)$$
$$+ (-2)(1)(3) + (1)(3)(1/2) + (1)(1)(2)$$
$$= 2.5.$$

Because $\text{Cov}(U, W) \neq 0$, it follows that U and W are *dependent*.

We now proceed with the proof of Theorem 5.12.

Proof The theorem consists of three parts, of which (a) follows directly from Theorems 5.7 and 5.8. To prove (b), we appeal to the definition of variance and write

$$V(U_1) = E[U_1 - E(U_1)]^2 = E\left[\sum_{i=1}^{n} a_i Y_i - \sum_{i=1}^{n} a_i \mu_i\right]^2$$

$$= E\left[\sum_{i=1}^{n} a_i (Y_i - \mu_i)\right]^2$$

$$= E\left[\sum_{i=1}^{n} a_i^2 (Y_i - \mu_i)^2 + \sum\sum_{i \neq j} a_i a_j (Y_i - \mu_i)(Y_j - \mu_j)\right]$$

$$= \sum_{i=1}^{n} a_i^2 E(Y_i - \mu_i)^2 + \sum\sum_{i \neq j} a_i a_j E\left[(Y_i - \mu_i)(Y_j - \mu_j)\right].$$

By the definitions of variance and covariance, we have

$$V(U_1) = \sum_{i=1}^{n} a_i^2 V(Y_i) + \sum\sum_{i \neq j} a_i a_j \, \text{Cov}(Y_i, Y_j).$$

Because $\text{Cov}(Y_i, Y_j) = \text{Cov}(Y_j, Y_i)$, we can write

$$V(U_1) = \sum_{i=1}^{n} a_i^2 V(Y_i) + 2 \sum\sum_{i<j} a_i a_j \, \text{Cov}(Y_i, Y_j).$$

Similar steps can be used to obtain (c). We have

$$\text{Cov}(U_1, U_2) = E\{[U_1 - E(U_1)][U_2 - E(U_2)]\}$$

$$= E\left[\left(\sum_{i=1}^{n} a_i Y_i - \sum_{i=1}^{n} a_i \mu_i\right)\left(\sum_{j=1}^{m} b_j X_j - \sum_{j=1}^{m} b_j \xi_j\right)\right]$$

$$= E\left\{\left[\sum_{i=1}^{n} a_i (Y_i - \mu_i)\right]\left[\sum_{j=1}^{m} b_j (X_j - \xi_j)\right]\right\}$$

$$= E\left[\sum_{i=1}^{n}\sum_{j=1}^{m} a_i b_j (Y_i - \mu_i)(X_j - \xi_j)\right]$$

$$= \sum_{i=1}^{n}\sum_{j=1}^{m} a_i b_j E\left[(Y_i - \mu_i)(X_j - \xi_j)\right]$$

$$= \sum_{i=1}^{n}\sum_{j=1}^{m} a_i b_j \, \text{Cov}(Y_i, X_j).$$

On observing that $\text{Cov}(Y_i, Y_i) = V(Y_i)$, we can see that (b) is a special case of (c).   ∎

EXAMPLE **5.26**   Refer to Examples 5.4 and 5.20. In Example 5.20, we were interested in $Y_1 - Y_2$, the proportional amount of gasoline remaining at the end of a week. Find the variance of $Y_1 - Y_2$.

Solution   Using Theorem 5.12, we have

$$V(Y_1 - Y_2) = V(Y_1) + V(Y_2) - 2\,\text{Cov}(Y_1,\ Y_2).$$

Because

$$f_1(y_1) = \begin{cases} 3y_1^2, & 0 \le y_1 \le 1 \\ 0, & \text{elsewhere} \end{cases}$$

and

$$f_2(y_2) = \begin{cases} (3/2)(1 - y_2^2), & 0 \le y_2 \le 1 \\ 0, & \text{elsewhere,} \end{cases}$$

it follows that

$$E(Y_1^2) = \int_0^1 3y_1^4\,dy_1 = \frac{3}{5}$$

$$E(Y_2^2) = \int_0^1 \frac{3}{2}y_2^2(1 - y_2^2)\,dy_2 = \frac{3}{2}\left[\frac{1}{3} - \frac{1}{5}\right] = \frac{1}{5}.$$

From Example 5.20, we have $E(Y_1) = 3/4$ and $E(Y_2) = 3/8$. Thus,

$$V(Y_1) = (3/5) - (3/4)^2 = .04 \quad \text{and} \quad V(Y_2) = (1/5) - (3/8)^2 = .06.$$

In Example 5.22 we determined that $\text{Cov}(Y_1, Y_2) = .02$. Therefore,

$$V(Y_1 - Y_2) = V(Y_1) + V(Y_2) - 2\,\text{Cov}(Y_1,\ Y_2)$$
$$= .04 + .06 - 2(.02) = .06.$$

The standard deviation of $Y_1 - Y_2$ is then $\sqrt{.06} = .245$.

EXAMPLE **5.27**   Let $Y_1, Y_2, \ldots, Y_n$ be independent random variables with $E(Y_i) = \mu$ and $V(Y_i) = \sigma^2$. (These variables may denote the outcomes of n independent trials of an experiment.) Define

$$\overline{Y} = \frac{1}{n}\sum_{i=1}^n Y_i$$

and show that $E(\overline{Y}) = \mu$ and $V(\overline{Y}) = \sigma^2/n$.

Solution   Notice that $\overline{Y}$ is a linear function of $Y_1, Y_2, \ldots, Y_n$ with all constants a_i equal to $1/n$. That is,

$$\overline{Y} = \left(\frac{1}{n}\right) Y_1 + \cdots + \left(\frac{1}{n}\right) Y_n.$$

By Theorem 5.12 (a),

$$E(\overline{Y}) = \sum_{i=1}^{n} a_i \mu_i = \sum_{i=1}^{n} a_i \mu = \mu \sum_{i=1}^{n} a_i = \mu \sum_{i=1}^{n} \frac{1}{n} = \frac{n\mu}{n} = \mu.$$

By Theorem 5.12 (b),

$$V(\overline{Y}) = \sum_{i=1}^{n} a_i^2 V(Y_i) + 2 \sum\sum_{i<j} a_i a_j \, \text{Cov}(Y_i, \, Y_j).$$

The covariance terms all are zero because the random variables are independent. Thus,

$$V(\overline{Y}) = \sum_{i=1}^{n} \left(\frac{1}{n}\right)^2 \sigma_i^2 = \sum_{i=1}^{n} \left(\frac{1}{n}\right)^2 \sigma^2 = \frac{1}{n^2} \sum_{i=1}^{n} \sigma^2 = \frac{n\sigma^2}{n^2} = \frac{\sigma^2}{n}.$$

EXAMPLE **5.28**   The number of defectives Y in a sample of $n = 10$ items selected from a manufacturing process follows a binomial probability distribution. An estimator of the fraction defective in the lot is the random variable $\hat{p} = Y/n$. Find the expected value and variance of $\hat{p}$.

Solution   The term $\hat{p}$ is a linear function of a single random variable Y, where $\hat{p} = a_1 Y$ and $a_1 = 1/n$. Then by Theorem 5.12,

$$E(\hat{p}) = a_1 E(Y) = \frac{1}{n} E(Y).$$

The expected value and variance of a binomial random variable are np and npq, respectively. Substituting for $E(Y)$, we obtain

$$E(\hat{p}) = \frac{1}{n}(np) = p.$$

Thus the expected value of the number of defectives Y, divided by the sample size, is p. Similarly

$$V(\hat{p}) = a_1^2 V(Y) = \left(\frac{1}{n}\right)^2 npq = \frac{pq}{n}.$$

EXAMPLE **5.29**   Suppose that an urn contains r red balls and $(N - r)$ black balls. A random sample of n balls is drawn without replacement and Y, the number of red balls in the sample, is observed. From Chapter 3 we know that Y has a hypergeometric probability distribution. Find the mean and variance of Y.

Solution   We will first observe some characteristics of sampling without replacement. Suppose that the sampling is done sequentially, and we observe outcomes for $X_1, X_2, \ldots, X_n$, where

$$X_i = \begin{cases} 1, & \text{if the } i\text{th draw results in a red ball} \\ 0, & \text{otherwise.} \end{cases}$$

Unquestionably, $P(X_1 = 1) = r/N$. But it is also true that $P(X_2 = 1) = r/N$ because

$$P(X_2 = 1) = P(X_1 = 1, \ X_2 = 1) + P(X_1 = 0, \ X_2 = 1)$$

$$= P(X_1 = 1)P(X_2 = 1|X_1 = 1) + P(X_1 = 0)P(X_2 = 1|X_1 = 0)$$

$$= \left(\frac{r}{N}\right)\left(\frac{r-1}{N-1}\right) + \left(\frac{N-r}{N}\right)\left(\frac{r}{N-1}\right) = \frac{r(N-1)}{N(N-1)} = \frac{r}{N}.$$

The same is true for X_k; that is,

$$P(X_k = 1) = \frac{r}{N}, \qquad k = 1, 2, \ldots, n.$$

Thus the (unconditional) probability of drawing a red ball on any draw is r/N.
   In a similar way it can be shown that

$$P(X_j = 1, \ X_k = 1) = \frac{r(r-1)}{N(N-1)}, \qquad j \neq k.$$

Now, observe that $Y = \sum_{i=1}^{n} X_i$, and hence

$$E(Y) = \sum_{i=1}^{n} E(X_i) = \sum_{i=1}^{n} \left(\frac{r}{N}\right) = n\left(\frac{r}{N}\right).$$

To find $V(Y)$ we need $V(X_i)$ and $\text{Cov}(X_i, X_j)$. Because X_i is 1 with probability r/N and 0 with probability $1 - (r/N)$, it follows that

$$V(X_i) = \frac{r}{N}\left(1 - \frac{r}{N}\right).$$

Also,

$$\text{Cov}(X_i, \ X_j) = E(X_i X_j) - E(X_i)E(X_j) = \frac{r(r-1)}{N(N-1)} - \left(\frac{r}{N}\right)^2$$

$$= -\frac{r}{N}\left(1 - \frac{r}{N}\right)\left(\frac{1}{N-1}\right)$$

because $X_i X_j = 1$ if and only if $X_i = 1$ and $X_j = 1$ and $X_i X_j = 0$ otherwise. From Theorem 5.12, we know that

$$V(Y) = \sum_{i=1}^{n} V(X_i) + 2 \sum\sum_{i<j} \text{Cov}(X_i, X_j)$$

$$= \sum_{i=1}^{n} \left(\frac{r}{N}\right)\left(1 - \frac{r}{N}\right) + 2\sum\sum_{i<j}\left[-\frac{r}{N}\left(1 - \frac{r}{N}\right)\left(\frac{1}{N-1}\right)\right]$$

$$= n\left(\frac{r}{N}\right)\left(1 - \frac{r}{N}\right) - n(n-1)\left(\frac{r}{N}\right)\left(1 - \frac{r}{N}\right)\left(\frac{1}{N-1}\right)$$

because the double summation contains $n(n-1)/2$ equal terms. A little algebra yields

$$V(Y) = n\left(\frac{r}{N}\right)\left(1 - \frac{r}{N}\right)\left(\frac{N-n}{N-1}\right).$$

To appreciate the usefulness of Theorem 5.12, notice that the derivations contained in Example 5.29 are much simpler than those outlined in Exercise 3.176, where the mean and variance were derived by using the probabilities associated with the hypergeometric distribution.

Exercises

5.86   A firm purchases two types of industrial chemicals. The type I chemical costs $3.00 per gallon, whereas type II costs $5.00 per gallon. The mean and variance for the number of gallons of type I chemical purchased, Y_1, are 40 and 4, respectively. The amount of type II chemical purchased, Y_2, has $E(Y_2) = 65$ gallons and $V(Y_2) = 8$. Assume that Y_1 and Y_2 are independent, and find the mean and variance of the total amount of money spent per week on the two chemicals.

5.87   Assume that Y_1, Y_2, and Y_3 are random variables, with

$E(Y_1) = 2$	$E(Y_2) = -1$	$E(Y_3) = 4$
$V(Y_1) = 4$	$V(Y_2) = 6$	$V(Y_3) = 8$
$\text{Cov}(Y_1, Y_2) = 1$	$\text{Cov}(Y_1, Y_3) = -1$	$\text{Cov}(Y_2, Y_3) = 0$

Find $E(3Y_1 + 4Y_2 - 6Y_3)$ and $V(3Y_1 + 4Y_2 - 6Y_3)$.

5.88   In Exercise 5.3, we determined that the joint probability distribution of Y_1, the number of married executives, and Y_2, the number of never-married executives, is given by

$$p(y_1, y_2) = \frac{\binom{4}{y_1}\binom{3}{y_2}\binom{2}{3 - y_1 - y_2}}{\binom{9}{3}}.$$

where y_1 and y_2 are integers, $0 \le y_1 \le 3, 0 \le y_2 \le 3$, and $1 \le y_1 + y_2 \le 3$.

a   Find $E(Y_1 + Y_2)$ and $V(Y_1 + Y_2)$ by first finding the probability distribution of $Y_1 + Y_2$.

b   In Exercise 5.76, we determined that $\text{Cov}(Y_1, Y_2) = -1/3$. Find $E(Y_1 + Y_2)$ and $V(Y_1 + Y_2)$ by using Theorem 5.12.

5.89 In Exercise 5.6, we established that

$$f(y_1, y_2) = \begin{cases} 4y_1y_2, & 0 \le y_1 \le 1, \quad 0 \le y_2 \le 1 \\ 0, & \text{elsewhere} \end{cases}$$

is a valid joint probability density function. In Exercise 5.44 we established that Y_1 and Y_2 are independent; in Exercise 5.64, we determined that $E(Y_1 - Y_2) = 0$ and found the value for $V(Y_1)$. Find $V(Y_1 - Y_2)$.

5.90 In Exercise 5.7, we determined that

$$f(y_1, y_2) = \begin{cases} 6(1 - y_2), & 0 \le y_1 \le y_2 \le 1 \\ 0, & \text{elsewhere} \end{cases}$$

is a valid joint probability density function. In Exercise 5.64, we derived the fact that $E(Y_1 - 3Y_2) = -5/4$; in Exercise 5.78, we proved that $\text{Cov}(Y_1, Y_2) = 1/40$. Find $V(Y_1 - 3Y_2)$.

5.91 In Exercise 5.10, we were given the following joint probability density function for the random variables Y_1 and Y_2, which were the proportions of two components in a sample from a mixture of insecticide:

$$f(y_1, y_2) = \begin{cases} 2, & 0 \le y_1 \le 1, \quad 0 \le y_2 \le 1; \quad 0 \le y_1 + y_2 \le 1 \\ 0, & \text{elsewhere.} \end{cases}$$

For the two chemicals under consideration, an important quantity is the total proportion $Y_1 + Y_2$ found in any sample. Find $E(Y_1 + Y_2)$ and $V(Y_1 + Y_2)$.

5.92 If Y_1 is the total time between a customer's arrival in the store and departure from the service window and if Y_2 is the time spent in line before reaching the window, the joint density of these variables was given in Exercise 5.13 to be

$$f(y_1, y_2) = \begin{cases} e^{-y_1}, & 0 \le y_2 \le y_1 \le \infty \\ 0, & \text{elsewhere.} \end{cases}$$

The random variable $Y_1 - Y_2$ represents the time spent at the service window. Find $E(Y_1 - Y_2)$ and $V(Y_1 - Y_2)$. Is it highly likely that a randomly selected customer would spend more than 4 minutes at the service window?

5.93 In Exercise 5.14, Y_1 and Y_2 denoted the proportions of time that employees I and II actually spent working on their assigned tasks during a workday. The joint density of Y_1 and Y_2 is given by

$$f(y_1, y_2) = \begin{cases} y_1 + y_2, & 0 \le y_1 \le 1, \quad 0 \le y_2 \le 1 \\ 0, & \text{elsewhere.} \end{cases}$$

In Exercise 5.68, we derived the mean of the productivity measure $30Y_1 + 25Y_2$. Find the variance of this measure of productivity. Give an interval in which you think the total productivity measures of the two employees should lie for at least 75% of the days in question.

5.94 In Exercise 5.16, Y_1 and Y_2 denoted the lengths of life, in hours, for components of types I and II, respectively, in an electronic system. The joint density of Y_1 and Y_2 is

$$f(y_1, y_2) = \begin{cases} (1/8)y_1 e^{-(y_1+y_2)/2}, & y_1 > 0, \ y_2 > 0 \\ 0, & \text{elsewhere.} \end{cases}$$

The cost C of replacing the two components depends upon their length of life at failure and is given by $C = 50 + 2Y_1 + 4Y_2$. Find $E(C)$ and $V(C)$.

5.95 A retail grocery merchant figures that her daily gain X from sales is a normally distributed random variable with $\mu = 50$ and $\sigma = 3$ (measurements in dollars). X can be negative if she is forced to dispose of enough perishable goods. Also, she figures daily overhead costs Y to have a gamma distribution with $\alpha = 4$ and $\beta = 2$. If X and Y are independent, find the expected value and variance of her net daily *gain*. Would you expect her net gain for tomorrow to rise above $70?

5.96 For the daily output of an industrial operation, let Y_1 denote the amount of sales and Y_2, the costs, in thousands of dollars. Assume that the density functions for Y_1 and Y_2 are given by

$$f_1(y_1) = \begin{cases} (1/6)y_1^3 e^{-y_1}, & y_1 > 0 \\ 0, & y_1 \leq 0 \end{cases} \quad \text{and} \quad f_2(y_2) = \begin{cases} (1/2)e^{-y_2/2}, & y_2 > 0 \\ 0, & y_2 \leq 0. \end{cases}$$

The daily profit is given by $U = Y_1 - Y_2$.
a Find $E(U)$.
b Assuming that Y_1 and Y_2 are independent, find $V(U)$.
c Would expect the daily profit to drop below zero very often? Why?

*5.97 A population of N alligators is to be sampled in order to obtain an approximate measure of the difference between the proportions of sexually mature males and sexually mature females. Obviously, this parameter has important implications for the future of the population. Assume that n animals are to be sampled without replacement. Let Y_1 denote the number of mature females and Y_2 the number of mature males in the sample. If the population contains proportions p_1 and p_2 of mature females and males, respectively (with $p_1 + p_2 < 1$), find expressions for

$$E\left(\frac{Y_1}{n} - \frac{Y_2}{n}\right) \quad \text{and} \quad V\left(\frac{Y_1}{n} - \frac{Y_2}{n}\right).$$

5.98 The total sustained load on the concrete footing of a planned building is the sum of the dead load plus the occupancy load. Suppose that the dead load X_1 has a gamma distribution with $\alpha_1 = 50$ and $\beta_1 = 2$, whereas the occupancy load X_2 has a gamma distribution with $\alpha_2 = 20$ and $\beta_2 = 2$. (Units are in kips.) Assume that X_1 and X_2 are independent.
a Find the mean and variance of the total sustained load on the footing.
b Find a value for the sustained load that will be exceeded with probability less than $1/16$.

5.9 The Multinomial Probability Distribution

Recall from Chapter 3 that a binomial random variable results from an experiment consisting of n trials with two possible outcomes per trial. Frequently we encounter similar situations in which the number of possible outcomes per trial is more than two. For example, experiments that involve blood typing typically have at least four possible outcomes per trial. Experiments that involve sampling for defectives may categorize the type of defects observed into more than two classes.

A multinomial experiment is a generalization of the binomial experiment.

DEFINITION 5.11   A *multinomial experiment* possesses the following properties:

1. The experiment consists of n identical trials.
2. The outcome of each trial falls into one of k classes or cells.

3. The probability that the outcome of a single trial falls into cell i, is p_i, $i = 1, 2, \ldots, k$ and remains the same from trial to trial. Notice that $p_1 + p_2 + p_3 + \cdots + p_k = 1$.

4. The trials are independent.

5. The random variables of interest are $Y_1, Y_2, \ldots, Y_k$, where Y_i equals the number of trials for which the outcome falls into cell i. Notice that $Y_1 + Y_2 + Y_3 + \cdots + Y_k = n$. ∎

The joint probability function for $Y_1, Y_2, \ldots, Y_k$ is given by

$$p(y_1, y_2, \ldots, y_k) = \frac{n!}{y_1! y_2! \cdots y_k!} p_1^{y_1} p_2^{y_2} \cdots p_k^{y_k}$$

where

$$\sum_{i=1}^{k} p_i = 1 \quad \text{and} \quad \sum_{i=1}^{k} y_i = n.$$

Finding the probability that the n trials in a multinomial experiment result in $(Y_1 = y_1, Y_2 = y_2, \ldots, Y_k = y_k)$ is an excellent application of the probabilistic methods of Chapter 2. We leave this problem as an exercise.

DEFINITION 5.12

Assume that $p_1, p_2, \ldots, p_k$ are such that $\sum_{i=1}^{k} p_i = 1$, and $p_i > 0$ for $i = 1, 2, \ldots, k$. The random variables $Y_1, Y_2, \ldots, Y_k$, are said to have a *multinomial distribution* with parameters n and $p_1, p_2, \ldots, p_k$ if the joint probability function of $Y_1, Y_2, \ldots, Y_k$ is given by

$$p(y_1, y_2, \ldots, y_k) = \frac{n!}{y_1! y_2! \cdots y_k!} p_1^{y_1} p_2^{y_2} \cdots p_k^{y_k},$$

where, for each i, $y_i = 0, 1, 2, \ldots, n$ and $\sum_{i=1}^{k} y_i = n$.

Many experiments involving classification are multinomial experiments. For example, classifying people into five income brackets results in an enumeration or count corresponding to each of five income classes. Or we might be interested in studying the reaction of mice to a particular stimulus in a psychological experiment. If the mice can react in one of three ways when the stimulus is applied, the experiment yields the number of mice falling into each reaction class. Similarly, a traffic study might require a count and classification of the types of motor vehicles using a section of highway. An industrial process might manufacture items that fall into one of three quality classes: acceptable, seconds, and rejects. A student of the arts might classify paintings into one of k categories according to style and period or we might wish to classify philosophical ideas of authors in a study of literature. The result of an advertising campaign might yield count data indicating a classification of consumer reactions. Many observations in the physical sciences are not amenable to measurement on a continuous scale and hence result in enumerative data that corresponds to the numbers of observations falling into various classes.

Notice that the binomial experiment is a special case of the multinomial experiment (when there are $k = 2$ classes).

EXAMPLE **5.30**   According to recent census figures, the proportions of adults (persons over 18 years of age) in the United States associated with five age categories are as given in the following table.

Age	Proportion
18–24	.18
25–34	.23
35–44	.16
45–64	.27
65↑	.16

If these figures are accurate and five adults are randomly sampled, find the probability that the sample contains one person between the ages of 18 and 24, two between the ages of 25 and 34, and two between the ages of 45 and 64.

Solution   We will number the five age classes 1, 2, 3, 4, and 5 from top to bottom and will assume that the proportions given are the probabilities associated with each of the classes. Then we wish to find

$$p(y_1, y_2, y_3, y_4, y_5) = \frac{n!}{y_1!\, y_2!\, y_3!\, y_4!\, y_5!} p_1^{y_1} p_2^{y_2} p_3^{y_3} p_4^{y_4} p_5^{y_5}$$

for $n = 5$ and $y_1 = 1$, $y_2 = 2$, $y_3 = 0$, $y_4 = 2$, and $y_5 = 0$. Substituting these values into the formula for the joint probability function, we obtain

$$p(1, 2, 0, 2, 0) = \frac{5!}{1!\, 2!\, 0!\, 2!\, 0!}(.18)^1(.23)^2(.16)^0(.27)^2(.16)^0$$

$$= 30(.18)(.23)^2(.27)^2 = .0208.$$

THEOREM **5.13**   If $Y_1, Y_2, \ldots, Y_k$ have a multinomial distribution with parameters n and $p_1, p_2, \ldots, p_k$, then

1. $E(Y_i) = np_i$, $V(Y_i) = np_i q_i$.
2. $\text{Cov}(Y_s, Y_t) = -np_s p_t$,   if $s \neq t$.

Proof   The marginal distribution of Y_i can be used to derive the mean and variance. Recall that Y_i may be interpreted as the number of trials falling into cell i. Imagine all of the cells, excluding cell i, combined into a single large cell. Then every trial will result in cell i, or in a cell other than cell i, with probabilities p_i and $1 - p_i$, respectively. Thus Y_i possesses a binomial marginal probability distribution. Consequently,

$$E(Y_i) = np_i \quad \text{and} \quad V(Y_i) = np_i q_i, \qquad \text{where } q_i = 1 - p_i.$$

The same results can be obtained by setting up the expectations and evaluating. For example,

$$E(Y_1) = \sum_{y_1} \sum_{y_2} \cdots \sum_{y_k} y_1 \frac{n!}{y_1! y_2! \cdots y_k!} p_1^{y_1} p_2^{y_2} \cdots p_k^{y_k}.$$

Because we have already derived the expected value and variance of Y_i, we leave the summation of this expectation to the interested reader.

The proof of part 2 uses Theorem 5.12. Think of the multinomial experiment as a sequence of n independent trials, and define, for $s \neq t$,

$$U_i = \begin{cases} 1, & \text{if trial } i \text{ results in class } s \\ 0, & \text{otherwise} \end{cases}$$

and

$$W_i = \begin{cases} 1, & \text{if trial } i \text{ results in class } t \\ 0, & \text{otherwise.} \end{cases}$$

Then

$$Y_s = \sum_{i=1}^{n} U_i \quad \text{and} \quad Y_t = \sum_{j=1}^{n} W_j.$$

(Because $U_i = 1$ or 0 depending upon whether the ith trial resulted in class s, Y_s is simply the sum of a series of 0s and 1s. A 1 occurs in the sum everytime we observe an item from class s, and a 0 occurs everytime we observe any other class. Thus Y_s is simply the number of times class s is observed. A similar interpretation applies to Y_t.)

Notice that U_i and W_i cannot both equal 1 (the ith item cannot simultaneously be in classes s and t). Thus, the product $U_i W_i$ always equals zero, and $E(U_i W_i) = 0$. The following results allow us to evaluate $\text{Cov}(Y_s, Y_t)$:

$$E(U_i) = p_s \qquad E(W_j) = p_t$$

$$\text{Cov}(U_i, W_j) = 0 \qquad \text{if } i \neq j \text{ because the trials are independent}$$

$$\text{Cov}(U_i, W_i) = E(U_i W_i) - E(U_i)E(W_i) = 0 - p_s p_t$$

From Theorem 5.12 we then have

$$\text{Cov}(Y_s, Y_t) = \sum_{i=1}^{n}\sum_{j=1}^{n} \text{Cov}(U_i, W_j)$$

$$= \sum_{i=1}^{n} \text{Cov}(U_i, W_i) + \sum\sum_{i\neq j} \text{Cov}(U_i, W_j)$$

$$= \sum_{i=1}^{n}(-p_s p_t) + \sum\sum_{i\neq j} 0 = -np_s p_t.$$

The covariance here is negative, which is to be expected because a large number of outcomes in cell s would force the number in cell t to be small.

Inferential problems associated with the multinomial experiment will be discussed later.

Exercises

5.99   A learning experiment requires a rat to run a maze (a network of pathways) until it locates one of three possible exits. Exit 1 presents a reward of food, but exits 2 and 3 do not. (If the rat eventually selects exit 1 almost every time, learning may have taken place.) Let Y_i denote the number of times exit i is chosen in successive runnings. For the following, assume that the rat chooses an exit at random on each run.

   a   Find the probability that $n = 6$ runs result in $Y_1 = 3$, $Y_2 = 1$, and $Y_3 = 2$.

   b   For general n, find $E(Y_1)$ and $V(Y_1)$.

   c   Find $\text{Cov}(Y_2, Y_3)$ for general n.

   d   To check for the rat's preference between exits 2 and 3, we may look at $Y_2 - Y_3$. Find $E(Y_2 - Y_3)$ and $V(Y_2 - Y_3)$ for general n.

5.100   A sample of size n is selected from a large lot of items in which a proportion p_1 contains exactly one defect and a proportion p_2 contains more than one defect (with $p_1 + p_2 < 1$). The cost of repairing the defective items in the sample is $C = Y_1 + 3Y_2$, where Y_1 denotes the number of items with one defect and Y_2 denotes the number with two or more defects. Find the expected value and variance of C.

5.101   Refer to Exercise 5.97. Suppose that the number N of alligators in the population is very large, with $p_1 = .3$ and $p_2 = .1$.

   a   Find the probability that, in a sample of five alligators, $Y_1 = 2$ and $Y_2 = 1$.

   b   If $n = 5$, find $E\left(\dfrac{Y_1}{n} - \dfrac{Y_2}{n}\right)$ and $V\left(\dfrac{Y_1}{n} - \dfrac{Y_2}{n}\right)$.

5.102   The weights of a population of mice fed on a certain diet since birth are assumed to be normally distributed with $\mu = 100$ and $\sigma = 20$ (measurement in grams). Suppose that a random sample of $n = 4$ mice is taken from this population.

   a   Find the probability that exactly two weigh between 80 and 100 grams and exactly one weighs more than 100 grams.

   b   Find the probability that all four mice weigh more than 100 grams.

5.103   The National Fire Incident Reporting Service stated that, among residential fires, 73% are in family homes, 20% are in apartments, and 7% are in other types of dwellings. If four residential fires are independently reported on a single day, what is the probability that two are in family homes, one is in an apartment, and one is in another type of dwelling?

5.104   The typical cost of damages caused by a fire in a family home is $20,000. Comparable costs for an apartment fire and for fire in other dwelling types are $10,000 and $2,000, respectively. If four fires are independently reported, use the information in Exercise 5.103 to

    a   find the expected total damage cost.

    b   find the variance of the total damage cost.

5.105   When commercial aircraft are inspected, wing cracks are reported as nonexistent, detectable, or critical. The history of a particular fleet indicates that 70% of the planes inspected have no wing cracks, 25% have detectable wing cracks, and 5% have critical wing cracks. Five planes are randomly selected. Find the probability that

    a   one has a critical crack, two have detectable cracks, and two have no cracks.

    b   at least one plane has critical cracks.

5.106   A large lot of manufactured items contains 10% with exactly one defect, 5% with more than one defect, and the remainder with no defects. Ten items are randomly selected from this lot for sale. If Y_1 denotes the number of items with one defect and Y_2, the number with more than one defect, the repair costs are $Y_1 + 3Y_2$. Find the mean and variance of the repair costs.

5.107   Refer to Exercise 5.106. Let Y denote the number of items among the ten that contain at least one defect. Find the probability that Y

    a   equals 2.

    b   is at least 1.

5.10 The Bivariate Normal Distribution (Optional)

No discussion of multivariate probability distributions would be complete without reference to the multivariate normal distribution, which is a keystone of much modern statistical theory. In general, the multivariate normal density function is defined for k continuous random variables, $Y_1, Y_2, \ldots, Y_k$. Because of its complexity, we will present only the bivariate density function ($k = 2$):

$$f(y_1, y_2) = \frac{e^{-Q/2}}{2\pi \sigma_1 \sigma_2 \sqrt{1 - \rho^2}}, \qquad -\infty < y_1 < \infty, \ -\infty < y_2 < \infty$$

where

$$Q = \frac{1}{1 - \rho^2}\left[\frac{(y_1 - \mu_1)^2}{\sigma_1^2} - 2\rho\frac{(y_1 - \mu_1)(y_2 - \mu_2)}{\sigma_1 \sigma_2} + \frac{(y_2 - \mu_2)^2}{\sigma_2^2}\right].$$

The bivariate normal density function is a function of five parameters: $\mu_1, \mu_2, \sigma_1^2,$ $\sigma_2^2,$ and ρ. The choice of symbols employed for these parameters is not coincidental.

In Exercise 5.108, you will show that the marginal distributions of Y_1 and Y_2 are normal distributions with means μ_1 and μ_2 and variances σ_1^2 and σ_2^2, respectively. With a bit of somewhat tedious integration, we can show that $\text{Cov}(Y_1,\ Y_2) = \rho\sigma_1\sigma_2$. If $\text{Cov}(Y_1,\ Y_2) = 0$—or, equivalently, if $\rho = 0$—then

$$f(y_1,\ y_2) = g(y_1)h(y_2)$$

where $g(y_1)$ is a nonnegative function of y_1 alone and $h(y_2)$ is a nonnegative function of y_2 alone. Therefore, if $\rho = 0$, Theorem 5.5 implies that Y_1 and Y_2 are independent. Recall that zero covariance for two random variables does not generally imply independence. However, if Y_1 and Y_2 have a bivariate normal distribution, they are independent if and only if their covariance is zero.

The expression for the joint density function, $k > 2$, is most easily expressed by using the matrix algebra. A discussion of the general case can be found in the references at the end of this chapter.

Exercises

*5.108   Let Y_1 and Y_2 have a bivariate normal distribution. Show that the marginal distribution of Y_1 is normal with mean μ_1 and variance σ_1^2. What is the marginal distribution of Y_2?

*5.109   Let Y_1 and Y_2 have a bivariate normal distribution. Show that the conditional distribution of Y_1 given that $Y_2 = y_2$ is a normal distribution with mean $\mu_1 + \rho\frac{\sigma_1}{\sigma_2}(y_2 - \mu_2)$ and variance $\sigma_1^2(1 - \rho^2)$.

*5.110   Let $Y_1,\ Y_2,\ \ldots,\ Y_n$ be independent random variables with $E(Y_i) = \mu$ and $V(Y_i) = \sigma^2$ for $i = 1,\ 2,\ \ldots,\ n$. Let

$$U_1 = \sum_{i=1}^{n} a_i Y_i \quad \text{and} \quad U_2 = \sum_{i=1}^{n} b_i Y_i,$$

where $a_1,\ a_2,\ \ldots,\ a_n$, and $b_1,\ b_2,\ \ldots,\ b_n$ are constants. U_1 and U_2 are said to be orthogonal if $\text{Cov}(U_1, U_2) = 0$.

a   Show that U_1 and U_2 are orthogonal if and only if $\sum_{i=1}^{n} a_i b_i = 0$.

b   Suppose, in addition, that $Y_1,\ Y_2,\ \ldots,\ Y_n$ have a multivariate normal distribution. Then U_1 and U_2 have a bivariate normal distribution. Show that U_1 and U_2 are independent if they are orthogonal.

*5.111   Let Y_1 and Y_2 be independent normally distributed random variables with means μ_1 and μ_2, respectively, and variances $\sigma_1^2 = \sigma_2^2 = \sigma^2$.

a   Show that Y_1 and Y_2 have a bivariate normal distribution with $\rho = 0$.

b   Consider $U_1 = Y_1 + Y_2$ and $U_2 = Y_1 - Y_2$. Use the result in Exercise 5.110 to show that U_1 and U_2 have a bivariate normal distribution and that U_1 and U_2 are independent.

*5.112   Refer to Exercise 5.111. What are the marginal distributions of U_1 and U_2?

5.11 Conditional Expectations

Section 5.3 contains a discussion of conditional probability functions and conditional density functions, which we will now relate to conditional expectations. Conditional expectations are defined in the same manner as univariate expectations except that the conditional distributions are used in place of the marginal distributions.

DEFINITION 5.13

If Y_1 and Y_2 are any two random variables, the *conditional expectation* of $g(Y_1)$, given that $Y_2 = y_2$, is defined to be

$$E(g(Y_1) \mid Y_2 = y_2) = \int_{-\infty}^{\infty} g(y_1) f(y_1 \mid y_2) \, dy_1$$

if Y_1 and Y_2 are jointly continuous and

$$E(g(Y_1) \mid Y_2 = y_2) = \sum_{y_1} g(y_1) p(y_1 \mid y_2)$$

if Y_1 and Y_2 are jointly discrete.

EXAMPLE 5.31

Refer to the random variables Y_1 and Y_2 of Example 5.8, where the joint density function is given by

$$f(y_1, y_2) = \begin{cases} 1/2, & 0 \le y_1 \le y_2 \le 2 \\ 0, & \text{elsewhere.} \end{cases}$$

Find the conditional expectation of the amount of sales, Y_1, given that $Y_2 = 1.5$.

Solution

In Example 5.8, we found that, if $0 < y_2 \le 2$,

$$f(y_1 \mid y_2) = \begin{cases} 1/y_2, & 0 < y_1 \le y_2 \\ 0, & \text{elsewhere.} \end{cases}$$

Thus, from Definition 5.13, for any value of y_2 such that $0 < y_2 \le 2$,

$$E(Y_1 \mid Y_2 = y_2) = \int_{-\infty}^{\infty} y_1 f(y_1 \mid y_2) \, dy_1$$

$$= \int_0^{y_2} y_1 \left(\frac{1}{y_2} \right) dy_1 = \frac{1}{y_2} \left(\frac{y_1^2}{2} \right]_0^{y_2} = \frac{y_2}{2}.$$

Because we are interested in the value $y_2 = 1.5$, it follows that $E(Y_1 \mid Y_2 = 1.5) = 1.5/2 = 0.75$. That is, if the soft-drink machine contains 1.5 gallons at the start of the day, the expected amount to be sold that day is 0.75 gallon.

In general, the conditional expectation of Y_1 given $Y_2 = y_2$ is a function of y_2. If we now let Y_2 range over all of its possible values, we can think of the conditional expectation $E(Y_1 \mid Y_2)$ as a function of the random variable Y_2. In Example 5.13, we obtained $E(Y_1 \mid Y_2 = y_2) = y_2/2$. It follows that $E(Y_1 \mid Y_2) = Y_2/2$. Because $E(Y_1 \mid Y_2)$ is a function of the random variable Y_2, it is itself a random variable; and as such, it has a mean and a variance. We consider the mean of this random variable in Theorem 5.14 and the variance in Theorem 5.15.

THEOREM 5.14

Let Y_1 and Y_2 denote random variables. Then

$$E(Y_1) = E\left[E(Y_1 \mid Y_2)\right]$$

where, on the right-hand side, the inside expectation is with respect to the conditional distribution of Y_1 given Y_2, and the outside expectation is with respect to the distribution of Y_2.

Proof

Suppose that Y_1 and Y_2 are jointly continuous with joint density function $f(y_1, y_2)$ and marginal densities $f_1(y_1)$ and $f_2(y_2)$, respectively. Then

$$E(Y_1) = \int_{-\infty}^{\infty} \int_{-\infty}^{\infty} y_1 f(y_1, y_2)\, dy_1\, dy_2$$

$$= \int_{-\infty}^{\infty} \int_{-\infty}^{\infty} y_1 f(y_1 \mid y_2) f_2(y_2)\, dy_1\, dy_2$$

$$= \int_{-\infty}^{\infty} \left\{ \int_{-\infty}^{\infty} y_1 f(y_1 \mid y_2)\, dy_1 \right\} f_2(y_2)\, dy_2$$

$$= \int_{-\infty}^{\infty} E(Y_1 \mid Y_2 = y_2) f_2(y_2)\, dy_2 = E\left[E(Y_1 \mid Y_2)\right].$$

The proof is similar for the discrete case. ∎

EXAMPLE 5.32

A quality control plan for an assembly line involves sampling $n = 10$ finished items per day and counting Y, the number of defectives. If p denotes the probability of observing a defective, then Y has a binomial distribution, assuming that a large number of items are produced by the line. But p varies from day to day and is assumed to have a uniform distribution on the interval from 0 to $1/4$. Find the expected value of Y.

Solution

From Theorem 5.14, we know that $E(Y) = E\left[E(Y \mid p)\right]$. For a given p, Y has a binomial distribution, and hence $E(Y \mid p) = np$. Thus,

$$E(Y) = E[E(Y \mid p)] = E(np) = nE(p) = n\left(\frac{1/4 - 0}{2}\right) = \frac{n}{8},$$

and for $n = 10$

$$E(Y) = 10/8 = 1.25.$$

In the long run, this inspection policy will average 1.25 defectives per day.

The conditional variance of Y_1 given $Y_2 = y_2$ is defined by analogy with an ordinary variance, again using the conditional distribution of Y_1 given $Y_2 = y_2$ in place of the ordinary marginal distribution of Y_1. That is,

$$V(Y_1 \mid Y_2 = y_2) = E(Y_1^2 \mid Y_2 = y_2) - [E(Y_1 \mid Y_2 = y_2)]^2.$$

As in the case of the conditional mean, the conditional variance is a function of y_2. Letting Y_2 range over all of its possible values, we can define $V(Y_1 \mid Y_2)$ as a random variable that is a function of Y_2. Specifically, if $g(y_2) = V(Y_1 \mid Y_2 = y_2)$ is a particular function of the observed value, y_2, then $g(Y_2) = V(Y_1 \mid Y_2)$ is the *same function* of the random variable, Y_2. The expected value of $V(Y_1 \mid Y_2)$ is useful in computing the variance of Y_1, as detailed in Theorem 5.15.

THEOREM 5.15   Let Y_1 and Y_2 denote random variables. Then

$$V(Y_1) = E\big[V(Y_1 \mid Y_2)\big] + V\big[E(Y_1 \mid Y_2)\big].$$

Proof   As previously indicated, $V(Y_1 \mid Y_2)$ is given by

$$V(Y_1 \mid Y_2) = E(Y_1^2 \mid Y_2) - \big[E(Y_1 \mid Y_2)\big]^2$$

and

$$E\big[V(Y_1 \mid Y_2)\big] = E\big[E(Y_1^2 \mid Y_2)\big] - E\left\{\big[E(Y_1 \mid Y_2)\big]^2\right\}.$$

By definition,

$$V\big[E(Y_1 \mid Y_2)\big] = E\left\{\big[E(Y_1 \mid Y_2)\big]^2\right\} - \big\{E\big[E(Y_1 \mid Y_2)\big]\big\}^2.$$

The variance of Y_1 is

$$\begin{aligned} V(Y_1) &= E\big[Y_1^2\big] - \big[E(Y_1)\big]^2 \\ &= E\big\{E[Y_1^2 \mid Y_2]\big\} - \big\{E\big[E(Y_1 \mid Y_2)\big]\big\}^2 \\ &= E\big\{E[Y_1^2 \mid Y_2]\big\} - E\left\{\big[E(Y_1 \mid Y_2)\big]^2\right\} + E\left\{\big[E(Y_1 \mid Y_2)\big]^2\right\} \\ &\quad - \big\{E\big[E(Y_1 \mid Y_2)\big]\big\}^2 \\ &= E\big[V(Y_1 \mid Y_2)\big] + V\big[E(Y_1 \mid Y_2)\big]. \end{aligned}$$

∎

EXAMPLE **5.33**   Refer to Example 5.32. Find the variance of Y.

Solution   From Theorem 5.15 we know that

$$V(Y_1) = E\big[V(Y_1 \mid Y_2)\big] + V\big[E(Y_1 \mid Y_2)\big].$$

For a given p, Y has a binomial distribution, and hence $E(Y \mid p) = np$ and $V(Y \mid p) = npq$. Thus,

$$V(Y) = E\big[V(Y \mid p)\big] + V\big[E(Y \mid p)\big] = E(npq) + V(np) = nE\,[p(1-p)] + n^2 V(p).$$

Because p is uniformly distributed on the interval $(0, 1/4)$, and $E(p^2) = V(p) + [E(p)]^2$ it follows that

$$E(p) = \frac{1}{8}, \qquad V(p) = \frac{(1/4 - 0)^2}{12} = \frac{1}{192}, \quad \text{and} \quad E(p^2) = \frac{1}{192} + \frac{1}{64} = \frac{1}{48}.$$

Thus,

$$V(Y) = nE\,[p(1-p)] + n^2 V(p) = n\left[E(p) - E(p^2)\right] + n^2 V(p)$$

$$= n\left(\frac{1}{8} - \frac{1}{48}\right) + n^2\left(\frac{1}{192}\right) = \frac{5n}{48} + \frac{n^2}{192}$$

and for $n = 10$

$$V(Y) = 50/48 + 100/192 = 1.5625.$$

Thus, the standard deviation of Y is $\sigma = \sqrt{1.5625} = 1.25$.

The mean and variance of Y calculated in Examples 5.32 and 5.33 could be checked by finding the unconditional distribution of Y and computing $E(Y)$ and $V(Y)$ directly. In doing so, we would need to find the joint distribution of Y and p. From this joint distribution, the marginal distribution of Y can be obtained and $E(Y)$ determined by evaluating $\sum_y yp(y)$. The variance can be determined in the usual manner, again using the marginal distribution of Y. In Examples 5.32 and 5.33 we avoided working directly with these joint and marginal distributions. Theorems 5.14 and 5.15 permitted a much quicker calculation of the desired mean and variance. As always, the mean and variance of a random variable can be used with Tchebysheff's theorem to provide bounds for probabilities when the distribution of the variable is unknown or difficult to derive.

In Examples 5.32 and 5.33, we encountered a situation where the distribution of a random variable ($Y =$ the number of defectives) was given *conditionally* for possible values of a quantity p that could vary from day to day. The fact that p varied was accommodated by assigning a probability distribution to this variable. This is an example of a *hierarchical* model. In such models, the distribution of a variable of interest, say, Y, is given, *conditional* on the value of a "parameter" θ. Uncertainty about the actual value of θ is modeled by assigning a probability distribution to it. Once we specify the conditional distribution of Y given θ *and* the marginal distribution of θ, the joint distribution of Y and θ is obtained by multiplying the conditional by the marginal. The marginal distribution of Y is then obtained from the joint distribution by integrating or summing over the possible values of θ. The results of this section can be used to find $E(Y)$ and $V(Y)$ *without* finding this marginal distribution. Other examples of hierarchical models are contained in Exercises 5.116 and 5.118.

Exercises

5.113 In Exercise 5.7, we determined that

$$f(y_1, y_2) = \begin{cases} 6(1 - y_2), & 0 \le y_1 \le y_2 \le 1 \\ 0, & \text{elsewhere} \end{cases}$$

is a valid joint probability density function.

a Find $E(Y_1|Y_2 = y_2)$.

b Use the answer derived in (a) to find $E(Y_1)$. (Compare this with the answer found in Exercise 5.65.)

5.114 In Examples 5.32 and 5.33, we determined that if Y is the number of defectives, $E(Y) = 1.25$ and $V(Y) = 1.5625$. Is it likely that, on any given day, Y will exceed 6?

5.115 In Exercise 5.35, we considered a quality control plan that calls for randomly selecting three items from the daily production (assumed large) of a certain machine and observing the number of defectives. The proportion p of defectives produced by the machine varies from day to day and has a uniform distribution on the interval $(0, 1)$.

a Find the expected number of defectives observed among the three sampled items.

b Find the variance of the number of defectives among the three sampled.

5.116 In Exercise 5.36, the number of defects per yard in a certain fabric, Y, was known to have a Poisson distribution with parameter λ. The parameter λ was assumed to be a random variable with a density function given by

$$f(\lambda) = \begin{cases} e^{-\lambda}, & \lambda \ge 0 \\ 0, & \text{elsewhere.} \end{cases}$$

a Find the expected number of defects per yard by first finding the conditional expectation of Y for given λ.

b Find the variance of Y.

c Is it likely that Y exceeds 9?

5.117 In Exercise 5.32 we assumed that Y_1, the weight of a bulk item stocked by a supplier, had a uniform distribution over the interval $(0, 1)$. The random variable Y_2 denoted the weight of the item sold and was assumed to have a uniform distribution over the interval $(0, y_1)$, where y_1 was a specific value of Y_1. If the supplier stocked 3/4 ton, what amount could be expected to be sold during the week?

5.118 Assume that Y denotes the number of bacteria per cubic centimeter in a particular liquid and that Y has a Poisson distribution with parameter λ. Further assume that λ varies from location to location and has a gamma distribution with parameters α and β, where α is a positive integer. If we randomly select a location,

a what is the expected number of bacteria per cubic centimeter?

b what is the standard deviation of the number of bacteria per cubic centimeter?

*5.119 If Y_1 and Y_2 are independent random variables, each having a normal distribution with mean 0 and variance 1, find the moment-generating function of $U = Y_1 Y_2$. Use this moment-generating function to find $E(U)$ and $V(U)$. Check the result by evaluating $E(U)$ and $V(U)$ directly from the density functions for Y_1 and Y_2.

5.12 Summary

The multinomial experiment (Section 5.9) and its associated multinomial probability distribution convey the theme of this chapter. Most experiments yield sample measurements, $y_1, y_2, \ldots, y_k$, which may be regarded as observations on k random variables. Inferences about the underlying structure that generates the observations—the probabilities of falling into cells $1, 2, \ldots, k$—are based on knowledge of the probabilities associated with various samples $(y_1, y_2, \ldots, y_k)$. Joint, marginal, and conditional distributions are essential concepts in finding the probabilities of various sample outcomes.

Generally we draw from a population a sample of n observations, which are specific values of $Y_1, Y_2, \ldots, Y_n$. Many times the random variables are independent and have the same probability distribution. As a consequence, the concept of independence is useful in finding the probability of observing the given sample.

The objective of this chapter has been to convey the ideas contained in the two preceding paragraphs. The numerous details contained in the chapter are essential in providing a solid background for a study of inference. At the same time, you should be careful to avoid overemphasis on details; be sure to keep the broader inferential objectives in mind.

References and Further Readings

Hoel, P. G. 1984. *Introduction to Mathematical Statistics*, 5th ed. New York: John Wiley.

Hogg, R. V., and A. T. Craig. 1995. *Introduction to Mathematical Statistics*, 5th ed. Upper Saddle River, N.J.: Prentice Hall.

Mood, A. M., F. A. Graybill, and D. Boes. 1974. *Introduction to the Theory of Statistics*, 3d ed. New York: McGraw-Hill.

Myers, R. H. 1990. *Classical and Modern Regression with Applications*, 2d ed. Boston: PWS-Kent.

Parzen, E. 1960. *Modern Probability Theory and Its Applications*. New York: John Wiley.

Supplementary Exercises

5.120 A target for a bomb is in the center of a circle with radius of 1 mile. A bomb falls at a randomly selected point inside that circle. If the bomb destroys everything within $1/2$ mile of its landing point, what is the probability that the target is destroyed?

5.121   Two friends are to meet at the library. Each independently and randomly selects an arrival time within the same 1-hour period. Each agrees to wait a maximum of 10 min for the other to arrive. What is the probability that they will meet?

5.122   A committee of three people is to be randomly selected from a group containing four Republicans, three Democrats, and two Independents. Let Y_1 and Y_2 denote numbers of Republicans and Democrats, respectively, on the committee.

    a   What is the joint probability distribution for Y_1 and Y_2?

    b   Find the marginal distributions of Y_1 and Y_2.

    c   Find $P(Y_1 = 1 | Y_2 \geq 1)$.

5.123   Let Y_1 and Y_2 have a joint density function given by

$$f(y_1, y_2) = \begin{cases} 3y_1, & 0 \leq y_2 \leq y_1 \leq 1 \\ 0, & \text{elsewhere.} \end{cases}$$

    a   Find the marginal density functions of Y_1 and Y_2.

    b   Find $P(Y_1 \leq 3/4 | Y_2 \leq 1/2)$.

    c   Find the conditional density function of Y_1 given $Y_2 = y_2$.

    d   Find $P(Y_1 \leq 3/4 | Y_2 = 1/2)$.

5.124   Refer to Exercise 5.123.

    a   Find $E(Y_2 | Y_1 = y_1)$.

    b   Use Theorem 5.14 to find $E(Y_2)$.

    c   Find $E(Y_2)$ directly from the marginal density of Y_2.

5.125   The lengths of life Y for a type of fuse has an exponential distribution with a density function given by

$$f(y) = \begin{cases} (1/\beta)e^{-y/\beta}, & y \geq 0 \\ 0, & \text{elsewhere.} \end{cases}$$

    a   If two such fuses have independent life lengths Y_1 and Y_2, find their joint probability density function.

    b   One fuse from (a) is in a primary system, and the other is in a backup system that comes into use only if the primary system fails. The total effective life length of the two fuses, therefore, is $Y_1 + Y_2$. Find $P(Y_1 + Y_2 \leq a)$, where $a > 0$.

5.126   In the production of a certain type of copper, two types of copper powder (types A and B) are mixed together and sintered (heated) for a certain length of time. For a fixed volume of sintered copper, the producer measures the proportion Y_1 of the volume due to solid copper (some pores will have to be filled with air) and the proportion Y_2 of the solid mass due to type A crystals. Assume that appropriate probability densities for Y_1 and Y_2 are

$$f_1(y_1) = \begin{cases} 6y_1(1 - y_1), & 0 \leq y_1 \leq 1 \\ 0, & \text{elsewhere} \end{cases}$$

$$f_2(y_2) = \begin{cases} 3y_2^2, & 0 \leq y_2 \leq 1 \\ 0, & \text{elsewhere.} \end{cases}$$

The proportion of the sample volume due to type A crystals is then $Y_1 Y_2$. Assuming that Y_1 and Y_2 are independent, find $P(Y_1 Y_2 \leq .5)$.

5.127 Suppose that the number of eggs laid by a certain insect has a Poisson distribution with mean λ. The probability that any one egg hatches is p. Assume that the eggs hatch independently of one another.

   a Find the expected value of Y, the total number of eggs that hatch.

   b Find the variance of Y.

5.128 In a clinical study of a new drug formulated to reduce the effects of rheumatoid arthritis, researchers found that the proportion p of patients who respond favorably to the drug is a random variable that varies from batch to batch of the drug. Assume that p has a probability density function given by

$$f(p) = \begin{cases} 12p^2(1-p), & 0 \le p \le 1 \\ 0, & \text{elsewhere.} \end{cases}$$

Suppose that n patients are injected with portions of the drug taken from the same batch. Let Y denote the number showing a favorable response.

   a Find the unconditional probability distribution of Y for general n.

   b Find $E(Y)$ for $n = 2$.

5.129 A forester studying diseased pine trees models the number of diseased trees per acre, Y, as a Poisson random variable with mean λ. However, λ changes from area to area, and its random behavior is modeled by a gamma distribution. That is, for some integer α,

$$f(\lambda) = \begin{cases} \dfrac{1}{\Gamma(\alpha)\beta^\alpha}\lambda^{\alpha-1}e^{-\lambda/\beta}, & \lambda > 0 \\ 0, & \text{elsewhere.} \end{cases}$$

Find the unconditional probability distribution for Y.

5.130 A coin has probability p of coming up heads when tossed. In n independent tosses of the coin, let $X_i = 1$ if the ith toss results in heads and $X_i = 0$ if the ith toss results in tails. Then Y, the number of heads in the n tosses, has a binomial distribution and can be represented as $Y = \sum_{i=1}^{n} X_i$. Find $E(Y)$ and $V(Y)$ using Theorem 5.12.

*5.131 The negative binomial random variable Y was defined in Section 3.6 as the number of the trial on which the rth success occurs, in a sequence of independent trials with constant probability p of success on each trial. Let X_i denote a random variable defined as the number of the trial on which the ith success occurs, for $i = 1, 2, \ldots, r$. Now define

$$W_i = X_i - X_{i-1}, \qquad i = 1, 2, \ldots, r$$

where X_0 is defined to be zero. Then we can write $Y = \sum_{i=1}^{r} W_i$. Notice that the random variables $W_1, W_2, \ldots, W_r$ have identical geometric distributions and are mutually independent. Use Theorem 5.12 to show that $E(Y) = r/p$ and $V(Y) = r(1-p)/p^2$.

5.132 A box contains four balls, numbered 1 through 4. One ball is selected at random from this box. Let

$$X_1 = 1 \text{ if ball number 1 or ball number 2 is drawn}$$

$$X_2 = 1 \text{ if ball number 1 or ball number 3 is drawn}$$

$$X_3 = 1 \text{ if ball number 1 or ball number 4 is drawn.}$$

The X_i values are zero otherwise. Show that any two of the random variables X_1, X_2, and X_3 are independent but that the three together are not.

5.133 Suppose that we are to observe two independent random samples: Y_1, Y_2, ..., Y_n denoting a random sample from a normal distribution with mean μ_1 and variance σ_1^2; and $X_1, X_2, \ldots, X_m$ denoting a random sample from another normal distribution with mean μ_2 and variance σ_2^2. An approximation for $\mu_1 - \mu_2$ is given by $\overline{Y} - \overline{X}$, the difference between the sample means. Find $E(\overline{Y} - \overline{X})$ and $V(\overline{Y} - \overline{X})$.

*5.134 Let X_1, X_2, and X_3 be random variables, either continuous or discrete. The joint moment-generating function of X_1, X_2, and X_3 is defined by

$$m(t_1, t_2, t_3) = E(e^{t_1 X_1 + t_2 X_2 + t_3 X_3})$$

a Show that $m(t, t, t)$ gives the moment-generating function of $X_1 + X_2 + X_3$.

b Show that $m(t, t, 0)$ gives the moment-generating function of $X_1 + X_2$.

c Show that

$$\frac{\partial^{k_1+k_2+k_3} m(t_1, t_2, t_3)}{\partial t_1^{k_1} \partial t_2^{k_2} \partial t_3^{k_3}} \Bigg]_{t_1=t_2=t_3=0} = E\left(X_1^{k_1} X_2^{k_2} X_3^{k_3}\right).$$

*5.135 Let X_1, X_2, and X_3 have a multinomial distribution with probability function

$$p(x_1, x_2, x_3) = \frac{n!}{x_1! x_2! x_3!} p_1^{x_1} p_2^{x_2} p_3^{x_3}, \qquad \sum_{i=1}^{n} x_i = n.$$

Use the results of Exercise 5.134 to do the following.

a Find the joint moment-generating function of X_1, X_2, and X_3.

b Use the answer to (a) to show that the marginal distribution of X_1 is binomial with parameter p_1.

c Use the joint moment-generating function to find $\text{Cov}(X_1, X_2)$.

*5.136 A box contains N_1 white balls, N_2 black balls, and N_3 red balls ($N_1 + N_2 + N_3 = N$). A random sample of n balls is selected from the box (without replacement). Let Y_1, Y_2, and Y_3 denote the number of white, black, and red balls, respectively, observed in the sample. Find the correlation coefficient for Y_1 and Y_2. (Let $p_i = N_i/N$, for $i = 1, 2, 3$.)

*5.137 Let Y_1 and Y_2 be jointly distributed random variables with finite variances.

a Show that $[E(Y_1 Y_2)]^2 \leq E(Y_1^2) E(Y_2^2)$. [*Hint:* Observe that $E[(tY_1 - Y_2)^2] \geq 0$ for any real number t or, equivalently,

$$t^2 E(Y_1^2) - 2t E(Y_1 Y_2) + E(Y_2^2) \geq 0.$$

This is a quadratic expression of the form $At^2 + Bt + C$; and because it is nonnegative, we must have $B^2 - 4AC \leq 0$. The preceding inequality follows directly.]

b Let ρ denote the correlation coefficient of Y_1 and Y_2. Using the inequality in (a), show that $\rho^2 \leq 1$.

CHAPTER **6**

Functions of Random Variables

6.1   Introduction

6.2   Finding the Probability Distribution of a Function of Random Variables

6.3   The Method of Distribution Functions

6.4   The Method of Transformations

6.5   The Method of Moment-Generating Functions

6.6   Multivariable Transformations Using Jacobians (Optional)

6.7   Order Statistics

6.8   Summary

References and Further Readings

6.1 Introduction

As we indicated in Chapter 1, the objective of statistics is to make inferences about a population based on information contained in a sample taken from that population. Any truly useful inference must be accompanied by an associated measure of goodness. Each of the topics discussed in the preceding chapters plays a role in the development of statistical inference. However, none of the topics discussed thus far pertains to the objective of statistics as closely as the study of the distributions of functions of random variables. This is because all quantities used to estimate population parameters or to make decisions about a population are functions of the n random observations that appear in a sample.

To illustrate, consider the problem of estimating a population mean, μ. Intuitively we draw a random sample of n observations, $y_1, y_2, \ldots, y_n$, from the population and employ the sample mean

$$\bar{y} = \frac{y_1 + y_2 + \cdots + y_n}{n} = \frac{1}{n} \sum_{i=1}^{n} y_i$$

as an estimate for μ. How good is this estimate? The answer depends upon the behavior of the random variables $Y_1, Y_2, \ldots, Y_n$ and their effect on the distribution of $\bar{Y} = (1/n) \sum_{i=1}^{n} Y_i$.

A measure of the goodness of an estimate is the *error of estimation*, the difference between the estimate and the parameter estimated (for our example, the difference between $\bar{y}$ and μ). Because $Y_1, Y_2, \ldots, Y_n$ are random variables, in repeated sampling $\bar{Y}$ is also a random variable (and a function of the n variables $Y_1, Y_2, \ldots, Y_n$). Therefore, we cannot be certain that the error of estimation will be less than a specific value, say, B. However, if we can determine the probability distribution of the estimator $\bar{Y}$, this probability distribution can be used to determine the *probability* that the error of estimation is less than or equal to B.

To determine the probability distribution for a function of n random variables, $Y_1, Y_2, \ldots, Y_n$, we must find the joint probability distribution for the random variables themselves. We generally assume that observations are obtained through random sampling, as defined in Section 2.12. We saw in Section 3.7 that random sampling from a finite population (sampling without replacement) results in dependent trials but that these trials become essentially independent if the population is large when compared to the size of the sample.

We will assume throughout the remainder of this text that populations are large in comparison to the sample size and consequently that the random variables obtained through a random sample are, in fact, independent of one another. Thus in the discrete case, the joint probability function for $Y_1, Y_2, \ldots, Y_n$, all sampled from the same population, is given by

$$p(y_1, y_2, \ldots, y_n) = p(y_1)p(y_2) \cdots p(y_n).$$

In the continuous case the joint density function is

$$f(y_1, y_2, \ldots, y_n) = f(y_1)f(y_2) \cdots f(y_n).$$

The statement "$Y_1, Y_2, \ldots, Y_n$ is a random sample from a population with density $f(y)$" will mean that the random variables are independent with common density function $f(y)$.

6.2 Finding the Probability Distribution of a Function of Random Variables

We will present three methods for finding the probability distribution for a function of random variables and a fourth method for finding the *joint* distribution of several functions of random variables. Any one of these may be employed to find the distribution a given function of the variables, but one of the methods usually leads to a simpler derivation than the others. The method that works "best" varies from one application to another. Hence acquaintance with the first three methods is desirable. The fourth method, presented in Section 6.7, is optional. Although the first three methods will be discussed separately in the next three sections, a brief summary of each of these methods is provided here.

Consider random variables $Y_1, Y_2, \ldots, Y_n$ and a function $U(Y_1, Y_2, \ldots, Y_n)$, denoted simply as U. Then three of the methods for finding the probability distribution of U are as follows:

1. The method of distribution functions: This method is typically used when the Y's have continuous distributions. First, find the distribution function for U, $F_U(u) = P(U \leq u)$ by using the methods that we discussed in Chapter 5. To do so, we must find the region in the $y_1, y_2, \ldots, y_n$ space for which $U \leq u$ and then find $P(U \leq u)$ by integrating $f(y_1, y_2, \ldots, y_n)$ over this region. The density function for U is then obtained by differentiating the distribution function, $F_U(u)$. A detailed account of this procedure will be presented in Section 6.3.

2. The method of transformations: If we are given the density function of a random variable Y, the method of transformations results in a general expression for the density of $U = h(Y)$ for an increasing or decreasing function $h(y)$. Then if Y_1 and Y_2 have a bivariate distribution, we can use the univariate result explained earlier to find the joint density of Y_1 and $U = h(Y_1, Y_2)$. By integrating over y_1, we find the marginal probability density function of U, which is our objective. This method will be illustrated in Section 6.4.

3. The method of moment-generating functions: This method is based on a uniqueness theorem, Theorem 6.1, which states that, if two random variables have identical moment-generating functions, the two random variables possess the same probability distributions. To use this method, we must find the moment-generating function for U and compare it with the moment-generating functions for the common discrete and continuous random variables derived in Chapters 3 and 4. If it is identical to one of these moment-generating functions, the probability distribution of U can be identified because of the uniqueness theorem. Applications of the method of moment-generating functions will be presented in Section 6.5. Probability-generating functions can be employed in a way similar to the method of moment-generating functions. If you are interested in their use, see the references at the end of the chapter.

6.3 The Method of Distribution Functions

We will illustrate the method of distribution functions with a simple univariate example. If Y has probability density function $f(y)$ and if U is some function of Y, then we can find $F_U(u) = P(U \leq u)$ directly by integrating $f(y)$ over the region for which $U \leq u$. The probability density function for U is found by differentiating $F_U(u)$. The following example illustrates the method.

EXAMPLE **6.1**   A process for refining sugar yields up to 1 ton of pure sugar per day, but the actual amount produced, Y, is a random variable because of machine breakdowns and other slowdowns. Suppose that Y has density function given by

$$f(y) = \begin{cases} 2y, & 0 \le y \le 1 \\ 0, & \text{elsewhere.} \end{cases}$$

The company is paid at the rate of $300 per ton for the refined sugar, but it also has a fixed overhead cost of $100 per day. Thus the daily profit, in hundreds of dollars, is $U = 3Y - 1$. Find the probability density function for U.

Solution   To employ the distribution function approach, we must find

$$F_U(u) = P(U \le u) = P(3Y - 1 \le u) = P\left(Y \le \frac{u+1}{3}\right).$$

If $u < -1$, then $(u+1)/3 < 0$ and, therefore, $F_U(u) = P(Y \le (u+1)/3) = 0$. Also, if $u > 2$, then $(u+1)/3 > 1$ and $F_U(u) = P(Y \le (u+1)/3) = 1$. However, if $-1 \le u \le 2$, the probability can be written as an integral of $f(y)$, and

$$P\left(Y \le \frac{u+1}{3}\right) = \int_{-\infty}^{(u+1)/3} f(y)dy = \int_{0}^{(u+1)/3} 2y \, dy = \left(\frac{u+1}{3}\right)^2.$$

(Notice that, as Y ranges from 0 to 1, U ranges from -1 to 2.) Thus, the distribution function of the random variable U is given by

$$F_U(u) = \begin{cases} 0, & u < -1 \\ \left(\dfrac{u+1}{3}\right)^2, & -1 \le u \le 2 \\ 1, & u > 2 \end{cases}$$

and the density function for U is

$$f_U(u) = \frac{dF_U(u)}{du} = \begin{cases} (2/9)(u+1), & -1 \le u < 2 \\ 0, & \text{elsewhere.} \end{cases}$$

In the bivariate situation, let Y_1 and Y_2 be random variables with joint density $f(y_1, y_2)$, and let $U = h(Y_1, Y_2)$ be a function of Y_1 and Y_2. Then for every point (y_1, y_2) there corresponds one and only one value of U. If we can find the region of values (y_1, y_2) such that $U \le u$, then the integral of the joint density function $f(y_1, y_2)$ over this region equals $P(U \le u) = F_U(u)$. As before, the density function for U can be obtained by differentiation.

We will illustrate these ideas with two examples.

EXAMPLE **6.2**   In Example 5.4, we considered the random variables Y_1 (the proportional amount of gasoline stocked at the beginning of a week) and Y_2 (the proportional amount of

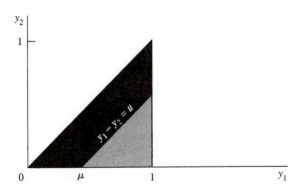

gasoline sold during the week). The joint density function of Y_1 and Y_2 is given by

$$f(y_1, y_2) = \begin{cases} 3y_1, & 0 \le y_2 \le y_1 \le 1 \\ 0, & \text{elsewhere.} \end{cases}$$

Find the probability density function for $U = Y_1 - Y_2$, the proportional amount of gasoline remaining at the end of the week. Use the density function of U to find $E(U)$.

Solution   The region over which $f(y_1, y_2)$ is not zero is sketched in Figure 6.1. Also shown there is the line $y_1 - y_2 = u$, for a value of u between 0 and 1. Notice that any point (y_1, y_2) such that $y_1 - y_2 \le u$ lies above the line $y_1 - y_2 = u$.
  If $u < 0$, the line $y_1 - y_2 = u$ has intercept $-u < 0$ and $F_U(u) = P(Y_1 - Y_2 \le u) = 0$. When $u > 1$, the line $y_1 - y_2 = u$ has intercept $-u < -1$ and $F_U(u) = 1$. For $0 \le u \le 1$, $F_U(u) = P(Y_1 - Y_2 \le u)$ is the integral over the dark shaded region above the line $y_1 - y_2 = u$. Because it is easier to integrate over the lower triangular region, we can write, for $0 \le u \le 1$,

$$F_U(u) = P(U \le u) = 1 - P(U \ge u)$$

$$= 1 - \int_u^1 \int_0^{y_1 - u} 3y_1 \, dy_2 \, dy_1$$

$$= 1 - \int_u^1 3y_1(y_1 - u) \, dy_1$$

$$= 1 - 3\left(\frac{y_1^3}{3} - \frac{uy_1^2}{2} \right) \Big]_u^1$$

$$= 1 - \left(1 - \frac{3}{2}(u) + \frac{u^3}{2} \right)$$

$$= \frac{1}{2}(3u - u^3).$$

FIGURE **6.2**
Distribution and
density functions
for Example 6.2

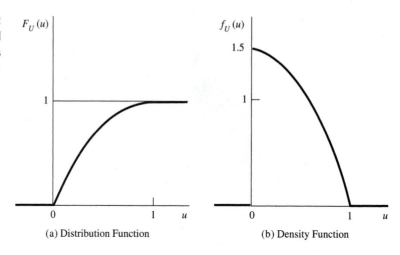

(a) Distribution Function   (b) Density Function

Summarizing,

$$F_U(u) = \begin{cases} 0, & u < 0 \\ (3u - u^3)/2, & 0 \le u \le 1 \\ 1, & u > 1. \end{cases}$$

A graph of $F_U(u)$ is given in Figure 6.2(a).
   It follows that

$$f_U(u) = \frac{dF_U(u)}{du} = \begin{cases} 3(1 - u^2)/2, & 0 \le u \le 1 \\ 0, & \text{elsewhere.} \end{cases}$$

The density function $f_U(u)$ is graphed in Figure 6.2(b).
   We can use this derived density function to find $E(U)$, because

$$E(U) = \int_0^1 u \left(\frac{3}{2}\right)(1 - u^2) \, du = \frac{3}{2}\left(\frac{u^2}{2} - \frac{u^4}{4}\right)\Bigg]_0^1 = \frac{3}{8},$$

which agrees with the value of $E(Y_1 - Y_2)$ found in Example 5.20 by using the
methods developed in Chapter 5 for finding the expected value of a linear function
of random variables.

EXAMPLE **6.3**   Let (Y_1, Y_2) denote a random sample of size $n = 2$ from the uniform distribution on
the interval $(0, 1)$. Find the probability density function for $U = Y_1 + Y_2$.

Solution   The density function for each Y_i is

$$f(y) = \begin{cases} 1, & 0 \le y \le 1 \\ 0, & \text{elsewhere.} \end{cases}$$

Therefore, because we have a random sample, Y_1 and Y_2 are independent, and

$$f(y_1, y_2) = f(y_1)f(y_2) = \begin{cases} 1, & 0 \leq y_1 \leq 1, \quad 0 \leq y_2 \leq 1 \\ 0, & \text{elsewhere.} \end{cases}$$

The random variables Y_1 and Y_2 have nonzero density over the unit square, as shown in Figure 6.3. We wish to find $F_U(u) = P(U \leq u)$. The first step is to find the points (y_1, y_2) that imply $y_1 + y_2 \leq u$. The easiest way to find this region is to locate the points that divide the regions $U \leq u$ and $U > u$. These points lie on the line $y_1 + y_2 = u$.

Graphing this relationship in Figure 6.3 and arbitrarily selecting y_2 as the dependent variable, we find that the line possesses a slope equal to -1 and a y_2 intercept equal to u. The points associated with $U < u$ are either above or below the line and can be determined by testing points on either side of the line. Suppose that $u = 1.5$. Let $y_1 = y_2 = 1/4$; then $y_1 + y_2 = 1/4 + 1/4 = 1/2$ and (y_1, y_2) satisfies the inequality $y_1 + y_2 < u$. Therefore, $y_1 = y_2 = 1/4$ falls in the shaded region below the line. Similarly, all points such that $y_1 + y_2 < u$ lie below the line $y_1 + y_2 = u$. Thus,

$$F_U(u) = P(U \leq u) = P(Y_1 + Y_2 \leq u) = \iint\limits_{y_1 + y_2 \leq u} f(y_1, y_2)\, dy_1\, dy_2.$$

If $u < 0$,

$$F_U(u) = P(U \leq u) = \iint\limits_{y_1 + y_2 \leq u} f(y_1, y_2)\, dy_1\, dy_2 = \iint\limits_{y_1 + y_2 \leq u} 0\, dy_1\, dy_2 = 0$$

and for $u > 2$,

$$F_U(u) = P(U \leq u) = \iint\limits_{y_1 + y_2 \leq u} f(y_1, y_2)\, dy_1\, dy_2 = \int_0^1 \int_0^1 (1)\, dy_1\, dy_2 = 1.$$

FIGURE **6.3**
The region of
integration for
Example 6.3

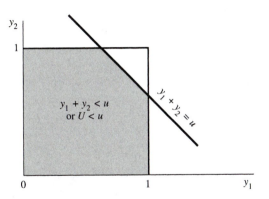

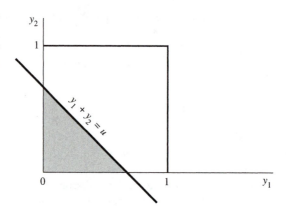

FIGURE **6.4**
The region
$y_1 + y_2 \le u$ for
$0 \le u \le 1$

For $0 \le u \le 2$, the limits of integration depend upon the particular value of u (where u is the y_2 intercept of the line $y_1 + y_2 = u$). Thus, the mathematical expression for $F_U(u)$ changes depending on whether $0 \le u \le 1$ or $1 < u \le 2$.

If $0 \le u \le 1$, the region $y_1 + y_2 \le u$, is the shaded area in Figure 6.4. Then for $0 \le u \le 1$, we have

$$
F_U(u) = \iint\limits_{y_1 + y_2 \le u} f(y_1, y_2)\, dy_1\, dy_2 = \int_0^u \int_0^{u - y_2} (1)\, dy_1\, dy_2 = \int_0^u (u - y_2)\, dy_2
$$

$$
= \left(u y_2 - \frac{y_2^2}{2} \right) \Bigg]_0^u = u^2 - \frac{u^2}{2} = \frac{u^2}{2}.
$$

The solution, $F_U(u), 0 \le u \le 1$, could have been acquired directly by using elementary geometry. The bivariate density $f(y_1, y_2) = 1$ is uniform over the unit square, $0 \le y_1 \le 1, 0 \le y_2 \le 1$. Hence $F_U(u)$ is the volume of a solid with height equal to $f(y_1, y_2) = 1$ and a triangular cross section, as shown in Figure 6.4. Hence

$$
F_U(u) = (\text{area of triangle}) \cdot (\text{height}) = \frac{u^2}{2}(1) = \frac{u^2}{2}.
$$

The distribution function can be acquired in a similar manner when u is defined over the interval $1 < u \le 2$. Although the geometric solution is easier, we will obtain $F_U(u)$ directly by integration. The region $y_1 + y_2 \le u, 1 \le u \le 2$ is the shaded area indicated in Figure 6.5.

The complement of the event $U \le u$ is the event that (Y_1, Y_2) falls in the region A of Figure 6.5. Then for $1 < u \le 2$,

$$
F_U(u) = 1 - \int_A \int f(y_1, y_2)\, dy_1\, dy_2
$$

$$
= 1 - \int_{u-1}^1 \int_{u-y_2}^1 (1)\, dy_1\, dy_2 = 1 - \int_{u-1}^1 \left(y_1 \Big]_{u-y_2}^1 \right) dy_2
$$

FIGURE **6.5**
The region
$y_1 + y_2 \leq u$,
$1 < u \leq 2$

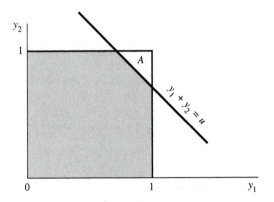

$$= 1 - \int_{u-1}^{1} (1 - u + y_2)\, dy_2 = 1 - \left[(1-u)y_2 + \frac{y_2^2}{2} \right]_{u-1}^{1}$$

$$= (-u^2/2) + 2u - 1.$$

To summarize,

$$F_U(u) = \begin{cases} 0, & u < 0 \\ u^2/2, & 0 \leq u \leq 1 \\ (-u^2/2) + 2u - 1, & 1 < u \leq 2 \\ 1, & u > 2. \end{cases}$$

The distribution function for U is shown in Figure 6.6(a).

The density function $f_U(u)$ can be obtained by differentiating $F_U(u)$. Thus,

$$f_U(u) = \frac{dF_U(u)}{du} \begin{cases} \frac{d}{du}(0) = 0, & u < 0 \\ \frac{d}{du}(u^2/2) = u, & 0 \leq u \leq 1 \\ \frac{d}{du}[(-u^2/2) + 2u - 1] = 2 - u, & 1 < u \leq 2 \\ \frac{d}{du}(1) = 0, & u > 2. \end{cases}$$

FIGURE **6.6**
Distribution and
density functions
for Example 6.3

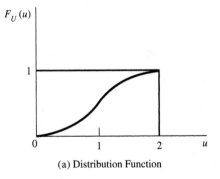

(a) Distribution Function

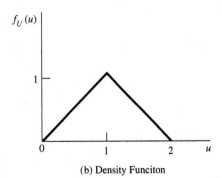

(b) Density Funciton

or, more simply,

$$f_U(u) = \begin{cases} u, & 0 \le u \le 1 \\ 2 - u, & 1 < u \le 2 \\ 0, & \text{otherwise.} \end{cases}$$

A graph of $f_U(u)$ is shown in Figure 6.6(b).

Summary of the Distribution Function Method

Let U be a function of the random variables $Y_1, Y_2, \ldots, Y_n$.

1. Find the region $U = u$ in the $(y_1, y_2, \ldots, y_n)$ space.
2. Find the region $U \le u$.
3. Find $F_U(u) = P(U \le u)$ by integrating $f(y_1, y_2, \ldots, y_n)$ over the region $U \le u$.
4. Find the density function $f_U(u)$ by differentiating $F_U(u)$. Thus, $f_U(u) = dF_U(u)/du$.

To illustrate, we will consider the case $U = h(Y) = Y^2$, where Y is a continuous random variable with distribution function $F_Y(y)$ and density function $f_Y(y)$. If $u \le 0$, $F_U(u) = P(U \le u) = P(Y^2 \le u) = 0$ and for $u > 0$ (see Figure 6.7),

$$F_U(u) = P(U \le u) = P(Y^2 \le u)$$

$$= P(-\sqrt{u} \le Y \le \sqrt{u})$$

$$= \int_{-\sqrt{u}}^{\sqrt{u}} f(y)\, dy = F_Y(\sqrt{u}) - F_Y(-\sqrt{u}).$$

FIGURE 6.7
The function
$h(y) = y^2$

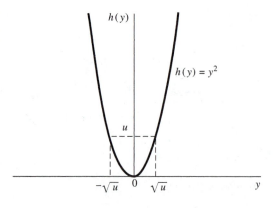

In general,

$$F_U(u) = \begin{cases} F_Y(\sqrt{u}) - F_Y(-\sqrt{u}), & u > 0 \\ 0, & \text{otherwise.} \end{cases}$$

On differentiating with respect to u, we see that

$$f_U(u) = \begin{cases} f_Y(\sqrt{u}) \left(\dfrac{1}{2\sqrt{u}}\right) + f_Y(-\sqrt{u}) \left(\dfrac{1}{2\sqrt{u}}\right), & u > 0 \\ 0, & \text{otherwise,} \end{cases}$$

or, more simply,

$$f_U(u) = \begin{cases} \dfrac{1}{2\sqrt{u}} \left[f_Y(\sqrt{u}) + f_Y(-\sqrt{u}) \right], & u > 0 \\ 0, & \text{otherwise.} \end{cases}$$

EXAMPLE 6.4   Let Y have probability density function given by

$$f_Y(y) = \begin{cases} \dfrac{y+1}{2}, & -1 \le y \le 1 \\ 0, & \text{elsewhere.} \end{cases}$$

Find the density function for $U = Y^2$.

Solution   We know that

$$f_U(u) = \begin{cases} \dfrac{1}{2\sqrt{u}} \left[f_Y(\sqrt{u}) + f_Y(-\sqrt{u}) \right], & u > 0 \\ 0, & \text{otherwise.} \end{cases}$$

and on substituting into this equation, we obtain

$$f_U(u) = \begin{cases} \dfrac{1}{2\sqrt{u}} \left(\dfrac{\sqrt{u}+1}{2} + \dfrac{-\sqrt{u}+1}{2} \right) = \dfrac{1}{2\sqrt{u}}, & 0 < u \le 1 \\ 0, & \text{elsewhere.} \end{cases}$$

Because Y has positive density only over the interval $-1 \le y \le 1$, it follows that $U = Y^2$ has positive density only over the interval $0 < u \le 1$.

In some instances, it is possible to find a transformation that, when applied to a random variable with a uniform distribution on the interval $(0, 1)$, results in a random variable with some other specified distribution function, say, $F(y)$. The next example illustrates a technique for achieving this objective. A brief discussion of one practical use of this transformation follows the example.

EXAMPLE 6.5 Let U be a uniform random variable on the interval $(0,1)$. Find a transformation $G(U)$ such that $G(U)$ possesses an exponential distribution with mean β.

Solution If U possesses a uniform distribution on the interval $(0,1)$, then the distribution function of U (see Exercise 4.28) is given by

$$F_U(u) = \begin{cases} 0, & u < 0 \\ u, & 0 \le u \le 1 \\ 1, & u > 1. \end{cases}$$

Let Y denote a random variable that has an exponential distribution with mean β. Then (see Section 4.6) Y has distribution function

$$F_Y(y) = \begin{cases} 0, & y < 0 \\ 1 - e^{-y/\beta}, & y \ge 0. \end{cases}$$

Notice that $F_Y(y)$ is strictly increasing on the interval $[0, \infty)$. Let $0 < u < 1$, and observe that there is a unique value y such that $F_Y(y) = u$. Thus, $F_Y^{-1}(u)$, $0 < u < 1$, is well defined. In this case $F_Y(y) = 1 - e^{-y/\beta} = u$ if and only if $y = -\beta \ln(1 - u) = F_Y^{-1}(u)$. Consider the random variable $F_Y^{-1}(U) = -\beta \ln(1 - U)$, and observe that if $y > 0$,

$$P(F_Y^{-1}(U) \le y) = P(-\beta \ln(1 - U) \le y)$$
$$= P(\ln(1 - U) \ge -y/\beta)$$
$$= P(U \le 1 - e^{-y/\beta})$$
$$= 1 - e^{-y/\beta}.$$

Also, $P(F_Y^{-1}(U) \le y) = 0$ if $y \le 0$. Thus, $F_Y^{-1}(U) = -\beta \ln(1 - U)$ possesses an exponential distribution with mean β, as desired.

Computer simulations are frequently used to evaluate proposed statistical techniques. Typically, these simulations require that we obtain observed values of random variables with a prescribed distribution. As noted in Section 4.4, most computer systems contain a subroutine that provides observed values of a random variable U that has a uniform distribution on the interval $(0, 1)$. How can the result of Example 6.5 be used to generate a set of observations from an exponential distribution with mean β? Simply use the computer's random number generator to produce values $u_1, u_2, \ldots, u_n$ from a uniform $(0, 1)$ distribution, and then calculate $y_i = -\beta \ln(1 - u_i)$, $i = 1, 2, \ldots, n$ to obtain values of random variables with the required exponential distribution.

As long as a prescribed distribution function $F(y)$ possesses a unique inverse $F^{-1}(\cdot)$, the preceding technique can be applied. In instances such as that illustrated in Example 6.5, we can readily write down the form of $F^{-1}(\cdot)$ and proceed as earlier. If the form of a distribution function cannot be written in an easily invertible form (recall that the distribution functions of normal, gamma, and beta distributed random variables are given in tables that were obtained by using numerical integration techniques), our task is more difficult. In these instances other methods are used to generate observations with the desired distribution.

In the following exercise set you will find problems that can be solved by using the techniques presented in this section. The exercises that involve finding $F^{-1}(U)$ for some specific distribution $F(y)$ focus on cases where $F^{-1}(\cdot)$ exists in a closed form.

Exercises

6.1 Let Y be a random variable with probability density function given by

$$f(y) = \begin{cases} 2(1-y), & 0 \le y \le 1 \\ 0, & \text{elsewhere.} \end{cases}$$

a Find the density function of $U_1 = 2Y - 1$.
b Find the density function of $U_2 = 1 - 2Y$.
c Find the density function of $U_3 = Y^2$.
d Find $E(U_1)$, $E(U_2)$, and $E(U_3)$ by using the derived density functions for these random variables.
e Find $E(U_1)$, $E(U_2)$, and $E(U_3)$ by the methods of Chapter 4.

6.2 Let Y be a random variable with a density function given by

$$f(y) = \begin{cases} (3/2)y^2, & -1 \le y \le 1 \\ 0, & \text{elsewhere.} \end{cases}$$

a Find the density function of $U_1 = 3Y$.
b Find the density function of $U_2 = 3 - Y$.
c Find the density function of $U_3 = Y^2$.

6.3 A supplier of kerosene has a weekly demand Y possessing a probability density function given by

$$f(y) = \begin{cases} y, & 0 \le y \le 1 \\ 1, & 1 < y \le 1.5 \\ 0, & \text{elsewhere} \end{cases}$$

with measurements in hundreds of gallons. (This problem was introduced in Exercise 4.7.) The supplier's profit is given by $U = 10Y - 4$.

a Find the probability density function for U.
b Use the answer to (a) to find $E(U)$.
c Find $E(U)$ by the methods of Chapter 4.

6.4 The amount of flour used per day by a bakery is a random variable Y that has an exponential distribution with mean equal to 4 tons. The cost of the flour is proportional to $U = 3Y + 1$.

a Find the probability density function for U.
b Use the answer in (a) to find $E(U)$.

6.5 The waiting time Y until delivery of a new component for an industrial operation is uniformly distributed over the interval from 1 to 5 days. The cost of this delay is given by $U = 2Y^2 + 3$. Find the probability density function for U.

6.6 The joint distribution of amount of pollutant emitted from a smokestack without a cleaning device (Y_1) and a similar smokestack with a cleaning device (Y_2) was given in Exercise 5.8 to be

$$f(y_1, y_2) = \begin{cases} 1, & 0 \le y_1 \le 2, \ 0 \le y_2 \le 1, \ 2y_2 \le y_1 \\ 0, & \text{elsewhere.} \end{cases}$$

The reduction in amount of pollutant due to the cleaning device is given by $U = Y_1 - Y_2$.

a Find the probability density function for U.

b Use the answer in (a) to find $E(U)$. Compare your results with those of Exercise 5.66(c).

6.7 Suppose that a unit of mineral ore contains a proportion Y_1 of metal A and a proportion Y_2 of metal B. Experience has shown that the joint probability density function of Y_1 and Y_2 is uniform over the region $0 \le y_1 \le 1, \ 0 \le y_2 \le 1, \ 0 \le y_1 + y_2 \le 1$. Let $U = Y_1 + Y_2$, the proportion of either metal A or B per unit.

a Find the probability density function for U.

b Find $E(U)$ by using the answer to (a).

c Find $E(U)$ by using only the marginal densities of Y_1 and Y_2.

6.8 The total time from arrival to completion of service at a fast-food outlet, Y_1, and the time spent waiting in line before arriving at the service window, Y_2, were given in Exercise 5.13 with joint density function

$$f(y_1, y_2) = \begin{cases} e^{-y_1}, & 0 \le y_2 \le y_1 < \infty \\ 0, & \text{elsewhere.} \end{cases}$$

Another random variable of interest is $U = Y_1 - Y_2$, the time spent at the service window.

a Find the probability density function for U.

b Find $E(U)$ and $V(U)$. Compare your answers with the results of Exercise 5.92.

6.9 Suppose that two electronic components in the guidance system for a missile operate independently and that each has a length of life governed by the exponential distribution with mean 1 (with measurements in hundreds of hours).

a Find the probability density function for the average length of life of the two components.

b Find the mean and variance of this average, using the answer in (a). Check your answer by computing the mean and variance, using Theorem 5.12.

6.10 In a process of sintering (heating) two types of copper powder (see Exercise 5.126) the density function for Y_1, the volume proportion of solid copper in a sample, was given by

$$f_1(y_1) = \begin{cases} 6y_1(1 - y_1), & 0 \le y_1 \le 1 \\ 0, & \text{elsewhere.} \end{cases}$$

The density function for Y_2, the proportion of type A crystals among the solid copper, was given as

$$f_2(y_2) = \begin{cases} 3y_2^2, & 0 \le y_2 \le 1 \\ 0, & \text{elsewhere.} \end{cases}$$

The variable $U = Y_1 Y_2$ gives the proportion of the sample volume due to type A crystals. If Y_1 and Y_2 are independent, find the probability density function for U.

6.11 Let Y have a distribution function given by

$$F(y) = \begin{cases} 0, & y < 0 \\ 1 - e^{-y^2}, & y \geq 0. \end{cases}$$

Find a transformation $G(U)$ such that, if U has a uniform distribution on the interval $(0, 1)$, $G(U)$ has the same distribution as Y.

6.12 In Exercise 4.9, we determined that

$$f(y) = \begin{cases} \dfrac{b}{y^2}, & y \geq b \\ 0, & \text{elsewhere} \end{cases}$$

is a bona fide probability density function for a random variable, Y. Assuming b is a known constant and U has a uniform distribution on the interval $(0, 1)$, transform U to obtain a random variable with the same distribution as Y.

6.13 A member of the power family of distributions has a distribution function given by

$$F(y) = \begin{cases} 0, & y < 0 \\ \left(\dfrac{y}{\theta}\right)^\alpha, & 0 \leq y \leq \theta \\ 1, & y > \theta \end{cases}$$

where $\alpha, \theta > 0$.

a Find the density function.

b For fixed values of α and θ, find a transformation $G(U)$ so that $G(U)$ has a distribution function of F when U possesses a uniform $(0, 1)$ distribution.

c Given that a random sample of size 5 from a uniform distribution on the interval $(0, 1)$ yielded the values .2700, .6901, .1413, .1523, and .3609, use the transformation derived in (b) to give values associated with a random variable with a power family distribution with $\alpha = 2, \theta = 4$.

6.14 A member of the Pareto family of distributions (often used in economics to model income distributions) has a distribution function given by

$$F(y) = \begin{cases} 0, & y < \beta \\ 1 - \left(\dfrac{\beta}{y}\right)^\alpha, & y \geq \beta, \end{cases}$$

where $\alpha, \beta > 0$.

a Find the density function.

b For fixed values of β and α, find a transformation $G(U)$ so that $G(U)$ has a distribution function of F when U has a uniform distribution on the interval $(0, 1)$.

c Given that a random sample of size 5 from a uniform distribution on the interval $(0, 1)$ yielded the values .0058, .2048, .7692, .2475 and .6078, use the transformation derived in (b) to give values associated with a random variable with a Pareto distribution with $\alpha = 2, \beta = 3$.

6.15 Refer to Exercises 6.13 and 6.14. If Y possesses a Pareto distribution with parameters α and β, prove that $X = 1/Y$ has a power family distribution with parameters α and $\theta = \beta^{-1}$.

6.16 Let the random variable Y possess a uniform distribution on the interval $(0, 1)$.
  a Derive the distribution of the random variable $W = Y^2$.
  b Derive the distribution of the random variable $W = \sqrt{Y}$.

*6.17 Suppose that Y is a random variable that takes on only integer values $1, 2, \ldots$. Let $F(y)$ denote the distribution function of this random variable. As discussed in Section 4.2, this distribution function is a step function and the magnitude of the step at each integer value is the probability that Y takes on that value. Let U be a continuous random variable that is uniformly distributed on the interval $(0, 1)$. Define a variable X such that $X = k$ if and only if $F(k - 1) < U \leq F(k)$, $k = 1, 2, \ldots$. Recall that $F(0) = 0$ because Y takes on only *positive* integer values. Show that $P(X = i) = F(i) - F(i - 1) = P(Y = i)$, $i = 1, 2, \ldots$. That is, X has the same distribution as Y. (*Hint:* Recall Exercise 4.1.)[1]

*6.18 Use the results derived in Exercises 4.2 and 6.17 to describe how to generate values of a geometrically distributed random variable.

6.4 The Method of Transformations

The transformation method for finding the probability distribution of a function of random variables is an offshoot of the distribution function method of Section 6.3. Through the distribution function approach we can arrive at a simple method of writing down the density function of $U = h(Y)$ provided that $h(y)$ is either decreasing or increasing. [By $h(y)$ increasing we mean that if $y_1 < y_2$, then $h(y_1) < h(y_2)$ for any real numbers y_1 and y_2.] The graph of an increasing function $h(y)$ appears in Figure 6.8.

Suppose that $h(y)$ is an increasing function of y and that $U = h(Y)$, where Y has density function $f_Y(y)$. Then $h^{-1}(u)$ is a increasing function of u: if $u_1 < u_2$, then $h^{-1}(u_1) = y_1 < y_2 = h^{-1}(u_2)$. We see from Figure 6.8 that the set of points y such that $h(y) \leq u_1$ is precisely the same as the set of points y such that $y \leq h^{-1}(u_1)$. Therefore (see Figure 6.8),

$$P(U \leq u) = P(h(Y) \leq u) = P(h^{-1}(h(Y)) \leq h^{-1}(u)) = P(Y \leq h^{-1}(u))$$

FIGURE 6.8
An increasing
function

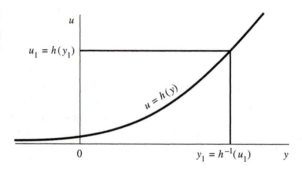

or

$$F_U(u) = F_Y(h^{-1}(u)).$$

Then differentiating with respect to u, we have

$$f_U(u) = \frac{dF_U(u)}{du} = \frac{dF_Y(h^{-1}(u))}{du} = f_Y(h^{-1}(u))\frac{d\left(h^{-1}(u)\right)}{du}.$$

To simplify notation, we will write dh^{-1}/du instead of $d\left(h^{-1}(u)\right)/du$ and

$$f_U(u) = f_Y(h^{-1}(u))\frac{dh^{-1}}{du}.$$

Thus we have acquired a new way to find $f_U(u)$ that evolved from the general method of distribution functions. To find $f_U(u)$, solve for y in terms of u; that is, find $y = h^{-1}(u)$ and substitute this expression into $f_Y(y)$. Then multiply this quantity by dh^{-1}/du. We will illustrate the procedure with an example.

EXAMPLE 6.6   In Example 6.1, we worked with a random variable Y (amount of sugar produced) with a density function given by

$$f_Y(y) = \begin{cases} 2y, & 0 \le y \le 1 \\ 0, & \text{elsewhere.} \end{cases}$$

We were interested in a new random variable (profit) given by $U = 3Y - 1$. Find the probability density function for U by the transformation method.

Solution   The function of interest here is $h(y) = 3y - 1$, which is increasing in y. If $u = 3y - 1$, then

$$y = h^{-1}(u) = \frac{u + 1}{3} \quad \text{and} \quad \frac{dh^{-1}}{du} = \frac{d\left(\frac{u+1}{3}\right)}{du} = \frac{1}{3}.$$

Thus,

$$f_U(u) = f_Y\left(h^{-1}(u)\right)\frac{dh^{-1}}{du}$$

$$= \begin{cases} 2\left(h^{-1}(u)\right)\dfrac{dh^{-1}}{du} = 2\left(\dfrac{u+1}{3}\right)\left(\dfrac{1}{3}\right), & 0 \le \dfrac{u+1}{3} \le 1 \\ 0, & \text{elsewhere,} \end{cases}$$

or, equivalently,

$$f_U(u) = \begin{cases} 2(u+1)/9, & -1 \le u \le 2 \\ 0, & \text{elsewhere.} \end{cases}$$

The range over which $f_U(u)$ is positive is simply the interval $0 \le y \le 1$ transformed to the u axis by the function $u = 3y - 1$. This answer agrees with that of Example 6.1.

If $h(y)$ is a decreasing function of y, then $h^{-1}(u)$ is a decreasing function of u. That is, if $u_1 < u_2$, then $h^{-1}(u_1) = y_1 > y_2 = h^{-1}(u_2)$. Also, as in Figure 6.9, the set of points y such that $h(y) \le u_1$ is the same as the set of points such that $y \ge h^{-1}(u_1)$.

It follows that, for $U = h(Y)$, as shown in Figure 6.9,

$$P(U \le u) = P\left(Y \ge h^{-1}(u)\right) \quad \text{or} \quad F_U(u) = 1 - F_Y\left(h^{-1}(u)\right).$$

If we differentiate with respect to u, we obtain

$$f_U(u) = -f_Y\left(h^{-1}(u)\right) \frac{d\left(h^{-1}(u)\right)}{du}.$$

If we again use the simplified notation dh^{-1}/du instead of $d\left(h^{-1}(u)\right)/du$, and recall that dh^{-1}/du is negative because $h^{-1}(u)$ is a decreasing function of u, the density of U is

$$f_U(u) = f_Y\left(h^{-1}(u)\right) \left| \frac{dh^{-1}}{du} \right|.$$

Actually, it is not necessary that $h(y)$ be increasing or decreasing (and hence invertable) for all values of y. The function $h(\cdot)$ need only be increasing or decreasing for the values of y such that $f_Y(y) > 0$. The set of points $\{y : f_Y(y) > 0\}$ is called the *support* of the density $f_Y(y)$. If $y = h^{-1}(u)$ is not in the support of the density, then $f_Y(h^{-1}(u)) = 0$. These results are combined in the following statement:

FIGURE 6.9
A decreasing function

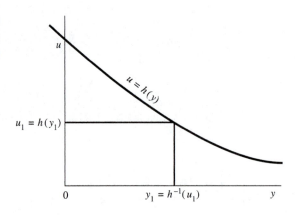

Let Y have probability density function $f_Y(y)$. If $h(y)$ is either increasing or decreasing for all y such that $f_Y(y) > 0$, then $U = h(Y)$ has density function

$$f_U(u) = f_Y\left(h^{-1}(u)\right)\left|\frac{dh^{-1}}{du}\right|, \quad \text{where} \quad \frac{dh^{-1}}{du} = \frac{d\left(h^{-1}(u)\right)}{du}.$$

EXAMPLE 6.7 Let Y have the probability density function given by

$$f_Y(y) = \begin{cases} 2y, & 0 \le y \le 1 \\ 0, & \text{elsewhere.} \end{cases}$$

Find the density function of $U = -4Y + 3$.

Solution In this example, the set of values of y such that $f_Y(y) > 0$ are the values $0 < y \le 1$. The function of interest, $h(y) = -4y + 3$, is decreasing for all y, and hence for all $0 < y \le 1$, if $u = -4y + 3$, then

$$y = h^{-1}(u) = \frac{3-u}{4} \quad \text{and} \quad \frac{dh^{-1}}{du} = -\frac{1}{4}.$$

Notice that $h^{-1}(u)$ is a decreasing function of u and that $dh^{-1}/du < 0$. Thus,

$$f_U(u) = f_Y\left(h^{-1}(u)\right)\left|\frac{dh^{-1}}{du}\right| = \begin{cases} 2\left(\frac{3-u}{4}\right)\left|-\frac{1}{4}\right|, & 0 \le \frac{3-u}{4} \le 1 \\ 0, & \text{elsewhere.} \end{cases}$$

Finally, some simple algebra gives

$$f_U(u) = \begin{cases} \dfrac{3-u}{8}, & -1 \le u \le 3 \\ 0, & \text{elsewhere.} \end{cases}$$

Direct application of the method of transformation requires that the function $h(y)$ be either increasing or decreasing for all y such that $f_Y(y) > 0$. If you want to use this method to find the distribution of $U = h(Y)$, you should be very careful to check that the function $h(\cdot)$ is either increasing or decreasing for all y in the support of $f_Y(y)$. If it is not, the method of transformations cannot be used, and you should instead use the method of distribution functions discussed in Section 6.3.

The transformation method can also be used in multivariate situations. The following example illustrates the bivariate case.

EXAMPLE 6.8   Let Y_1 and Y_2 have a joint density function given by

$$f(y_1, y_2) = \begin{cases} e^{-(y_1+y_2)}, & 0 \le y_1, \ 0 \le y_2 \\ 0, & \text{elsewhere.} \end{cases}$$

Find the density function for $U = Y_1 + Y_2$.

Solution   This problem must be solved in two stages: first, we will find the joint density of Y_1 and U; and second, we will find the marginal density of U. The approach is to let Y_1 be fixed at a value $y_1 \ge 0$. Then $U = y_1 + Y_2$, and we can consider the one-dimensional transformation problem in which $U = h(Y_2) = y_1 + Y_2$. Letting $g(y_1, u)$ denote the joint density of Y_1 and U, we have, with $y_2 = u - y_1 = h^{-1}(u)$,

$$g(y_1, u) = \begin{cases} f(y_1, h^{-1}(u)) \left| \dfrac{dh^{-1}}{du} \right| = e^{-(y_1+u-y_1)}(1), & 0 \le y_1, \quad 0 \le u - y_1 \\ 0, & \text{elsewhere.} \end{cases}$$

Simplifying, we obtain

$$g(y_1, u) = \begin{cases} e^{-u}, & 0 \le y_1 \le u \\ 0, & \text{elsewhere.} \end{cases}$$

(Notice that $Y_1 \le U$.) The marginal density of U is then given by

$$f_U(u) = \int_{-\infty}^{\infty} g(y_1, u) \, dy_1$$

$$= \begin{cases} \displaystyle\int_0^u e^{-u} \, dy_1 = ue^{-u}, & 0 \le u \\ 0, & \text{elsewhere.} \end{cases}$$

We will illustrate the use of the bivariate transformation with another example, this one involving the product of two random variables.

EXAMPLE 6.9   In Example 5.19, we considered a random variable Y_1, the proportion of impurities in a chemical sample, and Y_2, the proportion of type I impurities among all impurities in the sample. The joint density function was given by

$$f(y_1, y_2) = \begin{cases} 2(1 - y_1), & 0 \le y_1 \le 1, \quad 0 \le y_2 \le 1 \\ 0, & \text{elsewhere.} \end{cases}$$

We are interested in $U = Y_1 Y_2$, which is the proportion of type I impurities in the sample. Find the probability density function for U and use it to find $E(U)$.

Solution   Because we are interested in $U = Y_1 Y_2$, let us first fix Y_1 at a value $y_1, 0 < y_1 \le 1$, and think in terms of the univariate transformation $U = h(Y_2) = y_1 Y_2$. We can then determine the joint density function for Y_1 and U (with $y_2 = u/y_1 = h^{-1}(u)$) to be

$$g(y_1, u) = f(y_1, h^{-1}(u)) \left| \frac{dh^{-1}}{du} \right|$$

$$= \begin{cases} 2(1 - y_1) \left| \dfrac{1}{y_1} \right|, & 0 < y_1 \le 1, \quad 0 \le u/y_1 \le 1 \\ 0, & \text{elsewhere.} \end{cases}$$

Equivalently,

$$g(y_1, u) = \begin{cases} 2(1 - y_1) \left(\dfrac{1}{y_1} \right), & 0 \le u \le y_1 \le 1 \\ 0, & \text{elsewhere.} \end{cases}$$

(U also ranges between 0 and 1, but Y_1 always must be greater than or equal to U.)
Further,

$$f_U(u) = \int_{-\infty}^{\infty} g(y_1, u) \, dy_1$$

$$= \begin{cases} \displaystyle\int_u^1 2(1 - y_1) \left(\dfrac{1}{y_1} \right) dy_1, & 0 \le u \le 1 \\ 0, & \text{elsewhere.} \end{cases}$$

Because, for $0 \le u \le 1$,

$$\int_u^1 2(1 - y_1) \left(\frac{1}{y_1} \right) dy_1 = 2 \int_u^1 \left(\frac{1}{y_1} - 1 \right) dy_1$$

$$= 2 \left(\ln y_1 \big]_u^1 - y_1 \big]_u^1 \right) = 2 \left(-\ln u - 1 + u \right)$$

$$= 2(u - \ln u - 1),$$

we obtain

$$f_U(u) = \begin{cases} 2(u - \ln u - 1), & 0 \le u \le 1 \\ 0, & \text{elsewhere.} \end{cases}$$

(The symbol ln stands for natural logarithm.)
We now find $E(U)$:

$$E(U) = \int_{-\infty}^{\infty} u f_U(u) \, du = \int_0^1 2u(u - \ln u - 1) \, du$$

$$= 2 \left\{ \int_0^1 u^2 \, du - \int_0^1 u(\ln u) \, du - \int_0^1 u \, du \right\}$$

$$= 2\left\{ \left.\frac{u^3}{3}\right]_0^1 - \int_0^1 u(\ln u)\,du - \left.\frac{u^2}{2}\right]_0^1 \right\}.$$

The middle integral is most easily solved by using integration by parts, which yields

$$\int_0^1 u(\ln u)\,du = \left.\left(\frac{u^2}{2}\right)(\ln u)\right]_0^1 - \int_0^1 \left(\frac{u^2}{2}\right)\left(\frac{1}{u}\right)du = 0 - \left.\frac{u^2}{4}\right]_0^1 = -\frac{1}{4}.$$

Thus,

$$E(U) = 2\big[(1/3) - (-1/4) - (1/2)\big] = 2(1/12) = 1/6.$$

This answer agrees with the answer to Example 5.21, where $E(U) = E(Y_1 Y_2)$ was found by a different method.

Summary of the Transformation Method

Let $U = h(Y)$, where $h(y)$ is either an increasing or decreasing function of y for all y such that $f_Y(y) > 0$.

1. Find the inverse function, $y = h^{-1}(u)$.
2. Evaluate $\dfrac{dh^{-1}}{du} = \dfrac{d\left(h^{-1}(u)\right)}{du}$.
3. Find $f_U(u)$ by

$$f_U(u) = f_Y\left(h^{-1}(u)\right)\left|\frac{dh^{-1}}{du}\right|.$$

Exercises

6.19 In Exercise 6.1, we considered a random variable Y with probability density function given by

$$f(y) = \begin{cases} 2(1-y), & 0 \le y \le 1 \\ 0, & \text{elsewhere} \end{cases}$$

and used the method of distribution functions to find the density functions of

a $U_1 = 2Y - 1$.

b $U_2 = 1 - 2Y$.

c $U_3 = Y^2$.

Use the method of transformation to find the densities of $U_1, U_2,$ and U_3.

6.20 In Exercise 6.4 we considered a random variable Y that possessed an exponential distribution with mean 4 and used the method of distribution functions to derive the density function for $U = 3Y + 1$. Use the method of transformations to derive the density function for U.

6.21 In Exercise 6.9, we considered two electronic components that operate independently, each with life length governed by the exponential distribution with mean 1. We proceeded to use the method of distribution functions to obtain the distribution of the average length of life for the two components. Use the method of transformations to obtain the density function for the average life length of the two components.

6.22 The Weibull density function is given by

$$f(y) = \begin{cases} \dfrac{1}{\alpha} m y^{m-1} e^{-y^m/\alpha}, & y > 0 \\ 0, & \text{elsewhere} \end{cases}$$

where α and m are positive constants. This density function is often used as a model for the lengths of life of physical systems. Suppose Y has the Weibull density just given.

a Find the density function of $U = Y^m$.

b Find $E(Y^k)$ for any positive integer k.

6.23 Let Y have an exponential distribution with mean β.

a Prove that $W = \sqrt{Y}$ has a Weibull density with $\alpha = \beta$ and $m = 2$.

b Use the result in Exercise 6.22 (b) to give $E(Y^{k/2})$ for any positive integer k.

6.24 Let Y have a uniform $(0, 1)$ distribution. Show that $U = -2\ln(Y)$ has an exponential distribution with mean 2.

6.25 The speed of a molecule in a uniform gas at equilibrium is a random variable V whose density function is given by

$$f(v) = av^2 e^{-bv^2}, \qquad v > 0$$

where $b = m/2kT$ and k, T, and m denote Boltzmann's constant, the absolute temperature, and the mass of the molecule, respectively.

a Derive the distribution of $W = mV^2/2$, the kinetic energy of the molecule.

b Find $E(W)$.

6.26 A fluctuating electric current I may be considered a uniformly distributed random variable over the interval $(9, 11)$. If this current flows through a 2-ohm resistor, find the probability density function of the power $P = 2I^2$.

6.27 The joint distribution for the length of life of two different types of components operating in a system was given in Exercise 5.16 by

$$f(y_1, y_2) = \begin{cases} (1/8)y_1 e^{-(y_1+y_2)/2}, & y_1 > 0, \quad y_2 > 0 \\ 0, & \text{elsewhere.} \end{cases}$$

The relative efficiency of the two types of components is measured by $U = Y_2/Y_1$. Find the probability density function for U.

6.28 In Exercise 6.5, we considered a random variable Y that has a uniform distribution on the interval $[1, 5]$. The cost of delay is given by $U = 2Y^2 + 3$. Use the method of transformations to derive the density function of U.

6.29 The proportion of impurities in certain ore samples is a random variable Y with a density function given by

$$f(y) = \begin{cases} (3/2)y^2 + y, & 0 \le y \le 1 \\ 0, & \text{elsewhere.} \end{cases}$$

The dollar value of such samples is $U = 5 - (Y/2)$. Find the probability density function for U.

6.30 A density function sometimes used by engineers to model lengths of life of electronic components is the Rayleigh density, given by

$$f(y) = \begin{cases} \left(\dfrac{2y}{\theta}\right)e^{-y^2/\theta}, & y > 0 \\ 0, & \text{elsewhere.} \end{cases}$$

   a If Y has the Rayleigh density, find the probability density function for $U = Y^2$.

   b Use the result of (a) to find $E(Y)$ and $V(Y)$.

6.31 Let Y_1 and Y_2 be independent random variables, both uniformly distributed on $(0, 1)$. Find the probability density function for $U = Y_1 Y_2$.

6.32 Refer to Exercise 6.30. Let Y_1 and Y_2 be independent Rayleigh distributed random variables. Find the probability density function for $U = Y_1^2 + Y_2^2$. (*Hint:* Recall Example 6.8.)

6.5 The Method of Moment-Generating Functions

The moment-generating function method for finding the probability distribution of a function of random variables $Y_1, Y_2, \ldots, Y_n$ is based on the following uniqueness theorem.

THEOREM 6.1

Let $m_X(t)$ and $m_Y(t)$ denote the moment-generating functions of random variables X and Y, respectively. If both moment-generating functions exist and $m_X(t) = m_Y(t)$ for all values of t, then X and Y have the same probability distribution.

(The proof of Theorem 6.1 is beyond the scope of this text.)

If U is a function of n random variables, $Y_1, Y_2, \ldots, Y_n$, the first step in using Theorem 6.1 is to find the moment-generating function of U:

$$m_U(t) = E\left(e^{tU}\right).$$

Once the moment-generating function for U has been found, it is compared with the moment-generating functions for random variables with well-known distributions. If

$m_U(t)$ is identical to one of these, say, the moment-generating function for a random variable V, then, by Theorem 6.1, U and V possess identical probability distributions. The density functions, means, variances, and moment-generating functions for some frequently encountered random variables are presented in Appendix II. We will illustrate the procedure with a few examples.

EXAMPLE 6.10   Suppose that Y is a normally distributed random variable with mean μ and variance σ^2. Show that

$$Z = \frac{Y - \mu}{\sigma}$$

has a *standard normal* distribution, a normal distribution with mean 0 and variance 1.

Solution   We have seen in Example 4.16 that $Y - \mu$ has moment-generating function $e^{t^2\sigma^2/2}$. Hence

$$m_Z(t) = E(e^{tZ}) = E[e^{(t/\sigma)(Y-\mu)}] = m_{(Y-\mu)}\left(\frac{t}{\sigma}\right) = e^{(t/\sigma)^2(\sigma^2/2)} = e^{t^2/2}.$$

On comparing $m_Z(t)$ with the moment-generating function of a normal random variable, we see that Z must be normally distributed with $E(Z) = 0$ and $V(Z) = 1$.

EXAMPLE 6.11   Let Z be a normally distributed random variable with mean 0 and variance 1. Use the method of moment-generating functions to find the probability distribution of Z^2.

Solution   The moment-generating function for Z^2 is

$$m_{Z^2}(t) = E(e^{tZ^2}) = \int_{-\infty}^{\infty} e^{tz^2} f(z)\,dz = \int_{-\infty}^{\infty} e^{tz^2}\frac{e^{-z^2/2}}{\sqrt{2\pi}}\,dz$$

$$= \int_{-\infty}^{\infty} \frac{1}{\sqrt{2\pi}} e^{-(z^2/2)(1-2t)}\,dz.$$

This integral can be evaluated either by consulting a table of integrals or by noting that, if $1 - 2t > 0$ (equivalently, $t < 1/2$), the integrand

$$\frac{\exp\left[-\left(\frac{z^2}{2}\right)(1-2t)\right]}{\sqrt{2\pi}} = \frac{\exp\left[-\left(\frac{z^2}{2}\right)\Big/(1-2t)^{-1}\right]}{\sqrt{2\pi}}$$

is proportional to the density function of a normally distributed random variable with mean 0 and variance $(1 - 2t)^{-1}$. To make the integrand a normal density function (so that the definite integral is equal to 1), multiply the numerator and denominator by the standard deviation, $(1 - 2t)^{-1/2}$. Then

$$m_{Z^2}(t) = \frac{1}{(1 - 2t)^{1/2}} \int_{-\infty}^{\infty} \frac{1}{\sqrt{2\pi}(1 - 2t)^{-1/2}} \exp\left[-\left(\frac{z^2}{2}\right)\bigg/(1 - 2t)^{-1}\right] dz.$$

Because the integral equals 1, if $t < 1/2$,

$$m_{Z^2}(t) = \frac{1}{(1 - 2t)^{1/2}} = (1 - 2t)^{-1/2}.$$

A comparison of $m_{Z^2}(t)$ with the moment-generating functions in Appendix II shows that $m_{Z^2}(t)$ is identical to the moment-generating function for the gamma distributed random variable with $\alpha = 1/2$ and $\beta = 2$. Thus, using Definition 4.9, Z^2 has a chi-square distribution with $\nu = 1$ degree of freedom. It follows that the density function for $U = Z^2$ is given by

$$f_U(u) = \begin{cases} \dfrac{u^{-1/2}e^{-u/2}}{\Gamma(1/2)2^{1/2}}, & u \geq 0 \\ 0, & \text{elsewhere.} \end{cases}$$

The method of moment-generating functions is often very useful for finding the distributions of sums of independent random variables.

THEOREM 6.2 Let $Y_1, Y_2, \ldots, Y_n$ be independent random variables with moment-generating functions $m_{Y_1}(t), m_{Y_2}(t), \ldots, m_{Y_n}(t)$, respectively. If $U = Y_1 + Y_2 + \cdots + Y_n$, then

$$m_U(t) = m_{Y_1}(t) \times m_{Y_2}(t) \times \cdots \times m_{Y_n}(t).$$

Proof We know that, because the random variables $Y_1, Y_2, \ldots, Y_n$ are independent (see Theorem 5.9),

$$m_U(t) = E[e^{t(Y_1 + \cdots + Y_n)}] = E(e^{tY_1}e^{tY_2} \cdots e^{tY_n})$$
$$= E(e^{tY_1}) \times E(e^{tY_2}) \times \cdots \times E(e^{tY_n}).$$

Thus by the definition of moment-generating functions,

$$m_U(t) = m_{Y_1}(t) \times m_{Y_2}(t) \times \cdots \times m_{Y_n}(t). \qquad \blacksquare$$

EXAMPLE 6.12 The number of customer arrivals at a checkout counter in a given interval of time possesses approximately a Poisson probability distribution (see Section 3.8). If Y_1 denotes the time until the first arrival, Y_2 denotes the time between the first and second arrival, $\ldots$, Y_n denotes the time between the $(n - 1)$st and nth arrival, then it can be shown that $Y_1, Y_2, \ldots, Y_n$ are independent random variables, with the density function for Y_i given by

$$
f_{Y_i}(y_i) = \begin{cases} \dfrac{1}{\theta} e^{-y_i/\theta}, & y_i > 0 \\ 0, & \text{otherwise.} \end{cases}
$$

[Because the Y_i, for $i = 1, 2, \ldots, n$, are exponentially distributed, it follows that $E(Y_i) = \theta$; that is, θ is the average time between arrivals.] Find the probability density function for the waiting time from the opening of the counter until the nth customer arrives. (If $Y_1, Y_2, \ldots$ denote successive interarrival times, we want the density function of $U = Y_1 + Y_2 + \cdots + Y_n$.)

Solution   To use Theorem 6.2, we must first know $m_{Y_i}(t)$, $i = 1, 2, \ldots, n$. Because each of the Y_i's is exponentially distributed with mean θ, $m_{Y_i}(t) = (1 - \theta t)^{-1}$ and, by Theorem 6.2,

$$
m_U(t) = m_{Y_1}(t) \times m_{Y_1}(t) \times \cdots \times m_{Y_n}(t)
$$
$$
= (1 - \theta t)^{-1} \times (1 - \theta t)^{-1} \times \cdots \times (1 - \theta t)^{-1} = (1 - \theta t)^{-n}.
$$

This is the moment-generating function of a gamma distributed random variable with $\alpha = n$ and $\beta = \theta$. Theorem 6.1 implies that U actually has this gamma distribution and, therefore, that

$$
f_U(u) = \begin{cases} \dfrac{1}{\Gamma(n)\theta^n} (u^{n-1} e^{-u/\theta}), & u > 0 \\ 0, & \text{elsewhere.} \end{cases}
$$

The method of moment-generating functions can be used to establish some interesting and useful results about the distributions of functions of normally distributed random variables. Because these results will be used throughout Chapters 7, 8, and 9, we present them in the form of theorems.

THEOREM 6.3   Let $Y_1, Y_2, \ldots, Y_n$ be independent normally distributed random variables with $E(Y_i) = \mu_i$ and $V(Y_i) = \sigma_i^2$, for $i = 1, 2, \ldots, n$ and let $a_1, a_2, \ldots, a_n$ be constants. If

$$
U = \sum_{i=1}^{n} a_i Y_i = a_1 Y_1 + a_2 Y_2 + \cdots + a_n Y_n,
$$

then U is a normally distributed random variable with

$$
E(U) = \sum_{i=1}^{n} a_i \mu_i = a_1 \mu_1 + a_2 \mu_2 + \cdots + a_n \mu_n
$$

and

$$V(U) = \sum_{i=1}^{n} a_i^2 \sigma_i^2 = a_1^2 \sigma_1^2 + a_2^2 \sigma_2^2 + \cdots + a_n^2 \sigma_n^2.$$

Proof   Because Y_i is normally distributed with mean μ_i and variance σ_i^2, Y_i has moment-generating function given by

$$m_{Y_i}(t) = \exp\left(\mu_i t + \frac{\sigma_i^2 t^2}{2}\right).$$

[Recall that $\exp(\cdot)$ is a more convenient way to write $e^{(\cdot)}$ when the term in the exponent is long or complex.] Therefore, $a_i Y_i$ has moment-generating function given by

$$m_{a_i Y_i}(t) = E(e^{t a_i Y_i}) = m_{Y_i}(a_i t) = \exp\left(\mu_i a_i t + \frac{a_i^2 \sigma_i^2 t^2}{2}\right).$$

Because the random variables Y_i are independent, the random variables $a_i Y_i$ are independent, for $i = 1, 2, \ldots, n$, and Theorem 6.2 implies that

$$m_U(t) = m_{a_1 Y_1}(t) \times m_{a_2 Y_2}(t) \times \cdots \times m_{a_n Y_n}(t)$$

$$= \exp\left(\mu_1 a_1 t + \frac{a_1^2 \sigma_1^2 t^2}{2}\right) \times \cdots \times \exp\left(\mu_n a_n t + \frac{a_n^2 \sigma_n^2 t^2}{2}\right)$$

$$= \exp\left(t \sum_{i=1}^{n} a_i \mu_i + \frac{t^2}{2} \sum_{i=1}^{n} a_i^2 \sigma_i^2\right).$$

Thus U has a normal distribution with mean $\sum_{i=1}^{n} a_i \mu_i$ and variance $\sum_{i=1}^{n} a_i^2 \sigma_i^2$. ∎

THEOREM 6.4   Let $Y_1, Y_2, \ldots, Y_n$ be defined as in Theorem 6.3 and define Z_i by

$$Z_i = \frac{Y_i - \mu_i}{\sigma_i}, \quad i = 1, 2, \ldots, n.$$

Then $\sum_{i=1}^{n} Z_i^2$ has a χ^2 distribution with n degrees of freedom.

Proof   Because Y_i is normally distributed with mean μ_i and variance σ_i^2, the result of Example 6.10 implies that Z_i is normally distributed with mean 0 and variance 1. From Example 6.11 we then have that Z_i^2 is a χ^2-distributed random variable with 1 degree of freedom. Thus,

$$m_{Z_i^2}(t) = (1 - 2t)^{-1/2}$$

and from Theorem 6.2, with $V = \sum_{i=1}^{n} Z_i^2$,

$$m_V(t) = m_{Z_1^2}(t) \times m_{Z_2^2}(t) \times \cdots \times m_{Z_n^2}(t)$$

$$= (1 - 2t)^{-1/2} \times (1 - 2t)^{-1/2} \times \cdots \times (1 - 2t)^{-1/2} = (1 - 2t)^{-n/2}.$$

Because moment-generating functions are unique, V has a χ^2 distribution with n degrees of freedom.   ∎

Theorem 6.4 provides some clarification of the *degrees of freedom* associated with a χ^2 distribution. If n independent, standard normal random variables are squared and added together, the resulting sum has a χ^2 distribution with n degrees of freedom.

Summary of the Moment-Generating Function Method

Let U be a function of the random variables $Y_1, Y_2, \ldots, Y_n$.

1. Find the moment-generating function for U, $m_U(t)$.
2. Compare $m_U(t)$ with other well-known moment-generating functions. If $m_U(t) = m_V(t)$ for all values of t, Theorem 6.1 implies that U and V have identical distributions.

Exercises

6.33 In Exercises 6.9 and 6.21, we considered two electronic components that operate independently, each with a life length governed by the exponential distribution with mean 1. Use the method of moment-generating functions to obtain the density function for the average life length of the two components.

6.34 Suppose that Y_1 and Y_2 are independent, standard normal random variables. Find the density function of $U = Y_1^2 + Y_2^2$.

6.35 Let $Y_1, Y_2, \ldots, Y_n$ be independent, normal random variables, each with mean μ and variance σ^2. Let $a_1, a_2, \ldots, a_n$ denote known constants. Find the density function of the linear combination $U = \sum_{i=1}^{n} a_i Y_i$.

6.36 A type of elevator has a maximum weight capacity Y_1, which is normally distributed with mean 5000 pounds and standard deviation 300 pounds. For a certain building equipped with this type of elevator, the elevator's load, Y_2, is a normally distributed random variable with mean 4000 pounds and standard deviation 400 pounds. For any given time that the elevator is in use, find the probability that it will be overloaded, assuming that Y_1 and Y_2 are independent.

6.37 Refer to Exercise 6.35. Let $Y_1, Y_2, \ldots, Y_n$ be independent, normal random variables, each with mean μ and variance σ^2.

328 Chapter 6 Functions of Random Variables

a Find the density function of $\overline{Y} = \dfrac{1}{n}\displaystyle\sum_{i=1}^{n} Y_i$.

b If $\sigma^2 = 16$ and $n = 25$, what is the probability that the sample mean, $\overline{Y}$, takes on a value that is within one unit of the population mean, μ? That is, find $P\left(|\overline{Y} - \mu| \le 1\right)$.

c If $\sigma^2 = 16$, find $P\left(|\overline{Y} - \mu| \le 1\right)$ if $n = 36$, $n = 64$, and $n = 81$. Interpret the results of your calculations.

*6.38 The weight (in pounds) of "medium-size" watermelons is normally distributed with mean 15 and variance 4. A packing container for several melons has a nominal capacity of 140 pounds. What is the maximum number of melons that should be placed in a single packing container if the nominal weight limit is to be exceeded only 5% of the time? Give reasons for your answer.

6.39 The manager of a construction job needs to figure prices carefully before submitting a bid. He also needs to account for uncertainty (variability) in the amounts of products he might need. To oversimplify the real situation, suppose that a project manager treats the amount of sand, in yards, needed for a construction project as a random variable Y_1, which is normally distributed with mean 10 yards and standard deviation .5 yard. The amount of cement mix needed, in hundreds of pounds, is a random variable Y_2, which is normally distributed with mean 4 and standard deviation .2. The sand costs \$7 per yard, and the cement mix costs \$3 per hundred pounds. Adding \$100 for other costs, he computes his total cost to be

$$U = 100 + 7Y_1 + 3Y_2.$$

If Y_1 and Y_2 are independent, how much should the manager bid to ensure that the true costs will exceed the amount bid with a probability of only .01? Is the independence assumption reasonable here?

6.40 Suppose that Y has a gamma distribution with $\alpha = n/2$ for some positive integer n and β equal to some specified value. Use the method of moment-generating functions to show that $W = 2Y/\beta$ has a χ^2 distribution with n degrees of freedom.

6.41 A random variable Y has a gamma distribution with $\alpha = 3.5$ and $\beta = 4.2$. Use the result in Exercise 6.40 and the percentage points for the χ^2 distributions given in Table 6, Appendix III, to find $P(Y > 33.627)$.

6.42 In a missile-testing program, one random variable of interest is the distance between the point at which the missile lands and the center of the target at which the missile was aimed. If we think of the center of the target as the origin of a coordinate system, we can let Y_1 denote the north–south distance between the landing point and the target center and let Y_2 denote the corresponding east-west distance. (Assume that north and east define positive directions.) The distance between the landing point and the target center is then $U = \sqrt{Y_1^2 + Y_2^2}$. If Y_1 and Y_2 are independent, standard normal random variables, find the probability density function for U.

6.43 Let Y_1 be a binomial random variable with n_1 trials and probability of success given by p. Let Y_2 be another binomial random variable with n_2 trials and probability of success also given by p. If Y_1 and Y_2 are independent, find the probability function of $Y_1 + Y_2$.

6.44 Let Y_1 and Y_2 be independent Poisson random variables with means λ_1 and λ_2, respectively.

a Find the probability function of $Y_1 + Y_2$.

b Find the conditional probability function of Y_1 given $Y_1 + Y_2 = m$.

6.45 Let $Y_1, Y_2, \ldots, Y_n$ be independent binomial random variable with n_i trials and probability of success given by p_i, $i = 1, 2, \ldots, n$.

   a If all of the $n_i's$ are equal and all of the $p's$ are equal, find the distribution of $\sum_{i=1}^{n} Y_i$.

   b If all of the $n_i's$ are different and all of the $p's$ are equal, find the distribution of $\sum_{i=1}^{n} Y_i$.

   c If all of the $n_i's$ are different and all of the $p's$ are equal, find the conditional distribution Y_1 given $\sum_{i=1}^{n} Y_i = m$.

   d If all of the $n_i's$ are different and all of the $p's$ are equal, find the conditional distribution $Y_1 + Y_2$ given $\sum_{i=1}^{n} Y_i = m$.

   e If all of the $p's$ are different, does the method of moment-generating functions work well to find the distribution of $\sum_{i=1}^{n} Y_i$? Why?

6.46 Let $Y_1, Y_2, \ldots, Y_n$ be independent Poisson random variables with means $\lambda_1, \lambda_2, \ldots, \lambda_n$, respectively.

   a Find the probability function of $\sum_{i=1}^{n} Y_i$.

   b Find the conditional probability function of Y_1 given $\sum_{i=1}^{n} Y_i = m$.

   c Find the conditional probability function of $Y_1 + Y_2$ given $\sum_{i=1}^{n} Y_i = m$.

6.47 Customers arrive at a department store checkout counter according to a Poisson distribution with a mean of 7 per hour. In a given 2-hour period, what is the probability that 20 or more customers will arrive at the counter?

6.48 The length of time necessary to tune up a car is exponentially distributed with a mean of .5 hour. If two cars are waiting for a tune-up and the service times are independent, what is the probability that the total time for the two tune-ups will exceed 1.5 hours? (*Hint:* Recall the result of Example 6.12.)

6.49 Let $Y_1, Y_2, \ldots, Y_n$ be independent random variables such that each Y_i has a gamma distribution with parameters α_i and β. That is, the distributions of the Ys might have different α's, but all have the same value for β. Prove that $U = Y_1 + Y_2 + \cdots + Y_n$ has a gamma distribution with parameters $\alpha_1 + \alpha_2 + \cdots + \alpha_n$ and β.

6.50 We saw in Exercise 5.131 that the negative binomial random variable Y can be written as $Y = \sum_{i=1}^{r} W_i$, where $W_1, W_2, \ldots, W_r$ are independent geometric random variables with parameter p.

   a Use this fact to derive the moment-generating function for Y.

   b Use the moment-generating function to show that $E(Y) = r/p$ and $V(Y) = r(1-p)/p^2$.

   c Find the conditional probability function for W_1 given that $Y = W_1 + W_2 + \cdots + W_r = m$.

6.51 Show that if Y_1 has a χ^2 distribution with ν_1 degrees of freedom and Y_2 has a χ^2 distribution with ν_2 degrees of freedom, then $U = Y_1 + Y_2$ has a χ^2 distribution with $\nu_1 + \nu_2$ degrees of freedom, provided that Y_1 and Y_2 are independent.

*6.52 Let Y_1 and Y_2 be independent normal random variables, each with mean 0 and variance σ^2. Define $U_1 = Y_1 + Y_2$ and $U_2 = Y_1 - Y_2$. Show that U_1 and U_2 are independent normal random variables, each with mean 0 and variance $2\sigma^2$. [*Hint:* If (U_1, U_2) has a joint moment-generating function $m(t_1, t_2)$, then U_1 and U_2 are independent if and only if $m(t_1, t_2) = m_{U_1}(t_1)m_{U_2}(t_2)$.]

6.6 Multivariable Transformations Using Jacobians (Optional)

If Y is a random variable with density function $f_Y(y)$, the method of transformations (Section 6.4) can be used to find the density function for $U = h(Y)$ provided that $h(y)$ is either increasing or decreasing for all y such that $f_Y(y) > 0$. If $h(y)$ is increasing or decreasing for all y in the support of $f_Y(y)$, the function $h(\cdot)$ is one-to-one and there is an inverse function, $h^{-1}(\cdot)$ such that $u = h^{-1}(y)$. Further, the density function for U is given by

$$f_U(u) = f_Y(h^{-1}(u)) \left| \frac{dh^{-1}(u)}{du} \right|.$$

Suppose that Y_1 and Y_2 are jointly continuous random variables and that $U_1 = Y_1 + Y_2$ and $U_2 = Y_1 - Y_2$. How can we find the *joint density function* of U_1 and U_2?

For the rest of this section, we will write the joint density of Y_1 and Y_2 as $f_{Y_1,Y_2}(y_1, y_2)$. Extending the ideas of Section 6.4, the support of the joint density $f_{Y_1,Y_2}(y_1, y_2)$ is the set of all values of (y_1, y_2) such that $f_{Y_1,Y_2}(y_1, y_2) > 0$.

The Bivariate Transformation Method

Suppose that Y_1 and Y_2 are continuous random variables with joint density function $f_{Y_1,Y_2}(y_1, y_2)$ and that for all (y_1, y_2) such that $f_{Y_1,Y_2}(y_1, y_2) > 0$

$$u_1 = h_1(y_1, y_2) \quad \text{and} \quad u_2 = h_2(y_1, y_2)$$

is a one-to-one transformation from (y_1, y_2) to (u_1, u_2) with inverse

$$y_1 = h_1^{-1}(u_1, u_2) \quad \text{and} \quad y_2 = h_2^{-1}(u_1, u_2).$$

If $h_1^{-1}(u_1, u_2)$ and $h_2^{-1}(u_1, u_2)$ have continuous partial derivatives with respect to u_1 and u_2 and *Jacobian*

$$J = \det \begin{bmatrix} \dfrac{\partial h_1^{-1}}{\partial u_1} & \dfrac{\partial h_1^{-1}}{\partial u_2} \\[2ex] \dfrac{\partial h_2^{-1}}{\partial u_1} & \dfrac{\partial h_2^{-1}}{\partial u_2} \end{bmatrix} = \dfrac{\partial h_1^{-1}}{\partial u_1} \dfrac{\partial h_2^{-1}}{\partial u_2} - \dfrac{\partial h_2^{-1}}{\partial u_1} \dfrac{\partial h_1^{-1}}{\partial u_2} \neq 0$$

then the joint density of U_1 and U_2 is

$$f_{U_1,U_2}(u_1, u_2) = f_{Y_1,Y_2}\left(h_1^{-1}(u_1, u_2), h_2^{-1}(u_1, u_2) \right) |J|,$$

where $|J|$ is the absolute value of the Jacobian.

We will not prove this result, but it follows from calculus results used for change of variables in multiple integration. (Recall that sometimes double integrals are more easily calculated if we use polar coordinates instead of Euclidean coordinates; see Exercise 4.160.) The absolute value of the Jacobian, $|J|$, in the multivariate transformation is analogous to the quantity $\left|\frac{dh^{-1}(u)}{du}\right|$ that is used when making the one-variable transformation $U = h(Y)$.

A word of caution is in order. *Be sure that the bivariate transformation $u_1 = h_1(y_1, y_2)$, $u_2 = h_2(y_1, y_2)$ is a one-to-one transformation for all (y_1, y_2) such that $f_{Y_1,Y_2}(y_1, y_2) > 0$.* This step is easily overlooked. If the bivariate transformation is not one-to-one and this method is blindly applied, the resulting "density" function will not have the necessary properties of a valid density function. We illustrate the use of this method in the following examples.

EXAMPLE 6.13   Let Y_1 and Y_2 be independent standard normal random variables. If $U_1 = Y_1 + Y_2$ and $U_2 = Y_1 - Y_2$, both U_1 and U_2 are linear combinations of independent normally distributed random variables and Theorem 6.3 implies that U_1 is normally distributed with mean $0 + 0 = 0$ and variance $1 + 1 = 2$. Similarly, U_2 has a normal distribution with mean 0 and variance 2. What is the *joint density* of U_1 and U_2?

Solution   The density functions for Y_1 and Y_2 are

$$f_1(y_1) = \frac{e^{-\frac{1}{2}y_1^2}}{\sqrt{2\pi}}, \quad -\infty < y_1 < \infty \quad \text{and} \quad f_2(y_2) = \frac{e^{-\frac{1}{2}y_2^2}}{\sqrt{2\pi}}, \quad -\infty < y_2 < \infty$$

and the independence of Y_1 and Y_2 implies that their joint density is

$$f_{Y_1,Y_2}(y_1, y_2) = \frac{e^{-\frac{1}{2}y_1^2 - \frac{1}{2}y_2^2}}{2\pi}, \quad -\infty < y_1 < \infty, \; -\infty < y_2 < \infty.$$

In this case $f_{Y_1,Y_2}(y_1, y_2) > 0$ for all $-\infty < y_1 < \infty$, $-\infty < y_2 < \infty$, and we are interested in the transformation

$$u_1 = y_1 + y_2 = h_1(y_1, y_2) \quad \text{and} \quad u_2 = y_1 - y_2 = h_2(y_1, y_2)$$

with inverse transformation

$$y_1 = (u_1 + u_2)/2 = h_1^{-1}(u_1, u_2) \quad \text{and} \quad y_2 = (u_1 - u_2)/2 = h_2^{-1}(u_1, u_2).$$

Because $\frac{\partial h_1^{-1}}{\partial u_1} = 1/2$, $\frac{\partial h_1^{-1}}{\partial u_2} = 1/2$, $\frac{\partial h_2^{-1}}{\partial u_1} = 1/2$ and $\frac{\partial h_2^{-1}}{\partial u_2} = -1/2$, the Jacobian of this transformation is

$$J = \det \begin{bmatrix} 1/2 & 1/2 \\ 1/2 & -1/2 \end{bmatrix} = (1/2)(-1/2) - (1/2)(1/2) = -1/2$$

and the joint density of U_1 and U_2 is (with $\exp(\cdot) = e^{(\cdot)}$)

$$f_{U_1,U_2}(u_1, u_2) = \frac{\exp\left(-\frac{1}{2}\left(\frac{u_1+u_2}{2}\right)^2 - \frac{1}{2}\left(\frac{u_1-u_2}{2}\right)^2\right)}{2\pi}\left|-\frac{1}{2}\right|, \qquad \begin{array}{l} -\infty < \frac{u_1+u_2}{2} < \infty, \\ -\infty < \frac{u_1-u_2}{2} < \infty. \end{array}$$

A little algebra yields

$$-\frac{1}{2}\left(\frac{u_1+u_2}{2}\right)^2 - \frac{1}{2}\left(\frac{u_1-u_2}{2}\right)^2 = -\frac{1}{4}u_1^2 - \frac{1}{4}u_2^2$$

and

$$\{(u_1, u_2) : -\infty < (u_1 + u_2)/2 < \infty, -\infty < (u_1 - u_2)/2 < \infty\}$$
$$= \{(u_1, u_2) : -\infty < u_1 < \infty, -\infty < u_2 < \infty\}.$$

Finally, because $4\pi = \sqrt{2}\sqrt{2\pi}\sqrt{2}\sqrt{2\pi}$,

$$f_{U_1,U_2}(u_1, u_2) = \frac{e^{-u_1^2/4}}{\sqrt{2}\sqrt{2\pi}}\frac{e^{-u_2^2/4}}{\sqrt{2}\sqrt{2\pi}}, \qquad -\infty < u_1 < \infty, \qquad -\infty < u_2 < \infty.$$

Notice that U_1 and U_2 are *independent* and normally distributed, both with mean 0 and variance 2. The extra information provided by the joint distribution of U_1 and U_2 is that the two variables are independent!

The multivariable transformation method is also useful if we are interested in a single function of Y_1 and Y_2, say, $U_1 = h(Y_1, Y_2)$. Because we have only one function of Y_1 and Y_2, we can use the method of bivariate transformations to find the *joint* distribution of U_1 and another function $U_2 = h_2(Y_1, Y_2)$ and then find the desired marginal density of U_1 by integrating the joint density. Because we are really interested in only the distribution of U_1, we would typically choose the other function $U_2 = h_2(Y_1, Y_2)$ so that the bivariate transformation is easy to invert and the Jacobian is easy to work with. We illustrate this technique in the following example.

EXAMPLE 6.14   Let Y_1 and Y_2 be independent exponential random variables, both with mean $\beta > 0$. Find the density function of

$$U = \frac{Y_1}{Y_1 + Y_2}$$

Solution   The density functions for Y_1 and Y_2 are, again using $\exp(\cdot) = e^{(\cdot)}$,

$$f_1(y_1) = \begin{cases} \dfrac{1}{\beta}\exp(-y_1/\beta), & 0 < y_1 \\ 0, & \text{otherwise} \end{cases}$$

and

$$
f_2(y_2) \begin{cases} \dfrac{1}{\beta} \exp(-y_1/\beta), & 0 < y_1 \\ 0, & \text{otherwise.} \end{cases}
$$

and their joint density is

$$
f_{Y_1,Y_2}(y_1, y_2) = \begin{cases} \dfrac{1}{\beta^2} \exp(-(y_1 + y_2)/\beta), & 0 < y_1, \ 0 < y_2 \\ 0, & \text{otherwise.} \end{cases}
$$

because Y_1 and Y_2 are independent.

In this case $f_{Y_1,Y_2}(y_1, y_2) > 0$ for all (y_1, y_2) such that $0 < y_1, \ 0 < y_2$ and we are interested in the function $U_1 = \dfrac{Y_1}{Y_1 + Y_2}$. If we consider the function $u_1 = \dfrac{y_1}{y_1 + y_2}$, there are obviously many values for (y_1, y_2) that will give the same value for u_1. Let us define

$$
u_1 = \frac{y_1}{y_1 + y_2} = h_1(y_1, y_2) \quad \text{and} \quad u_2 = y_1 + y_2 = h_2(y_1, y_2).
$$

This choice of u_2 yields a convenient inverse transformation:

$$
y_1 = u_1 u_2 = h_1^{-1}(u_1, u_2) \quad \text{and} \quad y_2 = u_2(1 - u_1) = h_2^{-1}(u_1, u_2).
$$

The Jacobian of this transformation is

$$
J = \det \begin{bmatrix} u_2 & u_1 \\ -u_2 & 1 - u_1 \end{bmatrix} = u_2(1 - u_1) - (-u_2)(u_1) = u_2
$$

and the joint density of U_1 and U_2 is

$$
f_{U_1,U_2}(u_1, u_2)
$$
$$
= \begin{cases} \dfrac{1}{\beta^2} \exp\left(-[u_1 u_2 + u_2(1 - u_1)]/\beta \right) |u_2|, & 0 < u_1 u_2, \ 0 < u_2(1 - u_1) \\ 0, & \text{otherwise.} \end{cases}
$$

In this case, $f_{U_1,U_2}(u_1, u_2) > 0$ if u_1 and u_2 are such that $0 < u_1 u_2, \ 0 < u_2(1 - u_1)$. Notice that if $0 < u_1 u_2$ then

$$
0 < u_2(1 - u_1) = u_2 - u_1 u_2 \quad \Leftrightarrow \quad 0 < u_1 u_2 < u_2 \quad \Leftrightarrow \quad 0 < u_1 < 1.
$$

If $0 < u_1 < 1$, then $0 < u_2(1 - u_1)$ implies that $0 < u_2$. Therefore, the region of support for the joint density of U_1 and U_2 is $\{(u_1, u_2): 0 < u_1 < 1, \ 0 < u_2\}$ and the joint density of U_1 and U_2 is given by

$$
f_{U_1,U_2}(u_1, u_2) = \begin{cases} \dfrac{1}{\beta^2} u_2 e^{-u_2/\beta}, & 0 < u_1 < 1, \ 0 < u_2 \\ 0, & \text{otherwise.} \end{cases}
$$

Using Theorem 5.5 it is easily seen that U_1 and U_2 are independent. The marginal densities of U_1 and U_2 can be obtained by integrating the joint density derived earlier. In Exercise 6.53 you will show that U_1 is uniformly distributed over $(0, 1)$ and that U_2 has a gamma density with parameters $\alpha = 2$ and β.

The technique described in this section can be viewed to be a one-step version of the two-step process illustrated in Example 6.9.

In Example 6.14 it was more difficult to find the region of support (where the joint density is positive) than it was to find the equation of the joint density function. As you will see in the next example and in the exercises, this is often the case.

EXAMPLE 6.15 In Example 6.9, we considered a random variables Y_1 and Y_2 with joint density function

$$f_{Y_1,Y_2}(y_1, y_2) = \begin{cases} 2(1 - y_1), & 0 \le y_1 \le 1, \quad 0 \le y_2 \le 1 \\ 0, & \text{elsewhere} \end{cases}$$

and were interested in $U = Y_1 Y_2$. Find the probability density function for U by using the bivariate transformation method.

Solution In this case $f_{Y_1,Y_2}(y_1, y_2) > 0$ for all (y_1, y_2) such that $0 \le y_1 \le 1$, $0 \le y_2 \le 1$ and we are interested in the function $U_2 = Y_1 Y_2$. If we consider the function $u_2 = y_1 y_2$, this function alone is not a one-to-one function of the variables (y_1, y_2). Consider

$$u_1 = y_1 = h_1(y_1, y_2) \quad \text{and} \quad u_2 = y_1 y_2 = h_2(y_1, y_2).$$

For this choice of u_1, and $0 \le y_1 \le 1$, $0 \le y_2 \le 1$, the transformation from (y_1, y_2) to (u_1, u_2) is one-to-one and

$$y_1 = u_1 = h_1^{-1}(u_1, u_2) \quad \text{and} \quad y_2 = u_2/u_1 = h_2^{-1}(u_1, u_2).$$

The Jacobian is

$$J = \det \begin{bmatrix} 1 & 0 \\ -u_2/u_1^2 & 1/u_1 \end{bmatrix} = 1(1/u_1) - (-u_2/u_1^2)(0) = 1/u_1.$$

The original variable of interest is $U_2 = Y_1 Y_2$ and the joint density of U_1 and U_2 is

$$f_{U_1,U_2}(u_1, u_2) = \begin{cases} 2(1 - u_1)\left|\dfrac{1}{u_1}\right|, & 0 \le u_1 \le 1, \quad 0 \le u_2/u_1 \le 1 \\ 0, & \text{otherwise.} \end{cases}$$

Because

$$\{(u_1, u_2) : 0 \le u_1 \le 1, \ 0 \le u_2/u_1 \le 1\} = \{(u_1, u_2) : 0 \le u_2 \le u_1 \le 1\},$$

the joint density of U_1 and U_2 is

$$f_{U_1,U_2}(u_1, u_2) = \begin{cases} 2(1 - u_1)\dfrac{1}{u_1}, & 0 \le u_2 \le u_1 \le 1 \\ 0, & \text{otherwise.} \end{cases}$$

This joint density is exactly the same as the joint density obtained in Example 6.9 if we identify the variables Y_1 and U used in Example 6.9 with the variables U_1 and U_2, respectively, used here. With this identification, the marginal density of U_2 is precisely the density of U obtained in Example 6.9—that is,

$$f_2(u_2) = \begin{cases} 2(u_2 - \ln u_2 - 1), & 0 \le u_2 \le 1 \\ 0 & \text{elsewhere.} \end{cases}$$

If $Y_1, Y_2, \ldots, Y_k$ are jointly continuous random variables and

$$U_1 = h_1(Y_1, Y_2, \ldots, Y_k), \; U_2 = h_2(Y_1, Y_2, \ldots, Y_k), \ldots, \; U_k = h_k(Y_1, Y_2, \ldots, Y_k)$$

where the transformation

$$u_1 = h_1(y_1, y_2, \ldots, y_k), u_2 = h_2(y_1, y_2, \ldots, y_k), \ldots, u_k = h_k(y_1, y_2, \ldots, y_k)$$

is a one-to-one transformation from $(y_1, y_2, \ldots, y_k)$ to $(u_1, u_2, \ldots, u_k)$ with inverse

$$y_1 = h_1^{-1}(u_1, u_2, \ldots, u_k), \; y_2 = h_2^{-1}(u_1, u_2, \ldots, u_k), \ldots,$$
$$y_k = h_k^{-1}(u_1, u_2, \ldots, u_k)$$

and $h_1^{-1}(u_1, u_2, \ldots, u_k)$, $h_2^{-1}(u_1, u_2, \ldots, u_k)$, $\ldots$, $h_k^{-1}(u_1, u_2, \ldots, u_k)$ have continuous partial derivatives with respect to $u_1, u_2, \ldots, u_k$ and *Jacobian*

$$J = \det \begin{bmatrix} \dfrac{\partial h_1^{-1}}{\partial u_1} & \dfrac{\partial h_1^{-1}}{\partial u_2} & \cdots & \dfrac{\partial h_1^{-1}}{\partial u_k} \\[2mm] \dfrac{\partial h_2^{-1}}{\partial u_1} & \dfrac{\partial h_2^{-1}}{\partial u_2} & \cdots & \dfrac{\partial h_2^{-1}}{\partial u_k} \\[2mm] \vdots & \vdots & \ddots & \vdots \\[2mm] \dfrac{\partial h_k^{-1}}{\partial u_1} & \dfrac{\partial h_k^{-1}}{\partial u_2} & \cdots & \dfrac{\partial h_k^{-1}}{\partial u_k} \end{bmatrix} \neq 0,$$

then a result analogous to the one presented in this section can be used to find the joint density of $U_1, U_2, \ldots, U_k$. This requires the user to find the determinant of a $k \times k$ matrix, a skill that is not required in the rest of this text. For more details, see the references at the end of the chapter.

Exercises

*6.53 In Example 6.14, Y_1 and Y_2 were independent exponentially distributed random variables, both with mean β. We defined $U_1 = \frac{Y_1}{Y_1+Y_2}$ and $U_2 = Y_1 + Y_2$ and determined the joint density of (U_1, U_2) to be

$$f_{U_1,U_2}(u_1, u_2) = \begin{cases} \frac{1}{\beta^2} u_2 e^{-u_2/\beta}, & 0 < u_1 < 1, \quad 0 < u_2 \\ 0, & \text{otherwise.} \end{cases}$$

a Show that U_1 is uniformly distributed over the interval $(0, 1)$.
b Show that U_2 has a gamma density with parameters $\alpha = 2$ and β.
c Establish and that U_1 and U_2 are independent.

*6.54 Refer to Exercise 6.53 and Example 6.14. Suppose that Y_1 has a gamma distribution with parameters α_1 and β, that Y_1 is gamma distributed with parameters α_2 and β and that Y_1 and Y_2 are independent. Let $U_1 = \frac{Y_1}{Y_1+Y_2}$ and $U_2 = Y_1 + Y_2$.
a Derive the joint density function for U_1 and U_2.
b Show that the marginal distribution of U_1 is a beta distribution with parameters α_1 and α_2.
c Show that the marginal distribution of U_2 is a gamma distribution with parameters $\alpha = \alpha_1 + \alpha_2$ and β.
d Establish and that U_1 and U_2 are independent.

*6.55 The random variables Y_1 and Y_2 are independent, both with density

$$f(y) = \begin{cases} \frac{1}{y^2}, & 1 < y \\ 0, & \text{otherwise.} \end{cases}$$

Let $U_1 = \frac{Y_1}{Y_1+Y_2}$ and $U_2 = Y_1 + Y_2$.
a What is the joint density of Y_1 and Y_2?
b Show that the joint density of U_1 and U_2 is given by

$$f_{U_1,U_2}(u_1, u_2) = \begin{cases} \frac{1}{u_1^2(1-u_1)^2 u_2^3}, & \begin{array}{l} 1/u_1 < u_2, \quad 0 < u_1 < 1/2 \quad \text{and} \\ 1/(1-y_1) < u_2, \quad 1/2 \le u_1 \le 1 \end{array} \\ 0, & \text{otherwise.} \end{cases}$$

c Sketch the region where $f_{U_1,U_2}(u_1, u_2) > 0$.
d Show that the marginal density of U_1 is

$$f_{U_1}(u_1) = \begin{cases} \frac{1}{2(1-u_1)^2}, & 0 \le u_1 < 1/2 \\ \frac{1}{2u_1^2}, & 1/2 \le u_1 \le 1 \\ 0, & \text{otherwise.} \end{cases}$$

e Are U_1 and U_2 are independent? Why or why not?

*6.56 Suppose that Y_1 and Y_2 are independent and that both are uniformly distributed on the interval $(0, 1)$, and let $U_1 = Y_1 + Y_2$ and $U_2 = Y_1 - Y_2$.

a Show that the joint density of U_1 and U_2 is given by

$$f_{U_1,U_2}(u_1, u_2) = \begin{cases} 1/2, & -u_1 < u_2 < u_1, \quad 0 < u_1 < 1 \quad \text{and} \\ & u_1 - 2 < u_2 < 2 - u_1, \quad 1 \le u_1 < 2 \\ 0, & \text{otherwise.} \end{cases}$$

b Sketch the region where $f_{U_1,U_2}(u_1, u_2) > 0$.

c Show that the marginal density of U_1 is

$$f_{U_1}(u_1) = \begin{cases} u_1, & 0 < u_1 < 1 \\ 2 - u_1, & 1 \le u_1 < 2 \\ 0, & \text{otherwise.} \end{cases}$$

d Show that the marginal density of U_2 is

$$f_{U_2}(u_2) = \begin{cases} 1 + u_2, & -1 < u_2 < 0 \\ 1 - u_2, & 0 \le u_1 < 1 \\ 0, & \text{otherwise.} \end{cases}$$

e Are U_1 and U_2 are independent? Why or why not?

*6.57 Suppose that Y_1 and Y_2 are independent exponentially distributed random variables, both with mean β, and define $U_1 = Y_1 + Y_2$ and $U_2 = Y_1/Y_2$.

a Show that the joint density of (U_1, U_2) is

$$f_{U_1,U_2}(u_1, u_2) = \begin{cases} \dfrac{1}{\beta^2} u_1 e^{-u_1/\beta} \dfrac{1}{(1+u_2)^2}, & 0 < u_1, \ 0 < u_2 \\ 0, & \text{otherwise.} \end{cases}$$

b Are U_1 and U_2 are independent? Why?

6.7 Order Statistics

Many functions of random variables of interest in practice depend on the relative magnitudes of the observed variables. For instance, we may be interested in the fastest time in an automobile race or the heaviest mouse among those fed on a certain diet. Thus, we often order observed random variables according to their magnitudes. The resulting ordered variables are called *order statistics*.

Formally, let $Y_1, Y_2, \ldots, Y_n$ denote independent continuous random variables with distribution function $F(y)$ and density function $f(y)$. We denote the ordered random variables Y_i by $Y_{(1)}, Y_{(2)}, \ldots, Y_{(n)}$, where $Y_{(1)} \le Y_{(2)} \le \cdots \le Y_{(n)}$. (Because the random variables are continuous, the equality signs can be ignored.) Using this notation,

$$Y_{(1)} = \min(Y_1, Y_2, \ldots, Y_n)$$

is the minimum of the random variables Y_i, and

$$Y_{(n)} = \max(Y_1, Y_2, \ldots, Y_n)$$

is the maximum of the random variables Y_i.

The probability density functions for $Y_{(1)}$ and $Y_{(n)}$ can be found using the method of distribution functions. We will derive the density function of $Y_{(n)}$ first. Because $Y_{(n)}$ is the maximum of $Y_1, Y_2, \ldots, Y_n$, the event $(Y_{(n)} \leq y)$ will occur if and only if the events $(Y_i \leq y)$ occur, for every $i = 1, 2, \ldots, n$. That is,

$$P(Y_{(n)} \leq y) = P(Y_1 \leq y, Y_2 \leq y, \ldots, Y_n \leq y).$$

Because the Y_i are independent and $P(Y_i \leq y) = F(y)$ for $i = 1, 2, \ldots, n$, it follows that the distribution function of $Y_{(n)}$ is given by

$$F_{Y_{(n)}}(y) = P(Y_{(n)} \leq y) = P(Y_1 \leq y)P(Y_2 \leq y) \cdots P(Y_n \leq y) = [F(y)]^n.$$

Letting $g_{(n)}(y)$ denote the density function of $Y_{(n)}$, we see that, on taking derivatives of both sides,

$$g_{(n)}(y) = n[F(y)]^{n-1} f(y).$$

The density function for $Y_{(1)}$ can be found in a similar manner. The distribution function of $Y_{(1)}$ is

$$F_{Y_{(1)}}(y) = P(Y_{(1)} \leq y) = 1 - P(Y_{(1)} > y).$$

Because $Y_{(1)}$ is the minimum of $Y_1, Y_2, \ldots, Y_n$, it follows that the event $(Y_{(1)} > y)$ occurs if and only if the events $(Y_i > y)$ occur for $i = 1, 2, \ldots, n$. Because the Y_i are independent and $P(Y_i > y) = 1 - F(y)$ for $i = 1, 2, \ldots, n$, we see that

$$\begin{aligned}
F_{Y_{(1)}}(y) = P(Y_{(1)} \leq y) &= 1 - P(Y_{(1)} > y) \\
&= 1 - P(Y_1 > y, Y_2 > y, \ldots, Y_n > y) \\
&= 1 - \left[P(Y_1 > y)P(Y_2 > y) \cdots P(Y_n > y) \right] \\
&= 1 - \left[1 - F(y) \right]^n.
\end{aligned}$$

Thus, if $g_{(1)}(y)$ denotes the density function of $Y_{(1)}$, differentiation of both sides of the last expression yields

$$g_{(1)}(y) = n[1 - F(y)]^{n-1} f(y).$$

Let us now consider the case $n = 2$ and find the joint density function for $Y_{(1)}$ and $Y_{(2)}$. The event $(Y_{(1)} \leq y_1, Y_{(2)} \leq y_2)$ means that either $(Y_1 \leq y_1, Y_2 \leq y_2)$ or $(Y_2 \leq y_1, Y_1 \leq y_2)$. [Notice that $Y_{(1)}$ could be either Y_1 or Y_2, whichever is smaller.] Therefore, for $y_1 \leq y_2$, $P(Y_{(1)} \leq y_1, Y_{(2)} \leq y_2)$ is equal to the probability of the union of the two events $(Y_1 \leq y_1, Y_2 \leq y_2)$ and $(Y_2 \leq y_1, Y_1 \leq y_2)$. That is,

$$P(Y_{(1)} \leq y_1, Y_{(2)} \leq y_2) = P\left[(Y_1 \leq y_1, Y_2 \leq y_2) \cup (Y_2 \leq y_1, Y_1 \leq y_2) \right].$$

Using the additive law of probability and recalling that $y_1 \leq y_2$, we see that

$$P(Y_{(1)} \le y_1, Y_{(2)} \le y_2) = P(Y_1 \le y_1, Y_2 \le y_2) + P(Y_2 \le y_1, Y_1 \le y_2)$$
$$- P(Y_1 \le y_1, Y_2 \le y_1).$$

Because Y_1 and Y_2 are independent and $P(Y_i \le w) = F(w)$, for $i = 1, 2$, it follows that, for $y_1 \le y_2$,

$$P(Y_{(1)} \le y_1, Y_{(2)} \le y_2) = F(y_1)F(y_2) + F(y_2)F(y_1) - F(y_1)F(y_1)$$
$$= 2F(y_1)F(y_2) - [F(y_1)]^2.$$

If $y_1 > y_2$ (recall that $Y_{(1)} \le Y_{(2)}$),

$$P(Y_{(1)} \le y_1, Y_{(2)} \le y_2) = P(Y_{(1)} \le y_2, Y_{(2)} \le y_2)$$
$$= P(Y_1 \le y_2, Y_2 \le y_2) = [F(y_2)]^2 .$$

Summarizing, the joint distribution function of $Y_{(1)}$ and $Y_{(2)}$ is

$$F_{Y_{(1)}Y_{(2)}}(y_1, y_2) = \begin{cases} 2F(y_1)F(y_2) - [F(y_1)]^2, & y_1 \le y_2 \\ [F(y_2)]^2, & y_1 > y_2. \end{cases}$$

Letting $g_{(1)(2)}(y_1, y_2)$ denote the joint density of $Y_{(1)}$ and $Y_{(2)}$, we see that, on differentiating first with respect to y_2 and then with respect to y_1,

$$g_{(1)(2)}(y_1, y_2) = \begin{cases} 2f(y_1)f(y_2), & y_1 \le y_2 \\ 0, & \text{elsewhere.} \end{cases}$$

The same method can be used to find the joint density of $Y_{(1)}, Y_{(2)}, \ldots, Y_{(n)}$, which turns out to be

$$g_{(1)(2)\cdots(n)}(y_1, y_2, \ldots, y_n) = \begin{cases} n!f(y_1)f(y_2), \ldots, f(y_n), & y_1 \le y_2 \le \cdots \le y_n \\ 0, & \text{elsewhere.} \end{cases}$$

The marginal density function for any of the order statistics can be found from this joint density function, but we will not pursue this matter formally in this text.

EXAMPLE **6.16**   Electronic components of a certain type have a length of life Y, with probability density given by

$$f(y) = \begin{cases} (1/100)e^{-y/100}, & y > 0 \\ 0, & \text{elsewhere.} \end{cases}$$

(Length of life is measured in hours.) Suppose that two such components operate independently and in series in a certain system (hence, the system fails when either component fails). Find the density function for X, the length of life of the system.

Solution   Because the system fails at the first component failure, $X = \min(Y_1, Y_2)$, where Y_1 and Y_2 are independent random variables with the given density. Then, because

$F(y) = 1 - e^{-y/100}$, for $y \geq 0$,

$$f_X(y) = g_{(1)}(y) = n[1 - F(y)]^{n-1} f(y)$$

$$= \begin{cases} 2e^{-y/100}(1/100)e^{-y/100}, & y > 0 \\ 0, & \text{elsewhere} \end{cases}$$

and it follows that

$$f_X(y) = \begin{cases} (1/50)e^{-y/50}, & y > 0 \\ 0, & \text{elsewhere.} \end{cases}$$

Thus the minimum of two exponentially distributed random variables has an exponential distribution. Notice that the mean length of life for each component is 100 hours, whereas the mean length of life for the system is $E(X) = E(Y_{(1)}) = 50 = 100/2$.

EXAMPLE **6.17**   Suppose that the components in Example 6.13 operate in parallel (hence, the system does not fail until both components fail). Find the density function for X, the length of life of the system.

Solution   Now $X = \max(Y_1, Y_2)$ and

$$f_X(y) = g_{(2)}(y) = n[F(y)]^{n-1} f(y)$$

$$= \begin{cases} 2(1 - e^{-y/100})(1/100)e^{-y/100}, & y > 0 \\ 0, & \text{elsewhere} \end{cases}$$

and therefore

$$f_X(y) = \begin{cases} (1/50)(e^{-y/100} - e^{-y/50}), & y > 0 \\ 0, & \text{elsewhere.} \end{cases}$$

We see here that the maximum of two exponential random variables is not an exponential random variable.

Although a rigorous derivation of the density function of the kth-order statistic (k an integer, $1 < k < n$) is somewhat complicated, the resulting density function has an intuitively sensible structure. Once that structure is understood, the density can be written down with little difficulty. Think of the density function of a continuous random variable at a particular point as being proportional to the probability that the variable is "close" to that point. That is, if Y is a continuous random variable with density function $f(y)$, then

$$P(y \leq Y \leq y + dy) \approx f(y)\, dy.$$

Now consider the kth-order statistic, $Y_{(k)}$. If the kth-largest value is near y_k, then $k-1$ of the Ys must be less than y_k, one of the Ys must be near y_k, and the remaining $n-k$

of the Ys must be larger than y_k. Recall the multinomial distribution, Section 5.9. In the present case, we have three classes of values of Y:

Class 1: Ys that have values less than y_k.

Class 2: Ys that have values near y_k.

Class 3: Ys that have values larger than y_k.

The probabilities of each of these classes are, respectively, $p_1 = P(Y < y_k) = F(y_k)$, $p_2 = P(y_k \le Y \le y_k + dy_k) \approx f(y_k)dy_k$, and $p_3 = P(y > y_k) = 1 - F(y_k)$. Using the multinomial probabilities discussed earlier, we see that

$$P(y_k \le Y_{(k)} \le y_k + dy_k)$$

$$\approx P[(k - 1) \text{ from class 1, 1 from class 2, } (n - k) \text{ from class 3}]$$

$$\approx \binom{n}{k - 1 \quad 1 \quad n - k} p_1^{k-1} \, p_2^1 \, p_3^{n-k}$$

$$\approx \frac{n!}{(k - 1)! \, 1! \, (n - k)!} \left[(F(y_k))^{k-1} \, f(y_k)dy_k \, (1 - F(y_k))^{n-k} \right]$$

and

$$g_{(k)}(y_k)dy_k \approx \frac{n!}{(k - 1)! \, 1! \, (n - k)!} F^{k-1}(y_k) \, f(y_k). (1 - F(y_k))^{n-k} \, dy_k.$$

The density of the kth-order statistic and the joint density of two order statistics are given in the following theorem.

THEOREM 6.5   Let $Y_1, \ldots, Y_n$ be independent identically distributed continuous random variables with common distribution function $F(y)$ and common density function $f(y)$. If $Y_{(k)}$ denotes the kth-order statistic, then the density function of $Y_{(k)}$ is given by

$$g_{(k)}(y_k) = \frac{n!}{(k - 1)! \, (n - k)!} (F(y_k))^{k-1} \, [1 - F(y_k)]^{n-k} f(y_k),$$

$$-\infty < y_k < \infty.$$

If j and k are two integers such that $1 \le j < k \le n$, the joint density of $Y_{(j)}$ and $Y_{(k)}$ is given by

$$g_{(j)(k)}(y_j, y_k) = \frac{n!}{(j - 1)! \, (k - 1 - j)! \, (n - k)!} (F(y_j))^{j-1}$$

$$\times \left[F(y_k) - F(y_j) \right]^{k-1-j} \times \left[1 - F(y_k) \right]^{n-k} f(y_j) \, f(y_k),$$

$$-\infty < y_j < y_k < \infty.$$

The heuristic, intuitive derivation of the joint density given in Theorem 6.5 is similar to that given earlier for the density of a single order statistic. For $y_j < y_k$, the joint density can be interpreted as the probability that the jth largest observation is close to y_j *and* the kth-largest is close to y_k. Define five classes of values of Y:

Class 1: Ys that have values less than y_j need: $j - 1$
Class 2: Ys that have values near y_j need: 1
Class 3: Ys that have values between y_j and y_k need: $k - 1 - j$
Class 4: Ys that have values near y_k need: 1
Class 5: Ys that have values larger than y_k need: $n - k$.

Again, use the multinomial distribution to complete the heuristic argument.

EXAMPLE 6.18   Suppose that $Y_1, Y_2, \ldots, Y_5$ denotes a random sample from a uniform distribution defined on the interval $(0, 1)$. That is,

$$f(y) = \begin{cases} 1, & 0 \le y \le 1 \\ 0, & \text{elsewhere.} \end{cases}$$

Find the density function for the second-order statistic. Also, give the joint density function for the second- and fourth-order statistics.

Solution   The distribution function associated with each of the Y's is

$$F(y) = \begin{cases} 0, & y < 0 \\ y, & 0 \le y \le 1 \\ 1, & y > 1. \end{cases}$$

The density function of the second-order statistic, $Y_{(2)}$, can be obtained directly from Theorem 6.5 with $n = 5$, $k = 2$. Thus, with $f(y)$ and $F(y)$ as noted,

$$g_{(2)}(y_2) = \frac{5!}{(2-1)!\,(5-2)!} (F(y_2))^{2-1} [1 - F(y_2)]^{5-2} f(y_2), \quad -\infty < y_2 < \infty$$

$$= \begin{cases} 20 y_2 (1 - y_2)^3, & 0 \le y_2 \le 1 \\ 0, & \text{elsewhere.} \end{cases}$$

The preceding density is a beta density with $\alpha = 2$ and $\beta = 4$. In general, the kth-order statistic based on a sample of size n from a uniform $(0, 1)$ distribution has a beta density with $\alpha = k$ and $\beta = n - k + 1$.

The joint density of the second- and fourth-order statistics is readily obtained from the second result in Theorem 6.5. With $f(y)$ and $F(y)$ as before, $j = 2$, $k = 4$, and $n = 5$,

$$g_{(2)(4)}(y_2, y_4) = \frac{5!}{(2-1)!\,(4-1-2)!\,(5-4)!} (F(y_2))^{2-1} [F(y_4) - F(y_2)]^{4-1-2}$$

$$\times [1 - F(y_4)]^{5-4} f(y_2) f(y_4), \quad -\infty < y_2 < y_4 < \infty$$

$$= \begin{cases} 5!\, y_2(y_4 - y_2)(1 - y_4), & 0 \le y_2 < y_4 \le 1 \\ 0, & \text{elsewhere.} \end{cases}$$

Of course, this joint density can be used to evaluate joint probabilities about $Y_{(2)}$ and $Y_{(4)}$ or to evaluate the expected value of functions of these two variables.

Exercises

6.58 Let Y_1 and Y_2 be independent and uniformly distributed over the interval $(0, 1)$.
   a Find the probability density function of $U_1 = \min(Y_1, Y_2)$.
   b Find $E(U_1)$ and $V(U_1)$.

6.59 As in Exercise 5.58, let Y_1 and Y_2 be independent and uniformly distributed over the interval $(0, 1)$.
   a Find the probability density function of $U_2 = \max(Y_1, Y_2)$.
   b Find $E(U_2)$ and $V(U_2)$.

6.60 Let $Y_1, Y_2, \ldots, Y_n$ be independent, uniformly distributed random variables on the interval $[0, \theta]$.
   a Find the probability distribution function of $Y_{(n)} = \max(Y_1, Y_2, \ldots, Y_n)$.
   b Find the density function of $Y_{(n)}$.
   c Find the mean and variance of $Y_{(n)}$.

6.61 Refer to Exercise 6.60. Suppose that the number of minutes that you need to wait for a bus is uniformly distributed on the interval $[0, 15]$. If you take the bus five times, what is the probability that your longest wait is less than 10 minutes?

*6.62 Let $Y_1, Y_2, \ldots, Y_n$ be independent, uniformly distributed random variables on the interval $[0, \theta]$.
   a Find the density function of $Y_{(k)}$, the kth-order statistic, where k is an integer between 1 and n.
   b Use the result from (a) to find $E\left(Y_{(k)}\right)$.
   c Find $V\left(Y_{(k)}\right)$.
   d Use the result from (c) to find $E\left(Y_{(k)} - Y_{(k-1)}\right)$, the mean difference between two successive order statistics. Interpret this result.

*6.63 Let $Y_1, Y_2, \ldots, Y_n$ be independent, uniformly distributed random variables on the interval $[0, \theta]$.
   a Find the joint density function of $Y_{(j)}$ and $Y_{(k)}$ where j and k are integers $1 \le j < k \le n$.
   b Use the result from (a) to find $\text{Cov}\left(Y_{(j)}, Y_{(k)}\right)$ when j and k are integers $1 \le j < k \le n$.
   c Use the result from (b) and Exercise 6.62 to find $V\left(Y_{(k)} - Y_{(j)}\right)$, the variance of the difference between two order statistics.

6.64 Let $Y_1, Y_2, \ldots, Y_n$ be independent random variables, each with a beta distribution, with $\alpha = \beta = 2$.
   a Find the probability distribution function of $Y_{(n)} = \max(Y_1, Y_2, \ldots, Y_n)$.

b Find the density function of $Y_{(n)}$.

c Find $E\left(Y_{(n)}\right)$ when $n = 2$.

6.65 Let $Y_1, Y_2, \ldots, Y_n$ be independent, exponentially distributed random variables with mean β.

   a Show that $Y_{(1)} = \min(Y_1, Y_2, \ldots, Y_n)$ has an exponential distribution, with mean β/n.

   b If $n = 5$ and $\beta = 2$, find $P(Y_{(1)} \leq 3.6)$.

*6.66 Let $Y_1, Y_2, \ldots, Y_n$ be independent, exponentially distributed random variables with mean β.

   a Give the density function for $Y_{(k)}$, the kth-order statistic, where k is an integer between 1 and n.

   b Give the joint density function for $Y_{(j)}$ and $Y_{(k)}$ where j and k are integers $1 \leq j < k \leq n$.

6.67 The opening prices per share Y_1 and Y_2 of two similar stocks are independent random variables, each with a density function given by

$$f(y) = \begin{cases} (1/2)e^{-(1/2)(y-4)}, & y \geq 4 \\ 0, & \text{elsewhere.} \end{cases}$$

On a given morning an investor is going to buy shares of whichever stock is less expensive.

   a Find the probability density function for the price per share that the investor will pay.

   b Find the expected cost per share that the investor will pay.

6.68 Suppose that the length of time Y it takes a worker to complete a certain task has the probability density function given by

$$f(y) = \begin{cases} e^{-(y-\theta)}, & y > \theta \\ 0, & \text{elsewhere} \end{cases}$$

where θ is a positive constant that represents the minimum time until task completion. Let $Y_1, Y_2, \ldots, Y_n$ denote a random sample of completion times from this distribution.

   a Find the density function for $Y_{(1)} = \min(Y_1, Y_2, \ldots, Y_n)$.

   b Find $E(Y_{(1)})$.

*6.69 Let $Y_1, Y_2, \ldots, Y_n$ denote a random sample from the uniform distribution $f(y) = 1, 0 \leq y \leq 1$. Find the probability density function for the range $R = Y_{(n)} - Y_{(1)}$.

*6.70 Suppose that the number of occurrences of a certain event in time interval $(0, t)$ has a Poisson distribution. If we know that n such events have occurred in $(0, t)$, then the actual times, measured from 0, for the occurrences of the event in question form an ordered set of random variables, which we denote by $W_{(1)} \leq W_{(2)} \leq \cdots \leq W_{(n)}$. [$W_{(i)}$ actually is the waiting time from 0 until the occurrence of the ith event.] It can be shown that the joint density function for $W_{(1)}, W_{(2)}, \ldots, W_{(n)}$ is given by

$$f(w_1, w_2, \ldots, w_n) = \begin{cases} \dfrac{n!}{t^n}, & w_1 \leq w_2 \leq \cdots \leq w_n \\ 0, & \text{elsewhere.} \end{cases}$$

[This is the density function for an ordered sample of size n from a uniform distribution on the interval $(0, t)$.] Suppose that telephone calls coming into a switchboard follow a Poisson distribution with a mean of ten calls per minute. A slow period of 2 minutes' duration had only four calls.

a Find the probability that all four calls came in during the first minute; that is, find $P(W_{(4)} \leq 1)$.

b Find the expected waiting time from the start of the 2-minute period until the fourth call.

*6.71 Suppose that n electronic components, each having an exponentially distributed length of life with mean θ, are put into operation at the same time. The components operate independently and are observed until r have failed ($r \leq n$). Let W_j denote the length of time until the jth failure, with $W_1 \leq W_2 \leq \cdots \leq W_r$. Let $T_j = W_j - W_{j-1}$ for $j \geq 2$ and $T_1 = W_1$. Notice that T_j measures the time elapsed between successive failures.

a Show that T_j, for $j = 1, 2, \ldots, r$, has an exponential distribution with mean $\theta/(n-j+1)$.

b Show that

$$U_r = \sum_{j=1}^{r} W_j + (n-r)W_r = \sum_{j=1}^{r} (n-j+1)T_j$$

and hence that $E(U_r) = r\theta$. [U_r is called the *total observed life*, and we can use U_r/r as an approximation to (or "estimator" of) θ.]

6.8 Summary

This chapter has been concerned with finding probability distributions for functions of random variables. This is an important problem in statistics because estimators of population parameters are functions of random variables. Hence it is necessary to know something about the probability distributions of these functions (or estimators) in order to evaluate the goodness of our statistical procedures. A discussion of estimation will be presented in Chapters 8 and 9.

The methods for finding probability distributions for functions of random variables are the distribution function method (Section 6.3), the transformation method (Section 6.4), and the moment-generating-function method (Section 6.5). It should be noted that no particular method is best for all situations, because the method of solution depends a great deal upon the nature of the function involved. If U_1 and U_2 are two functions of the continuous random variables Y_1 and Y_2, the *joint* density function for U_1 and U_2 can be found using the Jacobian technique in Section 6.4. Facility for handling these methods can be achieved only through practice. The exercises at the end of each section and at the end of the chapter provide a good starting point.

The density functions of order statistics were presented in Section 6.7.

Some special functions of random variables that are particularly useful in statistical inference will be considered in Chapter 7.

References and Further Readings

Casella, G., and R. L. Berger. 1990. *Statistical Inference*. Pacific Grove, Calif.: Brooks Cole.

Hoel, P. G. 1984. *Introduction to Mathematical Statistics*, 5th ed. New York: John Wiley.

Hogg, R. V., and A. T. Craig. 1995. *Introduction to Mathematical Statistics*, 5th ed. Upper Saddle River, N.J.: Prentice Hall.

Mood, A. M., F. A. Graybill and D. Boes. 1974. *Introduction to the Theory of Statistics*, 3d ed. New York: McGraw-Hill.

Parzen, E. 1960. *Modern Probability Theory and Its Applications*. New York: John Wiley.

Supplementary Exercises

6.72 If Y_1 and Y_2 are independent and identically distributed normal random variables with mean μ and variance σ^2, find the probability density function for $U = (1/2)(Y_1 - 3Y_2)$.

6.73 When current I flows through resistance R, the power generated is given by $W = I^2 R$. Suppose that I has a uniform distribution over the interval $(0, 1)$ and that R has a density function given by

$$f(r) = \begin{cases} 2r, & 0 \leq r \leq 1 \\ 0, & \text{elsewhere.} \end{cases}$$

Find the probability density function for W. (Assume that I is independent of R.)

6.74 Two efficiency experts take independent measurements Y_1 and Y_2 on the length of time workers take to complete a certain task. Each measurement is assumed to have the density function given by

$$f(y) = \begin{cases} (1/4)ye^{-y/2}, & y > 0 \\ 0, & \text{elsewhere.} \end{cases}$$

Find the density function for the average $U = (1/2)(Y_1 + Y_2)$. [*Hint:* Use the method of moment-generating functions.]

6.75 Let Y_1 and Y_2 be independent and uniformly distributed over the interval $(0, 1)$. Find the probability density function of each of the following:

a $U_1 = Y_1^2$.

b $U_2 = Y_1/Y_2$.

c $U_3 = -\ln (Y_1 Y_2)$.

d $U_4 = Y_1 Y_2$.

*6.76 The length of time that a machine operates without failure is denoted by Y_1 and the length of time to repair a failure is denoted by Y_2. After a repair is made, the machine is assumed to operate like a new machine. Y_1 and Y_2 are independent and each has the density function

$$f(y) = \begin{cases} e^{-y}, & y > 0 \\ 0, & \text{elsewhere.} \end{cases}$$

Find the probability density function for $U = \frac{Y_1}{Y_1+Y_2}$, the proportion of time that the machine is in operation during any one operation–repair cycle.

*6.77 A parachutist wants to land at a target T, but she finds that she is equally likely to land at any point on a straight line (A, B), of which T is the midpoint. Find the probability density function of the distance between her landing point and the target. (*Hint:* Denote A by -1, B by $+1$, and T by 0. Then the parachutist's landing point has a coordinate X, which is uniformly distributed between -1 and $+1$. The distance between X and T is $|X|$.)

6.78 Two sentries are sent to patrol a road 1 mi long. The sentries are sent to points chosen independently and at random along the road. Find the probability that the sentries will be less than $1/2$ mile apart when they reach their assigned posts.

*6.79 Let Y_1 and Y_2 be independent, standard normal random variables. Find the probability density function of $U = Y_1/Y_2$.

6.80 Let Y_1 and Y_2 be independent random variables, each having the same geometric distribution.

    a Find $P(Y_1 = Y_2) = P(Y_1 - Y_2 = 0)$. (*Hint:* Your answer will involve evaluating an infinite geometric series. The results in Appendix A1.11 will be useful.)

    b Find $P(Y_1 - Y_2 = 1)$.

    *c If $U = Y_1 - Y_2$, find the (discrete) probability function for U. [*Hint:* Part (a) gives $P(U = 0)$ and part (b) gives $P(U = 1)$. Consider the positive and negative integer values for U separately.]

6.81 A random variable Y has a *beta distribution of the second kind*, if, for $\alpha > 0$ and $\beta > 0$, its density is

$$f_Y(y) = \begin{cases} \dfrac{y^{\alpha-1}}{B(\alpha, \beta)(1 + y)^{\alpha+\beta}}, & y > 0 \\ 0, & \text{elsewhere.} \end{cases}$$

Derive the density function of $U = \frac{1}{1+Y}$.

6.82 If Y is a continuous random variable with distribution function $F(y)$, find the probability density function of $U = F(Y)$.

6.83 Let Y be uniformly distributed over the interval $(-1, 3)$. Find the probability density function of $U = Y^2$.

6.84 If Y denotes the length of life of a component and $F(y)$ is the distribution function of Y, then $P(Y > y) = 1 - F(y)$ is called the *reliability* of the component. Suppose that a system consists of four components with identical reliability functions, $1 - F(y)$, operating as indicated in Figure 6.10. The system operates correctly if an unbroken chain of components is in operation

FIGURE **6.10**
Circuit diagram

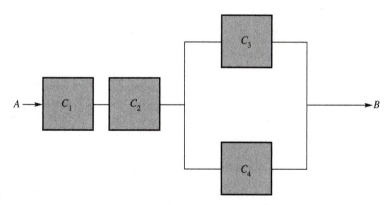

between A and B. If the four components operate independently, find the reliability of the system in terms of $F(y)$.

6.85   The percentage of alcohol in a certain compound is a random variable Y, with the following density function:

$$f(y) = \begin{cases} 20y^3(1 - y), & 0 < y < 1 \\ 0, & \text{otherwise.} \end{cases}$$

Suppose that the compound's selling price depends on its alcohol content. Specifically, if $1/3 < y < 2/3$, the compound sells for C_1 dollars per gallon; otherwise it sells for C_2 dollars per gallon. If the production cost is C_3 dollars per gallon, find the probability distribution of the price per gallon.

6.86   An engineer has observed that the gap times between vehicles passing a certain point on a highway have an exponential distribution with mean 10 seconds.

a   Find the probability that the next gap observed will be no longer than 1 minute.

b   Find the probability density function for the sum of the next four gap times to be observed. What assumptions are necessary for this answer to be correct?

6.87   If Y is a continuous random variable and m is the median of the distribution, then m is such that $P(Y \leq m) = P(Y \geq m) = 1/2$. If Y_1, Y_2, Y_3, and Y_4 are independent random variables with the same continuous distribution, what is the probability that the largest exceeds the median?

6.88   The time until failure of an electronic device has an exponential distribution with mean 15 months. If a random sample of five such devices are tested, what is the probability that the first failure among the five devices occurs

a   after 9 months?

b   before 12 months?

*6.89   If a random variable U is normally distributed with mean μ and variance σ^2 and if we write $U = \ln Y$, then Y is said to have a *log-normal* distribution. The log-normal distribution is often used in the biological and physical sciences to model sizes, by volume or weight, of various quantities, such as crushed coal particles, bacteria colonies, and individual animals. Let U and Y be as stated.

a   Show that the density function for Y is

$$f(y) = \begin{cases} \left(\dfrac{1}{y\sigma\sqrt{2\pi}}\right)e^{-(\ln y - \mu)^2/(2\sigma^2)}, & y > 0 \\ 0, & \text{elsewhere.} \end{cases}$$

b   Find $E(Y)$.

*6.90   A machine produces spherical containers whose radii vary according to the probability density function given by

$$f(r) = \begin{cases} 2r, & 0 \leq r \leq 1 \\ 0, & \text{elsewhere.} \end{cases}$$

Find the probability density function for the volume of the containers.

*6.91   Let v denote the volume of a three-dimensional figure. Let Y denote the number of particles observed in volume v, and assume that Y has a Poisson distribution with mean λv. The
particles might represent pollution particles in air, bacteria in water, or stars in the heavens.

    a   If a point is chosen at random within the volume v, show that the distance R to the nearest
particle has the probability density function given by

$$f(r) = \begin{cases} 4\lambda\pi r^2 e^{-(4/3)\lambda\pi r^3}, & r > 0 \\ 0, & \text{elsewhere.} \end{cases}$$

    b   If R is as in (a), show that $U = R^3$ has an exponential distribution.

CHAPTER 7

Sampling Distributions and the Central Limit Theorem

7.1 Introduction

7.2 Sampling Distributions Related to the Normal Distribution

7.3 The Central Limit Theorem

7.4 A Proof of the Central Limit Theorem (Optional)

7.5 The Normal Approximation to the Binomial Distribution

7.6 Summary

References and Further Readings

7.1 Introduction

In Chapter 6, we presented methods for finding the distributions of functions of random variables. Throughout this chapter we will be working with functions of the variables $Y_1, Y_2, \ldots, Y_n$ observed in a random sample selected from a population of interest. As discussed in Chapter 6, the variables $Y_1, Y_2, \ldots, Y_n$ are independent with a common distribution. Certain functions of the random variables observed in a sample are used to estimate or make decisions about unknown population parameters. For example, suppose that we want to estimate a population mean μ. If we obtain a random sample of n observations, $y_1, y_2, \ldots, y_n$, it seems reasonable to estimate μ with the sample mean

$$\overline{y} = \frac{1}{n} \sum_{i=1}^{n} y_i.$$

The goodness of this estimate depends on the behavior of the random variables $Y_1, Y_2, \ldots, Y_n$ and the effect that this behavior has on $\overline{Y} = (1/n) \sum_{i=1}^{n} Y_i$.

Notice that the random variable $\overline{Y}$ is a function of (only) the random variables $Y_1, Y_2, \ldots, Y_n$ and the (constant) sample size n. The random variable $\overline{Y}$ is therefore an example of a statistic.

DEFINITION 7.1

A *statistic* is a function of the observable random variables in a sample and known constants.

Statistics are used to make inferences (estimates or decisions) about unknown population parameters. Because a statistic is a function of the random variables observed in a sample, the statistic itself is a random variable. Consequently, using the methods of Chapter 6, we will derive its probability distribution, which we will call the *sampling distribution* of the statistic. From a practical point of view, the sampling distribution for a statistic provides a theoretical model of the relative frequency histogram of the possible values of the statistic that we would observe through repeated sampling.

The form of the theoretical sampling distribution of a statistic will depend upon the distribution of the observable random variables in the sample. In the next section we will derive the sampling distributions for some statistics used to make inferences about the parameters of a normal distribution.

7.2 Sampling Distributions Related to the Normal Distribution

We have already noted that many phenomena observed in the real world have relative frequency distributions that can be modeled adequately by a normal probability distribution. Thus in many applied problems it is reasonable to assume that the observable random variables in a random sample, $Y_1, Y_2, \ldots, Y_n$, are independent with a common, normal density function. In Exercise 6.37 you established that the statistic $\overline{Y} = (1/n)(Y_1 + Y_2 + \cdots + Y_n)$ actually has a normal distribution. Because this result is used so often in our subsequent discussions, we present it formally in the following theorem.

THEOREM 7.1

Let $Y_1, Y_2, \ldots, Y_n$ be a random sample of size n from a normal distribution with mean μ and variance σ^2. Then

$$\overline{Y} = \frac{1}{n}\sum_{i=1}^{n} Y_i$$

is normally distributed with mean $\mu_{\overline{Y}} = \mu$ and variance $\sigma_{\overline{Y}}^2 = \sigma^2/n$.

Proof

Because $Y_1, Y_2, \ldots, Y_n$ is a random sample from a normal distribution with mean μ and variance σ^2, the Y_i are independent, normally distributed variables, with $E(Y_i) = \mu$ and $V(Y_i) = \sigma^2$ for $i = 1, 2, \ldots, n$. Further,

$$\overline{Y} = \frac{1}{n}\sum_{i=1}^{n} Y_i = \frac{1}{n}(Y_1) + \frac{1}{n}(Y_2) + \cdots + \frac{1}{n}(Y_n)$$

$$= a_1 Y_1 + a_2 Y_2 + \cdots + a_n Y_n \qquad \text{where } a_i = 1/n, \quad i = 1, 2, \ldots, n.$$

Thus, $\overline{Y}$ is a linear combination of $Y_1, Y_2, \ldots, Y_n$ and Theorem 6.3 can be applied to conclude that $\overline{Y}$ is normally distributed with

$$E(\overline{Y}) = E\left[\frac{1}{n}(Y_1) + \cdots + \frac{1}{n}(Y_n)\right] = \frac{1}{n}(\mu) + \cdots + \frac{1}{n}(\mu) = \mu$$

and

$$V(\overline{Y}) = V\left[\frac{1}{n}(Y_1) + \cdots + \frac{1}{n}(Y_n)\right] = \frac{1}{n^2}(\sigma^2) + \cdots + \frac{1}{n^2}(\sigma^2)$$

$$= \frac{1}{n^2}(n\sigma^2) = \frac{\sigma^2}{n}.$$

That is, the sampling distribution of $\overline{Y}$ is normal with mean $\mu_{\overline{Y}} = \mu$ and variance $\sigma_{\overline{Y}}^2 = \sigma^2/n$. ∎

Notice that the variance of each of the random variables $Y_1, Y_2, \ldots, Y_n$ is σ^2 and the variance of the sampling distribution of the random variable $\overline{Y}$ is σ^2/n. In the discussions that follow, we will have occasion to refer to both of these variances. The notation σ^2 will be retained for the variance of the variables $Y_1, Y_2, \ldots, Y_n$, and $\sigma_{\overline{Y}}^2$ will be used to denote the variance of the sampling distribution of the random variable $\overline{Y}$.

Under the conditions of Theorem 7.1, $\overline{Y}$ is normally distributed with the mean $\mu_{\overline{Y}} = \mu$ and variance $\sigma_{\overline{Y}}^2 = \sigma^2/n$. It follows that

$$Z = \frac{\overline{Y} - \mu_{\overline{Y}}}{\sigma_{\overline{Y}}} = \frac{\overline{Y} - \mu}{\sigma/\sqrt{n}} = \sqrt{n}\left(\frac{\overline{Y} - \mu}{\sigma}\right)$$

has a standard normal distribution. We will illustrate the use of Theorem 7.1 in the following example.

EXAMPLE **7.1**   A bottling machine can be regulated so that it discharges an average of μ ounces per bottle. It has been observed that the amount of fill dispensed by the machine is normally distributed with $\sigma = 1.0$ ounce. A sample of $n = 9$ filled bottles is randomly selected from the output of the machine on a given day (all bottled with the same machine setting) and the ounces of fill measured for each. Find the probability that the sample mean will be within .3 ounce of the true mean μ for that particular setting.

Solution   If $Y_1, Y_2, \ldots, Y_9$ denote the ounces of fill to be observed, then we know that the Y_i are normally distributed with mean μ and variance $\sigma^2 = 1$ for $i = 1, 2, \ldots, 9$. Therefore, by Theorem 7.1, $\overline{Y}$ possesses a normal sampling distribution with mean $\mu_{\overline{Y}} = \mu$ and variance $\sigma_{\overline{Y}}^2 = \sigma^2/n = 1/9$. We want to find

$$P(|\overline{Y} - \mu| \leq .3) = P[-.3 \leq (\overline{Y} - \mu) \leq .3]$$

$$= P\left(-\frac{.3}{\sigma/\sqrt{n}} \leq \frac{\overline{Y} - \mu}{\sigma/\sqrt{n}} \leq \frac{.3}{\sigma/\sqrt{n}}\right).$$

Because $\dfrac{\overline{Y} - \mu_{\overline{Y}}}{\sigma_{\overline{Y}}} = \dfrac{\overline{Y} - \mu}{\sigma/\sqrt{n}}$ has a standard normal distribution, it follows that

$$P(|\overline{Y} - \mu| \leq .3) = P\left(-\frac{.3}{1/\sqrt{9}} \leq Z \leq \frac{.3}{1/\sqrt{9}}\right)$$

$$= P(-.9 \leq Z \leq .9).$$

Using Table 4, Appendix III, we find

$$P(-.9 \leq Z \leq .9) = 1 - 2P(Z > .9) = 1 - 2(.1841) = .6318.$$

Thus the chance is only .6318 that the sample mean will be within .3 ounce of the true population mean.

EXAMPLE 7.2   Refer to Example 7.1. How many observations should be included in the sample if we wish $\overline{Y}$ to be within .3 ounce of μ with probability .95?

Solution   Now we want

$$P(|\overline{Y} - \mu| \leq .3) = P[-.3 \leq (\overline{Y} - \mu) \leq .3] = .95.$$

Multiplying each term of the inequality by $\sqrt{n}/\sigma$ (recall $\sigma = 1$), we have

$$P\left[\frac{-.3\sqrt{n}}{\sigma} \leq \sqrt{n}\left(\frac{\overline{Y} - \mu}{\sigma}\right) \leq \frac{.3\sqrt{n}}{\sigma}\right] = P(-.3\sqrt{n} \leq Z \leq .3\sqrt{n}) = .95.$$

But using Table 4, Appendix III, we have that

$$P(-1.96 \leq Z \leq 1.96) = .95.$$

It must follow that

$$.3\sqrt{n} = 1.96$$

or

$$n = \left(\frac{1.96}{.3}\right)^2 = 42.68.$$

From a practical perspective, it is impossible to take a sample of size 42.68. Our solution indicates that a sample of size 42 is not quite large enough to reach our objective. If $n = 43$, $P(|\overline{Y} - \mu| \leq .3)$ slightly exceeds .95.

In succeeding chapters we will be interested in statistics that are functions of the squares of the observations in a random sample from a normal population. Theorem 7.2 establishes the sampling distribution of the sum of the squares of independent, standard normal random variables.

THEOREM 7.2

Let $Y_1, Y_2, \ldots, Y_n$ be defined as in Theorem 7.1. Then $Z_i = (Y_i - \mu)/\sigma$ are independent, standard normal random variables, where $i = 1, 2, \ldots, n$, and

$$\sum_{i=1}^{n} Z_i^2 = \sum_{i=1}^{n} \left(\frac{Y_i - \mu}{\sigma}\right)^2$$

has a χ^2 distribution with n degrees of freedom.

Proof

Because $Y_1, Y_2, \ldots, Y_n$ is a random sample from a normal distribution with mean μ and variance σ^2, Example 6.10 implies that $Z_i = (Y_i - \mu)/\sigma$ has a standard normal distribution for $i = 1, 2, \ldots, n$. Further, the random variables Z_i are independent because the random variables Y_i are independent, $i = 1, 2, \ldots, n$. The fact that $\sum_{i=1}^{n} Z_i^2$ has a χ^2 distribution with n degrees of freedom follows directly from Theorem 6.4. ∎

From Table 6, Appendix III, we can find values χ_α^2 so that

$$P(\chi^2 > \chi_\alpha^2) = \alpha$$

for random variables with χ^2 distributions (see Figure 7.1). For example, if the χ^2 random variable of interest has 10 degrees of freedom (d.f.), Table 6, Appendix III,

FIGURE 7.1
A χ^2 distribution showing upper-tail area α

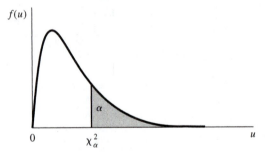

can be used to find $\chi^2_{.90}$. To do so, look in the row labeled 10 d.f. and the column headed $\chi^2_{.90}$ and read the value 4.86518. Therefore, if Y has a χ^2 distribution with 10 d.f., $P(Y > 4.86518) = .10$.

The following example illustrates the combined use of Theorem 7.2 and the χ^2 tables.

EXAMPLE 7.3   If $Z_1, Z_2, \ldots, Z_6$ denotes a random sample from the standard normal distribution, find a number b such that

$$P\left(\sum_{i=1}^{6} Z_i^2 \le b\right) = .95.$$

Solution   By Theorem 7.2, $\sum_{i=1}^{6} Z_i^2$ has a χ^2 distribution with 6 degrees of freedom. Looking at Table 6, Appendix III, in the row headed 6 d.f. and the column headed $\chi^2_{.05}$, we see the number 12.5916. Thus

$$P\left(\sum_{i=1}^{6} Z_i^2 > 12.5916\right) = .05, \quad \text{or, equivalently,} \quad P\left(\sum_{i=1}^{6} Z_i^2 \le 12.5916\right) = .95$$

and $b = 12.5916$.

The χ^2 distribution plays an important role in many inferential procedures. For example, suppose that we wish to make an inference about the population variance σ^2 based on a random sample $Y_1, Y_2, \ldots, Y_n$ from a normal population. As we will show in Chapter 8, a good estimator of σ^2 is the sample variance

$$S^2 = \frac{1}{n-1} \sum_{i=1}^{n} (Y_i - \overline{Y})^2.$$

The following theorem gives the probability distribution for a function of the statistic S^2.

THEOREM 7.3   Let $Y_1, Y_2, \ldots, Y_n$ be a random sample from a normal distribution with mean μ and variance σ^2. Then

$$\frac{(n-1)S^2}{\sigma^2} = \frac{1}{\sigma^2} \sum_{i=1}^{n} (Y_i - \overline{Y})^2$$

has a χ^2 distribution with $(n-1)$ degrees of freedom. Also, $\overline{Y}$ and S^2 are independent random variables.

Proof   The complete proof of Theorem 7.3 is outlined in Exercise 13.83. To make the general result more plausible, we will consider the case $n = 2$ and show that

$(n-1)S^2/\sigma^2$ has a χ^2 distribution with 1 degree of freedom. In the case $n = 2$,

$$\overline{Y} = (1/2)(Y_1 + Y_2)$$

and, therefore,

$$S^2 = \frac{1}{2-1} \sum_{i=1}^{2}(Y_i - \overline{Y})^2$$

$$= \left[Y_1 - \frac{1}{2}(Y_1 + Y_2) \right]^2 + \left[Y_2 - \frac{1}{2}(Y_1 + Y_2) \right]^2$$

$$= \left[\frac{1}{2}(Y_1 - Y_2) \right]^2 + \left[\frac{1}{2}(Y_2 - Y_1) \right]^2$$

$$= 2 \left[\frac{1}{2}(Y_1 - Y_2) \right]^2 = \frac{(Y_1 - Y_2)^2}{2}.$$

It follows that, when $n = 2$,

$$\frac{(n-1)S^2}{\sigma^2} = \frac{(Y_1 - Y_2)^2}{2\sigma^2} = \left(\frac{Y_1 - Y_2}{\sqrt{2\sigma^2}} \right)^2.$$

We will show that this quantity is equal to the square of a standard normal random variable, that is, it is a Z^2, which, as we have already shown in Example 6.11, possesses a χ^2 distribution with 1 degree of freedom.

Because $Y_1 - Y_2$ is a linear combination of independent, normally distributed random variables ($Y_1 - Y_2 = a_1 Y_1 + a_2 Y_2$ with $a_1 = 1$ and $a_2 = -1$), Theorem 6.3 tells us that $Y_1 - Y_2$ has a normal distribution with mean $1\mu - 1\mu = 0$ and variance $(1)^2\sigma^2 + (-1)^2\sigma^2 = 2\sigma^2$. Therefore,

$$Z = \frac{Y_1 - Y_2}{\sqrt{2\sigma^2}}$$

has a standard normal distribution. Because for $n = 2$

$$\frac{(n-1)S^2}{\sigma^2} = \left(\frac{Y_1 - Y_2}{\sqrt{2\sigma^2}} \right)^2 = Z^2,$$

it follows that $(n-1)S^2/\sigma^2$ has a χ^2 distribution with 1 degree of freedom.

In Example 6.13 we proved that $U_1 = (Y_1 + Y_2)/\sigma$ and $U_2 = (Y_1 - Y_2)/\sigma$ are independent random variables. Notice that, because $n = 2$,

$$\overline{Y} = \frac{Y_1 + Y_2}{2} = \frac{\sigma U_1}{2} \quad \text{and} \quad S^2 = \frac{(Y_1 - Y_2)^2}{2} = \frac{(\sigma U_2)^2}{2}.$$

> Because $\overline{Y}$ is a function of only U_1 and S^2 is a function of only U_2, the independence of U_1 and U_2 implies the independence of $\overline{Y}$ and S^2.   ∎

EXAMPLE 7.4   In Example 7.1 the ounces of fill from the bottling machine are assumed to have a normal distribution with $\sigma^2 = 1$. Suppose that we plan to select a random sample of ten bottles and measure the amount of fill in each bottle. If these ten observations are used to calculate S^2, it might be useful to specify an interval of values that will include S^2 with a high probability. Find numbers b_1 and b_2 such that

$$P(b_1 \leq S^2 \leq b_2) = .90.$$

Solution   Notice that

$$P(b_1 \leq S^2 \leq b_2) = P\left[\frac{(n-1)b_1}{\sigma^2} \leq \frac{(n-1)S^2}{\sigma^2} \leq \frac{(n-1)b_2}{\sigma^2}\right].$$

Because $\sigma^2 = 1$, it follows that $(n-1)S^2/\sigma^2 = (n-1)S^2$ has a χ^2 distribution with $(n-1)$ degrees of freedom. Therefore, we can use Table 6, Appendix III, to find two numbers a_1 and a_2 such that

$$P[a_1 \leq (n-1)S^2 \leq a_2] = .90.$$

One method of doing this is to find the value of a_2 that cuts off an area of .05 in the upper tail and the value of a_1 that cuts off .05 in the lower tail (.95 in the upper tail). Because there are $n - 1 = 9$ degrees of freedom, Table 6 gives $a_2 = 16.919$ and $a_1 = 3.325$. Consequently, values for b_1 and b_2 that satisfy our requirements are given by

$$3.325 = a_1 = \frac{(n-1)b_1}{\sigma^2} = 9b_1 \quad \text{or} \quad b_1 = \frac{3.325}{9} = .369 \quad \text{and}$$

$$16.919 = a_2 = \frac{(n-1)b_2}{\sigma^2} = 9b_2 \quad \text{or} \quad b_2 = \frac{16.919}{9} = 1.880.$$

Thus, if we wish to have an interval that will include S^2 with probability .90, one such interval is (.369, 1.880). Notice that this interval is fairly wide.

The result given in Theorem 7.1 provides the basis for development of inference-making procedures about the mean μ of a normal population with known variance σ^2. In that case Theorem 7.1 tells us that $\sqrt{n}(\overline{Y} - \mu)/\sigma$ has a standard normal distribution. When σ is unknown, it can be estimated by $S = \sqrt{S^2}$, and the quantity

$$\sqrt{n}\left(\frac{\overline{Y} - \mu}{S}\right)$$

provides the basis for developing methods for inferences about μ. We will show that $\sqrt{n}(\overline{Y} - \mu)/S$ has a distribution known as *Student's t distribution* with $n - 1$ degrees of freedom. The general definition of a random variable that possesses a Student's t distribution (or simply a t distribution) is as follows.

DEFINITION **7.2**

Let Z be a standard normal random variable, and let W be a χ^2-distributed variable with ν degrees of freedom. Then, if Z and W are independent,

$$T = \frac{Z}{\sqrt{W/\nu}}$$

is said to have a t *distribution* with ν degrees of freedom (d.f.).

If $Y_1, Y_2, \ldots, Y_n$ constitute a random sample from a normal population with mean μ and variance σ^2, Theorem 7.1 may be applied to show $Z = \sqrt{n}\,(\overline{Y} - \mu)/\sigma$ has a standard normal distribution. Theorem 7.3 tells us that $W = (n-1)S^2/\sigma^2$ has a χ^2 distribution with $\nu = n-1$ degrees of freedom and that Z and W are independent (because $\overline{Y}$ and S^2 are independent). Therefore, by Definition 7.2,

$$T = \frac{Z}{\sqrt{W/\nu}} = \frac{\sqrt{n}(\overline{Y} - \mu)/\sigma}{\sqrt{[(n-1)S^2/\sigma^2]/(n-1)}} = \sqrt{n}\left(\frac{\overline{Y} - \mu}{S}\right)$$

has a t distribution with $(n - 1)$ degrees of freedom.

The equation for the t density function will not be given here, but it can be found in Exercise 7.72, where hints about its derivation are given. Like the standard normal density function, the t density function is symmetric about zero. Further, for $\nu > 1$, $E(T) = 0$; and for $\nu > 2$, $V(T) = \nu/(\nu - 2)$. These results follow directly from results developed in Exercises 4.89 and 4.90 (see Exercise 7.14). Thus we see that a t-distributed random variable has the same expected value as a standard normal random variable. However, a standard normal random variable always has variance 1, whereas the variance of a variable with a t distribution always exceeds 1.

A standard normal density function and a t density function are sketched in Figure 7.2. Notice that both density functions are symmetric about the origin but that the t density has more probability mass in its tails.

Values of t_α such that $P(T > t_\alpha) = \alpha$ for $\alpha = .100, .050, .025, .010,$ and $.005$ are given in Table 5, Appendix III. For example, if a random variable has a t distribution with 21 d.f., $t_{.100}$ is found by looking in the row labeled 21 d.f. and the column

FIGURE **7.2**
A comparison of the standard normal and t density functions

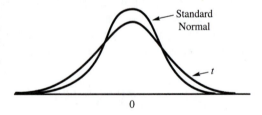

Standard Normal

t

0

headed $t_{.100}$. Using Table 5, we see that $t_{.100} = 1.323$. Therefore, for 21 d.f., $P(T > 1.323) = .100$.

EXAMPLE **7.5**   The tensile strength for a type of wire is normally distributed with unknown mean μ and unknown variance σ^2. Six pieces of wire were randomly selected from a large roll; and Y_i, the tensile strength for portion i, is measured for $i = 1, 2, \ldots, 6$. The population mean μ and variance σ^2 can be estimated by $\overline{Y}$ and S^2, respectively. Because $\sigma_{\overline{Y}}^2 = \sigma^2/n$, it follows that $\sigma_{\overline{Y}}^2$ can be estimated by S^2/n. Find the approximate probability that $\overline{Y}$ will be within $2S/\sqrt{n}$ of the true population mean μ.

Solution   We want to find

$$P\left[-\frac{2S}{\sqrt{n}} \leq (\overline{Y} - \mu) \leq \frac{2S}{\sqrt{n}}\right] = P\left[-2 \leq \sqrt{n}\left(\frac{\overline{Y} - \mu}{S}\right) \leq 2\right]$$

$$= P(-2 \leq T \leq 2)$$

where T has a t distribution with, in this case, $n - 1 = 5$ d.f. Looking at Table 5, Appendix III, we see that the upper-tail area to the right of 2.015 is .05. Hence

$$P(-2.015 \leq T \leq 2.015) = .90$$

and the probability that $\overline{Y}$ will be within two estimated standard deviations of μ is slightly less than .90.

Notice that, if σ^2 were known, the probability that $\overline{Y}$ will fall within $2\sigma_{\overline{Y}}$ of μ would be given by

$$P\left[-2\left(\frac{\sigma}{\sqrt{n}}\right) \leq (\overline{Y} - \mu) \leq 2\left(\frac{\sigma}{\sqrt{n}}\right)\right] = P\left[-2 \leq \sqrt{n}\left(\frac{\overline{Y} - \mu}{\sigma}\right) \leq 2\right]$$

$$= P(-2 \leq Z \leq 2) = .9544.$$

Suppose that we want to compare the variances of two normal populations based on information contained in independent random samples from the two populations. Samples of sizes n_1 and n_2 are taken from the two populations that have variances, σ_1^2 and σ_2^2, respectively. If we calculate S_1^2 from the observations in sample 1, then S_1^2 estimates σ_1^2. Similarly, S_2^2, calculated from the observations in the second sample, estimates σ_2^2. Thus, it seems intuitive that the ratio S_1^2/S_2^2 could be used to make inferences about the relative magnitudes of σ_1^2 and σ_2^2. If we divide each S_i^2 by σ_i^2, then the resulting ratio

$$\frac{S_1^2/\sigma_1^2}{S_2^2/\sigma_2^2} = \left(\frac{\sigma_2^2}{\sigma_1^2}\right)\left(\frac{S_1^2}{S_2^2}\right)$$

has an *F distribution* with $(n_1 - 1)$ numerator degrees of freedom and $(n_2 - 1)$ denominator degrees of freedom. The general definition of a random variable that possesses an *F* distribution appears next.

DEFINITION 7.3

Let W_1 and W_2 be *independent* χ^2-distributed random variables with ν_1 and ν_2 degrees of freedom, respectively. Then

$$F = \frac{W_1/\nu_1}{W_2/\nu_2}$$

is said to have an *F distribution* with ν_1 numerator degrees of freedom and ν_2 denominator degrees of freedom.

The density function for an *F*-distributed random variable is given in Exercise 7.73, where the method for its derivation is outlined. It can be shown (see Exercise 7.16) that if *F* possesses an *F* distribution with ν_1 numerator and ν_2 denominator degrees of freedom, then $E(F) = \nu_2/(\nu_2 - 2)$ if $\nu_2 > 2$. Also, if $\nu_2 > 4$, then $V(F) = \left[2\nu_2^2(\nu_1 + \nu_2 - 2)\right]/\left[\nu_1(\nu_2 - 2)^2(\nu_2 - 4)\right]$. Notice that the mean of an *F*-distributed random variable depends only on the number of denominator degrees of freedom, ν_2.

Considering once again the independent random samples from normal distributions, we know that $W_1 = (n_1 - 1)S_1^2/\sigma_1^2$ and $W_2 = (n_2 - 1)S_2^2/\sigma_2^2$ have independent χ^2 distributions with $\nu_1 = (n_1 - 1)$ and $\nu_2 = (n_2 - 1)$ degrees of freedom, respectively. Thus Definition 7.3 implies that

$$F = \frac{W_1/\nu_1}{W_2/\nu_2} = \frac{\left[(n_1 - 1)S_1^2/\sigma_1^2\right]/(n_1 - 1)}{\left[(n_2 - 1)S_2^2/\sigma_2^2\right]/(n_2 - 1)} = \frac{S_1^2/\sigma_1^2}{S_2^2/\sigma_2^2}$$

has an *F* distribution with $(n_1 - 1)$ numerator degrees of freedom and $(n_2 - 1)$ denominator degrees of freedom.

A typical *F* density function is sketched in Figure 7.3. Values of F_α such that $P(F > F_\alpha) = \alpha$ are given in Table 7, Appendix III, for values of $\alpha = .100, .050,$

FIGURE 7.3
A typical *F*
probability
density function

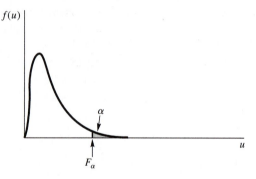

.025, .010, and .005. In Table 7, the column headings are the numerator degrees of freedom, whereas the denominator degrees of freedom are given in the main-row headings. Opposite each denominator degrees of freedom (row heading), the values of $\alpha = .100, .050, .025, 010,$ and $.005$ appear. For example, if the F variable of interest has 5 numerator degrees of freedom and 7 denominator degrees of freedom, then $F_{.100} = 2.88$, $F_{.050} = 3.97$, $F_{.025} = 5.29$, $F_{.010} = 7.46$, and $F_{.005} = 9.52$. Thus, if F has an F distribution with 5 numerator degrees of freedom and 7 denominator degrees of freedom, then $P(F > 7.46) = .01$.

EXAMPLE **7.6**   If we take independent samples of size $n_1 = 6$ and $n_2 = 10$ from two normal populations with equal population variances, find the number b such that

$$P\left(\frac{S_1^2}{S_2^2} \le b\right) = .95.$$

Solution   Because $n_1 = 6, n_2 = 10$ and the population variances are equal, then

$$\frac{S_1^2/\sigma_1^2}{S_2^2/\sigma_2^2} = \frac{S_1^2}{S_2^2}$$

has an F distribution with $v_1 = n_1 - 1 = 5$ numerator degrees of freedom and $v_2 = n_2 - 1 = 9$ denominator degrees of freedom. Also,

$$P\left(\frac{S_1^2}{S_2^2} \le b\right) = 1 - P\left(\frac{S_1^2}{S_2^2} > b\right).$$

Therefore, we want to find the number b cutting off an upper-tail area of .05 under the F density function with 5 numerator degrees of freedom and 9 denominator degrees of freedom. Looking in column 5 and row 9 in Table 7, we see that the appropriate value of b is 3.48.

Even when the population variances are equal, the probability that the ratio of the sample variances exceeds 3.48 is still .05 (assuming sample sizes of $n_1 = 6$ and $n_2 = 10$).

This section has been devoted to developing the sampling distributions of various statistics calculated by using the observations in a random sample from a normal population (or independent random samples from two normal populations). In particular, if $Y_1, Y_2, \ldots, Y_n$ represents a random sample from a normal population with mean μ and variance σ^2, we have seen that $\sqrt{n}(\overline{Y} - \mu)/\sigma$ has a standard normal distribution. Also, $(n - 1)S^2/\sigma^2$ has a χ^2 distribution, and $\sqrt{n}(\overline{Y} - \mu)/S$ has a t distribution (both with $n - 1$ degrees of freedom). If we have two independent random samples from normal populations with variances σ_1^2 and σ_2^2, then $F = (S_1^2/\sigma_1^2)/(S_2^2/\sigma_2^2)$ has an F distribution. These sampling distributions will enable us to evaluate the properties of inferential procedures in later chapters. In the next section we discuss approximations to certain sampling distributions. These approximations can be very useful when the exact form of the sampling distribution is unknown or when it is difficult or tedious to use the sampling distribution to compute probabilities.

Exercises

7.1   Refer to Example 7.1. The amount of fill dispensed by a bottling machine is normally distributed with $\sigma = 1$ ounce. If $n = 9$ bottles are randomly selected from the output of the machine, we found that the probability that the sample mean will be within .3 ounce of the true mean is .6318. Suppose that $\overline{Y}$ is to be computed using a sample of size n.

   a   If $n = 16$, what is $P(|\overline{Y} - \mu| \le .3)$?

   b   Find $P(|\overline{Y} - \mu| \le .3)$ when $\overline{Y}$ is to be computed using samples of sizes $n = 25$, $n = 36$, $n = 49$ and $n = 64$.

   c   What pattern do you observe among the values for $P(|\overline{Y} - \mu| \le .3)$ that you observed for the various values of n?

   d   Do the results that you obtained in part (b) seem to be consistent with the result obtained in Example 7.2?

7.2   Refer to Exercise 7.1. Assume now that the amount of fill dispensed by the bottling machine is normally distributed with $\sigma = 2$ ounce.

   a   If $n = 9$ bottles are randomly selected from the output of the machine, what is $P(|\overline{Y}-\mu| \le .3)$? Compare this with the answer obtained in Example 7.1.

   b   Find $P(|\overline{Y} - \mu| \le .3)$ when $\overline{Y}$ is to be computed using samples of sizes $n = 25$, $n = 36$, $n = 49$ and $n = 64$.

   c   What pattern do you observe among the values for $P(|\overline{Y} - \mu| \le .3)$ that you observed for the various values of n?

   d   How do the respective probabilities obtained in this problem (where $\sigma = 2$) compare to those obtained in Exercise 7.1 (where $\sigma = 1$)?

7.3   A forester studying the effects of fertilization on certain pine forests in the Southeast is interested in estimating the average basal area of pine trees. In studying basal areas of similar trees for many years, he has discovered these measurements (in square inches) to be normally distributed with standard deviation approximately 4 square inches. If the forester samples $n = 9$ trees, find the probability that the sample mean will be within 2 square inches of the population mean.

7.4   Suppose the forester in Exercise 7.3 would like the sample mean to be within 1 square inches of the population mean, with probability .90. How many trees must he measure in order to ensure this degree of accuracy?

7.5   The Environmental Protection Agency is concerned with the problem of setting criteria for the amounts of certain toxic chemicals to be allowed in freshwater lakes and rivers. A common measure of toxicity for any pollutant is the concentration of the pollutant that will kill half of the test species in a given amount of time (usually 96 hours for fish species). This measure is called the LC50 (lethal concentration killing 50% of the test species). In many studies, the values contained in the natural logarithm of LC50 measurements are normally distributed and hence the analysis is based on ln(LC50) data.

    Studies of the effects of copper on a certain species of fish (say, species A) show the variance of ln(LC50) measurements to be around .4 with concentration measurements in milligrams per liter. If $n = 10$ studies on LC50 for copper are to be completed, find the probability that the sample mean of ln(LC50) will differ from the true population mean by no more than .5.

7.6   If in Exercise 7.5 we want the sample mean to differ from the population mean by no more than .5 with probability .95, how many tests should be run?

7.7   Suppose that $X_1, X_2, \ldots, X_m$ and $Y_1, Y_2, \ldots, Y_n$ are independent random samples, with the variables X_i normally distributed with mean μ_1 and variance σ_1^2 and the variables Y_i normally distributed with mean μ_2 and variance σ_2^2. The difference between the sample means, $\overline{X} - \overline{Y}$, is then a linear combination of $m + n$ normally distributed random variables and, by Theorem 6.3, is itself normally distributed.

   a   Find $E(\overline{X} - \overline{Y})$.

   b   Find $V(\overline{X} - \overline{Y})$.

   c   Suppose that $\sigma_1^2 = 2$, $\sigma_2^2 = 2.5$, and $m = n$. Find the sample sizes so that $(\overline{X} - \overline{Y})$ will be within one unit of $(\mu_1 - \mu_2)$ with probability .95.

7.8   Referring to Exercise 7.5, suppose that the effects of copper on a second species (say, species B) of fish show the variance of ln(LC50) measurements to be .8. If the population means of ln(LC50) for the two species are equal, find the probability that, with random samples of ten measurements from each species, the sample mean for species A exceeds the sample mean for species B by at least one unit.

7.9   Ammeters produced by a manufacturer are marketed under the specification that the standard deviation of gauge readings is no larger than .2 amp. One of these ammeters was used to make ten independent readings on a test circuit with constant current. If the sample variance of these ten measurements is .065 and it is reasonable to assume that the readings are normally distributed, do the results suggest that the ammeter used does not meet the marketing specifications? (*Hint:* Find the approximate probability that the sample variance will exceed .065 if the true population variance is .04.)

7.10   **a**   If U has a χ^2 distribution with ν degrees of freedom, find $E(U)$ and $V(U)$.

     b   Using the results of Theorem 7.3, find $E(S^2)$ and $V(S^2)$ when $Y_1, Y_2, \ldots, Y_n$ is a random sample from a normal distribution with mean μ and variance σ^2.

7.11   Refer to Exercise 7.5. Suppose that $n = 20$ observations are to be taken on ln(LC50) measurements and that $\sigma^2 = 1.4$. Let S^2 denote the sample variance of the 20 measurements.

   a   Find a number b such that $P(S^2 \leq b) = .975$.

   b   Find a number a such that $P(a \leq S^2) = .975$.

   c   If a and b are as in the previous parts, what is $P(a \leq S^2 \leq b)$?

7.12   Refer to Exercise 7.3. Suppose that in the forest fertilization problem the population standard deviation of basal areas is not known and must be estimated from the sample. If a random sample of $n = 9$ basal areas is to be measured, find two statistics g_1 and g_2 such that $P[g_1 \leq (\overline{Y} - \mu) \leq g_2] = .90$.

7.13   If Y is a random variable that has an F distribution with ν_1 numerator and ν_2 denominator degrees of freedom, show that $U = 1/Y$ has an F distribution with ν_2 numerator and ν_1 denominator degrees of freedom.

*7.14   Suppose that Z has a standard normal distribution and that Y is an independent χ^2-distributed random variable with ν degrees. Then, according to Definition 7.2,

$$T = \frac{Z}{\sqrt{Y/\nu}}$$

has a t distribution with ν degrees of freedom.[1]

a If Z has a standard normal distribution, give $E(Z)$ and $E(Z^2)$. [*Hint:* For any random variable, $E(Z^2) = V(Z) + (E(Z))^2$.]

b According to the result derived in Exercise 4.90 (a), if Y has a χ^2 distribution with ν degrees of freedom, then

$$E\left(Y^a\right) = \frac{\Gamma\left([\nu/2] + a\right)}{\Gamma\left(\nu/2\right)} 2^a \qquad \text{if } \nu > -2a.$$

Use this result, the result from (a), and the structure of T to show the following. (*Hint:* Recall the independence of Z and Y.)

    i. $E(T) = 0$ if $\nu > 1$.

    ii. $V(T) = \nu/(\nu - 2)$ if $\nu > 2$.

7.15 Use the structures of T and F given in Definitions 7.2 and 7.3, respectively, to argue that if T has a t distribution with ν degrees of freedom, then $U = T^2$ has an F distribution with 1 numerator degree of freedom and ν denominator degrees of freedom.

*7.16 Suppose that W_1 and W_2 are independent χ^2-distributed random variables with ν_1 and ν_2 degrees of freedom, respectively. According to Definition 7.3,

$$F = \frac{W_1/\nu_1}{W_2/\nu_2}$$

has an F distribution with ν_1 and ν_2 numerator and denominator degrees of freedom, respectively. Use the preceding structure of F, the independence of W_1 and W_2, and the result summarized in Exercise 7.14(b) to show

a $E(F) = \nu_2/(\nu_2 - 2)$ if $\nu_2 > 2$.

b $V(F) = \left[2\nu_2^2(\nu_1 + \nu_2 - 2)\right] / \left[\nu_1(\nu_2 - 2)^2(\nu_2 - 4)\right]$ if $\nu_2 > 4$.

7.17 Refer to Exercise 7.16. Suppose that F has an F distribution with $\nu_1 = 50$ numerator degrees of freedom and $\nu_2 = 70$ denominator degrees of freedom. Notice that Table 7, Appendix III, does not contain entries for 50 numerator degrees of freedom and 70 denominator degrees of freedom.

a What is $E(F)$?

b Give $V(F)$.

c Is it likely that F will exceed 3? (*Hint:* Use Tchebysheff's theorem.)

*7.18 Let S_1^2 denote the sample variance for a random sample of ten ln(LC50) values for copper, and let S_2^2 denote the sample variance for a random sample of eight ln(LC50) values for lead, both samples using the same species of fish. The population variance for measurements on copper is assumed to be twice the corresponding population variance for measurements on lead. Assume S_1^2 to be independent of S_2^2.

---

[1]Exercises preceded by an asterisk are optional.

a Find a number b such that

$$P\left(\frac{S_1^2}{S_2^2} \le b\right) = .95$$

b Find a number a such that

$$P\left(a \le \frac{S_1^2}{S_2^2}\right) = .95$$

[*Hint:* Use the result of Exercise 7.13, and notice that $P(U_1/U_2 \le k) = P(U_2/U_1 \ge 1/k)$.]

c If a and b are as in the previous parts, find

$$P\left(a \le \frac{S_1^2}{S_2^2} \le b\right).$$

7.19 Let $Y_1, Y_2, \ldots, Y_5$ be a random sample of size 5 from a normal population with mean 0 and variance 1, and let $\overline{Y} = (1/5)\sum_{i=1}^{5} Y_i$. Let Y_6 be another independent observation from the same population.

a What is the distribution of $W = \sum_{i=1}^{5} Y_i^2$? Why?

b What is the distribution of $U = \sum_{i=1}^{5} (Y_i - \overline{Y})^2$? Why?

c What is the distribution of $\sum_{i=1}^{5} (Y_i - \overline{Y})^2 + Y_6^2$? Why?

7.20 Suppose that $Y_1, Y_2, \ldots, Y_5, Y_6, \overline{Y}, W,$ and U are as defined in Exercise 7.19.

a What is the distribution of $\sqrt{5}Y_6/\sqrt{W}$? Why?

b What is the distribution of $2Y_6/\sqrt{U}$? Why?

c What is the distribution of $2(5\overline{Y}^2 + Y_6^2)/U$? Why?

*7.21 Suppose that independent samples (of sizes n_i) are taken from each of k populations and that population i is normally distributed with mean μ_i and variance σ^2), $i = 1, 2, \ldots, k$. That is, all populations are normally distributed with the *same* variance, but with (possibly) different means. Let $\overline{X}_i$ and S_i^2, $i = 1, 2, \ldots, k$ be the respective sample means and variances. Let $\theta = c_1\mu_1 + c_2\mu_2 + \cdots + c_k\mu_k$ where $c_1, c_2, \ldots, c_k$ are given constants.

a Give the distribution of $\hat{\theta} = c_1\overline{X}_1 + c_2\overline{X}_2 + \cdots + c_k\overline{X}_k$. Provide reasons for any claims that you make.

b Give the distribution of

$$\frac{\text{SSE}}{\sigma^2} \quad \text{where} \quad \text{SSE} = \sum_{i=1}^{k} (n_i - 1)S_i^2.$$

Provide reasons for any claims that you make.

c Give the distribution of

$$\frac{\hat{\theta} - \theta}{\sqrt{\left(\frac{c_1^2}{n_1} + \frac{c_2^2}{n_2} + \cdots + \frac{c_k^2}{n_k}\right)\text{MSE}}}, \quad \text{where} \quad \text{MSE} = \frac{\text{SSE}}{n_1 + n_2 + \cdots + n_k - k}.$$

Provide reasons for any claims that you make.

7.3 The Central Limit Theorem

In Chapter 5 we showed that if Y_1, Y_2, ..., Y_n represents a random sample from *any* distribution with mean μ and variance σ^2, then $E(\overline{Y}) = \mu$ and $V(\overline{Y}) = \sigma^2/n$. In this section we will develop an approximation for the sampling distribution of $\overline{Y}$ that can be used regardless of the distribution of the population from which the sample is taken.

If we sample from a normal population, Theorem 7.1 tells us that $\overline{Y}$ has a normal sampling distribution. But what can we say about the sampling distribution of $\overline{Y}$ if the variables Y_i are not normally distributed? Fortunately, $\overline{Y}$ will have a sampling distribution that is approximately normal if the sample size is large. The formal statement of this result is called the *central limit theorem*. Before we state this theorem, however, we will look at some empirical investigations that demonstrate the sampling distribution of $\overline{Y}$.

A computer was used to generate random samples of size n from an exponential density function with mean 10—that is, from a population with density

$$f(y) = \begin{cases} (1/10)e^{-y/10}, & y > 0 \\ 0, & \text{elsewhere.} \end{cases}$$

A graph of this density function is given in Figure 7.4. The sample mean was computed for each sample and the relative frequency histogram for the values of the sample means for 1000 samples, each of size $n = 5$, is shown in Figure 7.5. Notice that Figure 7.5 portrays a histogram that is roughly mound-shaped, but the histogram is slightly skewed.

Figure 7.6 is a graph of a similar relative frequency histogram of the values of the sample mean for 1000 samples, each of size $n = 25$. In this case Figure 7.6 shows a mounded-shaped and nearly symmetric histogram, which can be approximated quite closely with a normal density function.

Recall from Chapter 5 that $E(\overline{Y}) = \mu_{\overline{Y}} = \mu$ and $V(\overline{Y}) = \sigma_{\overline{Y}}^2 = \sigma^2/n$. For the exponential density function used in the simulations, $\mu = E(Y_i) = 10$ and $\sigma^2 = V(Y_i) = (10)^2 = 100$. Thus, for this example, we see that

$$\mu_{\overline{Y}} = E(\overline{Y}) = \mu = 10 \quad \text{and} \quad \sigma_{\overline{Y}}^2 = V(\overline{Y}) = \frac{\sigma^2}{n} = \frac{100}{n}.$$

FIGURE 7.4
An exponential
density function

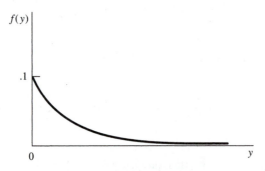

FIGURE **7.5**
Relative frequency
histogram: sample
means for 1000
samples ($n = 5$) from
an exponential
distribution

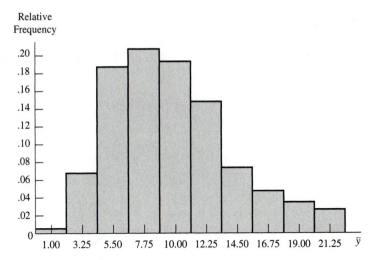

FIGURE **7.6**
Relative frequency
histogram: sample
means for 1000
samples ($n = 25$)
from an exponential
distribution

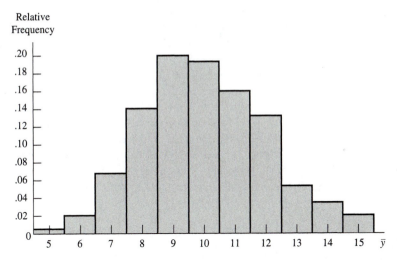

For each value of n (5 and 25), we calculated the average of the 1000 sample means generated in the study. The observed variance of the 1000 sample means was also calculated for each value of n. The results are shown in Table 7.1. In each empirical study ($n = 5$ and $n = 25$), the average of the observed sample means and the variance of the observed sample means are quite close to the theoretical values.

Table 7.1. Calculations for 1000 sample means

Sample Size	Average of 1000 Sample Means	$\mu_{\bar{Y}} = \mu$	Variance of 1000 Sample Means	$\sigma_{\bar{Y}}^2 = \sigma^2/n$
$n = 5$	9.86	10	19.63	20
$n = 25$	9.95	10	3.93	4

We now give a formal statement of the central limit theorem.

THEOREM 7.4   Let $Y_1, Y_2, \ldots, Y_n$ be independent and identically distributed random variables with $E(Y_i) = \mu$ and $V(Y_i) = \sigma^2 < \infty$. Define

$$U_n = \sqrt{n}\left(\frac{\overline{Y} - \mu}{\sigma}\right) \quad \text{where} \quad \overline{Y} = \frac{1}{n}\sum_{i=1}^{n} Y_i.$$

Then the distribution function of U_n converges to a standard normal distribution function as $n \to \infty$.

The conclusion of Theorem 7.4, stated slightly differently, says that

$$P(a \le U_n \le b) \to \int_a^b \left(\frac{1}{\sqrt{2\pi}}\right) e^{-u^2/2}\, du$$

as $n \to \infty$. That is, probability statements about U_n can be approximated by corresponding probabilities for the standard normal random variable if n is large. (Usually a value of n greater than 30 will ensure that the distribution of U_n can be closely approximated by a normal distribution.)

As a matter of convenience, the conclusion of the central limit theorem is often replaced with the simpler statement that $\overline{Y}$ is *asymptotically normally distributed* with mean μ and variance σ^2/n. The central limit theorem can be applied to a random sample $Y_1, Y_2, \ldots, Y_n$ from *any* distribution, so long as $E(Y_i) = \mu$ and $V(Y_i) = \sigma^2$ are both finite and the sample size is large.

We will give some examples of the use of the central limit theorem but will defer the proof until the next section (coverage of which is optional). The proof is not needed for an understanding of the applications of the central limit theorem that appear in this text.

EXAMPLE 7.7   Achievement test scores of all high school seniors in a state have mean 60 and variance 64. A random sample of $n = 100$ students from one large high school had a mean score of 58. Is there evidence to suggest that this high school is inferior? (Calculate the probability that the sample mean is at most 58 when $n = 100$.)

Solution   Let $\overline{Y}$ denote the mean of a random sample of $n = 100$ scores from a population with $\mu = 60$ and $\sigma^2 = 64$. We want to approximate $P(\overline{Y} \le 58)$. We know from Theorem 7.4 that $\sqrt{n}(\overline{Y} - \mu)/\sigma$ is approximately a standard normal random variable, which we denote by Z. Hence, using Table 4, Appendix III, we have

$$P(\overline{Y} \le 58) \approx P\left(Z \le \frac{\sqrt{100}(58 - 60)}{\sqrt{64}}\right) = P(Z \le -2.5) = .0062.$$

Because this probability is so small, it is unlikely that the sample from the school of interest can be regarded as a random sample from a population with $\mu = 60$ and

$\sigma^2 = 64$. The evidence suggests that the average score for this high school is lower than the overall average of $\mu = 60$.

This example illustrates the use of probability in the process of testing hypotheses, a common technique of statistical inference that will be further discussed in Chapter 10.

EXAMPLE **7.8**   The service times for customers coming through a checkout counter in a retail store are independent random variables with mean 1.5 minutes and variance 1.0. Approximate the probability that 100 customers can be served in less than 2 hours of total service time.

Solution   If we let Y_i denote the service time for the ith customer, then we want

$$P\left(\sum_{i=1}^{100} Y_i \leq 120\right) = P\left(\overline{Y} \leq \frac{120}{100}\right) = P(\overline{Y} \leq 1.20).$$

Because the sample size is large, the central limit theorem tells us that $\overline{Y}$ is approximately normally distributed with mean $\mu_{\overline{Y}} = \mu = 1.5$ and variance $\sigma_{\overline{Y}}^2 = \sigma^2/n = 1.0/100$. Therefore, using Table 4, Appendix III, we have

$$P(\overline{Y} \leq 1.20) = P\left(\frac{\overline{Y} - 1.50}{1/\sqrt{100}} \leq \frac{1.20 - 1.50}{1/\sqrt{100}}\right)$$

$$\approx P[Z \leq (1.2 - 1.5)\sqrt{100}] = P(Z \leq -3) = .0013.$$

Thus the probability that 100 customers can be served in less than 2 hours is approximately .0013. This small probability indicates that it is virtually impossible to serve 100 customers in only 2 hours.

Exercises

7.22   The fracture strength of tempered glass averages 14 (measured in thousands of pounds per square inch) and has standard deviation 2.

   a   What is the probability that the average fracture strength of 100 randomly selected pieces of this glass exceeds 14.5?

   b   Find an interval that includes, with probability 0.95, the average fracture strength of 100 randomly selected pieces of this glass.

7.23   An anthropologist wishes to estimate the average height of men for a certain race of people. If the population standard deviation is assumed to be 2.5 inches and if she randomly samples 100 men, find the probability that the difference between the sample mean and the true population mean will not exceed .5 inches.

7.24   Suppose that the anthropologist of Exercise 7.23 wants the difference between the sample mean and the population mean to be less than .4 inches, with probability .95. How many men should she sample to achieve this objective?

7.25 Workers employed in a large service industry have an average wage of $7.00 per hour with a standard deviation of $.50. The industry has 64 workers of a certain ethnic group. These workers have an average wage of $6.90 per hour. Is it reasonable to assume that the wage rate of the ethnic group is equivalent to that of a random sample of workers from those employed in the service industry? (*Hint:* Calculate the probability of obtaining a sample mean less than or equal to $6.90 per hour.)

7.26 The acidity of soils is measured by a quantity called the pH, which may range from 0 (high acidity) to 14 (high alkalinity). A soil scientist wants to estimate the average pH for a large field by randomly selecting n core samples and measuring the pH in each sample. Although the population standard deviation of pH measurements is not known, past experience indicates that most soils have a pH value of between 5 and 8. If the scientist selects $n = 40$ samples, find the approximate probability that the sample mean of the 40 pH measurements will be within .2 unit of the true average pH for the field. (*Hint:* See Exercise 1.15.)

7.27 Suppose that the scientist of Exercise 7.26 would like the sample mean to be within .1 of the true mean with probability .90. How many core samples should the scientist take?

7.28 An important aspect of a federal economic plan was that consumers would save a substantial portion of the money that they received from an income tax reduction. Suppose that early estimates of the portion of total tax saved, based on a random sampling of 35 economists, had mean 26% and standard deviation 12%.

   a What is the approximate probability that a sample mean estimate, based on a random sample of $n = 35$ economists, will lie within 1% of the mean of the population of the estimates of all economists?

   b Is it necessarily true that the mean of the population of estimates of all economists is equal to the percent tax saving that will actually be achieved?

7.29 The length of time required for the periodic maintenance of an automobile or other machine usually has a mound-shaped probability distribution. Because some occasional long service times will occur, the distribution tends to be skewed to the right. Suppose that the length of time required to run a 5000-mile check and to service a new automobile has mean 1.4 hours and standard deviation .7 hour. Suppose also that the service department plans to service 50 automobiles per 8-hour day and that, in order to do so, it can spend a maximum average service time of only 1.6 hours per automobile. On what proportion of all work days will the service department have to work overtime?

7.30 Shear strength measurements for spot welds have been found to have standard deviation 10 psi. If 100 test welds are to be measured, what is the approximate probability that the sample mean will be within 1 psi of the true population mean?

7.31 Refer to Exercise 7.30. If the standard deviation of shear strength measurements for spot welds is 10 psi, how many test welds should be sampled if we want the sample mean to be within 1 psi of the true mean with probability approximately .99?

7.32 Resistors to be used in a circuit have average resistance 200 ohms and standard deviation 10 ohms. Suppose 25 of these resistors are randomly selected to be used in a circuit.

   a What is the probability that the average resistance for the 25 resistors is between 199 and 202 ohms?

   b Find the probability that the *total* resistance does not exceed 5100 ohms. (*Hint:* see Example 7.8.)

7.33 One-hour carbon monoxide concentrations in air samples from a large city average 12 ppm (parts per million) with standard deviation 9 ppm.

    a Do you think that carbon monoxide concentrations in air samples from this city are normally distributed? Why or why not?

    b Find the probability that the average concentration in 100 randomly selected samples will exceed 14 ppm.

7.34 Unaltered bitumens, as commonly found in lead-zinc deposits have atomic hydrogen/carbon (H/C) ratios that average 1.4 with standard deviation .05. Find the probability that the average H/C ratio is less than 1.3 if we randomly select 25 bitumen samples.

7.35 The downtime per day for a computing facility has mean 4 hours and standard deviation .8 hour.

    a Suppose that we want to compute probabilities about the average daily downtime for a period of 30 days.

        i What assumptions must be true to use the result of Theorem 7.4 to obtain a valid approximation for probabilities about the average daily downtime?

        ii Under the assumptions described in part (i) of (a), what is the approximate probability the the average daily downtime for a period of 30 days is between 1 and 5 h?

    b Under the assumptions described in part (a), what is the approximate probability that the *total* downtime for a period of 30 days is less than 115 hours?

7.36 Many bulk products, such as iron ore, coal, and raw sugar, are sampled for quality by a method that requires many small samples to be taken periodically as the material is moving along a conveyor belt. The small samples are then combined and mixed to form one composite sample. Let Y_i denote the volume of the ith small sample from a particular lot, and suppose that $Y_1, Y_2, \ldots, Y_n$ constitute a random sample, with each Y_i value having mean μ (in cubic inches) and variance σ^2. The average volume μ of the samples can be set by adjusting the size of the sampling device. Suppose that the variance σ^2 of the volumes of the samples is known to be approximately 4. The total volume of the composite sample must exceed 200 cubic inches with probability approximately .95 when $n = 50$ small samples are selected. Determine a setting for μ that will allow the sampling requirements to be satisfied.

7.37 Twenty-five heat lamps are connected in a greenhouse so that when one lamp fails, another takes over immediately. (Only one lamp is turned on at any time.) The lamps operate independently, and each has a mean life of 50 hours and standard deviation of 4 hours. If the greenhouse is not checked for 1300 hours after the lamp system is turned on, what is the probability that a lamp will be burning at the end of the 1300-hour period?

7.38 Suppose that $X_1, X_2, \ldots, X_n$ and $Y_1, Y_2, \ldots, Y_n$ are independent random samples from populations with means μ_1 and μ_2 and variances σ_1^2 and σ_2^2, respectively. Show that the random variable

$$U_n = \frac{(\overline{X} - \overline{Y}) - (\mu_1 - \mu_2)}{\sqrt{(\sigma_1^2 + \sigma_2^2)/n}}$$

satisfies the conditions of Theorem 7.4 and thus that the distribution function of U_n converges to a standard normal distribution function as $n \to \infty$. [*Hint:* Consider $W_i = X_i - Y_i$, for $i = 1, 2, \ldots, n$.]

7.39 An experiment is designed to test whether operator A or operator B gets the job of operating a new machine. Each operator is timed on 50 independent trials involving the performance of a certain task using the machine. If the sample means for the 50 trials differ by more than 1 second, the operator with the smaller mean time gets the job. Otherwise, the experiment is considered to end in a tie. If the standard deviations of times for both operators are assumed to be 2 second, what is the probability that operator A will get the job even though both operators have equal ability?

7.40 The result in Exercise 7.38 holds even if the sample sizes differ. That is, if $X_1, X_2, \ldots, X_{n_1}$ and $Y_1, Y_2, \ldots, Y_{n_2}$ constitute independent random samples from populations with means μ_1 and μ_2 and variances σ_1^2 and σ_2^2, respectively, then $\overline{X} - \overline{Y}$ will be approximately normally distributed, for large n_1 and n_2, with mean $\mu_1 - \mu_2$ and variance $(\sigma_1^2/n_1) + (\sigma_2^2/n_2)$.

The flow of water through soil depends on, among other things, the porosity (volume proportion of voids) of the soil. To compare two types of sandy soil, $n_1 = 50$ measurements are to be taken on the porosity of soil A and $n_2 = 100$ measurements are to be taken on soil B. Assume that $\sigma_1^2 = .01$ and $\sigma_2^2 = .02$. Find the probability that the difference between the sample means will be within .05 unit of the difference between the population means $\mu_1 - \mu_2$.

7.41 Refer to Exercise 7.40. Suppose that $n_1 = n_2 = n$, and find the value of n that allows the difference between the sample means to be within .04 unit of $\mu_1 - \mu_2$ with probability .90.

7.42 The times that a cashier spends processing individual customer's order are independent random variables with mean 2.5 minutes and standard deviation 2 minutes. What is the approximate probability that it will take more than 4 hours to process the orders of 100 people?

7.43 Refer to Exercise 7.42. Find the number of customers n such that the probability that the orders of all n customers can be processed in less than 2 h is approximately .1.

7.4 A Proof of the Central Limit Theorem (Optional)

We will sketch a proof of the central limit theorem for the case in which the moment-generating functions exist for the random variables in the sample. The proof depends upon a fundamental result of probability theory, which cannot be proved here but that is stated in Theorem 7.5.

THEOREM 7.5

Let Y_n and Y be random variables with moment-generating functions $m_n(t)$ and $m(t)$, respectively. If

$$\lim_{n \to \infty} m_n(t) = m(t)$$

for all real t, then the distribution function of Y_n converges to the distribution function of Y as $n \to \infty$.

We now give the proof of the *central limit theorem*, Theorem 7.4.

Proof

In Chapters 3 and 4, we saw that if μ_k' denotes the kth moment of a random variable, W, then the moment-generating function of W can be written as

$$m_W(t) = E(e^{tW}) = 1 + t\mu_1' + \frac{t^2}{2!}\mu_2' + \frac{t^3}{3!}\mu_3' + \cdots.$$

Define the random variable Z_i by

$$Z_i = \frac{Y_i - \mu}{\sigma}.$$

Notice that $E(Z_i) = 0$ and $V(Z_i) = 1$. Thus, the first two moments of Z_i are 0 and 1, respectively, and the moment-generating function for Z_i can be written as

$$m_{Z_i}(t) = 1 + \frac{t^2}{2} + \frac{t^3}{3!} E(Z_i^3) + \cdots.$$

Further,

$$U_n = \sqrt{n} \left(\frac{\overline{Y} - \mu}{\sigma} \right) = \frac{1}{\sqrt{n}} \left(\frac{\sum\limits_{i=1}^{n} Y_i - n\mu}{\sigma} \right) = \frac{1}{\sqrt{n}} \sum_{i=1}^{n} Z_i.$$

Because the random variables Y_i are independent, it follows that the random variables Z_i are independent, for $i = 1, 2, \ldots, n$.

Recall that the moment-generating function of the sum of independent random variables is the product of their individual moment-generating functions. Hence,

$$m_n(t) = \left[m_Z \left(\frac{t}{\sqrt{n}} \right) \right]^n = \left(1 + \frac{t^2}{2n} + \frac{t^3}{3!n^{3/2}} k + \cdots \right)^n$$

where $k = E(Z_i^3)$.

Now take the limit of $m_n(t)$ as $n \to \infty$. One way to evaluate the limit is to consider the natural logarithm of $m_n(t)$, where

$$\ln \left[m_n(t) \right] = n \ln \left[1 + \left(\frac{t^2}{2n} + \frac{t^3 k}{6n^{3/2}} + \cdots \right) \right].$$

A standard series expansion for $\ln(1 + x)$ is

$$\ln(1 + x) = x - \frac{x^2}{2} + \frac{x^3}{3} - \frac{x^4}{4} + \cdots.$$

Letting

$$x = \left(\frac{t^2}{2n} + \frac{t^3 k}{6n^{3/2}} + \cdots \right)$$

we have

$$\ln m_n(t) = n \ln(1+x) = n\left(x - \frac{x^2}{2} + \cdots\right)$$

$$= n\left[\left(\frac{t^2}{2n} + \frac{t^3 k}{6n^{3/2}} + \cdots\right) - \frac{1}{2}\left(\frac{t^2}{2n} + \frac{t^3 k}{6n^{3/2}} + \cdots\right)^2 + \cdots\right]$$

where the succeeding terms in the expansion involve x^3, x^4, and so on. Multiplying through by n, we see that the first term, $t^2/2$, does not involve n, whereas all other terms have n to a positive power in the denominator. Thus it can be shown that

$$\lim_{n\to\infty} \ln[m_n(t)] = \frac{t^2}{2}$$

or

$$\lim_{n\to\infty} m_n(t) = e^{t^2/2}$$

the moment-generating function for a standard normal random variable. Applying Theorem 7.5, we conclude that U_n has a distribution function that converges to the distribution function of the standard normal random variable. ∎

7.5 The Normal Approximation to the Binomial Distribution

The central limit theorem also can be used to approximate probabilities for some discrete random variables when the exact probabilities are tedious to calculate. One useful example involves the binomial distribution for large values of the number of trials, n.

Suppose that Y has a binomial distribution with n trials and probability of success on any one trial denoted by p. If we want to find $P(Y \le b)$, we can use the binomial probability function to compute $P(Y = y)$ for each nonnegative integer y less than or equal to b and then sum these probabilities. Tables are available for some values of the sample size n, but direct calculation is cumbersome for large values of n for which tables may be unavailable.

Alternatively, we can view Y, the number of successes in n trials, as a sum of a sample consisting of 0s and 1as; that is,

$$Y = \sum_{i=1}^{n} X_i,$$

where

$$X_i = \begin{cases} 1, & \text{if the } i\text{th trial results in success} \\ 0, & \text{otherwise.} \end{cases}$$

The random variables X_i for $i = 1, 2, \ldots, n$ are independent (because the trials are independent), and it is easy to show that $E(X_i) = p$ and $V(X_i) = p(1 - p)$ for $i = 1, 2, \ldots, n$. Consequently, when n is large, the sample fraction of successes,

$$\frac{Y}{n} = \frac{1}{n} \sum_{i=1}^{n} X_i = \overline{X},$$

possesses an approximately normal sampling distribution with mean $E(X_i) = p$ and variance $V(X_i)/n = p(1 - p)/n$.

Thus, we have used Theorem 7.4 (the central limit theorem) to establish that if Y is a binomial random variable with parameters n and p and if n is large, then Y/n has approximately the same distribution as U, where U is normally distributed with mean $\mu_U = p$ and variance $\sigma_U^2 = p(1 - p)/n$. Equivalently, for large n, we can think of Y as having approximately the same distribution as W, where W is normally distributed with mean $\mu_W = np$ and variance $\sigma_W^2 = np(1 - p)$.

EXAMPLE 7.9   Candidate A believes that she can win a city election if she can earn at least 55% of the votes in precinct I. She also believes that about 50% of the city's voters favor her. If $n = 100$ voters show up to vote at precinct I, what is the probability that candidate A will receive at least 55% of their votes?

Solution   Let Y denote the number of voters at precinct I who vote for candidate A. We must approximate $P(Y/n \geq .55)$ when p is the probability that a randomly selected voter from precinct I favors candidate A. If we think of the $n = 100$ voters at precinct I as a random sample from the city, then Y has a binomial distribution with $n = 100$ and $p = .5$. We have seen that the fraction of voters who favor candidate A is

$$\frac{Y}{n} = \frac{1}{n} \sum_{i=1}^{n} X_i$$

where $X_i = 1$ if the ith voter favors candidate A and $X_i = 0$ otherwise.

Because it is reasonable to assume that $X_i, i = 1, 2, \ldots, n$ are independent, the central limit theorem implies that $\overline{X} = Y/n$ is approximately normally distributed with mean $p = .5$ and variance $pq/n = (.5)(.5)/100 = .0025$. Therefore,

$$P\left(\frac{Y}{n} \geq .55\right) = P\left(\frac{Y/n - .5}{\sqrt{.0025}} \geq \frac{.55 - .50}{.05}\right) \approx P(Z \geq 1) = .1587$$

from Table 4, Appendix III.

The normal approximation to binomial probabilities works well even for moderately large n, so long as p is not close to zero or one. A useful rule of thumb is that the normal approximation to the binomial distribution is appropriate when $p \pm 3\sqrt{pq/n}$

FIGURE 7.7
The normal
approximation to the
binomial distribution:
$n = 10$ and $p = .5$

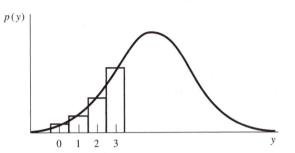

lies in the interval (0, 1)—that is, if

$$0 < p - 3\sqrt{pq/n} \quad \text{and} \quad p + 3\sqrt{pq/n} < 1.$$

In Exercise 7.44, you will show that a more convenient, but equivalent, criterion is that the normal approximation is adequate if

$$n > 9 \left(\frac{\text{larger of } p \text{ and } q}{\text{smaller of } p \text{ and } q} \right).$$

As you will see in Exercise 7.45, for some values of p, this criterion is sometimes met for moderate values of n. Especially for moderate values of n, substantial improvement in the approximation can be made by a slight adjustment on the boundaries used in the calculations. If we look at the segment of a binomial distribution graphed in Figure 7.7, we can see what happens when we try to approximate a discrete distribution represented by a histogram with a continuous density function.

If we want to find $P(Y \leq 3)$ by using the binomial distribution, we can find the total area in the four rectangles (above 0, 1, 2, and 3) illustrated in the binomial histogram (Figure 7.7). Notice that the total area in the rectangles can be approximated by an area under the normal curve. The area under the curve includes some areas not in the histogram and excludes the portion of the histogram that lies above the curve. If we want to approximate $P(Y \leq 3)$ by calculating an area under the density function, the area under the density function to the left of 3.5 provides a better approximation than does the area to the left of 3.0. The following example illustrates how close the normal approximation is for a case in which some exact binomial probabilities can be found.

EXAMPLE 7.10   Suppose that Y has a binomial distribution with $n = 25$ and $p = .4$. Find the exact probabilities that $Y \leq 8$ and $Y = 8$, and compare these to the corresponding values found by using the normal approximation.

Solution   From Table 1, Appendix III, we find that

$$P(Y \leq 8) = .274$$

and

$$P(Y = 8) = P(Y \le 8) - P(Y \le 7) = .274 - .154 = .120.$$

As previously stated, we can think of Y as having approximately the same distribution as W, where W is normally distributed with $\mu_W = np$ and $\sigma_W^2 = np(1-p)$. Because we want $P(Y \le 8)$, we look at the normal curve area to the left of 8.5. Thus,

$$P(Y \le 8) \approx P(W \le 8.5) = P\left(\frac{W - np}{\sqrt{np(1-p)}} \le \frac{8.5 - 10}{\sqrt{25(.4)(.6)}}\right)$$

$$= P(Z \le -.61) = .2709$$

from Table 4, Appendix III. This approximate value is close to the exact value for $P(Y \le 8) = .274$, obtained from the binomial tables.

To find the normal approximation to the binomial probability $p(8)$, we will find the area under the normal curve between the points 7.5 and 8.5, because this is the interval included in the histogram bar over $y = 8$ (see Figure 7.8).

Because Y has approximately the same distribution as W, where W is normally distributed with $\mu_W = np = 25(.4) = 10$ and $\sigma_W^2 = np(1-p) = 25(.4)(.6) = 6$, it follows that

$$P(Y = 8) \approx P(7.5 \le W \le 8.5)$$

$$= P\left(\frac{7.5 - 10}{\sqrt{6}} \le \frac{W - 10}{\sqrt{6}} \le \frac{8.5 - 10}{\sqrt{6}}\right)$$

$$= P(-1.02 \le Z \le -.61) = .2709 - .1539 = .1170.$$

Again we see that this approximate value is very close to the actual value, $P(Y = 8) = .120$, calculated earlier.

FIGURE 7.8
$P(Y = 8)$ for
binomial distribution
of Example 7.10

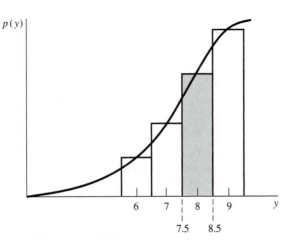

Exercises

7.44 In this section, we provided the "rule of thumb" that the normal approximation to the binomial distribution is adequate if $p \pm 3\sqrt{pq/n}$ lies in the interval $(0, 1)$, that is, if

$$0 < p - 3\sqrt{pq/n} \quad \text{and} \quad p + 3\sqrt{pq/n} < 1.$$

a Show that

$$p + 3\sqrt{pq/n} < 1 \quad \text{if and only if} \quad n > 9\,(p/q).$$

b Show that

$$0 < p - 3\sqrt{pq/n} \quad \text{if and only if} \quad n > 9\,(q/p).$$

c Combine the results from parts (a) and (b) to obtain that the normal approximation to the binomial is adequate if

$$n > 9\left(\frac{p}{q}\right) \quad \text{and} \quad n > 9\left(\frac{q}{p}\right), \quad \text{or, equivalently,} \quad n > 9\left(\frac{\text{larger of } p \text{ and } q}{\text{smaller of } p \text{ and } q}\right).$$

7.45 Refer to Exercise 7.44.

a For what values of n will the normal approximation to the binomial distribution be adequate if $p = .5$?

b Answer the question in part (a) if $p = .6$, $p = .4$, $p = .8$, $p = .2$, $p = .99$, and $p = .001$.

7.46 A machine is shut down for repairs if a random sample of 100 items selected from the daily output of the machine reveals at least 15% defectives. (Assume that the daily output is a large number of items.) If on a given day the machine is producing only 10% defective items, what is the probability that it will be shut down? (Use the .5 continuity correction.)

7.47 An airline finds that 5 percent of the persons who make reservations on a certain flight do not show up for the flight. If the airline sells 160 tickets for a flight with only 155 seats, what is the probability that a seat will be available for every person holding a reservation and planning to fly?

7.48 According to a survey conducted by the American Bar Association, 1 in every 410 Americans is a lawyer, but 1 in every 64 residents of Washington, D.C., is a lawyer.

a If you select a random sample of 1500 Americans, what is the approximate probability that the sample contains at least one lawyer?

b If the sample is selected from among the residents of Washington, D.C., what is the approximate probability that the sample contains more than 30 lawyers?

c If you stand on a Washington, D.C., street corner and interview the first 1000 persons who walked by and 30 say that they are lawyers, does this suggest that the density of lawyers passing the corner exceeds the density within the city? Explain.

7.49 A pollster believes that 20% of the voters in a certain area favor a bond issue. If 64 voters are randomly sampled from the large number of voters in this area, approximate the probability that the sampled fraction of voters favoring the bond issue will not differ from the true fraction by more than .06.

7.50 **a** Show that the variance of Y/n, where Y has a binomial distribution with n trials and a success probability of p, has a maximum at $p = .5$, for fixed n.

b A random sample of n items is to be selected from a large lot, and the number of defectives Y is to be observed. What value of n guarantees that Y/n will be within .1 of the true fraction of defectives, with probability .95.

7.51 The manager of a supermarket wants to obtain information about the proportion of customers who dislike a new policy on cashing checks. How many customers should he sample if he wants the sample fraction to be within .15 of the true fraction, with probability .98?

7.52 If the supermarket manager (Exercise 7.51) samples $n = 50$ customers and if the true fraction of customers who dislike the policy is approximately .9, find the probability that the sample fraction will be within .15 unit of the true fraction.

7.53 Suppose that a random sample of 25 items is selected from the machine of Exercise 7.46. If the machine produces 10% defectives, find the probability that the sample will contain at least two defectives, by the following methods:

a using the normal approximation to the binomial.

b using the exact binomial tables.

7.54 The median age of residents of the United States is 31 years. If a survey of 100 randomly selected United States residents is to be taken, what is the approximate probability that at least 60 will be under 31 years of age?

7.55 A lot acceptance sampling plan for large lots specifies that 50 items be randomly selected and that the lot be accepted if no more than 5 of the items selected do not conform to specifications.

a What is the approximate probability that a lot will be accepted if the true proportion of nonconforming items in the lot is .10?

b Answer the question in part (a) if the true proportion of nonconforming items in the lot is .20 and .30.

7.56 The quality of computer disks is measured by the number of missing pulses. Brand X is such that 80% of the disks have no missing pulses. If 100 disks of brand X are inspected, what is the probability that 15 or more contain missing pulses?

7.57 Vehicles entering an intersection from the east are equally likely to turn left, turn right, or proceed straight ahead. If 500 vehicles enter this intersection from the east, what is the approximate probability that

a 150 or fewer turn right?

b at least 350 turn?

7.58 Just as the difference between two sample means is normally distributed for large samples, so is the difference between two sample proportions. That is, if Y_1 and Y_2 are independent binomial random variables with parameters (n_1, p_1) and (n_2, p_2), respectively, then $(Y_1/n_1) - (Y_2/n_2)$ is approximately normally distributed for large values of n_1 and n_2.

a Find $E\left(\dfrac{Y_1}{n_1} - \dfrac{Y_2}{n_2}\right)$.

b Find $V\left(\dfrac{Y_1}{n_1} - \dfrac{Y_2}{n_2}\right)$.

7.59 As a check on the relative abundance of certain species of fish in two lakes, $n = 50$ observations are taken on results of net trapping in each lake. For each observation the experimenter merely records whether the desired species was present in the trap. Past experience has shown that this species appears in lake A traps approximately 10% of the time and in lake B traps approximately 20% of the time. Use these results to approximate the probability that the difference between the sample proportions will be within .1 of the difference between the true proportions.

7.60 An auditor samples 100 of a firm's travel vouchers to ascertain what percentage of the whole set of vouchers are improperly documented. What is the approximate probability that more than 30% of the sampled vouchers are improperly documented if, in fact, only 20% of all the vouchers are improperly documented. If you were the auditor and observed more than 30% with improper documentation, what would you conclude about the firm's claim that only 20% suffered from improper documentation? Why?

7.61 The times to process orders at the service counter of a pharmacy are exponentially distributed with mean 10 minutes. If 100 customers visit the counter in a 2-day period, what is the probability that at least half of them need to wait more than 10 minutes?

7.6 Summary

To make inferences about population parameters, we need to know the probability distributions for certain functions (statistics) of the observable random variables in the sample (or samples). These probability distributions provide models for the relative frequency behavior of the statistics in repeated sampling; consequently, they are referred to as *sampling distributions*. We have seen that the normal, χ^2, t, and F distributions provide models for the sampling distributions of statistics used to make inferences about the parameters associated with normal distributions.

When the sample size is large, the sample mean $\overline{Y}$ possesses an approximately normal distribution if the random sample is taken from *any* distribution with a finite mean μ and a finite variance σ^2. This result, known as the *central limit theorem*, also provides the justification for approximating the binomial probabilities with corresponding probabilities associated with the normal distribution.

The sampling distributions developed in this chapter will be utilized in the inference-making procedures presented in subsequent chapters.

References and Further Readings

Casella, G., and R. L. Berger. 1990. *Statistical Inference*. Pacific Grove, Calif.: Brooks Cole.

Hoel, P. G. 1984. *Introduction to Mathematical Statistics*, 5th ed. New York: John Wiley.

Hogg, R. V., and A. T. Craig. 1995. *Introduction to Mathematical Statistics*, 5th ed. Upper Saddle River, N.J.: Prentice Hall.

Mood, A. M., F. A. Graybill, and D. Boes. 1974. *Introduction to the Theory of Statistics*, 3d ed. New York: McGraw-Hill.

Parzen, E. 1960. *Modern Probability Theory and Its Applications*. New York: John Wiley.

Supplementary Exercises

7.62 The efficiency (in lumens per watt) of lightbulbs of a certain type has population mean 9.5 and standard deviation .5, according to production specifications. The specifications for a room in which eight of these bulbs are to be installed call for the average efficiency of the eight bulbs to exceed 10. Find the probability that this specification for the room will be met, assuming that efficiency measurements are normally distributed.

7.63 Refer to Exercise 7.62. What should be the mean efficiency per bulb if the specification for the room is to be met with a probability of approximately .80? (Assume that the variance of efficiency measurements remains at .5.)

7.64 Briggs and King developed the technique of nuclear transplantation, in which the nucleus of a cell from one of the later stages of an embryo's development is transplanted into a zygote (a single-cell, fertilized egg) to see if the nucleus can support normal development. If the probability that a single transplant from the early gastrula stage will be successful is .65, what is the probability that more than 70 transplants out of 100 will be successful?

7.65 A retail dealer sells three brands of automobiles. For brand A, her profit per sale, X, is normally distributed with parameters (μ_1, σ_1^2); for brand B, her profit per sale, Y, is normally distributed with parameters (μ_2, σ_2^2); for brand C, her profit per sale, W, is normally distributed with parameters (μ_3, σ_3^2). For the year, two-fifths of the dealer's sales are of brand A; one-fifth, of brand B; and the remaining two-fifths, of brand C. If you are given data on profits for n_1, n_2, and n_3 sales of brands A, B, and C, respectively, the quantity $U = .4\overline{X} + .2\overline{Y} + .4\overline{W}$ will approximate to the true average profit per sale for the year. Find the mean, variance, and probability density function for U. Assume that X, Y, and W are independent.

7.66 From each of two normal populations with identical means and with standard deviations of 6.40 and 7.20, independent random samples of 64 observations are drawn. Find the probability that the difference between the means of the samples exceeds .6 in absolute value.

7.67 If Y has an exponential distribution with mean θ, show that $U = 2Y/\theta$ has a χ^2 distribution with 2 degrees of freedom.

7.68 A plant supervisor is interested in budgeting weekly repair costs for a certain type of machine. Records over the past years indicate that these repair costs have an exponential distribution with mean 20 for each machine studied. Let Y_1, Y_2, $\ldots$, Y_5 denote the repair costs for five of these machines for the next week. Find a number c such that $P(\sum_{i=1}^{5} Y_i > c) = .05$, assuming that the machines operate independently. (*Hint:* Use the result given in Exercise 7.67.)

7.69 The *coefficient of variation* (C.V.) for a sample of values Y_1, Y_2, $\ldots$, Y_n is defined by

$$\text{C.V.} = S/\overline{Y}.$$

This quantity, which gives the standard deviation as a proportion of the mean, is sometimes informative. For example, the value $S = 10$ has little meaning unless we can compare it to something else. If S is observed to be 10 and $\overline{Y}$ is observed to be 1000, the amount of variation is small relative to the size of the mean. However, if S is observed to be 10 and $\overline{Y}$ is observed to be 5, the variation is quite large relative to the size of the mean. If we were studying the precision (variation in repeated measurements) of a measuring instrument, the first case (C.V. = 10/1000) might provide acceptable precision, but the second case (C.V. = 2) would be unacceptable.

Let $Y_1, Y_2, \ldots, Y_{10}$ denote a random sample of size ten from a normal distribution with mean 0 and variance σ^2. Use the following steps to find the number c such that

$$P\left(-c \leq \frac{S}{\overline{Y}} \leq c\right) = .95.$$

a   Use the result of Exercise 7.15 to find the distribution of $(10)\overline{Y}^2/S^2$.

b   Use the result of Exercise 7.13 to find the distribution of $S^2/[(10)\overline{Y}^2]$.

c   Use the answer to (b) to find the constant c.

7.70   Suppose that $Y_1, Y_2, \ldots, Y_{40}$ denote a random sample of measurements on the proportion of impurities in iron ore samples. Let each variable Y_i have a probability density function given by

$$f(y) = \begin{cases} 3y^2, & 0 \leq y \leq 1 \\ 0, & \text{elsewhere.} \end{cases}$$

The ore is to be rejected by the potential buyer if $\overline{Y}$ exceeds .7. Find $P(\overline{Y} > .7)$ for the sample of size 40.

*7.71   Let $X_1, X_2, \ldots, X_n$ be independent χ^2-distributed random variables, each with 1 degree of freedom. Define Y as

$$Y = \sum_{i=1}^{n} X_i.$$

It follows from Exercise 6.51 that Y has a χ^2 distribution with n degrees of freedom.

a   Use the preceding representation of Y as the sum of the Xs to show that $Z = (Y-n)/\sqrt{2n}$ has an asymptotic standard normal distribution.

b   A machine in a heavy-equipment factory produces steel rods of length Y, where Y is a normally distributed random variable with mean 6 inches and variance .2. The cost C of repairing a rod that is not exactly 6 inches in length is proportional to the square of the error and is given, in dollars, by $C = 4(Y - \mu)^2$. If 50 rods with independent lengths are produced in a given day, approximate the probability that the total cost for repairs for that day exceeds \$48.

*7.72   Suppose that T is defined as in Definition 7.2.

a   If W is fixed at w, then T is given by Z/c, where $c = \sqrt{w/\nu}$. Use this idea to find the conditional density of T for a fixed $W = w$.

b   Find the joint density of T and W, $f(t, w)$, by using $f(t, w) = f(t|w)f(w)$.

c   Integrate over w to show that

$$f(t) = \left\{ \frac{\Gamma[(\nu + 1)/2]}{\sqrt{\pi \nu}\,\Gamma(\nu/2)} \right\} \left(1 + \frac{t^2}{\nu}\right)^{-(\nu+1)/2}, \qquad -\infty < t < \infty.$$

*7.73   Suppose F is defined as in Definition 7.3.

a   If W_2 is fixed at w_2, then $F = W_1/c$, where $c = w_2\nu_1/\nu_2$. Find the conditional density of F for fixed $W_2 = w_2$.

b   Find the joint density of F and W_2.

c Integrate over w_2 to show that the probability density function of F, say, $g(y)$, is given by

$$g(y) = \frac{\Gamma[(v_1 + v_2)/2](v_1/v_2)^{v_1/2}}{\Gamma(v_1/2)\Gamma(v_2/2)} y^{(v_1/2)-1} \left(1 + \frac{v_1 y}{v_2}\right)^{-(v_1+v_2)/2}, \qquad 0 < y < \infty.$$

*7.74 Let X have a Poisson distribution with parameter λ.

    a Show that the moment-generating function of $Y = (X - \lambda)/\sqrt{\lambda}$ is given by

$$m_Y(t) = \exp(\lambda e^{t/\sqrt{\lambda}} - \sqrt{\lambda}t - \lambda).$$

    b Use the expansion

$$e^{t/\sqrt{\lambda}} = \sum_{i=0}^{\infty} \frac{[t/\sqrt{\lambda}]^i}{i!}$$

    to show that

$$\lim_{\lambda \to \infty} m_Y(t) = e^{t^2/2}.$$

    c Use Theorem 7.5 to show that the distribution function of Y converges to a standard normal distribution function as $\lambda \to \infty$.

*7.75 In the interest of pollution control, an experimenter wants to count the number of bacteria per small volume of water. Let X denote the bacteria count per cubic centimeter of water, and assume that X has a Poisson probability distribution with mean $\lambda = 100$. If the allowable pollution in a water supply is a count of 110 per cubic centimeter, approximate the probability that X will be at most 110. [*Hint:* Use the result in Exercise 7.74(c).]

*7.76 Y, the number of accidents per year at a given intersection, is assumed to have a Poisson distribution. Over the past few years, an average of 36 accidents per year have occurred at this intersection. If the number of accidents per year is at least 45, an intersection can qualify to be redesigned under an emergency program set up by the state. Approximate the probability that the intersection in question will come under the emergency program at the end of the next year.

*7.77 An experimenter is comparing two methods for removing bacteria colonies from processed luncheon meats. After treating some samples by method A and other identical samples by method B, the experimenter selects a 2-cm^3 subsample from each sample and makes bacteria colony counts on these subsamples. Let X denote the total count for the subsamples treated by method A, and let Y denote the total count for the subsamples treated by method B. Assume that X and Y are independent Poisson random variables with means λ_1 and λ_2, respectively. If X exceeds Y by more than 10, method B will be judged superior to A. Suppose that, in fact, $\lambda_1 = \lambda_2 = 50$. Find the approximate probability that method B will be judged superior to method A.

*7.78 Let Y_n be a binomial random variable with n trials and with success probability p. Suppose that n tends to infinity and p tends to zero in such a way that np remains fixed at $np = \lambda$. Use the result in Theorem 7.5 to prove that the distribution of Y_n converges to a Poisson distribution with mean λ.

7.79 If the probability that a person will suffer an adverse reaction from a medication is .001, use the result of Exercise 7.78 to approximate the probability that 2 or more persons will suffer an adverse reaction if the medication is administered to 1000 individuals.

CHAPTER **8**

Estimation

8.1   Introduction

8.2   The Bias and Mean Square Error of Point Estimators

8.3   Some Common Unbiased Point Estimators

8.4   Evaluating the Goodness of a Point Estimator

8.5   Confidence Intervals

8.6   Large-Sample Confidence Intervals

8.7   Selecting the Sample Size

8.8   Small-Sample Confidence Intervals for μ and $\mu_1 - \mu_2$

8.9   Confidence Intervals for σ^2

8.10 Summary

References and Further Readings

8.1 Introduction

As stated in Chapter 1, the purpose of statistics is to permit the user to make an inference about a population based on information contained in a sample. Because populations are characterized by numerical descriptive measures called *parameters*, the objective of many statistical investigations is to make an inference about one or more population parameters. Most statistical inference procedures involve either estimation or hypothesis testing. In this chapter we consider the general topic of estimation of population parameters. As you will see, the sampling distributions derived in Chapter 7 play an important role in the development of our estimation procedures.

Estimation has many practical applications. For example, a manufacturer of washing machines might be interested in estimating the proportion, p, of washers that can be expected to fail prior to the expiration of a 1-year guarantee time. Other important population parameters are the population mean, variance, and standard deviation. For example, we might wish to estimate the mean waiting time, μ, at a supermarket checkout station or the standard deviation of the error of measurement, σ, of an electronic instrument. To simplify our terminology, we will call the parameter of interest in the experiment the *target parameter*.

Suppose that we wish to estimate the average amount of mercury, μ, that a particular process can remove from 1 ounce of ore obtained at a particular geographic location. We could give our estimate in two distinct forms. First, we could use a single number, for instance .13 ounce, that we think is close to the unknown population mean μ. This type of estimate is called a *point estimate* because a single value, or point, is given as the estimate of μ. Second, we might say that μ will fall between two numbers, for example, between .07 and .19 ounce. In this second type of estimation procedure, the two values we give may be used to construct an interval (.07, .19) that is intended to enclose the parameter of interest; thus the estimate is called an *interval estimate*.

A point estimation procedure uses the information in the sample to arrive at a single number that is intended to be close to the true value of the target parameter. An interval estimation procedure uses the sample information to arrive at two numbers that are intended to enclose the parameter of interest. In either case the actual estimation is accomplished by using an *estimator* for the target parameter.

DEFINITION 8.1

> An *estimator* is a rule that tells how to calculate the value of an estimate based on the measurements contained in a sample.

An estimator is most frequently expressed as a formula. For example, the sample mean

$$\overline{Y} = \frac{1}{n} \sum_{i=1}^{n} Y_i$$

is one possible point estimator of the population mean μ. Clearly the expression for $\overline{Y}$ is both a rule and a formula. It tells us to sum the sample observations and divide by the sample size n.

An experimenter who wants an interval estimate of a parameter must use the sample data to calculate two values, chosen so that the interval formed by the two values possesses a specified probability of including the target parameter. Examples of interval estimators will be given in subsequent sections.

Many different estimators (rules for estimating) may be obtained for the same population parameter. This should not be surprising. Ten engineers, each assigned to estimate the cost of a large construction job, would most likely arrive at different estimates of the total cost. Such engineers, called *estimators* in the construction industry, base their estimates on specified fixed guidelines and intuition. Each estimator represents a unique human subjective rule for obtaining a single estimate. This brings us to a most important point: some estimators are considered *good*, and others, *bad*. The management of a construction firm must define *good* and *bad* as they relate to the estimation of the cost of a job. How can we establish criteria of goodness to compare statistical estimators? We will discuss these ideas in the following sections.

8.2 The Bias and Mean Square Error of Point Estimators

Point estimation is similar, in many respects, to firing a revolver at a target. The estimator, generating estimates, is analogous to the revolver, a particular estimate to one shot, and the parameter of interest to the bull's-eye. Drawing a single sample from the population and using it to compute an estimate for the value of the parameter corresponds to firing a single shot at the target.

Suppose that a man fires a single shot at a target and that the shot pierces the bull's-eye. Do we conclude that he is an excellent shot? Would you want to hold the target while a second shot is fired? Obviously, we would not decide that the man is an expert marksperson based on such a small amount of evidence. On the other hand, if 100 shots in succession hit the bull's-eye, we might acquire sufficient confidence in the marksperson to hold the target for the next shot if the compensation was adequate. The point is that we cannot evaluate the goodness of a point estimation procedure on the basis of the value of a single estimate; rather, we must observe the results when the estimation procedure is used many, many times. Because the estimates are numbers, we evaluate the goodness of the point estimator by constructing a frequency distribution of the values of the estimates obtained in repeated sampling and note how closely this distribution clusters about the target parameter.

Suppose that we wish to specify a point estimate for a population parameter that we will call θ. The estimator of θ will be indicated by the symbol $\hat{\theta}$, read as "θ hat". The "hat" indicates that we are estimating the parameter immediately beneath it. With the revolver-firing example in mind, we can say that it is highly desirable for the distribution of estimates—or, more properly, the sampling distribution of the estimator—to cluster about the target parameter, as shown in Figure 8.1. In other words, we would like the mean or expected value of the distribution of estimates to equal the parameter estimated; that is, $E(\hat{\theta}) = \theta$. Point estimators that satisfy this property are said to be *unbiased*. The sampling distribution for a positively biased point estimator, one for which $E(\hat{\theta}) > \theta$, is shown in Figure 8.2.

DEFINITION 8.2   Let $\hat{\theta}$ be a point estimator for a parameter θ. Then $\hat{\theta}$ is an *unbiased estimator* if $E(\hat{\theta}) = \theta$. Otherwise $\hat{\theta}$ is said to be *biased*.

DEFINITION 8.3   The *bias* of a point estimator $\hat{\theta}$ is given by $B(\hat{\theta}) = E(\hat{\theta}) - \theta$.

FIGURE 8.1
A distribution
of estimates

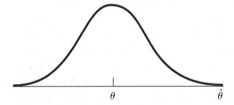

θ $\hat{\theta}$

FIGURE **8.2**
Sampling distribution
for a positively
biased estimator

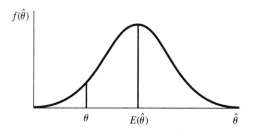

FIGURE **8.3**
Sampling
distributions for two
unbiased estimators:
(a) estimator with
large variation;
(b) estimator with
small variation

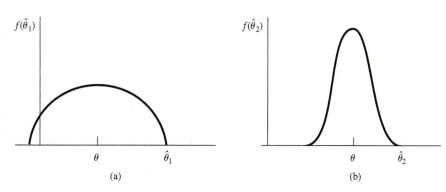

(a)   (b)

Figure 8.3 shows two possible sampling distributions for unbiased point estimators for a target parameter θ. We would prefer that our estimator have the type of distribution indicated in Figure 8.3(b) because the smaller variance guarantees that, in repeated sampling, a higher fraction of values of $\hat{\theta}_2$ will be "close" to θ. Thus, in addition to preferring unbiasedness, we want the variance of the distribution of the estimator $V(\hat{\theta})$ to be as small as possible. Given two unbiased estimators of a parameter θ, and all other things being equal, we would select the estimator with the smaller variance.

Rather than using the bias and variance of a point estimator to characterize its goodness, we might employ the expected value of $(\hat{\theta} - \theta)^2$, the average of the square of the distance between the estimator and its target parameter.

DEFINITION **8.4**
The *mean square error* of a point estimator $\hat{\theta}$ is the expected value of $(\hat{\theta} - \theta)^2$:

$$\text{MSE}(\hat{\theta}) = E\big((\hat{\theta} - \theta)^2\big).$$

The mean square error of an estimator $\hat{\theta}$, $\text{MSE}(\hat{\theta})$, is a function of both its variance and its bias. If $B(\hat{\theta})$ denotes the bias of the estimator $\hat{\theta}$, it can be shown that

$$\text{MSE}(\hat{\theta}) = V(\hat{\theta}) + \big(B(\hat{\theta})\big)^2.$$

We will leave the proof of this result as Exercise 8.1.

In this section we have defined properties of point estimators that are sometimes desirable. In particular, we often seek unbiased estimators with relatively small variances. In the next section we will consider some common and useful unbiased point estimators.

Exercises

8.1 Using the identity

$$(\hat{\theta} - \theta) = [\hat{\theta} - E(\hat{\theta})] + [E(\hat{\theta}) - \theta] = [\hat{\theta} - E(\hat{\theta})] + B(\hat{\theta}),$$

show that

$$\text{MSE}(\hat{\theta}) = E[(\hat{\theta} - \theta)^2] = V(\hat{\theta}) + (B(\hat{\theta}))^2.$$

8.2 Suppose that $E(\hat{\theta}_1) = E(\hat{\theta}_2) = \theta$, $V(\hat{\theta}_1) = \sigma_1^2$, and $V(\hat{\theta}_2) = \sigma_2^2$. Consider the estimator $\hat{\theta}_3 = a\hat{\theta}_1 + (1 - a)\hat{\theta}_2$.
 a Show that $\hat{\theta}_3$ is an unbiased estimator for θ.
 b If $\hat{\theta}_1$ and $\hat{\theta}_2$ are independent, how should the constant a be chosen in order to minimize the variance of $\hat{\theta}_3$?

8.3 Consider the situation described in Exercise 8.2. How should the constant a be chosen to minimize the variance of $\hat{\theta}_3$ if $\hat{\theta}_1$ and $\hat{\theta}_2$ are not independent but are such that $\text{Cov}(\hat{\theta}_1, \hat{\theta}_2) = c \neq 0$?

8.4 Suppose that Y_1, Y_2, Y_3 denote a random sample from an exponential distribution with density function

$$f(y) = \begin{cases} \left(\dfrac{1}{\theta}\right) e^{-y/\theta}, & y > 0 \\ 0, & \text{elsewhere.} \end{cases}$$

Consider the following five estimators of θ:

$$\hat{\theta}_1 = Y_1, \qquad \hat{\theta}_2 = \frac{Y_1 + Y_2}{2}, \qquad \hat{\theta}_3 = \frac{Y_1 + 2Y_2}{3}, \qquad \hat{\theta}_4 = \min(Y_1, Y_2, Y_3), \qquad \hat{\theta}_5 = \overline{Y}.$$

 a Which of these estimators are unbiased?
 b Among the unbiased estimators, which has the smallest variance?

8.5 Suppose that $Y_1, Y_2, \ldots, Y_n$ constitute a random sample from a population with probability density function

$$f(y) = \begin{cases} \left(\dfrac{1}{\theta + 1}\right) e^{-y/(\theta+1)}, & y > 0, \quad \theta > -1 \\ 0, & \text{elsewhere.} \end{cases}$$

Suggest a suitable statistic to use as an unbiased estimator for θ. (*Hint:* Consider $\overline{Y}$.)

8.6 The number of breakdowns per week for a type of minicomputer is a random variable Y with a Poisson distribution and mean λ. A random sample $Y_1, Y_2, \ldots, Y_n$ of observations on the weekly number of breakdowns is available.

a Suggest an unbiased estimator for λ.

b The weekly cost of repairing these breakdowns is $C = 3Y + Y^2$. Show that $E(C) = 4\lambda + \lambda^2$.

c Find a function of $Y_1, Y_2, \ldots, Y_n$ that is an unbiased estimator of $E(C)$. [*Hint:* Use what you know about $\overline{Y}$ and $(\overline{Y})^2$.]

8.7 Let $Y_1, Y_2, \ldots, Y_n$ denote a random sample of size n from a population with mean 3. Assume that $\hat{\theta}_2$ is an unbiased estimator of $E(Y^2)$ and that $\hat{\theta}_3$ is an unbiased estimator of $E(Y^3)$. Give an unbiased estimator for the third central moment of the underlying distribution.

8.8 The reading on a voltage meter connected to a test circuit is uniformly distributed over the interval $(\theta, \theta + 1)$, where θ is the true but unknown voltage of the circuit. Suppose that $Y_1, Y_2, \ldots, Y_n$ denotes a random sample of such readings.

a Show that $\overline{Y}$ is a biased estimator of θ, and compute the bias.

b Find a function of $\overline{Y}$ that is an unbiased estimator of θ.

c Find $\text{MSE}(\overline{Y})$ when $\overline{Y}$ is used as an estimator of θ.

8.9 We have seen that if Y has a binomial distribution with parameters n and p, then Y/n is an unbiased estimator of p. To estimate the variance of Y, we generally use $n(Y/n)(1 - Y/n)$.

a Show that the suggested estimator is a biased estimator of $V(Y)$.

b Modify $n(Y/n)(1 - Y/n)$ slightly to form an unbiased estimator of $V(Y)$.

8.10 Let $Y_1, Y_2, \ldots, Y_n$ denote a random sample of size n from a population whose density is given by

$$f(y) = \begin{cases} \alpha y^{\alpha-1}/\theta^\alpha, & 0 \le y \le \theta \\ 0, & \text{elsewhere} \end{cases}$$

where $\alpha > 0$ is a known, fixed value, but θ is unknown. (This is the power family distribution introduced in Exercise 6.13.) Consider the estimator $\hat{\theta} = \max(Y_1, Y_2, \ldots, Y_n)$.

a Show that $\hat{\theta}$ is a biased estimator for θ.

b Find a multiple of $\hat{\theta}$ that is an unbiased estimator of θ.

c Derive $\text{MSE}(\hat{\theta})$.

8.11 Let $Y_1, Y_2, \ldots, Y_n$ denote a random sample of size n from a population whose density is given by

$$f(y) = \begin{cases} 3\beta^3 y^{-4}, & \beta \le y \\ 0, & \text{elsewhere} \end{cases}$$

where $\beta > 0$ is unknown. (This is one of the Pareto distributions introduced in Exercise 6.14.) Consider the estimator $\hat{\beta} = \min(Y_1, Y_2, \ldots, Y_n)$.

a Derive the bias of the estimator $\hat{\beta}$.

b Derive $\text{MSE}(\hat{\beta})$.

*8.12 Suppose that $Y_1, Y_2, \ldots, Y_n$ constitute a random sample from a normal distribution with parameters μ and σ^2.[1]

a Show that $S = \sqrt{S^2}$ is a biased estimator of σ. [*Hint:* Recall the distribution of $(n - 1)S^2/\sigma^2$ and the result given in Exercise 4.90.]

b Adjust S to form an unbiased estimator of σ.

[1]Exercises preceded by an asterisk are optional.

   c Find an unbiased estimator of $\mu - z_\alpha \sigma$, the point that cuts off a lower-tail area of α under this normal curve.

8.13 If Y has a binomial distribution with parameters n and p, then $\hat{p}_1 = Y/n$ is an unbiased estimator of p. Another estimator of p is $\hat{p}_2 = (Y + 1)/(n + 2)$.

   a Derive the bias of $\hat{p}_2$.

   b Derive MSE($\hat{p}_1$) and MSE($\hat{p}_2$).

   c For what values of p is MSE($\hat{p}_1$) < MSE($\hat{p}_2$)?

8.14 Let $Y_1, Y_2, \ldots, Y_n$ denote a random sample of size n from a population with a uniform distribution on the interval $(0, \theta)$. Consider $Y_{(1)} = \min(Y_1, Y_2, \ldots, Y_n)$, the smallest-order statistic. Use the methods of Section 6.7 to derive $E(Y_{(1)})$. Find a multiple of $Y_{(1)}$ that is an unbiased estimator for θ.

8.15 Suppose that $Y_1, Y_2, \ldots, Y_n$ denotes a random sample of size n from a population with an exponential distribution whose density is given by

$$f(y) = \begin{cases} (1/\theta)e^{-y/\theta}, & y > 0 \\ 0, & \text{elsewhere.} \end{cases}$$

If $Y_{(1)} = \min(Y_1, Y_2, \ldots, Y_n)$ denotes the smallest-order statistic, show that $\hat{\theta} = nY_{(1)}$ is an unbiased estimator for θ and find MSE($\hat{\theta}$). (*Hint:* Recall the results of Exercise 6.64.)

*8.16 Suppose that Y_1, Y_2, Y_3, Y_4 denote a random sample of size four from a population with an exponential distribution whose density is given by

$$f(y) \begin{cases} (1/\theta)e^{-y/\theta}, & y > 0 \\ 0, & \text{elsewhere.} \end{cases}$$

   a Let $X = \sqrt{Y_1 Y_2}$. Find a multiple of X that is an unbiased estimator for θ. (*Hint:* Use your knowledge of the gamma distribution and the fact that $\Gamma(1/2) = \sqrt{\pi}$ to find $E(\sqrt{Y_1})$. Recall that the variables Y_i are independent.)

   b Let $W = \sqrt{Y_1 Y_2 Y_3 Y_4}$. Find a multiple of W that is an unbiased estimator for θ^2. Recall the hint for (a).

8.3 Some Common Unbiased Point Estimators

Some formal methods for deriving point estimators for target parameters are presented in Chapter 9. In this section we focus on some estimators that merit consideration on the basis of intuition. For example, it seems natural to use the sample mean $\overline{Y}$ to estimate the population mean μ and to use the sample proportion $\hat{p} = Y/n$ to estimate a binomial parameter p. How, then, would we estimate the difference between corresponding parameters for two different populations, such as the difference in means ($\mu_1 - \mu_2$) or the difference in two binomial parameters ($p_1 - p_2$), when the inference is to be based on independent random samples of n_1 and n_2 observations selected from the two populations? Again, our intuition suggests using the point estimators ($\overline{Y}_1 - \overline{Y}_2$), the difference in the sample means, to estimate ($\mu_1 - \mu_2$), and using ($\hat{p}_1 - \hat{p}_2$), the difference in the sample proportions, to estimate ($p_1 - p_2$).

   Because the four estimators $\overline{Y}$, $\hat{p}$, ($\overline{Y}_1 - \overline{Y}_2$), and ($\hat{p}_1 - \hat{p}_2$) are functions of the random variables observed in samples, we could find their expected values and variances by using the expectation theorems of Sections 5.6, 5.7, and 5.8. The standard

Table 8.1. Expected values and standard errors of some common point estimators

Target Parameter θ	Sample Size(s)	Point Estimator $\hat{\theta}$	$E(\hat{\theta})$	Standard Error $\sigma_{\hat{\theta}}$
μ	n	$\overline{Y}$	μ	$\dfrac{\sigma}{\sqrt{n}}$
p	n	$\hat{p} = \dfrac{Y}{n}$	p	$\sqrt{\dfrac{pq}{n}}$
$\mu_1 - \mu_2$	n_1 and n_2	$\overline{Y}_1 - \overline{Y}_2$	$\mu_1 - \mu_2$	$\sqrt{\dfrac{\sigma_1^2}{n_1} + \dfrac{\sigma_2^2}{n_2}}$ *†
$p_1 - p_2$	n_1 and n_2	$\hat{p}_1 - \hat{p}_2$	$p_1 - p_2$	$\sqrt{\dfrac{p_1 q_1}{n_1} + \dfrac{p_2 q_2}{n_2}}$ †

*σ_1^2 and σ_2^2 are the variances of populations 1 and 2, respectively.
†The two samples are assumed to be independent.

deviation of each of the estimators is simply the square root of the respective variance. Such an effort would show that, when random sampling has been employed, all four point estimators are unbiased and that they possess the standard deviations shown in Table 8.1. To facilitate communication we use the notation $\sigma_{\hat{\theta}}^2$ to denote the variance of the sampling distribution of the estimator $\hat{\theta}$. The standard deviation of the sampling distribution of the estimator $\hat{\theta}$, $\sigma_{\hat{\theta}} = \sqrt{\sigma_{\hat{\theta}}^2}$, is usually called the *standard error* of the estimator $\hat{\theta}$.

In Chapter 5, we did much of the derivation required for Table 8.1. In particular, we found the means and variances of $\overline{Y}$ and $\hat{p}$ in Examples 5.27 and 5.28, respectively. If the random samples are independent, these results and Theorem 5.12 imply that

$$E(\overline{Y}_1 - \overline{Y}_2) = E(\overline{Y}_1) - E(\overline{Y}_2) = \mu_1 - \mu_2$$

$$V(\overline{Y}_1 - \overline{Y}_2) = V(\overline{Y}_1) + V(\overline{Y}_2) = \frac{\sigma_1^2}{n_1} + \frac{\sigma_2^2}{n_2}.$$

The standard errors of the estimators are found by taking the square root of the respective variances. The expected value and standard error of $(\hat{p}_1 - \hat{p}_2)$, shown in Table 8.1, can be acquired similarly.

Although unbiasedness is often a desirable property for a point estimator, not all estimators are unbiased. In Chapter 1, we defined the sample variance

$$S^2 = \frac{\sum_{i=1}^{n}(Y_i - \overline{Y})^2}{n - 1}.$$

It probably seemed more natural to divide by n than by $n - 1$ in the preceding expression and to calculate

$$S'^2 = \frac{\sum_{i=1}^{n}(Y_i - \overline{Y})^2}{n}.$$

Example 8.1, however, shows that S'^2 is a biased estimator of the population variance σ^2, whereas S^2, defined with the divisor $n - 1$ instead of n, is an unbiased estimator of σ^2. This is the reason we initially identified S^2 as the *sample variance*.

EXAMPLE **8.1** Let $Y_1, Y_2, \ldots, Y_n$ be a random sample with $E(Y_i) = \mu$ and $V(Y_i) = \sigma^2$. Show that

$$S'^2 = \frac{1}{n} \sum_{i=1}^{n}(Y_i - \overline{Y})^2$$

is a biased estimator for σ^2 and that

$$S^2 = \frac{1}{n-1} \sum_{i=1}^{n}(Y_i - \overline{Y})^2$$

is an unbiased estimator for σ^2.

Solution It can be shown (see Exercise 1.7) that

$$\sum_{i=1}^{n}(Y_i - \overline{Y})^2 = \sum_{i=1}^{n}Y_i^2 - \frac{1}{n}\left(\sum_{i=1}^{n}Y_i\right)^2 = \sum_{i=1}^{n}Y_i^2 - n\overline{Y}^2.$$

Hence

$$E\left[\sum_{i=1}^{n}(Y_i - \overline{Y})^2\right] = E\left(\sum_{i=1}^{n}Y_i^2\right) - nE(\overline{Y}^2) = \sum_{i=1}^{n}E(Y_i^2) - nE\left(\overline{Y}^2\right).$$

Notice that $E(Y_i^2)$ is the same for $i = 1, 2, \ldots, n$. We use this and the fact that the variance of a random variable is given by $V(Y) = E(Y^2) - \mu^2$ to conclude that $E(Y_i^2) = V(Y_i) + (E(Y_i))^2 = \sigma^2 + \mu^2$, $E\left(\overline{Y}^2\right) = V(\overline{Y}) + \left(E(\overline{Y})\right)^2 = \sigma^2/n + \mu^2$, and that

$$E\left[\sum_{i=1}^{n}(Y_i - \overline{Y})^2\right] = \sum_{i=1}^{n}(\sigma^2 + \mu^2) - n\left(\frac{\sigma^2}{n} + \mu^2\right)$$

$$= n(\sigma^2 + \mu^2) - n\left(\frac{\sigma^2}{n} + \mu^2\right)$$

$$= n\sigma^2 - \sigma^2 = (n-1)\sigma^2.$$

It follows that

$$E(S'^2) = \frac{1}{n} E\left[\sum_{i=1}^{n}(Y_i - \overline{Y})^2\right] = \frac{1}{n}(n-1)\sigma^2 = \left(\frac{n-1}{n}\right)\sigma^2$$

and that S'^2 is biased because $E(S'^2) \neq \sigma^2$. However,

$$E(S^2) = \frac{1}{n-1} E\left[\sum_{i=1}^{n}(Y_i - \overline{Y})^2\right] = \frac{1}{n-1}(n-1)\sigma^2 = \sigma^2,$$

so we see that S^2 is an unbiased estimator for σ^2.

Two final comments can be made concerning the point estimators of Table 8.1. First, the expected values and standard errors shown in the table are valid regardless of the form of the population probability density function used. Second, all four estimators possess probability distributions that are approximately normal for large samples. The central limit theorem justifies this statement for $\overline{Y}$ and $\hat{p}$, and similar theorems for functions of sample means justify the assertion for $(\overline{Y}_1 - \overline{Y}_2)$ and $(\hat{p}_1 - \hat{p}_2)$. How large is "large"? For most populations the probability distribution of $\overline{Y}$ is mound-shaped even for relatively small samples (as low as $n = 5$), and will tend rapidly to normality as the sample size approaches $n = 30$ or larger. However, you sometimes will need to select larger samples from binomial populations because the required sample size depends on p. The binomial probability distribution is perfectly symmetrical about its mean when $p = 1/2$ and becomes more and more asymmetric as p tends to 0 or 1. As a rough rule you can assume that the distribution of $\hat{p}$ will be mound-shaped and approaching normality for sample sizes such that $p \pm 3\sqrt{pq/n}$ lies in the interval $(0, 1)$, or, as you demonstrated in Exercise 7.44, if $n > 9$(larger of p and q)/(smaller of p and q).

We know that $\overline{Y}$, $\hat{p}$, $(\overline{Y}_1 - \overline{Y}_2)$, and $(\hat{p}_1 - \hat{p}_2)$ are unbiased with near-normal (at least mound-shaped) sampling distributions for moderate-sized samples; now let us use this information to answer a couple of practical questions. If we use an estimator once and acquire a single estimate, how good will this estimate be? How much faith can we place in the validity of our inference? The answers to these questions are provided in the next section.

8.4   Evaluating the Goodness of a Point Estimator

One way to measure the goodness of any point estimation procedure is in terms of the distances between the estimates it generates and the target parameter. This quantity, which varies randomly in repeated sampling, is called the *error of estimation*. Naturally we would like the error of estimation to be as small as possible.

DEFINITION 8.5

The *error of estimation* ε is the distance between an estimator and its target parameter. That is, $\varepsilon = |\hat{\theta} - \theta|$.

FIGURE **8.4**
Sampling distribution
of a point estimator $\hat{\theta}$

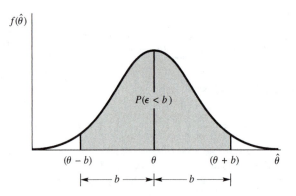

Because $\hat{\theta}$ is a random variable, the error of estimation is also a random quantity, and we cannot say how large or small it will be for a particular estimate. However, we can make probability statements about it. For example, suppose that $\hat{\theta}$, an unbiased estimator of θ, possesses a sampling distribution as shown in Figure 8.4. If we select two points, $(\theta - b)$ and $(\theta + b)$, located near the tails of the probability distribution, the probability that the error of estimation ε is less than b is represented by the shaded area in Figure 8.4. That is,

$$P(|\hat{\theta} - \theta| < b) = P[-b < (\hat{\theta} - \theta) < b] = P(\theta - b < \hat{\theta} < \theta + b).$$

We can think of b as a probabilistic bound on the error of estimation. Although we are not certain that a given error is less than b, we do know that $P(\varepsilon < b)$ is high. If b can be regarded from a practical point of view as small, then $P(\varepsilon < b)$ provides a measure of the goodness of a single estimate. This probability identifies the fraction of times, in repeated sampling, that the estimator $\hat{\theta}$ falls within b units of θ, the target parameter.

Suppose that we want to find the value of b so that $P(\varepsilon < b) = .90$. This is easy if we know the probability distribution of $\hat{\theta}$. Then we seek a value b such that

$$\int_{\theta - b}^{\theta + b} f(\hat{\theta}) \, d\hat{\theta} = .90.$$

But whether we know the probability distribution of $\hat{\theta}$ or not, if $\hat{\theta}$ is unbiased we can find an approximate bound on ε by expressing b as a multiple of the standard error of $\hat{\theta}$ (recall that the standard error of an estimator is simply a convenient alternative name for the standard deviation of the estimator). For example, for $k \geq 1$, if we let $b = k\sigma_{\hat{\theta}}$, we know from Tchebysheff's theorem that ε will be less than $k\sigma_{\hat{\theta}}$ with probability at least $1 - 1/k^2$. A convenient and often-used value of k is $k = 2$. Hence we know that ε will be less than $b = 2\sigma_{\hat{\theta}}$ with probability at least .75.

You will find that, with a probability in the vicinity of .95, most random variables observed in nature lie within two standard deviations of their mean. The probability that Y lies in the interval $(\mu \pm 2\sigma)$ is shown in Table 8.2 for the normal, uniform, and exponential probability distributions. The point is that $b = 2\sigma_{\hat{\theta}}$ is a good approximate bound on the error of estimation in most practical situations. According to Tchebysheff's theorem, the probability that the error of estimation will be less than this bound

Table 8.2. Probability that $(\mu - 2\sigma) < Y < (\mu + 2\sigma)$

Distribution	Probability
Normal	.9544
Uniform	1.0000
Exponential	.9502

is *at least* .75. As we have previously observed, the bounds for probabilities provided by Tchebysheff's theorem are usually very conservative: the actual probabilities usually exceed the Tchebysheff bounds by a considerable amount. In many situations, the probability that the error of estimation is less than $2\sigma_{\hat{\theta}}$ is near .95.

EXAMPLE 8.2   A sample of $n = 1000$ voters, randomly selected from a city, showed $y = 560$ in favor of candidate Jones. Estimate p, the fraction of voters in the population favoring Jones, and place a two-standard-error bound on the error of estimation.

Solution   We will use the estimator $\hat{p} = Y/n$ to estimate p. Hence the estimate of p, the fraction of voters favoring candidate Jones, is

$$\hat{p} = \frac{y}{n} = \frac{560}{1000} = .56.$$

How much faith can we place in this value? The probability distribution of $\hat{p}$ is very accurately approximated by a normal probability distribution for samples as large as $n = 1000$. Then when $b = 2\sigma_{\hat{p}}$, the probability that ε will be less than b is approximately .95.

From Table 8.1, the standard error of the estimator for p is given by $\sigma_{\hat{p}} = \sqrt{pq/n}$. Therefore,

$$b = 2\sigma_{\hat{p}} = 2\sqrt{\frac{pq}{n}}.$$

Unfortunately, to calculate b, we need to know p, and estimating p was the objective of our sampling. This apparent stalemate is not a handicap, however, because $\sigma_{\hat{p}}$ varies little for small changes in p. Hence, substitution of the estimate $\hat{p}$ for p produces little error in calculating the exact value of $b = 2\sigma_{\hat{p}}$. Then for our example, we have

$$b = 2\sigma_{\hat{p}} = 2\sqrt{\frac{pq}{n}} \approx 2\sqrt{\frac{(.56)(.44)}{1000}} = .03.$$

What is the significance of our calculations? The probability that the error of estimation is less than .03 is approximately .95. Consequently, we can be reasonably confident that our estimate, .56, is within .03 of the true value of p, the proportion of voters in the population who favor Jones.

EXAMPLE **8.3** A comparison of the durability of two types of automobile tires was obtained by road-testing samples of $n_1 = n_2 = 100$ tires of each type. The number of miles until wear-out was recorded, where wear-out was defined as the number of miles until the amount of remaining tread reached a prespecified small value. The measurements for the two types of tires were obtained independently and the following means and variances were computed:

$$\bar{y}_1 = 26,400 \text{ miles} \qquad \bar{y}_2 = 25,100 \text{ miles}$$

$$s_1^2 = 1,440,000 \qquad s_2^2 = 1,960,000.$$

Estimate the difference in mean miles to wear-out, and place a two-standard-error bound on the error of estimation.

Solution The point estimate of $(\mu_1 - \mu_2)$ is

$$(\bar{y}_1 - \bar{y}_2) = 26,400 - 25,100 = 1300 \text{ miles}$$

and the standard error of the estimator (see Table 8.1) is

$$\sigma_{(\bar{Y}_1 - \bar{Y}_2)} = \sqrt{\frac{\sigma_1^2}{n_1} + \frac{\sigma_2^2}{n_2}}.$$

We must know σ_1^2 and σ_2^2, or have good approximate values for them, in order to calculate $\sigma_{(\bar{Y}_1 - \bar{Y}_2)}$. Fairly accurate values of σ_1^2 and σ_2^2 often can be calculated from similar experimental data collected at some prior time, or they can be obtained from the current sample data by using the unbiased estimators

$$\hat{\sigma}_i^2 = S_i^2 = \frac{1}{n_i - 1} \sum_{j=1}^{n_i} (Y_{ij} - \bar{Y}_i)^2, \qquad i = 1, 2.$$

These estimates will be adequate if the sample sizes are reasonably large, say, $n_i \geq 30$, for $i = 1, 2$. The calculated values of S_1^2 and S_2^2, based on the two wear tests, are $s_1^2 = 1,440,000$ and $s_2^2 = 1,960,000$. Substituting these values for σ_1^2 and σ_2^2 in the formula for $\sigma_{(\bar{Y}_1 - \bar{Y}_2)}$, we have

$$\sigma_{(\bar{Y}_1 - \bar{Y}_2)} = \sqrt{\frac{\sigma_1^2}{n_1} + \frac{\sigma_2^2}{n_2}} \approx \sqrt{\frac{s_1^2}{n_1} + \frac{s_2^2}{n_2}} = \sqrt{\frac{1,440,000}{100} + \frac{1,960,000}{100}}$$

$$= \sqrt{34,000} = 184.4 \text{ miles.}$$

Consequently, we estimate the difference in mean wear to be 1300 miles, and we expect the error of estimation to be less than $2\sigma_{(\bar{Y}_1 - \bar{Y}_2)}$ or 368.8 miles, with a probability of approximately .95.

Exercises

8.17 An investigator is interested in the possibility of merging the capabilities of television and the Internet. A random sample of $n = 50$ Internet users yielded that the mean amount of time spent watching television per week was 11.5 hours and that the standard deviation was 3.5 hours. Estimate the population mean time that Internet users spend watching television and place a bound on the error of estimation.

8.18 An increase in the rate of consumer savings frequently is tied to a lack of confidence in the economy and is said to be an indicator of a recessional tendency in the economy. A random sampling of $n = 200$ savings accounts in a local community showed the mean increase in savings account values to be 7.2% over the past 12 months, with standard deviation 5.6%. Estimate the mean percentage increase in savings account values over the past 12 months for depositors in the community. Place a bound on your error of estimation.

8.19 The Environmental Protection Agency and the University of Florida recently cooperated in a large study of the possible effects of trace elements in drinking water on kidney stone disease. The accompanying table presents data on age, amount of calcium in home drinking water (measured in parts per million), and smoking activity. These data were obtained from individuals with recurrent kidney stone problems, all of whom lived in the Carolinas and the Rocky Mountain states.

	Carolinas	Rockies
Sample size	467	191
Mean age	45.1	46.4
Standard deviation of age	10.2	9.8
Mean calcium component (ppm)	11.3	40.1
Standard deviation of calcium	16.6	28.4
Proportion now smoking	.78	.61

a Estimate the average calcium concentration in drinking water for kidney stone patients in the Carolinas. Place a bound on the error of estimation.

b Estimate the difference in mean ages for kidney-stone patients in the Carolinas and in the Rockies. Place a bound on the error of estimation.

c Estimate and place a two-standard-deviation bound on the difference in proportions of kidney-stone patients from the Carolinas and Rockies who were smokers at the time of the study.

8.20 Results of a public opinion poll reported in a news magazine (*Time*, April 5, 1993)[2] indicated that 54% of respondents between the ages of 18 and 26 feel that religion is a "very important" part of their lives. The article states that 1013 individuals were interviewed and that the results have a sampling error of 3%. How was the 3% calculated, and how should it be interpreted? Can we conclude that a majority of the individuals in this age group feel that religion is a very important part of their lives?

8.21 A study was conducted to compare the mean number of police emergency calls per 8-hour shift in two districts of a large city. Samples of 100 8-hour shifts were randomly selected from the police records for each of the two regions, and the number of emergency calls was recorded for each shift. The sample statistics are given in the following table.

	Region	
	1	2
Sample size	100	100
Sample mean	2.4	3.1
Sample variance	1.44	2.64

Estimate the difference in the mean number of police emergency calls per 8-hour shift between the two districts in the city. Find a bound for the error of estimation.

8.22 Older Americans are more alienated politically than are young Americans, according to a poll conducted by the Times Mirror Center for the People and the Press (D. E. Rosenbaum, "Older More Alienated Politically than Young, Poll Shows," *Riverside Press-Enterprise* [July 8, 1992]). Two-thirds of respondents over age 50 said that new leaders were needed in Washington, "even if they were not as effective as experienced politicians." If $n = 1752$ of those polled were over age 50, what is the point estimate of the true proportion of Americans over age 50 who felt that new leaders were needed in Washington? Find a bound on the error of estimation.

8.23 A 1989 survey of college freshmen who would be graduating in 1992 showed that, relative to earlier graduating classes, more incoming college freshmen were smoking, their interest in business careers had declined, and they favored mandatory testing for AIDS and drugs. These findings were based on a survey by the American Council on Education (*New York Times*, January 9, 1989) that involved $n = 308,007$ freshmen entering 585 two- and four-year colleges and universities. Specifically, 67% agreed that the best way to control AIDS was through widespread mandatory testing, and 71% agreed that employers should be permitted to test employees or job applicants for drugs.

   a According to this study, the estimate for p, the proportion of all college freshmen who favored mandatory AIDS testing, was $\hat{p} = .67$. Find a bound for the error of estimation.

   b According to this study, the estimate for p, the proportion of all college freshmen who favored drug testing for employees and job applicants, was $\hat{p} = .71$. Find a bound for the error of estimation.

   c The results of the survey were reported to have a sampling error of no more than ± 2 percentage points. Is this the appropriate value for the margin of error?

8.24 In a study of the relationship between birth order and college success, an investigator found that 126 in a sample of 180 college graduates were firstborn or only children; in a sample of 100 nongraduates of comparable age and socioeconomic background, the number of firstborn or only children was 54. Estimate the difference in the proportions of first-born or only children for the two populations from which these samples were drawn. Give a bound for the error of estimation.

8.25 Surveys inquiring into the health habits of Americans have been conducted for more than 15 years by Louis Harris & Associates. The polling organization's recent findings were labeled "terribly disturbing" by Dr. Todd Davis, an executive vice president of the American Medical Association. The results of the surveys conducted in 1983 and 1992 are given in the accompanying table. The 1992 survey involved $n = 1251$ subjects. Although the size of the 1983 survey is not given in the article, assume that it involved $n = 1250$ individuals.

Health Issue	1983 Survey	1992 Survey
Ate recommended amount of fibrous food	.59	.53
Avoided fat	.55	.51
Avoided excess salt	.53	.46
Used seatbelts	.19	.70
Used smoke detectors	.67	.90

   a Give a point estimate for the difference in the proportions of Americans who ate the recommended amount of fibrous foods in 1983 and 1992. Provide a bound for the error of estimation.

b Based on your results from (a), do you think that there has been a demonstrable decrease in the proportion of Americans who ate the recommended amount of fibrous foods between 1983 and 1992? Explain.

8.26 Refer to Exercise 8.25. Repeat the instructions in (a) and (b) for the other two categories relating to dietary health habits. Do you agree with Dr. Davis's assessment that the results are "terribly disturbing"? Explain.

8.27 Refer to Exercise 8.25. The results of the 1992 Harris survey contain some good news about the proportion of individuals who regularly use seatbelts. Give the point estimate and a bound on the error of estimation for the proportion of individuals who regularly used seatbelts in 1992. Is it likely that the value of your estimate is off by as much as 10%? Why?

8.28 An auditor randomly samples 20 accounts receivable from among the 500 such accounts of a clients firm. The auditor lists the amount of each account and checks to see if the underlying documents comply with stated procedures. The data are recorded in the accompanying table (amounts are in dollars, Y = yes, N = no).

Account	Amount	Compliance	Account	Amount	Compliance
1	278	Y	11	188	N
2	192	Y	12	212	N
3	310	Y	13	92	Y
4	94	N	14	56	Y
5	86	Y	15	142	Y
6	335	Y	16	37	Y
7	310	N	17	186	N
8	290	Y	18	221	Y
9	221	Y	19	219	N
10	168	Y	20	305	Y

Estimate the total accounts receivable for the 500 accounts of the firm, and place a bound on the error of estimation. Do you think that the *average* account receivable for the firm exceeds $250? Why?

8.29 Refer to Exercise 8.28. From the data given on the compliance checks, estimate the proportion of the firm's accounts that fail to comply with stated procedures. Place a bound on the error of estimation. Do you think that the proportion of accounts that comply with stated procedures exceeds 80%? Why?

8.30 We can place a two-standard-deviation bound on the error of estimation with any estimator for which we can find a reasonable estimate of the standard error. Suppose that $Y_1, Y_2, \ldots, Y_n$ represent a random sample from a Poisson distribution with mean λ. We know that $V(Y_i) = \lambda$, and hence $E(\overline{Y}) = \lambda$ and $V(\overline{Y}) = \lambda/n$. How would you employ $Y_1, Y_2, \ldots, Y_n$ to estimate λ? How would you estimate the standard error of your estimator?

8.31 In polycrystalline aluminum, the number of grain nucleation sites per unit volume is modeled as having a Poisson distribution with mean λ. Fifty unit-volume test specimens subjected to annealing under regime A produced an average of 20 sites per unit volume. Fifty independently selected unit-volume test specimens subjected to annealing regime B produced an average of 23 sites per unit volume.

a Estimate the mean number λ_A of nucleation sites for regime A, and place a two-standard-error bound on the error of estimation.

b Estimate the difference in the mean numbers of nucleation sites $\lambda_A - \lambda_B$ for regimes A and B. Place a two-standard-error bound on the error of estimation. Would you say that regime B tends to produce a larger mean number of nucleation sites? Why?

8.32 If $Y_1, Y_2, \ldots, Y_n$ denotes a random sample from an exponential distribution with mean θ, then $E(Y_i) = \theta$ and $V(Y_i) = \theta^2$. Thus, $E(\overline{Y}) = \theta$ and $V(\overline{Y}) = \theta^2/n$, or $\sigma_{\overline{Y}} = \theta/\sqrt{n}$. Suggest an unbiased estimator for θ, and provide an estimate for the standard error of your estimator.

8.33 An engineer observes $n = 10$ independent length-of-life measurements on a type of electronic component. The average of these 10 measurements is 1020 hours. If these lengths of life come from an exponential distribution with mean θ, estimate θ and place a two-standard-error bound on the error of estimation.

8.34 The number of persons coming through a blood bank until the first person with type A blood is found is a random variable Y with a geometric distribution. If p denotes the probability that any one randomly selected person will possess type A blood, then $E(Y) = 1/p$ and $V(Y) = (1 - p)/p^2$. Find a function of Y that is an unbiased estimator of $V(Y)$. Then suggest how to form a two-standard-error bound on the error of estimation when Y is used to estimate $1/p$.

8.5 Confidence Intervals

An interval estimator is a rule specifying the method for using the sample measurements to calculate two numbers that form the endpoints of the interval. Ideally the resulting interval will have two properties: First, it will contain the target parameter θ; and second, it will be relatively narrow. One or both of the endpoints of the interval, being functions of the sample measurements, will vary randomly from sample to sample. Thus the length and location of the interval are random quantities, and we cannot be certain that the (fixed) target parameter θ will fall between the endpoints of any single interval calculated from a single sample. This being the case, our objective is to find an interval estimator capable of generating narrow intervals that have a high probability of enclosing θ.

Interval estimators are commonly called *confidence intervals*. The upper and lower endpoints of a confidence interval are called the *upper* and *lower confidence limits*, respectively. The probability that a confidence interval will enclose θ is called the *confidence coefficient*. From a practical point of view, the confidence coefficient identifies the fraction of the time, in repeated sampling, that the intervals constructed will contain the target parameter θ. If we know that the confidence coefficient associated with our estimator is high, we can be highly confident that any confidence interval, constructed by using the results from a single sample, will enclose θ.

Suppose that $\hat{\theta}_L$ and $\hat{\theta}_U$ are the (random) lower and upper confidences limits, respectively, for a parameter θ. Then, if

$$P\left(\hat{\theta}_L \leq \theta \leq \hat{\theta}_U\right) = 1 - \alpha$$

the probability $(1 - \alpha)$ is the *confidence coefficient*. The resulting random interval defined by $\left[\hat{\theta}_L, \hat{\theta}_U\right]$ is called a *two-sided confidence interval*.

It is also possible to form a *one-sided confidence interval* such that

$$P\left(\hat{\theta}_L \leq \theta\right) = 1 - \alpha.$$

Although only $\hat{\theta}_L$ is random in this case, the confidence interval is $[\hat{\theta}_L, \infty)$. Similarly, we could have an upper one-sided confidence interval such that

$$P(\theta \leq \hat{\theta}_U) = 1 - \alpha.$$

The implied confidence interval here is $(-\infty, \hat{\theta}_U]$.

One very useful method for finding confidence intervals is called the *pivotal method*. This method depends upon finding a pivotal quantity that possesses two characteristics:

1. It is a function of the sample measurements and the unknown parameter θ, where θ is the *only* unknown quantity.
2. Its probability distribution does not depend upon the parameter θ.

If the probability distribution of the pivotal quantity is known, the following logic can be used to form the desired interval estimate. If Y is any random variable, $c > 0$ is a constant, and $P(a \leq Y \leq b) = .7$, then certainly $P(ca \leq cY \leq cb) = .7$. Similarly, for any constant d, $P(a + d \leq Y + d \leq b + d) = .7$. That is, the probability of the event $(a \leq Y \leq b)$ is unaffected by a change of scale or a translation of Y. Thus if we know the probability distribution of a pivotal quantity, we may be able to use operations like these to form the desired interval estimator. We illustrate this method in the following examples.

EXAMPLE 8.4   Suppose that we are to obtain a single observation Y from an exponential distribution with mean θ. Use Y to form a confidence interval for θ with confidence coefficient .90.

Solution   The probability density function for Y is given by

$$f(y) = \begin{cases} \left(\dfrac{1}{\theta}\right) e^{-y/\theta}, & y \geq 0 \\ 0, & \text{elsewhere.} \end{cases}$$

By the transformation method of Chapter 6 we can see that $U = Y/\theta$ has the exponential density function given by

$$f_U(u) = \begin{cases} e^{-u}, & u > 0 \\ 0, & \text{elsewhere.} \end{cases}$$

The density function for U is graphed in Figure 8.5. $U = Y/\theta$ is a function of Y (the sample measurement) and θ, and the distribution of U does not depend upon θ. Thus we can use $U = Y/\theta$ as a pivotal quantity. Because we want an interval

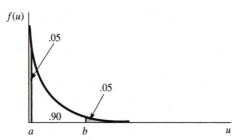

FIGURE 8.5
Density function for
U, Example 8.4

estimator with confidence coefficient equal to .90, we find two numbers a and b such that

$$P(a \leq U \leq b) = .90.$$

One way to do this is to choose a and b to satisfy

$$P(U < a) = \int_0^a e^{-u}\,du = .05 \quad \text{and} \quad P(U > b) = \int_b^\infty e^{-u}\,du = .05.$$

These equations yield

$$1 - e^{-a} = .05 \quad \text{and} \quad e^{-b} = .05 \quad \text{or, equivalently,} \quad a = .051,\ b = 2.996.$$

It follows that

$$.90 = P(.051 \leq U \leq 2.996) = P\left(.051 \leq \frac{Y}{\theta} \leq 2.996\right).$$

Because we seek an interval estimator for θ, let us manipulate the inequalities describing the event to isolate θ in the middle. Y has an exponential distribution, so $P(Y > 0) = 1$, and we maintain the direction of the inequalities if we divide through by Y. That is,

$$.90 = P\left(.051 \leq \frac{Y}{\theta} \leq 2.996\right) = P\left(\frac{.051}{Y} \leq \frac{1}{\theta} \leq \frac{2.996}{Y}\right).$$

Taking reciprocals (and, hence, reversing the direction of the inequalities), we obtain

$$.90 = P\left(\frac{Y}{.051} \geq \theta \geq \frac{Y}{2.996}\right) = P\left(\frac{Y}{2.996} \leq \theta \leq \frac{Y}{.051}\right).$$

Thus, we see that $Y/2.996$ and $Y/.051$ form the desired lower and upper confidence limits, respectively. To obtain numerical values for these limits, we must observe an actual value for Y and substitute that value into the given formulas for the confidence limits. We know that limits of the form $(Y/2.996, Y/.051)$ will include the true (unknown) values of θ for 90% of the values of Y we would obtain by repeatedly sampling from this exponential distribution.

EXAMPLE 8.5   Suppose that we take a sample of size $n = 1$ from a uniform distribution defined on the interval $[0, \theta]$, where θ is unknown. Find a 95% lower confidence bound for θ.

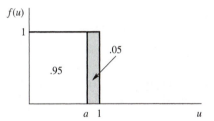

FIGURE 8.6
Density function for
U, Example 8.5

Solution   Because Y is uniform on $[0, \theta]$, the methods of Chapter 6 can be used to show that $U = Y/\theta$ is uniformly distributed over $[0, 1]$. That is,

$$f_U(u) = \begin{cases} 1, & 0 \le u \le 1 \\ 0, & \text{elsewhere.} \end{cases}$$

Figure 8.6 contains a graph of the density function for U. Again we see that U satisfies the requirements of a pivotal quantity. Because we seek a 95% lower confidence limit for θ, let us determine the value for a so that $P(U \le a) = .95$. That is,

$$\int_0^a (1)\, du = .95$$

or $a = .95$. Thus,

$$P(U \le .95) = P\left(\frac{Y}{\theta} \le .95\right) = P(Y \le .95\theta) = P\left(\frac{Y}{.95} \le \theta\right) = .95.$$

We see that $Y/.95$ is a lower confidence limit for θ, with confidence coefficient .95. Because any observed Y must be less than θ, it is intuitively reasonable to have the lower confidence limit for θ slightly larger than the observed value of Y.

The two preceding examples illustrate the use of the pivotal method for finding confidence limits for unknown parameters. In each instance the interval estimates were developed on the basis of a single observation from the distribution. These examples were introduced primarily for the illustrative purposes. In the remaining sections of this chapter we will utilize the pivotal method in conjunction with the sampling distributions presented in Chapter 7 to develop some interval estimates of greater practical importance.

Exercises

8.35   Suppose that the random variable Y has a gamma distribution with parameters $\alpha = 2$ and an unknown β. In Exercise 6.40, you used the method of moment-generating functions to prove a general result implying that $2Y/\beta$ has a χ^2 distribution with 4 degrees of freedom. Using $2Y/\beta$ as a pivotal quantity, derive a 90% confidence interval for β.

8.36   Suppose that the random variable Y is an observation from a normal distribution with unknown mean μ and variance 1.

    a   Find a 95% confidence interval for μ.

    b   Find a 95% upper confidence limit for μ.

    c   Find a 95% lower confidence limit for μ.

8.37   Suppose that Y is normally distributed with mean 0 and unknown variance σ^2. Then Y^2/σ^2 has a χ^2 distribution with 1 degree of freedom. Use the pivotal quantity Y^2/σ^2 to find:

    a   a 95% confidence interval for σ^2.

    b   a 95% upper confidence limit for σ^2.

    c   a 95% lower confidence limit for σ^2.

8.38   Use the answers from Exercise 8.37 to find:

    a   a 95% confidence interval for σ.

    b   a 95% upper confidence limit for σ.

    c   a 95% lower confidence limit for σ.

8.39   Let $Y_1, Y_2, \ldots, Y_n$ denote a random sample of size n from a population with a uniform distribution on the interval $(0, \theta)$. Let $Y_{(n)} = \max(Y_1, Y_2, \ldots, Y_n)$. Let $U = (1/\theta)Y_{(n)}$.

    a   Show that U has distribution function

$$F_U(u) = \begin{cases} 0, & u < 0 \\ u^n, & 0 \le u \le 1 \\ 1, & u > 1. \end{cases}$$

    b   Because the distribution of U does not depend on θ, U is a pivotal quantity. Find a 95% lower confidence bound for θ.

8.40   Let Y have probability density function

$$f_Y(y) = \begin{cases} \dfrac{2(\theta - y)}{\theta^2}, & 0 < y < \theta \\ 0, & \text{elsewhere.} \end{cases}$$

    a   Show that Y has distribution function

$$F_Y(y) = \begin{cases} 0, & y \le 0 \\ \dfrac{2y}{\theta} - \dfrac{y^2}{\theta^2}, & 0 < y < \theta \\ 1, & y \ge \theta. \end{cases}$$

    b   Show that Y/θ is a pivotal quantity.

    c   Use the pivotal quantity from part (b) to find a 90% lower confidence limit for θ.

8.41   Refer to Exercise 8.40.

    a   Use the pivotal quantity from Exercise 8.40(b) to find a 90% upper confidence limit for θ.

    b   If $\hat{\theta}_L$ is the lower confidence bound for θ obtained in Exercise 8.40(c) and $\hat{\theta}_U$ is the upper bound found in part (a), what is the confidence coefficient of the interval $(\hat{\theta}_L, \hat{\theta}_U)$?

8.6   Large-Sample Confidence Intervals

In Section 8.3 we presented some unbiased point estimators for the parameters μ, p, $\mu_1 - \mu_2$, and $p_1 - p_2$. As we indicated in that section, for large samples all these point estimators have approximately normal sampling distributions with standard errors as given in Table 8.1. That is, under the conditions of Section 8.3, if the target parameter θ is μ, p, $\mu_1 - \mu_2$, or $p_1 - p_2$, then for large samples

$$Z = \frac{\hat{\theta} - \theta}{\sigma_{\hat{\theta}}}$$

possesses approximately a standard normal distribution. Consequently, $Z = (\hat{\theta} - \theta)/\sigma_{\hat{\theta}}$ forms (at least approximately) a pivotal quantity, and the pivotal method can be employed to develop confidence intervals for the target parameter θ.

EXAMPLE **8.6**   Let $\hat{\theta}$ be a statistic that is normally distributed with mean θ and standard error $\sigma_{\hat{\theta}}$. Find a confidence interval for θ that possesses a confidence coefficient equal to $(1 - \alpha)$.

Solution   The quantity

$$Z = \frac{\hat{\theta} - \theta}{\sigma_{\hat{\theta}}}$$

has a standard normal distribution. Now select two values in the tails of this distribution, $z_{\alpha/2}$ and $-z_{\alpha/2}$, such that (see Figure 8.7)

$$P(-z_{\alpha/2} \leq Z \leq z_{\alpha/2}) = 1 - \alpha.$$

Substituting for Z in the probability statement, we have

$$P\left(-z_{\alpha/2} \leq \frac{\hat{\theta} - \theta}{\sigma_{\hat{\theta}}} \leq z_{\alpha/2}\right) = 1 - \alpha.$$

Multiplying by $\sigma_{\hat{\theta}}$, we obtain

$$P(-z_{\alpha/2}\sigma_{\hat{\theta}} \leq \hat{\theta} - \theta \leq z_{\alpha/2}\sigma_{\hat{\theta}}) = 1 - \alpha$$

FIGURE **8.7**
Location of $z_{\alpha/2}$
and $-z_{\alpha/2}$

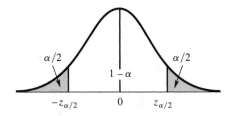

and subtracting $\hat{\theta}$ from each term of the inequality, we get

$$P(-\hat{\theta} - z_{\alpha/2}\sigma_{\hat{\theta}} \le -\theta \le -\hat{\theta} + z_{\alpha/2}\sigma_{\hat{\theta}}) = 1 - \alpha.$$

Finally, multiplying each term by -1 and, consequently, changing the direction of the inequalities, we have

$$P(\hat{\theta} - z_{\alpha/2}\sigma_{\hat{\theta}} \le \theta \le \hat{\theta} + z_{\alpha/2}\sigma_{\hat{\theta}}) = 1 - \alpha.$$

Thus, the endpoints for a $100(1-\alpha)\%$ confidence interval for θ are given by

$$\hat{\theta}_L = \hat{\theta} - z_{\alpha/2}\sigma_{\hat{\theta}} \quad \text{and} \quad \hat{\theta}_U = \hat{\theta} + z_{\alpha/2}\sigma_{\hat{\theta}}.$$

By analogous arguments, we can determine that $100(1-\alpha)\%$ one-sided confidence limits, often called upper and lower bounds, respectively, are given by

$$100(1-\alpha)\% \text{ lower bound for } \theta = \hat{\theta} - z_{\alpha}\sigma_{\hat{\theta}}$$

$$100(1-\alpha)\% \text{ upper bound for } \theta = \hat{\theta} + z_{\alpha}\sigma_{\hat{\theta}}.$$

Suppose that we compute both a $100(1-\alpha)\%$ lower bound and a $100(1-\alpha)\%$ upper bound for θ. We then decide to use both of these bounds to form a confidence interval for θ. What will be the confidence coefficient of this interval? A quick look at the preceding confirms that combining lower and upper bounds, each with confidence coefficient $1 - \alpha$, yields a two-sided interval with confidence coefficient $1 - 2\alpha$.

Under the conditions described in Section 8.3, the results given earlier in this section can be used to find large-sample confidence intervals (one-sided or two-sided) for μ, p, $(\mu_1 - \mu_2)$, and $(p_1 - p_2)$. The following examples illustrate applications of the general method developed in Example 8.6.

EXAMPLE 8.7   The shopping times of $n = 64$ randomly selected customers at a local supermarket were recorded. The average and variance of the 64 shopping times were 33 minutes and 256, respectively. Estimate μ, the true average shopping time per customer, with a confidence coefficient of $1 - \alpha = .90$.

Solution   In this case we are interested in the parameter $\theta = \mu$. Thus, $\hat{\theta} = \bar{y} = 33$ and $s^2 = 256$ for a sample of $n = 64$ shopping times. The population variance σ^2 is unknown, so (as in Section 8.3), we use s^2 as its estimated value. The confidence interval

$$\hat{\theta} \pm z_{\alpha/2}\sigma_{\hat{\theta}}$$

has the form

$$\bar{y} \pm z_{\alpha/2}\left(\frac{\sigma}{\sqrt{n}}\right) \approx \bar{y} \pm z_{\alpha/2}\left(\frac{s}{\sqrt{n}}\right).$$

From Table 4, Appendix III, $z_{\alpha/2} = z_{.05} = 1.645$; hence, the confidence limits are given by

$$\bar{y} - z_{\alpha/2}\left(\frac{s}{\sqrt{n}}\right) = 33 - 1.645\left(\frac{16}{8}\right) = 29.71$$

$$\bar{y} + z_{\alpha/2}\left(\frac{s}{\sqrt{n}}\right) = 33 + 1.645\left(\frac{16}{8}\right) = 36.29.$$

Thus, our confidence interval for μ is (29.71, 36.29). In repeated sampling, approximately 90% of all intervals of the form $\bar{Y} \pm 1.645(s/\sqrt{n})$ include μ, the true mean shopping time per customer. Although we do not know whether the particular interval (29.71, 36.29) contains μ, the procedure generating it yields intervals that do capture the true mean in approximately 95% of all instances where the procedure is used.

EXAMPLE **8.8**   Two brands of refrigerators, denoted A and B, are each guaranteed for 1 year. In a random sample of 50 refrigerators of brand A, 12 were observed to fail before the guarantee period ended. An independent random sample of 60 brand B refrigerators also revealed 12 failures during the guarantee period. Estimate the true difference $(p_1 - p_2)$ between proportions of failures during the guarantee period, with confidence coefficient .98.

Solution   The confidence interval

$$\hat{\theta} \pm z_{\alpha/2}\sigma_{\hat{\theta}}$$

now has the form

$$(\hat{p}_1 - \hat{p}_2) \pm z_{\alpha/2}\sqrt{\frac{p_1 q_1}{n_1} + \frac{p_2 q_2}{n_2}}.$$

Because $p_1, q_1, p_2,$ and q_2 are unknown, the exact value of $\sigma_{\hat{\theta}}$ cannot be evaluated. But as indicated in Section 8.3, we can get a good approximation for $\sigma_{\hat{\theta}}$ by substituting $\hat{p}_1, \hat{q}_1 = 1 - \hat{p}_1, \hat{p}_2,$ and $\hat{q}_2 = 1 - \hat{p}_2$ for $p_1, q_1, p_2,$ and q_2, respectively.

For this example, $\hat{p}_1 = .24, \hat{q}_1 = .76, \hat{p}_2 = .20, \hat{q}_2 = .80,$ and $z_{.01} = 2.33$. The desired 98% confidence interval is

$$(.24 - .20) \pm 2.33\sqrt{\frac{(.24)(.76)}{50} + \frac{(.20)(.80)}{60}}$$

$$.04 \pm .1851 \quad \text{or} \quad [-.1451, .2251].$$

Notice that this confidence interval contains zero. Thus, a zero value for the difference in proportions $(p_1 - p_2)$ is "believable" (at the 98% confidence level) on the basis of the observed data. However, the interval also includes the value .1. Thus, .1 represents another value of $(p_1 - p_2)$ that is "believable" on the basis of the data that we have analyzed.

We close this section with an empirical investigation of the performance of the large-sample interval estimation procedure for a single population mean μ, based on a sample of $Y_1, Y_2, \ldots, Y_n$. In this case $\theta = \mu$; therefore, $\hat{\theta} = \overline{Y}$ and $\sigma_{\hat{\theta}}^2 = \sigma_{\overline{Y}}^2 = \sigma^2/n$, where σ^2 is the population variance. The appropriate confidence limits then are

$$\hat{\theta}_L = \overline{Y} - z_{\alpha/2}\left(\frac{\sigma}{\sqrt{n}}\right) \quad \text{and} \quad \hat{\theta}_U = \overline{Y} + z_{\alpha/2}\left(\frac{\sigma}{\sqrt{n}}\right).$$

In the (large-sample) case when σ is unknown, it can be replaced (with no serious loss in accuracy) by the sample estimate $s = \sqrt{s^2}$ in these formulas.

One hundred independent random samples, each of size $n = 100$, were selected from an exponential probability distribution with mean $\mu = 10$. For each of the 100 samples, the sample mean $\overline{y}$ and the sample standard deviation s were calculated and used to form an approximate 95% confidence interval by using the formula $\overline{y} \pm 1.96s/\sqrt{100}$. (Notice that $z_{.025} = 1.96$.) For each of the 100 samples generated, Table 8.3 contains the sample mean, the sample standard deviation, and the resulting confidence limits. Some of the intervals (93, to be exact) contain 10, the true value of the population mean. The remaining 7 intervals do not contain the value 10 (these are marked with asterisks in Table 8.3). Therefore, 93% of the 100 intervals that we calculated actually captured 10, the true value of μ. Thus, the *percentage* of intervals capturing the true value of the population mean is close to .95, the nominal confidence coefficient.

Table 8.3. One hundred large-sample confidence intervals, each based on $n = 100$ observations

Sample	LCL	Mean	UCL	s
1	7.26064	8.6714	10.0822	7.1977
2	7.59523	9.5291	11.4630	9.8668
3	7.68938	9.3871	11.0849	8.6619
4	8.94686	11.1785	13.4100	11.3856
5	7.96793	10.1508	12.3336	11.1369
6	7.61223	10.0101	12.4080	12.2340
7	7.55475	9.2754	10.9961	8.7791
8	7.48558	9.1436	10.8016	8.4593
9	7.71919	9.5219	11.3245	9.1973
10	7.95952	9.7012	11.4428	8.8859
11	7.34959	8.7958	10.2420	7.3785
12	8.71580	10.8458	12.9759	10.8675
*13	6.41955	8.1029	9.7863	8.5885
14	7.27538	8.9598	10.6441	8.5938
15	8.30285	10.2902	12.2776	10.1398
16	9.48735	11.5540	13.6206	10.5441
17	8.35947	10.8521	13.3448	12.7176
18	7.63421	9.9196	12.2049	11.6599
19	7.97129	10.1196	12.2679	10.9608
20	8.86613	10.9223	12.9785	10.4906

Table 8.3. continued

Sample	LCL	Mean	UCL	s
21	7.16061	9.2142	11.2679	10.4778
22	6.99569	9.2782	11.5607	11.6454
23	8.13781	10.2701	12.4024	10.8791
24	7.70300	9.4910	11.2790	9.1226
25	9.61738	11.9111	14.2048	11.7026
26	7.23197	9.2257	11.2194	10.1720
27	7.85056	9.9053	11.9601	10.4836
28	8.82373	11.1849	13.5462	12.0470
29	8.36303	10.4008	12.4387	10.3970
30	8.74332	10.4288	12.1144	8.5996
31	7.80013	9.7064	11.6127	9.7259
32	7.88003	10.0706	12.2611	11.1763
33	7.01555	8.6994	10.3833	8.5913
34	8.31036	10.3074	12.3045	10.1891
35	7.91383	9.8344	11.7549	9.7986
36	8.80795	10.5208	12.2337	8.7390
37	7.27322	9.1728	11.0725	9.6919
38	7.47957	9.1333	10.7871	8.4375
39	7.77926	9.8034	11.8275	10.3272
40	7.09624	9.0955	11.0947	10.2001
41	8.32909	10.1639	11.9986	9.3610
42	8.23895	10.2593	12.2797	10.3080
43	8.44813	10.4284	12.4087	10.1034
44	7.11788	8.8708	10.6237	8.9436
45	8.14050	10.0992	12.0580	9.9936
46	7.53466	9.4198	11.3050	9.6183
47	8.93450	11.0165	13.0986	10.6227
48	7.43185	8.9012	10.3705	7.4965
49	7.83028	9.7367	11.6432	9.7268
50	7.61207	9.3698	11.1275	8.9680
51	7.97829	9.7442	11.5101	9.0098
52	8.97313	11.2890	13.6049	11.8158
*53	5.99726	7.7422	9.4871	8.9027
54	7.51538	9.4089	11.3025	9.6609
55	8.59019	10.4373	12.2843	9.4238
56	8.49077	10.3914	12.2920	9.6971
57	8.12970	10.4056	12.6814	11.6115
*58	6.85175	8.3456	9.8395	7.6217
59	8.87405	10.9482	13.0225	10.5827
60	8.70548	10.2597	11.8140	7.9298
61	6.37467	8.3475	10.3203	10.0653
62	8.64835	10.6703	12.6923	10.3161
63	8.37157	10.6196	12.8676	11.4696
64	7.76632	9.6919	11.6174	9.8241
65	9.19745	11.5565	13.9155	12.0359
66	7.54964	9.1818	10.8140	8.3275
67	8.57961	10.8918	13.2039	11.7967
68	8.23986	10.0088	11.7778	9.0254

Table 8.3. continued

Sample	LCL	Mean	UCL	s
69	8.08091	9.9399	11.7989	9.4846
70	7.52910	9.2224	10.9156	8.6391
71	7.52705	9.4949	11.4628	10.0401
72	8.61466	10.7728	12.9310	11.0110
73	9.13542	11.2654	13.3953	10.8670
*74	6.73171	8.1351	9.5386	7.1604
75	7.78546	10.1938	12.6021	12.2874
76	8.04925	9.6668	11.2843	8.2526
77	7.61919	9.6068	11.5943	10.1407
*78	5.91167	7.7383	9.5649	9.3194
79	8.36219	10.4003	12.4384	10.3984
80	8.78276	11.0946	13.4065	11.7953
81	7.75108	9.4840	11.2170	8.8415
82	8.41534	10.4168	12.4183	10.2117
83	7.56210	9.1461	10.7300	8.0815
*84	6.51658	8.0757	9.6349	7.9549
85	8.73332	10.7988	12.8643	10.5381
86	7.10357	8.9350	10.7664	9.3439
87	7.46386	9.5521	11.6403	10.6541
88	7.40553	9.0024	10.5993	8.1473
89	7.70464	9.6524	11.6002	9.9376
90	8.16395	10.3755	12.5870	11.2834
91	8.08567	9.8488	11.6120	8.9958
92	8.45424	10.2642	12.0741	9.2344
*93	6.78632	8.1950	9.6064	7.2010
94	8.56442	10.6626	12.7608	10.7050
95	7.49852	9.2474	10.9962	8.9226
96	8.01983	10.9016	13.7833	14.7028
97	7.40014	9.1321	10.8640	8.8363
98	7.80386	9.6444	11.4849	9.3903
99	8.47046	10.3403	12.2101	9.5399
100	8.08183	10.0133	11.9447	9.8542

*The intervals for these samples do not include 10, the true value of μ.

An analogy will help explain the meaning of the term *confident at the 95% level.* Suppose that you randomly select a battery from a group of 100 batteries, 95 of which are fully charged. The battery that you select is either fully charged or it is not. However, you are 95% confident that you have selected a fully charged battery because of the *composition of the group of batteries from which you made your selection.* The single interval that you compute on the basis of a random sample from the population of interest either contains the target parameter or it does not. You are "95% confident" that the interval contains the parameter because the interval was obtained by using a *procedure* that generates intervals that contain the parameter approximately 95% of the times the procedure is used.

In this section we have used the pivotal method to derive large-sample confidence intervals for the parameters μ, p, $\mu_1 - \mu_2$, and $p_1 - p_2$ under the conditions of

Section 8.3. The key formula is

$$\hat{\theta} \pm z_{\alpha/2}\sigma_{\hat{\theta}},$$

where the values of $\hat{\theta}$ and $\sigma_{\hat{\theta}}$ are as given in Table 8.1. When $\theta = \mu$ is the target parameter, then $\hat{\theta} = \overline{Y}$ and $\sigma_{\hat{\theta}}^2 = \sigma^2/n$, where σ^2 is the population variance. If the true value of σ^2 is known, this value should be used in calculating the confidence interval. If σ^2 is not known, there is no serious loss of accuracy if s^2 is substituted for σ^2 in the formula for the confidence interval. Similarly, if σ_1^2 and σ_2^2 are unknown, s_1^2 and s_2^2 can be substituted for these values in the formula for a large-sample confidence interval for $\theta = \mu_1 - \mu_2$. When $\theta = p$ is the target parameter, then $\hat{\theta} = \hat{p}$ and $\sigma_{\hat{\theta}} = \sqrt{pq/n}$. Because p is the unknown target parameter, $\sigma_{\hat{p}}$ cannot be evaluated. If we substitute $\hat{p}$ for p (and $\hat{q} = 1 - \hat{p}$ for q) in the formula for $\sigma_{\hat{p}}$, however, the resulting confidence interval will have approximately the stated confidence coefficient. Similar statements hold when $\hat{p}_1$ and $\hat{p}_2$ are used to estimate p_1 and p_2, respectively, in the formula for $\sigma_{\hat{p}_1 - \hat{p}_2}^2$. The theoretical justification for these substitutions will be provided in Section 9.3.

Exercises

8.42 "Athletes at major colleges graduated, on the whole, at virtually the same rate as other students," according the the National Collegiate Athletic Association (NCAA) ("Athletic Grad Rate Same as Others," *Riverside Press-Enterprise*, March 9, 1992). The survey involved students who had entered college before the NCAA's Proposition 48 took effect in 1985. (Proposition 48 requires that athletes maintain at least a 2.0 grade point average on a scale of 4.0.) Suppose that, in a new poll of 500 athletes at major colleges, the number graduating was 268.

    a Find a 98% confidence interval for p, the proportion of athletes at major colleges who graduate.

    b Does the interval include the value $p = .51$, the graduation rate prior to Proposition 48? What do you conclude?

8.43 Sometimes surveys provide meaningful information about issues that did not seem to be the focus of the survey initially. A *New York Times*/CBS poll ("Polls Say Clinton Keeps Lead Despite G.O.P. Fire," *New York Times*, November 1, 1992) was conducted to determine voter preferences in the 1992 presidential election. Telephone interviews were conducted of 2374 adults around the United States, excluding Alaska and Hawaii, and 1912 of the respondents said that they were registered voters. The sample of telephone exchanges called was computer selected from a complete list of exchanges in the country. For each exchange, the telephone numbers were formed by random digits, thus permitting access to both listed and unlisted numbers. Use the preceding data to form a 99% confidence interval for the proportion of adults in the continental United States who were registered voters in 1992. Interpret this interval.

8.44 The administrators for a hospital wished to estimate the average number of days required for in-patient treatment of patients between the ages of 25 and 34. A random sample of 500 hospital patients between these ages produced a mean and standard deviation equal to 5.4 and 3.1 days, respectively. Construct a 95% confidence interval for the mean length of stay for the population of patients from which the sample was drawn.

8.45 A *New York Times*/CBS survey of 224 children of ages 9 to 17 was conducted shortly after the January 28, 1986, space-shuttle disaster. Two-thirds of the children said that they would like to travel in space. (*Orlando Sentinel*, February 3, 1986).

a Construct a 90% confidence interval for the proportion of children of ages 9 to 17 in 1986 who would like to experience space travel.

b Do you think that "most" children in this age group in 1986 thought that they would like to experience space travel? Why?

8.46 Estimates of the earth's biomass—the total amount of vegetation held by the earth's forests— are important in determining the amount of unabsorbed carbon dioxide that we can expect to remain in the earths atmosphere (*Science News* 186, August 19, 1989). Suppose that a sample of 75 1-square-meter plots randomly chosen in North America's northern forests produced a mean biomass of 4.2 kilograms per square meter (kg/m^2) and a standard deviation of 1.5 kg/m^2. Give a 95% confidence interval for the average biomass for North America's northern forests.

8.47 A small amount of the trace element selenium, from 50 to 200 μg per day, is considered essential to good health. Suppose that independent random samples of $n_1 = n_2 = 30$ adults were selected from two regions of the United States, and a day's intake of selenium, from both liquids and solids, was recorded for each person. The mean and standard deviation of the selenium daily intakes for the 30 adults from Region 1 were $\bar{y}_1 = 167.1$ μg and $s_1 = 24.3$ μg, respectively. The corresponding statistics for the 30 adults from Region 2 were $\bar{y}_2 = 140.9$ μg and $s_2 = 17.6$ μg. Find a 95% confidence interval for the difference in the mean selenium intake for the two regions.

8.48 The following statistics are the result of an experiment conducted by P. I. Ward to investigate a theory concerning the moulting behavior of the male *Gammarus pulex*, a small crustacean ("*Gammarus pulex* Control Their Moult Timing to Secure Mates," *Animal Behaviour* 32 [1984]). If a male has to moult while paired with a female, he must release her and so loses her. The theory is that the male *Gammarus pulex* is able to postpone moulting, thereby reducing the possibility of losing his mate. Ward randomly assigned 100 pairs of males and females to two groups of 50 each. Pairs in the first group were maintained together (normal); those in the second group were separated (split). The length of time to moult was recorded for both males and females and the means, standard deviations, and sample sizes are shown in the accompanying table. (The number of crustaceans in each of the four samples is less than 50 because some in each group did not survive until moulting time.) Find a 99% confidence interval for the difference in mean moult time for "normal" males versus those "split" from their mates. Interpret the interval.

	Time to Moult (days)		
	Mean	s	n
Males			
Normal	24.8	7.1	34
Split	21.3	8.1	41
Females			
Normal	8.6	4.8	45
Split	11.6	5.6	48

8.49 Should professional athletes be tested for drugs? Reporting on this question, the *Orlando Sentinel* (October 31, 1985) noted that Donald Fehr, executive director of the Major League Baseball Player's Association, "criticized mandatory testing as being unnecessary." The public does not seem to agree. A *Washington Post*–ABC News survey found that 73% of 1506 people interviewed favored drug tests for professional athletes; 68% agreed that professional athletes discovered to be using drugs should be banned or suspended from professional sports on the first offense.

    a Find a 95% confidence interval for the percentage of the public that favors drug tests for professional athletes.

    b Upon what assumptions is your confidence interval in (a) based?

8.50 Refer to Exercise 8.25, where we reported on the results of Harris polls conducted in 1983 and 1992. The 1992 results summarized the responses of $n = 1251$ individuals, whereas we assumed that the results for the 1983 poll involved $n = 1250$ respondents. The data are summarized in the accompanying table for your convenience.

Health Issue	1983 Survey	1992 Survey
Ate recommended amount of fibrous food	.59	.53
Avoided fat	.55	.51
Avoided excess salt	.53	.46
Used seatbelts	.19	.70
Used smoke detectors	.67	.90

    a Construct a 98% confidence interval for the difference in the proportions of Americans who ate the recommended amounts of fibrous foods in 1983 and in 1992.

    b Based upon the interval you obtained in (a), do you think that the proportion of Americans who ate the recommended amount of fibrous foods decreased between 1983 and 1992? Explain.

8.51 Refer to Exercise 8.50. Construct a 90% confidence interval for the difference in the proportions of Americans who used seat belts in 1983 and in 1992. Do you think that a greater proportion of individuals used seat belts in 1992? Why?

8.52 Refer to Exercise 8.50. Construct a 99% confidence interval for comparing of the proportions of individuals who used smoke detectors in 1983 and in 1992.

8.53 For a comparison of the rates of defectives produced by two assembly lines, independent random samples of 100 items were selected from each line. Line *A* yielded 18 defectives in the sample, and line *B* yielded 12 defectives. Find a 98% confidence interval for the true difference in proportions of defectives for the two lines. Is there evidence here to suggest that one line produces a higher proportion of defectives than the other?

8.54 Alice Chang and her colleagues (A. Chang, T. L. Rosenthal, R. H. Bryant, R. H. Rosenthal, R. M. Heidlage, and B. K. Fritzler. "Comparing High School and College Students' Leisure Interests and Stress Ratings," *Behavioral Research Therapy* 31(2), 1993) tested 559 high school students using a leisure-interest checklist (LIC) survey. The means and standard deviations for each of the seven LIC factor scales are given in the following table for $n = 252$ male students and $n = 307$ female students.

    a Give a 95% confidence interval for the difference in the means for male and female students for the cultural activity scale. Interpret this interval.

LIC Factor	Males		Females	
Scale Activity	Mean	s	Mean	s
Hobbies	10.06	6.47	13.64	7.46
Social fun	22.05	5.12	25.96	5.07
Sports	13.65	4.82	9.88	4.41
Cultural	11.48	5.69	13.21	'5.31
Trips	6.90	3.41	6.49	2.97
Games	4.95	3.29	3.85	2.49
Church	2.15	2.10	3.00	2.26

b Find a 90% confidence interval for the difference in the means for male and female students for the social fun factor scale activity.

c Do the means for males and females for the two LIC activities considered in (a) and (b) appear to differ?

8.55 One suggested method for solving the electric power shortage in a region involves constructing floating nuclear power plants a few miles offshore in the ocean. Concern about the possibility of a ship collision with the floating (but anchored) plant has raised the need for an estimate of the density of ship traffic in the area. The number of ships passing within 10 mi of the proposed power-plant location per day, recorded for $n = 60$ days during July and August, possessed a sample mean and variance of $\bar{y} = 7.2$ and $s^2 = 8.8$.

a Find a 95% confidence interval for the mean number of ships passing within 10 mi of the proposed power-plant location during a 1-day time period.

b The density of ship traffic was expected to decrease during the winter months. A sample of $n = 90$ daily recordings of ship sightings for December, January, and February yielded a mean and variance of $\bar{y} = 4.7$ and $s^2 = 4.9$. Find a 90% confidence interval for the difference in mean density of ship traffic between the summer and winter months.

c What is the population associated with your estimate in (b)? What could be wrong with the sampling procedure for (a) and (b)?

*8.56 Suppose that (Y_1, Y_2, Y_3, Y_4) has a multinomial distribution with n trials and probabilities p_1, p_2, p_3, and p_4 for the four cells. Just as in the binomial case, any linear combination of Y_1, Y_2, Y_3, and Y_4 will be approximately normally distributed for large n.

a Determine the variance of $Y_1 - Y_2$. (*Hint:* Recall that the random variables Y_i are dependent.)

b A study of attitudes among residents of Florida with regard to policies for handling nuisance alligators in urban areas showed the following. Among 500 people sampled and presented with four management choices, 6% said the alligators should be completely protected, 16% said they should be destroyed by wildlife officers, 52% said they should be relocated live, and 26% said that a regulated commercial harvest should be allowed. Estimate the difference between the population proportion favoring complete protection and the population proportion favoring destruction by wildlife officers. Use a confidence coefficient of .95.

*8.57 The *Journal of Communication*, Winter 1978, reported on a study of the viewing of violence on TV. Samples from populations with low viewing rates (10–19 programs per week) and high viewing rates (40–49 programs per week) were divided into age groups, and Y, the number of persons watching a high number of violent programs, was recorded. The data for two age groups were as shown in the accompanying table, with n_i denoting the sample size for each cell. If Y_1, Y_2, Y_3, and Y_4 have independent binomial distributions with parameters

p_1, p_2, p_3, and p_4, respectively, find a 95% confidence interval for $(p_3 - p_1) - (p_4 - p_2)$. This function of the p_i values represents a comparison between the change in viewing habits for young adults and the corresponding change for older adults, as we move from those with low viewing rates to those with high viewing rates. (The data suggest that the rate of viewing violence may increase with young adults but decrease with older adults.)

Viewing Rate	Age Group			
	16–34		55 and Over	
Low	$y_1 = 20$	$n_1 = 31$	$y_2 = 13$	$n_2 = 30$
High	$y_3 = 18$	$n_3 = 26$	$y_4 = 7$	$n_4 = 28$

8.7 Selecting the Sample Size

The design of an experiment is essentially a plan for purchasing a quantity of information. Like any other commodity, information may be acquired at varying prices depending upon the manner in which the data are obtained. Some measurements contain a large amount of information about the parameter of interest; others may contain little or none. Research, scientific or otherwise, is done in order to obtain information. Obviously, we should seek to obtain information at minimum cost.

The sampling procedure—or *experimental design*, as it is usually called—affects the quantity of information per measurement. This, together with the sample size n, controls the total amount of relevant information in a sample. At this point in our study, we will be concerned with the simplest sampling situation: random sampling from a relatively large population. We will first devote our attention to selection of the sample size n.

A researcher makes little progress in planning an experiment before encountering the problem of selecting the sample size. Indeed, one of the most frequent questions asked of the statistician is: How many measurements should be included in the sample? Unfortunately, the statistician cannot answer this question without knowing how much information the experimenter wishes to obtain. Referring specifically to estimation, we would like to know how accurate the experimenter wishes the estimate to be. The experimenter can indicate the desired accuracy by specifying a bound on the error of estimation.

For instance, suppose that we wish to estimate the average daily yield μ of a chemical, and we wish the error of estimation to be less than 5 tons with probability .95. Because approximately 95% of the sample means will lie within $2\sigma_{\bar{Y}}$ of μ in repeated sampling, we are asking that $2\sigma_{\bar{Y}}$ equal 5 tons (see Figure 8.8). Then

$$\frac{2\sigma}{\sqrt{n}} = 5.$$

Solving for n, we obtain

$$n = \frac{4\sigma^2}{25}.$$

We cannot obtain an exact numerical value for n unless the population standard deviation σ is known. This is exactly what we would expect because the variability of the

FIGURE **8.8**
The approximate
distribution of $\overline{Y}$ for
large samples

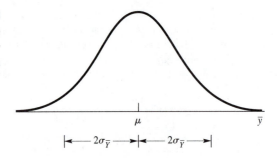

estimator $\overline{Y}$ depends upon the variability of the population from which the sample will be drawn.

Lacking an exact value for σ, we use the best approximation available, such as an estimate s obtained from a previous sample or knowledge of the range of the measurements in the population. Because the range is approximately equal to 4σ (recall the empirical rule), one-fourth of the range provides an approximate value of σ. For our example, suppose that the range of the daily yields is known to be approximately 84 tons. Then $\sigma \approx 84/4 = 21$ and

$$n = \frac{4\sigma^2}{25} \approx \frac{(4)(21)^2}{25} = 70.56$$

or

$$n = 71.$$

Using a sample size $n = 71$, we can be reasonably certain (with confidence coefficient approximately equal to .95) that our estimate will lie within 5 tons of the true average daily yield.

Actually we would expect the error of estimation to be much less than 5 tons. According to the empirical rule, the probability is approximately equal to .68 that the error of estimation will be less than $\sigma_{\overline{Y}} = 2.5$ tons. The probabilities, .95 and .68, used in these statements are inexact because σ was approximated. Although this method of choosing the sample size is only approximate for a specified accuracy of estimation, it is the best available and is certainly better than selecting the sample size intuitively.

The method of choosing the sample sizes for all the large-sample estimation procedures outlined in Table 8.1 is identical to that just described. The experimenter must specify a desired bound on the error of estimation and an associated confidence level $1 - \alpha$. For example, if the parameter is θ and the desired bound is B, we equate

$$z_{\alpha/2}\sigma_{\hat{\theta}} = B$$

where, as in Section 8.6,

$$P(Z > z_{\alpha/2}) = \frac{\alpha}{2}.$$

We will illustrate the use of this method in the following examples.

EXAMPLE **8.9**   The reaction of an individual to a stimulus in a psychological experiment may take one of two forms, A or B. If an experimenter wishes to estimate the probability p that a person will react in manner A, how many people must be included in the experiment? Assume that the experimenter will be satisfied if the error of estimation is less than .04 with probability equal to .90. Assume also that he expects p to lie somewhere in the neighborhood of .6.

Solution   Because we have specified that $1 - \alpha = .90$, α must equal .10 and $\alpha/2 = .05$. The z value corresponding to an area equal to .05 in the upper tail of the standard normal distribution is $z_{\alpha/2} = 1.645$. We then require that

$$1.645\sigma_{\hat{p}} = .04, \quad \text{or} \quad 1.645\sqrt{\frac{pq}{n}} = .04.$$

Because the standard error of $\hat{p}$ depends upon p, which is unknown, we could use the guessed value of $p = .6$ provided by the experimenter as an approximate value for n. Then

$$1.645\sqrt{\frac{(.6)(.4)}{n}} = .04$$

$$n = 406.$$

In this example, we assumed that $p \approx .60$. How would we proceed if we had no idea about the true value of p? In Exercise 7.38(a) we established that the *maximum* value for the variance of $\hat{p} = Y/n$ occurs when $p = .5$. If we did not know that $p \approx .6$, we would use $p = .5$, which would yield the maximum possible value for n: $n = 423$. No matter what the true value for p, $n = 423$ is large enough to provide an estimate that is within $B = .04$ of p with probability .90.

EXAMPLE **8.10**   An experimenter wishes to compare the effectiveness of two methods of training industrial employees to perform an assembly operation. The selected employees are to be divided into two groups of equal size, the first receiving training method 1 and the second receiving training method 2. After training, each employee will perform the assembly operation, and the length of assembly time will be recorded. The experimenter expects the measurements for both groups to have a range of approximately 8 minutes. If the estimate of the difference in mean assembly times is to be correct to within 1 minute with probability .95, how many workers must be included in each training group?

Solution   The manufacturer specified $1 - \alpha = .95$. Thus, $\alpha = .05$ and $z_{\alpha/2} = z_{.025} = 1.96$. Equating $1.96\sigma_{(\bar{Y}_1 - \bar{Y}_2)}$ to 1 minute, we obtain

$$1.96\sqrt{\frac{\sigma_1^2}{n_1} + \frac{\sigma_2^2}{n_2}} = 1.$$

Alternatively, because we desire n_1 to equal n_2, we may let $n_1 = n_2 = n$ and obtain the equation

$$1.96\sqrt{\frac{\sigma_1^2}{n} + \frac{\sigma_2^2}{n}} = 1.$$

As noted earlier, the variability of each method of assembly is approximately the same; hence, $\sigma_1^2 = \sigma_2^2 = \sigma^2$. Because the range, 8 minutes, is approximately equal to 4σ, we have

$$4\sigma \approx 8, \quad \text{or, equivalently} \quad \sigma \approx 2.$$

Substituting this value for σ_1 and σ_2 in the earlier equation, we obtain

$$1.96\sqrt{\frac{(2)^2}{n} + \frac{(2)^2}{n}} = 1.$$

Solving, we obtain $n = 30.73$. Therefore, each group should contain $n = 31$ members.

Exercises

8.58 Let Y be a binomial random variable with parameter p. Find the sample size necessary to estimate p to within .05 with probability .95 in the following situations:

a if p is thought to be approximately .9.

b if no information about p is known (use $p = .5$ in estimating the variance of $\hat{p}$).

8.59 If a wildlife service wishes to estimate the mean number of days of hunting per hunter for all hunters licensed in the state during a given season, with a bound on the error of estimation equal to 2 hunting days, how many hunters must be included in the survey? Assume that data collected in earlier surveys have shown σ to be approximately equal to 10.

8.60 A Louis Harris and Associates survey conducted in 1984 for Philip Morris, Inc., showed an increase over the decade in attendance at movies, concerts, and the theater. The poll was based on a telephone survey of 1504 adults, and the results were compared with similar polls conducted in 1973, 1975, and 1980 (*Wall Street Journal*, December 1, 1984).

a If the survey in 1973 was based on an equal number of interviews, how accurately could you estimate the difference in attendance proportions between 1973 and 1984? (*Hint:* Find the bound on the error of estimation based on a sample size of 1500 interviewees for each sample, and assume that both proportions are near .7 or .8, say .75.)

b Suppose that you were designing the surveys and wished to estimate the difference in a pair of proportions, correct to within .02, with probability .9. How many interviewees should be included in each sample?

8.61 Refer to Exercise 8.45. How many students should have been interviewed by the newspaper in order to estimate the proportion of children who would like to have traveled in space, correct to within .02, with probability .99? Use the proportion from the previous sample in approximating the standard error of the estimate.

8.62   Suppose that you want to estimate the mean pH of rainfalls in an area that suffers from heavy pollution due to the discharge of smoke from a power plant. Assume that σ is in the neighborhood of .5 pH and that you want your estimate to lie within .1 of μ with probability near .95. Approximately how many rainfalls must be included in your sample (one pH reading per rainfall)? Would it be valid to select all of your water specimens from a single rainfall? Explain.

8.63   Refer to Exercise 8.62. Suppose that you wish to estimate the difference between the mean acidity for rainfalls at two different locations, one in a relatively unpolluted area along the ocean and the other in an area subject to heavy air pollution. If you wish your estimate to be correct to the nearest .1 pH with probability near .90, approximately how many rainfalls (pH values) must you include in each sample? (Assume that the variance of the pH measurements is approximately .25 at both locations and that the samples are to be of equal size.)

8.64   Refer to the comparison of the daily adult intake of selenium in two different regions of the United States, in Exercise 8.47. Suppose that you wish to estimate the difference in the mean daily intake between the two regions, correct to within 5 μg, with probability .90. If you plan to select an equal number of adults from the two regions (that is, if $\mu_1 = \mu_2$), how large should n_1 and n_2 be?

8.65   Refer to Exercise 8.24. If the researcher wants to estimate the difference in proportions to within .05 with 90% confidence, how many graduates and nongraduates must be interviewed? (Assume that an equal number will be interviewed from each group.)

8.66   Refer to Exercise 8.53. How many items should be sampled from each line if a 95% confidence interval for the true difference in proportions is to have width .2? Assume that samples of equal size will be taken from each line.

8.67   Refer to Exercise 8.54.

  a   Another similar study is to be undertaken to compare the mean LIC inventory scores for males and females. The objective is to produce a 99% confidence interval for the true difference in the mean church activity scores. If we need to sample an equal number of males and females and want the width of the confidence interval to be 2 units, how many observations should be included in each group?

  b   Repeat the calculations from (a) if we are interested in comparing male and female students with respect to the activity scores for hobbies.

  c   What sample sizes should you use so that 99% confidence intervals to compare males and females on *all seven* of the activities have widths no larger than 2 units?

8.8   Small-Sample Confidence Intervals for μ and $\mu_1 - \mu_2$

The confidence intervals for a population mean μ that we will discuss in this section are based on the assumption that the experimenter's sample has been randomly selected from a normal population. The intervals are appropriate for samples of any size and the confidence coefficients of the intervals are close to the specified values even when the population is not normal, as long as the departure from normality is not excessive. We rarely know the form of the population frequency distribution before we sample. Consequently, if an interval estimator is to be of any value, it must work reasonably well even when the population is not normal. *Working well* means that the confidence coefficient should not be affected by modest departures from normality. For most mound-shaped population distributions, experimental studies indicate that these confidence intervals maintain confidence coefficients close to the nominal values used in their calculation.

We assume that Y_1, Y_2, ..., Y_n represent a random sample selected from a normal population, and we let $\overline{Y}$ and S^2 represent the sample mean and sample variance, respectively. We would like to construct a confidence interval for the population mean when $V(Y_i) = \sigma^2$ is unknown and the sample size is too small to permit us to to apply the large-sample techniques of the previous section. Under the assumptions just stated, Theorems 7.1 and 7.3 and Definition 7.2 imply that

$$T = \frac{\overline{Y} - \mu}{S/\sqrt{n}}$$

has a t distribution with $(n - 1)$ degrees of freedom. The quantity T serves as the pivotal quantity that we will use to form a confidence interval for μ. From Table 5, Appendix III, we can find values $t_{\alpha/2}$ and $-t_{\alpha/2}$ (see Figure 8.9) so that

$$P(-t_{\alpha/2} \leq T \leq t_{\alpha/2}) = 1 - \alpha.$$

The t distribution has a density function very much like the standard normal density, except that the tails are thicker (as illustrated in Figure 7.2). Recall that the values of $t_{\alpha/2}$ depend upon the degrees of freedom $(n - 1)$, as well as on the confidence coefficient $(1 - \alpha)$.

The confidence interval for μ is developed by manipulating the inequalities in the probability statement in a manner analogous to that used in the derivation presented in Example 8.6. In this case the resulting confidence interval for μ is

$$\overline{Y} \pm t_{\alpha/2} \left(\frac{S}{\sqrt{n}} \right).$$

Under the preceding assumptions, we can also obtain $100(1 - \alpha)\%$ *one-sided* confidence limits for μ. Notice that t_α, given in Table 5, Appendix III, is such that

$$P(T \leq t_\alpha) = 1 - \alpha.$$

Substituting T into this expression and manipulating the resulting inequality, we obtain

$$P(\overline{Y} - t_\alpha(S/\sqrt{n}) \leq \mu) = 1 - \alpha.$$

FIGURE 8.9
Location of $t_{\alpha/2}$
and $-t_{\alpha/2}$

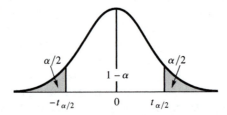

Thus, $\overline{Y} - t_\alpha(S/\sqrt{n})$ is a $100(1 - \alpha)\%$ *lower confidence bound* for μ. Analogously, $\overline{Y} + t_\alpha(S/\sqrt{n})$ is a $100(1-\alpha)\%$ *upper confidence bound* for μ. As in the large-sample case, if we determine *both* $100(1 - \alpha)\%$ lower and upper confidence bounds for μ and use the respective bounds as endpoints for a confidence interval, the resulting *two-sided* interval has confidence coefficient equal to $1 - 2\alpha$.

EXAMPLE **8.11**   A manufacturer of gunpowder has developed a new powder, which was tested in eight shells. The resulting muzzle velocities, in feet per second, were as follows:

$$3005 \quad 2925 \quad 2935 \quad 2965$$
$$2995 \quad 3005 \quad 2937 \quad 2905.$$

Find a 95% confidence interval for the true average velocity μ for shells of this type. Assume that muzzle velocities are approximately normally distributed.

Solution   If we assume that the velocities Y_i are normally distributed, the confidence interval for μ is

$$\overline{Y} \pm t_{\alpha/2}\left(\frac{S}{\sqrt{n}}\right)$$

where $t_{\alpha/2}$ is determined for $n - 1$ degrees of freedom. For the given data, $\overline{y} = 2959$ and $s = 39.1$. In this example, we have $n - 1 = 7$ degrees of freedom and, using Table 5, Appendix III, $t_{\alpha/2} = t_{.025} = 2.365$. Thus, we obtain

$$2959 \pm 2.365\left(\frac{39.1}{\sqrt{8}}\right), \quad \text{or} \quad 2959 \pm 32.7$$

as the observed confidence interval for μ.

Suppose that we are interested in comparing the means of two normal populations, one with mean μ_1 and variance σ_1^2 and the other with mean μ_2 and variance σ_2^2. If the samples are independent, confidence intervals for $\mu_1 - \mu_2$ based on a t-distributed random variable can be constructed if we assume that the two populations have a common, but unknown, variance, $\sigma_1^2 = \sigma_2^2 = \sigma^2$ (unknown).

If $\overline{Y}_1$ and $\overline{Y}_2$ are the respective sample means obtained from independent random samples from normal populations, the large-sample confidence interval for $(\mu_1 - \mu_2)$ is developed by using

$$Z = \frac{(\overline{Y}_1 - \overline{Y}_2) - (\mu_1 - \mu_2)}{\sqrt{\dfrac{\sigma_1^2}{n_1} + \dfrac{\sigma_2^2}{n_2}}}$$

as a pivotal quantity. Because we assumed that the sampled populations are both normally distributed, Z has a standard normal distribution, and using the assumption

$\sigma_1^2 = \sigma_2^2 = \sigma^2$, the quantity Z may be rewritten as

$$Z = \frac{(\overline{Y}_1 - \overline{Y}_2) - (\mu_1 - \mu_2)}{\sigma\sqrt{\dfrac{1}{n_1} + \dfrac{1}{n_2}}}.$$

Because σ is unknown, we need to find an estimator of the common variance σ^2 so that we can construct a quantity with a t distribution.

Let $Y_{11}, Y_{12}, \ldots, Y_{1n_1}$ denote the random sample of size n_1 from the first population, and let $Y_{21}, Y_{22}, \ldots, Y_{2n_2}$ denote an independent random sample of size n_2 from the second population. Then

$$\overline{Y}_1 = \frac{1}{n_1} \sum_{i=1}^{n_1} Y_{1i} \quad \text{and} \quad \overline{Y}_2 = \frac{1}{n_2} \sum_{i=1}^{n_2} Y_{2i}.$$

The usual unbiased estimator of the common variance σ^2 is obtained by pooling the sample data to obtain the *pooled estimator* S_p^2 :

$$S_p^2 = \frac{\sum\limits_{i=1}^{n_1}(Y_{1i} - \overline{Y}_1)^2 + \sum\limits_{i=1}^{n_2}(Y_{2i} - \overline{Y}_2)^2}{n_1 + n_2 - 2} = \frac{(n_1 - 1)S_1^2 + (n_2 - 1)S_2^2}{n_1 + n_2 - 2},$$

where S_i^2 is the sample variance from the ith sample, $i = 1, 2$. Notice that if $n_1 = n_2$, S_p^2 is simply the average of S_1^2 and S_2^2. If $n_1 \neq n_2$, S_p^2 is the *weighted* average of S_1^2 and S_2^2, with larger weight given to the sample variance associated with the larger sample size. Further,

$$W = \frac{(n_1 + n_2 - 2)S_p^2}{\sigma^2} = \frac{\sum\limits_{i=1}^{n_1}(Y_{1i} - \overline{Y}_1)^2}{\sigma^2} + \frac{\sum\limits_{i=1}^{n_2}(Y_{2i} - \overline{Y}_2)^2}{\sigma^2}$$

is the sum of two independent χ^2-distributed random variables with $(n_1 - 1)$ and $(n_2 - 1)$ degrees of freedom, respectively. Thus, W has a χ^2 distribution with $\nu = (n_1 - 1) + (n_2 - 1) = (n_1 + n_2 - 2)$ degrees of freedom. (See Theorems 7.2 and 7.3.) We now utilize the χ^2-distributed variable W and the *independent* standard normal quantity Z defined in the previous paragraph to form a pivotal quantity:

$$T = \frac{Z}{\sqrt{\dfrac{W}{\nu}}} = \left[\frac{(\overline{Y}_1 - \overline{Y}_2) - (\mu_1 - \mu_2)}{\sigma\sqrt{\dfrac{1}{n_1} + \dfrac{1}{n_2}}} \right] \Bigg/ \sqrt{\frac{(n_1 + n_2 - 2)S_p^2}{\sigma^2(n_1 + n_2 - 2)}}$$

$$= \frac{(\overline{Y}_1 - \overline{Y}_2) - (\mu_1 - \mu_2)}{S_p\sqrt{\dfrac{1}{n_1} + \dfrac{1}{n_2}}},$$

a quantity that, by construction, has a t distribution with $(n_1 + n_2 - 2)$ degrees of freedom.

Proceeding as we did earlier in this section, we see that the confidence interval for $(\mu_1 - \mu_2)$ has the form

$$(\overline{Y}_1 - \overline{Y}_2) \pm t_{\alpha/2} S_p \sqrt{\frac{1}{n_1} + \frac{1}{n_2}}$$

where $t_{\alpha/2}$ is determined from the t distribution with $(n_1 + n_2 - 2)$ degrees of freedom.

EXAMPLE 8.12   To reach maximum efficiency in performing an assembly operation in a manufacturing plant, new employees require approximately a 1-month training period. A new method of training was suggested, and a test was conducted to compare the new method with the standard procedure. Two groups of nine new employees each were trained for a period of 3 weeks, one group using the new method and the other following the standard training procedure. The length of time (in minutes) required for each employee to assemble the device was recorded at the end of the 3-week period. The resulting measurements are as shown in Table 8.4. Estimate the true mean difference $(\mu_1 - \mu_2)$, with confidence coefficient .95. Assume that the assembly times are approximately normally distributed, that the variances of the assembly times are approximately equal for the two methods, and that the samples are independent.

Solution   For the data in Table 8.4, with sample 1 denoting the standard procedure, we have

$$\overline{y}_1 = 35.22 \qquad\qquad \overline{y}_2 = 31.56$$

$$\sum_{i=1}^{9}(y_{1i} - \overline{y}_1)^2 = 195.56 \qquad \sum_{i=1}^{9}(y_{2i} - \overline{y}_2)^2 = 160.22$$

$$s_1^2 = 24.445 \qquad\qquad s_2^2 = 20.027.$$

Hence,

$$s_p^2 = \frac{8(24.445) + 8(20.027)}{9 + 9 - 2} = \frac{195.56 + 160.22}{16} = 22.236 \quad \text{and} \quad s_p = 4.716.$$

Notice that, because $n_1 = n_2 = 9$, s_p^2 is the simple average of s_1^2 and s_1^2. Also, $t_{.025} = 2.120$ for $(n_1 + n_2 - 2) = 16$ degrees of freedom. The observed confidence interval is, therefore,

$$(\overline{y}_1 - \overline{y}_2) \pm t_{\alpha/2} s_p \sqrt{\frac{1}{n_1} + \frac{1}{n_2}}$$

$$(35.22 - 31.56) \pm (2.120)(4.716)\sqrt{\frac{1}{9} + \frac{1}{9}}$$

$$3.66 \pm 4.71.$$

Table 8.4. Data for Example 8.12

Procedure	Measurements								
Standard	32	37	35	28	41	44	35	31	34
New	35	31	29	25	34	40	27	32	31

This confidence interval can be written in the form [−1.05, 8.37]. The interval is fairly wide and includes both positive and negative values. If $\mu_1 - \mu_2$ is positive, $\mu_1 > \mu_2$ and the standard procedure has a larger expected assembly time than the new procedure. If $\mu_1 - \mu_2$ is really negative, the reverse is true. Because the interval contains both positive and negative values, neither procedure can be said to differ from the other.

Summary of Small-Sample Confidence Intervals for Means of Normal Distributions with Unknown Variance(s)

Parameter	Confidence Interval ($v = $ degrees of freedom)
μ	$\overline{Y} \pm t_{\alpha/2}\left(\dfrac{S}{\sqrt{n}}\right), \qquad v = n-1$
$\mu_1 - \mu_2$	$(\overline{Y}_1 - \overline{Y}_2) \pm t_{\alpha/2} S_p \sqrt{\dfrac{1}{n_1} + \dfrac{1}{n_2}}$

where $\quad v = n_1 + n_2 - 2 \quad$ and

$$S_p^2 = \frac{(n_1 - 1)S_1^2 + (n_2 - 1)S_2^2}{n_1 + n_2 - 2}$$

(requires that the samples are independent and the assumption that $\sigma_1^2 = \sigma_2^2$)

As the sample size (or sizes) gets large, the number of degrees of freedom for the t distribution increases, and the t distribution can be approximated quite closely by the standard normal distribution. As a result, the small-sample confidence intervals of this section are nearly indistinguishable from the large-sample confidence intervals of Section 8.6 for large n (or large n_1 and n_2). The intervals are nearly equivalent when the degrees of freedom exceed 30.

The confidence intervals for a single mean and the difference in two means were developed under the assumptions that the populations of interest are normally distributed. There is considerable empirical evidence that these intervals maintain their nominal confidence coefficient as long as the populations sampled have roughly mound-shaped distributions. If $n_1 \approx n_2$, the intervals for $\mu_1 - \mu_2$ also maintain their nominal confidence coefficients as long as the population variances are roughly equal. The independence of the samples is the most crucial assumption in using the confidence intervals developed in this section to compare two population means.

Exercises

8.68 Although there are many treatments for *bulimia nervosa*, some subjects fail to benefit from treatment. In a study to determine which factors predict who will benefit from treatment, Wendy K. Baell and E. H. Wertheim ["Predictors of Outcome in the Treatment of Bulimia

Nervosa," *British Journal of Clinical Psychology* 31 (1992)] found that self-esteem was one of the important predictors. The mean and standard deviation of posttreatment self-esteem scores for $n = 21$ subjects were $\bar{y} = 26.6$ and $s = 7.4$, respectively. Find a 95% confidence interval for the true posttreatment self-esteem scores.

8.69 The carapace lengths of ten lobsters examined in a study of the infestation of the *Thenus orientalis* lobster by two types of barnacles, *Octolasmis tridens* and *O. lowei*, are given in the following table. Find a 95% confidence interval for the mean carapace length of *T. orientalis* lobsters caught in the seas in the vicinity of Singapore [W. B. Jeffries, H. K. Voris, and C. M. Yang, "Diversity and Distribution of the Pedunculate Barnacle *Octolasmis* Gray, 1825 Epizoic on the Scyllarid Lobster, *Thenus orientalis* (Lund 1793)," *Crustaceana* 46(3) (1984)].

Lobster field number	A061	A062	A066	A070	A067	A069	A064	A068	A065	A063
Carapace length (mm)	78	66	65	63	60	60	58	56	52	50

8.70 Scholastic Aptitude Test (SAT) scores, which have fallen slowly since the inception of the test, have now begun to rise. Originally, a score of 500 was intended to be average. The mean scores for 1991 were approximately 422 for the verbal test and 474 for the mathematics test. A random sample of the test scores of 20 seniors from a large urban high school produced the means and standard deviations listed in the accompanying table.

	Verbal	Mathematics
Sample mean	419	455
Sample standard deviation	57	69

a Find a 90% confidence interval for the mean verbal SAT scores for urban high school seniors.

b Does the interval that you found in (a) include the value 422, the true mean verbal SAT score for 1991? What can you conclude?

c Construct a 90% confidence interval for the mean mathematics SAT score for the urban high school seniors. Does the interval include 474, the true mean mathematics score for 1991? What can you conclude?

8.71 Chronic anterior compartment syndrome is a condition characterized by exercise-induced pain in the lower leg. Swelling and impaired nerve and muscle function also accompany the pain, which is relieved by rest. Susan Beckham and colleagues [S. J. Beckham, W. A. Grana, P. Buckley, J. E. Breasile, and P. L. Claypool, "A Comparison of Anterior Compartment Pressures in Competitive Runners and Cyclists," *American Journal of Sports Medicine* 21 (1) (1993)] conducted an experiment involving ten healthy runners and ten healthy cyclists to determine if pressure measurements within the anterior muscle compartment differ between runners and cyclists. The data—compartment pressure, in millimeters of mercury—are summarized in the following table.

	Runners		Cyclists	
Condition	Mean	s	Mean	s
Rest	14.5	3.92	11.1	3.98
80% maximal O_2 consumption	12.2	3.49	11.5	4.95

a Construct a 95% confidence interval for the difference in mean compartment pressures between runners and cyclists under the resting condition.

b Construct a 90% confidence interval for the difference in mean compartment pressures between runners and cyclists who exercise at 80% of maximal oxygen consumption.

c Consider the intervals constructed in (a) and (b). How would you interpret the results you obtained?

8.72 Organic chemists often purify organic compounds by a method known as fractional crystallization. An experimenter wanted to prepare and purify 4.85 g of aniline. Ten 4.85-g specimens of aniline were prepared and purified to produce acetanilide. The following dry yields were obtained:

$$3.85, \ 3.88, \ 3.90, \ 3.62, \ 3.72, \ 3.80, \ 3.85, \ 3.36, \ 4.01, \ 3.82.$$

Construct a 95% confidence interval for the mean number of grams of acetanilide that can be recovered from 4.85 g of aniline.

8.73 Two new drugs were given to patients with hypertension. The first drug lowered the blood pressure of 16 patients an average of 11 points, with a standard deviation of 6 points. The second drug lowered the blood pressure of 20 other patients an average of 12 points, with a standard deviation of 8 points. Determine a 95% confidence interval for the difference in the mean reductions in blood pressure, assuming that the measurements are normally distributed with equal variances.

8.74 The Environmental Protection Agency has collected data on LC50 measurements (concentrations that kill 50% of test animals) for certain chemicals likely to be found in freshwater rivers and lakes. (See Exercise 7.5 for additional details.) For certain species of fish, the LC50 measurements (in parts per million) for DDT in 12 experiments were as follows:

$$16, \ 5, \ 21, \ 19, \ 10, \ 5, \ 8, \ 2, \ 7, \ 2, \ 4, \ 9.$$

Estimate the true mean LC50 for DDT, with confidence coefficient .90. Assume that the LC50 measurements have an approximately normal distribution.

8.75 Refer to Exercise 8.74. Another common insecticide, Diazinon, yielded LC50 measurements in three experiments of 7.8, 1.6, and 1.3.

a Estimate the mean LC50 for Diazinon, with a 90% confidence interval.

b Estimate the difference between the mean LC50 for DDT and that for Diazinon, with a 90% confidence interval. What assumptions are necessary for the method you used to be valid?

8.76 Do SAT scores for high school students differ depending on the students' intended field of study? Fifteen students who intended to major in engineering were compared with 15 students who intended to major in language and literature. Given in the accompanying table are the means and standard deviations of the scores on the verbal and mathematics portion of the SAT for the two groups of students. (*Source:* "SAT Scores by Intended Field of Study," *Riverside Press Enterprise*, April 8, 1993.)

	Verbal		Math	
Engineering	$\bar{y} = 446$	$s = 42$	$\bar{y} = 548$	$s = 57$
Language/literature	$\bar{y} = 534$	$s = 45$	$\bar{y} = 517$	$s = 52$

a Construct a 95% confidence interval for the difference in average verbal scores of students majoring in engineering and of those majoring in language/literature.

b Construct a 95% confidence interval for the difference in average math scores of students majoring in engineering and of those majoring in language/literature.

c Interpret the results obtained in (a) and (b).

d What assumptions are necessary for the methods used previously to be valid?

8.77 Seasonal ranges (in hectares) for alligators were monitored on a lake outside Gainesville, Florida, by biologists from the Florida Game and Fish Commission. Five alligators monitored in the spring showed ranges of 8.0, 12.1, 8.1, 18.2, and 31.7. Four different alligators monitored in the summer showed ranges of 102.0, 81.7, 54.7, and 50.7. Estimate the difference between mean spring and summer ranges, with a 95% confidence interval. What assumptions did you make?

8.78 Solid copper produced by sintering (heating without melting) a powder under specified environmental conditions is then measured for porosity (the volume fraction due to voids) in a laboratory. A sample of $n_1 = 4$ independent porosity measurements have mean $\overline{y}_1 = .22$ and variance $s_1^2 = .0010$. A second laboratory repeats the same process on solid copper formed from an identical powder and gets $n_2 = 5$ independent porosity measurements with $\overline{y}_2 = 0.17$ and $s_2^2 = 0.0020$. Estimate the true difference between the population means $(\mu_1 - \mu_2)$ for these two laboratories, with confidence coefficient 0.95.

*8.79 A factory operates with two machines of type A and one machine of type B. The weekly repair costs X for type A machines are normally distributed with mean μ_1 and variance σ^2. The weekly repair costs Y for machines of type B are also normally distributed but with mean μ_2 and variance $3\sigma^2$. The expected repair cost per week for the factory is thus $2\mu_1 + \mu_2$. If you are given a random sample $X_1, X_2, \ldots, X_n$ on costs of type A machines and an independent random sample $Y_1, Y_2, \ldots, Y_m$ on costs for type B machines, show how you would construct a 95% confidence interval for $2\mu_1 + \mu_2$:

a if σ^2 is known.

b if σ^2 is not known.

8.80 Suppose that we obtain independent samples of sizes n_1 and n_2 from two normal populations with equal variances. Use the appropriate pivotal quantity from Section 8.8 to derive a $100(1 - \alpha)\%$ *upper confidence bound* for $\mu_1 - \mu_2$.

8.9 Confidence Intervals for σ^2

The population variance σ^2 quantifies the amount of variability in the population. Many times the actual value of σ^2 is unknown to an experimenter, and he or she must estimate σ^2. In Section 8.3 we proved that $S^2 = [1/(n - 1)] \sum_{i=1}^{n} (Y_i - \overline{Y})^2$ is an unbiased estimator for σ^2. Throughout our construction of confidence intervals for μ, we used S^2 to estimate σ^2 when σ^2 was unknown.

In addition to needing information about σ^2 in order to calculate confidence intervals for μ and $\mu_1 - \mu_2$, we may be interested in forming a confidence interval for σ^2. For example, if we performed a careful chemical analysis of tablets of a particular medication, we would be interested in the mean amount of active ingredient per tablet *and* the amount of tablet-to-tablet variability, as quantified by σ^2. Obviously, for a medication we desire a small amount of tablet-to-tablet variation and hence a small value for σ^2.

To proceed with our interval estimation procedure, we require the existence of a pivotal quantity. Again, assume that we have a random sample $Y_1, Y_2, \ldots, Y_n$ from a normal distribution with mean μ and variance σ^2, both unknown. We know from

Theorem 7.3 that

$$\frac{\sum_{i=1}^{n}(Y_i - \overline{Y})^2}{\sigma^2} = \frac{(n-1)S^2}{\sigma^2}$$

has a χ^2 distribution with $(n-1)$ degrees of freedom. We can then proceed, by the pivotal method, to find two numbers χ_L^2 and χ_U^2 such that

$$P\left[\chi_L^2 \leq \frac{(n-1)S^2}{\sigma^2} \leq \chi_U^2\right] = 1 - \alpha$$

for any confidence coefficient $(1 - \alpha)$. (The subscripts L and U stand for *lower* and *upper*, respectively.) The χ^2 density function is not symmetric, so we have some freedom in choosing χ_L^2 and χ_U^2. We would like to find the shortest interval that includes σ^2 with probability $(1 - \alpha)$. Generally, this is difficult and requires a trial-and-error search for the appropriate values of χ_L^2 and χ_U^2. We compromise by choosing points that cut off equal tail areas, as indicated in Figure 8.10. As a result, we obtain that

$$P\left[\chi_{1-\frac{\alpha}{2}}^2 \leq \frac{(n-1)S^2}{\sigma^2} \leq \chi_{1-(\alpha/2)}^2\right] = 1 - \alpha$$

and a reordering of the inequality in the probability statement gives

$$P\left[\frac{(n-1)S^2}{\chi_{1-(\alpha/2)}^2} \leq \sigma^2 \leq \frac{(n-1)S^2}{\chi_{1-\frac{\alpha}{2}}^2}\right] = 1 - \alpha.$$

The confidence interval for σ^2 is as follows.

A 100(1 − α)% Confidence Interval for σ^2

$$\left(\frac{(n-1)S^2}{\chi_{\alpha/2}^2}, \frac{(n-1)S^2}{\chi_{1-(\alpha/2)}^2}\right)$$

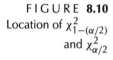

FIGURE **8.10**
Location of $\chi_{1-(\alpha/2)}^2$
and $\chi_{\alpha/2}^2$

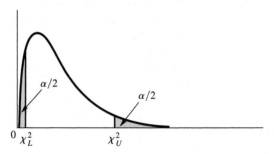

$\alpha/2$

$\alpha/2$

$0 \quad \chi_L^2 \qquad\qquad \chi_U^2$

EXAMPLE **8.13**   An experimenter wanted to check the variability of measurements obtained by using equipment designed to measure the volume of an audio source. Three independent measurements recorded by this equipment for the same sound were 4.1, 5.2, and 10.2. Estimate σ^2 with confidence coefficient .90.

Solution   If normality of the measurements recorded by this equipment can be assumed, the confidence interval just developed applies. For the data given, $s^2 = 10.57$. With $\alpha/2 = .05$ and $(n - 1) = 2$ degrees of freedom, Table 6, Appendix III, gives $\chi^2_{.95} = .103$ and $\chi^2_{.05} = 5.991$. Thus the 90% confidence interval for σ^2 is

$$\left(\frac{(n-1)s^2}{\chi^2_{.05}}, \frac{(n-1)s^2}{\chi^2_{.95}}\right) \quad \text{or} \quad \left(\frac{(2)(10.57)}{5.991}, \frac{(2)(10.57)}{.103}\right),$$

and finally, (3.53, 205.24).

Notice that this interval for σ^2 is very wide, primarily because n is quite small.

We have previously indicated that the confidence intervals developed in Section 8.8 for μ and $\mu_1 - \mu_2$ had confidence coefficients near the nominal level even if the underlying populations were not normally distributed. In contrast, the intervals for σ^2 presented in this section can have confidence coefficients that differ markedly from the nominal level if the sampled population is not normally distributed.

Exercises

8.81   Recently, the EPA set a maximum noise level for heavy trucks at 83 decibels (dB). The manner in which this limit is applied will greatly affect the trucking industry and the public. One way to apply the limit is to require all trucks to conform to the noise limit. A second, but less satisfactory, method is to require the truck fleet's mean noise level to be less than the limit. If the latter rule is adopted, variation in the noise level from truck to truck becomes important, because a large value of σ^2 would imply that many trucks exceed the limit, even if the mean fleet level were 83 dB. A random sample of six heavy trucks produced the following noise levels (in decibels):

$$85.4 \qquad 86.8 \qquad 86.1 \qquad 85.3 \qquad 84.8 \qquad 86.0.$$

Use these data to construct a 90% confidence interval for σ^2, the variance of the truck noise emission readings. Interpret your results.

8.82   In Exercise 8.69 we gave the carapace lengths of ten mature *Thenus orientalis* lobsters caught in the seas in the vicinity of Singapore. For your convenience, the data are reproduced here. Suppose that you wished to describe the variability of the carapace lengths of this population of lobsters. Find a 90% confidence interval for the population variance σ^2.

Lobster field number	A061	A062	A066	A070	A067	A069	A064	A068	A065	A063
Carapace length (mm)	78	66	65	63	60	60	58	56	52	50

8.83   Suppose that S^2 is the sample variance based on a sample of size n from a normal population with unknown mean and variance.

    a Derive a $100(1 - \alpha)\%$ upper confidence bound for σ^2.

    b Derive a $100(1 - \alpha)\%$ lower confidence bound for σ^2.

8.84 Given a random sample of size n from a normal population with unknown mean and variance, we developed a confidence interval for the population variance σ^2 in Section 8.9. What is the formula for a confidence interval for the population standard deviation σ?

8.85 In Exercise 8.83, you derived upper and lower confidence bounds, each with confidence coefficient $1 - \alpha$, for σ^2.

    a How would you construct a $100(1 - \alpha)\%$ upper confidence bound for σ?

    b How would you construct a $100(1 - \alpha)\%$ lower confidence bound for σ?

8.86 Industrial lightbulbs should have a mean life length acceptable to potential users and a relatively small variation in life length. If some bulbs fail too early in their life, users become annoyed and are likely to switch to bulbs produced by a different manufacturer. Large variations above the mean reduce replacement sales; and in general, variation in life lengths disrupts the user's replacement schedules. A random sample of 20 bulbs produced by a particular manufacturer produced the following lengths of life (in hours):

2100	2302	1951	2067	2415	1883	2101	2146	2278	2019
1924	2183	2077	2392	2286	2501	1946	2161	2253	1827

Set up a 99% upper confidence bound for the *standard deviation* of the lengths of life for the bulbs produced by this manufacturer. Is the true population standard deviation less than 150 hours? Why or why not?

8.87 In laboratory work it is desirable to run careful checks on the variability of readings produced on standard samples. In a study of the amount of calcium in drinking water undertaken as part of a water quality assessment, the same standard sample was run through the laboratory six times at random intervals. The six readings, in parts per million, were 9.54, 9.61, 9.32, 9.48, 9.70, and 9.26. Estimate the population variance σ^2 for readings on this standard, using a 90% confidence interval.

8.88 The ages of a random sample of five university professors are 39, 54, 61, 72, and 59. Using this information, find a 99% confidence interval for the population standard deviation of the ages of all professors at the university, assuming that the ages of university professors are normally distributed.

8.89 A precision instrument is guaranteed to read accurately to within 2 units. A sample of four instrument readings on the same object yielded the measurements 353, 351, 351, and 355. Find a 90% confidence interval for the population variance. What assumptions are necessary? Does the guarantee seem reasonable?

8.10 Summary

The objective of many statistical investigations is to make inferences about population parameters based on sample data. Often these inferences take the form of estimates—either point estimates or interval estimates. We like to have unbiased estimators with small variance. The goodness of an unbiased estimator $\hat{\theta}$ can be measured by $\sigma_{\hat{\theta}}$ because the error of estimation is generally smaller than $2\sigma_{\hat{\theta}}$ with high probability. The mean square error of an estimator, $\mathrm{MSE}(\hat{\theta}) = V(\hat{\theta}) + (B(\hat{\theta}))^2$, is small only if the estimator has small variance and small bias.

Interval estimates of many parameters, such as μ and p, can be derived from the normal distribution for large sample sizes because of the central limit theorem. If sample sizes are small, the normality of the population must be assumed, and the t distribution is used in deriving confidence intervals. However, the interval for a single mean is quite robust in relation to moderate departures from normality. That is, the actual confidence coefficient associated with intervals that have a nominal confidence coefficient of $100(1 - \alpha)\%$ is very close to the nominal level even if the population distribution differs moderately from normality. The confidence interval for a difference in two means is also robust in relation to moderate departures from normality and to the assumption of equal population variances if $n_1 \approx n_2$. As n_1 and n_2 become more dissimilar, the assumption of equal population variances becomes more crucial.

If sample measurements have been selected from a normal distribution, a confidence interval for σ^2 can be developed through use of the χ^2 distribution. These intervals are very sensitive to the assumption that the underlying population is normally distributed. Consequently, the actual confidence coefficient associated with the interval estimation procedure can differ markedly from the nominal value if the underlying population is not normally distributed.

References and Further Readings

Casella, G., and R. L. Berger. 1990. *Statistical Inference*. Pacific Grove, Calif.: Brooks Cole.

Hoel, P. G. 1984. *Introduction to Mathematical Statistics*, 5th ed. New York: John Wiley.

Hogg, R. V., and A. T. Craig. 1995. *Introduction to Mathematical Statistics*, 5th ed. Upper Saddle River, N.J.: Prentice Hall.

Mood, A. M., F. A. Graybill, and D. Boes. 1974. *Introduction to the Theory of Statistics*, 3d ed. New York: McGraw-Hill.

Supplementary Exercises

8.90 In a controlled pollination study involving *Phlox drummondii*, a spring-flowering annual plant common along roadsides in sandy fields in central Texas, Karen Pittman and Donald Levin ["Effects of Parental Identities and Environment on Components of Crossing Success on *Phlox drummondii*," *American Journal of Botany* 76(3) (1989)] found that seed survival rates were not affected by water or nutrition deprivation. In the experiment, flowers on plants were identified as males when they donated pollen and as females when they were pollinated by donor pollen in three treatment groups: control, low-water, and low-nutrient. The data in the following table reflect one aspect of the findings of the experiment: the number of seeds surviving to maturity for each of the three groups for both male and female parents.

Treament	Male		Female	
	n	Number Surviving	n	Number Surviving
Control	585	543	632	560
Low-water	578	522	510	466
Low-nutrient	568	510	589	546

a   Find a 99% confidence interval for the difference between survival proportions in the low-water group versus the low-nutrient group for male parents.

b   Find a 99% confidence interval for the difference between survival proportions in male and female parents subjected to low water.

8.91   Refer to Exercise 8.90. Suppose that you plan to estimate the difference in the survival rates of seeds for male parents in low-water and low-nutrient environments to within .03 with probability .95. If you plan to use an equal number of seeds from male parents in each environment (that is, $n_1 = n_2$), how large should n_1 and n_2 be?

8.92   A chemist who has prepared a product designed to kill 60% of a particular type of insect wants to evaluate the kill rate of her preparation. What sample size should she use if she wishes to be 95% confident that her experimental results fall within .02 of the true fraction of insects killed?

8.93   To estimate the proportion of unemployed workers in Panama, an economist selected at random 400 persons from the working class. Of these, 25 were unemployed.

a   Estimate the true proportion of unemployed workers, and place bounds on the error of estimation.

b   How many persons must be sampled to reduce the bound on the error of estimation to .02?

8.94   Past experience shows that the standard deviation of the yearly income of textile workers in a certain state is $400. How many textile workers would you need to sample if you wished to estimate the population mean to within $50.00, with probability .95?

8.95   How many voters must be included in a sample collected to estimate the fraction of the popular vote favorable to a presidential candidate in a national election if the estimate must be correct to within .005? Assume that the true fraction lies somewhere in the neighborhood of .5. Use a confidence coefficient of approximately .95.

8.96   In a poll taken among college students, 300 of 500 fraternity men favored a certain proposition, whereas 64 of 100 nonfraternity men favored it. Estimate the difference in the fractions favoring the proposition, and place a two-standard-deviation bound on the error of estimation.

8.97   Refer to Exercise 8.96. How many fraternity and nonfraternity men must be included in a poll if we wish to obtain an estimate, correct to within .05, for the difference in the proportions favoring the proposition? Assume that the groups will be of equal size and that $p = .6$ will suffice as an approximation of both fractions.

8.98   A chemical process has produced, on the average, 800 tons of chemical per day. The daily yields for the past week are 785, 805, 790, 793, and 802 tons. Estimate the mean daily yield, with confidence coefficient .90, from the data. What assumptions did you make?

8.99   Refer to Exercise 8.98. Find a 90% confidence interval for σ^2, the variance of the daily yields.

8.100   Do we lose our memory capacity as we get older? In a study of the effect of glucose on memory in elderly men and women, C. A. Manning and colleagues [C. A. Manning, J. L. Hall, and P. E. Gold, "Glucose Effects on Memory and Other Neuropsychological Tests in Elderly Humans," *Psychological Science* 1(5) (1990)] tested 16 volunteers (5 men and 11 women) for long-term memory, recording the number of words recalled from a list read to each person. Each person was reminded of the words missed and was asked to recall as many words as possible from the original list. The mean and standard deviation of the long-term word memory scores were $\bar{y} = 79.47$ and $s = 25.25$. Give a 99% confidence interval for the true long-term word memory scores for elderly men and women. Interpret this interval.

8.101 The annual main stem growth, measured for a sample of 17 4-year-old red pine trees, produced a mean of 11.3 inches and a standard deviation of 3.4 inches. Find a 90% confidence interval for the mean annual main stem growth of a population of 4-year-old red pine trees subjected to similar environmental conditions. Assume that the growth amounts are normally distributed.

8.102 Owing to the variability of trade-in allowance, the profit per new car sold by an automobile dealer varies from car to car. The profits per sale (in hundreds of dollars), tabulated for the past week, were 2.1, 3.0, 1.2, 6.2, 4.5, and 5.1. Find a 90% confidence interval for the mean profit per sale. What assumptions must be valid for the technique you used to be appropriate?

8.103 A mathematics test is given to a class of 50 students randomly selected from high school 1 and also to a class of 45 students randomly selected from high school 2. For the class at high school 1, the sample mean is 75 points and the sample standard deviation is 10 points. For the class at high school 2, the sample mean is 72 points and the sample standard deviation is 8 points. Construct a 95% confidence interval for the difference in the mean scores. What assumptions are necessary?

8.104 Two methods for teaching reading were applied to two randomly selected groups of elementary schoolchildren and were compared on the basis of a reading comprehension test given at the end of the learning period. The sample means and variances computed from the test scores are shown in the accompanying table. Find a 95% confidence interval for $(\mu_1 - \mu_2)$. What assumptions are necessary?

Statistic	Method 1	Method 2
Number of children in group	11	14
$\bar{y}$	64	69
s^2	52	71

8.105 A comparison of reaction times for two different stimuli in a psychological word association experiment produced the results (in seconds) shown in the accompanying table when applied to a random sample of 16 people. Obtain a 90% confidence interval for $(\mu_1 - \mu_2)$. What assumptions are necessary?

Stimulus 1		Stimulus 2	
1	2	4	1
3	1	2	2
2	3	3	3
1	2	3	3

8.106 The length of time between billing and receipt of payment was recorded for a random sample of 100 of a CPA firm's clients. The sample mean and standard deviation for the 100 accounts were 39.1 days and 17.3 days, respectively. Find a 90% confidence interval for the mean time between billing and receipt of payment for all of the CPA firm's accounts. Interpret the interval.

8.107 Television advertisers may mistakenly believe that most viewers understand most of the advertising that they see and hear. A recent research study asked 2300 viewers above age 13 to look at 30-second television advertising excerpts. Of these, 1914 of the viewers misunderstood all or part of the excerpt they saw. Find a 95% confidence interval for the proportion of all viewers (of which the sample is representative) who will misunderstand all or part of the television excerpts used in this study.

8.108 A survey of 415 corporate, government, and accounting executives of the Financial Accounting Foundation found that 278 rated cash flow (as opposed to earnings per share, etc.) as the most important indicator of a company's financial health. Assume that these 415 executives constitute a random sample from the population of all executives. Use the data to find a 95% confidence interval for the fraction of all corporate executives who consider cash flow the most important measure of a company's financial health.

8.109 Suppose that independent samples of sizes n_1 and n_2 are taken from two normally distributed populations with variances σ_1^2 and σ_2^2, respectively. If S_1^2 and S_2^2 denote the respective sample variances, Theorem 7.3 implies that $(n_1 - 1)S_1^2/\sigma_1^2$ and $(n_2 - 1)S_2^2/\sigma_2^2$ have χ^2 distributions with $n_1 - 1$ and $n_2 - 1$ degrees of freedom, respectively. Further, these χ^2-distributed random variables are independent because the samples were independently taken.

  a Use these quantities to construct a random variable that has an F distribution with $n_1 - 1$ numerator degrees of freedom and $n_2 - 1$ denominator degrees of freedom.

  b Use the F-distributed quantity from (a) as a *pivotal quantity*, and derive a formula for a $100(1 - \alpha)\%$ confidence interval for σ_2^2/σ_1^2.

8.110 A pharmaceutical manufacturer purchases raw material from two different suppliers. The mean level of impurities is approximately the same for both suppliers, but the manufacturer is concerned about the variability in the amount of impurities from shipment to shipment. If the level of impurities tends to vary excessively for one source of supply, this could affect the quality of the final product. To compare the variation in percentage impurities for the two suppliers, the manufacturer selects ten shipments from each supplier and measures the percentage of impurities in each shipment. The sample variances were $s_1^2 = .273$ and $s_2^2 = .094$, respectively. Form a 95% confidence interval for the ratio of the true population variances.

*8.111 We noted in Section 8.3 that, if

$$S'^2 = \frac{\sum\limits_{i=1}^{n}(Y_i - \overline{Y})^2}{n} \quad \text{and} \quad S^2 = \frac{\sum\limits_{i=1}^{n}(Y_i - \overline{Y})^2}{n - 1}$$

then S'^2 is a biased estimator of σ^2 but S^2 is an unbiased estimator of the same parameter. If we sample from a normal population,

  a find $V(S'^2)$.

  b show that $V(S^2) > V(S'^2)$.

*8.112 Exercise 8.111 suggests that S^2 is superior to S'^2 in regard to bias and that S'^2 is superior to S^2 because it possesses smaller variance. Which is the better estimator? (*Hint:* Compare the mean square errors.)

*8.113 Suppose that two independent random samples of n_1 and n_2 observations are selected from normal populations. Further, assume that the populations possess a common variance σ^2. Let

$$S_i^2 = \frac{\sum\limits_{j=1}^{n_i}(Y_{ij} - \overline{Y}_i)^2}{n_i - 1}, \qquad i = 1, 2.$$

a Show that S_p^2, the pooled estimator of σ^2 (which follows), is unbiased:

$$S_p^2 = \frac{(n_1 - 1)S_1^2 + (n_2 - 1)S_2^2}{n_1 + n_2 - 2}.$$

b Find $V(S_p^2)$.

*8.114 The small-sample confidence interval for μ, based on Student's t (Section 8.8), possesses a random width—in contrast to the large-sample confidence interval (Section 8.6), where the width is not random. Find the expected value of the interval width in the small-sample case.

*8.115 A confidence interval is *unbiased* if the expected value of the interval midpoint is equal to the estimated parameter. The expected value of the midpoint of the large-sample confidence interval (Section 8.6) is equal to the estimated parameter, and the same is true for the small-sample confidence intervals for μ and $(\mu_1 - \mu_2)$ (Section 8.8). For example, the midpoint of the interval $\overline{y} \pm ts/\sqrt{n}$ is $\overline{y}$, and $E(\overline{Y}) = \mu$. Now consider the confidence interval for σ^2. Show that the expected value of the midpoint of this confidence interval is not equal to σ^2.

*8.116 The sample mean $\overline{Y}$ is a good point estimator of the population mean μ. It can also be used to predict a future value of Y independently selected from the population. Assume that you have a sample mean $\overline{Y}$ and variance S^2 based on a random sample of n measurements from a normal population. Use Student's t to form a pivotal quantity to find a prediction interval for some new value of Y, say, Y_p, to be observed in the future. (*Hint:* Start with the quantity $Y_p - \overline{Y}$.) Notice the terminology: Parameters are *estimated*; values of random variables are *predicted*.

CHAPTER 9

Properties of Point Estimators and Methods of Estimation

9.1 Introduction

9.2 Relative Efficiency

9.3 Consistency

9.4 Sufficiency

9.5 The Rao–Blackwell Theorem and Minimum-Variance Unbiased Estimation

9.6 The Method of Moments

9.7 The Method of Maximum Likelihood

9.8 Some Large-Sample Properties of MLEs (Optional)

9.9 Summary

References and Further Readings

9.1 Introduction

In Chapter 8 we presented some intuitive estimators for parameters often of interest in practical problems. An estimator $\hat{\theta}$ for a target parameter θ is a function of the random variables observed in a sample and, therefore, is itself a random variable. Consequently, an estimator has a probability distribution, which we call the *sampling distribution* of the estimator. We noted in Section 8.2 that, if $E(\hat{\theta}) = \theta$, then the estimator has the (sometimes) desirable property of being unbiased.

In this chapter we undertake a more formal and detailed examination of some of the mathematical properties of point estimators—particularly the notions of efficiency, consistency, and sufficiency. We present a result, the Rao–Blackwell theorem, that provides a link between sufficient statistics and unbiased estimators for parameters. Generally speaking, an unbiased estimator with small variance is, or can be made to be, a function of a sufficient statistic. We also demonstrate a method that can sometimes be used to find minimum-variance unbiased estimators for parameters of interest. We then offer two other useful methods for deriving estimators: the method of moments and the method of maximum likelihood. Some properties of estimators derived by these methods are discussed as well.

9.2 Relative Efficiency

It usually is possible to obtain more than one unbiased estimator for the same target parameter θ. In Section 8.2 (Figure 8.3), we mentioned that if $\hat{\theta}_1$ and $\hat{\theta}_2$ denote two unbiased estimators for the same parameter θ, we prefer to use the estimator with the smaller variance. That is, if both estimators are unbiased, $\hat{\theta}_1$ is *relatively more efficient* than $\hat{\theta}_2$ if $V(\hat{\theta}_2) > V(\hat{\theta}_1)$. In fact, we use the ratio $V(\hat{\theta}_2)/V(\hat{\theta}_1)$ to define the *relative efficiency* of two unbiased estimators.

DEFINITION 9.1

Given two unbiased estimators, $\hat{\theta}_1$ and $\hat{\theta}_2$, of a parameter θ, with variances $V(\hat{\theta}_1)$ and $V(\hat{\theta}_2)$, respectively, then the *efficiency* of $\hat{\theta}_1$ relative to $\hat{\theta}_2$, denoted eff $(\hat{\theta}_1, \hat{\theta}_2)$, is defined to be the ratio

$$\text{eff}(\hat{\theta}_1, \hat{\theta}_2) = \frac{V(\hat{\theta}_2)}{V(\hat{\theta}_1)}.$$

If $\hat{\theta}_1$ and $\hat{\theta}_2$ are unbiased estimators for θ, the efficiency of $\hat{\theta}_1$ relative to $\hat{\theta}_2$, eff $(\hat{\theta}_1, \hat{\theta}_2)$, is greater than 1 only if $V(\hat{\theta}_2) > V(\hat{\theta}_1)$. In this case $\hat{\theta}_1$ is a better unbiased estimator than $\hat{\theta}_2$. For example, if eff $(\hat{\theta}_1, \hat{\theta}_2) = 1.8$, then $V(\hat{\theta}_2) = (1.8)V(\hat{\theta}_1)$ and $\hat{\theta}_1$ is preferred to $\hat{\theta}_2$. Similarly, if eff $(\hat{\theta}_1, \hat{\theta}_2)$ is less than 1, say, .73, then $V(\hat{\theta}_2) = (.73)V(\hat{\theta}_1)$ and $\hat{\theta}_2$ is preferred to $\hat{\theta}_1$. Let us consider an example involving two different estimators for a population mean. Suppose that we wish to estimate the mean of a normal population. Let $\hat{\theta}_1$ be the sample *median*, the middle observation when the sample measurements are ordered according to magnitude (n odd) or the average of the two middle observations (n even). Let $\hat{\theta}_2$ be the sample mean. Although proof is omitted, it can be shown that the variance of the sample median, for large n, is $V(\hat{\theta}_1) = (1.2533)^2(\sigma^2/n)$. Then the efficiency of the sample median relative to the sample mean is

$$\text{eff}(\hat{\theta}_1, \hat{\theta}_2) = \frac{V(\hat{\theta}_2)}{V(\hat{\theta}_1)} = \frac{\sigma^2/n}{(1.2533)^2\sigma^2/n} = \frac{1}{(1.2533)^2} = .6366.$$

Thus we see that the amount of variability associated with the sample mean is approximately 63% of the variability associated with the sample median. Therefore, we would prefer to use the sample mean as the estimator for the population mean.

EXAMPLE 9.1   Let $Y_1, Y_2, \ldots, Y_n$ denote a random sample from the uniform distribution on the interval $(0, \theta)$. Two unbiased estimators for θ are

$$\hat{\theta}_1 = 2\bar{Y} \quad \text{and} \quad \hat{\theta}_2 = \left(\frac{n+1}{n}\right) Y_{(n)}$$

where $Y_{(n)} = \max(Y_1, Y_2, \ldots, Y_n)$. Find the efficiency of $\hat{\theta}_1$ relative to $\hat{\theta}_2$.

Solution   Because each Y_i has a uniform distribution on the interval $(0, \theta)$, $\mu = E(Y_i) = \theta/2$ and $\sigma^2 = V(Y_i) = \theta^2/12$. Therefore,

$$E(\hat{\theta}_1) = E(2\bar{Y}) = 2E(\bar{Y}) = 2(\mu) = 2\left(\frac{\theta}{2}\right) = \theta$$

and $\hat{\theta}_1$ is unbiased, as claimed. Further,

$$V(\hat{\theta}_1) = V(2\bar{Y}) = 4V(\bar{Y}) = 4\left[\frac{V(Y_i)}{n}\right] = \left(\frac{4}{n}\right)\left(\frac{\theta^2}{12}\right) = \frac{\theta^2}{3n}.$$

To find the mean and variance of $\hat{\theta}_2$, recall (see Section 6.7) that the density function of $Y_{(n)}$ is given by

$$g_{(n)}(y) = n[F_Y(y)]^{n-1} f_Y(y) = \begin{cases} n\left(\dfrac{y}{\theta}\right)^{n-1}\left(\dfrac{1}{\theta}\right), & 0 \le y \le \theta \\ 0, & \text{elsewhere} \end{cases}$$

because

$$P(Y_i \le y) = F_Y(y) = \begin{cases} 0, & y < 0 \\ \dfrac{y}{\theta}, & 0 \le y \le \theta \\ 1, & y > \theta. \end{cases}$$

Thus

$$E(Y_{(n)}) = \frac{n}{\theta^n} \int_0^\theta y^n \, dy = \left(\frac{n}{n+1}\right)\theta$$

and it follows that $E\{[(n+1)/n]Y_{(n)}\} = \theta$; that is, $\hat{\theta}_2$ is an unbiased estimator for θ.
Because

$$E(Y_{(n)}^2) = \frac{n}{\theta^n} \int_0^\theta y^{n+1} \, dy = \left(\frac{n}{n+2}\right)\theta^2,$$

we obtain

$$V(Y_{(n)}) = E(Y_{(n)}^2) - [E(Y_{(n)})]^2 = \left[\frac{n}{n+2} - \left(\frac{n}{n+1}\right)^2\right]\theta^2$$

and

$$V(\hat{\theta}_2) = V\left[\left(\frac{n+1}{n}\right)Y_{(n)}\right] = \left(\frac{n+1}{n}\right)^2 V(Y_{(n)})$$

$$= \left[\frac{(n+1)^2}{n(n+2)} - 1\right]\theta^2 = \frac{\theta^2}{n(n+2)}.$$

Therefore, the efficiency of $\hat{\theta}_1$ relative to $\hat{\theta}_2$ is given by

$$\text{eff}(\hat{\theta}_1, \hat{\theta}_2) = \frac{V(\hat{\theta}_2)}{V(\hat{\theta}_1)} = \frac{\theta^2/[n(n+2)]}{\theta^2/3n} = \frac{3}{n+2}.$$

This efficiency is less than one if $n > 1$. That is, if $n > 1$, $\hat{\theta}_2$ has a smaller variance than $\hat{\theta}_1$, and therefore $\hat{\theta}_2$ is generally preferable to $\hat{\theta}_1$ as an estimator of θ.

We will present some methods for finding estimators with small variances later in this chapter. For now we wish only to point out that relative efficiency is one important criterion for comparing estimators.

Exercises

9.1 In Exercise 8.4 we considered a random sample of size three from an exponential distribution with a density function given by

$$f(y) = \begin{cases} (1/\theta)e^{-y/\theta}, & 0 < y \\ 0, & \text{elsewhere} \end{cases}$$

and determined that $\hat{\theta}_1 = Y_1$, $\hat{\theta}_2 = (Y_1 + Y_2)/2$, $\hat{\theta}_3 = (Y_1 + 2Y_2)/3$, and $\hat{\theta}_5 = \overline{Y}$ are all unbiased estimators for θ. Find the efficiency of $\hat{\theta}_1$ relative to $\hat{\theta}_5$, of $\hat{\theta}_2$ relative to $\hat{\theta}_5$, and of $\hat{\theta}_3$ relative to $\hat{\theta}_5$.

9.2 Let $Y_1, Y_2, \ldots, Y_n$ denote a random sample from a population with mean μ and variance σ^2. Consider the following three estimators for μ:

$$\hat{\mu}_1 = \frac{1}{2}(Y_1 + Y_2), \qquad \hat{\mu}_2 = \frac{1}{4}Y_1 + \frac{Y_2 + \cdots + Y_{n-1}}{2(n-2)} + \frac{1}{4}Y_n, \qquad \hat{\mu}_3 = \overline{Y}.$$

a Show that each of the three estimators is unbiased.
b Find the efficiency of $\hat{\mu}_3$ relative to $\hat{\mu}_2$ and $\hat{\mu}_1$, respectively.

9.3 Let $Y_1, Y_2, \ldots, Y_n$ denote a random sample from the uniform distribution on the interval $(\theta, \theta + 1)$. Let

$$\hat{\theta}_1 = \overline{Y} - \frac{1}{2} \quad \text{and} \quad \hat{\theta}_2 = Y_{(n)} - \frac{n}{n+1}.$$

    a Show that both $\hat{\theta}_1$ and $\hat{\theta}_2$ are unbiased estimators of θ.

    b Find the efficiency of $\hat{\theta}_1$ relative to $\hat{\theta}_2$.

9.4 Let $Y_1, Y_2, \ldots, Y_n$ denote a random sample of size n from a uniform distribution on the interval $(0, \theta)$. If $Y_{(1)} = \min(Y_1, Y_2, \ldots, Y_n)$, the result of Exercise 8.14 is that $\hat{\theta}_1 = (n+1)Y_{(1)}$ is an unbiased estimator for θ. If $Y_{(n)} = \max(Y_1, Y_2, \ldots, Y_n)$, the results of Example 9.1 imply that $\hat{\theta}_2 = [(n+1)/n]Y_{(n)}$ is another unbiased estimator for θ. Show that the efficiency of $\hat{\theta}_1$ to $\hat{\theta}_2$ is $1/n^2$. Notice that this implies that $\hat{\theta}_2$ is a markedly superior estimator.

9.5 Suppose that $Y_1, Y_2, \ldots, Y_n$ is a random sample from a normal distribution with mean μ and variance σ^2. Two unbiased estimators of σ^2 are

$$\hat{\sigma}_1^2 = S^2 = \frac{1}{n-1} \sum_{i=1}^{n}(Y_i - \overline{Y})^2 \quad \text{and} \quad \hat{\sigma}_2^2 = \frac{1}{2}(Y_1 - Y_2)^2.$$

Find the efficiency of $\hat{\sigma}_1^2$ relative to $\hat{\sigma}_2^2$.

9.6 Suppose that $Y_1, Y_2, \ldots, Y_n$ denote a random sample of size n from a Poisson distribution with mean λ. Consider $\hat{\lambda}_1 = (Y_1 + Y_2)/2$ and $\hat{\lambda}_2 = \overline{Y}$. Derive the efficiency of $\hat{\lambda}_1$ relative to $\hat{\lambda}_2$.

9.7 Suppose that $Y_1, Y_2, \ldots, Y_n$ denote a random sample of size n from an exponential distribution with density function given by

$$f(y) = \begin{cases} (1/\theta)e^{-y/\theta}, & 0 < y \\ 0, & \text{elsewhere.} \end{cases}$$

In Exercise 8.15, we determined that $\hat{\theta}_1 = nY_{(1)}$ is an unbiased estimator of θ with $\text{MSE}(\hat{\theta}_1) = \theta^2$. Consider the estimator $\hat{\theta}_2 = \overline{Y}$, and find the efficiency of $\hat{\theta}_1$ relative to $\hat{\theta}_2$.

*9.8 Let $Y_1, Y_2, \ldots, Y_n$ denote a random sample from a probability density function $f(y)$, which has unknown parameter θ. If $\hat{\theta}$ is an unbiased estimator of θ, then under very general conditions

$$V(\hat{\theta}) \geq I(\theta) \quad \text{where} \quad I(\theta) = 1 \left/ nE\left[-\frac{\partial^2 \ln f(Y)}{\partial \theta^2} \right] \right. .$$

(This is known as the Cramer–Rao inequality.) If $V(\hat{\theta}) = I(\theta)$, the estimator $\hat{\theta}$ is said to be *efficient*.[1]

    a Suppose that $f(y)$ is the normal density with mean μ and variance σ^2. Show that $\overline{Y}$ is an efficient estimator of μ.

    b This inequality also holds for discrete probability functions $p(y)$. Suppose that $p(y)$ is the Poisson probability function with mean λ. Show that $\overline{Y}$ is an efficient estimator of λ.

9.3 Consistency

Suppose that a coin, which has probability p of resulting in heads, is tossed n times. If the tosses are independent, then Y, the number of heads among the n tosses, has a binomial distribution. If the true value of p is unknown, the sample proportion Y/n

---

[1] Exercises preceded by an asterisk are optional.

is an estimator of p. What happens to this sample proportion as the number of tosses, n, increases? Our intuition leads us to believe that as n gets larger, Y/n should get closer to the true value of p. That is, our estimator should get closer to the quantity being estimated as the amount of information in the sample increases.

Because Y/n is a random variable, we may express this "closeness" to p in probabilistic terms. In particular, let us examine the probability that the distance between the estimator and the target parameter, $|(Y/n) - p|$, will be less than some arbitrary positive real number, ε. If our intuition is correct and n is large, this probability,

$$P\left(\left|\frac{Y}{n} - p\right| \leq \varepsilon\right),$$

should be close to 1. If this probability, in fact, does tend to 1 as $n \to \infty$, we then say that (Y/n) is a *consistent estimator* of p, or that (Y/n) "converges in probability to p."

DEFINITION 9.2

The estimator $\hat{\theta}_n$ is said to be a *consistent estimator* of θ if, for any positive number ε,

$$\lim_{n \to \infty} P(|\hat{\theta}_n - \theta| \leq \varepsilon) = 1,$$

or, equivalently,

$$\lim_{n \to \infty} P(|\hat{\theta}_n - \theta| > \varepsilon) = 0.$$

The notation $\hat{\theta}_n$ expresses that the estimator for θ is calculated by using a sample of size n. For example, $\overline{Y}_2$ is the average of two observations, whereas $\overline{Y}_{100}$ is the average of the 100 observations contained in a sample of size $n = 100$. If $\hat{\theta}_n$ is an unbiased estimator, the following theorem can often be used to prove that the estimator is consistent.

THEOREM 9.1

An unbiased estimator $\hat{\theta}_n$ for θ is a consistent estimator of θ if

$$\lim_{n \to \infty} V(\hat{\theta}_n) = 0.$$

Proof

If Y is any random variable with $E(Y) = \mu$ and $V(Y) = \sigma^2 < \infty$, and if k is any nonnegative constant, Tchebysheff's theorem (see Theorem 4.13) implies that

$$P(|Y - \mu| > k\sigma) \leq \frac{1}{k^2}.$$

Because $\hat{\theta}_n$ is an unbiased estimator for θ, it follows that $E(\hat{\theta}_n) = \theta$. Let $\sigma_{\hat{\theta}_n} = \sqrt{V(\hat{\theta}_n)}$ denote the standard error of the estimator $\hat{\theta}_n$. If we apply Tchebysheff's theorem for the random variable $\hat{\theta}_n$, we obtain

$$P\left(\left|\hat{\theta}_n - \theta\right| > k\sigma_{\hat{\theta}_n}\right) \le \frac{1}{k^2}.$$

Let n be any fixed sample size. For any positive number ε,

$$k = \frac{\varepsilon}{\sigma_{\hat{\theta}_n}}$$

is a positive number. Application of Tchebysheff's theorem for this fixed n and this choice of k shows that

$$P\left(\left|\hat{\theta}_n - \theta\right| > \varepsilon\right) = P\left(\left|\hat{\theta}_n - \theta\right| > \left[\frac{\varepsilon}{\sigma_{\hat{\theta}_n}}\right]\sigma_{\hat{\theta}_n}\right) \le \frac{1}{\left(\varepsilon/\sigma_{\hat{\theta}_n}\right)^2} = \frac{V(\hat{\theta}_n)}{\varepsilon^2}$$

or

$$P\left(\left|\hat{\theta}_n - \theta\right| > \varepsilon\right) \le \frac{V(\hat{\theta}_n)}{\varepsilon^2}.$$

Now we need to take the limit as $n \to \infty$ of the preceding sequence of probabilities. Notice that, for any fixed n,

$$0 \le P\left(\left|\hat{\theta}_n - \theta\right| > \varepsilon\right) \le \frac{V(\hat{\theta}_n)}{\varepsilon^2}.$$

If $\lim_{n\to\infty} V(\hat{\theta}_n) = 0$, then

$$\lim_{n\to\infty}(0) \le \lim_{n\to\infty} P\left(\left|\hat{\theta}_n - \theta\right| > \varepsilon\right) \le \lim_{n\to\infty} \frac{V(\hat{\theta}_n)}{\varepsilon^2} = 0.$$

Thus $\hat{\theta}_n$ is a consistent estimator for θ.   ∎

The consistency property given in Definition 9.2 and discussed in Theorem 9.1 involves a particular type of convergence of $\hat{\theta}_n$ to θ. For this reason the statement "$\hat{\theta}_n$ is a consistent estimator for θ" is sometimes replaced by the equivalent statement "$\hat{\theta}_n$ converges in probability to θ."

EXAMPLE 9.2   Let $Y_1, Y_2, \ldots, Y_n$ denote a random sample from a distribution with mean μ and variance $\sigma^2 < \infty$. Show that $\overline{Y}_n = \frac{1}{n}\sum_{i=1}^{n} Y_i$ is a consistent estimator of μ. (*Note:* We use the notation $\overline{Y}_n$ to explicitly indicate that $\overline{Y}$ is calculated by using a sample of size n.)

Solution   We know from earlier chapters that $E(\overline{Y}_n) = \mu$ and $V(\overline{Y}_n) = \sigma^2/n$. Because $\overline{Y}_n$ is unbiased for μ and $V(\overline{Y}_n) \to 0$ as $n \to \infty$, Theorem 9.1 establishes that $\overline{Y}_n$ is a consistent estimator of μ. Equivalently, we may say that $\overline{Y}_n$ converges in probability to μ.

The fact that $\overline{Y}_n$ is consistent for μ, or converges in probability to μ, is sometimes referred to as the *law of large numbers*. It provides the theoretical justification for the averaging process employed by many experimenters to obtain precision in measurements. For example, an experimenter may take the average of the weights of many animals to obtain a more precise estimate of the average weight of animals of this species. The experimenter's feeling, a feeling confirmed by Theorem 9.1, is that the average of many independently selected weights should be quite close to the true mean weight with high probability.

In Section 8.3 we considered an intuitive estimator for $\mu_1 - \mu_2$, the difference in the means of two populations. The estimator discussed at that time was $\overline{Y}_1 - \overline{Y}_2$, the difference in the means of independent random samples selected from two populations. The results of Theorem 9.2 will be very useful in establishing the consistency of such estimators.

THEOREM 9.2   Suppose that $\hat{\theta}_n$ converges in probability to θ and that $\hat{\theta}'_n$ converges in probability to θ'.

    a $\hat{\theta}_n + \hat{\theta}'_n$ converges in probability to $\theta + \theta'$.

    b $\hat{\theta}_n \times \hat{\theta}'_n$ converges in probability to $\theta \times \theta'$.

    c $\hat{\theta}_n / \hat{\theta}'_n$ converges in probability to θ/θ', provided that $\theta' \neq 0$.

    d   If $g(\cdot)$ is a real-valued function that is continuous at θ, then $g(\hat{\theta}_n)$ converges in probability to $g(\theta)$.

The proof of Theorem 9.2 closely resembles the corresponding proof in the case where $\{a_n\}$ and $\{b_n\}$ are sequences of real numbers converging to real limits a and b, respectively. For example, if $a_n \to a$ and $b_n \to b$ then

$$a_n + b_n \to a + b.$$

EXAMPLE 9.3   Suppose that $Y_1, Y_2, \ldots, Y_n$ represent a random sample such that $E(Y_i) = \mu$, $E(Y_i^2) = \mu'_2$, and $E(Y_i^4) = \mu'_4$ are all finite. Show that

$$S_n^2 = \frac{1}{n-1} \sum_{i=1}^{n} (Y_i - \overline{Y}_n)^2$$

is a consistent estimator of $\sigma^2 = V(Y_i)$. (*Note:* We use subscript n on both S^2 and $\overline{Y}$ to explicitly convey their dependence on the value of the sample size, n.)

Solution   We have seen in earlier chapters that S^2, now written as S_n^2, is

$$S_n^2 = \frac{1}{n-1}\left(\sum_{i=1}^{n} Y_i^2 - n\overline{Y}_n^2\right) = \left(\frac{n}{n-1}\right)\left(\frac{1}{n}\sum_{i=1}^{n} Y_i^2 - \overline{Y}_n^2\right).$$

The statistic $(1/n)\sum_{i=1}^{n} Y_i^2$ is the average of n independent and identically distributed random variables, with $E(Y_i^2) = \mu_2'$ and $V(Y_i^2) = \mu_4' - (\mu_2')^2 < \infty$. By the law of large numbers (Example 9.2), we know that $(1/n)\sum_{i=1}^{n} Y_i^2$ converges in probability to μ_2'.

Example 9.2 also implies that $\overline{Y}_n$ converges in probability to μ. Because the function $g(x) = x^2$ is continuous for all finite values of x, Theorem 9.2(d) implies that $\overline{Y}_n^2$ converges in probability to μ^2. It then follows from Theorem 9.2(a) that

$$\frac{1}{n}\sum_{i=1}^{n} Y_i^2 - \overline{Y}_n^2$$

converges in probability to $\mu_2' - \mu^2 = \sigma^2$. Because $n/(n-1)$ is a sequence of constants converging to 1 as $n \to \infty$, we can conclude that S_n^2 converges in probability to σ^2. Equivalently, S_n^2, the sample variance, is a consistent estimator for σ^2, the population variance.

In Section 8.6 we considered large-sample confidence intervals for some parameters of practical interest. In particular, if $Y_1, Y_2, \ldots, Y_n$ is a random sample from any distribution with mean μ and variance σ^2, we established that

$$\overline{Y} \pm z_{\alpha/2}\left(\frac{\sigma}{\sqrt{n}}\right)$$

is a valid large-sample confidence interval with confidence coefficient approximately equal to $(1-\alpha)$. If σ^2 is known, this interval can and should be calculated. However, if σ^2 is not known but the sample size is large, we recommended substituting S for σ in the calculation, because this entails no significant loss of accuracy. The following theorem provides the theoretical justification for these claims.

THEOREM 9.3

Suppose that U_n has a distribution function that converges to a standard normal distribution function as $n \to \infty$. If W_n converges in probability to 1, then the distribution function of U_n/W_n converges to a standard normal distribution function.

This result follows from a general result known as *Slutsky's theorem* (Serfling 1980). The proof of this result is beyond the scope of this text. However, the usefulness of the result is illustrated in the following example.

EXAMPLE **9.4**   Suppose that $Y_1, Y_2, \ldots, Y_n$ is a random sample of size n from a distribution with $E(Y_i) = \mu$ and $V(Y_i) = \sigma^2$. Define S_n^2 as

$$S_n^2 = \frac{1}{n-1} \sum_{i=1}^{n} (Y_i - \overline{Y}_n)^2.$$

Show that the distribution function of

$$\sqrt{n}\left(\frac{\overline{Y}_n - \mu}{S_n}\right)$$

converges to a standard normal distribution function.

Solution   In Example 9.3 we showed that S_n^2 converges in probability to σ^2. Notice that $g(x) = +\sqrt{x/c}$ is a continuous function of x if both x and c are positive. Hence it follows from Theorem 9.2(d) that $S_n/\sigma = +\sqrt{S_n^2/\sigma^2}$ converges in probability to 1. We also know from the central limit theorem (Theorem 7.4) that the distribution function of

$$U_n = \sqrt{n}\left(\frac{\overline{Y}_n - \mu}{\sigma}\right)$$

converges to a standard normal distribution function. Therefore, Theorem 9.3 implies that the distribution function of

$$\sqrt{n}\left(\frac{\overline{Y}_n - \mu}{\sigma}\right) \bigg/ (S_n/\sigma) = \sqrt{n}\left(\frac{\overline{Y}_n - \mu}{S_n}\right)$$

converges to a standard normal distribution function.

The result of Example 9.4 tells us that, when n is large, $\sqrt{n}(\overline{Y}_n - \mu)/S_n$ has approximately a standard normal distribution *whatever* is the form of the distribution from which the sample is taken. If the sample is taken from a *normal distribution*, the results of Chapter 7 imply that $t = \sqrt{n}(\overline{Y}_n - \mu)/S_n$ has a t distribution with $n-1$ degrees of freedom. Combining this information, we see that, if a large sample is taken from a normal distribution, the distribution function of $t = \sqrt{n}(\overline{Y}_n - \mu)/S_n$ can be approximated by a standard normal distribution function. That is, as n gets large, and hence as the number of degrees of freedom gets large, the t distribution function converges to the standard normal distribution function.

If we obtain a large sample from any distribution, we know from Example 9.4 that $\sqrt{n}(\overline{Y}_n - \mu)/S_n$ has approximately a standard normal distribution. Therefore, it follows that

$$P\left[-z_{\alpha/2} \le \sqrt{n}\left(\frac{\overline{Y}_n - \mu}{S_n}\right) \le z_{\alpha/2}\right] \approx 1 - \alpha.$$

If we manipulate the inequalities in the probability statement to isolate μ in the middle, we obtain

$$P\left[\overline{Y}_n - z_{\alpha/2}\left(\frac{S_n}{\sqrt{n}}\right) \leq \mu \leq \overline{Y}_n + z_{\alpha/2}\left(\frac{S_n}{\sqrt{n}}\right)\right] \approx 1 - \alpha.$$

Thus, $\overline{Y}_n \pm z_{\alpha/2}(S_n/\sqrt{n})$ forms a valid large-sample confidence interval for μ, with confidence coefficient approximately equal to $1 - \alpha$. Similarly, Theorem 9.3 can be applied to show that

$$\hat{p}_n \pm z_{\alpha/2}\sqrt{\frac{\hat{p}_n\hat{q}_n}{n}}$$

is a valid large-sample confidence interval for p with confidence coefficient approximately equal to $1 - \alpha$.

In this section we have seen that the property of consistency tells us something about the distance between an estimator and the quantity being estimated. We have seen that, when the sample size is large, $\overline{Y}_n$ is close to μ and S_n^2 is close to σ^2, with high probability. We will see other examples of consistent estimators in the exercises and later in the chapter.

In this section, we have used the notation $\overline{Y}_n$, S_n^2, $\hat{p}_n$, and, in general, $\hat{\theta}_n$ to explicitly convey the dependence of the estimators on the sample size, n. We needed to do so because we were interested in computing

$$\lim_{n\to\infty} P(|\hat{\theta}_n - \theta| \leq \varepsilon).$$

If this limit is 1, then $\hat{\theta}_n$ is a "consistent" estimator for θ (more precisely, $\hat{\theta}_n$ a consistent *sequence of estimators* for θ). Unfortunately, this notation makes our estimators look overly complicated. Henceforth, we will revert to the notation $\hat{\theta}$ as our estimator for θ and not explicitly display the dependence of the estimator on n. The dependence of $\hat{\theta}$ on the sample size, n, is always implicit and should be used whenever the consistency of the estimator is considered.

Exercises

9.9 Refer to Exercise 9.3. Show that both $\hat{\theta}_1$ and $\hat{\theta}_2$ are consistent estimators for θ.

9.10 Refer to Exercise 9.5. Is $\hat{\sigma}_2^2$ a consistent estimator of σ^2?

9.11 Suppose that $X_1, X_2, \ldots, X_n$ and $Y_1, Y_2, \ldots, Y_n$ are independent random samples from populations with means μ_1 and μ_2 and variances σ_1^2 and σ_2^2, respectively. Show that $\overline{X} - \overline{Y}$ is a consistent estimator of $\mu_1 - \mu_2$.

9.12 In Exercise 9.11, suppose that the populations are normally distributed with $\sigma_1^2 = \sigma_2^2 = \sigma^2$. Show that

$$\frac{\sum_{i=1}^{n}(X_i - \overline{X})^2 + \sum_{i=1}^{n}(Y_i - \overline{Y})^2}{2n - 2}$$

is a consistent estimator of σ^2.

9.13   Let $Y_1, Y_2, \ldots, Y_n$ denote a random sample from the probability density function

$$f(y) = \begin{cases} \theta y^{\theta-1}, & 0 < y < 1 \\ 0, & \text{elsewhere} \end{cases}$$

where $\theta > 0$. Show that $\overline{Y}$ is a consistent estimator of $\theta/(\theta + 1)$.

9.14   If Y has a binomial distribution with n trials and success probability p, show that Y/n is a consistent estimator of p.

9.15   Let $Y_1, Y_2, \ldots, Y_n$ be a random sample of size n from a normal population with mean μ and variance σ^2. Assuming $n = 2k$ for some integer k, one possible estimator for σ^2 is given by:

$$\hat{\sigma}^2 = \frac{1}{2k} \sum_{i=1}^{k} (Y_{2i} - Y_{2i-1})^2.$$

  a   Show that $\hat{\sigma}^2$ an unbiased estimator for σ^2.
  b   Show that $\hat{\sigma}^2$ is a consistent estimator for σ^2.

9.16   Refer to Exercise 9.15. Suppose that $Y_1, Y_2, \ldots, Y_n$ is a random sample of size n from a Poisson-distributed population with mean λ. Again, assume that $n = 2k$ for some integer k. Consider

$$\hat{\lambda} = \frac{1}{2k} \sum_{i=1}^{k} (Y_{2i} - Y_{2i-1})^2.$$

  a   Show that $\hat{\lambda}$ an unbiased estimator for λ.
  b   Show that $\hat{\lambda}$ is a consistent estimator for λ.

9.17   Refer to Exercise 9.15. Suppose that $Y_1, Y_2, \ldots, Y_n$ is a random sample of size n from a population for which the first four moments are finite. That is, $m_1' = E(Y_1) < \infty$, $m_2' = E(Y_1^2) < \infty$, $m_3' = E(Y_1^3) < \infty$, and $m_4' = E(Y_1^4) < \infty$. (*Note:* This assumption is valid for the normal and Poisson distributions in Exercise 9.15 and 9.16, respectively.) Again, assume that $n = 2k$ for some integer k. Consider

$$\hat{\sigma}^2 = \frac{1}{2k} \sum_{i=1}^{k} (Y_{2i} - Y_{2i-1})^2.$$

  a   Show that $\hat{\sigma}^2$ an unbiased estimator for σ^2.
  b   Show that $\hat{\sigma}^2$ is a consistent estimator for σ^2.
  c   Why did you need the assumption that $m_4' = E(Y_1^4) < \infty$?

9.18   Let $Y_1, Y_2, Y_3, \ldots$ be independent standard normal random variables.

  a   What is the distribution of $\sum_{i=1}^{n} Y_i^2$?

  b   Let $W_n = \frac{1}{n} \sum_{i=1}^{n} Y_i^2$. Does W_n *converge in probability* to some constant? If so, what is the value of the constant?

9.19   Suppose that $Y_1, Y_2, \ldots, Y_n$ denote a random sample of size n from a normal distribution with mean μ and variance 1. Consider the first observation Y_1 as an estimator for μ.

a Show that Y_1 is an unbiased estimator for μ.

b Find $P(|Y_1 - \mu| \le 1)$.

c Look at the basic definition of consistency given in Definition 9.2. Based on the result of (b), is Y_1 a consistent estimator for μ?

*9.20 It is sometimes relatively easy to establish consistency or lack of consistency by appealing directly to Definition 9.2, evaluating $P(|\hat{\theta}_n - \theta| \le \varepsilon)$ directly, and then showing that $\lim_{n \to \infty} P(|\hat{\theta}_n - \theta| \le \varepsilon) = 1$. Let $Y_1, Y_2, \ldots, Y_n$ denote a random sample of size n from a uniform distribution on the interval $(0, \theta)$. If $Y_{(n)} = \max(Y_1, Y_2, \ldots, Y_n)$, we showed in Exercise 6.59 that the probability distribution function of $Y_{(n)}$ is given by

$$F_{(n)}(y) = \begin{cases} 0, & y < 0 \\ (y/\theta)^n, & d0 \le y \le \theta \\ 1, & y > \theta. \end{cases}$$

a For each $n \ge 1$ and every $\varepsilon > 0$, it follows that $P(|Y_{(n)} - \theta| \le \varepsilon) = P(\theta - \varepsilon \le Y_{(n)} \le \theta + \varepsilon)$. If $\varepsilon > 0$, verify that $P(\theta - \varepsilon \le Y_{(n)} \le \theta + \varepsilon) = 1$ and that, for every positive $\varepsilon < \theta$, we obtain $P(\theta - \varepsilon \le Y_{(n)} \le \theta + \varepsilon) = 1 - [(\theta - \varepsilon)/\theta]^n$.

b Using the result from (a), show that $Y_{(n)}$ is a consistent estimator for θ by showing that, for every $\varepsilon > 0$, we have $\lim_{n \to \infty} P(|Y_{(n)} - \theta| \le \varepsilon) = 1$.

*9.21 Use the method described in Exercise 9.20 to show that, if $Y_{(1)} = \min(Y_1, Y_2, \ldots, Y_n)$ when $Y_1, Y_2, \ldots, Y_n$ are independent uniform random variables on the interval $(0, \theta)$, then $Y_{(1)}$ is *not* a consistent estimator for θ. (*Hint:* Based on the methods of Section 6.7, $Y_{(1)}$ has the distribution function

$$F_{(1)}(y) = \begin{cases} 0, & y < 0 \\ 1 - (1 - y/\theta)^n, & 0 \le y \le \theta \\ 1, & y > \theta.) \end{cases}$$

*9.22 Let $Y_1, Y_2, \ldots, Y_n$ denote a random sample of size n from a Pareto distribution (see Exercise 6.14). Then the methods of Section 6.7 imply that $Y_{(1)} = \min(Y_1, Y_2, \ldots, Y_n)$ has the distribution function given by

$$F_{(1)}(y) = \begin{cases} 0, & y \le \beta \\ 1 - (\beta/y)^{\alpha n}, & y > \beta. \end{cases}$$

Use the method described in Exercise 9.20 to show that $Y_{(1)}$ is a consistent estimator of β.

*9.23 Let $Y_1, Y_2, \ldots, Y_n$ denote a random sample of size n from a power family distribution (see Exercise 6.13). Then the methods of Section 6.7 imply that $Y_{(n)} = \max(Y_1, Y_2, \ldots, Y_n)$ has the distribution function given by

$$F_{(n)}(y) = \begin{cases} 0, & y < 0 \\ (y/\theta)^{\alpha n}, & 0 \le y \le \theta \\ 1, & y > \theta. \end{cases}$$

Use the method described in Exercise 9.20 to show that $Y_{(n)}$ is a consistent estimator of θ.

9.24 Let $Y_1, Y_2, \ldots, Y_n$ be independent random variables, each with probability density function

$$f(y) = \begin{cases} 3y^2, & 0 \le y \le 1 \\ 0, & \text{elsewhere.} \end{cases}$$

Show that $\overline{Y}$ converges in probability to some constant, and find the constant.

9.25   If $Y_1, Y_2, \ldots, Y_n$ denote a random sample from a gamma distribution with parameters α and β, show that $\overline{Y}$ converges in probability to some constant and find the constant.

9.26   Let $Y_1, Y_2, \ldots, Y_n$ denote a random sample from the probability density function

$$f(y) = \begin{cases} \dfrac{2}{y^2}, & y \geq 2 \\ 0, & \text{elsewhere.} \end{cases}$$

Does the law of large numbers apply to $\overline{Y}$ in this case? Why or why not?

9.27   An experimenter wishes to compare the numbers of bacteria of types A and B in samples of water. A total of n independent water samples are taken, and counts are made for each sample. Let X_i denote the number of type A bacteria and Y_i denote the number of type B bacteria for sample i. Assume that the two bacteria types are sparsely distributed within a water sample, so that $X_1, X_2, \ldots, X_n$ and $Y_1, Y_2, \ldots, Y_n$ can be considered independent random samples from Poisson distributions with means λ_1 and λ_2, respectively. Suggest an estimator of $\lambda_1/(\lambda_1 + \lambda_2)$. What properties does your estimator have?

9.28   Suppose that Y has a binomial distribution based on n trials and success probability p. Then $\hat{p}_n = Y/n$ is an unbiased estimator of p. Use Theorem 9.3 to prove that the distribution of $(\hat{p}_n - p)/\sqrt{\hat{p}_n \hat{q}_n/n}$ converges to a standard normal distribution. (*Hint:* Write Y as we did in Section 7.5.)

9.4   Sufficiency

Up to this point we have chosen estimators on the basis of intuition. Thus, we chose $\overline{Y}$ and S^2 as the estimators of the mean and variance, respectively, of the normal distribution. (It *seems* as though these estimators should be good estimators of the population parameters.) We have seen that it is sometimes desirable to use estimators that are unbiased. Indeed, $\overline{Y}$ and S^2 have been shown to be unbiased estimators of the population mean μ and variance σ^2, respectively. Notice that we have used the information in a sample of size n to calculate the value of two statistics that function as estimators for the parameters of interest. At this stage the actual sample values are no longer important; rather, we summarize the information in the sample that relates to the parameters of interest by using the statistics $\overline{Y}$ and S^2. Has this process of summarizing or reducing the data to the two statistics, $\overline{Y}$ and S^2, retained all the information about μ and σ^2 in the original set of n sample observations? Or has some information about these parameters been lost or obscured through the process of reducing the data? In this section we will present methods for finding statistics that, in a sense, summarize *all* the information in a sample about a target parameter. Such statistics are said to have the property of *sufficiency*; or more simply, they are called *sufficient statistics*. As we will see in the next section, "good" estimators are (or can be made to be) functions of any sufficient statistic. Indeed, sufficient statistics often can be used to develop estimators that have the minimum variance among all unbiased estimators.

To illustrate the notion of a sufficient statistic, let us consider the outcomes of n trials of a binomial experiment, $X_1, X_2, \ldots, X_n$, where

$$X_i = \begin{cases} 1, & \text{if the } i\text{th trial is a success} \\ 0, & \text{if the } i\text{th trial is a failure.} \end{cases}$$

If p is the probability of success on any trial then, for $i = 1, 2, \ldots, n$,

$$X_i = \begin{cases} 1, & \text{with probability } p \\ 0, & \text{with probability } q = 1 - p. \end{cases}$$

Suppose that we are given a value of $Y = \sum_{i=1}^{n} X_i$, the number of successes among the n trials. If we know the value of Y, can we gain any further information about p by looking at other functions of $X_1, X_2, \ldots, X_n$? One way to answer this question is to look at the conditional distribution of $X_1, X_2, \ldots, X_n$, given Y:

$$P(X_1 = x_1, \ldots, X_n = x_n | Y = y) = \frac{P(X_1 = x_1, \ldots, X_n = x_n, Y = y)}{P(Y = y)}.$$

The numerator on the right side of this expression is 0 if $\sum_{i=1}^{n} x_i \neq y$, and it is the probability of an independent sequence of 0s and 1s with a total of y 1s and $(n - y)$ 0s if $\sum_{i=1}^{n} x_i = y$. Also, the denominator is the binomial probability of exactly y successes in n trials. Therefore, if $y = 0, 1, 2, \ldots, n$,

$$P(X_1 = x_1, \ldots, X_n = x_n | Y = y) = \begin{cases} \dfrac{p^y(1-p)^{n-y}}{\binom{n}{y}p^y(1-p)^{n-y}} = \dfrac{1}{\binom{n}{y}}, & \text{if } \sum_{i=1}^{n} x_i = y \\ 0, & \text{otherwise.} \end{cases}$$

It is important to note that the conditional distribution of $X_1, X_2, \ldots, X_n$ given Y *does not* depend upon p. That is, once Y is known, no other function of $X_1, X_2, \ldots, X_n$ will shed additional light on the possible value of p. In this sense Y contains all the information about p. Therefore, the statistic Y is said to be *sufficient* for p. We generalize this idea in the following definition.

DEFINITION 9.3

Let $Y_1, Y_2, \ldots, Y_n$ denote a random sample from a probability distribution with unknown parameter θ. Then the statistic $U = g(Y_1, Y_2, \ldots, Y_n)$ is said to be *sufficient* for θ if the conditional distribution of $Y_1, Y_2, \ldots, Y_n$ given U does not depend on θ.

In many previous discussions we have considered the probability function $p(y)$ associated with a discrete random variable or the density function $f(y)$ for a continuous random variable to be functions of the argument y only. Our future discussions will be simplified if we adopt notation that will permit us to explicitly display the fact that the distribution associated with a random variable Y often depends on the value of a parameter θ. If Y is a discrete random variable that has a probability mass function that depends on the value of a parameter θ, instead of $p(y)$ we use the notation

$$p(y \mid \theta) = P(Y = y) \qquad \text{where } \theta \text{ denotes the value of the parameter.}$$

Similarly, we will indicate the explicit dependence of the form of a continuous density function on the value of a parameter θ by writing the density function as $f(y \mid \theta)$ instead of the previously used $f(y)$.

Definition 9.3 tells us how to check whether a statistic is sufficient, but it really does not tell us how to find a sufficient statistic. Recall that in the discrete case the joint distribution of discrete random variables $Y_1, Y_2, \ldots, Y_n$ is given by a probability function $p(y_1, y_2, \ldots, y_n)$. If this joint probability function depends explicitly on the value of a parameter θ, we write it as $p(y_1, y_2, \ldots, y_n \mid \theta)$. This function gives the probability or *likelihood* of observing the event $(Y_1 = y_1, Y_2 = y_2, \ldots, Y_n = y_n)$ when the value of the parameter is θ. In the continuous case when the joint distribution of $Y_1, Y_2, \ldots, Y_n$ depends on a parameter θ, we will write the joint density function as $f(y_1, y_2, \ldots, y_n \mid \theta)$. Henceforth, it will be convenient to have a single name for the function that defines the joint distribution of the variables $Y_1, Y_2, \ldots, Y_n$ observed in a sample.

DEFINITION **9.4**   Let $y_1, y_2, \ldots, y_n$ be sample observations taken on corresponding random variables $Y_1, Y_2, \ldots, Y_n$ whose distribution depends on a parameter θ. Then, if $Y_1, Y_2, \ldots, Y_n$ are discrete random variables, the *likelihood of the sample*, $L(y_1, y_2, \ldots, y_n \mid \theta)$, is defined to be the joint probability of $y_1, y_2, \ldots, y_n$. If $Y_1, Y_2, \ldots, Y_n$ are continuous random variables, the likelihood $L(y_1, y_2, \ldots, y_n \mid \theta)$ is defined to be the joint density evaluated at $y_1, y_2, \ldots, y_n$.

If the set of random variables $Y_1, Y_2, \ldots, Y_n$ denotes a random sample from a discrete distribution with probability function $p(y \mid \theta)$, then

$$L(y_1, y_2, \ldots, y_n \mid \theta) = P(Y_1 = y_1, Y_2 = y_2, \ldots, Y_n = y_n \mid \theta)$$
$$= p(y_1 \mid \theta) \times p(y_2 \mid \theta) \times \cdots \times p(y_n \mid \theta),$$

whereas if $Y_1, Y_2, \ldots, Y_n$ have a continuous distribution with density function $f(y \mid \theta)$, then

$$L(y_1, y_2, \ldots, y_n \mid \theta) = f(y_1, y_2, \ldots, y_n \mid \theta)$$
$$= f(y_1 \mid \theta) \times f(y_2 \mid \theta) \times \cdots \times f(y_n \mid \theta).$$

To simplify notation, we will sometimes denote the likelihood by $L(\theta)$ instead of by $L(y_1, y_2, \ldots, y_n \mid \theta)$.

The following theorem relates the property of sufficiency to the likelihood $L(\theta)$.

THEOREM **9.4**   Let U be a statistic based on the random sample $Y_1, Y_2, \ldots, Y_n$. Then U is a *sufficient statistic* for the estimation of a parameter θ if and only if the likelihood $L(\theta) = L(y_1, y_2, \ldots, y_n \mid \theta)$ can be factored into two nonnegative functions,

$$L(y_1, y_2, \ldots, y_n \mid \theta) = g(u, \theta) \times h(y_1, y_2, \ldots, y_n)$$

where $g(u, \theta)$ is a function only of u and θ and $h(y_1, y_2, \ldots, y_n)$ is not a function of θ.

Although the proof of Theorem 9.4 (also known as the *factorization criterion*) is beyond the scope of this book, we illustrate the usefulness of the theorem in the following example.

EXAMPLE 9.5   Let $Y_1, Y_2, \ldots, Y_n$ be a random sample in which Y_i possesses the probability density function

$$f(y) = \begin{cases} (1/\theta)e^{-y_i/\theta}, & 0 \le y < \infty \\ 0, & \text{elsewhere} \end{cases}$$

where $\theta > 0$, $i = 1, 2, \ldots, n$. Show that $\overline{Y}$ is a sufficient statistic for the estimation of θ.

Solution   The likelihood $L(\theta)$ of the sample is the joint density

$$
\begin{aligned}
L(y_1, y_2, \ldots, y_n \mid \theta) &= f(y_1, y_2, \ldots, y_n \mid \theta) \\
&= f(y_1 \mid \theta) \times f(y_2 \mid \theta) \times \cdots \times f(y_n \mid \theta) \\
&= \frac{e^{-y_1/\theta}}{\theta} \times \frac{e^{-y_2/\theta}}{\theta} \times \cdots \times \frac{e^{-y_n/\theta}}{\theta} = \frac{e^{-\sum y_i/\theta}}{\theta^n} = \frac{e^{-n\overline{y}/\theta}}{\theta^n}.
\end{aligned}
$$

Notice that $L(\theta)$ is a function only of θ and $\overline{y}$ and that if

$$g(\overline{y}, \theta) = \frac{e^{-n\overline{y}/\theta}}{\theta^n} \quad \text{and} \quad h(y_1, y_2, \ldots, y_n) = 1$$

then

$$L(y_1, y_2, \ldots, y_n \mid \theta) = g(\overline{y}, \theta) \times h(y_1, y_2, \ldots, y_n).$$

Hence Theorem 9.4 implies that $\overline{Y}$ is a sufficient statistic for the estimation of θ.

Theorem 9.4 can be used to show that there are many possible sufficient statistics for any one population parameter. First of all, according to Definition 9.3 or the factorization criterion (Theorem 9.4), the random sample itself is a sufficient statistic. Second, if $Y_1, Y_2, \ldots, Y_n$ denote a random sample from a distribution with a density function with parameter θ, then the set of order statistics $Y_{(1)} \le Y_{(2)} \le \cdots \le Y_{(n)}$, which is a function of $Y_1, Y_2, \ldots, Y_n$, is sufficient for θ. In Example 9.5, we decided that $\overline{Y}$ is a sufficient statistic for the estimation of θ. Theorem 9.4 could also have been used to show that $\sum_{i=1}^{n} Y_i$ is another sufficient statistic. Indeed, for the exponential distribution described in Example 9.5, any statistic that is a one–to–one function of $\overline{Y}$ is a sufficient statistic.

In our initial example of this section, involving the number of successes in n trials, $Y = \sum_{i=1}^{n} X_i$ reduces the data $X_1, X_2, \ldots, X_n$ to a single value that remains sufficient for p. Generally we would like to find a sufficient statistic that reduces the data in the sample as much as possible. Although many statistics are sufficient for the parameter θ associated with a specific distribution, application of the factorization criterion typically leads to a statistic that provides the "best" summary of the information in the data. In Example 9.5, this statistic is $\overline{Y}$ (or some one–to–one function of it). In the next section we show how these sufficient statistics can be used to develop unbiased estimators with minimum variance.

Exercises

9.29 Let $X_1, X_2, \ldots, X_n$ denote n independent and identically distributed random variables such that

$$P(X_i = 1) = p \quad \text{and} \quad P(X_i = 0) = 1 - p$$

for each $i = 1, 2, \ldots, n$. Show that $\sum_{i=1}^{n} X_i$ is sufficient for p by using the factorization criterion given in Theorem 9.4.

9.30 Let $Y_1, Y_2, \ldots, Y_n$ denote a random sample from a normal distribution with mean μ and variance σ^2.

    a If μ is unknown and σ^2 is known, show that $\overline{Y}$ is sufficient for μ.

    b If μ is known and σ^2 is unknown, show that $\sum_{i=1}^{n}(Y_i - \mu)^2$ is sufficient for σ^2.

    c If μ and σ^2 are both unknown, show that $\sum_{i=1}^{n} Y_i$ and $\sum_{i=1}^{n} Y_i^2$ are jointly sufficient for μ and σ^2. (Thus, it follows that $\overline{Y}$ and $\sum_{i=1}^{n}(Y_i - \overline{Y})^2$ or $\overline{Y}$ and S^2 are also jointly sufficient for μ and σ^2.)

9.31 Let $Y_1, Y_2, \ldots, Y_n$ denote a random sample from a Poisson distribution with parameter λ. Show by conditioning that $\sum_{i=1}^{n} Y_i$ is sufficient for λ.

9.32 Let $Y_1, Y_2, \ldots, Y_n$ denote a random sample from a Rayleigh distribution with parameter θ. (Refer to Exercise 6.30.) Show that $\sum_{i=1}^{n} Y_i^2$ is sufficient for θ.

9.33 Let $Y_1, Y_2, \ldots, Y_n$ denote a random sample from a Weibull distribution with known m and unknown α. (Refer to Exercise 6.22.) Show that $\sum_{i=1}^{n} Y_i^m$ is sufficient for α.

9.34 If $Y_1, Y_2, \ldots, Y_n$ denote a random sample from a geometric distribution with parameter p, show that $\overline{Y}$ is sufficient for p.

9.35 Let $Y_1, Y_2, \ldots, Y_n$ denote independent and identically distributed random variables from a power family distribution with parameters α and θ. Then, by the result in Exercise 6.13, if $\alpha, \theta > 0$,

$$f(y \mid \alpha, \theta) = \begin{cases} \alpha y^{\alpha-1}/\theta^\alpha, & 0 \le y \le \theta \\ 0, & \text{elsewhere.} \end{cases}$$

If θ is known, show that $\prod_{i=1}^{n} Y_i$ is sufficient for α.

9.36 Let $Y_1, Y_2, \ldots, Y_n$ denote independent and identically distributed random variables from a Pareto distribution with parameters α and β. Then, by the result in Exercise 6.14, if α, $\beta > 0$,

$$f(y \mid \alpha, \beta) = \begin{cases} \alpha \beta^\alpha y^{-(\alpha+1)}, & y \geq \beta \\ 0, & \text{elsewhere.} \end{cases}$$

If β is known, show that $\prod_{i=1}^n Y_i$ is sufficient for α.

9.37 Suppose that $Y_1, Y_2, \ldots, Y_n$ is a random sample from a probability density function in the (one-parameter) exponential family so that

$$f(y \mid \theta) = \begin{cases} a(\theta)b(y)e^{-[c(\theta)d(y)]}, & a \leq y \leq b \\ 0, & \text{elsewhere} \end{cases}$$

where a and b do not depend on θ. Show that $\sum_{i=1}^n d(Y_i)$ is sufficient for θ.

9.38 If $Y_1, Y_2, \ldots, Y_n$ denotes a random sample from an exponential distribution with mean β, show that $f(y \mid \beta)$ is in the exponential family and that $\overline{Y}$ is sufficient for β.

9.39 Refer to Exercise 9.35. If θ is known, show that the power family of distributions is in the exponential family. What is a sufficient statistic for α? Does this contradict your answer to Exercise 9.35?

9.40 Refer to Exercise 9.36. If β is known, show that the Pareto distribution is in the exponential family. What is a sufficient statistic for α? Argue that there is no contradiction between your answer to this exercise and the answer you found in Exercise 9.36.

*9.41 Let $Y_1, Y_2, \ldots, Y_n$ denote a random sample from the uniform distribution over the interval $(0, \theta)$. Show that $Y_{(n)} = \max(Y_1, Y_2, \ldots, Y_n)$ is sufficient for θ.

*9.42 Let $Y_1, Y_2, \ldots, Y_n$ denote a random sample from the uniform distribution over the interval (θ_1, θ_2). Show that $Y_{(1)} = \min(Y_1, Y_2, \ldots, Y_n)$ and $Y_{(n)} = \max(Y_1, Y_2, \ldots, Y_n)$ are jointly sufficient for θ_1 and θ_2.

*9.43 Let $Y_1, Y_2, \ldots, Y_n$ denote a random sample from the probability density function

$$f(y \mid \theta) = \begin{cases} e^{-(y-\theta)}, & y \geq \theta \\ 0, & \text{elsewhere.} \end{cases}$$

Show that $Y_{(1)} = \min(Y_1, Y_2, \ldots, Y_n)$ is sufficient for θ.

*9.44 Let $Y_1, Y_2, \ldots, Y_n$ be a random sample from a population with density function

$$f(y \mid \theta) = \begin{cases} \dfrac{3y^2}{\theta^3}, & 0 \leq y \leq \theta \\ 0, & \text{elsewhere.} \end{cases}$$

Show that $Y_{(n)} = \max(Y_1, Y_2, \ldots, Y_n)$ is sufficient for θ.

*9.45 Let $Y_1, Y_2, \ldots, Y_n$ be a random sample from a population with density function

$$f(y \mid \theta) = \begin{cases} \dfrac{2\theta^2}{y^3}, & \theta < y < \infty \\ 0, & \text{elsewhere.} \end{cases}$$

Show that $Y_{(1)} = \min(Y_1, Y_2, \ldots, Y_n)$ is sufficient for θ.

*9.46 Let $Y_1, Y_2, \ldots, Y_n$ denote independent and identically distributed random variables from a power family distribution with parameters α and θ. Then, as in Exercise 9.35, if $\alpha, \theta > 0$,

$$f(y \mid \alpha, \theta) = \begin{cases} \alpha y^{\alpha-1}/\theta^{\alpha}, & 0 \le y \le \theta \\ 0, & \text{elsewhere.} \end{cases}$$

Show that $\max(Y_1, Y_2, \ldots, Y_n)$ and $\prod_{i=1}^{n} Y_i$ are jointly sufficient for α and θ.

*9.47 Let $Y_1, Y_2, \ldots, Y_n$ denote independent and identically distributed random variables from a Pareto distribution with parameters α and β. Then, as in Exercise 9.36, if $\alpha, \beta > 0$,

$$f(y \mid \alpha, \beta) = \begin{cases} \alpha \beta^{\alpha} y^{-(\alpha+1)}, & y \ge \beta \\ 0, & \text{elsewhere.} \end{cases}$$

Show that $\prod_{i=1}^{n} Y_i$ and $\min(Y_1, Y_2, \ldots, Y_n)$ are jointly sufficient for α and β.

9.5 The Rao–Blackwell Theorem and Minimum-Variance Unbiased Estimation

Sufficient statistics play an important role in finding good estimators for parameters. If $\hat{\theta}$ is an unbiased estimator for θ and if U is a statistic that is sufficient for θ, then there is a function of U that is also an unbiased estimator for θ and has *no larger* variance than $\hat{\theta}$. If we seek unbiased estimators with small variances, we can restrict our search to estimators that are functions of sufficient statistics. The theoretical basis for the preceding remarks is provided in the following result, known as the *Rao–Blackwell theorem*.

THEOREM 9.5

The Rao–Blackwell Theorem

Let $\hat{\theta}$ be an unbiased estimator for θ such that $V(\hat{\theta}) < \infty$. If U is a sufficient statistic for θ, define $\hat{\theta}^* = E(\hat{\theta} \mid U)$. Then, for all θ,

$$E(\hat{\theta}^*) = \theta \quad \text{and} \quad V(\hat{\theta}^*) \le V(\hat{\theta}).$$

Proof

Because U is sufficient for θ, the conditional distribution of any statistic (including $\hat{\theta}$), given U, does not depend on θ. Thus, $\hat{\theta}^* = E(\hat{\theta} \mid U)$ is not a function of θ and is therefore a statistic.

Recall Theorems 5.14 and 5.15, where we considered how to find means and variances of random variables by using conditional means and variances. Because $\hat{\theta}$ is an unbiased estimator for θ, Theorem 5.14 implies that

$$E(\hat{\theta}^*) = E\big(E(\hat{\theta} \mid U)\big) = E(\hat{\theta}) = \theta.$$

Thus, $\hat{\theta}^*$ is an unbiased estimator for θ.

Theorem 5.15 implies that

$$V(\hat{\theta}) = V\big(E(\hat{\theta} \mid U)\big) + E\big(V(\hat{\theta} \mid U)\big)$$

$$= V(\hat{\theta}^*) + E\big(V(\hat{\theta} \mid U)\big).$$

Because $V(\hat{\theta} \mid U = u) \geq 0$ for all u, it follows that $E\big(V(\hat{\theta} \mid U)\big) \geq 0$ and therefore that $V(\hat{\theta}) \geq V(\hat{\theta}^*)$, as claimed.   ∎

Theorem 9.5 implies that an unbiased estimator for θ with a small variance is, or can be made to be, a function of a sufficient statistic. If we have an unbiased estimator for θ, we might be able to improve it by using the result in Theorem 9.5. It might initially seem that the Rao–Blackwell theorem could be applied once to get a better unbiased estimator and then reapplied to the resulting new estimator to get an even better unbiased estimator. If we apply the Rao–Blackwell theorem using the sufficient statistic U, then $\hat{\theta}^* = E(\hat{\theta} \mid U)$ will be a function of the statistic U, say, $\hat{\theta}^* = h(U)$. Suppose that we reapply the Rao–Blackwell theorem to $\hat{\theta}^*$ by using the same sufficient statistic U. Since, in general, $E(h(U) \mid U) = h(U)$, we see that by using the Rao–Blackwell theorem again, our "new" estimator is just $h(U) = \hat{\theta}^*$. That is, if we use the same sufficient statistic in successive applications of the Rao–Blackwell theorem, we gain nothing after the first application. The only way that successive applications can lead to better unbiased estimators is if we use a different sufficient statistic when the theorem is reapplied. Thus, it is unnecessary to use the Rao–Blackwell theorem successively if we use the right sufficient statistic in our initial application.

Because many statistics are sufficient for a parameter θ associated with a distribution, which sufficient statistic should we use when we apply this theorem? For the distributions that we discuss in this text, the factorization criterion typically identifies a statistic U that best summarizes the information in the data about the parameter θ. Such statistics are called *minimal sufficient statistics*. Exercise 9.58 introduces a method for determining a minimal sufficient statistic that might be of interest to some readers. In a few of the subsequent exercises, you will see that this method usually yields the same sufficient statistics as those obtained from the factorization criterion. In the cases we consider, these statistics possess another property (completeness) that guarantees that, if we apply Theorem 9.5 using U, we not only get an estimator with a smaller variance, but we actually obtain an unbiased estimator for θ with *minimum variance*. Such an estimator is called a *minimum variance unbiased estimator* (MVUE). See Casella and Berger (1990), Hogg and Craig (1995), or Mood, Graybill, and Boes (1974) for additional details.

Thus, if we start with an unbiased estimator for a parameter θ and the sufficient statistic obtained through the factorization criterion, application of the Rao–Blackwell theorem typically leads to an MVUE for the parameter. Direct computation of conditional expectations can be difficult. However, if U is the sufficient statistic that best summarizes the data and some function of U, say $h(U)$ can be found such that $E(h(U)) = \theta$, it follows that $h(U)$ is the MVUE for θ. We will illustrate this approach with several examples.

EXAMPLE 9.6   Let $Y_1, Y_2, \ldots, Y_n$ denote a random sample from a distribution where $P(Y_i = 1) = p$ and $P(Y_i = 0) = 1 - p$, with p unknown (such random variables are often called *Bernoulli* variables). Use the factorization criterion to find a sufficient statistic that best summarizes the data. Give an MVUE for p.

Solution   Notice that the preceding probability function can be written as

$$P(Y_i = y_i) = p^{y_i}(1 - p)^{1-y_i}, \qquad y_i = 0, 1.$$

Thus, the likelihood $L(p)$ is

$$L(y_1, y_2, \ldots, y_n \mid p) = p(y_1, y_2, \ldots, y_n \mid p)$$

$$= p^{y_1}(1 - p)^{1-y_1} \times p^{y_2}(1 - p)^{1-y_2} \times \cdots \times p^{y_n}(1 - p)^{1-y_n}$$

$$= \underbrace{p^{\sum y_i}(1 - p)^{n-\sum y_i}}_{g(\sum y_i,\, p)} \times \underbrace{1}_{h(y_1, y_2, \ldots, y_n)}.$$

According to the factorization criterion, $U = \sum_{i=1}^{n} Y_i$ is sufficient for p. This statistic best summarizes the information about the parameter p. Notice that $E(U) = np$, or, equivalently, $E(U/n) = p$. Thus, $U/n = \overline{Y}$ is an unbiased estimator for p. Because this estimator is a function of the sufficient statistic $\sum_{i=1}^{n} Y_i$, the estimator $\hat{p} = \overline{Y}$ is the MVUE for p.

EXAMPLE 9.7   Suppose that $Y_1, Y_2, \ldots, Y_n$ denote a random sample from the Weibull density function, given by

$$f(y \mid \theta) = \begin{cases} \left(\dfrac{2y}{\theta} \right) e^{-y^2/\theta}, & y > 0 \\ 0, & \text{elsewhere.} \end{cases}$$

Find an MVUE for θ.

Solution   We begin by using the factorization criterion to find the sufficient statistic that best summarizes the information about θ.

$$L(y_1, y_2, \ldots, y_n \mid \theta) = f(y_1, y_2, \ldots, y_n \mid \theta)$$

$$= \left(\frac{2}{\theta} \right)^n (y_1 \times y_2 \times \cdots \times y_n) \exp\left(-\frac{1}{\theta} \sum_{i=1}^{n} y_i^2 \right)$$

$$= \underbrace{\left(\frac{2}{\theta} \right)^n \exp\left(-\frac{1}{\theta} \sum_{i=1}^{n} y_i^2 \right)}_{g(\sum y_i^2,\, \theta)} \times \underbrace{(y_1 \times y_2 \times \cdots \times y_n)}_{h(y_1, y_2, \ldots, y_n)}.$$

Thus, $U = \sum_{i=1}^{n} Y_i^2$ is the minimal sufficient statistic for θ.

We now must find a function of this statistic that is unbiased for θ. Letting $W = Y_i^2$, we have

$$f_W(w) = f(\sqrt{w})\frac{d(\sqrt{w})}{dw} = \left(\frac{2}{\theta}\right)(\sqrt{w}\,e^{-w/\theta})\left(\frac{1}{2\sqrt{w}}\right) = \left(\frac{1}{\theta}\right)e^{-w/\theta}, \quad w > 0.$$

That is, Y_i^2 has an exponential distribution with parameter θ. Because

$$E(Y_i^2) = E(W) = \theta \quad \text{and} \quad E\left(\sum_{i=1}^{n} Y_i^2\right) = n\theta,$$

it follows that

$$\hat{\theta} = \frac{1}{n}\sum_{i=1}^{n} Y_i^2$$

is an unbiased estimator of θ that is a function of the sufficient statistic $\sum_{i=1}^{n} Y_i^2$. Therefore, $\hat{\theta}$ is an MVUE of the Weibull parameter θ.

The following example illustrates the use of this technique for estimating two unknown parameters.

EXAMPLE 9.8   Suppose $Y_1, Y_2, \ldots, Y_n$ denotes a random sample from a normal distribution with unknown mean μ and variance σ^2. Find the MVUEs for μ and σ^2.

Solution   Again, looking at the likelihood function, we have

$$L(y_1, y_2, \ldots, y_n \mid \mu, \sigma^2)$$

$$= f(y_1, y_2, \ldots, y_n \mid \mu, \sigma^2)$$

$$= \left(\frac{1}{\sigma\sqrt{2\pi}}\right)^n \exp\left\{-\frac{1}{2\sigma^2}\sum_{i=1}^{n}(y_i - \mu)^2\right\}$$

$$= \left(\frac{1}{\sigma\sqrt{2\pi}}\right)^n \exp\left\{-\frac{1}{2\sigma^2}\left(\sum_{i=1}^{n}y_i^2 - 2\mu\sum_{i=1}^{n}y_i + n\mu^2\right)\right\}$$

$$= \left(\frac{1}{\sigma\sqrt{2\pi}}\right)^n \exp\left\{\frac{-n\mu^2}{2\sigma^2}\right\}\exp\left\{-\frac{1}{2\sigma^2}\left(\sum_{i=1}^{n}y_i^2 - 2\mu\sum_{i=1}^{n}y_i\right)\right\}.$$

Thus, $\sum_{i=1}^{n} Y_i$ and $\sum_{i=1}^{n} Y_i^2$, jointly, are sufficient statistics for μ and σ^2. We know from past work that $\overline{Y}$ is unbiased for μ and

$$S^2 = \frac{1}{n-1}\sum_{i=1}^{n}(Y_i - \overline{Y})^2 = \frac{1}{n-1}\left[\sum_{i=1}^{n} Y_i^2 - n\overline{Y}^2\right]$$

is unbiased for σ^2. Because these estimators are functions of the statistics that best summarize the information about μ and σ^2, they are MVUEs for μ and σ^2.

The factorization criterion, together with the Rao–Blackwell theorem, can also be used to find minimum-variance unbiased estimators for functions of the parameters associated with a distribution. We illustrate the technique in the following example.

EXAMPLE 9.9   Let $Y_1, Y_2, \ldots, Y_n$ denote a random sample from the exponential density function given by

$$f(y \mid \theta) = \begin{cases} \left(\dfrac{1}{\theta}\right) e^{-y/\theta}, & y > 0 \\ 0, & \text{elsewhere.} \end{cases}$$

Find an MVUE of $V(Y_i)$.

Solution   In Chapter 4 we determined that $E(Y_i) = \theta$ and that $V(Y_i) = \theta^2$. The factorization criterion implies that $\sum_{i=1}^{n} Y_i$ is the best sufficient statistic for θ. In fact, $\overline{Y}$ is the MVUE of θ. Therefore, it is tempting to use $\overline{Y}^2$ as an estimator of θ^2. But

$$E(\overline{Y}^2) = V(\overline{Y}) + [E(\overline{Y})]^2 = \frac{\theta^2}{n} + \theta^2 = \left(\frac{n+1}{n}\right)\theta^2.$$

It follows that $\overline{Y}^2$ is a biased estimate for θ^2. However,

$$\left(\frac{n}{n+1}\right)\overline{Y}^2$$

is an MVUE of θ^2 because it is an unbiased estimator for θ^2, which is a function of the sufficient statistic. No other unbiased estimator of θ^2 will have a smaller variance than this one.

A sufficient statistic for a parameter θ often can be used to construct an exact confidence interval for θ, if the probability distribution of the statistic can be found. The resulting intervals generally are the shortest that can be found with a specified confidence coefficient. We illustrate the technique with an example involving the Weibull distribution.

EXAMPLE 9.10   The following data, with measurements in hundreds of hours, represent the lengths of life of ten identical electronic components operating in a guidance control system for missiles:

.637   1.531   .733   2.256   2.364
1.601   .152   1.826   1.868   1.126.

The length of life of a component of this type is assumed to follow a Weibull distribution with density function given by

$$f(y \mid \theta) = \begin{cases} \left(\dfrac{2y}{\theta}\right) e^{-y^2/\theta}, & y > 0 \\ 0, & \text{elsewhere.} \end{cases}$$

Use the data to construct a 95% confidence interval for θ.

Solution We saw in Example 9.7 that the sufficient statistic that best summarizes the information about θ is $\sum_{i=1}^{n} Y_i^2$. We will use this statistic to form a pivotal quantity for constructing the desired confidence interval.

Recall from Example 9.7 that $W_i = Y_i^2$ has an exponential distribution with mean θ. Now consider the transformation $T_i = 2W_i/\theta$. Then

$$f_T(t) = f_W \left(\frac{\theta t}{2}\right) \frac{d(\theta t/2)}{dt} = \left(\frac{1}{\theta}\right) e^{-(\theta t/2)/\theta} \left(\frac{\theta}{2}\right) = \left(\frac{1}{2}\right) e^{-t/2}, \qquad t > 0.$$

Thus, for each $i = 1, 2, \ldots, n$, T_i has a χ^2 distribution with 2 degrees of freedom. Further, because the variables Y_i are independent, the variables T_i are independent, for $i = 1, 2, \ldots, n$. The sum of independent χ^2 random variables has a χ^2 distribution, with degrees of freedom equal to the sum of the degrees of freedom of the variables in the sum. Therefore, the quantity

$$\sum_{i=1}^{10} T_i = \frac{2}{\theta} \sum_{i=1}^{10} W_i = \frac{2}{\theta} \sum_{i=1}^{10} Y_i^2$$

has a χ^2 distribution with 20 degrees of freedom. Thus,

$$\frac{2}{\theta} \sum_{i=1}^{10} Y_i^2$$

is a pivotal quantity, and we can use the pivotal method (Section 8.5) to construct the desired confidence interval.

From Table 6, Appendix III, we can find two numbers a and b such that

$$P \left(a \le \frac{2}{\theta} \sum_{i=1}^{10} Y_i^2 \le b \right) = .95.$$

Manipulating the inequality to isolate θ in the middle, we have

$$.95 = P \left(a \le \frac{2}{\theta} \sum_{i=1}^{10} Y_i^2 \le b \right) = P \left(\frac{1}{b} \le \frac{\theta}{2 \sum_{i=1}^{10} Y_i^2} \le \frac{1}{a} \right)$$

$$= P \left(\frac{2 \sum_{i=1}^{10} Y_i^2}{b} \le \theta \le \frac{2 \sum_{i=1}^{10} Y_i^2}{a} \right).$$

From Table 6, Appendix III, the value that cuts off an area of .025 in the lower tail of the χ^2 distribution with 20 degrees of freedom is $a = 9.591$. The value that cuts off an area of .025 in the upper tail of the same distribution is $b = 34.170$. For the preceding data, $\sum_{i=1}^{10} Y_i^2 = 24.643$. Therefore, the 95% confidence interval for the Weibull parameter θ is

$$\left(\frac{2(24.643)}{34.170}, \frac{2(24.643)}{9.591} \right), \quad \text{or} \quad (1.442, 5.139).$$

This is a fairly wide interval for θ, but it is based on only ten observations.

In this section we have seen that the Rao–Blackwell theorem implies that unbiased estimators with small variances are functions of sufficient statistics. Generally speaking, the factorization criterion presented in Section 9.4 can be applied to find sufficient statistics that best summarize the information contained in sample data about parameters of interest. For the distributions that we consider in this text, a minimum-variance unbiased estimator (MVUE) for a target parameter θ can be found if we find some function of the best sufficient statistic and this function is an unbiased estimator for θ. This method often works well. However, sometimes a best sufficient statistic is a fairly complicated function of the observable random variables in the sample. In cases like these it may be difficult to find a function of the sufficient statistic that is an unbiased estimator for the target parameter. For this reason two additional methods of finding estimators—the method of moments and the method of maximum likelihood—are presented in the next two sections. A third important method for estimation, the method of least squares, is the topic of Chapter 11.

Exercises

9.48 Refer to Exercise 9.30(b). Find an MVUE of σ^2.

9.49 Refer to Exercise 9.12. Is the estimator of σ^2 given there an MVUE of σ^2?

9.50 Refer to Exercise 9.32. Use $\sum_{i=1}^{n} Y_i^2$ to find an MVUE of θ.

9.51 The number of breakdowns Y per day for a certain machine is a Poisson random variable with mean λ. The daily cost of repairing these breakdowns is given by $C = 3Y^2$. If $Y_1, Y_2, \ldots, Y_n$ denote the observed number of breakdowns for n independently selected days, find an MVUE for $E(C)$.

9.52 Let $Y_1, Y_2, \ldots, Y_n$ denote a random sample from the probability density function

$$f(y \mid \theta) = \begin{cases} \theta y^{\theta-1}, & 0 < y < 1; \theta > 0 \\ 0, & \text{elsewhere.} \end{cases}$$

a Show that this density function is in the (one-parameter) exponential family and that $\sum_{i=1}^{n} -\ln(Y_i)$ is sufficient for θ. (See Exercise 9.37.)

b If $W_i = -\ln(Y_i)$, show that W_i has an exponential distribution with mean $1/\theta$.

c Use methods similar to those in Example 9.10 to show that $2\theta \sum_{i=1}^{n} W_i$ has a χ^2 distribution with $2n$ degrees of freedom.

d   Show that

$$E\left(\frac{1}{2\theta \sum_{i=1}^{n} W_i}\right) = \frac{1}{2(n-1)}.$$

(*Hint:* Recall Exercise 4.89.)

e   What is the MVUE for θ?

9.53   Refer to Exercise 9.41. Use $Y_{(n)}$ to find an MVUE of θ. (See Example 9.1.)

9.54   Refer to Exercise 9.43. Find a function of $Y_{(1)}$ that is a minimum variance unbiased estimator for θ.

9.55   Let $Y_1, Y_2, \ldots, Y_n$ be a random sample from a population with density function

$$f(y \mid \theta) = \begin{cases} \dfrac{3y^2}{\theta^3}, & 0 \le y \le \theta \\[2mm] 0, & \text{elsewhere.} \end{cases}$$

In Exercise 9.44 you showed that $Y_{(n)} = \max(Y_1, Y_2, \ldots, Y_n)$ is sufficient for θ.

a   Show that $Y_{(n)}$ has probability density function

$$f_{(n)}(y \mid \theta) = \begin{cases} \dfrac{3ny^{3n-1}}{\theta^{3n}}, & 0 \le y \le \theta \\[2mm] 0, & \text{elsewhere.} \end{cases}$$

b   Find the UMVUE of θ.

9.56   Let $Y_1, Y_2, \ldots, Y_n$ be a random sample from a normal distribution with mean μ and variance 1.

a   Show that the MVUE of μ^2 is $\widehat{\mu^2} = \bar{Y}^2 - 1/n$.

b   Derive the variance of $\widehat{\mu^2}$.

*9.57   In this exercise, we illustrate the direct use of the Rao–Blackwell theorem. Let $Y_1, Y_2, \ldots, Y_n$ be independent Bernoulli random variables with

$$p(y_i \mid p) = p^{y_i}(1-p)^{1-y_i}, \qquad y_i = 0, 1.$$

That is, $P(Y_i = 1) = p$ and $P(Y_i = 0) = 1 - p$. Find the MVUE of $p(1-p)$, which is a term in the variance of Y_i or $W = \sum_{i=1}^{n} Y_i$, by the following steps.

a   Let

$$T = \begin{cases} 1, & \text{if } Y_1 = 1 \text{ and } Y_2 = 0 \\ 0, & \text{otherwise.} \end{cases}$$

Show that $E(T) = p(1-p)$.

b   Show that

$$P(T = 1 \mid W = w) = \frac{w(n-w)}{n(n-1)}.$$

c Show that

$$E(T \mid W) = \frac{n}{n-1}\left[\frac{W}{n}\left(1 - \frac{W}{n}\right)\right] = \frac{n}{n-1}\overline{Y}(1 - \overline{Y})$$

and, hence, that $n\overline{Y}(1 - \overline{Y})/(n-1)$ is the MVUE of $p(1-p)$.

*9.58 The likelihood function $L(y_1, y_2, \ldots, y_n \mid \theta)$ takes on different values depending on the arguments $(y_1, y_2, \ldots, y_n)$. A method for deriving a *minimal* sufficient statistic developed by Lehmann and Scheffé uses the ratio of the likelihoods evaluated at two points, $(x_1, x_2, \ldots, x_n)$ and $(y_1, y_2, \ldots, y_n)$:

$$\frac{L(x_1, x_2, \ldots, x_n \mid \theta)}{L(y_1, y_2, \ldots, y_n \mid \theta)}.$$

Many times it is possible to find a function $g(x_1, x_2, \ldots, x_n)$ such that this ratio is free of the unknown parameter θ if and only if $g(x_1, x_2, \ldots, x_n) = g(y_1, y_2, \ldots, y_n)$. If such a function g can be found, then $g(Y_1, Y_2, \ldots, Y_n)$ is a minimal sufficient statistic for θ.

a Let $Y_1, Y_2, \ldots, Y_n$ be a random sample from a Bernoulli distribution (see Example 9.6 and Exercise 9.57) with p unknown.

  i. Show that

$$\frac{L(x_1, x_2, \ldots, x_n \mid p)}{L(y_1, y_2, \ldots, y_n \mid p)} = \left(\frac{p}{1-p}\right)^{\Sigma x_i - \Sigma y_i}.$$

  ii. Argue that for this ratio to be independent of p, we must have

$$\sum_{i=1}^{n} x_i - \sum_{i=1}^{n} y_i = 0 \quad \text{or} \quad \sum_{i=1}^{n} x_i = \sum_{i=1}^{n} y_i.$$

  iii. Using the method of Lehmann and Scheffé, what is a minimal sufficient statistic for p? How does this sufficient statistic compare to the sufficient statistic derived in Example 9.6 by using the factorization criterion?

b Consider the Weibull density discussed in Example 9.7.

  i. Show that

$$\frac{L(x_1, x_2, \ldots, x_n \mid \theta)}{L(y_1, y_2, \ldots, y_n \mid \theta)} = \left(\frac{x_1 x_2 \cdots x_n}{y_1 y_2 \cdots y_n}\right)\exp\left[-\frac{1}{\theta}\left(\sum_{i=1}^{n} x_i^2 - \sum_{i=1}^{n} y_i^2\right)\right].$$

  ii. Argue that $\sum_{i=1}^{n} Y_i^2$ is a minimal sufficient statistic for θ.

*9.59 Refer to Exercise 9.58. Suppose that a sample of size n is taken from a normal population with mean μ and variance σ^2. Show that $\sum_{i=1}^{n} Y_i$ and $\sum_{i=1}^{n} Y_i^2$, jointly, form minimal sufficient statistics for μ and σ^2.

*9.60 Suppose that a statistic U has a probability density function that is positive over the interval $a \leq U \leq b$, and suppose that the density depends on a parameter θ that can range over the interval $\alpha_1 \leq \theta \leq \alpha_2$. Suppose also that $g(u)$ is continuous for u in the interval $[a, b]$. If $E[g(U)] = 0$ for all θ in the interval $[\alpha_1, \alpha_2]$ implies that $g(u)$ is identically zero, then the family of density functions $f_U(u \mid \theta)$ is said to be *complete*. (All of the statistics that we employed in in Section 9.5 have complete families of density functions.) Suppose that U is a sufficient statistic for θ, and $g_1(U)$ and $g_2(U)$ are both unbiased estimators of θ. Show that, if the family of density functions for U is complete, $g_1(U)$ must equal $g_2(U)$, and thus there is a *unique* function of U that is an unbiased estimator of θ.

Coupled with the Rao–Blackwell theorem, the property of completeness of $f_U(u \mid \theta)$, along with the sufficiency of U, assures us that there is a unique minimum-variance unbiased estimator of θ (UMVUE).

9.6 The Method of Moments

In this section we will discuss one of the oldest methods for deriving point estimators: the method of moments. A more sophisticated method, the method of maximum likelihood, is the topic of Section 9.7.

The method of moments is a very simple procedure for finding an estimator for one or more population parameters. Recall that the kth moment of a random variable, taken about the origin, is

$$\mu_k' = E(Y^k).$$

The corresponding kth sample moment is the average

$$m_k' = \frac{1}{n} \sum_{i=1}^{n} Y_i^k.$$

The method of moments is based on the intuitively appealing idea that sample moments should provide good estimates of the corresponding population moments. That is, m_k' should be a good estimator of μ_k', for $k = 1, 2, \ldots$. Then because the population moments $\mu_1', \mu_2', \ldots, \mu_k'$ are functions of the population parameters, we can equate corresponding population and sample moments and solve for the desired estimators. Hence the method of moments can be stated as follows.

Method of Moments

Choose as estimates those values of the parameters that are solutions of the equations $\mu_k' = m_k'$, for $k = 1, 2, \ldots, t$, where t is the number of parameters to be estimated.

EXAMPLE 9.11   A random sample of n observations, $Y_1, Y_2, \ldots, Y_n$, is selected from a population in which Y_i, for $i = 1, 2, \ldots, n$, possesses a uniform probability density function over the interval $(0, \theta)$ where θ is unknown. Use the method of moments to estimate the parameter θ.

Solution   The value of μ_1' for a uniform random variable is

$$\mu_1' = \mu = \frac{\theta}{2}.$$

The corresponding first sample moment is

$$m_1' = \frac{1}{n}\sum_{i=1}^{n} Y_i = \overline{Y}.$$

Equating the corresponding population and sample moment, we obtain

$$\mu_1' = \frac{\theta}{2} = \overline{Y}.$$

The method of moments estimator for θ is the solution of the above equation. That is, $\hat{\theta} = 2\overline{Y}$.

For the distributions we consider in this text, the methods of Section 9.3 can be used to show that sample moments are consistent estimators of the corresponding population moments. Because the estimators obtained from the method of moments obviously are functions of the sample moments, estimators obtained using the method of moments are usually consistent estimators of their respective parameters.

EXAMPLE **9.12**   Show that the estimator $\hat{\theta} = 2\overline{Y}$, derived in Example 9.11, is a consistent estimator for θ.

Solution   In Example 9.1, we showed that $\hat{\theta} = 2\overline{Y}$ is an unbiased estimator for θ and that $V(\hat{\theta}) = \theta^2/3n$. Because $\lim_{n\to\infty} V(\hat{\theta}) = 0$, Theorem 9.1 implies that $\hat{\theta} = 2\overline{Y}$ is a consistent estimator for θ.

Although the estimator $\hat{\theta}$ derived in Example 9.11 is consistent, it is not necessarily the best estimator for θ. Indeed, the factorization criterion yields $Y_{(n)} = \max(Y_1, Y_2, \ldots, Y_n)$ to be the best sufficient statistic for θ. Thus, according to the Rao–Blackwell theorem, the method of moments estimator will have larger variance than an unbiased estimator based on $Y_{(n)}$. This, in fact, was shown to be the case in Example 9.1.

EXAMPLE **9.13**   A random sample of n observations, $Y_1, Y_2, \ldots, Y_n$, is selected from a population where Y_i, for $i = 1, 2, \ldots, n$, possesses a gamma probability density function with parameters α and β (see Section 4.6 for the gamma probability density function). Find method of moments estimators for the unknown parameters α and β.

Solution   Because we seek estimators for two parameters α and β, we must equate two pairs of population and sample moments.
The first two moments of the gamma distribution with parameters α and β are (see the inside of the back cover of the text, if necessary)

$$\mu_1' = \mu = \alpha\beta \quad \text{and} \quad \mu_2' = \sigma^2 + \mu^2 = \alpha\beta^2 + \alpha^2\beta^2.$$

Now equate these quantities to their corresponding sample moments, and solve for $\hat{\alpha}$ and $\hat{\beta}$. Thus,

$$\mu'_1 = \alpha\beta = m'_1 = \overline{Y}$$

$$\mu'_2 = \alpha\beta^2 + \alpha^2\beta^2 = m'_2 = \frac{1}{n}\sum_{i=1}^{n} Y_i^2.$$

From the first equation we obtain $\hat{\beta} = \overline{Y}/\hat{\alpha}$. Substituting into the second equation and solving for $\hat{\alpha}$, we obtain

$$\hat{\alpha} = \frac{\overline{Y}^2}{(\sum Y_i^2/n) - \overline{Y}^2} = \frac{n\overline{Y}^2}{\sum_{i=1}^{n}(Y_i - \overline{Y})^2}.$$

Substituting $\hat{\alpha}$ into the first equation, we obtain

$$\hat{\beta} = \frac{\overline{Y}}{\hat{\alpha}} = \frac{\sum_{i=1}^{n}(Y_i - \overline{Y})^2}{n\overline{Y}}.$$

The method of moments estimators $\hat{\alpha}$ and $\hat{\beta}$ in Example 9.13 are consistent. $\overline{Y}$ converges in probability to $E(Y_i) = \alpha\beta$, and $(1/n)\sum_{i=1}^{n} Y_i^2$ converges in probability to $E(Y_i^2) = \alpha\beta^2 + \alpha^2\beta^2$. Thus,

$$\hat{\alpha} = \frac{\overline{Y}^2}{\frac{1}{n}\sum_{i=1}^{n} Y_i^2 - \overline{Y}^2} \quad \text{is a consistent estimator of} \quad \frac{(\alpha\beta)^2}{\alpha\beta^2 + \alpha^2\beta^2 - (\alpha\beta)^2} = \alpha$$

and

$$\hat{\beta} = \frac{\overline{Y}}{\hat{\alpha}} \quad \text{is a consistent estimator of} \quad \frac{\alpha\beta}{\alpha} = \beta.$$

Using the factorization criterion, we can show $\sum_{i=1}^{n} Y_i$ and the product $\prod_{i=1}^{n} Y_i$ to be sufficient statistics for the gamma density function. Because the method of moments estimators $\hat{\alpha}$ and $\hat{\beta}$ are not functions of these sufficient statistics, we can find more efficient estimators for the parameters α and β. However, it is considerably more difficult to apply other methods to find estimators for these parameters.

To summarize, the method of moments finds estimators of unknown parameters by equating corresponding sample and population moments. The method is easy to employ and provides consistent estimators. However, the estimators derived by this method are often not functions of sufficient statistics. As a result, method of moments estimators are sometimes not very efficient. In many cases the method of moments estimators are biased. The primary virtues of this method are its ease of use and that it sometimes yields estimators with reasonable properties.

Exercises

9.61 Let $Y_1, Y_2, \ldots, Y_n$ denote a random sample from the probability density function

$$f(y \mid \theta) = \begin{cases} (\theta + 1)y^\theta, & 0 < y < 1; \ \theta > -1 \\ 0, & \text{elsewhere.} \end{cases}$$

Find an estimator for θ by the method of moments. Show that the estimator is consistent. Is the estimator a function of the sufficient statistic $-\sum_{i=1}^{n} \ln(Y_i)$ that we can obtain from the factorization criterion? What implications does this have?

9.62 Suppose that $Y_1, Y_2, \ldots, Y_n$ constitute a random sample from a Poisson distribution with mean λ. Find the method of moments estimator of λ.

9.63 If $Y_1, Y_2, \ldots, Y_n$ denote a random sample from the normal distribution with known mean $\mu = 0$ and unknown variance σ^2, find the method of moments estimator of σ^2.

9.64 If $Y_1, Y_2, \ldots, Y_n$ denote a random sample from the normal distribution with mean μ and variance σ^2, find the method of moments estimators of μ and σ^2.

9.65 An urn contains θ black balls and $N - \theta$ white balls. A sample of n balls is to be selected without replacement. Let Y denote the number of black balls in the sample. Show that $(N/n)Y$ is the method of moments estimator of θ.

9.66 Let $Y_1, Y_2, \ldots, Y_n$ constitute a random sample from the probability density function given by

$$f(y \mid \theta) = \begin{cases} \left(\dfrac{2}{\theta^2}\right)(\theta - y), & 0 \le y \le \theta \\ 0, & \text{elsewhere.} \end{cases}$$

a Find an estimator for θ by using the method of moments.

b Is this estimator a sufficient statistic for θ?

9.67 Let $Y_1, Y_2, \ldots, Y_n$ be a random sample from the probability density function given by

$$f(y \mid \theta) = \begin{cases} \dfrac{\Gamma(2\theta)}{[\Gamma(\theta)]^2}(y^{\theta-1})(1 - y)^{\theta-1}, & 0 \le y \le 1 \\ 0, & \text{elsewhere.} \end{cases}$$

Find the method of moments estimator for θ.

9.68 Let $X_1, X_2, X_3, \ldots$ be independent random variables such that $P(X_i = 1) = p$ and $P(X_i = 0) = 1 - p$ for each $i = 1, 2, 3, \ldots$. Let the random variable Y denote the number of trials necessary to obtain the first success—that is, the value of i for which $X_i = 1$ first occurs. Then Y has a geometric distribution with $P(Y = y) = (1 - p)^{y-1}p$, for $y = 1, 2, 3, \ldots$. Find the method of moments estimator of p based on this single observation Y.

9.69 Let $Y_1, Y_2, \ldots, Y_n$ denote independent and identically distributed uniform random variables on the interval $(0, 3\theta)$. Derive the method of moments estimator for θ.

9.70 Let $Y_1, Y_2, \ldots, Y_n$ denote independent and identically distributed random variables from a power family distribution with parameters α and $\theta = 3$. Then, as in Exercise 9.35, if $\alpha > 0$,

$$f(y|\alpha) = \begin{cases} \alpha y^{\alpha-1}/3^\alpha, & 0 \le y \le 3 \\ 0, & \text{elsewhere.} \end{cases}$$

Show that $E(Y_1) = 3\alpha/(\alpha+1)$, and derive the method of moments estimator for α.

*9.71 Let $Y_1, Y_2, \ldots, Y_n$ denote independent and identically distributed random variables from a Pareto distribution with parameters α and β, where β is known. Then, if $\alpha > 0$,

$$f(y|\alpha, \beta) = \begin{cases} \alpha\beta^\alpha y^{-(\alpha+1)}, & y \ge \beta \\ 0, & \text{elsewhere.} \end{cases}$$

Show that $E(Y_i) = \alpha\beta/(\alpha-1)$ if $\alpha > 1$ and $E(Y_i)$ is undefined if $0 < \alpha < 1$. Thus, the method of moments estimator for α is undefined.

9.7 The Method of Maximum Likelihood

In Section 9.5 we presented a method for deriving a minimum-variance unbiased estimator for a target parameter: using the factorization criterion together with the Rao–Blackwell theorem. The method requires that we find some function of a minimal sufficient statistic that is an unbiased estimator for the target parameter. Although we have a method for finding a sufficient statistic, the determination of the function of the minimal sufficient statistic that gives us an unbiased estimator can be largely a matter of hit or miss. Section 9.6 contained a discussion of the method of moments. The method of moments is intuitive and easy to apply but does not usually lead to the best estimators. In this section we present a method, the method of maximum likelihood, that often leads to minimum-variance unbiased estimators.

We will use an example to illustrate the logic upon which the method of maximum likelihood is based. Suppose that we are confronted with a box that contains three balls. We know that each of the balls may be red or white, but we do not know the total number of either color. However, we are allowed to randomly sample two of the balls, without replacement. If our random sample yields two red balls, what would be a good estimate of the total number of red balls in the box? Obviously, the number of red balls in the box must be two or three (if there were 0 or 1 red ball in the box, it would be impossible to obtain two red balls when sampling without replacement). If there are two red balls and one white ball in the box, the probability of randomly selecting two red balls is

$$\binom{2}{2}\binom{1}{0}\Big/\binom{3}{2} = \frac{1}{3}.$$

On the other hand, if there are three red balls in the box, the probability of randomly selecting two red balls is

$$\binom{3}{2}\Big/\binom{3}{2} = 1.$$

It should seem reasonable to choose three as the estimate of the number of red balls in the box, because this estimate *maximizes the probability* of obtaining the observed sample. Of course, it is possible for the box to contain only two red balls, but the observed outcome gives more credence to there being three red balls in the box.

This example illustrates a method for finding an estimator that can be applied to any situation. The technique, called the *method of maximum likelihood*, selects as estimates the values of the parameters that maximize the likelihood (the joint probability function or joint density function) of the observed sample. (See Definition 9.4.) Recall that we referred to this method of estimation in Chapter 3, where, in Examples 3.10 and 3.13 and Exercise 3.83, we found the maximum likelihood estimates of the parameter p based on single observations on binomial, geometric, and negative binomial random variables, respectively.

Method of Maximum Likelihood
Suppose that the likelihood function depends on k parameters $\theta_1, \theta_2, \ldots, \theta_k$. Choose as estimates those values of the parameters that maximize the likelihood $L(y_1, y_2, \ldots, y_n \mid \theta_1, \theta_2, \ldots, \theta_k)$.

To emphasize the fact that the likelihood function is a function of the parameters $\theta_1, \theta_2, \ldots, \theta_k$, we sometimes write the likelihood function as $L(\theta_1, \theta_2, \ldots, \theta_k)$. It is common to refer to maximum-likelihood estimators as MLEs. We will illustrate the method with an example.

EXAMPLE 9.14   A binomial experiment consisting of n trials resulted in observations $y_1, y_2, \ldots, y_n$, where $y_i = 1$ if the ith trial was a success and $y_i = 0$ otherwise. Find the maximum-likelihood estimator (MLE) of p, the probability of a success.

Solution   The likelihood of the observed sample is the probability of observing $y_1, y_2, \ldots, y_n$. Hence,

$$L(p) = L(y_1, y_2, \ldots, y_n \mid p) = p^y(1-p)^{n-y} \quad \text{where} \quad y = \sum_{i=1}^{n} y_i.$$

We now wish to find the value of p that maximizes $L(p)$. If $y = 0$, $L(p) = (1-p)^n$, and $L(p)$ is maximized when $p = 0$. Analogously, if $y = n$, $L(p) = p^n$, and $L(p)$ is maximized when $p = 1$. If $y = 1, 2, \ldots, n-1$, then $L(p) = p^y(1-p)^{n-y}$ is zero at $p = 0$ and $p = 1$ and is continuous for values of p between 0 and 1. Thus, for $y = 1, 2, \ldots, n-1$, we can find the value of p that maximizes $L(p)$ by setting the derivative $dL(p)/dp$ equal to zero and solving for p.

You will notice that $\ln[L(p)]$ is a monotonically increasing function of $L(p)$. Hence both $\ln[L(p)]$ and $L(p)$ are maximized for the same value of p. Because $L(p)$ is a product of functions of p, and finding the derivative of products is tedious, it is easier to find the value of p that maximizes $\ln[L(p)]$. We have

$$\ln[L(p)] = \ln\left[p^y(1-p)^{n-y}\right] = y \ln p + (n-y)\ln(1-p).$$

If $y = 1, 2, \ldots, n - 1$, the derivative of $\ln[L(p)]$ with respect to p, is

$$\frac{d \ln[L(p)]}{dp} = y\left(\frac{1}{p}\right) + (n - y)\left(\frac{-1}{1 - p}\right).$$

For $y = 1, 2, \ldots, n - 1$, the value of p that maximizes (or minimizes) $\ln[L(p)]$ is the solution of the equation

$$\frac{y}{\hat{p}} - \frac{n - y}{1 - \hat{p}} = 0.$$

Solving, we obtain the estimate $\hat{p} = y/n$. You can easily verify that this solution occurs when $\ln[L(p)]$ (and hence $L(p)$) achieves a maximum.

Because $L(p)$ is maximized at $p = 0$ when $y = 0$, at $p = 1$ when $y = n$ and at $p = y/n$ when $y = 1, 2, \ldots, n - 1$, whatever the observed value of y, $L(p)$ is maximized when $p = y/n$.

The maximum-likelihood *estimator* (MLE), $\hat{p} = Y/n$, is the fraction of successes in the total number of trials n. Hence, the MLE of p is is actually the intuitive estimator for p that we used throughout Chapter 8.

EXAMPLE 9.15   Let $Y_1, Y_2, \ldots, Y_n$ be a random sample from a normal distribution with mean μ and variance σ^2. Find the maximum-likelihood estimators (MLEs) of μ and σ^2.

Solution   Because $Y_1, Y_2, \ldots, Y_n$ are continuous random variables, $L(\mu, \sigma^2)$ is the joint density of the sample. Thus $L(\mu, \sigma^2) = f(y_1, y_2, \ldots, y_n \mid \mu, \sigma^2)$. In this case

$$L(\mu, \sigma^2) = f(y_1, y_2, \ldots, y_n \mid \mu, \sigma^2)$$
$$= f(y_1 \mid \mu, \sigma^2) \times f(y_2 \mid \mu, \sigma^2) \times \cdots \times f(y_n \mid \mu, \sigma^2)$$
$$= \left\{\frac{1}{\sigma\sqrt{2\pi}} \exp\left[\frac{-(y_1 - \mu)^2}{2\sigma^2}\right]\right\} \times \cdots \times \left\{\frac{1}{\sigma\sqrt{2\pi}} \exp\left[\frac{-(y_n - \mu)^2}{2\sigma^2}\right]\right\}$$
$$= \left(\frac{1}{2\pi\sigma^2}\right)^{n/2} \exp\left[\frac{-1}{2\sigma^2}\sum_{i=1}^{n}(y_i - \mu)^2\right]$$

[Recall that $\exp(w)$ is just another way of writing e^w.] Further,

$$\ln\left[L(\mu, \sigma^2)\right] = -\frac{n}{2}\ln\sigma^2 - \frac{n}{2}\ln 2\pi - \frac{1}{2\sigma^2}\sum_{i=1}^{n}(y_i - \mu)^2.$$

The MLEs of μ and σ^2 are the values that make $\ln\left[L(\mu, \sigma^2)\right]$ a maximum. Taking derivatives with respect to μ and σ^2, we obtain

$$\frac{\partial\left\{\ln\left[L(\mu, \sigma^2)\right]\right\}}{\partial\mu} = \frac{1}{\sigma^2}\sum_{i=1}^{n}(y_i - \mu)$$

and

$$\frac{\partial\left\{\ln\left[L(\mu, \sigma^2)\right]\right\}}{\partial\sigma^2} = -\left(\frac{n}{2}\right)\left(\frac{1}{\sigma^2}\right) + \frac{1}{2\sigma^4}\sum_{i=1}^{n}(y_i - \mu)^2.$$

Setting these derivatives equal to zero and solving simultaneously, we obtain from the first equation

$$\frac{1}{\hat{\sigma}^2} \sum_{i=1}^{n} (y_i - \hat{\mu}) = 0, \quad \text{or} \quad \sum_{i=1}^{n} y_i - n\hat{\mu} = 0, \quad \text{and} \quad \hat{\mu} = \frac{1}{n} \sum_{i=1}^{n} y_i = \overline{y}.$$

Substituting $\overline{y}$ for $\hat{\mu}$ in the second equation and solving for $\hat{\sigma}^2$, we have

$$-\left(\frac{n}{\hat{\sigma}^2}\right) + \frac{1}{\hat{\sigma}^4} \sum_{i=1}^{n} (y_i - \overline{y})^2 = 0 \quad \text{or} \quad \hat{\sigma}^2 = \frac{1}{n} \sum_{i=1}^{n} (y_i - \overline{y})^2.$$

Thus $\overline{Y}$ and $\hat{\sigma}^2$ are the maximum-likelihood estimators of μ and σ^2, respectively. Notice that $\overline{Y}$ is unbiased for μ. Although $\hat{\sigma}^2$ is not unbiased for σ^2, it can easily be adjusted to the unbiased estimator S^2 (see Example 8.1).

EXAMPLE 9.16   Let $Y_1, Y_2, \ldots, Y_n$ be a random sample of observations from a uniform distribution with probability density function $f(y_i \mid \theta) = 1/\theta$, for $0 \le y_i \le \theta$ and $i = 1, 2, \ldots, n$. Find the MLE of θ.

Solution   In this case the likelihood is given by

$$L(\theta) = f(y_1, y_2, \ldots, y_n \mid \theta) = f(y_1 \mid \theta) \times f(y_2 \mid \theta) \times \cdots \times f(y_n \mid \theta)$$

$$= \begin{cases} \dfrac{1}{\theta} \times \dfrac{1}{\theta} \times \cdots \times \dfrac{1}{\theta} = \dfrac{1}{\theta^n}, & \text{if all the } y_i \text{ values are between 0 and } \theta \\ 0, & \text{otherwise.} \end{cases}$$

Obviously, $L(\theta)$ is not maximized when $L(\theta) = 0$. You will notice that $1/\theta^n$ is a monotonically decreasing function of θ. Hence, nowhere in the interval $0 < \theta < \infty$ is $d[1/\theta^n]/d\theta$ equal to zero. However, $1/\theta^n$ increases as θ decreases, and $1/\theta^n$ is maximized by selecting θ to be as small as possible, subject to the constraint that all of the y_i values are between 0 and θ. The smallest value of θ that satisfies this constraint is the maximum observation in the set $y_1, y_2, \ldots, y_n$. That is, $\hat{\theta} = Y_{(n)} = \max(Y_1, Y_2, \ldots, Y_n)$ is the MLE for θ. This MLE for θ is not an unbiased estimator of θ, but it can be adjusted to be unbiased, as shown in Example 9.1.

We have seen that sufficient statistics that best summarize the data have desirable properties and often can be used to find an MVUE for parameters of interest. If U is *any* sufficient statistic for the estimation of a parameter θ, including the sufficient statistic obtained from the optimal use of the factorization criterion, the maximum-likelihood estimator is always some function of U. That is, the MLE depends on the sample observations only through the value of a sufficient statistic. To show this, we need only observe that if U is a sufficient statistic for θ, the factorization criterion (Theorem 9.4) implies that the likelihood can be factored as

$$L(\theta) = L(y_1, y_2, \ldots, y_n \mid \theta) = g(u, \theta)h(y_1, y_2, \ldots, y_n),$$

where $g(u, \theta)$ is a function of only u and θ and $h(y_1, y_2, \ldots, y_n)$ *does not depend on* θ. Therefore, it follows that

$$\ln[L(\theta)] = \ln[g(u, \theta)] + \ln[h(y_1, y_2, \ldots, y_n)].$$

Notice that $\ln[h(y_1, y_2, \ldots, y_n)]$ does not depend on θ and, therefore, that maximizing $\ln[L(\theta)]$ relative to θ is equivalent to maximizing $\ln[g(u, \theta)]$ relative to θ. Because $\ln[g(u, \theta)]$ depends on the data only through the value of the sufficient statistic U, it follows that $\hat{\theta}$ depends on the data only through the value of U. That is, if U is *any* sufficient statistic for estimating θ, the maximum-likelihood estimator (MLE) is always some function of U. Consequently, if a maximum-likelihood estimator for a parameter can be found and then adjusted by a constant to be unbiased, the resulting estimator often is an MVUE of the parameter in question. These realizations imply that the method of maximum likelihood can be a very useful tool in finding estimators with good properties.

Maximum-likelihood estimators (MLEs) have some additional properties that make this method of estimation particularly attractive. In Example 9.9, we considered estimation of θ^2, a function of the parameter θ. Functions of other parameters may also be of interest. For example, the variance of a binomial random variable is $np(1 - p)$, a function of the parameter p. If Y has a Poisson distribution with mean λ, it follows that $P(Y = 0) = e^{-\lambda}$; we may wish to estimate this function of λ. Generally, if θ is the parameter associated with a distribution, we are sometimes interested in estimating some function of θ, say, $t(\theta)$, rather than θ itself. In Exercise 9.86, you will prove that, if $t(\theta)$ is a one-to-one function of θ and if $\hat{\theta}$ is the MLE for θ, then the MLE of $t(\theta)$ is given by

$$\widehat{t(\theta)} = t(\hat{\theta}).$$

This result, sometimes referred to as the *invariance property* of MLEs, also holds for any function of a parameter of interest (not just one-to-one functions). See Casella and Berger (1990) for details.

EXAMPLE 9.17   In Example 9.14, we found that the MLE of the binomial proportion p is given by $\hat{p} = Y/n$. What is the MLE for the variance of Y?

Solution   The variance of a binomial random variable Y is given by $V(Y) = np(1 - p)$. Because $V(Y)$ is a function of the binomial parameter p, namely, $V(Y) = t(p)$ with $t(p) = np(1 - p)$, it follows that the MLE of $V(Y)$ is given by

$$\widehat{V(Y)} = \widehat{t(p)} = t(\hat{p}) = n\left(\frac{Y}{n}\right)\left(1 - \frac{Y}{n}\right).$$

This estimator is not unbiased. However, using the result in Exercise 9.57, we can easily adjust it to make it unbiased. Actually,

$$n\left(\frac{Y}{n}\right)\left(1 - \frac{Y}{n}\right)\left(\frac{n}{n-1}\right) = \left(\frac{n^2}{n-1}\right)\left(\frac{Y}{n}\right)\left(1 - \frac{Y}{n}\right)$$

is the UMVUE for $t(p) = np(1 - p)$.

In the next section (optional), we summarize some of the convenient and useful large-sample properties of maximum-likelihood estimators.

Exercises

9.72   Suppose that $Y_1, Y_2, \ldots, Y_n$ denote a random sample from the Poisson distribution with mean λ.

    a   Find the maximum-likelihood estimator $\hat{\lambda}$ for λ.

    b   Find the expected value and variance of $\hat{\lambda}$.

    c   Show that the estimator of (a) is consistent for λ.

    d   What is the MLE for $P(Y = 0) = e^{-\lambda}$?

9.73   Suppose that $Y_1, Y_2, \ldots, Y_n$ denote a random sample from an exponentially distributed population with mean θ. Find the MLE of the population variance θ^2. (*Hint:* Recall Example 9.9.)

9.74   Let $Y_1, Y_2, \ldots, Y_n$ denote a random sample from the density function given by

$$f(y \mid \theta) = \begin{cases} \left(\dfrac{1}{\theta}\right) r y^{r-1} e^{-y^r/\theta}, & \theta > 0, y > 0 \\ 0, & \text{elsewhere} \end{cases}$$

where r is a known positive constant.

    a   Find a sufficient statistic for θ.

    b   Find the maximum-likelihood estimator of θ.

    c   Is the estimator in part (b) an MVUE for θ?

9.75   Suppose that $Y_1, Y_2, \ldots, Y_n$ constitute a random sample from a uniform distribution with probability density function

$$f(y \mid \theta) = \begin{cases} \dfrac{1}{2\theta + 1}, & 0 \le y \le 2\theta + 1 \\ 0, & \text{otherwise.} \end{cases}$$

    a   Obtain the maximum-likelihood estimator of θ.

    b   Obtain the MLE for the *variance* of the underlying distribution.

9.76   A certain type of electronic component has a lifetime Y (in hours) with probability density function given by

$$f(y \mid \theta) = \begin{cases} \left(\dfrac{1}{\theta^2}\right) y e^{-y/\theta}, & y > 0 \\ 0, & \text{otherwise.} \end{cases}$$

That is, Y has a gamma distribution with parameters $\alpha = 2$ and θ. Let $\hat{\theta}$ denote the maximum-likelihood estimator of θ. Suppose that three such components, tested independently, had lifetimes of 120, 130, and 128 h.

    a   Find the maximum-likelihood estimate of θ.

    b   Find $E(\hat{\theta})$ and $V(\hat{\theta})$.

   c Suppose that θ actually equals 130. Give an approximate bound that you might expect for the error of estimation.

   d What is the MLE for the variance of Y?

9.77 Let $Y_1, Y_2, \ldots, Y_n$ denote a random sample from the density function given by

$$f(y \mid \alpha, \theta) = \begin{cases} \left(\dfrac{1}{\Gamma(\alpha)\theta^\alpha}\right) y^{\alpha-1} e^{-y/\theta}, & y > 0, \\ 0, & \text{elsewhere} \end{cases}$$

where $\alpha > 0$ is known.

   a Find the maximum-likelihood estimator $\hat{\theta}$ of θ.

   b Find the expected value and variance of $\hat{\theta}$.

   c Show that $\hat{\theta}$ is consistent for θ.

   d What is the best (minimal) sufficient statistic for θ in this problem?

   e Suppose that $n = 5$ and $\alpha = 2$. Use the minimal sufficient statistic to construct a 90% confidence interval for θ. (*Hint:* Transform to a χ^2 distribution.)

9.78 Suppose that $X_1, X_2, \ldots, X_m$, representing yields per acre for corn variety A, constitute a random sample from a normal distribution with mean μ_1 and variance σ^2. Also, $Y_1 Y_2, \ldots, Y_n$, representing yields for corn variety B, constitute a random sample from a normal distribution with mean μ_2 and variance σ^2. If the Xs and Ys are independent, find the maximum-likelihood estimator for the common variance σ^2. Assume that μ_1 and μ_2 are unknown.

9.79 A random sample of 100 voters selected from a large population revealed 30 favoring candidate A, 38 favoring candidate B, and 32 favoring candidate C. Find maximum-likelihood estimates for the proportions of voters in the population favoring candidates A, B, and C, respectively. Estimate the difference between the fractions favoring A and B, and place a two-standard-deviation bound on the error of estimation.

9.80 Let $Y_1, Y_2, \ldots, Y_n$ denote a random sample from the probability density function

$$f(y \mid \theta) = \begin{cases} (\theta + 1)y^\theta, & 0 < y < 1, \quad \theta > -1 \\ 0, & \text{elsewhere.} \end{cases}$$

Find the maximum-likelihood estimator for θ. Compare your answer to the method of moments estimator found in Exercise 9.61.

9.81 It is known that the probability p of tossing heads on an unbalanced coin is either 1/4 or 3/4. The coin is tossed twice and a value for Y, the number of heads, is observed. For each possible value of Y, which of the two values for p (1/4 or 3/4) maximizes the probability that $Y = y$? Depending on the value of y actually observed, what is the maximum-likelihood estimator of p?

9.82 A random sample of 100 men produced a total of 25 who favored a controversial local issue. An independent random sample of 100 women produced a total of 30 who favored the issue. Assume that p_M is the true underlying proportion of men who favor the issue and that p_W is the true underlying proportion of women who favor of the issue. If it actually is true that $p_W = p_M = p$, find the maximum-likelihood estimator of the common proportion p.

*9.83 Find the maximum-likelihood estimator of θ based on a random sample of size n from a uniform distribution on the interval $(0, 2\theta)$.

*9.84 Let $Y_1, Y_2, \ldots, Y_n$ be a random sample from a population with density function

$$f(y \mid \theta) = \begin{cases} \dfrac{3y^2}{\theta^3}, & 0 \le y \le \theta \\ 0, & \text{elsewhere.} \end{cases}$$

In Exercise 9.44 you showed that $Y_{(n)} = \max(Y_1, Y_2, \ldots, Y_n)$ is sufficient for θ.
a Find the MLE for θ. (*Hint:* See Example 9.16.)
b Find a function of the MLE in part (a) that is a pivotal quantity. (*Hint:* see Exercise 9.55.)
c Use the pivotal quantity from part (b) to find a $(1 - \alpha)100\%$ confidence interval for θ.

*9.85 Let $Y_1, Y_2, \ldots, Y_n$ be a random sample from a population with density function

$$f(y \mid \theta) = \begin{cases} \dfrac{2\theta^2}{y^3}, & \theta < y < \infty \\ 0, & \text{elsewhere.} \end{cases}$$

In Exercise 9.45 you showed that $Y_{(1)} = \min(Y_1, Y_2, \ldots, Y_n)$ is sufficient for θ.
a Find the MLE for θ. (*Hint:* See Example 9.16.)
b Find a function of the MLE in part (a) that is a pivotal quantity.
c Use the pivotal quantity from part (b) to find a $(1 - \alpha)100\%$ confidence interval for θ.

*9.86 Suppose that $\hat{\theta}$ is the maximum-likelihood estimator for a parameter θ. Let $t(\theta)$ be a function of θ that possesses a unique inverse [that is, if $\beta = t(\theta)$, then $\theta = t^{-1}(\beta)$]. Show that $t(\hat{\theta})$ is the maximum-likelihood estimator of $t(\theta)$.

*9.87 A random sample of n items is selected from the large number of items produced by a certain production line in one day. Find the maximum-likelihood estimator of the ratio R, the proportion of defective items divided by the proportion of good items.

9.88 Consider a random sample of size n from a normal population with mean μ and variance σ^2, both unknown. Derive the maximum-likelihood estimator of σ.

9.8 Some Large-Sample Properties of MLEs (Optional)

Maximum-likelihood estimators also have interesting large–sample properties. Suppose that $t(\theta)$ is a differentiable function of θ. In Section 9.7, we argued by the invariance property that, if $\hat{\theta}$ is the MLE of θ, then the MLE of $t(\theta)$ is given by $t(\hat{\theta})$. Under some conditions of regularity that hold for the distributions we will consider, $t(\hat{\theta})$ is a *consistent* estimator for $t(\theta)$. In addition, for large sample sizes,

$$Z = \frac{t(\hat{\theta}) - t(\theta)}{\sqrt{\left[\dfrac{\partial t(\theta)}{\partial \theta}\right]^2 \Big/ nE\left[-\dfrac{\partial^2 \ln f(Y \mid \theta)}{\partial \theta^2}\right]}}$$

has approximately a standard normal distribution. In this expression, the quantity $f(Y \mid \theta)$ in the denominator is the density function corresponding to the continuous distribution of interest, evaluated at the random value Y. In the discrete case, the analogous result holds with the probability function evaluated at the random value Y, $p(Y \mid \theta)$ substituted for the density $f(Y \mid \theta)$. If we desire a confidence interval for $t(\theta)$, we can use quantity Z as a pivotal quantity. If we proceed as in Section 8.6, we obtain the following approximate large sample $100(1 - \alpha)\%$ confidence interval for $t(\theta)$:

$$t(\hat{\theta}) \pm z_{\alpha/2} \sqrt{\left[\frac{\partial t(\theta)}{\partial \theta}\right]^2 \bigg/ nE\left[-\frac{\partial^2 \ln f(Y \mid \theta)}{\partial \theta^2}\right]}$$

$$\approx t(\hat{\theta}) \pm z_{\alpha/2} \sqrt{\left(\left[\frac{\partial t(\theta)}{\partial \theta}\right]^2 \bigg/ nE\left[-\frac{\partial^2 \ln f(Y \mid \theta)}{\partial \theta^2}\right]\right)}\bigg|_{\theta = \hat{\theta}}.$$

We will illustrate this with the following example.

EXAMPLE **9.18**   For random variable with a *Bernoulli* distribution, $p(y \mid p) = p^y(1 - p)^{1-y}$, for $y = 0$, 1. If Y_1, Y_2, $\ldots$, Y_n denote a random sample of size n from this distribution, derive a $100(1 - \alpha)\%$ confidence interval for $p(1 - p)$, the variance associated with this distribution.

Solution   As in Example 9.14, the MLE of the parameter p is given by $\hat{p} = W/n$ where $W = \sum_{i=1}^{n} Y_i$. It follows that the MLE for $t(p) = p(1 - p)$ is $t(p) = \hat{p}(1 - \hat{p})$. In this case,

$$t(p) = p(1 - p) = p - p^2 \quad \text{and}$$
$$\frac{\partial t(p)}{\partial p} = 1 - 2p.$$

Also,

$$p(y \mid p) = p^y(1 - p)^{1-y}$$
$$\ln[p(y \mid p)] = y(\ln p) + (1 - y)\ln(1 - p)$$
$$\frac{\partial \ln[p(y \mid p)]}{\partial p} = \frac{y}{p} - \frac{1 - y}{1 - p}$$
$$\frac{\partial^2 \ln[p(y \mid p)]}{\partial p^2} = -\frac{y}{p^2} - \frac{1 - y}{(1 - p)^2}$$
$$E\left[-\frac{\partial^2 \ln[p(Y \mid p)]}{\partial p^2}\right] = E\left[\frac{Y}{p^2} + \frac{1 - Y}{(1 - p)^2}\right]$$
$$= \frac{p}{p^2} + \frac{1 - p}{(1 - p)^2} = \frac{1}{p} + \frac{1}{1 - p} = \frac{1}{p(1 - p)}.$$

Substituting into the earlier formula for the confidence interval for $t(\theta)$, we obtain

$$t(\hat{p}) \pm z_{\alpha/2} \sqrt{\left(\left[\frac{\partial t(p)}{\partial p}\right]^2 \bigg/ nE\left[-\frac{\partial^2 \ln p(Y \mid p)}{\partial p^2}\right]\right)}\bigg|_{p=\hat{p}}$$

$$= \hat{p}(1-\hat{p}) \pm z_{\alpha/2} \sqrt{\left((1-2p)^2 \bigg/ n\left[\frac{1}{p(1-p)}\right]\right)}\bigg|_{p=\hat{p}}$$

$$= \hat{p}(1-\hat{p}) \pm z_{\alpha/2} \sqrt{\frac{\hat{p}(1-\hat{p})(1-2\hat{p})^2}{n}}$$

as the desired confidence interval for $p(1-p)$.

Exercises

*9.89 Consider the distribution discussed in Example 9.18. Use the method presented in Section 9.8 to derive a $100(1-\alpha)\%$ confidence interval for $t(p) = p$. Is the resulting interval familiar to you?

*9.90 Suppose that $Y_1, Y_2, \ldots, Y_n$ constitute a random sample of size n from an exponential distribution with mean θ. Find a $100(1-\alpha)\%$ confidence interval for $t(\theta) = \theta^2$.

*9.91 Let $Y_1, Y_2, \ldots, Y_n$ denote a random sample of size n from a Poisson distribution with mean λ. Find a $100(1-\alpha)\%$ confidence interval for $t(\lambda) = e^{-\lambda} = P(Y = 0)$.

9.9 Summary

In this chapter we continued and extended the discussion of estimation begun in Chapter 8. Good estimators are consistent and efficient when compared to other estimators. The most efficient estimators, those with the smallest variances, are functions of the sufficient statistics that best summarize all of the information about the parameter of interest.

Two methods of finding estimators—the method of moments and the method of maximum likelihood—were presented. Moment estimators are consistent but generally not very efficient. Maximum-likelihood estimators, on the other hand, are consistent and, if adjusted to be unbiased, often lead to minimum-variance unbiased estimators. Since maximum-likelihood estimators have many good properties, this method is a popular method of estimation.

References and Further Readings

Casella, G., and R. L. Berger. 1990. *Statistical Inference*. Pacific Grove, Calif.: Brooks Cole.

Cramer, H. 1963. *Mathematical Methods of Statistics*. Princeton, N.J.: Princeton University Press.

Hogg, R. V., and A. T. Craig. 1995. *Introduction to Mathematical Statistics*, 5th ed., Upper Saddle River, N.J.: Prentice Hall.

Lindgren, B. W. 1993. *Statistical Theory*, 4th ed. New York: Chapman and Hall.

Mood, A. M., F. A. Graybill, and D. Boes. 1974. *Introduction to the Theory of Statistics*, 3d ed. New York: McGraw-Hill.

Serfling, R. J. 1980. *Approximation Theorems of Mathematical Statistics*. New York: John Wiley.

Wilks, S. S. 1963. *Mathematical Statistics*. New York: John Wiley.

Supplementary Exercises

9.92 Suppose that $Y_1, Y_2, \ldots, Y_n$ constitute a random sample from the density function

$$f(y \mid \theta) = \begin{cases} e^{-(y-\theta)}, & y > \theta \\ 0, & \text{elsewhere} \end{cases}$$

where θ is an unknown, positive constant.

a Find an estimator $\hat{\theta}_1$ for θ by the method of moments.

b Find an estimator $\hat{\theta}_2$ for θ by the method of maximum likelihood.

c Adjust $\hat{\theta}_1$ and $\hat{\theta}_2$ so that they are unbiased. Find the efficiency of the adjusted $\hat{\theta}_1$ relative to the adjusted $\hat{\theta}_2$.

9.93 Refer to Exercise 9.30(b). Under the conditions outlined there, find the maximum-likelihood estimator of σ^2.

*9.94 Suppose that $Y_1, Y_2, \ldots, Y_n$ denote a random sample from a Poisson distribution with mean λ. Find the MVUE of $P(Y_i = 0) = e^{-\lambda}$. (*Hint:* Make use of the Rao–Blackwell theorem.]

9.95 Suppose that a random sample of length-of-life measurements, $Y_1, Y_2, \ldots, Y_n$, is to be taken of components whose length of life has an exponential distribution with mean θ. It is frequently of interest to estimate

$$\overline{F}(t) = 1 - F(t) = e^{-t/\theta},$$

the *reliability* at time t of such a component. For any fixed value of t, find the maximum-likelihood estimator of $\overline{F}(t)$.

*9.96 The maximum-likelihood estimator obtained in Exercise 9.95 is a function of the minimal sufficient statistic for θ, but it is not unbiased. Use the Rao–Blackwell theorem to find the MVUE of $e^{-t/\theta}$ by the following steps.

a Let

$$V = \begin{cases} 1, & Y_1 > t \\ 0, & \text{elsewhere.} \end{cases}$$

Show that V is an unbiased estimator of $e^{-t/\theta}$.

b Because $U = \sum_{i=1}^{n} Y_i$ is the minimal sufficient statistic for θ, show that the conditional density function for Y_1, given $U = u$, is

$$f_{Y_1 \mid U}(y_1 \mid u) = \begin{cases} \left(\dfrac{n-1}{u^{n-1}}\right)(u - y_1)^{n-2}, & 0 < y_1 < u \\ 0, & \text{elsewhere.} \end{cases}$$

   c Show that

$$E(V \mid U) = P(Y_1 > t \mid U) = \left(1 - \frac{t}{U}\right)^{n-1}.$$

This is the MVUE of $e^{-t/\theta}$ by the Rao–Blackwell theorem and by the fact that the density function for U is complete.

*9.97 Suppose that n integers are drawn at random and *with replacement* from the integers $1, 2, \ldots, N$. That is, each sampled integer has probability $1/N$ of taking on any of the values $1, 2, \ldots, N$, and the sampled values are independent.

   a Find the method of moments estimator $\hat{N}_1$ of N.

   b Find $E(\hat{N}_1)$ and $V(\hat{N}_1)$.

*9.98 Refer to Exercise 9.97.

   a Find the maximum-likelihood estimator $\hat{N}_2$ of N.

   b Show that $E(\hat{N}_2)$ is approximately $[n/(n+1)]N$. Adjust $\hat{N}_2$ to form an estimator $\hat{N}_3$ that is approximately unbiased for N.

   c Find an approximate variance for $\hat{N}_3$ by using the fact that, for large N, the variance of the largest sampled integer is approximately

$$\frac{nN^2}{(n+1)^2(n+2)}.$$

   d Show that for large N and $n > 1$, $V(\hat{N}_3) < V(\hat{N}_1)$.

*9.99 Suppose that enemy tanks have serial numbers $1, 2, \ldots, N$. A spy randomly observed five tanks (with replacement) with serial numbers 97, 64, 118, 210, and 57. Estimate N and place a bound on the error of estimation.

9.100 Let $Y_1, Y_2, \ldots, Y_n$ denote a random sample from a Poisson distribution with mean λ and define

$$W_n = \frac{\overline{Y} - \lambda}{\sqrt{\overline{Y}/n}}.$$

   a Show that the distribution of W_n converges to a standard normal distribution.

   b Use W_n and the result in part (a) to derive the formula for an approximate 95% confidence interval for λ.

Hypothesis Testing

10.1 Introduction

10.2 Elements of a Statistical Test

10.3 Common Large-Sample Tests

10.4 Calculating Type II Error Probabilities and Finding the Sample Size for the Z Test

10.5 Relationships Between Hypothesis-Testing Procedures and Confidence Intervals

10.6 Another Way to Report the Results of a Statistical Test: Attained Significance Levels or p-Values

10.7 Some Comments on the Theory of Hypothesis Testing

10.8 Small-Sample Hypothesis Testing for μ and $\mu_1 - \mu_2$

10.9 Testing Hypotheses Concerning Variances

10.10 Power of Tests and the Neyman–Pearson Lemma

10.11 Likelihood Ratio Tests

10.12 Summary

References and Further Readings

10.1 Introduction

Recall that the objective of statistics often is to make inferences about unknown population parameters based on information contained in sample data. These inferences are phrased in one of two ways: as estimates of the respective parameters or as tests of hypotheses about their values. Chapters 8 and 9 dealt with estimation. In this chapter we discuss the general topic of hypothesis testing.

In many ways the formal procedure for hypothesis testing is similar to the scientific method. The scientist observes nature, formulates a theory, and then tests this theory against observation. In our context the scientist poses a hypothesis concerning one or more population parameters—that they equal specified values. She then samples the population and compares her observations with the hypothesis. If the observations disagree with the hypothesis, the scientist rejects it. If not, the scientist concludes either that the hypothesis is true or that the sample did not detect the difference between the real and hypothesized values of the population parameters.

For example, a medical researcher may hypothesize that a new drug is more effective than another in combating a disease. To test her hypothesis, she randomly selects patients infected with the disease and randomly divides them into two groups. The new drug A is given to the patients in the first group and the old drug B is given to the patients in the second group. Then, based on the number of patients in each group who recover from the disease, the researcher must decide whether the new drug is more effective than the old.

Hypothesis tests are conducted in all fields in which theory can be tested against observation. A quality control engineer may hypothesize that a new assembly method produces only 5% defective items. An educator may claim that two methods of teaching reading are equally effective, or a political candidate may claim that a plurality of voters favor her election. All such hypotheses can be subjected to statistical verification by using observed sample data.

What is the role of statistics in testing hypotheses? Putting it more bluntly, of what value is statistics in this hypothesis-testing procedure? Testing a hypothesis requires making a decision when comparing the observed sample with theory. How do we decide whether the sample disagrees with the scientist's hypothesis? When should we reject the hypothesis, when should we accept it, and when should we withhold judgment? What is the probability that we will make the wrong decision and consequently be led to a loss? And, particularly, what function of the sample measurements should be employed to reach a decision? The answers to these questions are contained in a study of statistical hypothesis testing.

Chapter 8 introduced the general topic of estimation and presented some intuitive estimation procedures. Chapter 9 presented some properties of estimators and some formal methods for deriving estimators. We will use the same approach in our discussion of hypothesis testing. That is, we will introduce the topic, present some intuitive testing procedures, and then consider some formal methods for deriving statistical hypothesis testing procedures.

10.2   Elements of a Statistical Test

Many times the objective of a statistical test is to test a hypothesis concerning the values of one or more population parameters. We generally have a theory—a *research hypothesis*—about the parameter(s) that we wish to support. For example, suppose that a political candidate, Jones, claims that he will gain more than 50% of the votes in a city election and thereby emerge as the winner. If we do not believe Jones' claim, we might seek to support the research hypothesis that Jones is *not* favored by more than 50% of the electorate. Support for this research hypothesis, also called the *alternative hypothesis*, is obtained by showing (using the sample data as evidence) that the converse of the alternative hypothesis, called the *null hypothesis*, is false. Thus support for one theory is obtained by showing lack of support for its converse—in a sense, a proof by contradiction. Because we seek support for the alternative hypothesis that Jones' claim is false, our alternative hypothesis is that p, the probability of selecting a voter favoring Jones, is less than .5. If we can show that the data support rejection of the null hypothesis $p = .5$ (the minimum value needed for a plurality) in favor of the alternative hypothesis $p < .5$, we have achieved our research objective. Although it is common to speak of testing a null hypothesis, the research objective usually is to show support for the alternative hypothesis, if such support is warranted.

How do we utilize that data to decide between the null hypothesis and the alternative hypothesis? Suppose that $n = 15$ voters are randomly selected from the city and Y, the number favoring Jones, is recorded. If none in the sample favor Jones ($Y = 0$), what would you conclude about Jones' claim? If Jones is actually favored by more than 50% of the electorate, it is not *impossible* to observe $Y = 0$ favoring Jones in a sample of size $n = 15$, but it is highly *improbable*. It is much more likely that we would observe $Y = 0$ if the alternative hypothesis were true. Thus we would reject the null hypothesis ($p = .5$) in favor of the alternative hypothesis ($p < .5$). If we observed $Y = 1$ (or any small value of Y), the same type of reasoning would lead us to the same conclusion.

Any statistical test of hypotheses works in exactly the same way and is composed of the same essential elements.

The Elements of a Statistical Test

1. Null hypothesis, H_0.
2. Alternative hypothesis, H_a.
3. Test statistic.
4. Rejection region.

For our example the hypothesis to be tested, called the *null hypothesis* and denoted by H_0, is $p = .5$. The alternative (or research) hypothesis, denoted as H_a, is the hypothesis to be accepted in case H_0 is rejected. The alternative hypothesis usually is the hypothesis we seek to support on the basis of the information contained in the sample; thus, in our example, H_a is $p < .5$.

The functioning parts of a statistical test are the test statistic and an associated rejection region. The *test statistic* (like an estimator) is a function of the sample measurements (Y in our example) upon which the statistical decision will be based. The *rejection region*, which will henceforth be denoted by RR, specifies the values of the test statistic for which the null hypothesis is to be *rejected* in favor of the alternative hypothesis. If for a particular sample the computed value of the test statistic falls in the rejection region RR, we reject the null hypothesis H_0 and accept the alternative hypothesis H_a. If the value of the test statistic does not fall into the rejection region RR, we accept H_0. As previously indicated, for our example small values of Y would lead us to reject H_0. Therefore, one rejection region we might want to consider is the set of all values of Y less than or equal to 2. We will use the notation RR $= \{y : y \leq 2\}$—or, more simply, RR $= \{y \leq 2\}$—to denote this rejection region.

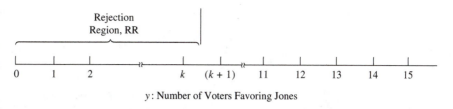

FIGURE **10.1**
Rejection region,
$RR = \{y \leq k\}$, for a
test of the hypothesis
$H_0: p = .5$ against
the alternative
$H_a: p < .5$

Finding a good rejection region for a statistical test is an interesting problem that merits further attention. It is clear that small values of Y, say, $y \leq k$ (see Figure 10.1), are contradictory to the hypothesis $H_0: p = .5$ but favorable to the alternative $H_a: p < .5$. So we intuitively choose the rejection region as $RR = \{y \leq k\}$. But what value should we choose for k? More generally, we seek some objective criteria for deciding which value of k specifies a good rejection region of the form $\{y \leq k\}$.

For any fixed rejection region (determined by a particular value of k), two types of errors can be made in reaching a decision. We can decide in favor of H_a when H_0 is true (make a *type I error*), or we can decide in favor of H_0 when H_a is true (make a *type II error*).

DEFINITION **10.1**

A *type I error* is made if H_0 is rejected when H_0 is true. The *probability of a type I error* is denoted by α. The value of α is called the *level* of the test.

A *type II error* is made if H_0 is accepted when H_a is true. The *probability of a type II error* is denoted by β.

For Jones' political poll, making a type I error—rejecting $H_0: p = .5$ (and thereby accepting $H_a: p < .5$) when in fact H_0 is true—means concluding that Jones will lose when, in fact, he is going to win. In contrast, making a type II error means accepting $H_0: p = .5$ when $p < .5$ and concluding that Jones will win, when, in fact, he will lose. For most real situations, incorrect decisions cost money, prestige, or time and imply a loss. Thus α and β, the probabilities of making these two types of errors, measure the risks associated with the two possible erroneous decisions that might result from a statistical test. As such, they provide a very practical way to measure the goodness of a test.

EXAMPLE **10.1**   For Jones' political poll, $n = 15$ voters were sampled. We wish to test $H_0: p = .5$ against the alternative, $H_a: p < .5$. The test statistic is Y, the number of sampled voters favoring Jones. Calculate α if we select $RR = \{y \leq 2\}$ as the rejection region.

Solution   By definition,

$$\alpha = P(\text{type I error}) = P(\text{rejecting } H_0 \text{ when } H_0 \text{ is true})$$

$$= P(\text{value of test statistic is in RR when } H_0 \text{ is true})$$

$$= P(Y \leq 2 \text{ when } p = .5).$$

Observing that Y is a binomial random variable with $n = 15$ and $p = .5$, we obtain

$$\alpha = \sum_{y=0}^{2} \binom{15}{y}(.5)^y(.5)^{15-y} = \binom{15}{0}(.5)^{15} + \binom{15}{1}(.5)^{15} + \binom{15}{2}(.5)^{15}.$$

Using Table I, Appendix III, to circumvent this computation, we find $\alpha = .004$. Thus, if we decide to use the rejection region RR $= \{y \leq 2\}$, we subject ourselves to a very small risk ($\alpha = .004$) of concluding that Jones will lose if, in fact, he is a winner.

EXAMPLE **10.2**   Refer to Example 10.1. Is our test equally good in protecting us from concluding that Jones is a winner if, in fact, he will lose? Suppose that he will receive 30% of the votes ($p = .3$). What is the probability β that the sample will erroneously lead us to conclude that H_0 is true and that Jones is going to win?

Solution   By definition,

$$\beta = P(\text{type II error}) = P(\text{accepting } H_0 \text{ when } H_a \text{ is true})$$

$$= P(\text{value of the test statistic is } not \text{ in RR when } H_a \text{ is true}).$$

Because we want to calculate β when $p = .3$ (a particular value of p that is in H_a),

$$\beta = P(Y > 2 \text{ when } p = .3) = \sum_{y=3}^{15} \binom{15}{y}(.3)^y(.7)^{15-y}.$$

Again consulting Table 1, Appendix III, we find that $\beta = .873$. In other words, if we use RR $= \{y \leq 2\}$, our test will usually lead us to conclude that Jones is a winner (with probability $\beta = .873$), even if p is as low as $p = .3$.

The value of β depends on the true value of the parameter p. The larger the difference is between p and the (null) hypothesized value of $p = .5$, the smaller is the likelihood that we will fail to reject the null hypothesis.

EXAMPLE **10.3**   Refer to Examples 10.1 and 10.2. Calculate the value of β if Jones will receive only 10% of the votes ($p = .1$).

Solution   In this case we want to calculate β when $p = .1$ (another particular value of p in H_a).

$$\beta = P(\text{type II error}) = P(\text{accepting } H_0 \text{ when } p = .1)$$

$$= P(\text{value of test statistic is } not \text{ in RR when } p = .1)$$

$$= P(Y > 2 \text{ when } p = .1) = \sum_{y=3}^{15} \binom{15}{y}(.1)^y(.9)^{15-y} = .184.$$

Consequently, if we use $\{y \le 2\}$ as the rejection region, the value of β when $p = .10$ is smaller than the value for β that we obtained in Example 10.2 with $p = .30$ (.184 versus .873). Nonetheless, in using this rejection region, we still have a fairly large probability of claiming that Jones is a winner if, in fact, he will receive only 10% of the votes.

Examples 10.1 through 10.3 show that the test using $RR = \{y \le 2\}$ guarantees a low risk of making a type I error ($\alpha = .004$), but it does not offer adequate protection against a type II error. How can we improve our test? One way is to balance α and β by changing the rejection region. If we enlarge RR into a new rejection region RR^* (that is, $RR \subset RR^*$), the test using RR^* will lead us to reject H_0 more often. If α^* and α denote the probabilities of type I errors (levels of the tests) when we use RR^* and RR as the rejection regions, respectively, then, because $RR \subset RR^*$,

$$\alpha^* = P(\text{test statistic is in } RR^* \text{ when } H_0 \text{ is true})$$

$$\ge P(\text{test statistic is in RR when } H_0 \text{ is true}) = \alpha.$$

Likewise, if we use the enlarged rejection region RR^*, the test procedure will lead us to accept H_0 less often. If β^* and β denote the probabilities of type II errors for the tests using RR^* and RR, respectively, then

$$\beta^* = P(\text{test statistic is } not \text{ in } RR^* \text{ when } H_a \text{ is true})$$

$$\le P(\text{test statistic is } not \text{ in RR when } H_a \text{ is true}) = \beta.$$

Hence, if we change the rejection region to increase α, then β will decrease. Similarly, if the change in rejection region results in a decrease in α, then β will increase. Thus, α and β are inversely related.

EXAMPLE **10.4**   Refer to the test discussed in Example 10.1. Now assume that $RR = \{y \le 5\}$. Calculate the level α of the test and calculate β if $p = .3$. Compare the results with the values obtained in Examples 10.1 and 10.2 (where we used $RR = \{y \le 2\}$).

Solution   In this case,

$$\alpha = P(\text{test statistic is in RR when } H_0 \text{ is true})$$

$$= P(Y \le 5 \text{ when } p = .5) = \sum_{y=0}^{5} \binom{15}{y}(.5)^{15} = .151.$$

When $p = .3$,

$$\beta = P(\text{test statistic is not in RR when } H_a \text{ is true and } p = .3)$$

$$= P(Y > 5 \text{ when } p = .3) = \sum_{y=6}^{15} \binom{15}{y}(.3)^y(.7)^{15-y} = .278.$$

Table 10.1. Comparison of α and β for two different rejection regions

Probabilities	RR	
of Error	$(0, 1, 2)$	$(0, 1, 2, \ldots, 5)$
α	.004	.151
β when $p = .3$	.873	.278

A comparison of the α and β calculated here with the results of Examples 10.1 and 10.2 shows that enlarging the rejection region from RR $= \{y \le 2\}$ to RR* $= \{y \le 5\}$ increased α and decreased β (see Table 10.1). Hence we have achieved a better balance between the risks of type I and type II errors, but both α and β remain disconcertingly large. *How can we reduce both α and β?* The answer is intuitively clear: shed more light on the true nature of the population by increasing the sample size. For almost all statistical tests, if α is fixed at some acceptably small value, β decreases as the sample size increases.

In this section we have defined the essential elements of any statistical test. We have seen that two possible types of error can be made when testing hypotheses: type I and type II errors. The probabilities of these errors serve as criteria for evaluating a testing procedure. In the next few sections we will utilize the sampling distributions derived in Chapter 7 to develop methods for testing hypotheses about parameters of frequent practical interest.

Exercises

10.1 Define α and β for a statistical test of hypotheses.

10.2 An experimenter has prepared a drug dosage level that she claims will induce sleep for 80% of people suffering from insomnia. After examining the dosage, we feel that her claims regarding the effectiveness of the dosage are inflated. In an attempt to disprove her claim, we administer her prescribed dosage to 20 insomniacs, and we observe Y, the number for which the drug dose induces sleep. We wish to test the hypothesis $H_0: p = .8$ versus the alternative, $H_a: p < .8$. Assume that the rejection region $\{y \le 12\}$ is used.

  a In terms of this problem, what is a type I error?
  b Find α.
  c In terms of this problem, what is a type II error?
  d Find β when $p = .6$.
  e Find β when $p = .4$.

10.3 Refer to Exercise 10.2.

  a Find the rejection region of the form $\{y \le c\}$ so that $\alpha \approx .01$.
  b For the rejection region in (a), find β when $p = .6$.
  c For the rejection region in (a), find β when $p = .4$.

10.4   Suppose that we wish to test the null hypothesis, H_0, that the proportion p of ledger sheets with errors is equal to .05 versus the alternative, H_a, that the proportion is larger than .05, by using the following scheme. Two ledger sheets are selected at random. If both are error-free, we reject H_0. If one or more contains an error, we look at a third sheet. If the third sheet is error-free, we reject H_0. In all other cases we accept H_0.

a   In terms of this problem, what is a type I error?

b   What is the value of α associated with this test?

c   In terms of this problem, what is a type II error?

d   Calculate $\beta = P(\text{type II error})$ as a function of p.

10.5   Let Y_1 and Y_2 be independent and identically distributed with a uniform distribution over the interval $(\theta, \theta + 1)$. For testing $H_0: \theta = 0$ versus $H_a: \theta > 0$, we have two competing tests:

Test 1: Reject H_0 if $Y_1 > 0.95$.

Test 2: Reject H_0 if $Y_1 + Y_2 > c$.

Find the value of c so that Test 2 has the same value for α as Test 1. [*Hint*: In Example 6.3 we derived the density and distribution function of the sum of two independent random variables that are uniformly distributed on the interval $(0, 1)$.]

*10.6   A two-stage clinical trial is planned for testing $H_0: p = 0.10$ versus $H_a: p > 0.10$, where p is the proportion of responders among patients who were treated by the protocol treatment. At the first stage, 15 patients are accrued and treated. If 4 or more responders are observed among the (first) 15 patients, H_0 is rejected, the study is terminated, and no more patients are accrued. Otherwise, another 15 patients will be accrued and treated in the second stage. If a total of 6 or more responders are observed among the 30 patients accrued in the two stages (15 in the first stage and 15 more in the second stage), then H_0 is rejected. For example, if 5 responders are found among the first stage patients, H_0 is rejected and the study is over. However, if 2 responders are found among the first stage patients, 15 second-stage patients are accrued, and if an additional 4 or more responders (for a total of 6 or more among the 30) are identified H_0 is rejected and the study is over.[1]

a   Use the binomial table to find the numerical value of α for this testing procedure.

b   Use the binomial table to find the probability of rejecting the null hypothesis when using this rejection region if $p = 0.30$.

c   For the rejection region defined above, find β if $p = 0.30$.

10.3   Common Large-Sample Tests

Suppose that we want to test a set of hypotheses concerning a parameter θ based on a random sample $Y_1, Y_2, \ldots, Y_n$. In this section we will develop hypothesis-testing procedures that are based on an estimator $\hat{\theta}$ that has an (approximately) normal sampling distribution with mean θ and standard error $\sigma_{\hat{\theta}}$. The large-sample estimators of Chapter 8 (Table 8.1), such as $\overline{Y}$ and $\hat{p}$ used for estimating a population mean μ and proportion p, respectively, satisfy these requirements. So do the estimators for the comparison of two means $(\mu_1 - \mu_2)$ and for the comparison of two binomial parameters $(p_1 - p_2)$.

---

[1] Exercises preceded by an asterisk are optional.

FIGURE **10.2**
Sampling
distributions of
the estimator $\hat{\theta}$ for
various values of θ

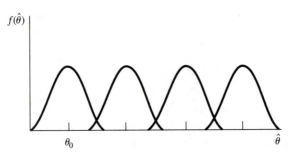

If θ_0 is a specific value of θ, we may wish to test $H_0 : \theta = \theta_0$ versus $H_a : \theta > \theta_0$. Figure 10.2 contains a graph illustrating the sampling distributions of $\hat{\theta}$ for various values of θ. If $\hat{\theta}$ is close to θ_0, it seems reasonable to accept H_0. If, in reality, $\theta > \theta_0$, however, $\hat{\theta}$ is more likely to be large. Consequently, large values of $\hat{\theta}$ (values larger than θ_0 by a suitable amount) favor rejection of $H_0 : \theta = \theta_0$ and acceptance of $H_a : \theta > \theta_0$. That is, the null and alternative hypotheses, the test statistic, and the rejection region are as follows:

$H_0 : \theta = \theta_0$.

$H_a : \theta > \theta_0$.

Test statistic: $\hat{\theta}$.

Rejection region: RR $= \{\hat{\theta} > k\}$ for some choice of k.

The actual value of k in the rejection region RR is determined by fixing the type I error probability α (the level of the test) and choosing k accordingly (see Figure 10.3). If H_0 is true, $\hat{\theta}$ has an approximately normal distribution with mean θ_0 and standard error $\sigma_{\hat{\theta}}$. Therefore, if we desire an α-level test,

$$k = \theta_0 + z_\alpha \sigma_{\hat{\theta}}$$

is the appropriate choice for k [if Z has a standard normal distribution, then z_α is such that $P(Z > z_\alpha) = \alpha$]. Because

FIGURE **10.3**
Large-sample
rejection region for
$H_0 : \theta = \theta_0$ versus
$H_a : \theta > \theta_0$

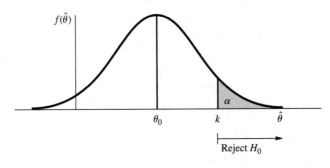

$$\text{RR} = \{\hat{\theta} : \hat{\theta} > \theta_0 + z_\alpha \sigma_{\hat{\theta}}\} = \left\{ \hat{\theta} : \frac{\hat{\theta} - \theta_0}{\sigma_{\hat{\theta}}} > z_\alpha \right\}$$

if $Z = (\hat{\theta} - \theta_0)/\sigma_{\hat{\theta}}$ is used as the test statistic, the rejection region may also be written as RR $= \{z > z_\alpha\}$. Notice that Z measures the number of standard errors between the estimator for θ and θ_0, the value of θ specified in H_0. Thus an equivalent form of the test of hypothesis, with level α, is as follows:

$H_0 : \theta = \theta_0$.

$H_a : \theta > \theta_0$.

Test statistic: $Z = \dfrac{\hat{\theta} - \theta_0}{\sigma_{\hat{\theta}}}$.

Rejection region: $\{z > z_\alpha\}$.

H_0 is rejected if Z falls far enough into the upper tail of the standard normal distribution. The alternative hypothesis $H_a : \theta > \theta_0$ is called an *upper-tail* alternative, and RR $= \{z > z_\alpha\}$ is referred to as an *upper-tail rejection region*. Notice that the preceding formula for Z is simply

$$Z = \frac{\text{estimator for the parameter} - \text{value for the parameter given by } H_0}{\text{standard error of the estimator}}.$$

EXAMPLE 10.5   A vice president in charge of sales for a large corporation claims that salespeople are averaging no more than 15 sales contacts per week. (He would like to increase this figure.) As a check on his claim, $n = 36$ salespeople are selected at random, and the number of contacts made by each is recorded for a single randomly selected week. The mean and variance of the 36 measurements were 17 and 9, respectively. Does the evidence contradict the vice president's claim? Use a test with level $\alpha = .05$.

Solution   We are interested in the research hypothesis that the vice president's claim is incorrect. This can be formally written as $H_a : \mu > 15$, where μ is the mean number of sales contacts per week. Thus, we are interested in testing

$$H_0 : \mu = 15 \quad \text{against} \quad H_a : \mu > 15.$$

We know that for large enough n, the sample mean $\overline{Y}$ is a point estimator of μ that is approximately normally distributed with $\mu_{\overline{Y}} = \mu$ and $\sigma_{\overline{Y}} = \sigma/\sqrt{n}$. Hence, our test statistic is

$$Z = \frac{\overline{Y} - \mu_0}{\sigma_{\overline{Y}}} = \frac{\overline{Y} - \mu_0}{\sigma/\sqrt{n}}.$$

The rejection region, with $\alpha = .05$, is given by $\{z > z_{.05} = 1.645\}$ (see Table 4, Appendix III). The population variance σ^2 is not known, but it can be estimated very accurately (because $n = 36$ is sufficiently large) by the sample variance $s^2 = 9$.

Thus, the observed value of the test statistic is approximately

$$z = \frac{\bar{y} - \mu}{s/\sqrt{n}} = \frac{17 - 15}{3/\sqrt{36}} = 4.$$

Because the observed value of Z lies in the rejection region (because $z = 4$ exceeds $z_{.05} = 1.645$), we reject $H_0 : \mu = 15$. Thus, at the $\alpha = .05$ level of significance, the evidence is sufficient to indicate that the vice president's claim is incorrect and that the average number of sales contacts per week exceeds 15.

EXAMPLE 10.6   A machine in a factory must be repaired if it produces more than 10% defectives among the large lot of items it produces in a day. A random sample of 100 items from the day's production contains 15 defectives, and the supervisor says that the machine must be repaired. Does the sample evidence support his decision? Use a test with level .01.

Solution   If Y denotes the number of observed defectives, then Y is a binomial random variable, with p denoting the probability that a randomly selected item is defective. Hence, we want to test the null hypothesis

$$H_0 : p = .10 \quad \text{against the alternative} \quad H_a : p > .10.$$

The test statistic, which is based on $\hat{p} = Y/n$ (the unbiased point estimator of p), is given by

$$Z = \frac{\hat{p} - p_0}{\sigma_{\hat{p}}} = \frac{\hat{p} - p_0}{\sqrt{\dfrac{p_0(1 - p_0)}{n}}}.$$

We could have used $\sqrt{\hat{p}(1 - \hat{p})/n}$ to approximate the standard error of $\hat{p}$, but because we are considering the distribution of Z under H_0, it is more appropriate to use $\sqrt{p_0(1 - p_0)/n}$, the true value of the standard error of $\hat{p}$ when H_0 is true.

From Table 4, Appendix III, we see that $P(Z > 2.33) = .01$. Hence we take $\{z > 2.33\}$ as the rejection region. The observed value of the test statistic is given by

$$z = \frac{\hat{p} - p_0}{\sqrt{\dfrac{p_0(1 - p_0)}{n}}} = \frac{.15 - .10}{\sqrt{\dfrac{(.1)(.9)}{100}}} = \frac{5}{3} = 1.667.$$

Because the observed value of Z is not in the rejection region, we cannot reject $H_0 : p = .10$ in favor of $H_a : p > .10$. In terms of this application, we conclude that, at the $\alpha = .01$ level of significance, the evidence does not support the supervisor's decision.

Is the supervisor wrong? We can not make a statistical judgment about this until we have evaluated the probability of accepting H_0 when H_a is true—that is, until we have calculated β. The method for calculating β is presented in Section 10.4.

Testing $H_0 : \theta = \theta_0$ against $H_a : \theta < \theta_0$ is done in an analogous manner, except that we now reject H_0 for values of $\hat{\theta}$ that are much smaller than θ_0. The test statistic remains

$$Z = \frac{\hat{\theta} - \theta_0}{\sigma_{\hat{\theta}}}$$

but for a fixed level α we reject the null hypothesis when $z < -z_\alpha$. Because we reject H_0 in favor of H_a when z falls far enough into the lower tail of the standard normal distribution, we call $H_a : \theta < \theta_0$ a *lower-tail* alternative and RR: $\{z < -z_\alpha\}$ a *lower-tail rejection region*.

In testing $H_0 : \theta = \theta_0$ against $H_a : \theta \neq \theta_0$, we reject H_0 if $\hat{\theta}$ is either much smaller or much larger than θ_0. The test statistic is still Z, as before, but the rejection region is located symmetrically in the two tails of the probability distribution for Z. Thus, we reject H_0 if either $z < -z_{\alpha/2}$ or $z > z_{\alpha/2}$. Equivalently, we reject H_0 if $|z| > z_{\alpha/2}$. This test is called a *two-tailed test*, as opposed to the *one-tailed tests* used for the alternatives $\theta < \theta_0$ and $\theta > \theta_0$. The rejection regions for the lower-tail alternative, $H_a : \theta < \theta_0$, and the two-sided alternative, $H_a : \theta \neq \theta_0$, are displayed in Figure 10.4.

A summary of the large-sample α-level hypothesis tests developed so far is given next.

FIGURE **10.4**
Rejection regions for testing $H_0 : \theta = \theta_0$ versus (a) $H_a : \theta < \theta_0$, and (b) $H_a : \theta \neq \theta_0$, based on $Z = \frac{\hat{\theta} - \theta_0}{\sigma_{\hat{\theta}}}$

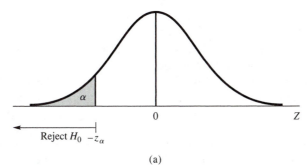

(a)

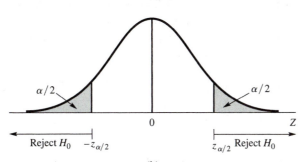

(b)

> **Large-Sample α-Level Hypothesis Tests**
>
> H_0: $\theta = \theta_0$.
>
> H_a: $\begin{cases} \theta > \theta_0 & \textit{(upper-tail alternative).} \\ \theta < \theta_0 & \textit{(lower-tail alternative).} \\ \theta \neq \theta_0 & \textit{(two-tailed alternative).} \end{cases}$
>
> Test statistic: $Z = \dfrac{\hat{\theta} - \theta_0}{\sigma_{\hat{\theta}}}$.
>
> Rejection region: $\begin{cases} \{z > z_\alpha\} & \textit{(upper-tail} \text{ RR).} \\ \{z < -z_\alpha\} & \textit{(lower-tail} \text{ RR).} \\ \{|z| > z_{\alpha/2}\} & \textit{(two-tailed} \text{ RR).} \end{cases}$

In any particular test, only one of the listed alternatives H_a is appropriate. Whatever alternative hypothesis we choose, we must be sure to use the corresponding rejection region.

How do we decide which alternative to use for a test? The answer depends on the hypothesis that we seek to support. If we are interested only in detecting an increase in the percentage of defectives (Example 10.6), we should locate the rejection region in the upper tail of the standard normal distribution. On the other hand, if we wish to detect a change in p either above or below $p = .10$, we should locate the rejection region in both tails of the standard normal distribution and employ a two-tailed test. The following example illustrates a situation in which a two-tailed test is appropriate.

EXAMPLE 10.7   A psychological study was conducted to compare the reaction times of men and women to a stimulus. Independent random samples of 50 men and 50 women were employed in the experiment. The results are shown in Table 10.2. Do the data present sufficient evidence to suggest a difference between true mean reaction times for men and women? Use $\alpha = .05$.

Solution   Let μ_1 and μ_2 denote the true mean reaction times for men and women, respectively. If we wish to test the hypothesis that the means differ, we must test H_0: $(\mu_1 - \mu_2) = 0$ against H_a: $(\mu_1 - \mu_2) \neq 0$. The two-sided alternative permits us to detect either the case $\mu_1 > \mu_2$ or the reverse case $\mu_2 > \mu_1$; in either case, H_0 is false.

The point estimator of $(\mu_1 - \mu_2)$ is $(\overline{Y}_1 - \overline{Y}_2)$. As we discussed in Sections 8.3 and 8.6, because the samples are independent and both are large, this estimator satisfies

Table 10.2. Data for Example 10.7

Men	Women
$n_1 = 50$	$n_2 = 50$
$\overline{y}_1 = 3.6$ seconds	$\overline{y}_2 = 3.8$ seconds
$s_1^2 = .18$	$s_2^2 = .14$

the assumptions necessary to develop a large-sample test. Hence, if we desire to test $H_0: \mu_1 - \mu_2 = D_0$ (where D_0 is some fixed value) versus any alternative, the test statistic is given by

$$Z = \frac{(\overline{Y}_1 - \overline{Y}_2) - D_0}{\sqrt{\dfrac{\sigma_1^2}{n_1} + \dfrac{\sigma_2^2}{n_2}}}$$

where σ_1^2 and σ_2^2 are the respective population variances. In this application we want to use a two-tailed test. Thus, for $\alpha = .05$, we reject H_0 for $|z| > z_{\alpha/2} = z_{.025} = 1.96$.

For large samples (say, $n_i > 30$) the sample variances provide good estimates of their corresponding population variances. Substituting these values, along with $\overline{y}_1, \overline{y}_2, n_1, n_2$, and $D_0 = 0$, into the formula for the test statistic, we have

$$z = \frac{\overline{y}_1 - \overline{y}_2 - 0}{\sqrt{\dfrac{\sigma_1^2}{n_1} + \dfrac{\sigma_2^2}{n_2}}} \approx \frac{3.6 - 3.8}{\sqrt{\dfrac{.18}{50} + \dfrac{.14}{50}}} = -2.5.$$

This value is less than $-z_{\alpha/2} = -1.96$ and, therefore, falls in the rejection region. Hence, at the $\alpha = .05$ level, we conclude that sufficient evidence exists to permit us to conclude that mean reaction times differ for men and women.

In this section we have described the general procedure for implementing large-sample test of hypotheses for some parameters of frequent practical interest. We will discuss in Section 10.4 how to calculate β, the probability of a type II error, for these large-sample tests. Constructing confidence intervals for these parameters and implementing formal tests of hypotheses are remarkably similar. Both procedures use the estimators of the respective parameters, the standard errors of these estimators, and quantities obtained from the table of the standard normal distribution. In Section 10.5 we will explicitly point out a correspondence between large sample testing procedures and large sample confidence intervals.

Exercises

10.7 Many computer buyers have discovered that they can save a considerable amount by purchasing a personal computer (PC) from a mail-order company—an average of $900 by their estimates ("Who's Tops in Service?" *PC World*, November 1992). In a test of this claim, a random sample of 35 customers who recently purchased a PC through a mail-order company were contacted and asked to estimate the amount that they had saved by purchasing by mail. The mean and standard deviation of these 35 estimates were $885 and $50, respectively. Is there sufficient evidence to indicate that the average savings differs from the $900 claimed by mail-order PC buyers?

a   State the null and alternative hypotheses.

b   What is the appropriate rejection region for an $\alpha = .01$ level test?

c   Calculate the observed value of the appropriate test statistic.

d   What is your conclusion?

10.8   The hourly wages in a particular industry are normally distributed with mean $13.20 and standard deviation $2.50. A company in this industry employs 40 workers, paying them an average of $12.20 per hour. Can this company be accused of paying substandard wages? Use an $\alpha = .01$ level test.

10.9   The output voltage for a certain electric circuit is specified to be 130. A sample of 40 independent readings on the voltage for this circuit gave a sample mean of 128.6 and a standard deviation of 2.1. Test the hypothesis that the average output voltage is 130 against the alternative that it is less than 130. Use a test with level .05.

10.10   The Rockwell hardness index for steel is determined by pressing a diamond point into the steel and measuring the depth of penetration. For 50 specimens of a certain type of steel, the Rockwell hardness index averaged 62 with standard deviation 8. The manufacturer claims that this type of steel has an average hardness index of at least 64. Is there sufficient evidence to refute the manufacturer's claim at the 1% significance level?

10.11   Shear strength measurements derived from unconfined compression tests for two types of soils gave the results shown in the following table (measurements in tons per square foot).

Soil Type I	Soil Type II
$n_1 = 30$	$n_2 = 35$
$\bar{y}_1 = 1.65$	$\bar{y}_2 = 1.43$
$s_1 = 0.26$	$s_2 = 0.22$

Do the soils appear to differ with respect to average shear strength, at the 1% significance level?

10.12   In Exercise 8.54, we examined the results of a 1993 study by Alice Chang and her colleagues comparing the leisure interest checklist (LIC) inventory scores for various activities. The results were given for samples of 252 boys and 307 girls. One of the activities studied was sports.

a   Without looking at the data, would you expect that boys would have a greater or lesser interest in sports as a leisure time activity? Based on your conjecture, what alternative hypothesis would you choose to test versus the null hypothesis that there is no difference in the mean LIC scores for boys and girls?

b   Does the alternative hypothesis you posed in (a) correspond to a one-tailed or a two-tailed statistical test?

c   The mean and standard deviation of the LIC inventory scores for boys were 13.65 and 4.82, respectively. For girls, the corresponding mean and standard deviation were 9.88 and 4.41. Do the data provide support for the conjecture that the mean LIC scores for sports is higher for boys than for girls? Test using $\alpha = .01$.

10.13   Charles Dickey ("A Strategy for Big Bucks," *Field and Stream*, October 1980) discusses studies of the habits of white-tailed deer that indicate that these deer live and feed within very limited ranges, approximately 150 to 205 acres. To determine whether the ranges of deer located in two different geographical areas differ, researchers caught, tagged, and fitted 40 deer with small radio transmitters. Several months later, the deer were tracked and identified and the distance y from the release point was recorded. The mean and standard deviation of the distances from the release point were as given in the accompanying table.

	Location	
	1	2
Sample size	40	40
Sample mean	2980 ft	3205 ft
Sample standard deviation	1140 ft	963 ft
Population mean	μ_1	μ_2

a   If you have no preconceived reason for believing one population mean to be larger than the other, what would you choose for your alternative hypothesis? Your null hypothesis?

b   Would your alternative hypothesis in (a) imply a one-tailed or a two-tailed test? Explain.

c   Do the data provide sufficient evidence to indicate that the mean distances differ for the two geographical locations? Test using $\alpha = .10$.

10.14   In a recent Associated Press article ("College Campuses Graying: 1 in 4 Students 30 or Older," *Riverside Press-Enterprise*, December 1, 1991), the columnist claimed that one in four college students is 30 years old or older, indicating that older Americans are returning to school at a higher rate than in previous years. A random sample of 300 students in community colleges found 98 who were 30 or older.

a   Test whether the percentage claimed by the columnist is applicable to the group of community college students from which this sample was selected. Use $\alpha = .05$.

b   Do the results of this study invalidate the columnists claim? Why?

10.15   According to a recent *New York Times*/CBS News poll (*Gainesville Sun*, November 3, 1994), 60% of the 1429 adults interviewed were unable to name an elected official whom they admired. Is there sufficient evidence to claim that a majority of adults are unable to name an elected official whom they admire? Use a .01 level test.

10.16   An article in the March 16, 1993, *Washington Post* stated that nearly 45% of all Americans have brown eyes. A random sample of 80 people found 32 with brown eyes. Is there sufficient evidence at the .01 level to indicate that the proportion of brown-eyed people in the region where the study was performed differs from the value reported in the *Washington Post*?

10.17   A survey conducted by the National Association of College Admission Counselors examined enrollment in colleges and universities for the years 1991 and 1992 (*USA Today*, August 18, 1992). The 1992 survey was based on 1232 questionnaires; we will assume that the number of responses in 1991 was 1225. The 1992 results indicated that 71% of the colleges reported an increase in applications, compared with 50% in 1991. Do the data indicate a significant difference in the proportions of colleges reporting an increase in applications for the years 1991 and 1992? Use $\alpha = .01$.

10.18   For several years, aspirin-based products lost market share to substitute analgesics such as ibuprofen and acetaminophen. As a result, large companies such as Bayer, which had previously produced only aspirin-based products, began marketing acetaminophen-based pain relievers (*Los Angeles Times*, March 27, 1993). The following table gives the proportions of individuals who preferred each of the three major pain relievers for the years 1986 and 1991.

Pain Reliever	1986	1991
Aspirin	45%	34%
Acetaminophen	41%	41%
Ibuprofen	14%	26%

Assume that these results were obtained on the basis of two independent samples, each of size 1000.

a Does statistically significant evidence exist to suggest that the proportions of aspirin users differ for 1986 and 1991? Use a .05 level test.

b Is there sufficient evidence to indicate that ibuprofen has increased its market share between 1986 and 1991? Use $\alpha = .05$.

c Are the tests performed in (a) and (b) related? Explain

10.19 A manufacturer of automatic washers offers a particular model in one of three colors: A, B, or C. Of the first 1000 washers sold, 400 were of color A. Would you conclude that customers have a preference for color A? Justify your answer.

10.20 A manufacturer claimed that at least 20% of the public preferred her product. A sample of 100 persons is taken to check her claim. With $\alpha = .05$, how small would the sample percentage have to be before the claim could legitimately be refuted? (Notice that this would involve a one-tailed test of the hypothesis.)

10.21 What conditions must be met in order for the Z test to be used to test a hypothesis concerning a population mean μ?

10.22 The results on a *New York Times*/CBS News poll involving 1429 adults indicated that 56% of those interviewed approved of the performance of their representatives in Congress but only 33% thought that their representatives deserved reelection (*Gainesville Sun*, November 3, 1994).

a Is there sufficient evidence to permit us to claim that the majority of adults approved of the performance of their congressional representatives? Use $\alpha = .01$.

b Can we claim that the majority of adults felt that their congressional representatives deserved reelection? Test using $\alpha = .05$.

10.23 A political researcher believes that the fraction p_1 of Republicans strongly in favor of the death penalty is greater than the fraction p_2 of Democrats strongly in favor of the death penalty. He acquired independent random samples of 200 Republicans and 200 Democrats and found 46 Republicans and 34 Democrats strongly favoring the death penalty. Does this evidence provide statistical support for the researcher's belief? Use $\alpha = .05$.

10.24 Exercise 8.44 stated that a random sample of 500 measurements on the length of stay in hospitals had sample mean 5.4 days and sample standard deviation 3.1 days. A federal regulatory agency hypothesizes that the average length of stay is in excess of 5 days. Do the data support this hypothesis? Use $\alpha = .05$.

10.25 Michael Sosin ("Homeless and Vulnerable Meal Program Users: A Comparison Study," *Social Problems* 39(2), 1992) recently investigated determinants that account for individuals' making a transition from having a home (domiciled) but utilizing meal programs to becoming homeless. The following table contains the data obtained in the study.

	Homeless Men	Domiciled Men
Sample size	112	260
Number currently working	34	98

Is there sufficient evidence to indicate that a greater proportion of domiciled men than of homeless men are currently working? Use $\alpha = .01$.

*10.26   Refer to Exercise 8.56(b). Is there evidence of a difference between the proportion of residents favoring complete protection of alligators and the proportion favoring their destruction? Use $\alpha = .01$.

10.4 Calculating Type II Error Probabilities and Finding the Sample Size for the Z Test

Calculating β can be very difficult for some statistical tests, but it is easy for the tests developed in Section 10.3. Consequently, we can use the Z test to demonstrate both the calculation of β and the logic employed in selecting the sample size for a test.

For the test $H_0 : \theta = \theta_0$ versus $H_a : \theta > \theta_0$, we can calculate type II error probabilities only for specific values for θ in H_a. Suppose that the experimenter has in mind a specific alternative, say, $\theta = \theta_a$ (where $\theta_a > \theta_0$). Because the rejection region is of the form

$$\mathrm{RR} = \{\hat{\theta} : \hat{\theta} > k\}$$

the probability β of a type II error is

$$\beta = P(\hat{\theta} \text{ is not in RR when } H_a \text{ is true})$$

$$\beta = P(\hat{\theta} \leq k \text{ when } \theta = \theta_a) = P\left(\frac{\hat{\theta} - \theta_a}{\sigma_{\hat{\theta}}} \leq \frac{k - \theta_a}{\sigma_{\hat{\theta}}} \text{ when } \theta = \theta_a \right).$$

If θ_a is the true value of θ, then $(\hat{\theta} - \theta_a)/\sigma_{\hat{\theta}}$ has approximately a standard normal distribution. Consequently, the probability β can be determined (approximately) by finding a corresponding area under a standard normal curve.

For a fixed sample of size n, the size of β depends upon the distance between θ_a and θ_0. If θ_a is close to θ_0, the true value of θ (θ_0 or θ_a) is difficult to detect, and the probability of accepting H_0 when H_a is true tends to be large. If θ_a is far from θ_0, the true value is relatively easy to detect, and β is considerably smaller. As we saw in Section 10.2, for a specified value of α, β can both be made smaller by choosing a large sample size n.

EXAMPLE **10.8**   Suppose that the vice president in Example 10.5 wants to be able to detect a difference equal to one call in the mean number of customer calls per week. That is, he wishes to test $H_0 : \mu = 15$ against $H_a : \mu = 16$. With the data as given in Example 10.5, find β for this test.

FIGURE **10.5**
Rejection region for
Example 10.8
($k = 15.8225$)

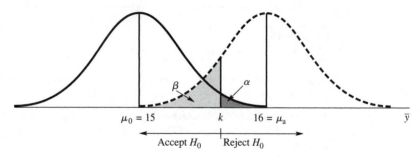

Solution   In Example 10.5, we had $n = 36$, $\bar{y} = 17$, and $s^2 = 9$. The rejection region for a .05 level test was given by

$$z = \frac{\bar{y} - \mu_0}{\sigma/\sqrt{n}} > 1.645$$

which is equivalent to

$$\bar{y} - \mu_0 > 1.645 \left(\frac{\sigma}{\sqrt{n}} \right) \quad \text{or} \quad \bar{y} > \mu_0 + 1.645 \left(\frac{\sigma}{\sqrt{n}} \right).$$

Substituting $\mu_0 = 15$ and $n = 36$, and using s to approximate σ, we find the rejection region to be

$$\bar{y} > 15 + 1.645 \left(\frac{3}{\sqrt{36}} \right), \quad \text{or, equivalently,} \quad \bar{y} > 15.8225.$$

This rejection region is shown in Figure 10.5. Then, by definition, $\beta = P(\bar{Y} \le 15.8225$ when $\mu = 16)$ is the shaded area under the dashed curve to the left of $k = 15.8225$ in Figure 10.5. Thus, for $\mu_a = 16$,

$$\beta = P \left(\frac{\bar{Y} - \mu_a}{\sigma/\sqrt{n}} \le \frac{15.8225 - 16}{3/\sqrt{36}} \right) = P(Z \le -.36) = .3594.$$

The large value of β tells us that samples of size $n = 36$ frequently will fail to detect a difference of one unit from the hypothesized means. We can reduce the value of β by increasing the sample size n.

The preceding example suggests the procedure an experimenter employs when choosing the sample size(s) for an experiment. Suppose that you want to test $H_0 : \mu = \mu_0$ versus $H_a : \mu > \mu_0$. If you specify the desired values of α and β (where β is evaluated when $\mu = \mu_a$ and $\mu_a > \mu_0$), any further adjustment of the test must involve two remaining quantities: the sample size, n, and the point at which the rejection region begins, k. Because α and β can be written as probabilities involving n and k, we have two equations in two unknowns, which can be solved simultaneously for n. Thus,

$$\alpha = P(\overline{Y} > k \text{ when } \mu = \mu_0)$$

$$= P\left(\frac{\overline{Y} - \mu_0}{\sigma/\sqrt{n}} > \frac{k - \mu_0}{\sigma/\sqrt{n}} \text{ when } \mu = \mu_0\right) = P(Z > z_\alpha)$$

$$\beta = P(\overline{Y} \le k \text{ when } \mu = \mu_a)$$

$$= P\left(\frac{\overline{Y} - \mu_a}{\sigma/\sqrt{n}} \le \frac{k - \mu_a}{\sigma/\sqrt{n}} \text{ when } \mu = \mu_a\right) = P(Z \le -z_\beta).$$

(See Figure 10.5.)

From the previous equations for α and β, we have

$$\frac{k - \mu_0}{\sigma/\sqrt{n}} = z_\alpha \quad \text{and} \quad \frac{k - \mu_a}{\sigma/\sqrt{n}} = -z_\beta.$$

Eliminating k from these two equations gives

$$\mu_0 + z_\alpha \left(\frac{\sigma}{\sqrt{n}}\right) = \mu_a - z_\beta \left(\frac{\sigma}{\sqrt{n}}\right).$$

Thus,

$$(z_\alpha + z_\beta) \left(\frac{\sigma}{\sqrt{n}}\right) = \mu_a - \mu_0, \quad \text{or, equivalently,} \quad \sqrt{n} = \frac{(z_\alpha + z_\beta)\sigma}{(\mu_a - \mu_0)}.$$

Sample Size for an Upper-Tail α-Level Test

$$n = \frac{(z_\alpha + z_\beta)^2 \sigma^2}{(\mu_a - \mu_0)^2}$$

Exactly the same solution would be obtained for a one-tailed alternative, H_a : $\mu = \mu_a$ with $\mu_a < \mu_0$. The method just employed can be used to develop a similar formula for sample size for any one-tailed, hypothesis-testing problem that satisfies the conditions of Section 10.3.

EXAMPLE 10.9   Suppose that the vice president of Example 10.5 wants to test $H_0 : \mu = 15$ against $H_a : \mu = 16$ with $\alpha = \beta = .05$. Find the sample size that will ensure this accuracy. Assume that σ^2 is approximately 9.

Solution   Because $\alpha = \beta = .05$, it follows that $z_\alpha = z_\beta = z_{.05} = 1.645$. Then

$$n = \frac{(z_\alpha + z_\beta)^2 \sigma^2}{(\mu_a - \mu_0)^2} = \frac{(1.645 + 1.645)^2 (9)}{(16 - 15)^2} = 97.4.$$

Hence, $n = 98$ observations should be used to meet the requirement that $\alpha \approx \beta \approx .05$ for the vice president's test.

Exercises

10.27 Refer to Exercise 10.9. If the voltage falls as low as 128, serious consequences may result. For testing $H_0 : \mu = 130$ versus $H_a : \mu = 128$, find the probability β of a type II error for the rejection region used in Exercise 10.9.

10.28 Refer to Exercise 10.10. The steel is sufficiently hard to meet usage requirements if the mean Rockwell hardness measure does not drop below 60. Using the rejection region found in Exercise 10.10, find β for the specific alternative $\mu_a = 60$.

10.29 Refer to Exercise 10.20. Calculate the value of β for the alternative $p_a = .15$.

10.30 Refer to Exercise 10.23. The political researcher should have designed a test for which β is tolerably low when p_1 exceeds p_2 by a meaningful amount. For example, find a common sample size n for a test with $\alpha = .05$ and $\beta \le .20$ when, in fact, p_1 exceeds p_2 by .1. [*Hint:* The maximum value of $p(1 - p)$ is .25.]

10.31 Refer to Exercise 10.24. Using the rejection region found there, calculate β when $\mu_a = 5.5$.

10.32 In Exercises 10.24 and 10.31, how large should the sample size be if we require that $\alpha = .01$ and $\beta = .05$ when $\mu_a = 5.5$?

10.33 A random sample of 37 second-graders who participated in sports had manual dexterity scores with mean 32.19 and standard deviation 4.34. An independent sample of 37 second-graders who did not participate in sports had manual dexterity scores with mean 31.68 and standard deviation 4.56.

  a Test to see whether sufficient evidence exists to indicate that second-graders who participate in sports have a higher mean dexterity score. Use $\alpha = .05$.

  b For the rejection region used in (a), calculate β when $\mu_1 - \mu_2 = 3$.

10.34 Refer to Exercise 10.33. Find the sample sizes that give $\alpha = .05$ and $\beta = .05$ when $\mu_1 - \mu_2 = 3$. (Assume equal-size samples for each group.)

10.5 Relationships Between Hypothesis-Testing Procedures and Confidence Intervals

Thus far, we have considered two large-sample procedures for making inferences about a target parameter θ. In Section 8.6 we observed that if $\hat{\theta}$ is an estimator for θ that has an approximately normal sampling distribution, a two-sided confidence interval for θ with confidence coefficient $1 - \alpha$ is given by

$$\hat{\theta} \pm z_{\alpha/2}\sigma_{\hat{\theta}}.$$

In this expression, $\sigma_{\hat{\theta}}$ is the standard error of the estimator $\hat{\theta}$ (the standard deviation of the sampling distribution of $\hat{\theta}$) and $z_{\alpha/2}$ is a number read out of the standard normal table such that $P(Z > z_{\alpha/2}) = \alpha/2$. For large samples, if we were interested in an α-level test of $H_0 : \theta = \theta_0$ versus the two-sided alternative $H_a : \theta \neq \theta_0$, the results of the previous section indicate that we would use a Z test based on the test statistic

$$Z = \frac{\hat{\theta} - \theta_0}{\sigma_{\hat{\theta}}}$$

and would reject H_0 if the value of Z fell in the rejection region $\{|z| > z_{\alpha/2}\}$. Both of these procedures make heavy use of the estimator $\hat{\theta}$, its standard error $\sigma_{\hat{\theta}}$, and the table value $z_{\alpha/2}$. Let us explore these two procedures more fully.

The complement of the rejection region associated with any test is sometimes called the *acceptance region* for the test. For any of our large-sample two-tailed α-level tests, the acceptance region is given by $\overline{RR} = \{-z_{\alpha/2} \leq z \leq z_{\alpha/2}\}$. That is, we do not reject $H_0 : \theta = \theta_0$ in favor of the two-tailed alternative if

$$-z_{\alpha/2} \leq \frac{\hat{\theta} - \theta_0}{\sigma_{\hat{\theta}}} \leq z_{\alpha/2}.$$

Restated, the null hypothesis is not rejected (is "accepted") at level α if

$$\hat{\theta} - z_{\alpha/2}\sigma_{\hat{\theta}} \leq \theta_0 \leq \hat{\theta} + z_{\alpha/2}\sigma_{\hat{\theta}}.$$

Notice that the quantities on the far left and far right of the previous string of inequalities are the lower and upper endpoints, respectively, of a $100(1-\alpha)\%$ two-sided confidence interval for θ. Thus, a duality exists between our large-sample procedures for constructing a $100(1 - \alpha)\%$ two-sided confidence interval and for implementing a two-sided hypothesis test with level α. Do not reject $H_0 : \theta = \theta_0$ in favor of $H_a : \theta \neq \theta_0$ if the value θ_0 lies inside a $100(1-\alpha)\%$ confidence interval for θ. Reject H_0 if θ_0 lies outside the interval. Equivalently, a $100(1 - \alpha)\%$ two-sided confidence interval can be interpreted as the set of all values of θ_0 for which $H_0 : \theta = \theta_0$ is "acceptable" at level α. Notice that *any* value inside the confidence interval is an acceptable value of the parameter. There is not *one acceptable value* for the parameter, but many (indeed, the infinite number of values inside the interval). For this reason, we usually do not *accept* the null hypothesis that θ takes on a particular value θ_0, even if the value θ_0 falls inside our confidence interval. Because we recognize that many values of θ are acceptable, we refrain from accepting a single θ value as being *the true value*. Additional comments regarding hypothesis testing are contained in Section 10.7.

Our previous discussion focused on the duality between two-sided confidence intervals and two-sided hypothesis tests. In the exercises that follow this section, you will be asked to demonstrate the correspondence between large-sample one-sided hypothesis tests of level α and the construction of the appropriate upper or lower bounds with confidence coefficients $1-\alpha$. If you desire an α-level test of $H_0 : \theta = \theta_0$ versus $H_a : \theta > \theta_0$ (an upper-tail test), you should accept the alternative hypothesis if θ_0 is less than a $100(1 - \alpha)\%$ *lower confidence bound* for θ. If the appropriate alternative hypothesis is $H_a : \theta < \theta_0$ (a lower-tail test), you should reject $H_0 : \theta = \theta_0$ if in favor of H_a if θ_0 is larger than a $100(1 - \alpha)\%$ *upper confidence bound* for θ.

Exercises

10.35   Refer to Exercise 10.7. Construct a 99% confidence interval for the true average savings if computers are purchased through the mail. Is the value $\mu_0 = 900$ inside or outside this interval? Based on this interval, should the null hypothesis discussed in Exercise 10.7 be rejected? Why? How does the conclusion you reached compare with your answer to Exercise 10.7?

10.36   A large-sample α-level test of hypothesis for $H_0 : \theta = \theta_0$ versus $H_a : \theta > \theta_0$ rejects the null hypothesis if

$$\frac{\hat{\theta} - \theta_0}{\sigma_{\hat{\theta}}} > z_\alpha.$$

Show that this is equivalent to rejecting H_0 if θ_0 is less than the large-sample $100(1 - \alpha)\%$ lower confidence bound for θ.

10.37   Refer to Exercise 10.15. Construct a 99% lower confidence bound for the proportion of voters who are unable to name an elected official whom they admire. How does the value $p = .5$ compare to this lower bound? Based on this lower bound should the alternative hypothesis of Exercise 10.15 be accepted? Is there any conflict between this answer and your answer to Exercise 10.15?

10.38   A large-sample α-level test of hypothesis for $H_0 : \theta = \theta_0$ versus $H_a : \theta < \theta_0$ rejects the null hypothesis if

$$\frac{\hat{\theta} - \theta_0}{\sigma_{\hat{\theta}}} < -z_\alpha.$$

Show that this is equivalent to rejecting H_0 if θ_0 is greater than the large-sample $100(1 - \alpha)\%$ upper confidence bound for θ.

10.39   Refer to Exercise 10.9. Construct a 95% upper confidence bound for the average voltage reading. How does the value $\mu = 130$ compare to this upper bound? Based on this upper bound, should the alternative hypothesis of Exercise 10.9 be accepted? Is there any conflict between this answer and your answer to Exercise 10.9?

10.6   Another Way to Report the Results of a Statistical Test: Attained Significance Levels or *p*-Values

As previously indicated, the probability α of a type I error is often called the *significance level*, or, more simply, the *level* of the test. These terms originated in the following way. The probability of the observed value of the test statistic, or some value even more contradictory to the null hypothesis, in a sense measures the weight of evidence favoring rejection of the null hypothesis. Although small values of α are recommended, the actual value of α to use in an analysis is chosen somewhat arbitrarily. One experimenter may choose to implement a test with $\alpha = .05$, whereas another experimenter might prefer $\alpha = .01$. It is possible, therefore, for two persons to analyze the same data and reach opposite conclusions—one concluding that the null hypothesis should be rejected at the $\alpha = .05$ significance level, and the other deciding that the null hypothesis should not be rejected with $\alpha = .01$. Further, α values of .05 or .01 often are used out of habit or for the sake of convenience rather than as a result of careful consideration of the ramifications of making a type I error.

Once a test statistic (Y in our polling example, or one of the Zs of Section 10.3) is decided upon, it is often possible to report the p-value or attained significance level associated with a test. This quantity is a statistic representing the smallest value of α for which the null hypothesis can be rejected.

DEFINITION **10.2**

If W is a test statistic, the *p-value*, or *attained significance level*, is the smallest level of significance α for which the observed data indicate that the null hypothesis should be rejected.

The smaller the p-value becomes, the more compelling is the evidence that the null hypothesis should be rejected. Many scientific journals require researchers to report p-values associated with statistical tests because these values provide the reader with *more information* than is contained in a statement that the null hypothesis was rejected or not rejected for some value of α chosen by the researcher. If the p-value is small enough to be convincing to you, you should reject the null hypothesis. If an experimenter has a value of α in mind, the p-value can be used to implement an α-level test. The p-value is the *smallest* value of α for which the null hypothesis can be rejected. Thus, if the desired value of α is greater than or equal to the p-value, the null hypothesis is rejected for that value of α. Indeed, the null hypothesis should be rejected for any value of α *down to and including* the p-value. Otherwise, if α is less than the p-value, the null hypothesis cannot be rejected. In a sense, the p-value allows the reader of published research to evaluate the extent to which the observed data disagree with the null hypothesis. Particularly, the p-value permits each reader to use his or her own choice for α in deciding whether the observed data should lead to rejection of the null hypothesis.

The procedures for finding p-values for the tests we have discussed thus far are presented in the following examples.

EXAMPLE **10.10**   Recall our discussion of the political poll (see Examples 10.1 through 10.4), where $n = 15$ voters were sampled. If we wish to test $H_0 : p = .5$ versus $H_a : p < .5$, using Y = the number of voters favoring Jones as our test statistic, what is the p-value if $Y = 3$? Interpret the result.

Solution   In previous discussions we noted that H_0 should be rejected for small values of Y. Thus the p-value for this test is given by p-value $= P\{Y \leq 3\}$, where Y has a binomial distribution with $n = 15$ and $p = .5$ (the shaded area in the binomial distribution of Figure 10.6). Using Table 1, Appendix III, we find that the p-value is .018.

Because the p-value $= .018$ represents the smallest value of α for which the null hypothesis is rejected, an experimenter who specifies any value of $\alpha \geq .018$ would

FIGURE **10.6**
Illustration of p-value
for Example 10.10

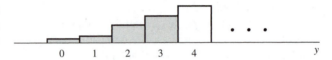

be led to reject H_0 and to conclude that Jones does *not* have a plurality of the vote. If the experimenter chose a value of α less than .018, however, the null hypothesis could *not* be rejected.

This example illustrates that the reporting of p-values is particularly beneficial where the appropriate test statistic possesses a *discrete* distribution. In situations like these, one often cannot find any rejection region that yields an α value of a particular magnitude. For example, in this instance, no rejection region of the form $\{y \leq a\}$ can be found for which α is equal to .05. In such cases, reporting the p-value is usually preferable to limiting oneself to values of α that can be obtained on the basis of the discrete distribution of the test statistic.

Example 10.10 also indicates the general method of computing p-values. If we were to reject H_0 in favor of H_a for small values of a test statistic W, say, RR: $\{w \leq k\}$, the p-value associated with an observed value w_0 of W would be given by

$$p\text{-value} = P(W \leq w_0, \text{ when } H_0 \text{ is true}).$$

Analogously, if we were to reject H_0 in favor of H_a for large values of W, say, RR: $\{w \geq k\}$, the p-value associated with the observed value w_0 would be given by

$$p\text{-value} = P(W \geq w_0, \text{ when } H_0 \text{ is true}).$$

Calculation of a p-value for a two-tailed alternative is illustrated in the following example.

EXAMPLE 10.11 Find the p-value for the statistical test of Example 10.7.

Solution Example 10.7 presents a test of the null hypothesis $H_0 : \mu_1 - \mu_2 = 0$ versus the alternative hypothesis $H_a : \mu_1 - \mu_2 \neq 0$. The value of the test statistic, computed from the observed data, was $z = -2.5$. Because this test is two-tailed, the p-value is the probability that either $Z \leq -2.5$ or $Z \geq 2.5$ (the shaded area in Figure 10.7). From Table 4, Appendix III, we find that $P(Z \geq 2.5) = P(Z \leq -2.5) = .0062$. Because this is a two-tailed test, the p-value $= 2(.0062) = .0124$. Thus, if $\alpha = .05$

FIGURE **10.7**
Shaded area
gives p-value for
Example 10.11

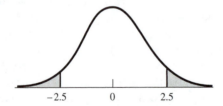

(a value larger than .0124), we reject H_0 in favor of H_a and, in agreement with the conclusion of Example 10.7, conclude that evidence of a difference in mean reaction time for men and women exists. However, if $\alpha = .01$ (or any value of $\alpha < .0124$) were chosen, we could *not* legitimately claim to have detected a difference in mean reaction times for the two sexes.

For the statistical tests we have developed thus far, the experimenter can compute exact p-values by using the binomial and Z tables in Appendix III. Tables of distributions for some of the test statistics that we encounter in later sections give critical values only for largely differential values of α (for example, .10, .05, .025, .01, and .005). Consequently, such tables cannot be used to compute exact p-values. However, the tables provided in the appendix for the F, t, and χ^2 (and some other) distributions do permit us to determine a region of values, inside which the p-value is known to lie. For example, if a test result is statistically significant for $\alpha = .05$ but *not* for $\alpha = .025$, we will report that $.025 \leq p\text{-value} \leq .05$. Thus, for any $\alpha \geq .05$, we reject the null hypothesis; for $\alpha < .025$, we do *not* reject the null hypothesis; and for values of α that fall between .025 and .05, we need to seek more complete tables of the appropriate distribution before reaching a conclusion. In the last case, we nevertheless have very useful information about the p-value available to us. It should be noted that p-values are reported by most statistical software that can be used to implement tests of hypotheses. Many calculators are also capable of computing exact p-values.

The recommendation that a researcher report the p-value for a test and leave its interpretation to a reader does not violate the traditional (decision theoretic) statistical testing procedures described in the preceding sections. The reporting of a p-value simply leaves the decision regarding whether to reject the null hypothesis (with the associated potential of committing type I or type II errors) to the reader. Thus, the responsibility of choosing α and, possibly, the problem of evaluating the probability β of making a type II error are shifted to the reader.

Exercises

10.40 High airline occupancy rates on scheduled flights are essential for profitability. Suppose that a scheduled flight must average at least 60% occupancy to be profitable and that an examination of the occupancy rates for 120 10:00 A.M. flights from Atlanta to Dallas showed mean occupancy rate per flight of 58% and standard deviation 11%. Test to see if sufficient evidence exists to support a claim that the flight is unprofitable. Find the p-value associated with the test. What would you conclude if you wished to implement the test at the $\alpha = .10$ level?

10.41 Two sets of elementary schoolchildren were taught to read by using different methods, 50 by each method. At the conclusion of the instructional period, a reading test yielded the results $\bar{y}_1 = 74$, $\bar{y}_2 = 71$, $s_1 = 9$, and $s_2 = 10$. What is the attained significance level if you wish to see whether evidence indicates a difference between the two population means? What would you conclude if you desired an α value of .05?

10.42   Does auto safety education increase the use of auto safety devices? Researchers at the University of California at San Francisco gave 78 couples randomly selected from a group of 136 expectant couples a special 30-minute lecture on the value of child safety seats (*USA Today*, March 12, 1985). Contacted 4 to 6 weeks after the birth of their children, the researchers found that 96% of those who had attended the lecture claimed that they were using the safety seats, as did 78% of the 58 couples who did not attend the lecture.

  a   Is this difference in percentages large enough to imply that the lecture was effective in increasing the percentage of couples who say that they use child safety seats? Test using $\alpha = .05$.

  b   Find the *p*-value for the test.

10.43   How would you like to live to be 200 years old? For centuries, humankind has sought the key to the mystery of aging. What causes aging? How can aging be slowed? Recent studies focus on *biomarkers*, physical or biological changes that occur at a predictable time in a person's life. The theory is that, if ways can be found to delay the occurrence of these biomarkers, human life can be extended. A key biomarker, according to scientists, is forced vital capacity (FVC), the volume of air that a person can expel after taking a deep breath. A study of 5209 men and women aged 30 to 62 showed that FVC declined, on the average, 3.8 deciliters per decade for men and 3.1 deciliters per decade for women (T. Boddé, "Biomarkers of Aging: Key to a Younger Life," *Bioscience* 31(8), 1981, pp. 566–67). Suppose that you wished to determine whether a physical fitness program for men and women aged 50 to 60 would delay aging; to do so, you measured the FVC for 30 men and 30 women participating in the fitness program at the beginning and end of the 50-to-60-year age interval and recorded the drop in FVC for each person. A summary of the data appears in the accompanying table.

	Men	Women
Sample size	30	30
Sample average drop in FVC	3.6	2.7
Sample standard deviation	1.1	1.2
Population mean drop in FVC	μ_1	μ_2

  a   Do the data provide sufficient evidence to indicate that the decrease in the mean FVC over the decade for the men on the physical fitness program is less than 3.8 deciliters? Find the attained significance level for the test.

  b   Refer to (a). If you choose $\alpha = .05$, do the data support the contention that the mean decrease in FVC is less than 3.8 deciliters?

  c   Test to determine whether the FVC drop for women on the physical fitness program was less than 3.1 deciliters for the decade. Find the attained significance level for the test.

  d   Refer to (c). If you choose $\alpha = .05$, do the data support the contention that the mean decrease in FVC is less than 3.1 deciliters?

10.44   Do you believe that an exceptionally high percentage of the executives of large corporations are right-handed? Although 85% of the general public is right-handed, a survey of 300 chief executive officers of large corporations found that 96% were right-handed.

  a   Is this difference in percentages statistically significant? Test using $\alpha = .01$.

  b   Find the *p*-value for the test, and explain what it means.

10.45   A check-cashing service found that approximately 5% of all checks submitted to the service were bad. After instituting a check-verification system to reduce its losses, the service found that only 45 checks were bad in a random sample of 1124 that were cashed. Does sufficient evidence exist to affirm that the check verification system reduced the proportion of bad checks? What attained significance level is associated with the test? What would you conclude at the $\alpha = .01$ level?

10.46 A pharmaceutical company conducted an experiment to compare the mean times (in days) necessary to recover from the effects and complications that follow the onset of the common cold. This experiment compared persons on a daily dose of 500 mg of vitamin C to those who were not given a vitamin supplement. Thirty-five adults were randomly selected for each treatment category, and the mean recovery times and standard deviations for the two groups were found to be as given in the accompanying table. Do the data indicate that the use of vitamin C reduces the mean time required to recover? Find the attained significance level. What would the company conclude at the $\alpha = .05$ level?

	Treatment	
	No Supplement	500 mg Vitamin C
Sample size	35	35
Sample mean	6.9	5.8
Sample standard deviation	2.9	1.2

10.47 A publisher of a newsmagazine has found through past experience that 60% of subscribers renew their subscriptions. In a random sample of 200 subscribers, 108 indicated that they planned to renew their subscriptions. What is the p-value associated with the test that the current rate of renewals differs from the rate previously experienced?

10.48 In a study to assess various effects of using a female model in automobile advertising, each of 100 male subjects was shown photographs of two automobiles matched for price, color, and size, but of different makes. Fifty of the subjects (group A) were shown automobile 1 with a female model and automobile 2 with no model. Both automobiles were shown without the model to the other 50 subjects (group B). In group A, automobile 1 (shown with the model) was judged to be more expensive by 37 subjects. In group B, automobile 1 was judged to be more expensive by 23 subjects. Do these results indicate that using a female model increases the perceived cost of an automobile? Find the associated p-value, and indicate your conclusion for an $\alpha = .05$ level test.

10.7   Some Comments on the Theory of Hypothesis Testing

As previously indicated, we can choose between implementing a one-tailed or a two-tailed test for a given situation. This choice is dictated by the practical aspects of the problem and depends upon the alternative value of the parameter θ that the experimenter is trying to detect. Thus, if we stood to suffer a large financial loss if θ were greater than θ_0 but not if it were less, we would concentrate our attention on detecting values of θ greater than θ_0. Hence, we would reject in the upper tail of the distribution for the test statistics previously discussed. On the other hand, if we were equally interested in detecting values of θ less than or greater than θ_0, we would employ a two-tailed test.

The theory of statistical tests of hypotheses (outlined in Section 10.2 and utilized in Section 10.3) is a very clearcut procedure that enables the researcher either to reject or to accept the null hypothesis, with measured risk α or β. Unfortunately, this theoretical framework does not suffice for all practical situations.

For any statistical test, the probability α of a type I error depends upon the value of the parameter specified in the null hypothesis. This probability can be calculated, at least approximately, for each of the testing procedures discussed in this text. For the procedures discussed thus far, the probability, β, of a type II error can be calculated only after a *specific* value of the parameter of interest has been singled out for consideration. The selection of a practically meaningful value for this parameter is often difficult. And even if a meaningful alternative can be identified, the actual calculation of β can be quite tedious. Specification of a meaningful alternative hypothesis is even more difficult for some of the testing procedures we will consider in subsequent chapters.

Of course, we do not want to ignore the possibility of committing a type II error. Later in this chapter we will determine methods for selecting tests with the smallest possible value of β for tests where α, the probability of a type I error, is a fixed value selected by the researcher. Even in these situations, however, the smallest possible value of β can be quite large.

These obstacles do not invalidate the use of statistical tests; rather, they urge us to be cautious about drawing conclusions where insufficient evidence is available to permit rejection of the null hypothesis. If a truly meaningful (and believable) value for β can be calculated, we should feel justified in accepting H_0 if the value of β is small. In the more typical situation where a truly meaningful value for β is unavailable, we will modify our procedure as follows. When the value of the test statistic is not in the rejection region, we will "fail to reject" rather than "accept" the null hypothesis. In the polling example discussed in Example 10.1, we tested $H_0: p = .5$ versus $H_a: p < .5$. If our observed value of Y falls into the rejection region, we reject H_0 and say that the evidence supports the research hypothesis that Jones will lose. In this situation, we will have demonstrated support for the hypothesis we wanted to support—the research hypothesis. If, however, Y does not fall in the rejection region, and we can determine no specific value of p in H_a that is of direct interest, we simply state that we will *not* reject H_0 and must seek additional information before reaching a conclusion. Alternatively, we could report the p-value associated with the statistical test and leave the interpretation to the reader.

If H_0 is rejected for a "small" value of α (or for a small p-value), this occurrence does *not* imply that the null hypothesis is "wrong by a large amount." It *does mean* that the null hypothesis can be rejected based on a procedure that incorrectly rejects the null hypothesis (when H_0 is true) with a small probability (that is, with a small probability of a type I error). We also must refrain from equating *statistical* with *practical* significance. If we consider the experiment described and analyzed in Examples 10.7 and 10.11, the p-value of .0124 is "small" and the result is statistically significant for any choice of $\alpha \geq .0124$. However, the difference between the mean reaction times for the two samples is only .2 seconds, a result that may or may not be *practically* significant. To assess the practical significance of such a difference, you may wish to form a confidence interval for $\mu_1 - \mu_2$ by using the methods of Section 8.6.

Finally, some comments are in order regarding the choice of the null hypotheses that we have used, particularly in the one-sided tests. For example, in Example 10.1, we identified the appropriate alternative hypothesis as $H_a : p < .5$ and used $H_0 : p = .5$ as our null hypothesis. The test statistic was $Y =$ the number of voters who favored Jones in a sample of size $n = 15$. One rejection region that we considered was $\{y \le 2\}$. You might wonder why we did not use $H_0^* : p \ge .5$ as the null hypothesis. This makes a lot of sense, because *every* possible value of p is either in $H_0^* : p \ge .5$ or in $H_a : p < .5$. So why did we use $H_0 : p = .5$? The brief answer is that what we really care about is the *alternative hypothesis* $H_a : p < .5$; the null hypothesis is not our primary concern. As previously discussed, we usually do not actually accept the null hypothesis anyway, regardless of its form. In addition, $H_0 : p = .5$ is easier to deal with and *leads to exactly the same conclusions at the same α value* without requiring us to develop additional theory to deal with the more complicated $H_0^* : p \ge .5$. When we used $H_0 : p = .5$ as our null hypotheses, calculating the level α of the test was relatively simple: we just found $P(Y \le 2$ when $p = .5)$. If we had used $H_0^* : p \ge .5$ as the null hypothesis, our previous definition of α would have been inadequate because the value of $P(Y \le 2)$ is actually a function of p for $p \ge .5$. In cases like these, α is defined to be the *maximum* (over all values of $p \ge .5$) value of $P(Y \le 2)$. Although we will not derive this result here, $\max_{p \ge .5} P(Y \le 2)$ occurs when $p = .5$, the "boundary" value of p in $H_0^* : p \ge .5$. Thus, we get the "right" value of α if we use the simpler null hypothesis $H_0 : p = .5$. Similar statements are true for all of the tests that we have considered thus far and that we will consider in future discussions. That is, if we consider $H_a : \theta > \theta_0$ to be the appropriate research hypothesis, $\alpha = \max_{\theta \le \theta_0} P(\text{test statistic in RR})$ typically occurs when $\theta = \theta_0$, the "boundary" value of θ. Similarly, if $H_a : \theta < \theta_0$ is the appropriate research hypothesis $\alpha = \max_{\theta \ge \theta_0} P(\text{test statistic in RR})$ typically occurs when $\theta = \theta_0$. Thus, using $H_0 : \theta = \theta_0$ instead of $H_0^* : \theta \ge \theta_0$ leads to the correct testing procedure and the correct calculation of α without needlessly raising additional considerations.

10.8 Small-Sample Hypothesis Testing for μ and $\mu_1 - \mu_2$

In Section 10.3, we discussed large-sample hypothesis-testing procedures that, like the interval-estimation procedures developed in Section 8.6, are useful for large samples. For these procedures to be applicable, the sample size must be large enough that $Z = (\hat{\theta} - \theta_0)/\sigma_{\hat{\theta}}$ has approximately a standard normal distribution. Section 8.8 contains procedures based on the t distribution for constructing confidence intervals for μ (the mean of a single normal population) and $\mu_1 - \mu_2$ (the difference in the means of two normal populations with equal variances). In this section we will develop formal procedures for testing hypotheses about μ and $\mu_1 - \mu_2$, procedures that are appropriate for small samples from normal populations.

We assume that $Y_1, Y_2, \ldots, Y_n$ denote a random sample of size n from a normal distribution with unknown mean μ and unknown variance σ^2. If $\overline{Y}$ and S denote the sample mean and sample standard deviation, respectively, and if $H_0 : \mu = \mu_0$ is true, then

$$T = \frac{\overline{Y} - \mu_0}{S/\sqrt{n}}$$

has a t distribution with $n - 1$ degrees of freedom (see Section 8.8).

Because the t distribution is symmetric and mound-shaped, the rejection region for a small-sample test of the hypothesis $H_0 : \mu = \mu_0$ must be located in the tails of the t distribution and should be determined in exactly the same way as for the large-sample Z statistic. By analogy with the Z test developed in Section 10.3, the proper rejection region for the upper-tail alternative $H_a : \mu > \mu_0$ is given by

$$RR = \{t > t_\alpha\}$$

where t_α is such that $P\{T > t_\alpha\} = \alpha$ for a t distribution with $n - 1$ degrees of freedom (see Table 5, Appendix III).

A summary of the tests for μ based on the t distribution, known as t *tests*, is as follows.

A Small-Sample Test for μ

Assumptions: $Y_1, Y_2, \ldots, Y_n$ constitute a random sample from a normal distribution with $E(Y_i) = \mu$.

H_0: $\mu = \mu_0$.

H_a:
$\begin{cases} \mu > \mu_0 & \text{(upper-tail alternative).} \\ \mu < \mu_0 & \text{(lower-tail alternative).} \\ \mu \neq \mu_0 & \text{(two-tailed alternative).} \end{cases}$

Test statistic: $T = \dfrac{\overline{Y} - \mu_0}{S/\sqrt{n}}$.

Rejection region:
$\begin{cases} t > t_\alpha & \text{(upper-tail RR).} \\ t < -t_\alpha & \text{(lower-tail RR).} \\ |t| > t_{\alpha/2} & \text{(two-tailed RR).} \end{cases}$

(See Table 5, Appendix III, for values of t_α, with $\nu = n - 1$ degrees of freedom.)

EXAMPLE **10.12**   Example 8.11 gives muzzle velocities of eight shells tested with a new gunpowder, along with the sample mean and sample standard deviation, $\overline{y} = 2959$ and $s = 39.1$. The manufacturer claims that the new gunpowder produces an average velocity of not less than 3000 feet per second. Do the sample data provide sufficient evidence to contradict the manufacturer's claim at the .025 level of significance?

Solution   Assuming that muzzle velocities are approximately normally distributed, we can use the test just outlined. We want to test $H_0 : \mu = 3000$ versus the alternative, $H_a :$

$\mu < 3000$. The rejection region is given by $t < -t_{.025} = -2.365$, where t possesses $\nu = (n-1) = 7$ degrees of freedom. Computing, we find that the observed value of the test statistic is

$$t = \frac{\bar{y} - \mu_0}{s/\sqrt{n}} = \frac{2959 - 3000}{39.1/\sqrt{8}} = -2.966.$$

This value falls in the rejection region (that is, $t = -2.966$ is less than -2.365); hence the null hypothesis is rejected at the $\alpha = .025$ level of significance. So we conclude that sufficient evidence exists to contradict the manufacturer's claim, and we conclude that the true mean muzzle velocity is less than 3000 feet per second.

EXAMPLE 10.13   What is the p-value associated with the statistical test in Example 10.12?

Solution   Because the null hypothesis should be rejected if t is "small," the smallest value of α for which the null hypothesis can be rejected is p-value $= P(T < -2.966)$, where T has a t distribution with $n - 1 = 7$ degrees of freedom.

Unlike the table of areas under the normal curve (Table 4, Appendix III), Table 5 in Appendix III does not give areas corresponding to many values of t. Rather, it gives the values of t corresponding to upper-tail areas equal to .10, .05, .025, .010, and .005. Because the t distribution is symmetric about 0, we can use these upper-tail areas to provide corresponding lower-tail areas. In this instance, the t statistic is based on 7 degrees of freedom; hence, we consult the d.f. = 7 row of Table 5 and find that -2.966 falls between $-t_{.025} = -2.365$ and $-t_{.01} = -2.998$. These values are indicated in Figure 10.8. Because the observed value of T (-2.966) is less than $-t_{.025} = -2.376$ but not less than $-t_{.01} = -2.998$, we reject H_0 for $\alpha = .025$ but not for $\alpha = .01$. Thus the p-value for the test satisfies $.01 \leq p$-value $\leq .025$.

A second application of the t distribution is in constructing a small-sample test to compare the means of two normal populations that possess equal variances. Suppose that independent random samples are selected from each of two normal populations: $Y_{11}, Y_{12}, \ldots, Y_{1n_1}$ from the first, and $Y_{21}, Y_{22}, \ldots, Y_{2n_2}$ from the second, where the mean and variance of the ith population are μ_i and σ^2, for $i = 1, 2$. Further, assume that $\bar{Y}_i$ and S_i^2, for $i = 1, 2$, are the corresponding sample means and variances.

FIGURE 10.8
Determination of the p-value for Example 10.13

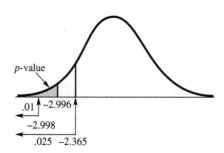

When these assumptions are satisfied, we showed in Section 8.8 that if

$$S_p^2 = \frac{(n_1 - 1)S_1^2 + (n_2 - 1)S_2^2}{n_1 + n_2 - 2}$$

is the pooled estimator for σ^2, then

$$T = \frac{(\overline{Y}_1 - \overline{Y}_2) - (\mu_1 - \mu_2)}{S_p \sqrt{\dfrac{1}{n_1} + \dfrac{1}{n_2}}}$$

has a Student's t distribution with $n_1 + n_2 - 2$ degrees of freedom. If we want to test the null hypothesis $H_0 : \mu_1 - \mu_2 = D_0$ for some fixed value D_0, it follows that, if H_0 is true, then

$$T = \frac{\overline{Y}_1 - \overline{Y}_2 - D_0}{S_p \sqrt{\dfrac{1}{n_1} + \dfrac{1}{n_2}}}$$

has a Student's t distribution with $n_1 + n_2 - 2$ degrees of freedom. Notice that this small-sample test statistic resembles its large-sample counterpart, the Z statistic of Section 10.3. Tests of the hypothesis $H_0 : \mu_1 - \mu_2 = D_0$ versus upper-tail, lower-tail, and two-tailed alternatives are conducted in the same manner as in the large-sample test, except that we employ the t statistic and tables of the t distribution to reach our conclusions. A summary of the small-sample testing procedures for $\mu_1 - \mu_2$ follows.

Small-Sample Tests for Comparing Two Population Means

Assumptions: Independent samples from normal distributions with $\sigma_1^2 = \sigma_2^2$.

H_0: $\mu_1 - \mu_2 = D_0$

H_a: $\begin{cases} \mu_1 - \mu_2 > D_0 & \text{(upper-tail alternative).} \\ \mu_1 - \mu_2 < D_0 & \text{(lower-tail alternative).} \\ \mu_1 - \mu_2 \neq D_0 & \text{(two-tailed alternative).} \end{cases}$

Test statistic: $T = \dfrac{\overline{Y}_1 - \overline{Y}_2 - D_0}{S_p \sqrt{\dfrac{1}{n_1} + \dfrac{1}{n_2}}}$, where $S_p = \sqrt{\dfrac{(n_1 - 1)S_1^2 + (n_2 - 1)S_2^2}{n_1 + n_2 - 2}}$.

Rejection region: $\begin{cases} t > t_\alpha & \text{(upper-tail RR).} \\ t < -t_\alpha & \text{(lower-tail RR).} \\ |t| > t_{\alpha/2} & \text{(two-tailed RR).} \end{cases}$

Here $P(T > t_\alpha) = \alpha$; and degrees of freedom $\nu = n_1 + n_2 - 2$. (See Table 5, Appendix III.)

Table 10.3. Data for Example 10.14

	Standard Procedure		New Procedure
	$n_1 = \quad 9$		$n_2 = \quad 9$
	$\bar{y}_1 = \quad 35.22$ seconds		$\bar{y}_2 = \quad 31.56$ seconds
	$\sum_{i=1}^{9}(y_{1i} - \bar{y}_1)^2 = 195.56$		$\sum_{i=1}^{9}(y_{2i} - \bar{y}_2)^2 = 160.22$

EXAMPLE **10.14**   Example 8.12 gives data on the length of time required to complete an assembly procedure using each of two different training methods. The sample data are as shown in Table 10.3. Is there sufficient evidence to indicate a difference in true mean times for the two methods? Test at the $\alpha = .05$ level of significance.

Solution   We are testing $H_0 : (\mu_1 - \mu_2) = 0$ against the alternative $H_a : (\mu_1 - \mu_2) \neq 0$. Consequently, we must use a two-tailed test. The test statistic is

$$T = \frac{(\bar{Y}_1 - \bar{Y}_2) - D_0}{S_p\sqrt{\dfrac{1}{n_1} + \dfrac{1}{n_2}}}$$

with $D_0 = 0$, and the rejection region for $\alpha = .05$ is $|t| > t_{\alpha/2} = t_{.025}$. In this case, $t_{.025} = 2.120$ because t is based on $(n_1 + n_2 - 2) = 9 + 9 - 2 = 16$ degrees of freedom.

The observed value of the test statistic is found by first computing

$$s_p = \sqrt{s_p^2} = \sqrt{\frac{195.56 + 160.22}{9 + 9 - 2}} = \sqrt{22.24} = 4.716.$$

Then

$$t = \frac{\bar{y}_1 - \bar{y}_2}{s_p\sqrt{\dfrac{1}{n_1} + \dfrac{1}{n_2}}} = \frac{35.22 - 31.56}{4.716\sqrt{\dfrac{1}{9} + \dfrac{1}{9}}} = 1.65.$$

This value does not fall in the rejection region ($|t| > 2.120$); hence, the null hypothesis is not rejected. There is insufficient evidence to indicate a difference in the mean assembly times for the two training periods at the $\alpha = .05$ level of significance.

Notice that, in line with the comments of Section 10.7, we have not accepted $H_0 : \mu_1 - \mu_2 = 0$. Rather, we have stated that we lack sufficient evidence to reject H_0 and to accept the alternative $H_a : \mu_1 - \mu_2 \neq 0$.

EXAMPLE **10.15**   Find the p-value for the statistical test of Example 10.14.

FIGURE **10.9**
Shaded area is
the *p*-value for
Example 10.15

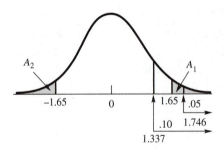

Solution   The observed value of the test statistic for this two-tailed test was $t = 1.65$. The *p*-value for this test is thus the probability that $T > 1.65$ or $T < -1.65$, the area shaded in Figure 10.9, that is, $A_1 + A_2$.

Because this test statistic is based on $n_1 + n_2 - 2 = 16$ d.f., we consult Table 5, Appendix III, to find $t_{.05} = 1.746$ and $t_{.10} = 1.337$. Thus, $A_1 = P(T > 1.65)$ lies between .05 and .10; that is, $.05 < A_1 < .1$. Similarly, $.05 < A_2 < .1$. Because the *p*-value $= A_1 + A_2$, it follows that $.1 < p\text{-value} < .2$.

In this case, we see that, if we select $\alpha = .05$, we cannot reject the null hypothesis, which is the same conclusion we reached in Example 10.14.

The test of Example 10.12 is based on the assumption that the muzzle velocity measurements have been randomly selected from a normal population. In most cases it is impossible to verify this assumption. We might ask how this predicament affects the validity of our conclusions.

Empirical studies of the test statistic

$$\frac{\overline{Y} - \mu}{S/\sqrt{n}}$$

have been conducted by sampling from many populations with nonnormal distributions. Such investigations have shown that moderate departures from normality in the distribution of the population have little effect on the probability distribution of the test statistic. This result, coupled with the common occurrence of near-normal distributions of data in nature, makes the *t* test of a population mean extremely useful. Statistical tests that lack sensitivity to departures from the assumptions upon which they are based possess wide applicability. Because of their insensitivity to assumptions, they have been called *robust statistical tests*.

Like the *t* test for a single mean, the *t* test for comparing two population means (often called the *two-sample t test*) is robust relative to the assumption of normality. It is also robust relative to the assumption that $\sigma_1^2 = \sigma_2^2$ when n_1 and n_2 are equal (or nearly equal).

Finally, the duality between tests and confidence intervals that we considered in Section 10.6 holds for the tests based on the *t* distributions that we considered in this section and the confidence intervals presented in Section 8.8.

Exercises

10.49   Why is the Z test usually inappropriate as a test procedure when the sample size is small?

10.50   What assumptions are made when a Student's t test is employed to test a hypothesis involving a population mean?

10.51   A chemical process has produced, on the average, 800 tons of chemical per day. The daily yields for the past week are 785, 805, 790, 793, and 802 tons. Do these data indicate that the average yield is less than 800 tons and hence that something is wrong with the process? Test at the 5% level of significance. What assumptions must be satisfied in order for the procedure you used to analyze these data to be valid? Give bounds for the associated p-value.

10.52   A coin-operated soft drink machine was designed to discharge, on the average, 7 ounces of beverage per cup. In a test of the machine, ten cupfuls of beverage were drawn from the machine and measured. The mean and standard deviation of the ten measurements were 7.1 ounces and .12 ounce, respectively. Do these data present sufficient evidence to indicate that the mean discharge differs from 7 ounces? What can be said about the attained significance level for this test? What is the appropriate decision if $\alpha = .10$?

10.53   Operators of gasoline-fueled vehicles complain about the price of gasoline in gas stations. The federal gas tax per gallon is constant (14.45¢ as of January 1, 1993 [*Time*, February 8, 1992]), but state and local taxes vary from 0.5¢ to 25.45¢ for $n = 18$ key metropolitan areas around the country. The total tax per gallon for gasoline at each of these 18 locations is given next. Suppose that these measurements constitute a random sample of size 18:

42.89	53.91	48.55	47.90	47.73	46.61
40.45	39.65	38.65	37.95	36.80	35.95
35.09	35.04	34.95	33.45	28.99	27.45.

a   Is there sufficient evidence to claim that the average per gallon gas tax is less than 45¢? Give the p-value associated with the test.

b   Construct a 95% confidence interval for the average per gallon gas tax in the United States.

10.54   Researchers have shown that cigarette smoking has a deleterious effect on lung function. In their study of the effect of cigarette smoking on the carbon monoxide diffusing capacity (DL) of the lung, Ronald Knudson, Walter Kaltenborn, and Benjamin Burrows (*American Review of Respiratory Diseases* 140, 1989, pp. 645–51) found that current smokers had DL readings significantly lower than either ex-smokers or nonsmokers. The carbon monoxide diffusing capacity for a random sample of current smokers was as follows:

103.768	88.602	73.003	123.086	91.052
92.295	61.675	90.677	84.023	76.014
100.615	88.017	71.210	82.115	89.222
102.754	108.579	73.154	106.755	90.479.

Do these data indicate that the mean DL reading for current smokers is lower than 100, the average DL reading for nonsmokers? Test at the $\alpha = .01$ level. What is the p-value associated with this test?

10.55   Researchers are calling for closer scrutiny of milk after reports have shown that vitamin D levels often failed to reach or far exceeded federal standards (*Riverside Press-Enterprise*, April 30, 1992, p. A16). Vitamin D deficiency causes rickets in the young; among the elderly, it leads to weakened bones that may fracture. In a test of 42 containers of 13 different brands of milk, 62% fell at least 20% below the 400 units required by the U.S. Food and Drug Administration; another 10% had levels at least 20% above the required 400 units. Suppose that, in an independent study of 20 containers from a particular distributor, the average and standard deviation of the number of units of vitamin D were $\bar{y} = 365$ and $s = 46$.

    a Use the data from the independent study to determine whether sufficient evidence exists to permit you to conclude that the mean number of units of vitamin D is not 400, the amount required by the Food and Drug Administration. Give the p-value for your test.

    b Construct a 95% confidence interval for the true mean number of units of vitamin D in the milk provided by the distributor.

    c On the basis of the interval you obtained in (b), what would you conclude about the claim that the mean number of units of vitamin D is 400? How does your conclusion here compare with your conclusion in (a), where you conducted a formal test of hypothesis?

10.56 What assumptions are made about the populations from which independent random samples are obtained when the t distribution is used to make small-sample inferences concerning the differences in population means?

10.57 Two methods for teaching reading were applied to two randomly selected groups of elementary schoolchildren and then compared on the basis of a reading comprehension test given at the end of the learning period. The sample means and variances computed from the test scores are shown in the accompanying table. Do the data present sufficient evidence to indicate a difference in the mean scores for the populations associated with the two teaching methods? What can be said about the attained significance level? What assumptions are required? What would you conclude at the $\alpha = .05$ level of significance?

	Method I	Method II
Number of children in group	11	14
$\bar{y}$	64	69
s^2	52	71

10.58 A study was conducted by the Florida Game and Fish Commission to assess the amounts of chemical residues found in the brain tissue of brown pelicans. In a test for DDT, random samples of $n_1 = 10$ juveniles and $n_2 = 13$ nestlings produced the results shown in the accompanying table (measurements in parts per million). Test the hypothesis that mean amounts of DDT found in juveniles and nestlings do not differ versus the alternative, that the juveniles have a larger mean. Use $\alpha = .05$. (This test has important implications regarding the accumulation of DDT over time.)

Juveniles	Nestlings
$n_1 = 10$	$n_2 = 13$
$\bar{y}_1 = .041$	$\bar{y}_2 = .026$
$s_1 = .017$	$s_2 = .006$

10.59 Refer to Exercise 10.58. Is there evidence that the mean for juveniles exceeds that for nestlings by more than .01 parts per million? Give the p-value.

10.60 The strength of concrete depends, to some extent, on the method used for drying it. Two different drying methods yielded the results shown in the accompanying table for independently tested specimens (measurements in psi).

Method I	Method II
$n_1 = 7$	$n_2 = 10$
$\bar{y}_1 = 3250$	$\bar{y}_2 = 3240$
$s_1 = 210$	$s_2 = 190$

Do the methods appear to produce concrete with different mean strengths? Use $\alpha = .05$. What is the attained significance level?

10.61 In Exercise 8.71 we presented some data collected in a 1993 study by Susan Beckham and her colleagues. In this study, measurements were made of anterior compartment pressure (in millimeters of mercury) for 10 healthy runners and 10 healthy cyclists. The data summary is repeated here for your convenience.

Condition	Runners Mean	s	Cyclists Mean	s
Rest	14.5	3.92	11.1	3.98
80% maximal O_2 consumption	12.2	3.49	11.5	4.95

a Is there sufficient evidence to justify claiming that a difference exists in the mean compartment pressure between runners and cyclists who are resting? Use $\alpha = .05$. What can be said about the associated p-value?

b Does sufficient evidence exist to permit us to identify a difference in the mean compartment pressure between runners and cyclists at 80% maximal O_2 consumption? Use $\alpha = .05$. What can be said about the associated p-value?

10.62 Refer to Exercise 8.74. A report from a testing laboratory claims that, for these species of fish, the average LC50 measurement is 6 parts per million. Use the data of Exercise 8.74 to determine whether sufficient evidence exists to indicate that the average LC50 measurement is less than 6 parts per million. Use $\alpha = .05$.

10.63 The tremendous growth of the Florida lobster (called *spiny lobster*) industry over the past 20 years has made it the state's second most valuable fishery industry. A recent declaration by the Bahamian government that prohibits U.S. lobsterers from fishing on the Bahamian portion of the continental shelf is expected to reduce dramatically the landings in pounds per lobster trap. According to the records, the prior mean landings per trap was 30.31 pounds. A random sampling of 20 lobster traps since the Bahamian fishing restriction went into effect gave the following results (in pounds):

17.4	18.9	39.6	34.4	19.6
33.7	37.2	43.4	41.7	27.5
24.1	39.6	12.2	25.5	22.1
29.3	21.1	23.8	43.2	24.4.

Do these landings provide sufficient evidence to support the contention that the mean landings per trap has decreased since imposition of the Bahamian restrictions? Test using $\alpha = .05$.

10.64 Jan Lindhe, D.M.D., conducted a study ("Clinical Assessment of Antiplaque Agents," *Compendium of Continuing Education in Dentistry*, supl. no. 5, 1984) on the effect of an oral antiplaque rinse on plaque buildup on teeth. Fourteen subjects, whose teeth were thoroughly cleaned and polished, were randomly assigned to two groups of seven subjects each. Both groups were assigned to use oral rinses (no brushing) for a 2-week period. Group 1 used a rinse that contained an antiplaque agent. Group 2, the control group, received a similar rinse except that, unknown to the subjects, the rinse contained no antiplaque agent. A plaque index y, a measure of plaque buildup, was recorded at 4, 7, and 14 days. The mean and standard deviation for the 14-day plaque measurements for the two groups are given in the following table.

	Control Group	Antiplaque Group
Sample size	7	7
Mean	1.26	.78
Standard deviation	.32	.32

a State the null and alternative hypotheses that should be used to test the effectiveness of the antiplaque oral rinse.

b Do the data provide sufficient evidence to indicate that the oral antiplaque rinse is effective? Test using $\alpha = .05$.

c Find the p-value for the test.

10.65 In Exercise 8.76, we presented a summary of data (*Riverside Press-Enterprise*, April 8, 1993) regarding SAT scores (verbal and math) for high school students who intended to major in engineering or in language and literature. The data are summarized in the following table.

Prospective Major	Verbal		Math	
Engineering ($n = 15$)	$\bar{y} = 446$	$s = 42$	$\bar{y} = 548$	$s = 57$
Language/literature ($n = 15$)	$\bar{y} = 534$	$s = 45$	$\bar{y} = 517$	$s = 52$

a Is there sufficient evidence to indicate a difference in the mean verbal SAT scores between high school students intending to major in engineering and in language/literature? Give the associated p-value. What would you conclude at the $\alpha = .05$ significance level?

b Are the results you obtained in (a) consistent with those you obtained in Exercise 8.76(a)?

c Answer the questions posed in (a) in relation to the mean math SAT scores for the two groups of students.

d Are the results you obtained in (c) consistent with those you obtained in Exercise 8.76(b)?

10.9 Testing Hypotheses Concerning Variances

We again assume that we have a random sample $Y_1, Y_2, \ldots, Y_n$ from a normal distribution with unknown mean μ and unknown variance σ^2. In Section 8.9, we used the pivotal method to construct a confidence interval for the parameter σ^2. In this section we consider the problem of testing $H_0: \sigma^2 = \sigma_0^2$ for some fixed value σ_0^2 versus various alternative hypotheses. If H_0 is true and $\sigma^2 = \sigma_0^2$, Theorem 7.3 implies that

$$\chi^2 = \frac{(n-1)S^2}{\sigma_0^2}$$

has a χ^2 distribution with $n-1$ degrees of freedom. If we desire to test $H_0 : \sigma^2 = \sigma_0^2$ versus $H_a : \sigma^2 > \sigma_0^2$, we can use $\chi^2 = (n-1)S^2/\sigma_0^2$ as our test statistic, but how should we select the rejection region RR?

If H_a is true and the actual value of σ^2 is larger than σ_0^2, we would expect S^2 (which estimates the true value of σ^2) to be larger than σ_0^2. The larger S^2 is relative to σ_0^2, the stronger is the evidence to support $H_a : \sigma^2 > \sigma_0^2$. Notice that S^2 is large relative to σ_0^2 if and only if $\chi^2 = (n-1)S^2/\sigma_0^2$ is large. Thus we see that a rejection region of the form $RR = \{\chi^2 > k\}$ for some constant k is appropriate for testing $H_0 : \sigma^2 = \sigma_0^2$ versus $H_a : \sigma^2 > \sigma_0^2$. If we desire a test for which the probability of a type I error is α, we use the rejection region

$$RR = \{\chi^2 > \chi_\alpha^2\}$$

where $P(\chi^2 > \chi_\alpha^2) = \alpha$. (Values of χ_α^2 can be found in Table 6, Appendix III.) An illustration of this rejection region is found in Figure 10.10(a).

If we want to test $H_0 : \sigma^2 = \sigma_0^2$ versus $H_a : \sigma^2 < \sigma_0^2$ (a lower-tail alternative), analogous reasoning leads to a rejection region located in the lower tail of the χ^2 distribution. Alternatively, we can test $H_0 : \sigma^2 = \sigma_0^2$ versus $H_a : \sigma^2 \neq \sigma_0^2$ (a two-tailed test) by using a two-tailed rejection region. Graphs illustrating these rejection regions are given in Figure 10.10.

FIGURE **10.10**
Rejection regions
RR for testing
$H_0 : \sigma^2 = \sigma_0^2$ versus
(a) $H_a : \sigma^2 > \sigma_0^2$;
(b) $H_a : \sigma^2 < \sigma_0^2$;
and (c) $H_a : \sigma^2 \neq \sigma_0^2$

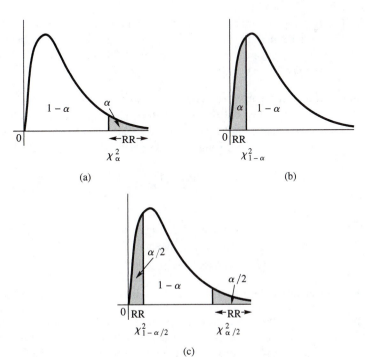

Test of Hypotheses Concerning a Population Variance

Assumptions: $Y_1, Y_2, \ldots, Y_n$ constitute a random sample from a normal distribution with

$$E(Y_i) = \mu \quad \text{and} \quad V(Y_i) = \sigma^2.$$

$H_0: \sigma^2 = \sigma_0^2.$

$$H_a: \begin{cases} \sigma^2 > \sigma_0^2 & \text{(upper-tail alternative)}. \\ \sigma^2 < \sigma_0^2 & \text{(lower-tail alternative)}. \\ \sigma^2 \neq \sigma_0^2 & \text{(two-tailed alternative)}. \end{cases}$$

Test statistic: $\chi^2 = \dfrac{(n-1)S^2}{\sigma_0^2}.$

$$\text{Rejection region: } \begin{cases} \chi^2 > \chi_\alpha^2 & \text{(upper-tail RR)}. \\ \chi^2 < \chi_{1-\alpha}^2 & \text{(lower-tail RR)}. \\ \chi^2 > \chi_{\alpha/2}^2 \text{ or } \chi^2 < \chi_{1-\alpha/2}^2 & \text{(two-tailed RR)}. \end{cases}$$

Notice that χ_α^2 is chosen so that, for $\nu = n - 1$ degrees of freedom, $P(\chi^2 > \chi_\alpha^2) = \alpha$. (See Table 6, Appendix III.)

EXAMPLE **10.16**   A company produces machined engine parts that are supposed to have a diameter variance no larger than .0002 (diameters measured in inches). A random sample of 10 parts gave a sample variance of .0003. Test, at the 5% level, $H_0 : \sigma^2 = .0002$ against $H_a : \sigma^2 > .0002$.

Solution   If it is reasonable to assume that the measured diameters are normally distributed, the appropriate test statistic is $\chi^2 = (n-1)S^2/\sigma_0^2$. Because we have posed an upper-tail test, we reject H_0 for values of this statistic larger than $\chi_{.05}^2 = 16.919$ (based on 9 d.f.). The observed value of the test statistic is

$$\frac{(n-1)s^2}{\sigma_0^2} = \frac{(9)(.0003)}{.0002} = 13.5.$$

Thus, H_0 is not rejected. There is not sufficient evidence to indicate that σ^2 exceeds .0002 at the 5% level of significance.

EXAMPLE **10.17**   Determine the p-value associated with the statistical test of Example 10.16.

Solution   The p-value is the probability that a χ^2 random variable with 9 d.f. is larger than the observed value of 13.5. The area corresponding to this probability is shaded in Figure 10.11. By examining the row corresponding to 9 d.f. in Table 6, Appendix III,

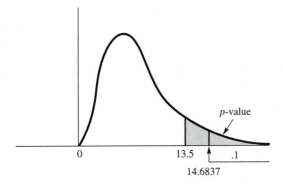

FIGURE **10.11**
Illustration of the
p-value for Example
10.17 (χ^2 density
with 9 d.f.)

we find that $\chi^2_{.1} = 14.6837$. As Figure 10.11 indicates, the shaded area exceeds .1 and thus p-value $> .1$. That is, for any value of $\alpha < .1$, the null hypothesis cannot be rejected. This agrees with the conclusion of Example 10.16.

EXAMPLE **10.18**   An experimenter was convinced that the variability in his measuring equipment yielded a standard deviation of 2. Sixteen measurements resulted in a value of $s^2 = 6.1$. Do the data disagree with his claim? Determine the p-value for the test. What would you conclude if you chose $\alpha = .05$?

Solution   We require a test of $H_0 : \sigma^2 = 4$ versus $H_a : \sigma^2 \neq 4$, a two-tailed test. The value of the test statistic is $\chi^2 = 15(6.1)/4 = 22.875$. Referring to Table 6, we see that, for 15 degrees of freedom, $\chi^2_{.05} = 24.9958$ and $\chi^2_{.10} = 22.3072$. Thus, the portion of the p-value that falls in the upper tail is between .05 and .10. Because we need to account for a corresponding equal area in the lower tail (this area is also between .05 and .10), it follows that $.1 < p\text{-value} < .2$. The chosen value of $\alpha = .05$ is smaller than even the smallest possible value for the p-value; therefore, we cannot reject the experimenters claim at the $\alpha = .05$ level.

Sometimes we wish to compare the variances of two normal distributions, particularly by testing to determine whether they are equal. These problems are encountered in comparing the precision of two measuring instruments, the variation in quality characteristics of a manufactured product, or the variation in scores for two testing procedures. For example, suppose that $Y_{11}, Y_{12}, \ldots, Y_{1n_1}$ and $Y_{21}, Y_{22}, \ldots, Y_{2n_2}$ are independent random samples from normal distributions with unknown means and that $V(Y_{1i}) = \sigma^2_1$ and $V(Y_{2i}) = \sigma^2_2$, where σ^2_1 and σ^2_2 are unknown. Suppose that we want to test the null hypothesis $H_0 : \sigma^2_1 = \sigma^2_2$ against the alternative $H_a : \sigma^2_1 > \sigma^2_2$.

Because the sample variances S^2_1 and S^2_2 estimate the respective population variances σ^2_1 and σ^2_2, we will reject H_0 in favor of H_a if S^2_1 is much larger than S^2_2. That is, we will use a rejection region RR of the form

$$\text{RR} = \left\{ \frac{S^2_1}{S^2_2} > k \right\}$$

FIGURE **10.12**
Rejection region
RR for testing
$H_0: \sigma_1^2 = \sigma_2^2$ versus
$H_a: \sigma_1^2 > \sigma_2^2$

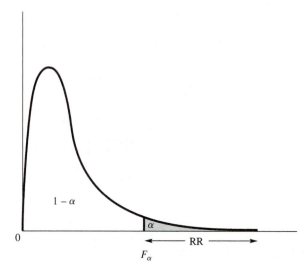

where k is chosen so that the probability of a type I error is α. The appropriate value of k depends upon the probability distribution of the statistic S_1^2/S_2^2. Notice that $(n_1 - 1)S_1^2/\sigma_1^2$ and $(n_2 - 1)S_2^2/\sigma_2^2$ are independent chi-square random variables. From Definition 7.3, it follows that

$$F = \frac{(n_1 - 1)S_1^2}{\sigma_1^2(n_1 - 1)} \left/ \frac{(n_2 - 1)S_2^2}{\sigma_2^2(n_2 - 1)} \right. = \frac{S_1^2\sigma_2^2}{S_2^2\sigma_1^2}$$

has an F distribution with $(n_1 - 1)$ numerator degrees of freedom and $(n_2 - 1)$ denominator degrees of freedom. Under the null hypothesis that $\sigma_1^2 = \sigma_2^2$, it follows that $F = S_1^2/S_2^2$ and the rejection region RR given earlier is equivalent to RR = $\{F > k\} = \{F > F_\alpha\}$, where $k = F_\alpha$ is the value of the F distribution with $\nu_1 = (n_1 - 1)$ and $\nu_2 = (n_2 - 1)$ such that $P(F > F_\alpha) = \alpha$. Values of F_α are given in Table 7, Appendix III. This rejection region is shown in Figure 10.12.

EXAMPLE **10.19**   Suppose that we wish to compare the variation in diameters of parts produced by the company in Example 10.16 with the variation in diameters of parts produced by a competitor. Recall that the sample variance for our company, based on $n = 10$ diameters, was $s_1^2 = .0003$. In contrast, the sample variance of the diameter measurements for 20 of the competitor's parts was $s_2^2 = .0001$. Do the data provide sufficient information to indicate a smaller variation in diameters for the competitor? Test with $\alpha = .05$.

Solution   We are testing $H_0: \sigma_1^2 = \sigma_2^2$ against the alternative $H_a: \sigma_1^2 > \sigma_2^2$. The test statistic, $F = (S_1^2/S_2^2)$, is based on $\nu_1 = 9$ numerator and $\nu_2 = 19$ denominator degrees of freedom, and we reject H_0 for values of F larger than $F_{.05} = 2.42$. (See Table 7,

Appendix III.) Because the observed value of the test statistic is

$$F = \frac{s_1^2}{s_2^2} = \frac{.0003}{.0001} = 3$$

we see that $F > F_{.05}$; therefore, at the $\alpha = .05$ level, we reject $H_0 : \sigma_1^2 = \sigma_2^2$ in favor of $H_a : \sigma_1^2 > \sigma_2^2$ and conclude that the competing company produces parts with smaller variation in their diameters.

EXAMPLE 10.20   Give bounds for the p-value associated with the data of Example 10.19.

Solution   The calculated F value for this upper-tail test is $F = 3$. Because this value is based on $\nu_1 = 9$ and $\nu_2 = 19$ numerator and denominator degrees of freedom, respectively, Table 7, Appendix III, can be used to determine that $F_{.025} = 2.88$, whereas $F_{.01} = 3.52$. Thus, the observed value, $F = 3$, would lead to rejection of the null hypothesis for $\alpha = .025$ but not for $\alpha = .01$. Hence, $.01 < p\text{-value} < .025$.

Suppose that, for Example 10.19, our research hypothesis was $H_a : \sigma_1^2 < \sigma_2^2$. How would we proceed? We are at liberty to identify either population as population 1. Therefore, if we simply interchange the arbitrary labels of 1 and 2 on the two populations (and the corresponding identifiers on sample sizes, sample variances, etc.), our alternative hypothesis becomes $H_a : \sigma_1^2 > \sigma_2^2$, and we can proceed as before. That is, if the research hypothesis is that the variance of one population is larger than the variance of another population, we identify the population with the *hypothesized larger* variance as population 1 and proceed as indicated in the solution to Example 10.19.

Test of the Hypothesis $\sigma_1^2 = \sigma_2^2$

Assumptions: Independent samples from normal populations.

$H_0: \sigma_1^2 = \sigma_2^2$.

$H_a: \sigma_1^2 > \sigma_2^2$.

Test statistic: $F = \dfrac{S_1^2}{S_2^2}$.

Rejection region: $F > F_\alpha$, where F_α is chosen so that $P(F > F_\alpha) = \alpha$ when F has $\nu_1 = n_1 - 1$ numerator degrees of freedom and $\nu_2 = n_2 - 1$ denominator degrees of freedom. (See Table 7, Appendix III.)

If we wish to test $H_0: \sigma_1^2 = \sigma_2^2$ versus $H_a: \sigma_1^2 \neq \sigma_2^2$ with type I error probability α, we can employ $F = S_1^2/S_2^2$ as a test statistic and reject H_0 in favor of H_a if the calculated F value is in either the upper or the lower $\alpha/2$ tail of the F distribution.

The upper-tail critical values can be determined directly from Table 7, Appendix III; but how do we determine the lower-tail critical values?

Notice that $F = S_1^2/S_2^2$ and $F^{-1} = S_2^2/S_1^2$ both have F distributions, but the numerator and denominator degrees of freedom are interchanged (the process of inversion switches the roles of numerator and denominator). Let F_b^a denote a random variable with an F distribution having a and b numerator and denominator degrees of freedom, respectively, and let $F_{b,\alpha/2}^a$ be such that

$$P(F_b^a > F_{b,\alpha/2}^a) = \alpha/2.$$

Then

$$P[(F_b^a)^{-1} < (F_{b,\alpha/2}^a)^{-1}] = \alpha/2$$

and, therefore,

$$P[F_a^b < (F_{b,\alpha/2}^a)^{-1}] = \alpha/2.$$

That is, the value that cuts off a lower-tail area of $\alpha/2$ for an F_a^b distribution can be found by inverting $F_{b,\alpha/2}^a$. Thus, if we use $F = S_1^2/S_2^2$ as a test statistic for testing $H_0 : \sigma_1^2 = \sigma_2^2$ versus $H_a : \sigma_1^2 \neq \sigma_2^2$, the appropriate rejection region is

$$\text{RR:} \left\{ F > F_{n_2-1,\alpha/2}^{n_1-1} \quad \text{or} \quad F < (F_{n_1-1,\alpha/2}^{n_2-1})^{-1} \right\}.$$

An equivalent test (see Exercise 10.69) is obtained as follows. Let n_L and n_S denote the sample sizes associated with the larger and smaller sample variances, respectively. Place the larger sample variance in the numerator and the smaller sample variance in the denominator of the F statistic, and reject $H_0 : \sigma_1^2 = \sigma_2^2$ in favor of $H_a : \sigma_1^2 \neq \sigma_2^2$ if $F > F_{\alpha/2}$, where $F_{\alpha/2}$ is determined for $\nu_1 = n_L - 1$ and $\nu_2 = n_S - 1$ numerator and denominator degrees of freedom, respectively.

EXAMPLE **10.21** An experiment to explore the pain thresholds to electrical shocks for males and females resulted in the data summary given in Table 10.4. Do the data provide sufficient evidence to indicate a significant difference in the variability of pain thresholds for men and women? Use $\alpha = .10$. What can be said about the p-value?

Solution Let us assume that the pain thresholds for men and women are approximately normally distributed. We desire to test $H_0 : \sigma_M^2 = \sigma_F^2$ versus $H_a : \sigma_M^2 \neq \sigma_F^2$, where σ_M^2

Table 10.4. Data for Example 10.21

	Males	Females
n	14	10
$\bar{y}$	16.2	14.9
s^2	12.7	26.4

and σ_F^2 are the variances of pain thresholds for men and women, respectively. The larger S^2 is 26.4 (the S^2 for women), and the sample size associated with the larger S^2 is $n_L = 10$. The smaller S^2 is 12.7 (the S^2 for men), and $n_S = 14$ (the number of men in the sample). Therefore, we compute

$$F = \frac{26.4}{12.7} = 2.079$$

and we compare this value to $F_{\alpha/2} = F_{.05}$ with $v_1 = 10-1 = 9$ and $v_2 = 14-1 = 13$ numerator and denominator degrees of freedom, respectively. Because $F_{.05} = 2.71$ and because 2.079 is not larger than the critical value (2.71), insufficient evidence exists to support a claim that the variability of pain thresholds for men and women differs.

The p-value associated with the observed value of F for this two-tailed test can be found as follows. Referring to Table 7, with $v_1 = 9$, $v_2 = 13$ numerator and denominator degrees of freedom, respectively, we find $F_{.10} = 2.16$. Thus, p-value $> 2(.10) = .20$. Unless we were willing to work with a very large value of α (some value greater than .2), these results would *not* allow us to conclude that the variances of pain thresholds for men and women differ.

Although we used the notation F in Example 10.21 to denote the ratio with the larger S^2 in the numerator and the smaller S^2 in the denominator, this ratio *does not* have an F distribution (notice that the ratio defined in this way *must* be greater than or equal to 1). Nevertheless, the tables of the F distribution can be used to determine the rejection region for an α-level test (see Exercise 10.69).

Both the χ^2 tests and the F tests presented in this section are *very sensitive* to departures from the assumption of normality of the underlying population(s). Thus, unlike the t tests of Section 10.8, these tests are *not robust* if the normality assumption is violated.

Exercises

10.66 A manufacturer of hard safety hats for construction workers is concerned about the mean and the variation of the forces its helmets transmit to wearers when subjected to a standard external force. The manufacturer desires the mean force transmitted by helmets to be 800 pounds (or less), well under the legal 1000-pound limit, and desires σ to be less than 40. Tests were run on a random sample of $n = 40$ helmets, and the sample mean and variance were found to be equal to 825 pounds and 2350 pounds^2, respectively.

    a If $\mu = 800$ and $\sigma = 40$, is it likely that any helmet subjected to the standard external force will transmit a force to a wearer in excess of 1000 pounds? Explain.

    b Do the data provide sufficient evidence to indicate that when subjected to the standard external force, the helmets transmit a mean force exceeding 800 pounds?

    c Do the data provide sufficient evidence to indicate that σ exceeds 40?

10.67 The manufacturer of a machine to package soap powder claimed that her machine could load cartons at a given weight with a range of no more than .4 ounce. The mean and variance of a sample of eight 3-pound boxes were found to equal 3.1 and .018, respectively. Test the hypothesis that the variance of the population of weight measurements is $\sigma^2 = .01$ against the alternative that it is $\sigma^2 > .01$. Use an $\alpha = .05$ level of significance. What assumptions are required for this test? What can be said about the attained significance level?

10.68   Under what assumptions may the F distribution be used in making inferences about the ratio of population variances?

10.69   From two normal populations with respective variances σ_1^2 and σ_2^2, we observe independent sample variances S_1^2 and S_2^2, with corresponding degrees of freedom $\nu_1 = n_1 - 1$ and $\nu_2 = n_2 - 1$. We wish to test $H_0 : \sigma_1^2 = \sigma_2^2$ versus $H_a : \sigma_1^2 \neq \sigma_2^2$.

a   Show that the rejection region given by

$$\{F > F_{\nu_2,\alpha/2}^{\nu_1} \quad \text{or} \quad F < (F_{\nu_1,\alpha/2}^{\nu_2})^{-1}\},$$

where $F = S_1^2/S_2^2$, is the same as the rejection region given by

$$\{S_1^2/S_2^2 > F_{\nu_2,\alpha/2}^{\nu_1} \quad \text{or} \quad S_2^2/S_1^2 > F_{\nu_1,\alpha/2}^{\nu_2}\}.$$

b   Let S_L^2 denote the larger of S_1^2 and S_2^2, and let S_S^2 denote the smaller of S_1^2 and S_2^2. Let ν_L and ν_S denote the degrees of freedom associated with S_L^2 and S_S^2, respectively. Use (a) to show that, under H_0,

$$P(S_L^2/S_S^2 > F_{\nu_S,\alpha/2}^{\nu_L}) = \alpha.$$

Notice that this gives an equivalent method for testing the equality of two variances.

10.70   Exercises 8.71 and 10.61 presented some data collected in a 1993 study by Susan Beckham and her colleagues. In this study, measurements of anterior compartment pressure (in millimeters of mercury) were taken for 10 healthy runners and 10 healthy cyclists. The researchers also obtained pressure measurements for the runners and cyclists at maximal O_2 consumption. The data summary is given in the accompanying table.

	Runners		Cyclists	
Condition	Mean	s	Mean	s
Rest	14.5	3.92	11.1	3.98
80% maximal O_2 consumption	12.2	3.49	11.5	4.95
Maximal O_2 consumption	19.1	16.9	12.2	4.67

a   Is there sufficient evidence to support a claim that the variability of compartment pressure between runners and cyclists who are resting differs? Use $\alpha = .05$. What can be said about the associated p-value?

b   Is there sufficient evidence to support a claim that the variability in compartment pressure between runners and cyclists at maximal O_2 consumption differs? Use $\alpha = .05$. What can be said about the associated p-value?

10.71   The manager of a dairy is in the market for a new bottle-filling machine and is considering machines manufactured by companies A and B. If ruggedness, cost, and convenience are comparable in the two machines, the deciding factor will be the variability of fills (the machine producing fills with the smaller variance being preferable). Let σ_1^2 and σ_2^2 be the fill variances for machines produced by companies A and B, respectively. Now consider various tests of the null hypothesis $H_0: \sigma_1^2 = \sigma_2^2$. Obtaining samples of fills from the two machines and utilizing the test statistic S_1^2/S_2^2, we could set up as the rejection region an upper-tail area, a lower-tail area, or a two-tailed area of the F distribution, depending on the interests to be served. Identify the type of rejection region that would be most favored by the following persons, and explain why:

a the manager of the dairy.

b a salesperson for company A.

c a salesperson for company B.

10.72 The closing prices of two common stocks were recorded for a period of 16 days. The means and variances were

$$\bar{y}_1 = 40.33 \quad \bar{y}_2 = 42.54$$
$$s_1^2 = 1.54 \quad s_2^2 = 2.96.$$

Do these data present sufficient evidence to indicate a difference in variability of closing prices of the two stocks for the populations associated with the two samples? Give bounds for the attained significance level. What would you conclude, with $\alpha = .02$?

10.73 A precision instrument is guaranteed to be accurate to within 2 units. A sample of four instrument readings on the same object yielded the measurements 353, 351, 351, and 355. Give the attained significance level for testing the null hypothesis that $\sigma = .7$ versus the alternative that $\sigma > .7$.

10.74 Aptitude tests should produce scores with a large amount of variation so that an administrator can distinguish between persons with low aptitude and persons with high aptitude. The standard test used by a certain industry has been producing scores with a standard deviation of 10 points. A new test is given to 20 prospective employees and produces a sample standard deviation of 12 points. Are scores from the new test significantly more variable than scores from the standard? Use $\alpha = .01$.

10.75 Refer to Exercise 10.58. Is there sufficient evidence, at the 5% significance level, to support concluding that the variance in measurements of DDT levels is greater for juveniles than it is for nestlings?

10.10 Power of Tests and the Neyman–Pearson Lemma

In the remaining sections of this chapter we move from practical examples of statistical tests to a theoretical discussion of their properties. We have suggested specific tests for a number of practical hypothesis-testing situations, but you may wonder why we chose those particular tests. How did we decide on the test statistics that were presented, and how did we know that we had selected the best rejection regions?

The goodness of a test is measured by α and β, the probabilities of type I and type II errors, respectively, where α is chosen in advance and determines the location of the rejection region. A related but more useful concept for evaluating the performance of a test is called the *power* of the test. Basically, the power of a test is the probability that the test will lead to rejection of the null hypothesis.

DEFINITION **10.3**

Suppose that W is the test statistic and RR is the rejection region for a test of a hypothesis involving the value of a parameter θ. Then the *power* of the test, denoted by power(θ), is the probability that the test will lead to rejection of H_0 when the actual parameter value is θ. That is,

$$\text{power}(\theta) = P(W \text{ in RR when the parameter value is } \theta).$$

Suppose that we want to test the null hypothesis $H_0 : \theta = \theta_0$ and that θ_a is a particular value for θ chosen from H_a. The power of the test at $\theta = \theta_0$, power(θ_0), is equal to the probability of rejecting H_0 when H_0 is true. That is, power(θ_0) $= \alpha$, the probability of a type I error. For any value of θ from H_a, the power of a test measures the test's ability to detect that the null hypothesis is false. That is, for $\theta = \theta_a$,

$$\text{power}(\theta_a) = P(\text{rejecting } H_0 \text{ when } \theta = \theta_a).$$

If we express the probability β of a type II error when $\theta = \theta_a$ as $\beta(\theta_a)$, then

$$\beta(\theta_a) = P(\text{accepting } H_0 \text{ when } \theta = \theta_a).$$

It follows that the power of the test at θ_a and the probability of a type II error are related as follows.

Relationship Between Power and β

If θ_a is a value of θ in the alternative hypothesis H_a, then

$$\text{power}(\theta_a) = 1 - \beta(\theta_a).$$

A typical *power curve*, a graph of power(θ), is shown in Figure 10.13.

FIGURE **10.13**
A typical power curve for the test of $H_0 : \theta = \theta_0$ against the alternative $H_a : \theta \neq \theta_0$

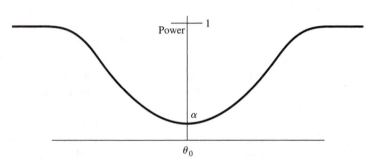

FIGURE **10.14**
Ideal power curve for
the test of $H_0: \theta = \theta_0$
versus $H_a: \theta \neq \theta_0$

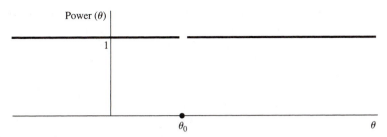

FIGURE **10.14**
Ideal power curve for
the test of $H_0: \theta = \theta_0$
versus $H_a: \theta \neq \theta_0$

Ideally a test would detect a departure from $H_0 : \theta = \theta_0$ with certainty; that is, power(θ_a) would be 1 for all θ_a in H_a (see Figure 10.14). Because, for a fixed sample size, α and β both cannot be made arbitrarily small, this is clearly not possible. Therefore, for a fixed sample size n, we adopt the procedure of selecting a (small) value for α and finding a rejection region RR to *minimize* $\beta(\theta_a)$ at each θ_a in H_a. That is, we choose RR to maximize power(θ) for θ in H_a. From among all tests with a significance level of α, we seek the test whose power function comes closest to the ideal power function (Figure 10.14) if such a test exists. How do we find such a testing procedure?

Before we proceed, we must define *simple* and *composite* hypotheses. Suppose that $Y_1, Y_2, \ldots, Y_n$ constitute a random sample from an exponential distribution with parameter λ; that is, $f(y) = (1/\lambda)e^{-y/\lambda}$, $y > 0$. Then the hypothesis $H : \lambda = 2$ uniquely specifies the distribution from which the sample is taken as having density function $f(y) = (1/2)e^{-y/2}$, $y > 0$. The hypothesis $H : \lambda = 2$ is, therefore, an example of a *simple* hypothesis. In contrast, the hypothesis $H : \lambda > 2$ is a *composite* hypothesis because under H the density function, $f(y)$ is not uniquely determined. The form of the density is exponential, but the parameter λ could be 3 or 15 or any value greater than 2.

DEFINITION **10.4**

> If a random sample is taken from a distribution with parameter θ, a hypothesis is said to be a *simple hypothesis* if that hypothesis uniquely specifies the distribution of the population from which the sample is taken. Any hypothesis that is not a simple hypothesis is called a *composite hypothesis*.

If $Y_1, Y_2, \ldots, Y_n$ represent a random sample from a normal distribution with known variance $\sigma^2 = 1$, then $H : \mu = 5$ is a simple hypothesis because, if H is true, the density function is uniquely specified to be a normal density function with $\mu = 5$ and $\sigma^2 = 1$. If, on the other hand, σ^2 is not known, the hypothesis $H : \mu = 5$ determines the mean of the normal distribution but does not determine the value of the variance. Therefore, if σ^2 is not known, $H : \mu = 5$ is a composite hypothesis.

Suppose that we would like to test a *simple* null hypothesis $H_0 : \theta = \theta_0$ versus a *simple* alternative hypothesis $H_a : \theta = \theta_a$. Because we are concerned only with two particular values of θ (θ_0 and θ_a), we would like to choose a rejection region, RR, so that $\alpha = $ power (θ_0) is a fixed value and $\beta(\theta_a) = 1-$ power(θ_a) is as small as possible. Equivalently, we seek a *most powerful* test, that is, a test with maximum power: power(θ_a). The following theorem provides the methodology for deriving the most powerful test for testing simple H_0 versus simple H_a. [*Note:* In Definition 9.4, we defined the likelihood function $L(\theta) = L(y_1, y_2, \ldots, y_n \mid \theta)$. At that time we indicated that the likelihood function depends on $y_1, y_2, \ldots, y_n$ and on θ. As before, we use the notation $L(\theta)$ to emphasize the dependence of the likelihood function on the value of the parameter θ.]

THEOREM 10.1

The Neyman–Pearson Lemma.

Suppose that we wish to test the simple null hypothesis $H_0 : \theta = \theta_0$ versus the simple alternative hypothesis $H_a : \theta = \theta_a$, based on a random sample $Y_1, Y_2, \ldots, Y_n$ from a distribution with parameter θ. Let $L(\theta)$ denote the likelihood of the sample when the value of the parameter is θ. Then, for a given α, the test that maximizes the power at θ_a has a rejection region, RR, determined by

$$\frac{L(\theta_0)}{L(\theta_a)} < k.$$

The value of k is chosen so that the test has the desired value for α. Such a test is a most powerful α−level test for H_0 versus H_a.

The proof of Theorem 10.1 is not given here, but it can be found in most of the texts listed in the references at the end of the chapter. We illustrate the application of the theorem with the following example.

EXAMPLE 10.22

Suppose that Y represents a single observation from the probability density function given by

$$f(y \mid \theta) = \begin{cases} \theta y^{\theta-1}, & 0 < y < 1 \\ 0, & \text{elsewhere.} \end{cases}$$

Find the most powerful test with significance level $\alpha = .05$ to test $H_0 : \theta = 2$ versus $H_a : \theta = 1$.

Solution Because both of the hypotheses are simple, Theorem 10.1 can be applied to derive the required test. In this case,

$$\frac{L(\theta_0)}{L(\theta_a)} = \frac{f(y \mid \theta_0)}{f(y \mid \theta_a)} = \frac{2y}{(1)y^0} = 2y \qquad \text{for } 0 < y < 1$$

and the form of the rejection region for the most powerful test is

$$2y < k.$$

Equivalently, the rejection region, RR, is $\{y < k/2\}$. Or because $k/2 = k^*$, a constant, the rejection region is RR: $\{y < k^*\}$.

Because $\alpha = .05$ is specified, the value of k^* is determined by

$$.05 = P(Y \text{ in RR when } \theta = 2) = P(Y < k^* \text{ when } \theta = 2) = \int_0^{k^*} 2y \, dy = (k^*)^2.$$

Therefore, $(k^*)^2 = .05$, and the rejection region of the most powerful test is

$$\text{RR}: \{y < \sqrt{.05} = .2236\}.$$

Among all tests for H_0 versus H_a based on a sample size of 1 and with α fixed at .05, this test has the smallest type II error probability when $\beta(\theta_a)$ is evaluated at $\theta_a = 1$. That is, it is the most powerful test (at $\theta_a = 1$) with significance level $\alpha = .05$.

What is the actual value for power(θ) when $\theta = 1$?

$$\text{power}(1) = P(Y \text{ in RR when } \theta = 1) = P(Y < .2236 \text{ when } \theta = 1)$$

$$= \int_0^{.2236} (1) \, dy = .2236.$$

Even though the rejection region $\{y < .2236\}$ gives the *maximum* value for power(1) among all tests with $\alpha = .05$, we see that $\beta(1) = 1 - .2236 = .7764$ is still very large.

Notice that the forms of the test statistic and of the rejection region depend upon both H_0 and H_a. If the alternative is changed to $H_a : \theta = 4$, the most powerful test is based upon Y^2, and we reject H_0 in favor of H_a if $Y^2 > k'$, for some constant k'. Also notice that the Neyman–Pearson lemma gives the *form* of the rejection region; the actual rejection region depends upon the specified value for α.

For discrete distributions it is not always possible to find a test whose significance level is exactly equal to some predetermined value of α. In such cases we specify the test to be the one for which the probability of a type I error is closest to the predetermined value of α, without exceeding it.

Suppose that we sample from a population whose distribution is completely specified except for the value of a single parameter θ. If we desire to test $H_0 : \theta = \theta_0$ (simple) versus $H_a : \theta > \theta_0$ (composite), no general theorem comparable to Theorem 10.1 is applicable if either hypothesis is composite. However, Theorem 10.1 can be applied to obtain a most powerful test for $H_0 : \theta = \theta_0$ versus $H_a : \theta = \theta_a$ for any single value θ_a, where $\theta_a > \theta_0$. In many situations the actual rejection region for the most powerful test depends only on the value of θ_0 (and does not depend upon the particular choice of θ_a). When a test obtained by Theorem 10.1 actually maximizes the power for every value of θ greater than θ_0, it is said to be a *uniformly most powerful* test for $H_0 : \theta = \theta_0$ versus $H_a : \theta > \theta_0$. Analogous remarks apply to the derivation of tests for $H_0 : \theta = \theta_0$ versus $H_a : \theta < \theta_0$. We illustrate these ideas in the following example.

EXAMPLE **10.23**   Suppose that $Y_1, Y_2, \ldots, Y_n$ constitute a random sample from a normal distribution with unknown mean μ and known variance σ^2. We wish to test $H_0 : \mu = \mu_0$ against

$H_a : \mu > \mu_0$ for a specified constant μ_0. Find the uniformly most powerful test with significance level α.

Solution   We begin by looking for the most powerful α-level test of $H_0 : \mu = \mu_0$ versus $H_a^* : \mu = \mu_a$ for one fixed value of μ_a that is larger than μ_0. Because

$$f(y \mid \mu) = \left(\frac{1}{\sigma \sqrt{2\pi}} \right) \exp \left[\frac{-(y - \mu)^2}{2\sigma^2} \right], \qquad -\infty < y < \infty$$

we have

$$L(\mu) = f(y_1 \mid \mu) f(y_2 \mid \mu) \cdots f(y_n \mid \mu) = \left(\frac{1}{\sigma \sqrt{2\pi}} \right)^n \exp \left[-\sum_{i=1}^{n} \frac{(y_i - \mu)^2}{2\sigma^2} \right].$$

[Recall that $\exp(w)$ is simply e^w in another form.] Appealing to Theorem 10.1, we see that the most powerful test of $H_0 : \mu = \mu_0$ versus $H_a^* : \mu = \mu_a$ is given by

$$\frac{L(\mu_0)}{L(\mu_a)} < k,$$

which in this case is equivalent to

$$\frac{\left(\dfrac{1}{\sigma \sqrt{2\pi}} \right)^n \exp \left[-\sum_{i=1}^{n} \dfrac{(y_i - \mu_0)^2}{2\sigma^2} \right]}{\left(\dfrac{1}{\sigma \sqrt{2\pi}} \right)^n \exp \left[-\sum_{i=1}^{n} \dfrac{(y_i - \mu_a)^2}{2\sigma^2} \right]} < k.$$

This inequality can be rearranged as follows:

$$\exp \left\{ -\frac{1}{2\sigma^2} \left[\sum_{i=1}^{n} (y_i - \mu_0)^2 - \sum_{i=1}^{n} (y_i - \mu_a)^2 \right] \right\} < k.$$

Taking natural logarithms and simplifying, we have

$$-\frac{1}{2\sigma^2} \left[\sum_{i=1}^{n} (y_i - \mu_0)^2 - \sum_{i=1}^{n} (y_i - \mu_a)^2 \right] < \ln(k)$$

$$\sum_{i=1}^{n} (y_i - \mu_0)^2 - \sum_{i=1}^{n} (y_i - \mu_a)^2 > -2\sigma^2 \ln(k)$$

$$\sum_{i=1}^{n} y_i^2 - 2n\bar{y}\mu_0 + n\mu_0^2 - \sum_{i=1}^{n} y_i^2 + 2n\bar{y}\mu_a - n\mu_a^2 > -2\sigma^2 \ln(k)$$

$$\bar{y}(\mu_a - \mu_0) > \frac{-2\sigma^2 \ln(k) - n\mu_0^2 + n\mu_a^2}{2n}$$

or, since $\mu_a > \mu_0$,

$$\overline{y} > \frac{-2\sigma^2 \ln(k) - n\mu_0^2 + n\mu_a^2}{2n(\mu_a - \mu_0)}.$$

Because σ^2, n, μ_0 and μ_a are all known constants, the quantity on the right-hand side of this inequality is a constant, call it k'. Therefore, the most powerful test of $H_0 : \mu = \mu_0$ versus $H_a^* : \mu = \mu_a$ has the rejection region given by

$$\text{RR} = \{\overline{y} > k'\}.$$

The precise value of k' is determined by fixing α and noting that

$$\alpha = P(\overline{Y} \text{ in RR when } \mu = \mu_0)$$
$$= P(\overline{Y} > k' \text{ when } \mu = \mu_0)$$
$$= P\left(\frac{\overline{Y} - \mu_0}{\sigma/\sqrt{n}} > \frac{k' - \mu_0}{\sigma/\sqrt{n}}\right)$$
$$= P\left(Z > \sqrt{n}(k' - \mu_0)/\sigma\right).$$

Because, under H_0, Z has a standard normal distribution, $P(Z > z_\alpha) = \alpha$ and the required value for k' must satisfy

$$\sqrt{n}(k' - \mu_0)/\sigma = z_\alpha, \quad \text{or, equivalently,} \quad k' = \mu_0 + z_\alpha \sigma/\sqrt{n}.$$

Thus, our test is based on the statistic $\overline{Y}$ and the rejection region is $\{\overline{y} > \mu_0 + z_\alpha \sigma/\sqrt{n}\}$. We now observe that neither the test statistic nor the rejection region for an α-level test depends upon the particular value assigned to μ_a. That is, any value of μ_a greater than μ_0 would lead to exactly the same rejection region. Thus, we have found the uniformly most powerful test for $H_0 : \mu = \mu_0$ versus $H_a : \mu > \mu_0$. This is exactly the test we considered in Section 10.3.

Again consider the situation where the random sample is taken from a distribution that is completely specified except for the value of a single parameter θ. If we wish to derive a test for $H_0 : \theta \leq \theta_0$ versus $H_a : \theta > \theta_0$ (so that both H_0 and H_a are composite hypotheses), how do we proceed? Suppose that we use the method illustrated in Example 10.23 to find a uniformly most powerful test for $H_0' : \theta = \theta_0$ versus $H_a : \theta > \theta_0$. If θ_1 is a fixed value of θ that is less than θ_0 and we use the same test for $H_0'' : \theta = \theta_1$ versus H_a, typically α will decrease and $\beta(\theta_a)$ will remain unchanged for all θ_a in H_a. In other words, if we have a good test for discriminating between H_0' and H_a, the same test will be even better for discriminating between H_0'' and H_a. For tests with composite null hypotheses of the form $H_0 : \theta \leq \theta_0$ (or $H_0 : \theta \geq \theta_0$), we define the significance level α to be the probability of a type I error when $\theta = \theta_0$, that is, $\alpha = \text{power}(\theta_0)$. Generally this value for α is the maximum value of the power function for $\theta \leq \theta_0$ (or $\theta \geq \theta_0$). Using this methodology we can show that the test derived in Example 10.23 for testing $H_0 : \theta = \theta_0$ versus

$H_a : \theta > \theta_0$ is also the uniformly most powerful $\alpha-$level test for testing $H_0 : \theta \le \theta_0$ versus $H_a : \theta > \theta_0$.

In Example 10.23, we derived the uniformly most powerful test for $H_0 : \mu = \mu_0$ versus $H_a : \mu > \mu_0$ and found it to have rejection region $\{\bar{y} > \mu_0 + z_\alpha \sigma / \sqrt{n}\}$. That is, for any other test with the same α value and with any value of μ_a such that $\mu_a > \mu_0$, the test based on this rejection region gives the smallest value for $\beta(\mu_a)$. If we wished to test $H_0 : \mu = \mu_0$ versus $H_a : \mu < \mu_0$, analogous calculations would lead us to $\{\bar{y} < \mu_0 - z_\alpha \sigma / \sqrt{n}\}$ as the rejection region for the test that is uniformly most powerful for all $\mu_a < \mu_0$. Therefore, if we wish to test $H_0 : \mu = \mu_0$ versus $H_a : \mu \ne \mu_0$, no single rejection region yields the most powerful test for all values of $\mu_a \ne \mu_0$. Although there are some special exceptions, in most instances there do not exist uniformly most powerful two-tailed tests. Thus, there are many null and alternative hypotheses for which uniformly most powerful tests do not exist.

The Neyman–Pearson lemma is useless if we wish to test a hypothesis about a single parameter θ when the sampled distribution contains other unspecified parameters. For example, we might want to test $H_0 : \mu = \mu_0$ when the sample is taken from a normal distribution with unknown variance σ^2. In this case $H_0 : \mu = \mu_0$ does not uniquely determine the form of the distribution (since σ^2 could be any nonnegative number), and it is therefore *not* a simple hypothesis. The next section will present a very general and widely used method for developing tests of hypotheses. The method is particularly useful when unspecified parameters (called *nuisance parameters*) are present.

Exercises

10.76 Refer to Exercise 10.2. Find the power of the test for each alternative in (a)–(d).

    a $p = .4$.

    b $p = .5$.

    c $p = .6$.

    d $p = .7$.

    e Sketch a graph of the power function.

10.77 Refer to Exercise 10.5. Find the power of test 1 for each alternative in (a)–(e).

    a $\theta = .1$.

    b $\theta = .4$.

    c $\theta = .7$.

    d $\theta = 1$.

    e Sketch a graph of the power function.

*10.78 Refer to Exercise 10.5.

    a Find the power of test 2 for each of the following alternatives: $\theta = .1, \theta = .4, \theta = .7$, and $\theta = 1$.

    b Sketch a graph of the power function.

    c Compare the power function in part (b) with the power function that you found in Exercise 10.77 (this is the power function for test 1, Exercise 10.5). What can you conclude about the power of test 2 compared to the power of test 1 *for all* $\theta \ge 0$?

10.79   Let $Y_1, Y_2, \ldots, Y_{20}$ be a random sample of size $n = 20$ from a normal distribution with unknown mean μ and known variance $\sigma^2 = 5$. We wish to test $H_0 : \mu = 7$ versus $H_a : \mu > 7$.
   a   Find the uniformly most powerful test with significance level .05.
   b   For the test in (a), find the power at each of the following alternative values for μ: $\mu_a = 7.5$, $\mu_a = 8.0$, $\mu_a = 8.5$, and $\mu_a = 9.0$.
   c   Sketch a graph of the power function.

10.80   Consider the situation described in Exercise 10.79. What is the smallest sample size such that an $\alpha = .05$ level test has power at least .80 when $\mu = 8$?

10.81   For a normal distribution with mean μ and variance $\sigma^2 = 25$, an experimenter wishes to test $H_0 : \mu = 10$ versus $H_a : \mu = 5$. Find the sample size n for which the most powerful test will have $\alpha = \beta = .025$.

10.82   Suppose that $Y_1, Y_2, \ldots, Y_n$ constitute a random sample from a normal distribution with *known* mean μ and unknown variance σ^2. Find the most powerful α-level test of $H_0 : \sigma^2 = \sigma_0^2$ versus $H_a : \sigma^2 = \sigma_1^2$ where $\sigma_1^2 > \sigma_0^2$. Show that this test is equivalent to a χ^2 test. Is the test uniformly most powerful for $H_a : \sigma^2 > \sigma_0^2$?

10.83   Suppose that we have a random sample of four observations from the density function

$$f(y \mid \theta) = \begin{cases} \left(\dfrac{1}{2\theta^3} \right) y^2 e^{-y/\theta}, & y > 0 \\ 0, & \text{elsewhere.} \end{cases}$$

   a   Find the rejection region for the most powerful test of $H_0 : \theta = \theta_0$ against $H_a : \theta = \theta_a$, assuming that $\theta_a > \theta_0$. (*Hint:* Make use of the χ^2 distribution.)
   b   Is the test given in (a) uniformly most powerful for the alternative $\theta > \theta_0$?

10.84   Suppose Y is a random sample of size 1 from a population with density function

$$f(y \mid \theta) = \begin{cases} \theta y^{\theta-1}, & 0 \le y \le 1 \\ 0, & \text{elsewhere,} \end{cases}$$

where $\theta > 0$.
   a   Sketch the power function of the test with rejection region: $Y > .5$.
   b   Based on the single observation Y, find a uniformly most powerful test of size α for testing $H_0 : \theta = 1$ vs. $H_a : \theta > 1$.

*10.85   Let $Y_1, Y_2, \ldots, Y_n$ be independent and identically distributed random variables with discrete probability function given by:

	y		
	1	2	3
$p(y \mid \theta)$	θ^2	$2\theta(1-\theta)$	$(1-\theta)^2$

where $0 < \theta < 1$. Let N_i denote the number of observations equal to i for $i = 1, 2, 3$.
   a.   Derive the likelihood function, $L(\theta)$ as a function of N_1, N_2 and N_3.
   b.   Find the most powerful test for testing $H_0 : \theta = \theta_0$ versus $H_a : \theta = \theta_a$ where $\theta_a > \theta_0$. Show that your test specifies that H_0 be rejected for certain values of $2N_1 + N_2$.
   c.   How do you determine the value of k so that the test has nominal level α? You need not do the actual computation. A clear description of how to determine k is adequate.

d. Is the test derived in the previous parts uniformly most powerful for testing $H_0 : \theta = \theta_0$ versus $H_a : \theta > \theta_0$? Why or why not?

10.86 Let $Y_1, \ldots, Y_n$ be a random sample from the probability density function given by

$$f(y \mid \theta) = \begin{cases} \left(\dfrac{1}{\theta}\right) m y^{m-1} e^{-y^m/\theta}, & y > 0 \\ 0, & \text{elsewhere,} \end{cases}$$

with m denoting a known constant.

a Find the uniformly most powerful test for testing $H_0 : \theta = \theta_0$ against $H_a : \theta > \theta_0$.

b If the test in (a) is to have $\theta_0 = 100$, $\alpha = .05$, and $\beta = .05$ when $\theta_a = 400$, find the appropriate sample size and critical region.

10.87 Let $Y_1, Y_2, \ldots, Y_n$ denote a random sample from a population having a Poisson distribution with mean λ.

a Find the form of the rejection region for a most powerful test of $H_0 : \lambda = \lambda_0$ against $H_a : \lambda = \lambda_a$, where $\lambda_a > \lambda_0$.

b Recall that $\sum_{i=1}^{n} Y_i$ has a Poisson distribution with mean $n\lambda$. Indicate how this information can be used to find any constants associated with the rejection region derived in (a).

c Is the test derived in (a) uniformly most powerful for testing $H_0 : \lambda = \lambda_0$ against $H_a : \lambda > \lambda_0$? Why?

d Find the form of the rejection region for a most powerful test of $H_0 : \lambda = \lambda_0$ against $H_a : \lambda = \lambda_a$, where $\lambda_a < \lambda_0$.

10.88 Let $Y_1, Y_2, \ldots, Y_n$ denote a random sample from a population having a Poisson distribution with mean λ_1. Let $X_1, X_2, \ldots, X_m$ denote an independent random sample from a population having a Poisson distribution with mean λ_2. Derive the most powerful test for testing $H_0 : \lambda_1 = \lambda_2 = 2$ versus $H_a : \lambda_1 = 1/2$, $\lambda_2 = 3$.

10.89 Suppose that $Y_1, Y_2, \ldots, Y_n$ denote a random sample from a population having an exponential distribution with mean θ.

a Derive the most powerful test for $H_0 : \theta = \theta_0$ against $H_a : \theta = \theta_a$ where $\theta_a < \theta_0$.

b Is the test derived in (a) uniformly most powerful for testing $H_0 : \theta = \theta_0$ against $H_a : \theta < \theta_0$?

10.90 Let $Y_1, Y_2, \ldots, Y_n$ denote a random sample from a Bernoulli-distributed population with parameter p. That is,

$$p(y_i \mid p) = p^{y_i} (1 - p)^{1 - y_i}, \qquad y_i = 0, 1.$$

a Suppose that we are interested in testing $H_0 : p = p_0$ versus $H_a : p = p_a$, where $p_0 < p_a$.

   i. Show that

$$\frac{L(p_0)}{L(p_a)} = \left(\frac{p_0(1 - p_a)}{(1 - p_0)p_a}\right)^{\sum y_i} \left(\frac{1 - p_0}{1 - p_a}\right)^n.$$

   ii. Argue that $L(p_0)/L(p_a) < k$ if and only if $\sum_{i=1}^{n} y_i > k^*$ for some constant k^*.

   iii. Give the rejection region for the most powerful test of H_0 versus H_a.

b Recall that $\sum_{i=1}^{n} Y_i$ has a binomial distribution with parameters n and p. Indicate how to determine the values of any constants contained in the rejection region derived in (a(iii)).

c Is the test derived in (a) uniformly most powerful for testing $H_0: p = p_0$ versus $H_a: p > p_0$? Why or why not?

*10.91 Let $Y_1, Y_2, \ldots, Y_n$ denote a random sample from a uniform distribution over the interval $(0, \theta)$.

a Find the most powerful α-level test for testing $H_0: \theta = \theta_0$ against $H_a: \theta = \theta_a$, where $\theta_a < \theta_0$.

b Is the test in (a) uniformly most powerful for testing $H_0: \theta = \theta_0$ against $H_a: \theta < \theta_0$?

*10.92 Refer to the random sample of Exercise 10.91.

a Find the most powerful α-level test for testing $H_0: \theta = \theta_0$ against $H_a: \theta = \theta_a$ where $\theta_a > \theta_0$.

b Is the test in (a) uniformly most powerful for testing $H_0: \theta = \theta_0$ against $H_a: \theta > \theta_0$?

c Is the most powerful α-level test you found in (a) unique?

10.11 Likelihood Ratio Tests

Theorem 10.1 provides a method of constructing most powerful tests for simple hypotheses when the distribution of the observations is known except for the value of a single unknown parameter. This method can sometimes be utilized to find uniformly most powerful tests for composite hypotheses that involve a single parameter. In many cases, in fact, the distribution of concern has more than one unknown parameter. In this section we present a very general method that can be used to derive tests of hypotheses. The procedure works for simple or composite hypotheses, and whether or not nuisance parameters with unknown values are present.

Suppose that a random sample is selected from a distribution and that the likelihood function $L(y_1, y_2, \ldots, y_n \mid \theta_1, \theta_2, \ldots, \theta_k)$ is a function of k parameters, $\theta_1, \theta_2, \ldots, \theta_k$. To simplify notation, let Θ denote the vector of all k parameters, that is, $\Theta = (\theta_1, \theta_2, \ldots, \theta_k)$, and write the likelihood function as $L(\Theta)$. It may be the case that we are interested in testing hypotheses only about one of the parameters, say, θ_1. For example, if, as in Example 10.24, we take a sample from a normally distributed population with unknown mean μ and unknown variance σ^2, then the likelihood function depends on the *two* parameters μ and σ^2 and $\Theta = (\mu, \sigma^2)$. If we are interested in testing hypotheses about only the mean μ then σ^2, a parameter not of particular interest to us, is called a *nuisance parameter*. Thus, the likelihood function may be a function with both unknown nuisance parameters and a parameter of interest.

Suppose that the null hypothesis specifies that Θ (may be a vector) lies in a particular set of possible values, say, Ω_0, and that the alternative hypothesis specifies that Θ lies in another set of possible values Ω_a, which does not overlap Ω_0. For example, if we sample from a population with an exponential distribution with mean λ (in this case λ is the only parameter of the distribution and $\Theta = \lambda$), we might be interested in testing $H_0: \lambda = \lambda_0$ versus $H_a: \lambda \neq \lambda_0$. In this exponential example, Ω_0 contains only the single value λ_0 and $\Omega_a = \{\lambda > 0 : \lambda \neq \lambda_0\}$. Denote the union of the two sets, Ω_0 and Ω_a, by Ω; that is, $\Omega = \Omega_0 \cup \Omega_a$. In the exponential example $\Omega = \{\lambda_0\} \cup \{\lambda > 0 \,:\, \lambda \neq \lambda_0\} = \{\lambda \,:\, \lambda > 0\}$, the set of all possible values for λ. Either or both of the hypotheses H_0 and H_a can be composite because they might contain multiple values of the parameter of interest or because other unknown parameters may be present.

Let $L(\hat{\Omega}_0)$ denote the maximum (actually the supremum) of the likelihood function for all $\Theta \in \Omega_0$. That is, $L(\hat{\Omega}_0) = \max_{\Theta \in \Omega_0} L(\Theta)$. Notice that $L(\hat{\Omega}_0)$ represents the best explanation for the observed data for all $\Theta \in \Omega_0$ and can be found by using methods similar to those used in Section 9.7. Similarly, $L(\hat{\Omega}) = \max_{\Theta \in \Omega} L(\Theta)$ represents the best explanation for the observed data for all $\Theta \in \Omega = \Omega_0 \cup \Omega_a$. If $L(\hat{\Omega}_0) = L(\hat{\Omega})$, then a best explanation for the observed data can be found inside Ω_0 and we should not reject the null hypothesis $H_0 : \Theta \in \Omega_0$. However, if $L(\hat{\Omega}_0) < L(\hat{\Omega})$, then the best explanation for the observed data can be found inside Ω_a, and we should consider rejecting H_0 in favor of H_a. A likelihood ratio test is based on the ratio $L(\hat{\Omega}_0)/L(\hat{\Omega})$.

A Likelihood Ratio Test

Define λ by

$$\lambda = \frac{L(\hat{\Omega}_0)}{L(\hat{\Omega})} = \frac{\max\limits_{\Theta \in \Omega_0} L(\Theta)}{\max\limits_{\Theta \in \Omega} L(\Theta)}.$$

A likelihood ratio test of $H_0 : \Theta \in \Omega_0$ versus $H_a : \Theta \in \Omega_a$ employs λ as a test statistic, and the rejection region is determined by $\lambda \leq k$.

It can be shown that $0 \leq \lambda \leq 1$. A value of λ close to zero indicates that the likelihood of the sample is smaller under H_0 than it is under H_a. Therefore, the data suggest favoring H_a over H_0. The actual value of k is chosen so that α achieves the desired value. We will illustrate the mechanics of this method with the following example.

EXAMPLE **10.24**   Suppose that $Y_1, Y_2, \ldots, Y_n$ constitute a random sample from a normal distribution with unknown mean μ and unknown variance σ^2. We want to test $H_0 : \mu = \mu_0$ versus $H_a : \mu > \mu_0$. Find the appropriate likelihood ratio test.

Solution   In this case, $\Theta = (\mu, \sigma^2)$. Notice that Ω_0 is the set $\{(\mu_0, \sigma^2) : \sigma^2 > 0\}$, $\Omega_a = \{(\mu, \sigma^2) : \mu > \mu_0, \sigma^2 > 0\}$, and hence that $\Omega = \Omega_0 \cup \Omega_a = \{(\mu, \sigma^2) : \mu \geq \mu_0 \ \sigma^2 > 0\}$. The constant σ^2 is completely unspecified. We must now find $L(\hat{\Omega}_0)$ and $L(\hat{\Omega})$.

For the normal distribution, we have

$$L(\Theta) = L(\mu, \sigma^2) = \left(\frac{1}{\sqrt{2\pi}}\right)^n \left(\frac{1}{\sigma^2}\right)^{n/2} \exp\left[-\sum_{i=1}^{n} \frac{(y_i - \mu)^2}{2\sigma^2}\right].$$

Restricting μ to Ω_0 implies that $\mu = \mu_0$, and hence we can find $L(\hat{\Omega}_0)$ if we determine the value of σ^2 that maximizes $L(\mu, \sigma^2)$ subject to the constraint that $\mu = \mu_0$. From Example 9.15 we see that, when $\mu = \mu_0$, the value of σ^2 that maximizes $L(\mu_0, \sigma^2)$ is

$$\hat{\sigma}_0^2 = \frac{1}{n} \sum_{i=1}^{n} (y_i - \mu_0)^2.$$

Thus, $L(\hat{\Omega}_0)$ is obtained by replacing μ with μ_0 and σ^2 with $\hat{\sigma}_0^2$ in $L(\mu, \sigma^2)$, which gives

$$L(\hat{\Omega}_0) = \left(\frac{1}{\sqrt{2\pi}}\right)^n \left(\frac{1}{\hat{\sigma}_0^2}\right)^{n/2} e^{-n/2}.$$

We now turn to finding $L(\hat{\Omega})$. As in Example 9.15, it is easier to look at $\ln L(\mu, \sigma^2)$,

$$\ln \left[L(\mu, \sigma^2) \right] = -\frac{n}{2} \ln \sigma^2 - \frac{n}{2} \ln 2\pi - \frac{1}{2\sigma^2} \sum_{i=1}^{n} (y_i - \mu)^2.$$

Taking derivatives with respect to μ and σ^2, we obtain

$$\frac{\partial \left\{ \ln \left[L(\mu, \sigma^2) \right] \right\}}{\partial \mu} = \frac{1}{\sigma^2} \sum_{i=1}^{n} (y_i - \mu),$$

$$\frac{\partial \left\{ \ln \left[L(\mu, \sigma^2) \right] \right\}}{\partial \sigma^2} = -\left(\frac{n}{2\sigma^2}\right) + \frac{1}{2\sigma^4} \sum_{i=1}^{n} (y_i - \mu)^2.$$

We need to find the maximum of $L(\mu, \sigma^2)$ over the set $\Omega = \{(\mu, \sigma^2) : \mu \geq \mu_0, \sigma^2 > 0\}$. Notice that

$$\partial L(\mu, \sigma^2)/\partial \mu < 0 \quad \text{if} \quad \mu > \bar{y}$$
$$\partial L(\mu, \sigma^2)/\partial \mu = 0 \quad \text{if} \quad \mu = \bar{y}$$
$$\partial L(\mu, \sigma^2)/\partial \mu > 0 \quad \text{if} \quad \mu < \bar{y}$$

Thus, over the set $\Omega = \{(\mu, \sigma^2) : \mu \geq \mu_0 \ \sigma^2 > 0\}$, $\ln L(\mu, \sigma^2)$ (and also $L(\mu, \sigma^2)$) is maximized at $\hat{\mu}$ where

$$\hat{\mu} = \begin{cases} \bar{y}, & \text{if } \bar{y} > \mu_0 \\ \mu_0, & \text{if } \bar{y} \leq \mu_0. \end{cases}$$

Just as earlier, the value of σ^2 in Ω that maximizes $L(\mu, \sigma^2)$ is

$$\hat{\sigma}^2 = \frac{1}{n} \sum_{i=1}^{n} (y_i - \hat{\mu})^2.$$

$L(\hat{\Omega})$ is obtained by replacing μ with $\hat{\mu}$ and σ^2 with $\hat{\sigma}^2$, which yields

$$L(\hat{\Omega}) = \left(\frac{1}{\sqrt{2\pi}}\right)^n \left(\frac{1}{\hat{\sigma}^2}\right)^{n/2} e^{-n/2}.$$

Thus,

$$\lambda = \frac{L(\hat{\Omega}_0)}{L(\hat{\Omega})} = \left(\frac{\hat{\sigma}^2}{\hat{\sigma}_0^2}\right)^{n/2}$$

$$= \begin{cases} \left[\dfrac{\displaystyle\sum_{i=1}^{n}(y_i - \bar{y})^2}{\displaystyle\sum_{i=1}^{n}(y_i - \mu_0)^2}\right]^{n/2}, & \text{if } \bar{y} > \mu_0 \\[2em] 1, & \text{if } \bar{y} \leq \mu_0. \end{cases}$$

Notice that λ is always less than or equal to 1. Thus, "small" values of λ are those less than some $k < 1$. Because

$$\sum_{i=1}^{n}(y_i - \mu_0)^2 = \sum_{i=1}^{n}[(y_i - \bar{y}) + (\bar{y} - \mu_0)]^2$$

$$= \sum_{i=1}^{n}(y_i - \bar{y})^2 + n(\bar{y} - \mu_0)^2$$

if $k < 1$, it follows that the rejection region, $\lambda \leq k$, is equivalent to

$$\frac{\displaystyle\sum_{i=1}^{n}(y_i - \bar{y})^2}{\displaystyle\sum_{i=1}^{n}(y_i - \mu_0)^2} < k^{2/n} = k'$$

$$\frac{\displaystyle\sum_{i=1}^{n}(y_i - \bar{y})^2}{\displaystyle\sum_{i=1}^{n}(y_i - \bar{y})^2 + n(\bar{y} - \mu_0)^2} < k'$$

$$\frac{1}{1 + \dfrac{n(\bar{y} - \mu_0)^2}{\displaystyle\sum_{i=1}^{n}(y_i - \bar{y})^2}} < k'.$$

This inequality, in turn, is equivalent to

$$\frac{n(\bar{y} - \mu_0)^2}{\displaystyle\sum_{i=1}^{n}(y_i - \bar{y})^2} > \frac{1}{k'} - 1 = k''$$

$$\frac{n(\overline{y}-\mu_0)^2}{\dfrac{1}{n-1}\displaystyle\sum_{i=1}^{n}(y_i-\overline{y})^2} > (n-1)k''$$

or, because $\overline{y} > \mu_0$ when $\lambda < k < 1$,

$$\frac{\sqrt{n}(\overline{y}-\mu_0)}{s} > \sqrt{(n-1)k''},$$

where

$$s^2 = \frac{1}{n-1}\sum_{i=1}^{n}(y_i-\overline{y})^2.$$

Notice that $\sqrt{n}(\overline{Y}-\mu_0)/S$ is the t statistic employed in previous sections. Consequently, the likelihood ratio test is equivalent to the t test of Section 10.8.

Situations in which the likelihood ratio test assumes a well-known form are not uncommon. In fact, all the tests of Sections 10.8 and 10.9 can be obtained by the likelihood ratio method. For most practical problems, the likelihood ratio method produces the best possible test, in terms of power.

Unfortunately, the likelihood ratio method does not always produce a test statistic with a known probability distribution, such as the t statistic of Example 10.24. If the sample size is large, however, we can obtain an approximation to the distribution of λ if some reasonable "regularity conditions" are satisfied by the underlying population distribution(s). These are general conditions that hold for most (but not all) of the distributions we have considered. The regularity conditions mainly involve the existence of derivatives, with respect to the parameters, of the likelihood function. Another key condition is that the region over which the likelihood function is positive cannot depend on unknown parameter values.

THEOREM **10.2**

Let $Y_1, Y_2, \ldots, Y_n$ have joint likelihood function $L(\Theta)$. Let r_0 denote the number of free parameters that are specified by $H_0: \Theta \in \Omega_0$, and let r denote the number of free parameters specified by the statement $\Theta \in \Omega$. Then, for large n, $-2\ln\lambda$ has approximately a χ^2 distribution with $r_0 - r$ degrees of freedom.

The proof of this result is beyond the scope of this text. Theorem 10.2 allows us to use the table of the χ^2 distribution to find rejection regions with fixed α, when n is large. Notice that $-2\ln(\lambda)$ is a decreasing function of λ. Because the likelihood ratio test specifies that we use RR: $\{\lambda < k\}$, this rejection may be rewritten as RR:$\{-2\ln(\lambda) > -2\ln(k) = k^*\}$. For large sample sizes, if we desire an α level test, Theorem 10.2 implies that $k^* \approx \chi^2_\alpha$. That is, a large-sample likelihood ratio test has rejection region given by

$$-2\ln\lambda > \chi^2_\alpha, \qquad \text{where } \chi^2_\alpha \text{ is based on } r_0 - r \text{ degrees for freedom.}$$

The size of the sample necessary for a "good" approximation varies from application to application. It is important to realize that large-sample likelihood ratio tests are based on $-2\ln\lambda$, where λ is the *original* likelihood ratio, $\lambda = L(\hat{\Omega}_0)/L(\hat{\Omega})$.

EXAMPLE **10.25** Suppose that an engineer wishes to compare the number of complaints per week filed by union stewards for two different shifts at a manufacturing plant. One hundred independent observations on the number of complaints gave means $\bar{x} = 20$ for shift 1 and $\bar{y} = 22$ for shift 2. Assume that the number of complaints per week on the ith shift has a Poisson distribution with mean θ_i, for $i = 1, 2$. Use the likelihood ratio method to test $H_0: \theta_1 = \theta_2$ versus $H_a: \theta_1 \neq \theta_2$ with $\alpha \approx .01$.

Solution The likelihood of the sample is now the joint density of all Xs and Ys and is given by

$$L(\theta_1, \theta_2) = \left(\frac{1}{k}\right) \theta_1^{\sum x_i} e^{-n\theta_1} \theta_2^{\sum y_i} e^{-n\theta_2}$$

where $k = x_1! \cdots x_n! y_1! \cdots y_n!$, and $n = 100$. In this example $\Theta = (\theta_1, \theta_2)$ and $\Omega_0 = \{(\theta_1, \theta_2) : \theta_1 = \theta_2 = \theta\}$, where θ is unknown. Hence, under H_0 the likelihood function is a function of the single parameter θ and

$$L(\theta) = \left(\frac{1}{k}\right) \theta^{\sum x_i + \sum y_i} e^{-2n\theta}.$$

Notice that, for $\Theta \in \Omega_0$, $L(\theta)$ is maximized when θ is equal to its maximum likelihood estimate,

$$\hat{\theta} = \frac{1}{2n}\left(\sum_{i=1}^{n} x_i + \sum_{i=1}^{n} y_i\right) = \frac{1}{2}(\bar{x} + \bar{y}).$$

In this example $\Omega_a = \{(\theta_1, \theta_2) : \theta_1 \neq \theta_2\}$ and $\Omega = \{(\theta_1, \theta_2) : \theta_1 > 0, \theta_2 > 0\}$. Using the general likelihood $L(\theta_1, \theta_2)$, a function of both θ_1 and θ_2, we see that $L(\theta_1, \theta_2)$ is maximized when $\hat{\theta}_1 = \bar{x}$ and $\hat{\theta}_2 = \bar{y}$, respectively. That is, $L(\theta_1, \theta_2)$ is maximized when both θ_1 and θ_2 are replaced by their maximum likelihood estimates. Thus,

$$\lambda = \frac{L(\hat{\Omega}_0)}{L(\hat{\Omega})} = \frac{k^{-1}(\hat{\theta})^{n\bar{x}+n\bar{y}} e^{-2n\hat{\theta}}}{k^{-1}(\hat{\theta}_1)^{n\bar{x}}(\hat{\theta}_2)^{n\bar{y}} e^{-n\hat{\theta}_1 - n\hat{\theta}_2}} = \frac{(\hat{\theta})^{n\bar{x}+n\bar{y}}}{(\bar{x})^{n\bar{x}}(\bar{y})^{n\bar{y}}}.$$

Notice that λ is a complicated function of $\bar{x}$ and $\bar{y}$. The observed value of $\hat{\theta}$ is $(1/2)(\bar{x} + \bar{y}) = (1/2)(20 + 22) = 21$. The observed value of λ is

$$\lambda = \frac{21^{(100)(20+22)}}{20^{(100)(20)} 22^{(100)(22)}}$$

and hence

$$-2\ln\lambda = -(2)[4200\ln(21) - 2000\ln(20) - 2200\ln(22)] = 9.53.$$

In this application, the number of free parameters in $\Omega = \{(\theta_1, \theta_2) : \theta_1 > 0, \theta_2 > 0\}$ is $k = 2$. In $\Omega_0 = \{(\theta_1, \theta_2) : \theta_1 = \theta_2 = \theta\}$, $r_0 = 1$ of these free parameters is fixed. In the set Ω, $r = 0$ of the parameters are fixed. Theorem 10.2 implies that $-2 \ln \lambda$ has an approximately χ^2 distribution with $r_0 - r = 1$ degree of freedom. Small values of λ correspond to large values of $-2 \ln \lambda$, so the rejection region for a test at approximately the $\alpha = .01$ level contains the values of $-2 \ln \lambda$ that exceed $\chi_{.01}^2 = 6.635$, the value that cuts off an area of .01 in the right-hand tail of a χ^2 density with 1 degree of freedom.

Because the observed value of $-2 \ln \lambda$ is larger than $\chi_{.01}^2$, we reject $H_0 : \theta_1 = \theta_2$. We conclude, at approximately the $\alpha = .01$ level of significance, that the mean numbers of complaints filed by the union stewards do differ.

Exercises

10.93 Let $Y_1, Y_2, \ldots, Y_n$ denote a random sample from a normal distribution with mean μ (unknown) and variance σ^2. For testing $H_0 : \sigma^2 = \sigma_0^2$ against $H_a : \sigma^2 > \sigma_0^2$, show that the likelihood ratio test is equivalent to the χ^2 test given in Section 10.9.

10.94 A survey of voter sentiment was conducted in four midcity political wards to compare the fraction of voters favoring candidate A. Random samples of 200 voters were polled in each of the four wards, with the results as shown in the accompanying table. The numbers of voters favoring A in the four samples can be regarded as four independent binomial random variables. Construct a likelihood ratio test of the hypothesis that the fractions of voters favoring candidate A are the same in all four wards. Use $\alpha = .05$.

	Ward				
Opinion	1	2	3	4	Total
Favor A	76	53	59	48	236
Do not favor A	124	147	141	152	564
Total	200	200	200	200	800

10.95 Let S_1^2 and S_2^2 denote, respectively, the variances of independent random samples of sizes n and m selected from normal distributions with means μ_1 and μ_2 and common variance σ^2. If μ_1 and μ_2 are unknown, construct a likelihood ratio test of $H_0 : \sigma^2 = \sigma_0^2$ against $H_a : \sigma^2 = \sigma_a^2$, assuming that $\sigma_a^2 > \sigma_0^2$.

10.96 Suppose that $X_1, X_2, \ldots, X_{n_1}, Y_1, Y_2, \ldots, Y_{n_2}$, and $W_1, W_2, \ldots, W_{n_3}$ are independent random samples from normal distributions with respective unknown means μ_1, μ_2, and μ_3 and variances σ_1^2, σ_2^2, and σ_3^2.

a Find the likelihood ratio test for $H_0 : \sigma_1^2 = \sigma_2^2 = \sigma_3^2$ against the alternative of at least one inequality.

b Find an approximate critical region for the test in (a) if n_1, n_2, and n_3 are large and $\alpha = .05$.

*10.97 Let $X_1, X_2, \ldots, X_m$ denote a random sample from the exponential density with mean θ_1, and let $Y_1, Y_2, \ldots, Y_n$ denote an independent random sample from an exponential density with mean θ_2.

    a Find the likelihood ratio criterion for testing $H_0: \theta_1 = \theta_2$ versus $H_a: \theta_1 \neq \theta_2$.

    b Show that the test in (a) is equivalent to an exact F test (*Hint:* Transform $\sum X_i$ and $\sum Y_i$ to χ^2 random variables.)

*10.98 Show that a likelihood ratio test depends on the data only through the value of a sufficient statistic. (*Hint:* Use the factorization criterion.)

10.99 Suppose that we are interested in testing the *simple* null hypothesis $H_0: \theta = \theta_0$ versus the *simple* alternative hypothesis $H_a: \theta = \theta_a$. According to the Neyman–Pearson lemma, the test that maximizes the power at θ_a has a rejection region determined by

$$\frac{L(\theta_0)}{L(\theta_a)} < k.$$

In the context of a likelihood ratio test, if we are interested in the *simple* H_0 and H_a, as stated, then $\Omega_0 = \{\theta_0\}$, $\Omega_a = \{\theta_a\}$, and $\Omega = \{\theta_0, \theta_a\}$.

    a Show that the likelihood ratio λ is given by

$$\lambda = \frac{L(\theta_0)}{\max\{L(\theta_0), L(\theta_a)\}} = \frac{1}{\max\left\{1, \dfrac{L(\theta_a)}{L(\theta_0)}\right\}}.$$

    b Argue that $\lambda < k$ if and only if, for some constant k',

$$\frac{L(\theta_0)}{L(\theta_a)} < k'.$$

    c What do the results in (a) and (b) imply about likelihood ratio tests when both the null and alternative hypotheses are simple?

10.100 Suppose that independent random samples of sizes n_1 and n_2 are to be selected from normal populations with means μ_1 and μ_2, respectively, and common variance σ^2. For testing $H_0: \mu_1 = \mu_2$ versus $H_a: \mu_1 - \mu_2 > 0$ (σ^2 unknown), show that the likelihood ratio test reduces to the two-sample t test presented in Section 10.8.

10.101 Refer to Exercise 10.100. Show that, in testing of $H_0: \mu_1 = \mu_2$ versus $H_a: \mu_1 \neq \mu_2$ (σ^2 unknown), the likelihood ratio test reduces to the two-sample t test.

*10.102 Refer to Exercise 10.101. Suppose that another independent random sample of size n_3 is selected from a third normal population with mean μ_3 and variance σ^2. Find the likelihood ratio test for testing $H_0: \mu_1 = \mu_2 = \mu_3$ versus the alternative that there is at least one inequality. Show that this test is equivalent to an exact F test.

10.12 Summary

In Chapters 8, 9, and 10 we have presented the basic concepts associated with two methods for making inferences: estimation and tests of hypotheses. Philosophically, estimation (Chapters 8 and 9) focuses on the question, What is the numerical value of a parameter θ? In contrast, a test of a hypothesis attempts to answer the question, Is θ equal to a specific numerical value θ_0? Often the inferential method you employ for a given situation depends on how you, the experimenter, prefer to phrase your inference. Sometimes this decision is taken out of your hands. That is, the practical question clearly implies that either an estimation or a hypothesis-testing procedure be used. For example, acceptance or rejection of incoming supplies of outgoing products in a manufacturing process clearly requires a decision, or a statistical test. We have seen that a duality exists between these two inference-making procedures. A two-sided confidence interval with confidence coefficient $1 - \alpha$ may be viewed as the set of all values of θ_0 that are "acceptable" null hypothesis values for θ if we use a two-sided α-level test. Similarly, a two-sided α-level test for $H_0 : \theta = \theta_0$ can be implemented by constructing a two-sided confidence interval (with confidence coefficient $1 - \alpha$) and rejecting H_0 if the value θ_0 *falls outside* the confidence interval.

Associated with both methods for making inferences are measures of their goodness. Thus the expected width of a confidence interval and the confidence coefficient both measure the goodness of the estimation procedure. Likewise, the goodness of a statistical test is measured by the probabilities α and β of type I and type II errors. These measures of goodness enable us to compare one statistical test with another and to develop a theory for acquiring statistical tests with desirable properties. The ability to evaluate the goodness of an inference is one of the major contributions of statistics to the analysis of experimental data. Of what value is an inference if you have no measure of its validity?

In this chapter we have investigated the elements of a statistical test and discussed how a test works. Some useful tests are given to show how they can be used in practical situations, and you will see other interesting applications in the chapters that follow.

Many of the testing procedures developed in this chapter were presented from an intuitive perspective. However, we have also illustrated the use of the Neyman–Pearson lemma in deriving most powerful procedures for testing a simple null hypothesis versus a simple alternative hypothesis. In addition, we have seen how the Neyman–Pearson method can sometimes be used to find uniformly most powerful tests for composite null and alternative hypotheses if the underlying distribution is specified except for the value of a single parameter. The likelihood ratio procedure provides a general method for developing a statistical test. The likelihood ratio tests can be derived whether or not nuisance parameters are present. In general, the likelihood ratio tests possess desirable properties. The Neyman–Pearson and likelihood ratio procedures both require that the distribution of the sampled population(s) must be known, except for the values of some parameters. Otherwise, the likelihood functions cannot be determined and the methods cannot be applied.

References and Further Readings

Casella, G., and R. L. Berger 1990. *Statistical Inference*. Pacific Grove, Calif: Brooks Cole, 1990.

Cramer, H. 1963. *Mathematical Methods of Statistics*. Princeton, N.J.: Princeton University Press.

Hoel, P. G. 1984. *Introduction to Mathematical Statistics*, 5th ed. New York: John Wiley.

Hogg, R. V., and A. T. Craig. 1995. *Introduction to Mathematical Statistics*, 5th ed., Upper Saddle River, N.J.: Prentice Hall.

Lehmann, E. L. 1997. *Testing Statistical Hypotheses*, 2d ed. New York: Springer.

Mood, A. M., F. A. Graybill, and D. Boes. 1974. *Introduction to the Theory of Statistics*, 3d ed. New York: McGraw-Hill.

Wilks, S. S. 1963. *Mathematical Statistics*. New York: John Wiley.

Supplementary Exercises

10.103 Lord Rayleigh was one of the earliest scientists to study the density of nitrogen. In his studies, he noticed something peculiar. The nitrogen densities produced from chemical compounds tended to be smaller than the densities of nitrogen produced from the air. Lord Rayleigh's measurements (*Proceedings, Royal Society* (London) 55 (1894), pp. 340–44) are given in the following table. These measurements correspond to the mass of nitrogen filling a flask of specified volume under specified temperature and pressure.

Chemical	Atmospheric
2.30143	2.31017
2.29890	2.30986
2.29816	2.31010
2.30182	2.31001
2.29869	2.31024
2.29940	2.31010
2.29849	2.31028
2.29889	2.31163
2.30074	2.30956
2.30054	

a For the measurements from the chemical compound, $\bar{y} = 2.29971$ and $s = .001310$; for the measurements from the atmosphere, $\bar{y} = 2.310217$ and $s = .000574$. Is there sufficient evidence to indicate a difference in the mean mass of nitrogen per flask for chemical compounds and air? What can be said about the p-value associated with your test?

b Find a 95% confidence interval for the difference in mean mass of nitrogen per flask for chemical compounds and air.

c Based on your answer to (b), at the $\alpha = .05$ level of significance, is there sufficient evidence to indicate a difference in mean mass of nitrogen per flask for measurements from chemical compounds and air?

d Is there any conflict between your conclusions in (a) and (b)? Although the difference in these mean nitrogen masses is small, Lord Rayleigh emphasized this difference rather than ignoring it, and this led to the discovery of inert gases in the atmosphere.

10.104 The effect of alcohol consumption on the body appears to be much greater at higher altitudes. To test this theory, a scientist randomly selected 12 subjects and divided them into two groups of 6 each. One group was transported to an altitude of 12,000 feet and each member in the group ingested 100cm^3 of alcohol. The members of the second group were taken to sea level and given the same amount of alcohol. After 2 hours, the amount of alcohol in the blood of each subject was measured (measurements in grams per 100 cm^3). The data are given in the following table. Is there sufficient evidence to indicate that retention of alcohol is greater at 12,000 feet than at sea level? Test at the $\alpha = .10$ level of significance.

Sea Level	12,000 feet
.07	.13
.10	.17
.09	.15
.12	.14
.09	.10
.13	.14

10.105 Currently 20% of potential customers buy soap of brand A. To increase sales, the company will conduct an extensive advertising campaign. At the end of the campaign a sample of 400 potential customers will be interviewed to determine whether the campaign was successful.

a State H_0 and H_a in terms of p, the probability that a customer prefers soap brand A.

b The company decides to conclude that the advertising campaign was a success if at least 92 of the 400 customers interviewed prefer brand A. Find α. (Use the normal approximation to the binomial distribution to evaluate the desired probability.)

10.106 In the past a chemical plant has produced an average of 1100 pounds of chemical per day. The records for the past year, based on 260 operating days, show the following:

$$\bar{y} = 1060 \text{ pounds/day}, \qquad s = 340 \text{ pounds/day}.$$

We wish to test whether the average daily production has dropped significantly over the past year.

a Give the appropriate null and alternative hypotheses.

b If Z is used as a test statistic, determine the rejection region corresponding to a level of significance of $\alpha = .05$.

c Do the data provide sufficient evidence to indicate a drop in average daily production?

10.107 The braking ability of two types of automobiles was compared. Random samples of 64 automobiles were tested for each type. The recorded measurement was the distance required to stop when the brakes were applied at 40 miles per hour. The computed sample means and variances were as follows:

$$\bar{y}_1 = 118 \quad \bar{y}_2 = 109$$
$$s_1^2 = 102 \quad s_2^2 = \;\; 87.$$

Do the data provide sufficient evidence to indicate a difference in the mean stopping distances of the two types of automobiles? Give the attained significance level.

10.108 The stability of measurements of the characteristics of a manufactured product is important in maintaining product quality. In fact, it is sometimes better to obtain small variation in the measured value of some important characteristic of a product and have the process mean slightly off target than to get wide variation with a mean value that perfectly fits requirements. The latter situation may produce a higher percentage of defective product than the former. A manufacturer of lightbulbs suspected that one of his production lines was producing bulbs with a high variation in length of life. To test this theory, he compared the lengths of life of $n = 50$ bulbs randomly sampled from the suspect line and $n = 50$ from a line that seemed to be in control. The sample means and variances for the two samples were as shown in the following table.

Suspect Line	Line in Control
$\bar{y}_1 = 1{,}520$	$\bar{y}_2 = 1{,}476$
$s_1^2 = 92{,}000$	$s_2^2 = 37{,}000$

a Do the data provide sufficient evidence to indicate that bulbs produced by the suspect line possess a larger variance in length of life than those produced by the line that is assumed to be in control? Use $\alpha = .05$.

b Find the approximate observed significance level for the test, and interpret its value.

10.109 A pharmaceutical manufacturer purchases a particular material from two different suppliers. The mean level of impurities in the raw material is approximately the same for both suppliers, but the manufacturer is concerned about the variability of the impurities from shipment to shipment. If the level of impurities tends to vary excessively for one source of supply, it could affect the quality of the pharmaceutical product. To compare the variation in percentage impurities for the two suppliers, the manufacturer selects ten shipments from each of the two suppliers and measures the percentage of impurities in the raw material for each shipment. The sample means and variances are shown in the accompanying table.

Supplier A	Supplier B
$\bar{y}_1 = 1.89$	$\bar{y}_2 = 1.85$
$s_1^2 = .273$	$s_2^2 = .094$
$n_1 = 10$	$n_2 = 10$

a Do the data provide sufficient evidence to indicate a difference in the variability of the shipment impurity levels for the two suppliers? Test using $\alpha = .10$. Based on the results of your test, what recommendation would you make to the pharmaceutical manufacturer?

b Find a 90% confidence interval for σ_B^2, and interpret your results.

10.110 The data in the following table give readings in foot-pounds of the impact strength of two kinds of packaging material. Determine whether the data suggests a difference in mean strength between the two kinds of material. Test at the $\alpha = .10$ level of significance.

A	B
1.25	.89
1.16	1.01
1.33	.97
1.15	.95
1.23	.94
1.20	1.02
1.32	.98
1.28	1.06
1.21	.98
$\sum y_i = 11.13$	$\sum y_i = 8.80$
$\bar{y} = 1.237$	$\bar{y} = .978$
$\sum y_i^2 = 13.7973$	$\sum y_i^2 = 8.6240$

10.111 How much combustion efficiency should a homeowner expect from an oil furnace? The EPA states that 80% or higher is excellent, 75 to 79% is good, 70 to 74% is fair, and below 70% is poor. A home-heating contractor who sells two makes of oil heaters (call them A and B) decided to compare their mean efficiencies by analyzing the efficiencies of eight heaters of type A and six of type B. The resulting efficiency ratings in percentages for the fourteen heaters are shown in the accompanying table.

Type A	Type B
72	78
78	76
73	81
69	74
75	82
74	75
69	
75	

a Do the data provide sufficient evidence to indicate a difference in mean efficiencies for the two makes of home heaters? Find the approximate p-value for the test, and interpret its value.

b Find a 90% confidence interval for $(\mu_A - \mu_B)$, and interpret the result.

10.112 Suppose that $X_1, X_2, \ldots, X_{n_1}, Y_1, Y_2, \ldots, Y_{n_2}$, and $W_1, W_2, \ldots, W_{n_3}$ are independent random samples from normal distributions with respective unknown means μ_1, μ_2, and μ_3 and common variances $\sigma_1^2 = \sigma_2^2 = \sigma_3^2 = \sigma^2$. Suppose that we want to estimate a linear function of the means: $\theta = a_1\mu_1 + a_2\mu_2 + a_3\mu_3$. Because the maximum-likelihood estimator of a function of parameters is the function of the maximum-likelihood estimators of the parameters, the maximum likelihood estimator of θ is $\hat{\theta} = a_1\overline{X} + a_2\overline{Y} + a_3\overline{W}$.

a What is the standard error of the estimator $\hat{\theta}$?

b What is the distribution of the estimator $\hat{\theta}$?

c If the sample variances are given by S_1^2, S_2^2, and S_3^2, respectively, consider

$$S_p^2 = \frac{(n_1 - 1)S_1^2 + (n_2 - 1)S_2^2 + (n_3 - 1)S_3^2}{n_1 + n_2 + n_3 - 3}.$$

What is the distribution of $(n_1 + n_2 + n_3 - 3)S_p^2/\sigma^2$?
What is the distribution of

$$T = \frac{\hat{\theta} - \theta}{S_p\sqrt{\dfrac{a_1^2}{n_1} + \dfrac{a_2^2}{n_2} + \dfrac{a_3^2}{n_3}}}?$$

d Give a confidence interval for θ with confidence coefficient $1 - \alpha$.

e Develop a test for $H_0: \theta = \theta_0$ versus $H_a: \theta \neq \theta_0$.

10.113 A merchant figures her weekly profit to be a function of three variables: retail sales (denoted by X), wholesale sales (denoted by Y), and overhead costs (denoted by W). The variables X, Y, and W are regarded as independent, normally distributed random variables with means μ_1, μ_2, and μ_3 and variances $\sigma^2, a\sigma^2$, and $b\sigma^2$, respectively, for known constants a and b but unknown σ^2. The merchant's expected profit per week is $\mu_1 + \mu_2 - \mu_3$. If the merchant has made independent observations of X, Y, and W for the past n weeks, construct a test of

$H_0: \mu_1 + \mu_2 - \mu_3 = k$ against the alternative $H_a: \mu_1 + \mu_2 - \mu_3 \neq k$, for a given constant k. You may specify $\alpha = .05$.

10.114   A reading exam is given to the sixth-graders at three large elementary schools. The scores on the exam at each school are regarded as having normal distributions with unknown means μ_1, μ_2, and μ_3, respectively, and unknown common variance $\sigma^2 (\sigma_1^2 = \sigma_2^2 = \sigma_3^2 = \sigma^2)$. Using the data in the accompanying table on independent random samples from each school, test to see if evidence exists of a difference between μ_1 and μ_2. Use $\alpha = .05$.

School I	School II	School III
$n_1 = 10$	$n_2 = 10$	$n_3 = 10$
$\sum x_i^2 = 36{,}950$	$\sum y_i^2 = 25{,}850$	$\sum w_i^2 = 49{,}900$
$\bar{x} = 60$	$\bar{y} = 50$	$\bar{w} = 70$

*10.115   Suppose that $Y_1, Y_2, \ldots, Y_n$ denotes a random sample from the probability density function given by

$$f(y \mid \theta_1, \theta_2) = \begin{cases} \left(\dfrac{1}{\theta_1}\right) e^{-(y-\theta_2)/\theta_1}, & y > \theta_2 \\ 0, & \text{elsewhere.} \end{cases}$$

Find the likelihood ratio test for testing $H_0 : \theta_1 = \theta_{1,0}$ versus $H_a : \theta_1 > \theta_{1,0}$, with θ_2 unknown.

*10.116   Refer to Exercise 10.115. Find the likelihood ratio test for testing $H_0 : \theta_2 = \theta_{2,0}$ versus $H_a : \theta_2 > \theta_{2,0}$, with θ_1 unknown.

CHAPTER **11**

Linear Models and Estimation by Least Squares

11.1 Introduction

11.2 Linear Statistical Models

11.3 The Method of Least Squares

11.4 Properties of the Least Squares Estimators: Simple Linear Regression

11.5 Inferences Concerning the Parameters β_i

11.6 Inferences Concerning Linear Functions of the Model Parameters: Simple Linear Regression

11.7 Predicting a Particular Value of Y Using Simple Linear Regression

11.8 Correlation

11.9 Some Practical Examples

11.10 Fitting the Linear Model by Using Matrices

11.11 Linear Functions of the Model Parameters: Multiple Linear Regression

11.12 Inferences Concerning Linear Functions of the Model Parameters: Multiple Linear Regression

11.13 Predicting a Particular Value of Y Using Multiple Regression

11.14 A Test for $H_0 : \beta_{g+1} = \beta_{g+2} = \cdots = \beta_k = 0$

11.15 Summary and Concluding Remarks

References and Further Readings

11.1 Introduction

In Chapter 9 we considered several methods for finding estimators of parameters, including the methods of moments and maximum likelihood and also methods based upon sufficient statistics. Another method of estimation, the method of least squares, is the topic of this chapter.

In all our previous discussions of statistical inference, we assumed that the observable random variables $Y_1, Y_2, \ldots, Y_n$ were independent and identically distributed. One of the implications of this assumption is that the expected value of Y_i, $E(Y_i)$, is constant (if it exists). That is, $E(Y_i) = \mu$ does not depend on the value of any other variable. Obviously this assumption is unrealistic in many inferential problems. For example, the mean stopping distance for a particular type of automobile will depend upon the speed the automobile is traveling; the mean potency of an antibiotic depends upon the amount of time that the antibiotic has been stored; the mean amount of elongation observed in a particular metal alloy depends upon the force applied and the temperature of the alloy. In this chapter we undertake a study of inferential procedures that can be used when a random variable Y, called the *dependent variable*, has a mean that is a function of one or more nonrandom variables $x_1, x_2, \ldots, x_k$, called *independent variables*. (In this context the terms *independent* and *dependent* are used in their mathematical sense. There is no relationship with the probabilistic concept of independent random variables.)

Many different types of mathematical functions can be used to model a response that is a function of one or more independent variables. These can be classified into two categories: deterministic and probabilistic models. For example, suppose that y and x are related according to the equation

$$y = \beta_0 + \beta_1 x$$

(where β_0 and β_1 are unknown parameters). This model is called a *deterministic mathematical model* because it does not allow for any error in predicting y as a function of x. We imply that y always takes the value $\beta_0 + \beta_1(5.5)$ whenever $x = 5.5$.

Suppose that we collect a sample of n values of y corresponding to n different settings of the independent variable, x, and that a plot of the data is as shown in Figure 11.1. It is quite clear from the figure that the expected value of Y may increase as a linear function of x but that a deterministic model is far from an adequate description of reality. Repeated experiments when $x = 5.5$ would yield values of Y that vary in a random manner. This tells us that the deterministic model is not an exact represen-

FIGURE **11.1**
Plot of data

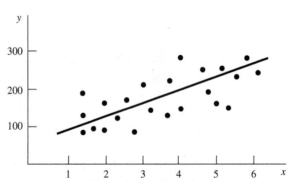

tation of the relationship between the two variables. Further, if the model were used to predict Y when $x = 5.5$, the prediction would be subject to some unknown error. This, of course, leads us to the use of statistical methods. Predicting Y for a given value of x is an inferential process, and we need to know the properties of the error of prediction if the prediction is to be of value in real life.

In contrast to the deterministic model, statisticians use *probabilistic* models. For example, we might represent the responses of Figure 11.1 by the model

$$E(Y) = \beta_0 + \beta_1 x$$

or, equivalently,

$$Y = \beta_0 + \beta_1 x + \varepsilon$$

where ε is a random variable possessing a specified probability distribution with mean zero. We think of Y as the sum of a deterministic component $E(Y)$ and a random component ε. This model accounts for the random behavior of Y exhibited in Figure 11.1 and provides a more accurate description of reality than the deterministic model. Further, the properties of the error of prediction for Y can be derived for many probabilistic models.

Figure 11.2 presents a graphical representation of the probabilistic model $Y = \beta_0 + \beta_1 x + \varepsilon$. When $x = 5.5$, there is a *population* of possible values of Y. The distribution of this population is indicated on the main portion of the graph and is centered on the line $E(Y) = \beta_0 + \beta_1 x$ at the point $x = 5.5$. This population has a distribution with mean $\beta_0 + \beta_1(5.5)$ and variance σ^2, as shown in the magnified

FIGURE **11.2**
Graph of the
probabilistic model
$Y = \beta_0 + \beta_1 x + \varepsilon$

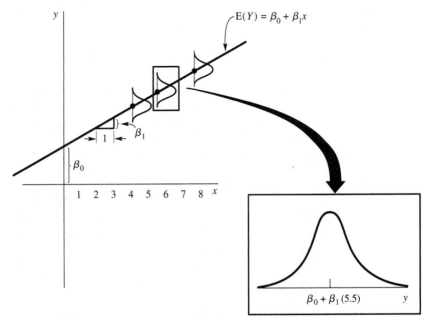

version of the distribution that is boxed in Figure 11.2. When $x = 7$, there is another *population* of possible values for Y. The distribution of this population has the same form as the distribution of Y-values when $x = 5.5$ and has the same variance σ^2, but when $x = 7$, the distribution of Y has mean $\beta_0 + \beta_1(7)$. The same is true for each possible value of the independent variable x. That is, in a regression model, a *separate population* of response values exists for each possible setting of the independent variable(s). These populations all have the same variance, and the distributions of the populations are all the same (see Figure 11.2); however, the mean of each population depends on the setting of the independent variable(s) through the regression model.

Scientific and mathematical textbooks are filled with deterministic models of reality. Indeed, many of the mathematical functions that appear in calculus and physics books are deterministic mathematical models of nature. For example, Newton's law relating the force of a moving body to its mass and acceleration,

$$F = ma,$$

is a deterministic model that, for practical purposes, predicts with little error. In contrast, other models—such as functions graphically represented in scientific journals and texts—are often poor. The spatter of points that would give graphic evidence of their inadequacies, similar to the random behavior of the points in Figure 11.1, has been deemphasized, which leads novice scientists to accept the corresponding "laws" and theories as an exact description of nature.

If deterministic models can be used to predict with negligible error, for all practical purposes, we use them. If not, we seek a probabilistic model, which will not be an exact characterization of nature but which will enable us to assess the validity of our inferences.

11.2 Linear Statistical Models

Although infinitely many different functions can be used to model the mean value of the response variable Y as a function of one or more independent variables, we will concentrate on a set of models called *linear statistical models*. If Y is the response variable and x is a single independent variable, it may be reasonable in some situations to use the model $E(Y) = \beta_0 + \beta_1 x$ for unknown parameter values β_0 and β_1. Notice that, in this model, $E(Y)$ is a linear function of x (for a given β_0 and β_1) and also a linear function of β_0 and β_1 [because $E(Y) = c\beta_0 + d\beta_1$ with $c = 1$ and $d = x$]. In the model $E(Y) = \beta_0 + \beta_1 x^2$, $E(Y)$ is not a linear function of x, but it is a linear function of β_0 and β_1 [because $E(Y) = c\beta_0 + d\beta_1$ with $c = 1$ and $d = x^2$]. When we say we have a linear statistical model for Y, we mean that $E(Y)$ is a linear function of the unknown parameters β_0 and β_1 and *not* necessarily a linear function of x. Thus $Y = \beta_0 + \beta_1(\ln x) + \varepsilon$ is a linear model (because $\ln x$ is a known constant).

If the model relates $E(Y)$ as a linear function of β_0 and β_1 only, the model is called a *simple* linear regression model. If more than one independent variable, say,

FIGURE **11.3**

Plot of $E(Y) = \beta_0 + \beta_1 x_1 + \beta_2 x_2$

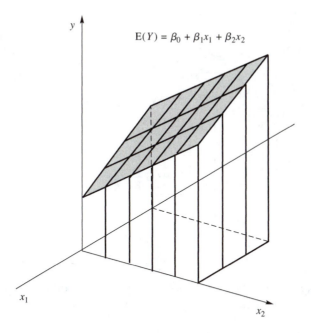

$x_1, x_2, \ldots, x_k$, are of interest and we model $E(Y)$ by

$$E(Y) = \beta_0 + \beta_1 x_1 + \cdots + \beta_k x_k,$$

the model is called a *multiple* linear regression model. Because $x_1, x_2, \ldots, x_k$ are regarded as known constants, they are assumed to be measured without error in an experiment. For example, if you think that the mean yield $E(Y)$ is a function of the variable t, the temperature of a chemical process, you might let $x_1 = t$ and $x_2 = e^t$ and then model $E(Y)$ as $E(Y) = \beta_0 + \beta_1 x_1 + \beta_2 x_2$ or, equivalently, $E(Y) = \beta_0 + \beta_1 t + \beta_2 e^t$. Or, if $E(Y)$ is a function of two variables x_1 and x_2, you might choose a planar approximation to the true mean response, using the linear model $E(Y) = \beta_0 + \beta_1 x_1 + \beta_2 x_2$. Thus, $E(Y)$ is a linear function of β_0, β_1, and β_2 and represents a plane in the y, x_1, x_2 space (see Figure 11.3). Similarly,

$$E(Y) = \beta_0 + \beta_1 x + \beta_2 x^2$$

is a linear statistical model, where $E(Y)$ is a second-order polynomial function of the independent variable x, with $x_1 = x$ and $x_2 = x^2$. This model would be appropriate for a response that traces a segment of a parabola over the experimental region.

The expected percentage $E(Y)$ of water in paper during its manufacture could be represented as a second-order function of the temperature of the dryer, x_1, and the speed of the paper machine, x_2. Thus,

$$E(Y) = \beta_0 + \beta_1 x_1 + \beta_2 x_2 + \beta_3 x_1 x_2 + \beta_4 x_1^2 + \beta_5 x_2^2$$

where $\beta_0, \beta_1, \ldots, \beta_5$ are unknown parameters in the model. Geometrically, $E(Y)$ traces a second-order (conic) surface over the x_1, x_2 plane (see Figure 11.4).

FIGURE 11.4
Plot of $E(Y) =$
$\beta_0 + \beta_1 x_1 + \beta_2 x_2 +$
$\beta_3 x_1 x_2 + \beta_4 x_1^2 + \beta_5 x_2^2$

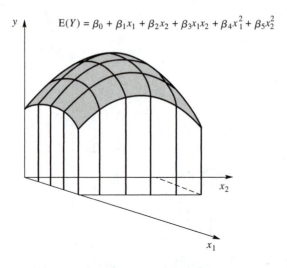

$$E(Y) = \beta_0 + \beta_1 x_1 + \beta_2 x_2 + \beta_3 x_1 x_2 + \beta_4 x_1^2 + \beta_5 x_2^2$$

DEFINITION 11.1

A *linear statistical* model relating a random response Y to a set of independent variables $x_1, x_2, \ldots, x_k$ is of the form

$$Y = \beta_0 + \beta_1 x_1 + \beta_2 x_2 + \cdots + \beta_k x_k + \varepsilon$$

where $\beta_0, \beta_1, \ldots, \beta_k$ are unknown parameters, ε is a random variable, and $x_1, x_2, \ldots, x_k$ are known constants. We will assume that $E(\varepsilon) = 0$ and hence that

$$E(Y) = \beta_0 + \beta_1 x_1 + \beta_2 x_2 + \cdots + \beta_k x_k.$$

Consider the physical interpretation of the linear model Y. It says that Y is equal to an expected value, $\beta_0 + \beta_1 x_1 + \beta_2 x_2 + \cdots + \beta_k x_k$ (a function of the independent variables $x_1, x_2, \ldots, x_k$), plus a random error ε. From a practical point of view, ε acknowledges our inability to provide an exact model for nature. In repeated experimentation, Y varies about $E(Y)$ in a random manner because we have failed to include in our model all the many variables that may affect Y. Fortunately, the net effect of these unmeasured, and most often unknown, variables is to cause Y to vary in a manner that may be adequately approximated by an assumption of random behavior.

In this chapter we will use the *method of least squares* to derive estimators for the parameters $\beta_0, \beta_1, \ldots, \beta_k$ in a linear regression model. In many applications, one or more of these parameters will have meaningful interpretations. For this reason, we develop inferential methods for an individual β parameter and for sets of β parameters. If we estimate the parameters $\beta_0, \beta_1, \ldots, \beta_5$ in the model expressing the expected percentage $E(Y)$ of water in paper as a second-order polynomial in x_1 (the dryer temperature) and x_2 (the dryer speed), we will be able to develop methods for estimating and forming confidence intervals for the value of $E(Y)$ when x_1 and x_2 take on specific values. Similarly, we can develop methods for predicting a future value of Y when the independent variables are set at particular values. Sections 11.3

through 11.9 focus on the *simple* linear regression model, whereas the later sections deal with *multiple* linear regression models.

11.3   The Method of Least Squares

A procedure for estimating the parameters of any linear model—the method of least squares—can be illustrated simply by fitting a straight line to a set of data points. Suppose that we wish to fit the model

$$E(Y) = \beta_0 + \beta_1 x$$

to the set of data points shown in Figure 11.5. [The independent variable x could be w^2 or $(w)^{1/2}$ or $\ln w$, and so on, for some other independent variable w.] That is, we postulate that $Y = \beta_0 + \beta_1 x + \varepsilon$, where ε possesses some probability distribution with $E(\varepsilon) = 0$. If $\hat{\beta}_0$ and $\hat{\beta}_1$ are estimators of the parameters β_0 and β_1, then $\hat{Y} = \hat{\beta}_0 + \hat{\beta}_1 x$ is clearly an estimator of $E(Y)$.

The least squares procedure for fitting a line through a set of n data points is similar to the method we might use if we fit a line by eye; that is, we want the deviations between the observed values and corresponding points on the fitted line to be "small" in some sense. A convenient way to accomplish this, and one that yields estimators with good properties, is to minimize the sum of squares of the vertical deviations from the fitted line (see the deviations indicated in Figure 11.5). Thus, if

$$\hat{y}_i = \hat{\beta}_0 + \hat{\beta}_1 x_i$$

is the predicted value of the ith y value (when $x = x_i$), then the deviation of the observed value of y from the $\hat{y}$ line (sometimes called the *error*) is

$$y_i - \hat{y}_i$$

and the sum of squares of deviations to be minimized is

$$\text{SSE} = \sum_{i=1}^{n} (y_i - \hat{y}_i)^2 = \sum_{i=1}^{n} [y_i - (\hat{\beta}_0 + \hat{\beta}_1 x_i)]^2.$$

FIGURE **11.5**
Fitting a straight
line through a
set of data points

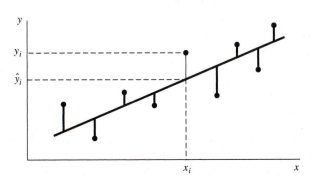

The quantity SSE is also called the *sum of squares for error*, for reasons that will subsequently become apparent.

If SSE possesses a minimum, it will occur for values of β_0 and β_1 that satisfy the equations, $\partial \text{SSE}/\partial \hat{\beta}_0 = 0$ and $\partial \text{SSE}/\partial \hat{\beta}_1 = 0$. Taking the partial derivatives of SSE with respect to $\hat{\beta}_0$ and $\hat{\beta}_1$ and setting them equal to zero, we obtain

$$\frac{\partial \text{SSE}}{\partial \hat{\beta}_0} = \frac{\partial \left\{ \sum_{i=1}^{n} [y_i - (\hat{\beta}_0 + \hat{\beta}_1 x_i)]^2 \right\}}{\partial \hat{\beta}_0} = -\sum_{i=1}^{n} 2[y_i - (\hat{\beta}_0 + \hat{\beta}_1 x_i)]$$

$$= -2 \left(\sum_{i=1}^{n} y_i - n\hat{\beta}_0 - \hat{\beta}_1 \sum_{i=1}^{n} x_i \right) = 0$$

and

$$\frac{\partial \text{SSE}}{\partial \hat{\beta}_1} = \frac{\partial \left\{ \sum_{i=1}^{n} [y_i - (\hat{\beta}_0 + \hat{\beta}_1 x_i)]^2 \right\}}{\partial \hat{\beta}_1} = -\sum_{i=1}^{n} 2[y_i - (\hat{\beta}_0 + \hat{\beta}_1 x_i)]x_i$$

$$= -2 \left(\sum_{i=1}^{n} x_i y_i - \hat{\beta}_0 \sum_{i=1}^{n} x_i - \hat{\beta}_1 \sum_{i=1}^{n} x_i^2 \right) = 0.$$

The equations $\partial \text{SSE}/\hat{\beta}_0 = 0$ and $\partial \text{SSE}/\hat{\beta}_1 = 0$ are called the *least squares equations* for estimating the parameters of a line.

The least squares equations are linear in $\hat{\beta}_0$ and $\hat{\beta}_1$ and hence can be solved simultaneously. You can verify that the solutions are

$$\hat{\beta}_1 = \frac{\sum_{i=1}^{n} (x_i - \bar{x})(y_i - \bar{y})}{\sum_{i=1}^{n} (x_i - \bar{x})^2} = \frac{\sum_{i=1}^{n} x_i y_i - \frac{1}{n} \sum_{i=1}^{n} x_i \sum_{i=1}^{n} y_i}{\sum_{i=1}^{n} x_i^2 - \frac{1}{n} \left(\sum_{i=1}^{n} x_i \right)^2}$$

$$\hat{\beta}_0 = \bar{y} - \hat{\beta}_1 \bar{x}.$$

Further, it can be shown that the simultaneous solution for the two least squares equations yields values of $\hat{\beta}_0$ and $\hat{\beta}_1$ that minimize SSE. We leave this for you to prove.

The expressions

$$\sum_{i=1}^{n} (x_i - \bar{x})(y_i - \bar{y}) \quad \text{and} \quad \sum_{i=1}^{n} (x_i - \bar{x})^2$$

that are used to calculate $\hat{\beta}_1$ are often encountered in the development of simple linear regression models. The first of these is calculated by summing products of *x*-values minus their mean and *y*-values minus their mean. In all subsequent discussions, we will denote this quantity by S_{xy}. Similarly, we will denote the second quantity by S_{xx}, because it is calculated by summing products that involve only the x-values.

Least Squares Estimators for the Simple Linear Regression Model

1. $\hat{\beta}_1 = \dfrac{S_{xy}}{S_{xx}}$ where $S_{xy} = \displaystyle\sum_{i=1}^{n}(x_i - \overline{x})(y_i - \overline{y})$ and

   $S_{xx} = \displaystyle\sum_{i=1}^{n}(x_i - \overline{x})^2.$

2. $\hat{\beta}_0 = \overline{y} - \hat{\beta}_1\overline{x}.$

We will illustrate the use of the preceding equations with a simple example.

EXAMPLE **11.1**   Use the method of least squares to fit a straight line to the $n = 5$ data points given in Table 11.1.

Solution   We commence by constructing Table 11.2 to compute the coefficients in the least squares equations. Then we have

Table 11.1.
Data, Example 11.1

x	y
-2	0
-1	0
0	1
1	1
2	3

Table 11.2. Calculations for finding the coefficients

x_i	y_i	$x_i y_i$	x_i^2
-2	0	0	4
-1	0	0	1
0	1	0	0
1	1	1	1
2	3	6	4
$\displaystyle\sum_{i=1}^{n} x_i = 0$	$\displaystyle\sum_{i=1}^{n} y_i = 5$	$\displaystyle\sum_{i=1}^{n} x_i y_i = 7$	$\displaystyle\sum_{i=1}^{n} x_i^2 = 10$

FIGURE **11.6**
Plot of data points
and least squares line
for Example 11.1

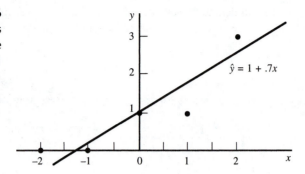

$$\hat{\beta}_1 = \frac{S_{xy}}{S_{xx}} = \frac{\displaystyle\sum_{i=1}^{n} x_i y_i - \frac{1}{n}\sum_{i=1}^{n} x_i \sum_{i=1}^{n} y_i}{\displaystyle\sum_{i=1}^{n} x_i^2 - \frac{1}{n}\left(\sum_{i=1}^{n} x_i\right)^2} = \frac{7 - \frac{1}{5}(0)(5)}{10 - \frac{1}{5}(0)^2} = .7$$

$$\hat{\beta}_0 = \bar{y} - \hat{\beta}_1 \bar{x} = \frac{5}{5} - (.7)(0) = 1$$

and the fitted line is

$$\hat{y} = 1 + .7x.$$

The five points and the fitted line are shown in Figure 11.6.

In this section we have determined the least squares estimators for the parameters β_0 and β_1 in the model $E(Y) = \beta_0 + \beta_1 x$. The simple example used here will reappear in future sections to illustrate other calculations. Exercises of a more realistic nature are presented at the ends of the sections, and two examples involving data from actual experiments are presented and analyzed in Section 11.9. In the next section, we develop the statistical properties of the least squares estimators for $\hat{\beta}_0$ and $\hat{\beta}_1$. Subsequent sections are devoted to using these estimators for a variety of inferential purposes.

Exercises

11.1 Fit a straight line to the five data points in the accompanying table. Give the estimates of β_0 and β_1. Plot the points and sketch the fitted line as a check on the calculations.

y	x
3	−2
2	−1
1	0
1	1
.5	2

11.2 Auditors are often required to compare the audited (or current) value of an inventory item with the book (or listed) value. If a company is keeping its inventory and books up to date, there should be a strong linear relationship between the audited and book values. A sample of ten inventory items from a certain company gave the data in the accompanying table on audited and book values. Fit the model $Y = \beta_0 + \beta_1 x + \varepsilon$ to these data. What is your estimate for the expected change in audited value for a one-unit change in book value? If the book value is $x = 100$, what would you use to estimate the audited value?

Item	Audit Value (y_i)	Book Value (x_i)
1	9	10
2	14	12
3	7	9
4	29	27
5	45	47
6	109	112
7	40	36
8	238	241
9	60	59
10	170	167

11.3 The median sale prices for new single-family houses over an 8-year period are given in the accompanying table. Letting Y denote the median sales price and x the year (using integers $1, 2, \ldots, 8$), fit the model $Y = \beta_0 + \beta_1 x + \varepsilon$. What can you conclude from the results?

Year	Median Sales Price ($\times 1000$)
1972 (1)	$27.6
1973 (2)	$32.5
1974 (3)	$35.9
1975 (4)	$39.3
1976 (5)	$44.2
1977 (6)	$48.8
1978 (7)	$55.7
1979 (8)	$62.9

Source: Adapted from *Time* (July 23, 1979), p. 67; the 1979 figure is for only the first quarter.

11.4 Laboratory experiments designed to measure LC50 values for the effect of certain toxicants on fish are run by two different methods. One method has water continuously flowing through laboratory tanks, and the other method has static water conditions. For purposes of establishing criteria for toxicants, the Environmental Protection Agency (EPA) wants to adjust all results to the flow-through condition. Thus a model is needed to relate the two types of observations. Observations on certain toxicants examined under both static and flow-through conditions yielded the data in the accompanying table (measurements in parts per million). Fit the model

$Y = \beta_0 + \beta_1 x + \varepsilon$. What interpretation can you give to the results? Estimate the flow-through value for a toxicant with an LC50 static value of $x = 12$ parts per million.

Toxicant	LC50 Flow-Through (y)	LC50 Static (x)
1	23.00	39.00
2	22.30	37.50
3	9.40	22.20
4	9.70	17.50
5	.15	.64
6	.28	.45
7	.75	2.62
8	.51	2.36
9	28.00	32.00
10	.39	.77

11.5   The 1980 EPA 49-state (all except California) combined mileage rating and engine volume are listed in the following table for nine standard-transmission, four-cylinder, gasoline-fueled, subcompact cars. The engine sizes are in total cubic inches of cylinder volume.

Car	Cylinder Volume (x)	Combined mpg (y)
VW Rabbit	97	24
Datsun 210	85	29
Chevrolet Chevette	98	26
Dodge Omni	105	24
Mazda 626	120	24
Oldsmobile Starfire	151	22
Mercury Capri	140	23
Toyota Celica	134	23
Datsun 810	146	21

a   Plot the data points on graph paper.

b   Find the least squares line for the data.

c   Graph the least squares line to see how well it fits the data.

d   Use the least squares line to estimate the mean miles per gallon for a subcompact automobile which has 125 cubic inches of engine volume.

11.6   Suppose that we have postulated the model

$$Y_i = \beta_1 x_i + \varepsilon_i \qquad i = 1, 2, \ldots, n,$$

where the ε_i are independent and identically distributed random variables with $E(\varepsilon_i) = 0$. Then $\hat{y}_i = \hat{\beta}_1 x_i$ is the predicted value of y when $x = x_i$ and $\text{SSE} = \sum_{i=1}^{n} [y_i - \hat{\beta}_1 x_i]^2$. Find the least squares estimator of β_1. (Notice that the equation $y = \beta x$ describes a straight line passing through the origin. The model just described often is called the *no-intercept* model.)

11.7   Some data obtained by Carlos E. Marcellari (1984) on the height x and diameter y of shells appear in the following table. If we consider the model

$$E(Y) = \beta_1 x,$$

then the slope β_1 is the ratio of the mean diameter to the height. Use the following data and the result of Exercise 11.6 to obtain the least squares estimate of the mean diameter to height ratio.

Specimen	Diameter (y)	Height (x)
OSU 36651	185	78
OSU 36652	194	65
OSU 36653	173	77
OSU 36654	200	76
OSU 36655	179	72
OSU 36656	213	76
OSU 36657	134	75
OSU 36658	191	77
OSU 36659	177	69
OSU 36660	199	65

11.8 Processors usually preserve cucumbers by fermenting them in a low-salt brine (6% to 9% sodium chloride) and then storing them in a high-salt brine until they are used by processors to produce various types of pickles. The high-salt brine is needed to retard softening of the pickles and to prevent freezing when they are stored outside in northern climates. Data showing the reduction in firmness of pickles stored over time in a low-salt brine (2% to 3%) are given in the accompanying table.

	Weeks in Storage at 72°F				
	0	4	14	32	52
Firmness (y) in pounds	19.8	16.5	12.8	8.1	7.5

Source: R. W. Buescher, J. M. Hudson, J. R. Adams, and D. H. Wallace, "Calcium Makes It Possible to Store Cucumber Pickles in Low-Salt Brine," *Arkansas Farm Research* 30(4) (July–August 1981).

a Fit a least squares line to the data.

b As a check on your calculations, plot the five data points and graph the line. Does the line appear to provide a good fit to the data points?

c Use the least squares line to estimate the mean firmness of pickles stored for 20 weeks.

11.9 The accompanying table gives the catches of Peruvian anchovies (in millions of metric tons) and the prices of fish meal (in current dollars per ton) for the years 1965 to 1978.

Variable	1965	1966	1967	1968	1969	1970	1971
Price of fish meal (y)	190	160	134	129	172	197	167
Anchovy catch (x)	7.23	8.53	9.82	10.26	8.96	12.27	10.28

Variable	1972	1973	1974	1975	1976	1977	1978
Price of fish meal (y)	239	542	372	245	376	454	410
Anchovy catch (x)	4.45	1.78	4.0	3.3	4.3	0.8	0.5

Source: John E. Bardach, and Regina M. Santerre, "Climate and the Fish in the Sea," *BioScience* 31(3) (March 1981): 206ff. Copyright ©1981 by the American Institute of Biological Sciences.

    a Find the least squares line appropriate for these data.

    b Plot the points and graph the line as a check on your calculations.

11.10 Matis and Wehrly (1979) report the following table of data on the proportion of green sunfish that survive a fixed level of thermal pollution for varying lengths of time.

    a Fit the linear model $Y = \beta_0 + \beta_1 x + \varepsilon$. Give your interpretation.

    b Plot the points and graph the result of (a). Does the line fit through the points?

Proportion of Survivors (y)	Scaled Time (x)
1.00	.10
.95	.15
.95	.20
.90	.25
.85	.30
.70	.35
.65	.40
.60	.45
.55	.50
.40	.55

11.4 Properties of the Least Squares Estimators: Simple Linear Regression

To employ the least squares estimators for making statistical inferences, we need to determine their statistical properties. In this section we show that the least squares estimators $\hat{\beta}_0$ and $\hat{\beta}_1$ for the parameters in the simple linear model

$$Y = \beta_0 + \beta_1 x + \varepsilon$$

are *unbiased* estimators of their respective parameter values. We also derive the variances of these estimators and, under the assumption that the error term ε is normally distributed, show that $\hat{\beta}_0$ and $\hat{\beta}_1$ have normal sampling distributions. Corresponding results applicable to the multiple linear regression model are presented without proof in Section 11.11.

Recall that ε was previously assumed to be a random variable with $E(\varepsilon) = 0$. We now add the assumption that $V(\varepsilon) = \sigma^2$. That is, we are assuming that the difference between the random variable Y and $E(Y) = \beta_0 + \beta_1 x$ is distributed about zero with a variance that does not depend on x. Notice that $V(Y) = V(\varepsilon) = \sigma^2$, because the other terms in the linear model are constants. (An unbiased estimator for the variance σ^2 of the error term in the model is also provided in this section.)

Assume that n independent observations are to be made on this model so that, before sampling, we have n independent random variables of the form

$$Y_i = \beta_0 + \beta_1 x_i + \varepsilon_i.$$

From Section 11.3, we know that

$$\hat{\beta}_1 = \frac{S_{xy}}{S_{xx}} = \frac{\sum\limits_{i=1}^{n}(x_i - \bar{x})(Y_i - \bar{Y})}{\sum\limits_{i=1}^{n}(x_i - \bar{x})^2}$$

which can be written as

$$\hat{\beta}_1 = \frac{\sum\limits_{i=1}^{n}(x_i - \bar{x})Y_i - \bar{Y}\sum\limits_{i=1}^{n}(x_i - \bar{x})}{S_{xx}}.$$

Then because $\sum_{i=1}^{n}(x_i - \bar{x}) = 0$, we have

$$\hat{\beta}_1 = \frac{\sum\limits_{i=1}^{n}(x_i - \bar{x})Y_i}{S_{xx}}.$$

Because all summations in the following discussion will be summed from $i = 1$ to n, we will simplify our notation by omitting the variable of summation and its index. Now let us find the expected value and variance of $\hat{\beta}_1$.

From the expectation theorems developed in Section 5.8, we have

$$E(\hat{\beta}_1) = E\left[\frac{\sum(x_i - \bar{x})Y_i}{S_{xx}}\right] = \frac{\sum(x_i - \bar{x})E(Y_i)}{S_{xx}}$$

$$= \frac{\sum(x_i - \bar{x})(\beta_0 + \beta_1 x_i)}{S_{xx}}$$

$$= \beta_0\frac{\sum(x_i - \bar{x})}{S_{xx}} + \beta_1\frac{\sum(x_i - \bar{x})x_i}{S_{xx}}.$$

Because $\sum(x_i - \bar{x}) = 0$ and $S_{xx} = \sum(x_i - \bar{x})^2 = \sum(x_i - \bar{x})x_i$, we have

$$E(\hat{\beta}_1) = 0 + \beta_1\frac{S_{xx}}{S_{xx}} = \beta_1.$$

Thus, $\hat{\beta}_1$ is an unbiased estimator of β_1.

To find $V(\hat{\beta}_1)$, we use Theorem 5.12. Recall that $Y_1, Y_2, \ldots, Y_n$ are independent and, therefore,

$$V(\hat{\beta}_1) = V\left[\frac{\sum(x_i - \bar{x})Y_i}{S_{xx}}\right] = \left[\frac{1}{S_{xx}}\right]^2\sum V[(x_i - \bar{x})Y_i]$$

$$= \left[\frac{1}{S_{xx}}\right]^2\sum(x_i - \bar{x})^2 V(Y_i).$$

Because $V(Y_i) = \sigma^2$, for $i = 1, 2, \ldots, n$,

$$V(\hat{\beta}_1) = \frac{\sigma^2}{S_{xx}}.$$

Now let us find the expected value and variance of $\hat{\beta}_0$, where $\hat{\beta}_0 = \overline{Y} - \hat{\beta}_1\overline{x}$. From Theorem 5.12 we have

$$V(\hat{\beta}_0) = V(\overline{Y}) + \overline{x}^2 V(\hat{\beta}_1) - 2\overline{x}\,\text{Cov}(\overline{Y}, \hat{\beta}_1).$$

Consequently, we must find $V(\overline{Y})$ and $\text{Cov}(\overline{Y}, \hat{\beta}_1)$ in order to obtain $V(\hat{\beta}_0)$. Because $Y_i = \beta_0 + \beta_1 x_i + \varepsilon_i$, we see that

$$\overline{Y} = \frac{1}{n}\sum Y_i = \beta_0 + \beta_1\overline{x} + \overline{\varepsilon}.$$

Thus,

$$E(\overline{Y}) = \beta_0 + \beta_1\overline{x} + E(\overline{\varepsilon}) = \beta_0 + \beta_1\overline{x}$$

and

$$V(\overline{Y}) = V(\overline{\varepsilon}) = \left(\frac{1}{n}\right)V(\varepsilon_i) = \frac{\sigma^2}{n}.$$

To find $\text{Cov}(\overline{Y}, \hat{\beta}_1)$, rewrite the expression for $\hat{\beta}_1$ as

$$\hat{\beta}_1 = \sum c_i Y_i,$$

where

$$c_i = \frac{x_i - \overline{x}}{S_{xx}}.$$

(Notice that $\sum c_i = 0$.) Then

$$\text{Cov}(\overline{Y}, \hat{\beta}_1) = \text{Cov}\left[\sum\left(\frac{1}{n}\right)Y_i, \sum c_i Y_i\right]$$

and, using Theorem 5.12,

$$\text{Cov}(\overline{Y}, \hat{\beta}_1) = \sum\left(\frac{c_i}{n}\right)V(Y_i) + \sum\sum_{i\neq j}\left(\frac{c_j}{n}\right)\text{Cov}(Y_i, Y_j).$$

Because Y_i and Y_j, where $i \neq j$, are independent, $\text{Cov}(Y_i, Y_j) = 0$. Also, $V(Y_i) = \sigma^2$, and hence

$$\text{Cov}(\overline{Y}, \hat{\beta}_1) = \frac{\sigma^2}{n}\sum c_i = 0.$$

Returning to our original task of finding the expected value and variance of $\hat{\beta}_0 = \overline{Y} - \hat{\beta}_1\overline{x}$, we apply expectation theorems to obtain

$$E(\hat{\beta}_0) = E(\overline{Y}) - E(\hat{\beta}_1)\overline{x} = \beta_0 + \beta_1\overline{x} - \beta_1\overline{x} = \beta_0.$$

Thus, we have shown that both $\hat{\beta}_0$ and $\hat{\beta}_1$ are unbiased estimators of their respective parameters.

Because we now have $V(\overline{Y})$, $V(\hat{\beta}_1)$, and $\text{Cov}(\overline{Y}, \hat{\beta}_1)$, we are ready to find $V(\hat{\beta}_0)$, because

$$V(\hat{\beta}_0) = V(\overline{Y}) + \overline{x}^2 V(\hat{\beta}_1) - 2\overline{x}\,\text{Cov}(\overline{Y}, \hat{\beta}_1).$$

Substituting the values for $V(\overline{Y})$, $V(\hat{\beta}_1)$, and $\text{Cov}(\overline{Y}, \hat{\beta}_1)$, we obtain

$$V(\hat{\beta}_0) = \frac{\sigma^2}{n} + \overline{x}^2\left[\frac{\sigma^2}{S_{xx}}\right] - 0$$

$$= \sigma^2\left[\frac{1}{n} + \frac{\overline{x}^2}{S_{xx}}\right] = \frac{\sigma^2 \sum x_i^2}{n S_{xx}}.$$

Further (see Exercise 11.17), Theorem 5.12 can be employed to show that

$$\text{Cov}(\hat{\beta}_0, \hat{\beta}_1) = \frac{-\overline{x}\sigma^2}{S_{xx}}.$$

Notice that $\hat{\beta}_0$ and $\hat{\beta}_1$ are correlated (and therefore dependent) unless $\overline{x} = 0$.

All the quantities necessary to determine the values of the variances and covariances, above, have already been calculated in the course of obtaining the values for $\hat{\beta}_0$ and $\hat{\beta}_1$.

EXAMPLE 11.2   Find the variances of the estimators $\hat{\beta}_0$ and $\hat{\beta}_1$ for Example 11.1.

Solution   In Example 11.1 (see the calculations for the denominator of $\hat{\beta}_1$), we found that

$$n = 5, \qquad \sum x_i = 0, \qquad \sum x_i^2 = 10, \quad \text{and} \quad S_{xx} = 10.$$

It follows that $\overline{x} = 0$ and that

$$V(\hat{\beta}_0) = \frac{\sigma^2 \sum x_i^2}{n S_{xx}} = \frac{\sigma^2(10)}{5(10)} = (1/5)\sigma^2$$

and

$$V(\hat{\beta}_1) = \frac{\sigma^2}{S_{xx}} = (1/10)\sigma^2.$$

Notice that $\text{Cov}(\hat{\beta}_0, \hat{\beta}_1) = 0$ in this case since $\sum x_i = 0$.

The preceding expressions give the variances for the least squares estimators in terms of σ^2, the variance of the error term ε. Usually the value of σ^2 will be unknown, and we will need to make use of the sample observations to estimate σ^2. If

$\overline{Y}$ is used to estimate the mean, we previously used

$$\left(\frac{1}{n-1}\right)\sum_{i=1}^{n}(Y_i - \overline{Y})^2$$

to estimate the population variance σ^2. Because we are now using $\hat{Y}_i$ to estimate $E(Y_i)$, it seems natural to base an estimate of σ^2 upon SSE $= \sum_{i=1}^{n}(Y_i - \hat{Y}_i)^2$. Indeed, we will show that

$$S^2 = \left(\frac{1}{n-2}\right)\sum_{i=1}^{n}(Y_i - \hat{Y}_i)^2 = \left(\frac{1}{n-2}\right)\text{SSE}$$

provides an unbiased estimator for σ^2. Notice that the 2 occurring in the denominator of S^2 corresponds to the number of β parameters estimated in the model.

Because

$$E(S^2) = E\left[\left(\frac{1}{n-2}\right)\text{SSE}\right] = \left(\frac{1}{n-2}\right)E(\text{SSE}),$$

it is necessary to find $E(\text{SSE})$ in order to verify that $E(S^2) = \sigma^2$.

Notice that

$$E(\text{SSE}) = E\left[\sum(Y_i - \hat{Y}_i)^2\right] = E\left[\sum(Y_i - \hat{\beta}_0 - \hat{\beta}_1 x_i)^2\right]$$

$$= E\left[\sum(Y_i - \overline{Y} + \hat{\beta}_1\overline{x} - \hat{\beta}_1 x_i)^2\right]$$

$$= E\left[\sum[(Y_i - \overline{Y}) - \hat{\beta}_1(x_i - \overline{x})]^2\right]$$

$$= E\left[\sum(Y_i - \overline{Y})^2 + \hat{\beta}_1^2\sum(x_i - \overline{x})^2 - 2\hat{\beta}_1\sum(x_i - \overline{x})(Y_i - \overline{Y})\right].$$

Because $\sum(x_i - \overline{x})(Y_i - \overline{Y}) = \sum(x_i - \overline{x})^2\hat{\beta}_1$, the last two terms in the expectation combine to give $-\hat{\beta}_1^2\sum(x_i - \overline{x})^2$. Also,

$$\sum(Y_i - \overline{Y})^2 = \sum Y_i^2 - n\overline{Y}^2$$

and, therefore,

$$E\left[\sum(Y_i - \hat{Y}_i)^2\right] = E\left[\sum Y_i^2 - n\overline{Y}^2 - \hat{\beta}_1^2 S_{xx}\right]$$

$$= \sum E(Y_i^2) - nE(\overline{Y}^2) - S_{xx}E(\hat{\beta}_1^2).$$

Noting that, for any random variable U, $E(U^2) = V(U) + [E(U)]^2$, we see that

$$E\left[\sum(Y_i - \hat{Y}_i)^2\right] = \sum\{V(Y_i) + [E(Y_i)]^2\} - n\{V(\overline{Y}) + [E(\overline{Y})]^2\}$$

$$- S_{xx}\{V(\hat{\beta}_1) + [E(\hat{\beta}_1)]^2\}$$

$$= n\sigma^2 + \sum (\beta_0 + \beta_1 x_i)^2 - n \left[\frac{\sigma^2}{n} + (\beta_0 + \beta_1 \bar{x})^2 \right]$$

$$- S_{xx} \left[\frac{\sigma^2}{S_{xx}} + \beta_1^2 \right].$$

This expression simplifies to $(n-2)\sigma^2$. Thus, we find that an unbiased estimator of σ^2 is given by

$$S^2 = \left(\frac{1}{n-2} \right) \sum (Y_i - \hat{Y}_i)^2 = \left(\frac{1}{n-2} \right) \text{SSE}.$$

One task remains, finding an easy way to calculate $\sum (y_i - \hat{y}_i)^2 = \text{SSE}$. In Exercise 11.11, you will show that a computing formula for SSE is given by

$$\text{SSE} = \sum_{i=1}^{n} (y_i - \bar{y})^2 - \hat{\beta}_1 \sum_{i=1}^{n} (x_i - \bar{x})(y_i - \bar{y})$$

$$= S_{yy} - \hat{\beta}_1 S_{xy}, \qquad \text{where } S_{yy} = \sum_{i=1}^{n} (y_i - \bar{y})^2.$$

EXAMPLE 11.3   Estimate σ^2 from the data given in Example 11.1.

Solution   For these data, $n = 5$ and we have already determined that

$$\sum y_i = 5, \qquad S_{xy} = 7, \quad \text{and} \quad \hat{\beta}_1 = .7.$$

It is easily determined that $\sum y_i^2 = 11$ and that

$$S_{yy} = \sum (y_i - \bar{y})^2 = \sum y_i^2 - n(\bar{y})^2 = 11 - 5(1)^2 = 6.0.$$

Therefore,

$$\text{SSE} = S_{yy} - \hat{\beta}_1 S_{xy} = 6.0 - (.7)(7) = 1.1$$

and

$$s^2 = \frac{\text{SSE}}{n-2} = \frac{1.1}{5-2} = \frac{1.1}{3} = .367.$$

These derivations establish the means and variances of the estimators $\hat{\beta}_0$ and $\hat{\beta}_1$ and show that $S^2 = \text{SSE}/(n-2)$ is an unbiased estimator for the parameter σ^2. Thus far, the only assumptions we have made about the error term ε in the model

$Y = \beta_0 + \beta_1 x + \varepsilon$ is that $E(\varepsilon) = 0$ and that $V(\varepsilon) = \sigma^2$, independent of x. The form of the sampling distributions for $\hat{\beta}_0$ and $\hat{\beta}_1$ depends upon the distribution of the error term ε. Because of the common occurrence of the normal distribution in nature, it is often reasonable to assume that ε is normally distributed with mean zero and variance σ^2. If this assumption of normality is warranted, it follows that Y_i is normally distributed with mean $\beta_0 + \beta_1 x_i$ and variance σ^2. Because both $\hat{\beta}_0$ and $\hat{\beta}_1$ are *linear functions* of $Y_1, Y_2, \ldots, Y_n$, the estimators are normally distributed, with means and variances as previously derived. Further, if the assumption of normality is warranted, it follows that

$$\frac{(n-2)S^2}{\sigma^2} = \frac{SSE}{\sigma^2}$$

has a χ^2 distribution with $n - 2$ degrees of freedom. (The proof of this result is omitted.)

As you will subsequently see, the assumption of normality of the distribution of the error term ε and the resulting normal distributions for $\hat{\beta}_0$ and $\hat{\beta}_1$ will allow us to develop tests and confidence intervals based upon the t distribution. The results of this section are summarized here because of their importance to discussions in subsequent sections. Notice that $V(\hat{\beta}_0)$, $V(\hat{\beta}_1)$, and $\text{Cov}(\hat{\beta}_0, \hat{\beta}_1)$ are all constant multiples of σ^2. Because $V(\hat{\beta}_i) = \text{Cov}(\hat{\beta}_i, \hat{\beta}_i)$, we will unify notation and provide consistency with the later sections of this chapter if we use the notation $V(\hat{\beta}_0) = c_{00}\sigma^2$, $V(\hat{\beta}_1) = c_{11}\sigma^2$, and $\text{Cov}(\hat{\beta}_0, \hat{\beta}_1) = c_{01}\sigma^2$.

Properties of the Least Squares Estimators—Simple Linear Regression

1. The estimators $\hat{\beta}_0$ and $\hat{\beta}_1$ are unbiased—that is, $E(\hat{\beta}_i) = \beta_i$, for $i = 0, 1$.

2. $V(\hat{\beta}_0) = c_{00}\sigma^2$, where $c_{00} = \dfrac{\sum x_i^2}{n S_{xx}}$.

3. $V(\hat{\beta}_1) = c_{11}\sigma^2$, where $c_{11} = \dfrac{1}{S_{xx}}$.

4. $\text{Cov}(\hat{\beta}_0, \hat{\beta}_1) = c_{01}\sigma^2$, where $c_{01} = \dfrac{-\bar{x}}{S_{xx}}$.

5. An unbiased estimator of σ^2 is $S^2 = SSE/(n-2)$, where $SSE = S_{yy} - \hat{\beta}_1 S_{xy}$ and $S_{yy} = \sum (y_i - \bar{y})^2$.

If, in addition, the ε_i, for $i = 1, 2, \ldots, n$ are normally distributed:

6. Both $\hat{\beta}_0$ and $\hat{\beta}_1$ are normally distributed.

7. The random variable $\dfrac{(n-2)S^2}{\sigma^2}$ has a χ^2 distribution with $n-2$ degrees of freedom.

8. The statistic S^2 is independent of both $\hat{\beta}_0$ and $\hat{\beta}_1$.

Exercises

11.11 Derive the following identity:

$$\text{SSE} = \sum_{i=1}^{n}(y_i - \hat{y}_i)^2 = \sum_{i=1}^{n}(y_i - \hat{\beta}_0 - \hat{\beta}_1 x_i)^2$$

$$= \sum_{i=1}^{n}(y_i - \overline{y})^2 - \hat{\beta}_1 \sum_{i=1}^{n}(x_i - \overline{x})(y_i - \overline{y}) = S_{yy} - \hat{\beta}_1 S_{xy}.$$

Notice that this provides an easier computational method of finding SSE.

11.12 An experiment was conducted to observe the effect of an increase in temperature on the potency of an antibiotic. Three 1-ounce portions of the antibiotic were stored for equal lengths of time at each of the following Fahrenheit temperatures: 30°, 50°, 70°, and 90°. The potency readings observed at the end of the experimental period were as shown in the following table.

Potency readings (y)	38, 43, 29	32, 26, 33	19, 27, 23	14, 19, 21
Temperature (x)	30°	50°	70°	90°

  a Find the least squares line appropriate for this data.
  b Plot the points and graph the line as a check on your calculations.
  c Calculate S^2.

11.13 **a** Calculate SSE and S^2 for Exercise 11.3.
  b It is sometimes convenient, for computational purposes, to have x values spaced symmetrically and equally about zero. The x values can be rescaled (or coded) in any convenient manner, with no loss of information in the statistical analysis. Refer to Exercise 11.3. Code the x values (originally given on a scale of 1 to 8) by using the formula

$$x^* = \frac{x - 4.5}{.5}.$$

Then fit the model $Y = \beta_0^* + \beta_1^* x^* + \varepsilon$. Calculate SSE. (Notice that the x^* values are integers symmetrically spaced about zero.) Compare the SSE with the value obtained in (a).

11.14 **a** Calculate SSE and S^2 for Exercise 11.4.
  b Refer to Exercise 11.4. Code the x values in a convenient manner and fit a simple linear model to the LC50 measurements presented there. Compute SSE and compare your answer to the result of (a).

11.15 A study was conducted to determine the effects of sleep deprivation on subjects' ability to solve simple problems. The amount of sleep deprivation varied over 8, 12, 16, 20, and 24 hours without sleep. A total of ten subjects participated in the study, two at each sleep-deprivation level. After his or her specified sleep-deprivation period, each subject was administered a set of simple addition problems and the number of errors recorded. The results shown in the following table were obtained.

Number of errors (y)	8, 6	6, 10	8, 14	14, 12	16, 12
Number of hours without sleep (x)	8	12	16	20	24

  a Find the least squares line appropriate to these data.

    b Plot the points and graph the least squares line as a check on your calculations.

    c Calculate S^2.

11.16 Suppose that $Y_1, Y_2, \ldots, Y_n$ are independent normal random variables with $E(Y_i) = \beta_0 + \beta_1 x_i$ and $V(Y_i) = \sigma^2$, for $i = 1, 2, \ldots, n$. Show that the maximum-likelihood estimators of β_0 and β_1 are the same as the least squares estimators of Section 11.3.

11.17 Under the assumptions of Exercise 11.16, find $\text{Cov}(\hat{\beta}_0, \hat{\beta}_1)$. Use this answer to show that $\hat{\beta}_0$ and $\hat{\beta}_1$ are independent if $\sum_{i=1}^{n} x_i = 0$. [*Hint:* $\text{Cov}(\hat{\beta}_0, \hat{\beta}_1) = \text{Cov}(\overline{Y} - \hat{\beta}_1\overline{x}, \hat{\beta}_1)$. Use Theorem 5.12 and the results of this section.]

11.18 Under the assumptions of Exercise 11.16, find the maximum-likelihood estimator of σ^2.

11.5 Inferences Concerning the Parameters β_i

Suppose that an engineer has fit the model

$$Y = \beta_0 + \beta_1 x + \varepsilon,$$

where Y is the strength of concrete after 28 days and x is the water/cement ratio used in the concrete. If, in reality, the strength of concrete does not change with the water/cement ratio, then $\beta_1 = 0$. Thus the engineer may wish to test $H_0 : \beta_1 = 0$ versus $H_a : \beta_1 \neq 0$ in order to assess whether the independent variable has an influence on the dependent variable. Or the engineer may wish to estimate the mean rate β_1 of change in $E(Y)$ for a one-unit change in the water/cement ratio x.

    In general, for any linear regression model, if the random error ε is normally distributed, we have established that $\hat{\beta}_i$ is an unbiased, normally distributed estimator of β_i with

$$V(\hat{\beta}_0) = c_{00}\sigma^2, \quad \text{with} \quad c_{00} = \frac{\sum x_i^2}{n S_{xx}}$$

and

$$V(\hat{\beta}_1) = c_{11}\sigma^2, \quad \text{with} \quad c_{11} = \frac{1}{S_{xx}}.$$

That is, the variances of both estimators are constant multiples of σ^2, the variance of the error term in the model. Using this information, we can construct a test of the hypothesis $H_0: \beta_i = \beta_{i0}$ (β_{i0} is a specified value of β_i), using the test statistic

$$Z = \frac{\hat{\beta}_i - \beta_{i0}}{\sigma \sqrt{c_{ii}}},$$

where

$$c_{00} = \frac{\sum x_i^2}{n S_{xx}} \quad \text{and} \quad c_{11} = \frac{1}{S_{xx}}.$$

The rejection region for a two-tailed test is given by

$$|z| \geq z_{\alpha/2}.$$

As in the case of the simple normal deviation tests studied in Chapter 10, to compute either of the preceding statistics Z, we must either know σ or possess a good estimate based upon an adequate number of degrees of freedom. (What would be adequate is a debatable point. We suggest that the estimate be based upon 30 or more degrees of freedom.) When this estimate is unavailable (which usually is the case), an estimate of σ may be calculated from the experimental data (in accordance with the procedure of Section 11.4) and substituted for σ in the Z statistic. If we estimate σ with $S = \sqrt{\frac{SSE}{n-2}}$, the resulting quantity,

$$T = \frac{\hat{\beta}_i - \beta_{i0}}{S\sqrt{c_{ii}}}$$

can be shown to possess a Student's t distribution with $n-2$ degrees of freedom.

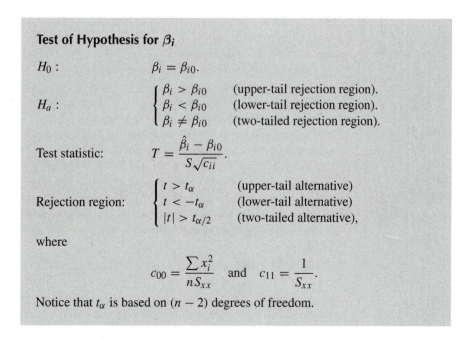

Test of Hypothesis for β_i

$H_0:$ $\beta_i = \beta_{i0}.$

$H_a:$ $\begin{cases} \beta_i > \beta_{i0} & \text{(upper-tail rejection region).} \\ \beta_i < \beta_{i0} & \text{(lower-tail rejection region).} \\ \beta_i \neq \beta_{i0} & \text{(two-tailed rejection region).} \end{cases}$

Test statistic: $T = \dfrac{\hat{\beta}_i - \beta_{i0}}{S\sqrt{c_{ii}}}.$

Rejection region: $\begin{cases} t > t_\alpha & \text{(upper-tail alternative)} \\ t < -t_\alpha & \text{(lower-tail alternative)} \\ |t| > t_{\alpha/2} & \text{(two-tailed alternative),} \end{cases}$

where

$$c_{00} = \frac{\sum x_i^2}{n\,S_{xx}} \quad \text{and} \quad c_{11} = \frac{1}{S_{xx}}.$$

Notice that t_α is based on $(n-2)$ degrees of freedom.

EXAMPLE **11.4**   Do the data of Example 11.1 present sufficient evidence to indicate that the slope differs from 0? Test using $\alpha = .05$ and give bounds to the attained significance level.

Solution   The preceding question assumes that the probabilistic model is a realistic description of the true response and implies a test of hypothesis $H_0 : \beta_1 = 0$ versus $H_a : \beta_1 \neq 0$ in the linear model $Y = \beta_0 + \beta_1 x + \varepsilon$. For these data, we determined in Example 11.1 that $\hat{\beta}_1 = .7$ and that $S_{xx} = 10$. Example 11.3 yielded that $s^2 = \frac{SSE}{n-2} = .367$ and, thus, that $s = \sqrt{.367} = .606$.

(Note: SSE is based upon $n - 2 = 3$ degrees of freedom.)

Because we are interested in the parameter β_1, we need the value

$$c_{11} = \frac{1}{S_{xx}} = \frac{1}{10} = .1.$$

Then

$$t = \frac{\hat{\beta}_1 - 0}{s\sqrt{c_{11}}} = \frac{.7 - 0}{.606\sqrt{.1}} = 3.65.$$

If we take $\alpha = .05$, the value of $t_{\alpha/2} = t_{.025}$ for 3 degrees of freedom is 3.182, and the rejection region would be

$$\text{reject if } |t| \geq 3.182.$$

Because the absolute value of the calculated value of t is larger than 3.182, we reject the null hypothesis that $\beta_1 = 0$ at the $\alpha = .05$ level of significance. Because the test is two-tailed, p-value $= 2P(t > 3.65)$, where t has a t distribution with 3 degrees of freedom. Using Table 5, Appendix III, we find that $.01 < P(t > 3.65) < .025$. Thus, we conclude that $.02 < p$-value $< .05$. Hence, we would reject the null hypothesis for any value of $\alpha \geq .05$. For values of $\alpha \leq .02$, we would fail to reject the null hypothesis. If we had chosen $.02 < \alpha < .05$, a more complete t table would be required before a decision could be reached. Again we notice the agreement between the conclusions reached by the formal (fixed α) test procedure and the proper interpretation of the attained significance level.

As a further step in the analysis, we could look at the width of a confidence interval for β_1 to see whether it is short enough to detect a departure from zero that would be of practical significance. We will show that the confidence interval for β_1 is quite wide, suggesting that the experimenter needs to collect more data before reaching a decision.

Based on the t statistic given earlier, we can follow the procedures of Chapter 10 to show that a confidence interval for β_i, with confidence coefficient $1 - \alpha$, is as follows.

A $(1 - \alpha)100\%$ Confidence Interval for β_i

$$\hat{\beta}_i \pm t_{\alpha/2} S \sqrt{c_{ii}}$$

where

$$c_{00} = \frac{\sum x_i^2}{n S_{xx}} \quad \text{and} \quad c_{11} = \frac{1}{S_{xx}}.$$

EXAMPLE **11.5**   Calculate a 95% confidence interval for the parameter β_1 of Example 11.4.

Solution   The tabulated value for $t_{.025}$, based upon 3 degrees of freedom, is 3.182. Then the 95% confidence interval for β_1 is

$$\hat{\beta}_1 \pm t_{.025} S \sqrt{c_{11}}.$$

Substituting, we get

$$.7 \pm (3.182)(.606)\sqrt{0.1}, \quad \text{or} \quad .7 \pm .610.$$

If we wish to estimate β_1 correct to within .15 unit, it is obvious that the confidence interval is too wide and that the sample size must be increased.

Exercises

11.19   Refer to Exercise 11.1.

 a   Do the data present sufficient evidence to indicate that the slope β_1 differs from zero? (Test at the 5% significance level.)

 b   Find a 95% confidence interval for β_1.

11.20   Refer to Exercise 11.9. Do the data present sufficient evidence to indicate that the size of the anchovy catch x contributes information for the prediction of price y of the fish meal? Give bounds on the attained significance level. What would you conclude at the $\alpha = .10$ level of significance?

11.21   Do the data in Exercise 11.15 present sufficient evidence to indicate that the number of errors is linearly related to the number of hours without sleep?

 a   Give bounds on the attained significance level.

 b   What would you conclude at the $\alpha = .05$ level of significance?

 c   Would you expect the relationship between y and x to be linear if x were varied over a wider range, say from $x = 4$ to $x = 48$?

 d   Give a 95% confidence interval for the slope. Provide a practical interpretation for this interval estimate.

11.22   Most sophomore physics students are required to conduct an experiment verifying Hooke's law. Hooke's law states that, when a force is applied to a body that is long in comparison to its cross-sectional area, the change y in its length is proportional to the force x; that is,

$$y = \beta_1 x,$$

where β_1 is a constant of proportionality. The results of an actual physics student's laboratory experiment are shown in the following table. Six lengths of steel wire, .34 millimeters (mm) in diameter and 2 meters (m) long, were used to obtain the six force-length change measurements.

Force x (kg)	Change in Length (y) (mm)
29.4	4.25
39.2	5.25
49.0	6.50
58.8	7.85
68.6	8.75
78.4	10.00

a   Fit the model, $Y = \beta_0 + \beta_1 x + \varepsilon$, to the data, using the method of least squares.

b   Find a 95% confidence interval for the slope of the line.

c   According to Hooke's law, the line should pass through the point $(0, 0)$; that is, β_0 should equal 0. Test the hypothesis that $E(Y) = 0$ when $x = 0$. Give bounds for the attained significance level. What would you conclude at the $\alpha = .05$ level?

11.23   Use the properties of the least squares estimators given in Section 11.4 to perform the following.

a   Show that, under the null hypothesis $H_0: \beta_i = \beta_{i0}$,

$$T = \frac{\hat{\beta}_i - \beta_{i0}}{S\sqrt{c_{ii}}}$$

possesses a t distribution with $n - 2$ degrees of freedom, where $i = 1, 2$.

b   Derive the confidence intervals for β_i given in this section.

11.24   Suppose that $Y_1, Y_2, \ldots, Y_n$ are independent, normally distributed random variables with $E(Y_i) = \beta_0 + \beta_1 x_i$ and $V(Y_i) = \sigma^2$, for $i = 1, 2, \ldots, n$. Show that the likelihood ratio test of $H_0: \beta_1 = 0$ versus $H_a: \beta_1 \neq 0$ is equivalent to the t test given in this section.

*11.25   Let $Y_1, Y_2, \ldots, Y_n$ be as given in Exercise 11.24. Suppose that we have an additional set of independent random variables $W_1, W_2, \ldots, W_m$, where W_i is normally distributed with $E(W_i) = \gamma_0 + \gamma_1 c_i$ and $V(W_i) = \sigma^2$, for $i = 1, 2, \ldots, m$. Construct a test of $H_0: \beta_1 = \gamma_1$ against the $H_a: \beta_1 \neq \gamma_1$.[1]

11.26   The octane number Y of refined petroleum is related to the temperature x of the refining process, but it is also related to the particle size of the catalyst. An experiment with a small-particle catalyst gave a fitted least squares line of

$$\hat{y} = 9.360 + .155x$$

with $n = 31$, $V(\hat{\beta}_1) = (.0202)^2$, and SSE $= 2.04$. An independent experiment with a large-particle catalyst gave

$$\hat{y} = 4.265 + .190x,$$

with $n = 11$, $V(\hat{\beta}_1) = (.0193)^2$, and SSE $= 1.86$.[2]

a   Test the hypotheses that the slopes are significantly different from zero, with each test at the significance level of .05.

*b   Test, at the .05 significance level, that the two types of catalyst produce the same slope in the relationship between octane number and temperature. (Use the test in Exercise 11.25.)

---

[1] Exercises preceded by an asterisk are optional.

[2] *Source:* Gweyson and Cheasley, *Petroleum Refiner* (August 1959): 135.

11.27   The accompanying table lists the number of cases of tuberculosis (per 100,000 population) for the years 1967 through 1976 in Florida. Is there sufficient evidence to indicate that the rate of tuberculosis is decreasing over this period? Use $\alpha = .05$. (The years may be coded in any convenient manner.)

Year	Cases
1967	26.3
1968	26.1
1969	24.7
1970	22.8
1971	22.1
1972	20.4
1973	19.0
1974	17.7
1975	19.3
1976	17.5

11.28   Refer to Exercises 11.3 and 11.13.

   a   Is there sufficient evidence to indicate that the median sales price for new single-family houses increased over the period from 1972 through 1979 at the .01 level of significance?

   b   Estimate the expected yearly increase in median sale price by constructing a 99% confidence interval.

11.29   Refer to Exercise 11.4 and 11.14. Is there evidence of a linear relationship between flow-through and static LC50s? Test at the .05 significance level.

11.30   Refer to Exercise 11.4 and 11.14. Is there evidence of a linear relationship between flow-through and static LC50s? Give bounds for the attained significance level.

11.6   Inferences Concerning Linear Functions of the Model Parameters: Simple Linear Regression

In addition to making inferences about a single β_i, we frequently are interested in making inferences about linear functions of the model parameters β_0 and β_1. For example, we might wish to estimate $E(Y)$, given by

$$E(Y) = \beta_0 + \beta_1 x,$$

where $E(Y)$ might represent the mean yield of a chemical process for the settings of controlled process variable x or the mean mileage rating of four-cylinder gasoline engines with cylinder volume x. Properties of estimators of such linear functions will appear in this section.

Suppose that we wish to make an inference about the linear function

$$\theta = a_0 \beta_0 + a_1 \beta_1,$$

where a_0 and a_1 are constants (one of which may equal zero). Then the same linear function of the parameter estimators,

$$\hat{\theta} = a_0\hat{\beta}_0 + a_1\hat{\beta}_1,$$

is an unbiased estimator of θ because, by Theorem 5.12,

$$E(\hat{\theta}) = a_0 E(\hat{\beta}_0) + a_1 E(\hat{\beta}_1) = a_0\beta_0 + a_1\beta_1 = \theta.$$

Applying the same theorem, we find the variance of $\hat{\theta}$ to be

$$V(\hat{\theta}) = a_0^2 V(\hat{\beta}_0) + a_1^2 V(\hat{\beta}_1) + 2a_0a_1\text{Cov}(\hat{\beta}_0, \hat{\beta}_1),$$

where $V(\hat{\beta}_i) = c_{ii}\sigma^2$ and $\text{Cov}(\hat{\beta}_0, \hat{\beta}_1) = c_{01}\sigma^2$. Using the expressions

$$c_{00} = \frac{\sum x_i^2}{nS_{xx}}, \qquad c_{11} = \frac{1}{S_{xx}}, \qquad \text{and} \quad c_{01} = \frac{-\bar{x}}{S_{xx}}$$

and some routine algebraic manipulations, we obtain

$$V(\hat{\theta}) = \left(\frac{a_0^2\frac{\sum x_i^2}{n} + a_1^2 - 2a_0a_1\bar{x}}{S_{xx}}\right)\sigma^2.$$

Finally, recalling that $\hat{\beta}_0$ and $\hat{\beta}_1$ are normally distributed in repeated sampling (Section 11.4), it is clear that $\hat{\theta}$ is a linear function of normally distributed random variables and hence itself is normally distributed in repeated sampling.

Thus, we conclude that

$$Z = \frac{\hat{\theta} - \theta}{\sigma_{\hat{\theta}}}$$

has a standard normal distribution and could be employed to test a hypothesis

$$H_0 : \theta = \theta_0$$

when θ_0 is some specified value of $\theta = a_0\beta_0 + a_1\beta_1$. Likewise, a large-sample $(1 - \alpha)100\%$ confidence interval for $\theta = a_0\beta_0 + a_1\beta_1$ is

$$\hat{\theta} \pm z_{\alpha/2}\sigma_{\hat{\theta}}.$$

We notice that, in both the Z statistic and the confidence interval above, $\sigma_{\hat{\theta}} = \sqrt{V(\hat{\theta})}$ is a constant (depending in the values of the x's and the a's) multiple of σ. If we substitute S for σ in the expression for Z, the resulting expression (which we identify as T) possesses a Student's t distribution in repeated sampling, with $n - 2$ degrees of freedom, and hence provides a test statistic to test hypotheses concerning $\theta = a_0\beta_0 + a_1\beta_1$.

Appropriate tests are summarized as follows.

A Test for $\theta = a_0\beta_0 + a_1\beta_1$

H_0 :

$$\theta = \theta_0.$$

H_a :

$$\begin{cases} \theta > \theta_0. \\ \theta < \theta_0. \\ \theta \neq \theta_0. \end{cases}$$

Test statistic:

$$T = \frac{\hat{\theta} - \theta_0}{S\sqrt{\left(\dfrac{a_0^2\dfrac{\sum x_i^2}{n} + a_1^2 - 2a_0a_1\bar{x}}{S_{xx}}\right)}}.$$

Rejection region:

$$\begin{cases} t > t_\alpha. \\ t < -t_\alpha. \\ |t| > t_{\alpha/2}. \end{cases}$$

Here t_α is based upon $n - 2$ degrees of freedom.

The corresponding $(1 - \alpha)100\%$ confidence interval for $\theta = a_0\beta_0 + a_1\theta_1$ is as follows.

A $(1 - \alpha)100\%$ Confidence Interval for $\theta = a_0\beta_0 + a_1\beta_1$

$$\hat{\theta} \pm t_{\alpha/2}S\sqrt{\left(\frac{a_0^2\dfrac{\sum x_i^2}{n} + a_1^2 - 2a_0a_1\bar{x}}{S_{xx}}\right)}$$

where the tabulated $t_{\alpha/2}$ is based upon $n - 2$ degrees of freedom.

One useful application of the hypothesis-testing and confidence interval techniques just presented is to the problem of estimating the mean value of Y, $E(Y)$, for a fixed value of the independent variable x. In particular, if x^* denotes a specific value of x that is of interest, then

$$E(Y) = \beta_0 + \beta_1 x^*.$$

Notice that $E(Y)$ is a special case of $a_0\beta_0 + a_1\beta_1$ with $a_0 = 1$ and $a_1 = x^*$. Thus, an inference about $E(Y)$ when $x = x^*$ can be made by using the techniques developed earlier for a general linear combination of the β's.

In the context of estimating the mean value for Y, $E(Y) = \beta_0 + \beta_1 x^*$ when the independent variable x takes on the value x^*, it can be shown (see Exercise 11.31) that, with $a_0 = 1$, $a_1 = x^*$,

$$\left(\frac{a_0^2\dfrac{\sum x_i^2}{n} + a_1^2 - 2a_0a_1\bar{x}}{S_{xx}}\right) = \frac{1}{n} + \frac{(x^* - \bar{x})^2}{S_{xx}}.$$

A confidence interval for the mean value of Y when $x = x^*$, a particular value of x, is as follows.

A $(1 - \alpha)100\%$ Confidence Interval for $E(Y) = \beta_0 + \beta_1 x^*$

$$\hat{\beta}_0 + \hat{\beta}_1 x^* \pm t_{\alpha/2} S \sqrt{\frac{1}{n} + \frac{(x^* - \bar{x})^2}{S_{xx}}}$$

where the tabulated $t_{\alpha/2}$ is based upon $n - 2$ degrees of freedom.

This formula makes it easy to see that for a fixed value of n and for given x values, the shortest confidence interval for $E(Y)$ is obtained when $x^* = \bar{x}$, the average of the x values used in the experiment. If our objective is to plan an experiment that yields short confidence intervals for $E(Y)$, n should be large, S_{xx} should be large (if possible), and x^* should be near $\bar{x}$. The physical interpretation of a large S_{xx} is that, when possible, the values of x used in the experiment should be *spread out* as much as possible.

EXAMPLE 11.6   For the data of Example 11.1, find a 90% confidence interval for $E(Y)$ when $x = 1$.

Solution   For the model of Example 11.1,

$$E(Y) = \beta_0 + \beta_1 x.$$

To estimate $E(Y)$ for any fixed value $x = x^*$, we use the unbiased estimator $\widehat{E(Y)} = \hat{\beta}_0 + \hat{\beta}_1 x^*$. Then

$$\hat{\beta}_0 + \hat{\beta}_1 x^* = 1 + .7x^*.$$

For this case $x^* = 1$; and because $n = 5, \bar{x} = 0$, and $S_{xx} = 10$, it follows that

$$\frac{1}{n} + \frac{(x^* - \bar{x})^2}{S_{xx}} = \frac{1}{5} + \frac{(1 - 0)^2}{10} = .3.$$

In Example 11.3, we found s^2 to be .367, or $s = .606$, for these data. The value of $t_{.05}$ with $n - 2 = 3$ degrees of freedom is 2.353.

The confidence interval for $E(Y)$ when $x = 1$ is

$$\hat{\beta}_0 + \hat{\beta}_1 x^* \pm t_{\alpha/2} S \sqrt{\frac{1}{n} + \frac{(x^* - \bar{x})^2}{S_{xx}}}$$

$$[(1 + (.7)(1)] \pm (2.353)(.606)\sqrt{.3}$$

$$1.7 \pm .781.$$

Exercises 561

That is, we are 90% confident that, when the independent variable takes on the value x = 1, the mean value E(Y) of the dependent variable is between .919 and 2.481. This interval obviously is very wide, but remember that it is based on only five data points and was used solely for purposes of illustration. We will show you some practical applications of regression analyses in Section 11.9.

Exercises

11.31 For the simple linear regression model $Y = \beta_0 + \beta_1 x + \varepsilon$ with $E(\varepsilon) = 0$ and $V(\varepsilon) = \sigma^2$, use the expression for $V(a_0\hat{\beta}_0 + a_1\hat{\beta}_1)$ derived in this section to show that

$$V(\hat{\beta}_0 + \hat{\beta}_1 x^*) = \left[\frac{1}{n} + \frac{(x^* - \bar{x})^2}{S_{xx}}\right]\sigma^2.$$

For what value of x^* does the confidence interval for $E(Y)$ achieve its minimum length?

11.32 Refer to Exercise 11.9 and 11.20. Find the 90% confidence interval for the mean price per ton of fish meal if the anchovy catch is 5 million metric tons.

11.33 Using the model fit to the data of Exercise 11.4, construct a 95% confidence interval for the mean value of flow-through LC50 measurements for a toxicant that has a static LC50 of 12 parts per million. (Also see Exercise 11.14.)

11.34 Refer to Exercise 11.1. Find a 90% confidence interval for $E(Y)$ when $x^* = 0$. Then find 90% confidence intervals for $E(Y)$ when $x^* = -2$ and $x^* = +2$. Compare the lengths of these intervals. Plot these confidence limits on the graph you constructed for Exercise 11.1.

11.35 Refer to Exercise 11.12. Find a 95% confidence interval for the mean potency of a 1-ounce portion of antibiotic stored at 65°F.

11.36 Refer to Exercise 11.10. Find a 90% confidence interval for the expected proportion of survivors at time period .30.

*11.37 Refer to Exercise 11.2. Suppose that the sample given there came from a large, but finite, population of inventory items. We wish to estimate the population mean of the audited values, using the fact that book values are known for every item on inventory. If the population contains N items and

$$E(Y_i) = \mu_i = \beta_0 + \beta_1 x_i,$$

then the population mean is given by

$$\mu_Y = \frac{1}{N}\sum_{i=1}^{N}\mu_i = \beta_0 + \beta_1\left(\frac{1}{N}\right)\sum_{i=1}^{N}x_i = \beta_0 + \beta_1\mu_x.$$

a Using the least squares estimators of β_0 and β_1, show that μ_Y can be estimated by

$$\hat{\mu}_Y = \bar{y} + \hat{\beta}_1(\mu_x - \bar{x}).$$

(Notice that $\bar{y}$ is adjusted up or down, depending upon whether $\bar{x}$ is larger or smaller than μ_x.)

b Using the data of Exercise 11.2 and the fact that $\mu_x = 74.0$, estimate μ_Y, the mean of the audited values, and place a 2-standard-deviation bound on the error of estimation. (Regard the x_i values as constants when computing the variance of $\hat{\mu}_Y$.)

11.7 Predicting a Particular Value of Y Using Simple Linear Regression

Suppose that, for a fixed pressure, the yield Y for a chemical experiment is a function of the temperature x at which the experiment is run. Assume that a linear model of the form

$$Y = \beta_0 + \beta_1 x + \varepsilon$$

adequately represents the response function traced by Y over the experimental region of interest. In Section 11.6, we discussed methods for estimating $E(Y)$ for a given temperature, say, x^*. That is, we know how to estimate the mean yield $E(Y)$ of the process at the setting $x = x^*$.

Now consider a different problem. Instead of estimating the mean yield at x^*, we wish to *predict* the particular response Y that we will observe if the experiment is run at some time in the future (such as next Monday). This situation would occur if, for some reason, the response next Monday held a special significance to us. Prediction problems frequently occur in business, where we may be interested in a particular gain associated with the investment we intend to make next month rather than the mean gain over a long series of investments.

Notice that Y is a random variable, not a parameter; predicting its value therefore represents a departure from our previous objective of making inferences about population parameters. If it is reasonable to assume that ε is normally distributed with mean 0 and variance σ^2, it follows that Y is normally distributed with mean $\beta_0 + \beta_1 x$ and variance σ^2. If the distribution of a random variable Y is known and a single value of Y is then selected, how would you predict the observed value? We contend that you would select a value of Y near the *center* of the distribution—in particular, the expected value of Y. If we are interested in the value of Y when $x = x^*$, call it Y^*, we could employ $\widehat{Y^*} = \hat{\beta}_0 + \hat{\beta}_1 x^*$ as a predictor of a particular value of Y^* and as an estimator of $E(Y)$, as well.

If $x = x^*$, the error of predicting a particular value of Y^*, using $\widehat{Y^*}$ as the predictor, is the difference between the actual value of Y^* and the predicted value:

$$\text{error} = Y^* - \widehat{Y^*}.$$

Let us now investigate the properties of this error in repeated sampling.

Because both Y^* and $\widehat{Y^*}$ are normally distributed random variables, their difference (the error) is also normally distributed.

Applying Theorem 5.12, which gives the formulas for the expected value and variance of a linear function of random variables, we obtain

$$E(\text{error}) = E(Y^* - \widehat{Y^*}) = E(Y^*) - E(\widehat{Y^*})$$

and, because $E(\widehat{Y^*}) = \beta_0 + \beta_1 x^* = E(Y^*)$,

$$E(\text{error}) = 0.$$

Likewise,

$$V(\text{error}) = V(Y^* - \widehat{Y^*}) = V(Y^*) + V(\widehat{Y^*}) - 2\text{Cov}(Y^*, \widehat{Y^*}).$$

Because we are predicting a future value Y^* that is not employed in the computation of $\widehat{Y^*}$, it follows that Y^* and $\widehat{Y^*}$ are independent and hence that $\text{Cov}(Y^*, \widehat{Y^*}) = 0$. Then

$$V(\text{error}) = V(Y^*) + V(\widehat{Y^*}) = \sigma^2 + V(\hat{\beta}_0 + \hat{\beta}_1 x^*)$$

$$= \sigma^2 + \left(\frac{1}{n} + \frac{(x^* - \bar{x})^2}{S_{xx}} \right) \sigma^2$$

$$= \sigma^2 \left[1 + \frac{1}{n} + \frac{(x^* - \bar{x})^2}{S_{xx}} \right].$$

We have shown that the error of predicting a particular value of Y is normally distributed with mean 0 and variance as given in the preceding equation. It follows that

$$Z = \frac{Y^* - \widehat{Y^*}}{\sigma \sqrt{1 + \dfrac{1}{n} + \dfrac{(x^* - \bar{x})^2}{S_{xx}}}}$$

has a standard normal distribution. Furthermore, if S is substituted for σ, it can be shown that

$$T = \frac{Y^* - \widehat{Y^*}}{S \sqrt{1 + \dfrac{1}{n} + \dfrac{(x^* - \bar{x})^2}{S_{xx}}}}$$

possesses a Student's t distribution with $n - 2$ degrees of freedom. We will use this result to place a bound on the error of prediction; in doing so, we will construct a prediction interval for the random variable Y^*. The procedure employed will be similar to that used to construct the confidence intervals presented in the preceding chapters.

We begin by observing that

$$P(-t_{\alpha/2} < T < t_{\alpha/2}) = 1 - \alpha.$$

Substituting for T, we obtain

$$P\left[-t_{\alpha/2} < \frac{Y^* - \widehat{Y^*}}{S\sqrt{1 + \dfrac{1}{n} + \dfrac{(x^* - \overline{x})^2}{S_{xx}}}} < t_{\alpha/2} \right] = 1 - \alpha.$$

In other words, in repeated sampling the inequality within the brackets will hold with a probability equal to $(1 - \alpha)$. Furthermore, the inequality will continue to hold with the same probability if each term is multiplied by the same factor or if the same quantity is added to each term of the inequality. Multiply each term by

$$S\sqrt{1 + \frac{1}{n} + \frac{(x^* - \overline{x})^2}{S_{xx}}}$$

and then add $\widehat{Y^*}$ to each. The result,

$$P\left[\widehat{Y^*} - t_{\alpha/2}S\sqrt{1 + \frac{1}{n} + \frac{(x^* - \overline{x})^2}{S_{xx}}} < Y^* < \widehat{Y^*} \right.$$

$$\left. + t_{\alpha/2}S\sqrt{1 + \frac{1}{n} + \frac{(x^* - \overline{x})^2}{S_{xx}}} \right] = 1 - \alpha,$$

places an interval about $\widehat{Y^*}$ that constitutes a $(1 - \alpha)100\%$ prediction interval for Y^*.

A $(1 - \alpha)100\%$ Prediction Interval for Y when $x = x^*$

$$\hat{\beta}_0 + \hat{\beta}_1 x^* \pm t_{\alpha/2}S\sqrt{1 + \frac{1}{n} + \frac{(x^* - \overline{x})^2}{S_{xx}}}.$$

In attempting to place a bound on the error of predicting Y, we would expect the error to be less in absolute value than

$$t_{\alpha/2}S\sqrt{1 + \frac{1}{n} + \frac{(x^* - \overline{x})^2}{S_{xx}}}$$

with probability equal to $(1 - \alpha)$.

Notice that the length of a *confidence interval* for $E(Y)$ when $x = x^*$ is given by

$$2 \times t_{\alpha/2}S\sqrt{\frac{1}{n} + \frac{(x^* - \overline{x})^2}{S_{xx}}},$$

whereas the length of a *prediction interval for an actual value* of Y when $x = x^*$ is given by

$$2 \times t_{\alpha/2} S \sqrt{1 + \frac{1}{n} + \frac{(x^* - \bar{x})^2}{S_{xx}}}.$$

Thus, we observe that *prediction intervals* for the actual value of Y are longer than *confidence intervals* for $E(Y)$ if both are determined for the same value of x.

EXAMPLE 11.7 Suppose that the experiment that generated the data of Example 11.1 is to be run again with $x = 2$. Predict the particular value of Y with $1 - \alpha = .90$.

Solution From Example 11.1, we have

$$\hat{\beta}_0 = 1 \quad \text{and} \quad \hat{\beta}_1 = .7$$

so the predicted value of Y with $x = 2$ is

$$\hat{\beta}_0 + \hat{\beta}_1 x^* = 1 + (.7)(2) = 2.4.$$

Further, with $x^* = 2$,

$$\frac{1}{n} + \frac{(x^* - \bar{x})^2}{S_{xx}} = \frac{1}{5} + \frac{(2 - 0)^2}{10} = .6.$$

From Example 11.3, we know that $s = .606$. The $t_{.05}$ value with 3 degrees of freedom is 2.353. Thus the prediction interval becomes

$$\hat{\beta}_0 + \hat{\beta}_1 x^* \pm t_{\alpha/2} s \sqrt{1 + \frac{1}{n} + \frac{(x^* - \bar{x})^2}{S_{xx}}}$$

$$2.4 \pm (2.353)(.606) \sqrt{1 + .6}$$

$$2.4 \pm 1.804.$$

Figure 11.7 represents some hypothetical data and the estimated regression line fitted to those data that indicates the estimated value of $E(Y)$ when $x = 8$. Also shown on this graph are *confidence bands* for $E(Y)$. For each value of x, we computed

$$\hat{\beta}_0 + \hat{\beta}_1 x \pm t_{\alpha/2} S \sqrt{\frac{1}{n} + \frac{(x - \bar{x})^2}{S_{xx}}}.$$

Thus, for each value of x we obtain a confidence interval for $E(Y)$. The confidence interval for $E(Y)$ when $x = 7$ is displayed on the Y axis in the figure. Notice that the distance between the confidence bands is smallest when $x = \bar{x}$, as expected. Using the same approach, we computed prediction bands for the prediction of an actual Y value for each setting of x. As discussed earlier, for each fixed value of x, the prediction interval is wider than the corresponding confidence interval. The result is

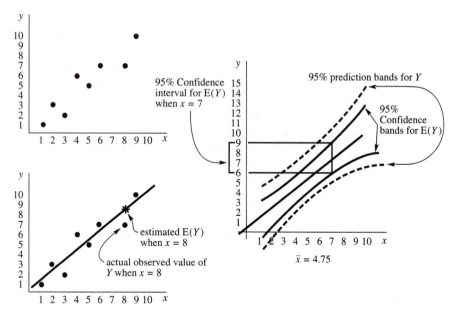

that the prediction bands fall uniformly farther from the prediction line than do the confidence bands. The prediction bands are also closest together when $x = \bar{x}$.

Exercises

11.38 Suppose that the model $Y = \beta_0 + \beta_1 x + \varepsilon$ is fit to the n data points $(y_1, x_1), \ldots, (y_n, x_n)$. At what value of x will the length of the prediction interval for Y be minimized?

11.39 Refer to Exercises 11.3 and 11.13. Use the data and model given there to construct a 95% prediction interval for the median sale price in 1980.

11.40 Refer to Exercise 11.39. Find a 95% prediction interval for the median sale price for the year 1981. Repeat for 1982. Would you feel comfortable in using this model and the data of Exercise 11.3 to predict the median sale price for the year 1988?

11.41 Refer to Exercises 11.4 and 11.14. Find a 95% prediction interval for a flow-through LC50 if the static LC50 is observed to be 12 parts per million. Compare the length of this interval to that of the interval found in Exercise 11.33.

11.42 Refer to Exercise 11.12. Find a 95% prediction interval for the potency of a 1-ounce portion of antibiotic stored at 65°F. Compare this interval to that calculated in Exercise 11.35.

11.43 Refer to Exercise 11.10. Find a 95% prediction interval for the proportion of survivors at time $x = .60$.

11.8 Correlation

The previous sections of this chapter dealt with modeling a response Y as a linear function of a nonrandom variable x, so that appropriate inferences could be made concerning the expected value of Y, or a future value of Y, for a given value of x. These models are useful in two quite different practical situations.

First, the variable x may be completely controlled by the experimenter. This occurs, for example, if x is the temperature setting and Y is the yield in a chemical experiment. Then x is merely the point at which the temperature dial is set when the experiment is run. Of course, x could vary from experiment to experiment, but it is under the complete control, practically speaking, of the experimenter. The linear model

$$Y = \beta_0 + \beta_1 x + \varepsilon$$

then implies that

$$E(Y) = \beta_0 + \beta_1 x$$

or that the average yield is a linear function of the temperature setting.

Second, the variable x may be an observed value of a random variable X. For example, we may want to relate the volume of usable timber Y in a tree to the circumference X of the base. If a functional relationship could be established, then in the future we could predict the amount of timber in any tree simply by measuring the circumference of the base. For this situation we use the model

$$Y = \beta_0 + \beta_1 x + \varepsilon$$

to imply that

$$E(Y|X = x) = \beta_0 + \beta_1 x.$$

That is, we are assuming that the conditional expectation of Y for a fixed value of X is a linear function of the x value. We generally assume that the vector random variable, (X, Y), has a bivariate normal distribution with $E(X) = \mu_X$, $E(Y) = \mu_Y$, $V(X) = \sigma_X^2$, $V(Y) = \sigma_Y^2$, and correlation coefficient ρ (see Section 5.10), in which case it can be shown that

$$E(Y|X = x) = \beta_0 + \beta_1 x, \quad \text{where} \quad \beta_1 = \frac{\sigma_Y}{\sigma_X}\rho.$$

The statistical theory for making inferences about the parameters β_0 and β_1 is exactly the same for both of these cases, but the differences in model interpretation should be kept in mind.

For the case where (X, Y) has a bivariate distribution, the experimenter may not always be interested in the linear relationship defining $E(Y|X)$. He or she may want to know only whether the random variables X and Y are *independent*. If (X, Y) has a bivariate normal distribution (see Section 5.10), then testing for independence is equivalent to testing whether the correlation coefficient ρ is equal to zero. Recall from Section 5.7 that ρ is positive if X and Y tend to increase together, and ρ is negative if Y decreases as X increases.

Let $(X_1, Y_1), (X_2, Y_2), \ldots, (X_n, Y_n)$ denote a random sample from a bivariate normal distribution. The maximum-likelihood estimator of ρ is given by the sample correlation coefficient:

$$r = \frac{\sum\limits_{i=1}^{n}(X_i - \overline{X})(Y_i - \overline{Y})}{\sqrt{\sum\limits_{i=1}^{n}(X_i - \overline{X})^2 \sum\limits_{i=1}^{n}(Y_i - \overline{Y})^2}}.$$

Notice that we can express r in terms of familiar quantities:

$$r = \frac{S_{xy}}{\sqrt{S_{xx}S_{yy}}} = \hat{\beta}_1 \sqrt{\frac{S_{xx}}{S_{yy}}}.$$

It follows that r and $\hat{\beta}_1$ have the same sign.

In the case where (X, Y) has a bivariate normal distribution, we have indicated that

$$E(Y|X = x) = \beta_0 + \beta_1 x, \quad \text{where} \quad \beta_1 = \frac{\sigma_Y}{\sigma_X}\rho.$$

Thus, for example, testing $H_0 : \rho = 0$ versus $H_a : \rho > 0$ is equivalent to testing $H_0 : \beta_1 = 0$ versus $H_a : \beta_1 > 0$. Similarly, $H_a : \rho < 0$ is equivalent to $H_a : \beta_1 < 0$, and $H_a : \rho \neq 0$ is equivalent to $H_a : \beta_1 \neq 0$. Tests for each of these sets of hypotheses involving β_1 can be based (see Section 11.5) on the statistic

$$t = \frac{\hat{\beta}_1 - 0}{S/\sqrt{S_{xx}}},$$

which possesses a t distribution with $n - 2$ degrees of freedom. In fact (see Exercise 11.47), this statistic can be rewritten in terms of r as follows:

$$t = \frac{r\sqrt{n-2}}{\sqrt{1-r^2}}.$$

Because the preceding two t statistics are algebraic equivalents, both possess the same distribution: the t distribution with $n - 2$ degrees of freedom.

It would seem natural to use r as a test statistic to test more general hypotheses about ρ, but the probability distribution for r is difficult to obtain. The difficulty can be overcome, for moderately large samples, by using the fact that $(1/2) \ln[(1 + r)/(1 - r)]$ is approximately normally distributed with mean $(1/2) \ln[(1 + \rho)/(1 - \rho)]$ and variance $1/(n - 3)$. Thus, for testing the hypothesis $H_0 : \rho = \rho_0$, we can employ a Z test in which

$$Z = \frac{\left(\frac{1}{2}\right) \ln \left(\frac{1+r}{1-r}\right) - \left(\frac{1}{2}\right) \ln \left(\frac{1+\rho_0}{1-\rho_0}\right)}{\frac{1}{\sqrt{n-3}}}.$$

If α is the desired probability of a type I error, the form of the rejection region depends upon the alternative hypothesis. The various alternatives of most frequent interest and the corresponding rejection regions are as follows:

$$H_a: \rho > \rho_0 \qquad \text{RR: } z > z_\alpha$$

$$H_a: \rho < \rho_0 \qquad \text{RR: } z < -z_\alpha$$

$$H_a: \rho \neq \rho_0 \qquad \text{RR: } |z| > z_{\alpha/2}.$$

We illustrate with an example.

EXAMPLE **11.8**   The data in Table 11.3 represent a sample of mathematics achievement test scores and calculus grades for ten independently selected college freshmen. From this evidence, would you say that the achievement test scores and calculus grades are independent? Use $\alpha = .05$. Identify the corresponding attained significance level.

Solution   We state as the null hypothesis that X and Y are independent; or, assuming that (X, Y) has a bivariate normal distribution, we test $H_0 : \rho = 0$ versus $H_a : \rho \neq 0$. Because we are focusing on $\rho = 0$, the test can be based on the statistic $t = (r\sqrt{n-2})/\sqrt{1-r^2}$. Denoting achievement test scores by x and calculus grades by y, we calculate

Table 11.3. Data for Example 11.8

Student	Mathematics Achievement Test Score	Final Calculus Grade
1	39	65
2	43	78
3	21	52
4	64	82
5	57	92
6	47	89
7	28	73
8	75	98
9	34	56
10	52	75

$$\sum x_i = 460 \qquad \sum x_i^2 = 23{,}634 \qquad S_{xx} = 2{,}474$$

$$\sum y_i = 760 \qquad \sum y_i^2 = 59{,}816 \qquad S_{yy} = 2{,}056$$

$$\sum x_i y_i = 36{,}854 \qquad\qquad\qquad S_{xy} = 1{,}894.$$

Thus,

$$r = \frac{S_{xy}}{\sqrt{S_{xx}S_{yy}}} = \frac{1894}{\sqrt{(2474)(2056)}} = .8398.$$

The value of the test statistic is

$$t = \frac{r\sqrt{n-2}}{\sqrt{1-r^2}} = \frac{(.8398)\sqrt{8}}{\sqrt{1-.7053}}$$

$$= 4.375.$$

Because t is based on $n - 2 = 8$ degrees of freedom, $t_{\alpha/2} = t_{.025} = 2.306$; and the observed value of our test statistic lies in the rejection region. Thus, the evidence strongly suggests that achievement test scores and calculus grades are dependent. Notice that $\alpha = .05$ is the probability that our test statistic will fall in the rejection region when H_0 is true. Hence, we are fairly confident that we have made a correct decision.

Because we are implementing a two-tailed test, p-value $= 2P(t > 4.375)$. From the values contained in Table 5, Appendix III, it follows that $P(t > 4.375) < .005$. Thus, p-value $< 2(.005) = .010$, and for any value of α greater than .01 (including $\alpha = .05$, as used in the initial part of this analysis), we would conclude that $\rho \neq 0$.

Exercises

11.44   The accompanying table gives the peak power load for a power plant and the daily high temperature for a random sample of 10 days. Test the hypothesis that the population correlation coefficient ρ between peak power load and high temperature is zero versus the alternative that it is positive. Use $\alpha = .05$. Identify the attained significance level.

Day	High Temperature	Peak Load
1	95°F	214
2	82°F	152
3	90°F	156
4	81°F	129
5	99°F	254
6	100°F	266
7	93°F	210
8	95°F	204
9	93°F	213
10	87°F	150

11.45   In Exercise 11.4 both the flow-through and static LC50 values could be considered random variables. Using the data of Exercise 11.4, test to see whether the correlation between static and flow-through values significantly differs from zero. Use $\alpha = .01$. Identify the associated p-value.

11.46   Suppose that we seek an intuitive estimator for

$$\rho = \frac{\text{Cov}(X, Y)}{\sigma_X \sigma_Y}.$$

a   The method of moments estimator of $\text{Cov}(X, Y) = E[(X - \mu_X)(Y - \mu_Y)]$ is

$$\widehat{\text{Cov}(X, Y)} = \frac{1}{n} \sum_{i=1}^{n} (X_i - \overline{X})(Y_i - \overline{Y}).$$

Show that the method of moments estimators for the standard deviations of X and Y are

$$\hat{\sigma}_X = \sqrt{\frac{1}{n} \sum_{i=1}^{n} (X_i - \overline{X})^2} \quad \text{and} \quad \hat{\sigma}_Y = \sqrt{\frac{1}{n} \sum_{i=1}^{n} (Y_i - \overline{Y})^2}.$$

b   Substitute the estimators for their respective parameters in the definition of ρ, and obtain the method of moments estimator for ρ. Compare your estimator to r, the maximum-likelihood estimator for ρ presented in this section.

11.47   Consider the simple linear regression model based on normal theory. If we are interested in testing $H_0 \colon \beta_1 = 0$ versus various alternatives, the statistic

$$T = \frac{\hat{\beta}_1 - 0}{S/\sqrt{S_{xx}}}$$

possesses a t distribution with $n - 2$ degrees of freedom if the null hypothesis is true. Show that the equation for T can also be written as

$$T = \frac{r\sqrt{n-2}}{\sqrt{1-r^2}}.$$

*11.48   Refer to Exercises 11.4 and 11.45. Suppose that independent tests, with the same toxicants and species but in a different laboratory, showed $r = .85$ with $n = 20$. Test the hypothesis that the two correlation coefficients between static and flow-through LC50 measurements are equal. Use $\alpha = .05$.

11.9   Some Practical Examples

In this section we present two examples illustrating the applicability of previously developed techniques to real data. Most of the methods are illustrated somewhere in the course of the discussions. We make no attempt to implement every method for each example.

Table 11.4. Data for Example 11.9

Water/Cement Ratio	Strength (100 ft/lb)
1.21	1.302
1.29	1.231
1.37	1.061
1.46	1.040
1.62	.803
1.79	.711

Source: Data adapted from Hamid Behbahani, "Econocrete—Design and Properties" (Ph.D. thesis, University of Florida, 1977), p. 95.

EXAMPLE 11.9   In his Ph.D. thesis, H. Behbahani examined the effect of varying the water/cement ratio on the strength of concrete that had been aged 28 days. For concrete with a cement content of 200 pounds per cubic yard, he obtained the data presented in Table 11.4. Let Y denote the strength and x denote the water/cement ratio.

a   Fit the model $E(Y) = \beta_0 + \beta_1 x$.

b   Test $H_0: \beta_1 = 0$ versus $H_a: \beta_1 < 0$ with $\alpha = .05$. (Notice that, if H_0 is rejected, we conclude that $\beta_1 < 0$ and that the strength tends to decrease with an increase in water/cement ratio.) Identify the corresponding attained significance level.

c   Find a 90% confidence interval for the expected strength of concrete when the water/cement ratio is 1.5. What will happen to the confidence interval if we try to estimate mean strengths for water/cement ratios of .3 or 2.7?

Solution   **a**   Using the formulas developed in Section 11.3, we obtain

$$S_{xy} = \sum_{i=1}^{n} x_i y_i - \frac{1}{n} \sum_{i=1}^{n} x_i \sum_{i=1}^{n} y_i = 8.709 - \frac{1}{6}(8.74)(6.148) = -.247$$

$$S_{xx} = \sum_{i=1}^{n} x_i^2 - \frac{1}{n} \left(\sum_{i=1}^{n} x_i \right)^2 = 12.965 - \frac{1}{6}(8.74)^2 = .234$$

$$S_{yy} = \sum_{i=1}^{n} y_i^2 - \frac{1}{n} \left(\sum_{i=1}^{n} y_i \right)^2 = 6.569 - \frac{1}{6}(6.148)^2 = .269$$

$$\hat{\beta}_1 = \frac{S_{xy}}{S_{xx}} = \frac{-0.247}{0.234} = -1.056$$

and

$$\hat{\beta}_0 = \bar{y} - \hat{\beta}_1 \bar{x} = \frac{6.148}{6} - (-1.056)\left(\frac{8.74}{6}\right) = 2.563.$$

(Throughout this example, all calculations are carried out to three decimal places.)

Thus the straight-line model that best fits the data is

$$\hat{y} = 2.563 - 1.056x.$$

b Because we desire to test whether there is evidence that $\beta_1 < 0$ with $\alpha = .05$, the appropriate test statistic is

$$t = \frac{\hat{\beta}_1 - 0}{S\sqrt{c_{11}}} \quad \text{or} \quad t = \frac{\hat{\beta}_1 - 0}{S\sqrt{\dfrac{1}{S_{xx}}}}.$$

For a simple linear regression model,

$$\text{SSE} = S_{yy} - \hat{\beta}_1 S_{xy} = .269 - (-1.056)(-.247) = .008$$

and hence

$$s = \sqrt{s^2} = \sqrt{\frac{\text{SSE}}{n-2}} = \sqrt{\frac{.008}{4}} = .045.$$

Thus, the value of the appropriate test statistic for testing $H_0 : \beta_1 = 0$ versus $H_a : \beta_1 < 0$ is

$$t = \frac{-1.056 - 0}{.045\sqrt{1/(.234)}} = -11.355.$$

Because this statistic is based on $n - 2 = 4$ degrees of freedom and the appropriate rejection region is $t < -t_{.05} = -2.132$, we reject H_0 in favor of H_a at the $\alpha = .05$ level of significance. Because the appropriate test is a lower-tail test, p-value $= P(t < -11.355)$, where t has a t distribution with 4 degrees of freedom. Thus, p-value $< .005$. In fact, more extensive tables of the t distribution indicate that the p-value is considerably less than .005. Hence, for most commonly used values of α, we conclude that there is evidence to indicate that strength decreases with an increase in the water/cement ratio *on the region where the experiment was conducted.* From a practical point of view, the water/cement ratio must be large enough to moisten the cement, sand, and other components that make up concrete. But if the water/cement ratio gets too large, the concrete will be useless.

c Because the model we are using is a simple linear regression model, the confidence interval can be obtained from the formula

$$\hat{\beta}_0 + \hat{\beta}_1 x^* \pm t_{\alpha/2} S\sqrt{\frac{1}{n} + \frac{(x^* - \bar{x})^2}{S_{xx}}}.$$

We want a confidence interval when $x = 1.5$; therefore, $x^* = 1.5$ and

$$\hat{\beta}_0 + \hat{\beta}_1 x^* = 2.563 - (1.056)(1.5) = .979.$$

Using calculations from (a) and (b), we obtain the desired 90% confidence interval:

$$.979 \pm (2.132)(.045)\sqrt{\frac{1}{6} + \frac{(1.5 - 1.457)^2}{.234}}$$

or

$$(.938, \ 1.020).$$

Thus we would estimate the mean strength of concrete with a water/cement ratio of 1.5 to be between .938 and 1.020.

We can see from the variance expression that the confidence interval gets wider as x^* gets farther from $\bar{x} = 1.457$. Also, the values $x^* = .3$ and $x^* = 2.7$ are far from the values that were used in the experiment. Considerable caution should be used before constructing a confidence interval for $E(Y)$ when the values of x^* are far removed from the experimental region. Water/cement ratios of .3 and 2.7 would probably yield concrete that is utterly useless!

In many real-world situations, the most appropriate deterministic component of a model is not linear. For example, many populations of plants or animals tend to grow at exponential rates. If Y_t denotes the size of the population at time t, we might employ the model

$$E(Y_t) = \alpha_0 e^{\alpha_1 t}.$$

Although this expression is not linear in the parameters α_0 and α_1, it can be linearized by taking natural logarithms. If Y_t can be observed for various values of t, we can write the model as

$$\ln Y_t = \ln \alpha_0 + \alpha_1 t + \varepsilon$$

and estimate $\ln \alpha_0$ and α_1 by least squares.

Other basic models can also be linearized. In the biological sciences it is sometimes possible to relate the weight (or volume) of an organism to some linear measurement such as length (or weight). If W denotes weight and l length, the model

$$E(W) = \alpha_0 l^{\alpha_1}$$

for unknown α_0 and α_1 is often applicable. (This model is known as an *allometric equation*.) If we want to relate the weight of randomly selected organisms to observable fixed lengths, we can take logarithms and obtain the linear model

$$\ln W = \ln \alpha_0 + \alpha_1 \ln l + \varepsilon = \beta_0 + \beta_1 x + \varepsilon$$

with $x = \ln l$. Then $\beta_0 = \ln \alpha_0$ and $\beta_1 = \alpha_1$ can be estimated by the method of least squares. The following example illustrates such a model.

Table 11.5. Data for Example 11.10

Alligator	$x = \ln l$	$y = \ln W$
1	3.87	4.87
2	3.61	3.93
3	4.33	6.46
4	3.43	3.33
5	3.81	4.38
6	3.83	4.70
7	3.46	3.50
8	3.76	4.50
9	3.50	3.58
10	3.58	3.64
11	4.19	5.90
12	3.78	4.43
13	3.71	4.38
14	3.73	4.42
15	3.78	4.25

EXAMPLE **11.10** In the data set of Table 11.5, W denotes the weight (in pounds) and l the length (in inches) for 15 alligators captured in central Florida. Because l is easier to observe (perhaps from a photograph) than W for alligators in their natural habitat, we want to construct a model relating weight to length. Such a model can then be used to predict the weights of alligators of specified lengths. Fit the model

$$\ln W = \ln \alpha_0 + \alpha_1 \ln l + \varepsilon = \beta_0 + \beta_1 x + \varepsilon$$

to the data. Find a 90% prediction interval for W if $\ln l$ is observed to be 4.00.

Solution We begin by calculating the quantities that have routine application throughout our solution:

$$S_{xy} = \sum_{i=1}^{n} x_i y_i - \frac{1}{n} \sum_{i=1}^{n} x_i \sum_{i=1}^{n} y_i = 251.9757 - \frac{1}{15}(56.37)(66.27) = 2.933$$

$$S_{xx} = \sum_{i=1}^{n} x_i^2 - \frac{1}{n}\left(\sum_{i=1}^{n} x_i\right)^2 = 212.6933 - \frac{1}{15}(56.37)^2 = 0.8548$$

$$S_{yy} = \sum_{i=1}^{n} y_i^2 - \frac{1}{n}\left(\sum_{i=1}^{n} y_i\right)^2 = 303.0409 - \frac{1}{15}(66.27)^2 = 10.26$$

$$\hat{\beta}_1 = \frac{S_{xy}}{S_{xx}} = \frac{2.933}{0.8548} = 3.4312$$

and

$$\hat{\beta}_0 = \bar{y} - \hat{\beta}_1 \bar{x} = \frac{66.27}{15} - (3.4312)\left(\frac{56.37}{15}\right) = -8.476.$$

We can now estimate α_0 by

$$\hat{\alpha}_0 = e^{\hat{\beta}_0} = e^{-8.476} = .0002$$

and α_1 by $\hat{\alpha}_1 = \hat{\beta}_1$ to arrive at the estimated model

$$\hat{w} = \hat{\alpha}_0 l^{\hat{\alpha}_1} = (.0002)l^{3.4312}.$$

(In many cases α_1 will be close to 3 because weight or volume is generally proportional to the cube of a linear measurement.)

For these data, SSE $= .1963$, $n = 15$, and $s = \sqrt{\text{SSE}/(n-2)} = .123$. The calculations leading to these numerical values are completely analogous to the calculations of Example 11.9.

To find a prediction interval for W where $x = \ln l = 4$, we must first form a prediction interval for $Y = \ln W$. As before, the prediction interval is

$$\hat{\beta}_0 + \hat{\beta}_1 x^* \pm t_{.05} S \sqrt{1 + \frac{1}{n} + \frac{(x^* - \bar{x})^2}{S_{xx}}},$$

where $t_{.05}$ is based on $n - 2 = 13$ degrees of freedom. Therefore, $t_{.05} = 1.771$ and the 90% prediction interval for $Y = \ln W$ is

$$-8.476 + 3.4312(4) \pm 1.771(.123)\sqrt{1 + \frac{1}{15} + \frac{(4 - 3.758)^2}{.8548}}$$

$$5.2488 \pm .2321$$

or

$$(5.0167, \ 5.4809).$$

Because $\hat{Y} = \ln \hat{W}$, we can predict W by $e^{\hat{Y}} = e^{5.2488} = 190.3377$. The observed 90% prediction interval for W is

$$\left(e^{5.0167}, \ e^{5.4809}\right), \quad \text{or} \quad (150.9125, \ 240.0627).$$

When $x = \ln l = 4$, then $l = e^4 = 54.598$. Thus, for an alligator of length 54.598 inches, we predict that its weight will fall between 150.91 and 240.06 pounds. The relatively narrow interval on the natural logarithm scale becomes a rather wide interval when transformed to the original scale.

The data presented and analyzed in this section are examples from real experiments; methods developed in previous sections of this chapter were applied to produce answers of actual interest to experimenters. Through Example 11.10, we have demonstrated how the theory of linear models sometimes can be applied after transformation of the scale of the original variables. Of course, not all models can be linearized, but numerous techniques for nonlinear least squares estimation are available.

Exercises

11.49 Refer to Example 11.9. Find a 90% prediction interval for the strength of concrete when the water/cement ratio is 1.5.

11.50 Refer to Example 11.10. Calculate the correlation coefficient r between the variables $\ln W$ and $\ln l$.

*11.51 It is well known that large bodies of water have a mitigating effect on the temperature of the surrounding land masses. On a cold night in central Florida, temperatures were recorded at equal distances along a transect running downwind from a large lake. The resulting data are given in the accompanying table. Notice that the temperatures drop rapidly and then level off as we move away from the lake. The suggested model for these data is

$$E(Y) = \alpha_0 e^{-\alpha_1 x}.$$

a Linearize the model and estimate the parameters by the method of least squares.

b Find a 90% confidence interval for α_0. Give an interpretation of the result.

Site (x)	Temperature (y)
1	37.00°F
2	36.25°F
3	35.41°F
4	34.92°F
5	34.52°F
6	34.45°F
7	34.40°F
8	34.00°F
9	33.62°F
10	33.90°F

*11.52 Refer to Exercise 11.10. One model proposed for these data on the proportion of survivors of thermal pollution is

$$E(Y) = \exp(-\alpha_0 x^{\alpha_1}).$$

Linearize this model and estimate the parameters by using the method of least squares and the data of Exercise 11.10. (Omit the observation with $y = 1.00$.)

*11.53 In the biological and physical sciences, a common model for proportional growth over time is

$$E(Y) = 1 - e^{-\beta t},$$

where Y denotes a proportion and t denotes time. Y might represent the proportion of eggs that hatch, the proportion of an organism filled with diseased cells, the proportion of patients reacting to a drug, or the proportion of a liquid that has passed through a porous medium. With n observations of the form (y_i, t_i), outline how you would estimate and then form a confidence interval for β.

11.10 Fitting the Linear Model by Using Matrices

Thus far in this chapter, we have dealt almost exclusively with *simple* linear regression models that have enabled us to express our derivations and results by using ordinary algebraic expressions. The only practical way to handle analogous derivations and results for *multiple* linear regression models is through the use of matrix algebra. In this section we use matrices to reexpress some of our previous results and to extend these results to the multiple linear regression model.

Suppose that we have the linear model

$$Y = \beta_0 + \beta_1 x_1 + \cdots + \beta_k x_k + \varepsilon$$

and we make n independent observations, $y_1, y_2, \ldots, y_n$, on Y. We can write the observation y_i as

$$y_i = \beta_0 + \beta_1 x_{i1} + \beta_2 x_{i2} + \cdots + \beta_k x_{ik} + \varepsilon_i$$

where x_{ij} is the setting of the jth independent variable for the ith observation, $i = 1, 2, \ldots, n$. We now define the following matrices, with $x_0 = 1$:

$$\mathbf{Y} = \begin{bmatrix} y_1 \\ y_2 \\ \vdots \\ y_n \end{bmatrix} \qquad \mathbf{X} = \begin{bmatrix} x_0 & x_{11} & x_{12} & \cdots & x_{1k} \\ x_0 & x_{21} & x_{22} & \cdots & x_{2k} \\ \vdots & \vdots & \vdots & & \vdots \\ x_0 & x_{n1} & x_{n2} & \cdots & x_{nk} \end{bmatrix}$$

$$\beta = \begin{bmatrix} \beta_0 \\ \beta_1 \\ \vdots \\ \beta_k \end{bmatrix} \qquad \epsilon = \begin{bmatrix} \varepsilon_1 \\ \varepsilon_2 \\ \vdots \\ \varepsilon_n \end{bmatrix}.$$

Thus the n equations representing y_i as a function of the x's, β's, and ε's can be simultaneously written as

$$\mathbf{Y} = \mathbf{X}\beta + \epsilon.$$

(See Appendix I for a discussion of matrix operations.)

For n observations from a simple linear model of the form

$$Y = \beta_0 + \beta_1 x + \varepsilon,$$

we have

$$\mathbf{Y} = \begin{bmatrix} y_1 \\ y_2 \\ \vdots \\ y_n \end{bmatrix}, \qquad \mathbf{X} = \begin{bmatrix} 1 & x_1 \\ 1 & x_2 \\ \vdots & \vdots \\ 1 & x_n \end{bmatrix}, \qquad \epsilon = \begin{bmatrix} \varepsilon_1 \\ \varepsilon_2 \\ \vdots \\ \varepsilon_n \end{bmatrix}, \qquad \beta = \begin{bmatrix} \beta_0 \\ \beta_1 \end{bmatrix}.$$

(We suppress the second subscript on x because only one x variable is involved.) The least squares equations for β_0 and β_1 were given in Section 11.3 as

$$n\hat{\beta}_0 + \hat{\beta}_1 \sum_{i=1}^{n} x_i = \sum_{i=1}^{n} y_i$$

$$\hat{\beta}_0 \sum_{i=1}^{n} x_i + \hat{\beta}_1 \sum_{i=1}^{n} x_i^2 = \sum_{i=1}^{n} x_i y_i.$$

Because

$$\mathbf{X'X} = \begin{bmatrix} 1 & 1 & \cdots & 1 \\ x_1 & x_2 & \cdots & x_n \end{bmatrix} \begin{bmatrix} 1 & x_1 \\ 1 & x_2 \\ \vdots & \vdots \\ 1 & x_n \end{bmatrix} = \begin{bmatrix} n & \sum_{i=1}^{n} x_i \\ \sum_{i=1}^{n} x_i & \sum_{i=1}^{n} x_i^2 \end{bmatrix}$$

and

$$\mathbf{X'Y} = \begin{bmatrix} \sum_{i=1}^{n} y_i \\ \sum_{i=1}^{n} x_i y_i \end{bmatrix}$$

we see that the least squares equations are given by

$$(\mathbf{X'X})\hat{\boldsymbol{\beta}} = \mathbf{X'Y}$$

where

$$\hat{\boldsymbol{\beta}} = \begin{bmatrix} \hat{\beta}_0 \\ \hat{\beta}_1 \end{bmatrix}.$$

Hence

$$\hat{\boldsymbol{\beta}} = (\mathbf{X'X})^{-1}\mathbf{X'Y}.$$

Although we have shown that this result holds only for a simple case, it can be shown that, in general, the least squares equations and solutions presented in matrix notation are as follows.

Least Squares Equations and Solutions for a General Linear Model

Equations: $(\mathbf{X'X})\hat{\boldsymbol{\beta}} = \mathbf{X'Y}$

Solutions: $\hat{\boldsymbol{\beta}} = (\mathbf{X'X})^{-1}\mathbf{X'Y}$

EXAMPLE 11.11   Solve Example 11.1 by using matrix operations.

Solution   From the data given in Example 11.1, we see that

$$
Y = \begin{bmatrix} 0 \\ 0 \\ 1 \\ 1 \\ 3 \end{bmatrix}, \quad \text{and} \quad X = \begin{matrix} x_0 \quad x_1 \\ \begin{bmatrix} 1 & -2 \\ 1 & -1 \\ 1 & 0 \\ 1 & 1 \\ 1 & 2 \end{bmatrix} \end{matrix}.
$$

It follows that

$$
X'X = \begin{bmatrix} 5 & 0 \\ 0 & 10 \end{bmatrix}, \quad X'Y = \begin{bmatrix} 5 \\ 7 \end{bmatrix}, \quad (X'X)^{-1} = \begin{bmatrix} 1/5 & 0 \\ 0 & 1/10 \end{bmatrix}.
$$

Thus,

$$
\hat{\beta} = (X'X)^{-1}X'Y = \begin{bmatrix} 1/5 & 0 \\ 0 & 1/10 \end{bmatrix}\begin{bmatrix} 5 \\ 7 \end{bmatrix} = \begin{bmatrix} 1 \\ .7 \end{bmatrix},
$$

or $\hat{\beta}_0 = 1$ and $\hat{\beta}_1 = .7$. Thus,

$$
\hat{y} = 1 + .7x,
$$

just as in Example 11.1.

EXAMPLE 11.12   Fit a parabola to the data of Example 11.1 using the model

$$
Y = \beta_0 + \beta_1 x + \beta_2 x^2 + \varepsilon.
$$

Solution   The X matrix for this example differs from that of Example 11.11 only by the addition of a third column corresponding to x^2. (Notice that $x_1 = x$, $x_2 = x^2$, and $k = 2$ in the notation of the general linear model.) Thus,

$$
Y = \begin{bmatrix} 0 \\ 0 \\ 1 \\ 1 \\ 3 \end{bmatrix}, \quad X = \begin{matrix} x_0 \quad x \quad x^2 \\ \begin{bmatrix} 1 & -2 & 4 \\ 1 & -1 & 1 \\ 1 & 0 & 0 \\ 1 & 1 & 1 \\ 1 & 2 & 4 \end{bmatrix} \end{matrix}.
$$

(The three variables, x_0, x, and x^2, are shown above their respective columns in the X matrix.) Thus, for the first measurement, $y = 0$, $x_0 = 1$, $x = -2$, and $x^2 = 4$; and for the second measurement, $y = 0$, $x_0 = 1$, $x = -1$, and $x^2 = 1$. Succeeding rows of the Y and X matrices are obtained in a similar manner.

The matrix products, $\mathbf{X'X}$ and $\mathbf{X'Y}$, are

$$\mathbf{X'X} = \begin{bmatrix} 1 & 1 & 1 & 1 & 1 \\ -2 & -1 & 0 & 1 & 2 \\ 4 & 1 & 0 & 1 & 4 \end{bmatrix} \begin{bmatrix} 1 & -2 & 4 \\ 1 & -1 & 1 \\ 1 & 0 & 0 \\ 1 & 1 & 1 \\ 1 & 2 & 4 \end{bmatrix} = \begin{bmatrix} 5 & 0 & 10 \\ 0 & 10 & 0 \\ 10 & 0 & 34 \end{bmatrix}$$

$$\mathbf{X'Y} = \begin{bmatrix} 1 & 1 & 1 & 1 & 1 \\ -2 & -1 & 0 & 1 & 2 \\ 4 & 1 & 0 & 1 & 4 \end{bmatrix} \begin{bmatrix} 0 \\ 0 \\ 1 \\ 1 \\ 3 \end{bmatrix} = \begin{bmatrix} 5 \\ 7 \\ 13 \end{bmatrix}.$$

We omit the process of inverting $\mathbf{X'X}$ and simply state that the inverse matrix is equal to

$$(\mathbf{X'X})^{-1} = \begin{bmatrix} 17/35 & 0 & -1/7 \\ 0 & 1/10 & 0 \\ -1/7 & 0 & 1/14 \end{bmatrix}.$$

[You may verify that $(\mathbf{X'X})^{-1}\mathbf{X'X} = \mathbf{I}$.]
   Finally,

$$\hat{\beta} = (\mathbf{X'X})^{-1}\mathbf{X'Y}$$

$$= \begin{bmatrix} 17/35 & 0 & -1/7 \\ 0 & 1/10 & 0 \\ -1/7 & 0 & 1/14 \end{bmatrix} \begin{bmatrix} 5 \\ 7 \\ 13 \end{bmatrix} = \begin{bmatrix} 4/7 \\ 7/10 \\ 3/14 \end{bmatrix} \approx \begin{bmatrix} .571 \\ .700 \\ .214 \end{bmatrix}.$$

Hence, $\hat{\beta}_0 = .571$, $\hat{\beta}_1 = .7$, and $\hat{\beta}_2 = .214$, and the prediction equation is

$$\hat{y} = .571 + .7x + .214x^2.$$

A graph of this parabola on Figure 11.6 will indicate a good fit to the data points.

The expressions for $V(\hat{\beta}_0)$, $V(\hat{\beta}_1)$, $\text{Cov}(\hat{\beta}_0, \hat{\beta}_1)$, and SSE that we derived in Section 11.4 for the *simple* linear regression model can be expressed conveniently in terms of matrices. We have seen that for the linear model $Y = \beta_0 + \beta_1 x + \varepsilon$, $\mathbf{X'X}$ is given by

$$\mathbf{X'X} = \begin{bmatrix} n & \sum x_i \\ \sum x_i & \sum x_i^2 \end{bmatrix}.$$

It can be shown that

$$(\mathbf{X'X})^{-1} = \begin{bmatrix} \dfrac{\sum x_i^2}{nS_{xx}} & -\dfrac{\overline{x}}{S_{xx}} \\ -\dfrac{\overline{x}}{S_{xx}} & \dfrac{1}{S_{xx}} \end{bmatrix} = \begin{bmatrix} c_{00} & c_{01} \\ c_{10} & c_{11} \end{bmatrix}.$$

By checking the variances and covariances derived in Section 11.4, you can see that

$$V(\hat{\beta}_i) = c_{ii}\sigma^2, \qquad i = 0, 1$$

and

$$\text{Cov}(\hat{\beta}_0, \hat{\beta}_1) = c_{01}\sigma^2 = c_{10}\sigma^2.$$

Recall that an unbiased estimator for σ^2, the variance of the error term ε, is given by $S^2 = \text{SSE}/(n-2)$. A bit of matrix algebra will show that $\text{SSE} = \sum(y_i - \hat{y}_i)^2$ can be expressed as

$$\text{SSE} = \mathbf{Y'Y} - \hat{\boldsymbol{\beta}}'\mathbf{X'Y}.$$

(Notice that $\mathbf{Y'Y} = \sum_i Y_i^2$.)

EXAMPLE 11.13   Find the variances of the estimators $\hat{\beta}_0$ and $\hat{\beta}_1$ for Example 11.11, and provide an estimator for σ^2.

Solution   In Example 11.11, we found that

$$(\mathbf{X'X})^{-1} = \begin{bmatrix} 1/5 & 0 \\ 0 & 1/10 \end{bmatrix}.$$

Hence,

$$V(\hat{\beta}_0) = c_{00}\sigma^2 = (1/5)\sigma^2$$

and

$$V(\hat{\beta}_1) = c_{11}\sigma^2 = (1/10)\sigma^2.$$

As before, $\text{Cov}(\hat{\beta}_0, \hat{\beta}_1) = 0$ in this case, because $\sum x_i = 0$. For these data,

$$\mathbf{Y} = \begin{bmatrix} 0 \\ 0 \\ 1 \\ 1 \\ 3 \end{bmatrix}, \qquad \mathbf{X} = \begin{bmatrix} 1 & -2 \\ 1 & -1 \\ 1 & 0 \\ 1 & 1 \\ 1 & 2 \end{bmatrix}, \qquad \hat{\boldsymbol{\beta}} = \begin{bmatrix} 1 \\ .7 \end{bmatrix}.$$

Hence,

$$\text{SSE} = \mathbf{Y'Y} - \hat{\boldsymbol{\beta}}'\mathbf{X'Y}$$

$$= [0 \quad 0 \quad 1 \quad 1 \quad 3] \begin{bmatrix} 0 \\ 0 \\ 1 \\ 1 \\ 3 \end{bmatrix} - [1 \quad .7] \begin{bmatrix} 1 & 1 & 1 & 1 & 1 \\ -2 & -1 & 0 & 1 & 2 \end{bmatrix} \begin{bmatrix} 0 \\ 0 \\ 1 \\ 1 \\ 3 \end{bmatrix}$$

$$= 11 - [1 \quad .7] \begin{bmatrix} 5 \\ 7 \end{bmatrix} = 11 - 9.9 = 1.1.$$

Then

$$s^2 = \frac{\text{SSE}}{n-2} = \frac{1.1}{5-2} = \frac{1.1}{3} = .367.$$

Notice the agreement with the results that were obtained in Examples 11.2 and 11.3.

Exercises

11.54 Refer to Exercise 11.1. Fit the model suggested there by use of matrices.

11.55 Use the matrix approach to fit a straight line to the data in the accompanying table, plot the points, and then sketch the fitted line as a check on the calculations. The data points are the same as for Exercises 11.1 and 11.54 except that they are translated one unit in the positive direction along the x axis. What effect does symmetric spacing of the x values about $x = 0$ have on the form of the $(\mathbf{X'X})$ matrix and the resulting calculations?

y	x
3	−1
2	0
1	1
1	2
.5	3

11.56 Fit the quadratic model $Y = \beta_0 + \beta_1 x + \beta_2 x^2 + \varepsilon$ to the data points in the following table. Plot the points and sketch the fitted parabola as a check on the calculations.

y	x
1	−3
0	−2
0	−1
−1	0
−1	1
0	2
0	3

11.57 Florida morbidity statistics for the decade ending in 1976 show that infectious hepatitis had the incidence rates shown in the accompanying table (in cases per 100,000 population).

x	y
1967	10.5
1968	18.5
1969	22.6
1970	27.2
1971	31.2
1972	33.0
1973	44.9
1974	49.4
1975	35.0
1976	27.6

a Letting Y denote incidence rate and x denote the coded year (-9 for 1967, -7 for 1968, through 9 for 1976), fit the model $Y = \beta_0 + \beta_1 x + \varepsilon$.

b For the same data, fit the model $Y = \beta_0 + \beta_1 x + \beta_2 x^2 + \varepsilon$.

11.58 a Calculate SSE and S^2 for Exercise 11.2. Use the matrix approach.

b Fit the model suggested in Exercise 11.2 for the relationship between audited values and book values by using matrices. We can simplify the computations by defining

$$x_i^* = x_i - \overline{x}$$

and fitting the model $Y = \beta_0^* + \beta_1^* x^* + \varepsilon$. Fit this latter model and calculate SSE. Compare your answer with the SSE calculation in (a).

11.11 Linear Functions of the Model Parameters: Multiple Linear Regression

All the theoretical results of Section 11.4 can be extended to the multiple linear regression model,

$$Y_i = \beta_0 + \beta_1 x_{i1} + \cdots + \beta_k x_{ik} + \varepsilon_i, \qquad i = 1, 2, \ldots, n.$$

Suppose that $\varepsilon_1, \varepsilon_2, \ldots, \varepsilon_n$ are independent random variables with $E(\varepsilon_i) = 0$ and $V(\varepsilon_i) = \sigma^2$. Then the least squares estimators are given by

$$\hat{\beta} = (\mathbf{X}'\mathbf{X})^{-1}\mathbf{X}'\mathbf{Y}$$

provided that $(\mathbf{X}'\mathbf{X})^{-1}$ exists. The properties of these estimators are as follows (proof omitted).

Properties of the Least Squares Estimators—Multiple Linear Regression

1. $E(\hat{\beta}_i) = \beta_i, i = 0, 1, \ldots, k.$
2. $V(\hat{\beta}_i) = c_{ii}\sigma^2$, where c_{ij} (or, in this special case, c_{ii}) is the element in row i and column j (here, column i) of $(\mathbf{X'X})^{-1}$. (Recall that this matrix has a row and column numbered 0.)
3. $\text{Cov}(\hat{\beta}_i, \hat{\beta}_j) = c_{ij}\sigma^2.$
4. An unbiased estimator of σ^2 is $S^2 = \text{SSE}/[n - (k + 1)]$, where $\text{SSE} = \mathbf{Y'Y} - \hat{\boldsymbol{\beta}}'\mathbf{X'Y}$. (Notice that there are $k + 1$ unknown β_i values in the model.)

   If, in addition, the ε_i, for $i = 1, 2, \ldots, n$ are normally distributed:

5. Each $\hat{\beta}_i$ is normally distributed.
6. The random variable

$$\frac{[n - (k + 1)]S^2}{\sigma^2}$$

   has a χ^2 distribution with $n - (k + 1)$ degrees of freedom.
7. The statistic S^2 and $\hat{\beta}_i$, for $i = 0, 1, 2, \ldots, k$ are independent.

11.12 Inferences Concerning Linear Functions of the Model Parameters: Multiple Linear Regression

As discussed in Sections 11.5 and 11.6, we might be interested in making inferences about a single β_i or about linear combinations of the model parameters $\beta_0, \beta_1, \ldots, \beta_k$. For example, we might wish to estimate $E(Y)$, given by

$$E(Y) = \beta_0 + \beta_1 x_1 + \cdots + \beta_k x_k,$$

where $E(Y)$ represents the mean yield of a chemical process for settings of controlled process variables $x_1, x_2, \ldots, x_k$; or the mean profit of a corporation for various investment expenditures $x_1, x_2, \ldots, x_k$. Properties of estimators of such linear functions are given in this section.

Suppose that we wish to make an inference about the linear function

$$a_0\beta_0 + a_1\beta_1 + a_2\beta_2 + \cdots + a_k\beta_k,$$

where $a_0, a_1, a_2, \ldots, a_k$ are constants (some of which may equal zero). Defining the $(k + 1) \times 1$ matrix,

$$\mathbf{a} = \begin{bmatrix} a_0 \\ a_1 \\ a_2 \\ \vdots \\ a_k \end{bmatrix},$$

it follows that a linear combination of the $\beta_0, \beta_1, \ldots, \beta_k$ corresponding to a_0, $a_1, \ldots, a_k$ may be expressed as

$$\mathbf{a}'\beta = a_0\beta_0 + a_1\beta_1 + a_2\beta_2 + \cdots + a_k\beta_k.$$

From now on, we will refer to such linear combinations in their matrix form. Because $\mathbf{a}'\beta$ is a linear combination of the model parameters, an unbiased estimator for $\mathbf{a}'\beta$ is given by the same linear combination of the parameter estimators. That is, by Theorem 5.12, if

$$\widehat{\mathbf{a}'\beta} = a_0\hat{\beta}_0 + a_1\hat{\beta}_1 + a_2\hat{\beta}_2 + \cdots + a_k\hat{\beta}_k = \mathbf{a}'\hat{\beta},$$

then

$$E(\mathbf{a}'\hat{\beta}) = E(a_0\hat{\beta}_0 + a_1\hat{\beta}_1 + a_2\hat{\beta}_2 + \cdots + a_k\hat{\beta}_k)$$
$$= a_0\beta_0 + a_1\beta_1 + a_2\beta_2 + \cdots + a_k\beta_k = \mathbf{a}'\beta.$$

Applying the same theorem, we find the variance of $\mathbf{a}'\hat{\beta}$:

$$\begin{aligned} V(\mathbf{a}'\hat{\beta}) &= V(a_0\hat{\beta}_0 + a_1\hat{\beta}_1 + a_2\hat{\beta}_2 + \cdots + a_k\hat{\beta}_k) \\ &= a_0^2 V(\hat{\beta}_0) + a_1^2 V(\hat{\beta}_1) + a_2^2 V(\hat{\beta}_2) + \cdots + a_k^2 V(\hat{\beta}_k) \\ &\quad + 2a_0a_1\text{Cov}(\hat{\beta}_0, \hat{\beta}_1) + 2a_0a_2\text{Cov}(\hat{\beta}_0, \hat{\beta}_2) \\ &\quad + \cdots + 2a_1a_2\text{Cov}(\hat{\beta}_1, \hat{\beta}_2) + \cdots + 2a_{k-1}a_k\text{Cov}(\hat{\beta}_{k-1}, \hat{\beta}_k) \end{aligned}$$

where $V(\hat{\beta}_i) = c_{ii}\sigma^2$ and $\text{Cov}(\hat{\beta}_i, \hat{\beta}_j) = c_{ij}\sigma^2$. You may verify that $V(\mathbf{a}'\hat{\beta})$ is given by

$$V(\mathbf{a}'\hat{\beta}) = [\mathbf{a}'(\mathbf{X}'\mathbf{X})^{-1}\mathbf{a}]\sigma^2.$$

Finally, recalling that $\hat{\beta}_0, \hat{\beta}_1, \hat{\beta}_2, \ldots, \hat{\beta}_k$ are normally distributed in repeated sampling (Section 11.11), it is clear that $\mathbf{a}'\hat{\beta}$ is a linear function of normally distributed random variables and hence itself is normally distributed in repeated sampling.

Because $\mathbf{a}'\hat{\beta}$ is normally distributed with

$$E(\mathbf{a}'\hat{\beta}) = \mathbf{a}'\beta$$

and $V(\mathbf{a}'\hat{\beta}) = [\mathbf{a}'(\mathbf{X}'\mathbf{X})^{-1}\mathbf{a}]\sigma^2$, we conclude that

$$Z = \frac{\mathbf{a}'\hat{\beta} - \mathbf{a}'\beta}{\sqrt{V(\mathbf{a}'\hat{\beta})}} = \frac{\mathbf{a}'\hat{\beta} - \mathbf{a}'\beta}{\sigma\sqrt{\mathbf{a}'(\mathbf{X}'\mathbf{X})^{-1}\mathbf{a}}}$$

has a standard normal distribution and could be employed to test a hypothesis

$$H_0: \mathbf{a}'\boldsymbol{\beta} = (\mathbf{a}'\boldsymbol{\beta})_0$$

when $(\mathbf{a}'\boldsymbol{\beta})_0$ is some specified value. Likewise, a large-sample $(1 - \alpha)100\%$ confidence interval for $\mathbf{a}'\boldsymbol{\beta}$ would be

$$\mathbf{a}'\hat{\boldsymbol{\beta}} \pm z_{\alpha/2}\sigma\sqrt{\mathbf{a}'(\mathbf{X}'\mathbf{X})^{-1}\mathbf{a}}.$$

Furthermore, as we might suspect, if we substitute S for σ, the quantity

$$T = \frac{\mathbf{a}'\hat{\boldsymbol{\beta}} - \mathbf{a}'\boldsymbol{\beta}}{S\sqrt{\mathbf{a}'(\mathbf{X}'\mathbf{X})^{-1}\mathbf{a}}}$$

possesses a Student's t distribution in repeated sampling, with $[n - (k + 1)]$ degrees of freedom, and hence provides a test statistic to test the hypothesis

$$H_0: \mathbf{a}'\boldsymbol{\beta} = (\mathbf{a}'\boldsymbol{\beta})_0.$$

A Test for $\mathbf{a}'\boldsymbol{\beta}$

H_0: $\mathbf{a}'\boldsymbol{\beta} = (\mathbf{a}'\boldsymbol{\beta})_0.$

H_a: $\begin{cases} \mathbf{a}'\boldsymbol{\beta} > (\mathbf{a}'\boldsymbol{\beta})_0. \\ \mathbf{a}'\boldsymbol{\beta} < (\mathbf{a}'\boldsymbol{\beta})_0. \\ \mathbf{a}'\boldsymbol{\beta} \neq (\mathbf{a}'\boldsymbol{\beta})_0. \end{cases}$

Test statistic: $T = \dfrac{\mathbf{a}'\hat{\boldsymbol{\beta}} - (\mathbf{a}'\boldsymbol{\beta})_0}{S\sqrt{\mathbf{a}'(\mathbf{X}'\mathbf{X})^{-1}\mathbf{a}}}.$

Rejection region: $\begin{cases} t > t_\alpha. \\ t < -t_\alpha. \\ |t| > t_{\alpha/2}. \end{cases}$

Here t_α is based upon $[n - (k + 1)]$ degrees of freedom.

The corresponding $(1 - \alpha)100\%$ confidence interval for $\mathbf{a}'\boldsymbol{\beta}$ is as follows.

A $(1 - \alpha)100\%$ Confidence Interval for $\mathbf{a}'\boldsymbol{\beta}$

$$\mathbf{a}'\hat{\boldsymbol{\beta}} \pm t_{\alpha/2}S\sqrt{\mathbf{a}'(\mathbf{X}'\mathbf{X})^{-1}\mathbf{a}}.$$

As earlier, the tabulated $t_{\alpha/2}$ in this formula is based upon $[n - (k + 1)]$ degrees of freedom.

Although we usually do not think of a single β_i as a linear combination of $\beta_0, \beta_1, \ldots, \beta_k$, if we choose

$$a_j = \begin{cases} 1, & \text{if } j = i \\ 0, & \text{if } j \neq i, \end{cases}$$

then $\beta_i = \mathbf{a}'\boldsymbol{\beta}$ for this choice of $\mathbf{a}$. In Exercise 11.59, you will show that with this choice of $\mathbf{a}$, $\mathbf{a}'(\mathbf{X}'\mathbf{X})^{-1}\mathbf{a} = c_{ii}$, where c_{ii} is the element in row i and column i of $(\mathbf{X}'\mathbf{X})^{-1}$. This fact greatly simplifies the form of both the test statistic and confidence intervals that can be used to make inferences about an individual β_i.

As previously indicated, one useful application of the hypothesis-testing and confidence interval techniques just presented is to the problem of estimating the mean value of Y, $E(Y)$, for fixed values of the independent variables $x_1, x_2, \ldots, x_k$. In particular, if x_i^* denotes a specific value of x_i, for $i = 1, 2, \ldots, k$, then

$$E(Y) = \beta_0 + \beta_1 x_1^* + \beta_2 x_2^* + \cdots + \beta_k x_k^*.$$

Notice that $E(Y)$ is a special case of $a_0 \beta_0 + a_1 \beta_1 + \cdots + a_k \beta_k = \mathbf{a}'\boldsymbol{\beta}$ with $a_0 = 1$ and $a_i = x_i^*$, for $i = 1, 2, \ldots, k$. Thus an inference about $E(Y)$ when $x_i = x_i^*$, for $i = 1, 2, \ldots, k$, can be made by using the techniques developed earlier for a general linear combination of the β's.

We illustrate with two examples.

EXAMPLE 11.14   Do the data of Example 11.1 present sufficient evidence to indicate curvature in the response function? Test using $\alpha = .05$ and give bounds to the attained significance level.

Solution   The preceding question assumes that the probabilistic model is a realistic description of the true response and implies a test of the hypothesis $H_0: \beta_2 = 0$ versus $H_a: \beta_2 \neq 0$ in the linear model $Y = \beta_0 + \beta_1 x + \beta_2 x^2 + \varepsilon$ that was fit to the data in Example 11.12. (If $\beta_2 = 0$, the quadratic term will not appear and the expected value of Y will represent a straight-line function of x.) The first step in the solution is to calculate SSE and s^2:

$$\text{SSE} = \mathbf{Y}'\mathbf{Y} - \hat{\boldsymbol{\beta}}'\mathbf{X}'\mathbf{Y} = 11 - [.571 \quad .700 \quad .214] \begin{bmatrix} 5 \\ 7 \\ 13 \end{bmatrix}$$

$$= 11 - 10.537 = .463$$

so then

$$s^2 = \frac{\text{SSE}}{n-3} = \frac{.463}{2} = .232 \quad \text{and} \quad s = .48.$$

(Notice that the model contains three parameters and hence SSE is based upon $n-3 = 2$ d.f.) The parameter β_2 is a linear combination of β_0, β_1 and β_2 with $a_0 = 0$, $a_1 = 0$, and $a_2 = 1$. For this choice of $\mathbf{a}$, we have $\beta_2 = \mathbf{a}'\boldsymbol{\beta}$ and $\mathbf{a}'(\mathbf{X}'\mathbf{X})^{-1}\mathbf{a} = c_{22}$.

The calculations in Example 11.12 yielded $\hat{\beta}_2 = 3/14 \approx .214$ and $c_{22} = 1/14$. The appropriate test statistic can therefore be written as

$$t = \frac{\hat{\beta}_2 - 0}{s\sqrt{c_{22}}} = \frac{.214}{.48\sqrt{1/14}} = 1.67.$$

If we take $\alpha = .05$, the value of $t_{\alpha/2} = t_{.025}$ for 2 degrees of freedom is 4.303, and the rejection region is

$$\text{reject if } |t| \geq 4.303.$$

Because the absolute value of the calculated value of t is less than 4.303, we cannot reject the null hypothesis that $\beta_2 = 0$. We do not accept $H_0 : \beta_2 = 0$, because we would need to know the probability of making a type II error—that is, the probability of falsely accepting H_0 for a specified alternative value of β_2—before we could make a statistically sound decision to accept. Because the test is two-tailed, p-value $= 2P(t > 1.67)$, where t has a t distribution with 2 degrees of freedom. Using Table 5, Appendix III, we find that $P(t > 1.67) > .10$. Thus, we conclude that p-value $> .2$. Unless we are willing to work with a relatively large value of α (greater than .2), we cannot reject H_0. Again we notice the agreement between the conclusions reached by the formal (fixed α) test procedure and the proper interpretation of the attained significance level.

As a further step in the analysis, we could look at the width of a confidence interval for β_2 to see whether it is short enough to detect a departure from zero that would be of practical significance. The resulting 95% confidence interval for β_2 is

$$\hat{\beta}_2 \pm t_{.025} S\sqrt{c_{22}}.$$

Substituting, we get

$$.214 \pm (4.303)(.48)\sqrt{1/14}, \quad \text{or} \quad .214 \pm .552.$$

Thus the confidence interval for β_2 is quite wide, suggesting that the experimenter needs to collect more data before reaching a decision.

EXAMPLE **11.15**   For the data of Example 11.1, find a 90% confidence interval for $E(Y)$ when $x = 1$.

Solution   For the model of Example 11.1,

$$E(Y) = \beta_0 + \beta_1 x = \mathbf{a}'\boldsymbol{\beta} \quad \text{with} \quad \mathbf{a} = \begin{bmatrix} a_0 \\ a_1 \end{bmatrix} = \begin{bmatrix} 1 \\ x \end{bmatrix}.$$

The desired confidence interval is given by

$$\mathbf{a}'\hat{\boldsymbol{\beta}} \pm t_{\alpha/2} S\sqrt{\mathbf{a}'(\mathbf{X}'\mathbf{X})^{-1}\mathbf{a}}.$$

In Example 11.11, we determined that

$$\hat{\beta} = \begin{bmatrix} 1 \\ .7 \end{bmatrix} \quad \text{and} \quad (\mathbf{X}'\mathbf{X})^{-1} = \begin{bmatrix} 1/5 & 0 \\ 0 & 1/10 \end{bmatrix}.$$

Because we are interested in $x = 1$,

$$\mathbf{a} = \begin{bmatrix} 1 \\ 1 \end{bmatrix} \qquad \mathbf{a}'\hat{\beta} = \begin{bmatrix} 1 & 1 \end{bmatrix} \begin{bmatrix} 1 \\ .7 \end{bmatrix} = 1.7$$

$$\mathbf{a}'(\mathbf{X}'\mathbf{X})^{-1}\mathbf{a} = \begin{bmatrix} 1 & 1 \end{bmatrix} \begin{bmatrix} 1/5 & 0 \\ 0 & 1/10 \end{bmatrix} \begin{bmatrix} 1 \\ 1 \end{bmatrix} = .3.$$

In Example 11.13 we found s^2 to be .367, or $s = .606$ for these data. The value of $t_{.05}$ with $n - 2 = 3$ degrees of freedom is 2.353, and the required 90% confidence interval for $E(Y)$ is given by

$$1.7 \pm (2.353)(.606)\sqrt{.3}, \quad \text{or} \quad 1.7 \pm .781.$$

Our answer here is the same as that obtained in Example 11.6 without the use of matrices.

Exercises

11.59   Consider the general linear model

$$Y = \beta_0 + \beta_1 x_1 + \beta_2 x_2 + \cdots + \beta_k x_k + \varepsilon,$$

where $E(\varepsilon) = 0$ and $V(\varepsilon) = \sigma^2$. Notice that $\beta_i = \mathbf{a}'\hat{\beta}$, where the vector $\mathbf{a}$ is defined by

$$a_j = \begin{cases} 1, & \text{if } j = i \\ 0, & \text{if } j \neq i. \end{cases}$$

Use this to verify that $E(\hat{\beta}_i) = \beta_i$ and $V(\hat{\beta}_i) = c_{ii}\sigma^2$, where c_{ii} is the element in row i and column i of $(\mathbf{X}'\mathbf{X})^{-1}$.

11.60   Refer to Exercise 11.57.

   a   Is there evidence of a quadratic effect in the relationship between Y and x? (Test H_0 : $\beta_2 = 0$.) Use $\alpha = .10$.

   b   Find a 90% confidence interval for β_2.

11.61   The experimenter who collected the data in Exercise 11.56 claims that the *minimum* value of $E(Y)$ occurs at $x = 1$. Test this claim at the 5% significance level. [*Hint:* $E(Y) = \beta_0 + \beta_1 x + \beta_2 x^2$ has its minimum at the point x_0, which satisfies the equation $\beta_1 + 2\beta_2 x_0 = 0$.]

11.62   An experiment was conducted to investigate the effect of four factors—temperature T_1, pressure P, catalyst C, and temperature T_2—on the yield Y of a chemical.

   a   The values (or levels) of the four factors used in the experiment are shown in the accompanying table. If each of the four factors is coded to produce the four variables x_1, x_2, x_3, and x_4, respectively, give the transformation relating each coded variable to its corresponding original.

T_1	x_1	P	x_2	C	x_3	T_2	x_4
50	-1	10	-1	1	-1	100	-1
70	1	20	1	2	1	200	1

b Fit the linear model

$$Y = \beta_0 + \beta_1 x_1 + \beta_2 x_2 + \beta_3 x_3 + \beta_4 x_4 + \varepsilon$$

to the following table of data.

				x_4			
				+1		−1	
				x_3		x_3	
				−1	1	−1	1
x_1	−1	x_2	−1	22.2	24.5	24.4	25.9
			1	19.4	24.1	25.2	28.4
	+1	x_2	−1	22.1	19.6	23.5	16.5
			1	14.2	12.7	19.3	16.0

c Do the data present sufficient evidence to indicate that T_1 contributes information for the estimation of Y? Does P? Does C? Does T_2? (Test the hypotheses, respectively, that $\beta_1 = 0$, $\beta_2 = 0$, $\beta_3 = 0$, and $\beta_4 = 0$.) Give bounds for the p-value associated with each test. What would you conclude if you used $\alpha = .01$ in each case?

11.63 Refer to Exercise 11.62. Find a 90% confidence interval for the expected yield, given that $T_1 = 50$, $P = 20$, $C = 1$, and $T_2 = 200$.

11.13   Predicting a Particular Value of Y Using Multiple Regression

In Section 11.7, we considered predicting an actual observed value of Y in the *simple* linear regression, setting the single independent variable $x = x^*$. The solution was based heavily on the properties of

$$\text{error} = Y^* - \widehat{Y^*},$$

where $\widehat{Y^*} = \hat{\beta}_0 + \hat{\beta}_1 x^*$ was observed to be a predictor of the actual value of Y and an estimator for $E(Y)$ as well. The same method will be used in this section to provide the corresponding solution in the *multiple* linear regression case. Suppose that we have fit a multiple linear regression model

$$Y = \beta_0 + \beta_1 x_1 + \beta_2 x_2 + \cdots + \beta_k x_k + \varepsilon$$

and that we are interested in predicting the value of Y^* when $x_1 = x_1^*, x_2 = x_2^*, \ldots, x_k = x_k^*$. We predict the value of Y^* with

$$\widehat{Y^*} = \hat{\beta}_0 + \hat{\beta}_1 x_1^* + \hat{\beta}_2 x_2^* + \cdots + \hat{\beta}_k x_k^* = \mathbf{a}'\hat{\boldsymbol{\beta}}$$

where

$$
\mathbf{a} = \begin{bmatrix} 1 \\ x_1^* \\ x_2^* \\ \vdots \\ x_k^* \end{bmatrix}.
$$

As in Section 11.7, we focus on the difference between the variable Y^* and the predicted value:

$$
\text{error} = Y^* - \widehat{Y^*}.
$$

Because both Y^* and $\widehat{Y^*}$ are normally distributed, the error is normally distributed; and using Theorem 5.12 and the results of Section 11.11, we find that

$$
E(\text{error}) = 0 \quad \text{and} \quad V(\text{error}) = \sigma^2[1 + \mathbf{a}'(\mathbf{X}'\mathbf{X})^{-1}\mathbf{a}]
$$

and that

$$
Z = \frac{Y^* - \widehat{Y^*}}{\sigma\sqrt{1 + \mathbf{a}'(\mathbf{X}'\mathbf{X})^{-1}\mathbf{a}}} x
$$

has a standard normal distribution. Furthermore, if S is substituted for σ, it can be shown that

$$
T = \frac{Y^* - \widehat{Y^*}}{S\sqrt{1 + \mathbf{a}'(\mathbf{X}'\mathbf{X})^{-1}\mathbf{a}}}
$$

possesses a Student's t distribution with $[n - (k + 1)]$ degrees of freedom.

Proceeding as in Section 11.7, we obtain the following $(1 - \alpha)100\%$ prediction interval for Y.

A $(1 - \alpha)100\%$ Prediction Interval for Y when $x_1 = x_1^*$, $x_2 = x_2^*, \ldots, x_k = x_k^*$

$$
\mathbf{a}'\hat{\boldsymbol{\beta}} \pm t_{\alpha/2}S\sqrt{1 + \mathbf{a}'(\mathbf{X}'\mathbf{X})^{-1}\mathbf{a}},
$$

where $\mathbf{a}' = [1, x_1^*, x_2^*, \ldots, x_k^*]$.

EXAMPLE **11.16**   Suppose that the experiment that generated the data of Example 11.11 is to be run again with $x = 2$. Predict the particular value of Y with $1 - \alpha = .90$.

Solution   In Example 11.11, we determined that

$$\hat{\beta} = \begin{bmatrix} 1 \\ .7 \end{bmatrix} \quad \text{and} \quad (\mathbf{X}'\mathbf{X})^{-1} = \begin{bmatrix} 1/5 & 0 \\ 0 & 1/10 \end{bmatrix}.$$

Because we are interested in $x = 2$, the desired prediction interval is given by

$$\mathbf{a}'\hat{\beta} \pm t_{\alpha/2} S \sqrt{1 + \mathbf{a}'(\mathbf{X}'\mathbf{X})^{-1}\mathbf{a}}$$

with

$$\mathbf{a} = \begin{bmatrix} 1 \\ 2 \end{bmatrix}, \quad \mathbf{a}'\hat{\beta} = \begin{bmatrix} 1 & 2 \end{bmatrix} \begin{bmatrix} 1 \\ .7 \end{bmatrix} = 2.4,$$

$$\mathbf{a}'(\mathbf{X}'\mathbf{X})^{-1}\mathbf{a} = \begin{bmatrix} 1 & 2 \end{bmatrix} \begin{bmatrix} 1/5 & 0 \\ 0 & 1/10 \end{bmatrix} \begin{bmatrix} 1 \\ 2 \end{bmatrix} = .6.$$

As before, $s = .606$ for these data and the value of $t_{.05}$ with $n - 2 = 3$ degrees of freedom is 2.353. The 90% prediction interval for a future observation on Y when $x = 2$ is, therefore,

$$2.4 \pm (2.353)(.606)\sqrt{1 + .6}, \quad \text{or} \quad 2.4 \pm 1.804.$$

Notice the agreement with the answer provided in Example 11.7, where we used ordinary algebra rather than the matrix approach in the solution.

Exercises

11.64   Refer to Exercise 11.57. Find a 98% prediction interval for the incidence rate of infectious hepatitis in 1977. Use the quadratic model.

11.65   Refer to Exercises 11.62 and 11.63. Find a 90% prediction interval for Y, given that $T_1 = 50$, $P = 20$, $C = 1$, and $T_2 = 200$.

11.14 A Test for $H_0: \beta_{g+1} = \beta_{g+2} = \cdots = \beta_k = 0$

In seeking an intuitively appealing test statistic to test a hypothesis concerning a set of parameters of the linear model, we are led to a consideration of the sum of squares of deviations SSE. Suppose, for example, that we were to fit a model involving only a subset of the independent variables under consideration, that is, fit a *reduced* model of the form

$$\text{model R: } Y = \beta_0 + \beta_1 x_1 + \beta_2 x_2 + \cdots + \beta_g x_g + \varepsilon$$

to the data and then were to calculate the sum of squares of deviations between the observed and predicted values of Y, SSE_R. Having done this, we might fit the linear model with *all* candidate independent variables present (the *complete* model):

$$\text{model C: } Y = \beta_0 + \beta_1 x_1 + \beta_2 x_2 + \cdots + \beta_g x_g + \beta_{g+1} x_{g+1} + \cdots + \beta_k x_k + \varepsilon.$$

This model contains all the terms of the reduced model, model R, plus the terms involving $x_{g+1}, x_{g+2}, \ldots, x_k$ (notice that $k > g$). Then we calculate the sum of squares of deviations for this model, SSE_C. Finally, let us suppose that $x_{g+1}, x_{g+2}, \ldots, x_k$ contribute a substantial quantity of information for the prediction of Y that is not contained in the variables $x_1, x_2, \ldots, x_g$ (that is, at least one of the parameters $\beta_{g+1}, \beta_{g+2}, \ldots, \beta_k$ differs from zero). What would be the relation between SSE_R and SSE_C? Intuitively, we see that, if $x_{g+1}, x_{g+2}, \ldots, x_k$ are important information-contributing variables, model C, the *complete* model, which contains all the variables in model R, the *reduced* model, plus the additional terms $x_{g+1}, x_{g+2}, \ldots, x_k$, should predict with a *smaller* error of prediction than model R. Thus SSE_C should be less than SSE_R. The greater the difference $(\text{SSE}_R - \text{SSE}_C)$, the stronger will be the evidence to support the alternative hypothesis that $x_{g+1}, x_{g+2}, \ldots, x_k$ contribute information for the prediction of Y and to reject the null hypothesis,

$$H_0 : \beta_{g+1} = \beta_{g+2} = \cdots = \beta_k = 0.$$

The decrease in the sum of squares of deviations $(\text{SSE}_R - \text{SSE}_C)$ is called the *sum of squares associated with the variables* $x_{g+1}, x_{g+2}, \ldots, x_k$, *adjusted for the variables* $x_1, x_2, x_3, \ldots, x_g$.

We indicated that large values of $(\text{SSE}_R - \text{SSE}_C)$ would lead us to reject the hypothesis

$$H_0 : \beta_{g+1} = \beta_{g+2} = \cdots = \beta_k = 0.$$

How large is "large"? We will develop a test statistic that is a function of $(\text{SSE}_R - \text{SSE}_C)$ for which we know the distribution when H_0 is true.

To acquire this test statistic, let us *assume* that the null hypothesis is true and then examine the quantities that we have calculated. Particularly, notice that

$$\text{SSE}_R = \text{SSE}_C + (\text{SSE}_R - \text{SSE}_C).$$

In other words, as indicated in Figure 11.8, we have partitioned SSE_R into two parts: SSE_C and the difference $(\text{SSE}_R - \text{SSE}_C)$. Although we omit the proof, if H_0 is true, then

$$\chi_3^2 = \frac{\text{SSE}_R}{\sigma^2}$$

$$\chi_2^2 = \frac{\text{SSE}_C}{\sigma^2}$$

$$\chi_1^2 = \frac{\text{SSE}_R - \text{SSE}_C}{\sigma^2}$$

FIGURE **11.8**
Partitioning SSE_R

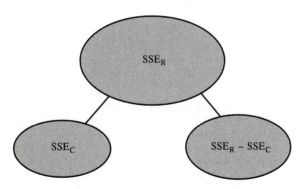

possess chi-square probability distributions in repeated sampling, with $(n - [g + 1])$, $(n - [k + 1])$, and $(k - g)$ degrees of freedom, respectively. Further, it can be shown that χ_2^2 and χ_1^2 are statistically independent.

The definition of a random variable with an F distribution is given in Definition 7.3. Consider the ratio

$$F = \frac{\chi_1^2/(k - g)}{\chi_2^2/(n - [k + 1])} = \frac{(SSE_R - SSE_C)/(k - g)}{(SSE_C)/(n - [k + 1])}.$$

If $H_0: \beta_{g+1} = \beta_{g+2} = \cdots = \beta_k = 0$ is true, then F possesses an F distribution with $\nu_1 = k - g$ numerator degrees of freedom and $\nu_2 = n - (k+1)$ denominator degrees of freedom. We have previously argued that large values of $(SSE_R - SSE_C)$ lead us to reject the null hypothesis. Thus, we see that large values of F favor rejection of H_0; if we desire a test with a type I error probability equal to α, it follows that

$$F > F_\alpha$$

is the appropriate rejection region. (See Table 7, Appendix III.)

EXAMPLE **11.17**   Do the data of Example 11.12 provide sufficient evidence to indicate that the second-order model

$$Y = \beta_0 + \beta_1 x + \beta_2 x^2 + \varepsilon$$

contributes information for the prediction of Y? That is, test the hypothesis $H_0: \beta_1 = \beta_2 = 0$ against the alternative hypothesis H_a: at least one of the parameters β_1, β_2, differs from 0. Use $\alpha = .05$. Give bounds for the attained significance level.

Solution   For the complete model, we determined in Example 11.14 that $SSE_C = .463$. Because we want to test $H_0: \beta_1 = \beta_2 = 0$, the appropriate reduced model is

$$Y = \beta_0 + \varepsilon$$

for which

$$\mathbf{Y} = \begin{bmatrix} 0 \\ 0 \\ 1 \\ 1 \\ 3 \end{bmatrix} \quad \text{and} \quad \mathbf{X} = \overset{x_0}{\begin{bmatrix} 1 \\ 1 \\ 1 \\ 1 \\ 1 \end{bmatrix}}.$$

Because $\mathbf{X}'\mathbf{X} = 5$, then $(\mathbf{X}'\mathbf{X})^{-1} = 1/5$, and $\hat{\beta} = (\mathbf{X}'\mathbf{X})^{-1}\mathbf{X}'\mathbf{Y} = (1/5)\sum_{i=1}^{5} y_i = \bar{y} = 5/5 = 1$. Thus,

$$\text{SSE}_R = \mathbf{Y}'\mathbf{Y} - \hat{\beta}'\mathbf{X}'\mathbf{Y}$$

$$= \sum_{i=1}^{5} y_i^2 - \bar{y}\left(\sum_{i=1}^{n} y_i\right) = \sum_{i=1}^{5} y_i^2 - \frac{1}{n}\left(\sum_{i=1}^{5} y_i\right)^2$$

$$= 11 - (1/5)(5)^2 = 11 - 5 = 6.$$

Notice that in this case the number of independent variables in the complete model is $k = 2$, whereas the number of independent variables in the reduced model is $g = 0$. Thus,

$$F = \frac{(\text{SSE}_R - \text{SSE}_C)/(k - g)}{(\text{SSE}_C)/(n - [k + 1])} = \frac{(6 - .463)/(2 - 0)}{.463/(5 - 3)} = 11.959.$$

The tabulated F value for $\alpha = .05$ with $v_1 = k - g = 2$ numerator degrees of freedom and $v_2 = n - (k + 1) = 2$ denominator degrees of freedom is 19.00. Hence the observed value of the test statistic does not fall in the rejection region and we conclude that, at the $\alpha = .05$ level, there is not enough evidence to support a claim that either β_1 or β_2 differs from zero. Because the proper form of the rejection region is $F > F_\alpha$, the p-value is given by $P(F > 11.959)$ when F is based on 2 numerator and 2 denominator degrees of freedom. Using Table 7, Appendix III, you can see that $.05 < p\text{-value} < .10$. Thus, if we chose $\alpha = .05$ (in agreement with the previous discussion), there is not enough evidence to support a claim that either β_1 or β_2 differs from zero. However, if an α value of .10 were selected, we could claim that either $\beta_1 \neq 0$ or $\beta_2 \neq 0$. Notice that the little additional effort required to place bounds on the p-value provides a considerable amount of additional information.

Consider the situation where we have fit a model with k independent variables and wish to test the null hypothesis

$$H_0: \beta_1 = \beta_2 = \cdots = \beta_k = 0$$

that *none* of the independent variables in the model contribute substantial information for the prediction of Y. This is exactly what was done in Example 11.17. An examination of the solution of that example will convince you that the appropriate *reduced* model is then of the form

$$Y = \beta_0 + \varepsilon.$$

This reduced model contains $g = 0$ independent variables and is such that $\text{SSE}_R = S_{yy}$. Thus, a test for

$$H_0: \beta_1 = \beta_2 = \cdots = \beta_k = 0$$

can be based on the statistic

$$F = \frac{(\text{SSE}_R - \text{SSE}_C)/(k - g)}{(\text{SSE}_C)/(n - [k + 1])} = \frac{(S_{yy} - \text{SSE}_C)/k}{(\text{SSE}_C)/(n - [k + 1])}$$

which possesses an F distribution with $\nu_1 = k$ and $\nu_2 = n - (k + 1)$ numerator and denominator degrees of freedom, respectively.

Another application of the general method for comparing complete and reduced models is given in the following example.

EXAMPLE 11.18   It is desired to relate abrasion resistance of rubber (Y) to the amount of silica filler x_1' and the amount of coupling agent x_2'. Fine-particle silica fibers are added to rubber to increase strength and resistance to abrasion. The coupling agent chemically bonds the filler to the rubber polymer chains and thus increases the efficiency of the filler. The unit of measurement for x_1' and x_2' is parts per 100 parts of rubber, which is denoted phr. For computational simplicity the actual amounts of silica filler and coupling agent are rescaled by the equations

$$x_1 = \frac{x_1' - 50}{6.7} \quad \text{and} \quad x_2 = \frac{x_2' - 4}{2}.$$

(Such rescaling of the independent variables does not affect the analysis or conclusions, but it does simplify computations.)

The data from an actual experiment are given in Table 11.6. Notice that five levels of both x_1 and x_2 are used, with the ($x_1 = 0, x_2 = 0$) point repeated three times. Let

Table 11.6. Data for Example 11.18

y	x_1	x_2
83	1	−1
113	1	1
92	−1	1
82	−1	−1
100	0	0
96	0	0
98	0	0
95	0	1.5
80	0	−1.5
100	1.5	0
92	−1.5	0

Source: Ronald Suich and G. C. Derringer, *Technometrics*, 19(2) (May 1977): 214.

us fit the second-order model

$$Y = \beta_0 + \beta_1 x_1 + \beta_2 x_2 + \beta_3 x_1^2 + \beta_4 x_2^2 + \beta_5 x_1 x_2 + \varepsilon$$

to these data. This model represents a conic surface over the (x_1, x_2) plane. Fit the second-order model and test $H_0: \beta_3 = \beta_4 = \beta_5 = 0$. (We are testing that the surface is actually a plane versus the alternative that it is a conic surface.) Give bounds for the attained significance level and indicate the proper conclusion if we choose $\alpha = .05$.

Solution We will first use matrix equations to fit the complete model, as indicated earlier. (With models of this size it is best to use a computer to do the computations.) For the data in Table 11.6, we have

$$
Y = \begin{bmatrix} 83 \\ 113 \\ 92 \\ 82 \\ 100 \\ 96 \\ 98 \\ 95 \\ 80 \\ 100 \\ 92 \end{bmatrix}, \quad
X = \begin{bmatrix}
1 & 1 & -1 & 1 & 1 & -1 \\
1 & 1 & 1 & 1 & 1 & 1 \\
1 & -1 & 1 & 1 & 1 & -1 \\
1 & -1 & -1 & 1 & 1 & 1 \\
1 & 0 & 0 & 0 & 0 & 0 \\
1 & 0 & 0 & 0 & 0 & 0 \\
1 & 0 & 0 & 0 & 0 & 0 \\
1 & 0 & 1.5 & 0 & 2.25 & 0 \\
1 & 0 & -1.5 & 0 & 2.25 & 0 \\
1 & 1.5 & 0 & 2.25 & 0 & 0 \\
1 & -1.5 & 0 & 2.25 & 0 & 0
\end{bmatrix}
$$

with columns labeled x_1, x_2, x_1^2, x_2^2, $x_1 x_2$.

and

$$
(X'X)^{-1} = \begin{bmatrix}
.33 & 0 & 0 & -.15 & -.15 & 0 \\
0 & 0.12 & 0 & 0 & 0 & 0 \\
0 & 0 & 0.12 & 0 & 0 & 0 \\
-.15 & 0 & 0 & .15 & .05 & 0 \\
-.15 & 0 & 0 & .05 & .15 & 0 \\
0 & 0 & 0 & 0 & 0 & .25
\end{bmatrix}
$$

These matrices yield

$$
\hat{\beta} = (X'X)^{-1}X'Y = \begin{bmatrix} 98.00 \\ 4.00 \\ 7.35 \\ -.88 \\ -4.66 \\ 5.00 \end{bmatrix}
$$

or the fitted second-order model,

$$\hat{y} = 98.00 + 4.00x + 7.35x_2 - .88x_1^2 - 4.66x_2^2 + 5.00x_1 x_2.$$

For this model, $SSE_C = Y'Y - \hat{\beta}'X'Y = 77.948.$

To test the hypothesis of interest ($H_0 : \beta_3 = \beta_4 = \beta_5 = 0$), we must fit the reduced model

$$Y = \beta_0 + \beta_1 x_1 + \beta_2 x_2 + \varepsilon.$$

By deleting the columns for x_1^2, x_2^2, and $x_1 x_2$ in the **X** matrix, we have

$$\hat{\beta} = (\mathbf{X}'\mathbf{X})^{-1}\mathbf{X}'\mathbf{Y} = \begin{bmatrix} 93.73 \\ 4.00 \\ 7.35 \end{bmatrix}$$

and the fitted planar model is

$$\hat{y} = 93.73 + 4.00x_1 + 7.35x_2.$$

(Notice that we cannot simply set $\hat{\beta}_3$, $\hat{\beta}_4$, and $\hat{\beta}_5$ equal to zero to produce the fitted model in the reduced case.) For the reduced model, $\text{SSE}_R = 326.623$.

We now test the hypothesis $H_0 : \beta_3 = \beta_4 = \beta_5 = 0$ by calculating F (notice that $k = 5$, $g = 2$, and $n = 11$):

$$F = \frac{(\text{SSE}_R - \text{SSE}_C)/(k - g)}{\text{SSE}_C/[n - (k + 1)]} = \frac{(326.623 - 77.948)/3}{77.948/5} = 5.32.$$

Because this statistic is based on $\nu_1 = (k - g) = 3$ numerator degrees of freedom and $\nu_2 = n - (k + 1) = 5$ denominator degrees of freedom, the p-value is given by $P(F > 5.32)$. Thus, $.05 < p\text{-value} < .10$. If we choose $\alpha = .05$, there is insufficient evidence to support a claim that the second-order model fits the data significantly better than does the planar model. Notice that we have tested whether the *group* of variables x_1^2, x_2^2, $x_1 x_2$ contributed to a significantly better fit of the model to the data.

Exercises

11.66   Refer to Exercise 11.27. Answer the question on the decreasing rate of tuberculosis by constructing an F test.

11.67   In Exercise 11.66, you used an F test to test the same hypothesis that was tested in Exercise 11.27 via a t test. Consider the general simple linear regression case and the F and t statistics that can be used to implement the test of $H_0 : \beta_1 = 0$ versus $H_a : \beta_1 \neq 0$. Show that, in general, $F = t^2$. Compare the value of F obtained in Exercise 11.66 to the corresponding value of t obtained in Exercise 11.27.

11.68   Refer to Exercise 11.57.

    a   For the quadratic model carry out an F test of $H_0 : \beta_2 = 0$, using $\alpha = .05$. Compare the result to the result of the test in Exercise 11.60.

    b   Test $H_0 : \beta_1 = \beta_2 = 0$ at the 5% significance level.

11.69   Refer to Exercise 11.62. Test the hypothesis, at the 5% level of significance, that neither T_1 nor T_2 affects the yield.

11.70 Utility companies, which must plan the operation and expansion of electricity generation, are vitally interested in predicting customer demand over both short and long periods of time. A short-term study was conducted to investigate the effect of each month's mean daily temperature, x_1, and of cost per kilowatt-hour, x_2, on the mean daily consumption (in kW·h) per household. The company officials expected the demand for electricity to rise in cold weather (due to heating), fall when the weather was moderate, and rise again when the temperature rose and there was a need for air conditioning. They expected demand to decrease as the cost per kilowatt-hour increased, reflecting greater attention to conservation. Data were available for 2 years, a period during which the cost per kilowatt-hour, x_2, increased due to the increasing costs of fuel. The company officials fitted the model

$$Y = \beta_0 + \beta_1 x_1 + \beta_2 x_1^2 + \beta_3 x_2 + \beta_4 x_1 x_2 + \beta_5 x_1^2 x_2 + \varepsilon$$

to the data in the following table and obtained $\hat{y} = 325.606 - 11.383x_1 + .113x_1^2 - 21.699x_2 + .873x_1x_2 - .009x_1^2x_2$ with SSE $= 152.177$.

Price per kW·h (x_2)		Mean Daily Consumption (kW·h) per Household					
8¢	Mean daily °F temperature (x_1)	31	34	39	42	47	56
	Mean daily consumption (y)	55	49	46	47	40	43
10¢	Mean daily °F temperature (x_1)	32	36	39	42	48	56
	Mean daily consumption (y)	50	44	42	42	38	40
8¢	Mean daily °F temperature (x_1)	62	66	68	71	75	78
	Mean daily consumption (y)	41	46	44	51	62	73
10¢	Mean daily °F temperature (x_1)	62	66	68	72	75	79
	Mean daily consumption (y)	39	44	40	44	50	55

When the model $Y = \beta_0 - \beta_1 x_1 + \beta_2 x_1^2 + \varepsilon$ was fit, the prediction equation was $\hat{y} = 130.009 - 3.302x_1 + .033x_1^2$ with SSE $= 465.134$. Test whether the terms involving x_2 (x_2, x_1x_2, $x_1^2x_2$) contribute to a significantly better fit of the model to the data. Give bounds for the attained significance level.

11.71 Refer to Example 11.18. Using the reduced model, construct a 95% confidence interval for the expected abrasion resistance of rubber when $x_1 = 1$ and $x_2 = -1$.

11.72 Refer to Example 11.18. Construct individual tests of the three hypotheses $H_0 : \beta_3 = 0$, $H_0 : \beta_4 = 0$, and $H_0 : \beta_5 = 0$. Use a 1% level of significance on each test. (If multiple tests are to be conducted on the same set of data, it is wise to use a very small α level on each test.)

11.15 Summary and Concluding Remarks

In this chapter we have used the method of least squares to fit a linear model to an experimental response. We assumed the expected value of Y to be a function of a set of variables $x_1, x_2, \ldots, x_k$, where the function is linear in a set of unknown parameters. We used the expression

$$Y = \beta_0 + \beta_1 x_1 + \beta_2 x_2 + \cdots + \beta_k x_k + \varepsilon$$

to denote a linear statistical model.

Inferential problems associated with the linear statistical model include estimation and tests of hypotheses relating to the model parameters $\beta_0, \beta_1, \ldots, \beta_k$ and—even more important—estimation of $E(Y)$, the expected response for a particular setting, and the prediction of some future value of Y. Experiments for which the least squares theory is appropriate include both controlled experiments and those where $x_1, x_2, \ldots, x_k$ are observed values of random variables.

Why use the method of least squares to fit a linear model to a set of data? Where the assumptions about the random errors ε hold [normality, independence, $V(\varepsilon) = \sigma^2$ for all values of $x_1, x_2, \ldots, x_k$], it can be shown that the least squares procedure gives the best *linear* unbiased estimators for $\beta_0, \beta_1, \ldots, \beta_k$. That is, if we estimate the parameters $\beta_0, \beta_1, \ldots, \beta_k$, using linear functions of $y_1, y_2, \ldots, y_k$, the least squares estimators have minimum variance. Some other nonlinear estimators for the parameters may possess a smaller variance than the least squares estimators, but if such estimators exist, they are not known at this time. Again, why use least squares estimators? They are easy to use and we know they possess good properties for many situations.

As you might imagine, the methodology presented in this chapter is employed widely in business and in all the sciences for exploring the relationship between a response and a set of independent variables. Estimation of $E(Y)$ or prediction of Y usually is the experimental objective.

Whole textbooks are devoted to the topic of regression. Our purpose has been to introduce many of the theoretical considerations associated with simple and multiple linear regression. Although the method of least squares can be utilized to estimate model parameters in general situations, the formal inference-making techniques that we presented (based on the t and F distributions) are valid only under the extra assumptions that we presented. Key assumptions include that the error terms in the model are normally distributed and that the variance of the error terms does not depend on the value of any independent variable(s). In practical applications, these assumptions may not be valid. Generally, assessments of the validity of model assumptions are based on analyses of the *residuals*, the differences between the observed and predicted (using the model) values of the response variable. Examination of the residuals, including plots of the residuals versus the independent variable(s) and plots of the residuals against their normal theory expected values, permits assessments of whether the assumptions are reasonable for a particular data set. Data points with unusually large residuals may be *outliers* that indicate that something went wrong when the corresponding observation was made. Some individual data points may have an unusually large impact on the fitted regression model in the sense that the model fitted with these data points included differs considerably from the model fitted with them excluded (such points are often called *high-influence points*). A regression model might suffer from *lack of fit*, indicating that the selected model is not adequate to model the response. In such cases, it might be necessary to fit a more complicated model to obtain any predictive precision. An important consideration in multiple regression models is that of *multicollinearity*, where some of the independent variables in the model are highly correlated with one another. We cannot do justice to these topics in a single introductory chapter on linear and multiple regression. We have focused on the general concept of least squares as a method for estimating model parameters and have provided the theoretical foundations for analyses based on the classical normal theory. The other issues described in this section are discussed in the supplemental references.

References and Further Readings

Draper, N. R., and H. Smith. 1998. *Applied Regression Analysis,* 3d ed. New York: John Wiley.

Graybill, F. 1976. *Theory and Application of the Linear Model.* Boston: Duxbury Press.

Marcellari, C. E. 1984. "Revision of Serpulids of the Genus Rotularia (Annelida) at Seymour Island (Antarctic Peninsula) and Their Value in Stratigraphy," *Journal of Paleontology* 58(4).

Matis, J. H., and T. E. Wehrly. 1979. "Stochastic Models of Compartmental Systems," *Biometrics* 35(1): 199–220.

Meyers, R. H. 1990. *Classical and Modern Regression with Applications* 2d ed. Boston: PWS-Kent.

Meyers, R. H., and J. S. Milton. 1998. *A First Course in the Theory of Linear Statistical Models.* New York: McGraw-Hill, Primis Custom Pub.

Montgomery, D. C., and E. A. Peck. 1992. *Introduction to Linear Regression Analysis.* New York: John Wiley.

Supplementary Exercises

11.73 At temperatures approaching absolute zero ($-273°C$), helium exhibits traits that defy many laws of conventional physics. An experiment has been conducted with helium in solid form at various temperatures near absolute zero. The solid helium is placed in a dilution refrigerator along with a solid impure substance, and the fraction (in weight) of the impurity passing through the solid helium is recorded. (The phenomenon of solids passing directly through solids is known as *quantum tunneling.*) The data are given in the following table.

°C Temperature (x)	Proportion of Impurity Passing Through Helium (y)
−262.0	.315
−265.0	.202
−256.0	.204
−267.0	.620
−270.0	.715
−272.0	.935
−272.4	.957
−272.7	.906
−272.8	.985
−272.9	.987

a Fit a least squares line to the data.

b Test the null hypothesis $H_0: \beta_1 = 0$ against the alternative hypothesis $H_a: \beta_1 < 0$, at the $\alpha = 0.01$ level of significance.

c Find a 95% prediction interval for the percentage of the solid impurity passing through solid helium at $-273°C$. (This value of x is outside the experimental region, where use of the model for prediction may be dangerous.)

11.74 A study was conducted to determine whether a linear relationship exists between the breaking strength y of wooden beams and the specific gravity x of the wood. Ten randomly selected beams of the same cross-sectional dimensions were stressed until they broke. The breaking strengths and the density of the wood are shown in the accompanying table for each of the ten beams.

a Fit the model $Y = \beta_0 + \beta_1 x + \varepsilon$.

b Test $H_0: \beta_1 = 0$ against the alternative hypothesis, $H_a: \beta_1 \neq 0$.

c Estimate the mean strength for beams with specific gravity .590, using a 90% confidence interval.

Beam	Specific Gravity (x)	Strength (y)
1	.499	11.14
2	.558	12.74
3	.604	13.13
4	.441	11.51
5	.550	12.38
6	.528	12.60
7	.418	11.13
8	.480	11.70
9	.406	11.02
10	.467	11.41

11.75 A response Y is a function of three independent variables x_1, x_2, and x_3 that are related as follows:

$$Y = \beta_0 + \beta_1 x_1 + \beta_2 x_2 + \beta_3 x_3 + \varepsilon.$$

a Fit this model to the $n = 7$ data points shown in the accompanying table.

y	x_1	x_2	x_3
1	−3	5	−1
0	−2	0	1
0	−1	−3	1
1	0	−4	0
2	1	−3	−1
3	2	0	−1
3	3	5	1

b Predict Y when $x_1 = 1, x_2 = -3, x_3 = -1$. Compare with the observed response in the original data. Why are these two not equal?

c Do the data present sufficient evidence to indicate that x_3 contributes information for the prediction of Y? (Test the hypothesis $H_0: \beta_3 = 0$, using $\alpha = .05$.)

d Find a 95% confidence interval for the expected value of Y, given $x_1 = 1, x_2 = -3, x_3 = -1$.

e Find a 95% prediction interval for Y, given $x_1 = 1, x_2 = -3, x_3 = -1$.

11.76 If values of independent variables are equally spaced, what is the advantage of coding to new variables that represent symmetric spacing about the origin?

11.77 Suppose that you wish to fit a straight line to a set of n data points, where n is an even integer, and that you can select the n values of x in the interval $-9 \le x \le 9$. How should you select the values of x so as to minimize $V(\hat{\beta}_1)$?

11.78 Refer to Exercise 11.77. It is common to employ equal spacing in selecting the values of x. Suppose that $n = 10$. Find the relative efficiency of the estimator $\hat{\beta}_1$ based on equal spacing versus the same estimator based on the spacing of Exercise 11.74. Assume that $-9 \le x \le 9$.

11.79 The data in the accompanying table come from the comparison of the growth rates for bacteria types A and B. The growth Y recorded at five equally spaced (and coded) points of time is shown in the table.

Bacteria Type	Time				
	-2	-1	0	1	2
A	8.0	9.0	9.1	10.2	10.4
B	10.0	10.3	12.2	12.6	13.9

a Fit the linear model

$$Y = \beta_0 + \beta_1 x_1 + \beta_2 x_2 + \beta_3 x_1 x_2 + \varepsilon$$

to the $n = 10$ data points. Let $x_1 = 1$ if the point refers to bacteria type B, and let $x_1 = 0$ if the point refers to type A. Let $x_2 = $ coded time.

b Plot the data points and graph the two growth lines. Notice that β_3 is the difference between the slopes of the two lines and represents time–bacteria interaction.

c Predict the growth of bacteria type A at time $x_2 = 0$, and compare the answer with the graph. Repeat the process for bacteria type B.

d Do the data present sufficient evidence to indicate a difference in the rates of growth for the two types of bacteria?

e Find a 90% confidence interval for the expected growth for bacteria type B at time $x_2 = 1$.

f Find a 90% prediction interval for the growth Y of bacteria type B at time $x_2 = 1$.

11.80 The following model was proposed for testing whether there was evidence of salary discrimination against women in a state university system:

$$Y = \beta_0 + \beta_1 x_1 + \beta_2 x_2 + \beta_3 x_1 x_2 + \beta_4 x_2^2 + \varepsilon$$

where $Y = $ annual salary (in thousands of dollars)

$$x_1 = \begin{cases} 1, & \text{if female} \\ 0, & \text{if male} \end{cases}$$

$$x_2 = \text{amount of experience (in years)}.$$

When this model was fit to data obtained from the records of 200 faculty members, SSE $= 783.90$. The reduced model $Y = \beta_0 + \beta_1 x_2 + \beta_2 x_2^2 + \varepsilon$ was also fit and produced a value of SSE = 795.23. Do the data provide sufficient evidence to support the claim that the mean salary depends on the gender of the faculty members? Use $\alpha = .05$.

11.81 Show that the least squares prediction equation

$$\hat{y} = \hat{\beta}_0 + \hat{\beta}_1 x_1 + \cdots + \hat{\beta}_k x_k$$

passes through the point $(\bar{x}_1, \bar{x}_2, \ldots, \bar{x}_k, \bar{y})$.

11.82 An experiment was conducted to determine the effect of pressure and temperature on the yield of a chemical. Two levels of pressure (in pounds per square inch) and three of temperature were used:

Pressure (psi)	Temperature (°F)
50	100
80	200
	300

One run of the experiment at each temperature–pressure combination gave the data listed in the following table.

Yield	Pressure (psi)	Temperature (°F)
21	50	100
23	50	200
26	50	300
22	80	100
23	80	200
28	80	300

a Fit the model $Y = \beta_0 + \beta_1 x_1 + \beta_2 x_2 + \beta_3 x_2^2 + \varepsilon$, where $x_1 = $ pressure and $x_2 = $ temperature.

b Test to see whether β_3 differs significantly from zero, with $\alpha = .05$.

c Test the hypothesis that temperature does not affect the yield, with $\alpha = .05$.

*11.83 the bivariate normal distribution. A test of $H_0 : \rho = 0$ against $H_a : \rho \neq 0$ can be derived as follows.

a Let $S_{yy} = \sum_{i=1}^{n} (y_i - \bar{y})^2$ and $S_{xx} = \sum_{i=1}^{n} (x_i - \bar{x})^2$. Show that

$$\hat{\beta}_1 = r \sqrt{\frac{S_{yy}}{S_{xx}}}.$$

b Conditional on $X_i = x_i$, for $i = 1, 2, \ldots, n$, show that, under $H_0 : \rho = 0$,

$$\frac{\hat{\beta}_1 \sqrt{(n-2) S_{xx}}}{\sqrt{S_{yy}(1 - r^2)}}$$

has a t distribution with $(n - 2)$ degrees of freedom.

c Conditional on $X_i = x_i$, for $i = 1, 2, \ldots, n$, conclude that

$$T = \frac{r \sqrt{n-2}}{\sqrt{1 - r^2}}$$

has a t distribution with $(n-2)$ degrees of freedom, under $H_0 : \rho = 0$. Hence conclude that T has the same distribution unconditionally.

11.84 Labor and material costs are two basic components in the cost of construction. Changes in the component costs, of course, lead to changes in total construction costs. The accompanying table tracks changes in construction cost and cost of all construction materials for eight consecutive months.

Month	Construction Cost (y)	Index of All Construction Materials (x)
January	193.2	180.0
February	193.1	181.7
March	193.6	184.1
April	195.1	185.3
May	195.6	185.7
June	198.1	185.9
July	200.9	187.7
August	202.7	189.6

Do the data provide sufficient evidence to indicate a nonzero correlation between the monthly construction costs and indexes of all construction materials? Give the attained significance level.

11.85 The data in the following table give the miles per gallon obtained by a test automobile when using gasolines of varying octane levels.

Miles per Gallon (y)	Octane (x)
13.0	89
13.2	93
13.0	87
13.6	90
13.3	89
13.8	95
14.1	100
14.0	98

a Calculate the value of r.

b Do the data provide sufficient evidence to indicate that octane level and miles per gallon are dependent? Give the attained significance level, and indicate your conclusion if you wish to implement an $\alpha = .05$ level test.

CHAPTER **12**

Considerations in Designing Experiments

12.1   The Elements Affecting the Information in a Sample

12.2   Designing Experiments to Increase Accuracy

12.3   The Matched Pairs Experiment

12.4   Some Elementary Experimental Designs

12.5   Summary

References and Further Readings

12.1   The Elements Affecting the Information in a Sample

A meaningful measure of the information in a sample that is available to make an inference about a population parameter is provided by the width (or half-width) of the confidence interval that could be constructed from the sample data. Recall that a 95% large-sample confidence interval for a population mean is

$$\overline{Y} \pm 1.96\left(\frac{\sigma}{\sqrt{n}}\right).$$

The widths of many of the commonly employed confidence intervals, like the confidence interval for a population mean, are dependent on the population variance σ^2 and the sample size n. The less variation in the population, measured by σ^2, the shorter the confidence interval will be. Similarly, the width of the confidence interval decreases as n increases. This interesting phenomenon would lead us to believe that two factors affect the quantity of information in a sample pertinent to a parameter: namely, the variation of the data and the sample size n. We will find this deduction to be slightly oversimplified but essentially true.

In previous chapters, when we were interested in comparing two population means or fitting a simple linear regression, we assumed that independent random samples were taken from the populations of interest. If we wish to compare two populations based on a total of n observations, how many observations should be taken from each population? If we have decided to fit a simple linear regression model and wish to maximize the information in the resulting data, how should we choose the values of the independent variable? These questions are addressed in the next section.

Generally, the design of experiments is a very broad subject concerned with methods of sampling to reduce the variation in an experiment and thereby to acquire a specified quantity of information at minimum cost. If the objective is to make a comparison of two population means, the *matched pairs experiment* often suffices. After considering the matched pairs experiment in Section 12.3, the remainder of the chapter presents some of the important considerations in the design of good experiments.

12.2 Designing Experiments to Increase Accuracy

As we will see, for the same total number of observations, some methods of data collection (*designs*) provide more information concerning specific population parameters than others. No single design is best in acquiring information concerning all types of population parameters. Indeed, the problem of finding the best design for focusing information on a specific population parameter has been solved in only a few specific cases. The purpose of this section is not to present a general theory but rather to present two examples that illustrate the principles involved.

Consider the problem of estimating the difference between a pair of population means, $\mu_1 - \mu_2$, based on independent random samples. If the experimenter has resources sufficient to sample a total of n observations, how many observations should she select from populations 1 and 2, say, n_1 and n_2 ($n_1 + n_2 = n$), respectively, to maximize the information in the data pertinent to $\mu_1 - \mu_2$? If $n = 10$, should she select $n_1 = n_2 = 5$ observations from each population, or would an allocation of $n_1 = 4$ and $n_2 = 6$ be better?

If the random samples are independently drawn, we estimate $\mu_1 - \mu_2$ with $\overline{Y}_1 - \overline{Y}_2$, which has standard error

$$\sigma_{(\overline{Y}_1-\overline{Y}_2)} = \sqrt{\frac{\sigma_1^2}{n_1} + \frac{\sigma_2^2}{n_2}}.$$

The smaller $\sigma_{(\overline{Y}_1-\overline{Y}_2)}$ is, the smaller will be the corresponding error of estimation and the greater will be the quantity of information in the sample pertinent to $\mu_1 - \mu_2$. If, as we frequently assume, $\sigma_1^2 = \sigma_2^2 = \sigma^2$, then

$$\sigma_{(\overline{Y}_1-\overline{Y}_2)} = \sigma\sqrt{\frac{1}{n_1} + \frac{1}{n_2}}.$$

You can verify that this quantity is a minimum when $n_1 = n_2$ and consequently that the sample contains a maximum of information about $\mu_1 - \mu_2$ when the n experimental units are equally divided between the two treatments. A more general case is considered in Example 12.1.

EXAMPLE 12.1   If n observations are to be used to estimate $\mu_1 - \mu_2$, based upon independent random samples from the two populations of interest, find n_1 and n_2 so that $V(\overline{Y}_1 - \overline{Y}_2)$ is minimized (assume that $n_1 + n_2 = n$).

Solution   Let b denote the fraction of the n observations assigned to the sample from population 1; that is, $n_1 = bn$ and $n_2 = (1 - b)n$. Then

$$V(\overline{Y}_1 - \overline{Y}_2) = \frac{\sigma_1^2}{bn} + \frac{\sigma_2^2}{(1-b)n}.$$

To find the fraction b that minimizes this variance, we set the first derivative, with respect to b, equal to zero. This process yields

$$-\frac{\sigma_1^2}{n}\left(\frac{1}{b^2}\right) + \frac{\sigma_2^2}{n}\left(\frac{1}{1-b}\right)^2 = 0.$$

Solving for b, we obtain

$$b = \frac{\sigma_1}{\sigma_1 + \sigma_2} \quad \text{and} \quad 1 - b = \frac{\sigma_2}{\sigma_1 + \sigma_2}.$$

Thus, $V(\overline{Y}_1 - \overline{Y}_2)$ is minimized when

$$n_1 = \left(\frac{\sigma_1}{\sigma_1 + \sigma_2}\right)n \quad \text{and} \quad n_2 = \left(\frac{\sigma_2}{\sigma_1 + \sigma_2}\right)n,$$

that is, when sample sizes are allocated proportionally to sizes of the standard deviations. Notice that $n_1 = n/2 = n_2$ if $\sigma_1 = \sigma_2$.

As a second example, consider the problem of fitting a straight line through a set of n points by using the least squares method of Chapter 11 (see Figure 12.1). Further, suppose that we are primarily interested in the slope β_1 of the line in the linear model

$$Y = \beta_0 + \beta_1 x + \varepsilon.$$

If we have the option of selecting the n values of x for which y will be observed, which values of x will maximize the quantity of information on β_1? We have one quantitative independent variable x, and our problem is to decide on the values $x_1, x_2, \ldots, x_n$ to employ, as well as the number of observations to take at each of these values.

FIGURE **12.1**
Fitting a straight line
by the method
of least squares

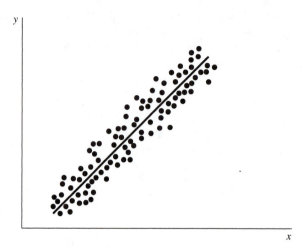

The best design for estimating the slope β_1 can be determined by considering the standard deviation of $\hat{\beta}_1$:

$$\sigma_{\hat{\beta}_1} = \frac{\sigma}{\sqrt{S_{xx}}} = \frac{\sigma}{\sqrt{\displaystyle\sum_{i=1}^{n}(x_i - \overline{x})^2}}.$$

The larger S_{xx}, the sum of squares of deviations of $x_1, x_2, \ldots, x_n$ about their mean, the smaller the standard deviation of $\hat{\beta}_1$. That is, we obtain a better estimator for the slope if the values of x are spread farther apart. In some cases, the experimenter has some experimental region, say, $x_1 < x < x_2$, over which he or she wishes to observe Y, and this range is frequently selected prior to experimentation. Then the smallest value for $\sigma_{\hat{\beta}_1}$ occurs when the n data points are equally divided, with half located at the lower boundary x_1 of the region and half at the upper boundary x_2. (The proof is omitted.) An experimenter who wished to fit a line by using $n = 10$ data points in the interval $2 \leq x \leq 6$ would select five data points at $x = 2$ and five at $x = 6$. Before concluding discussion of this example, you should notice that observing all values of Y at only two values of x will not provide information on curvature of the response curve in case the assumption of linearity in the relation of $E(Y)$ and x is incorrect. It is frequently safer to select a few points (as few as one or two) somewhere near the middle of the experimental region to detect curvature if it should be present (see Figure 12.2). A further comment is in order. One of the assumptions that we have made regarding the simple linear regression model is that the variance of the error term ε does not depend on the value of the independent variable x. If the x values are more spread out, the validity of this assumption may become more questionable.

To summarize, we have given optimal designs (allocation of experimental units per population and selection of settings for the independent variable x) for comparing a pair of means and fitting a straight line. These two simple designs illustrate how information in an experiment can be increased or decreased, depending on where observations are made and on the allocation of sample sizes. In the next section, we consider a method for controlling the amount of inherent variability in an experiment.

FIGURE **12.2**
A good design for
fitting a straight
line ($n = 10$)

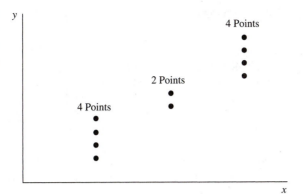

Exercises

12.1   Suppose that you wish to compare the means for two populations and that $\sigma_1^2 = 9$, $\sigma_2^2 = 25$, and $n = 90$. What allocation of $n = 90$ to the two samples will result in the maximum amount of information about $(\mu_1 - \mu_2)$?

12.2   Refer to Exercise 12.1. Suppose that you allocate $n_1 = n_2$ observations to each sample. How large must n_1 and n_2 be in order to obtain the same amount of information as that implied by the solution to Exercise 12.1?

12.3   Suppose, as in Exercise 12.1, that two populations have respective variances $\sigma_1^2 = 9$ and $\sigma_2^2 = 25$. Find the smallest sample size and the corresponding sample allocation that will yield a 95% confidence interval for $\mu_1 - \mu_2$ that is two units in length.

12.4   Refer to Exercise 12.3. How many observations are needed for a 95% confidence interval to be two units in length if $n_1 = n_2$?

12.5   Suppose that we wish to study the effect of the stimulant digitalis on the blood pressure Y of rats over a dosage range of $x = 2$ to $x = 5$ units. The response is expected to be linear over the region; that is, $Y = \beta_0 + \beta_1 x + \varepsilon$. Six rats are available for the experiment and each rat can receive only one dose. What dosages of digitalis should be employed in the experiment, and how many rats should be run at each dosage to maximize the quantity of information in the experiment relative to the slope β_1?

12.6   Refer to Exercise 12.5. Consider two methods for selecting the dosages. Method 1 assigns three rats to the dosage $x = 2$ and three rats to $x = 5$. Method 2 equally spaces the dosages between $x = 2$ and $x = 5$ ($x = 2, 2.6, 3.2, 3.8, 4.4, 5.0$). Suppose that σ is known and that the relationship between $E(Y)$ and x is truly linear (see Chapter 11). If we use the data from both methods to construct confidence intervals for the slope β_1, which method will yield the longer interval? How much longer is the longer interval? If we use method 2, approximately how many observations will be required to obtain an interval the same length as that obtained by the optimal assignment of method 1?

12.7   Refer to Exercise 12.5. Why might it be advisable to assign one or two points at $x = 3.5$?

12.8   The standard error of the estimator $\hat{\beta}_1$ in a simple linear regression model gets smaller as S_{xx} increases, that is, as the x values become more spread out. Why don't we always spread the x values out as much as possible?

12.3 The Matched Pairs Experiment

In Chapters 8 and 10 we considered methods for comparing the means of two populations based on independent samples from each. In the previous section, we examined how to determine the sizes of the samples from the two populations so that the standard error of the estimator $\overline{Y}_1 - \overline{Y}_2$ is minimized. In many experiments, the samples are paired rather than independent. A commonly occurring situation is one where repeated observations are made on the same sampling unit, such as weighing the same individual before and after he or she participated in a weight-loss program. In a medical experiment we might pair individuals who are of similar weight and age and are of the same gender. One individual from each pair is randomly selected to receive one of two competing medications to control hypertension, whereas the other individual from the same pair receives the other medication.

Comparing two populations on the basis of paired data can be a very effective experimental design that can control for extraneous sources of variability and result in decreasing the standard error of the estimator $\overline{Y}_1 - \overline{Y}_2$ for the difference in the population means $\mu_1 - \mu_2$. Let (Y_{1i}, Y_{2i}), for $i = 1, 2, \ldots, n$, denote a random sample of paired observations. Assume that

$$E(Y_{1i}) = \mu_1, \qquad \mathrm{Var}(Y_{1i}) = \sigma_1^2, \qquad E(Y_{2i}) = \mu_2, \qquad \mathrm{Var}(Y_{2i}) = \sigma_2^2,$$

$$\text{and} \quad \mathrm{Cov}(Y_{1i}, Y_{2i}) = \rho\sigma_1\sigma_2$$

where ρ is the common correlation coefficient of the variables within each pair (see Section 5.7). Define $D_i = Y_{1i} - Y_{2i}$, for $i = 1, 2, \ldots, n$, the differences between the observations within each pair. Because the pairs of observations were assumed to be independent and identically distributed, the D_i values, for $i = 1, 2, \ldots, n$, are independent and identically distributed; using Theorem 5.12, we see that

$$\mu_D = E(D_i) = E(Y_{1i}) - E(Y_{2i}) = \mu_1 - \mu_2$$

$$\sigma_D^2 = \mathrm{Var}(D_i) = \mathrm{Var}(Y_{1i}) + \mathrm{Var}(Y_{2i}) - 2\mathrm{Cov}(Y_{1i}, Y_{2i})$$

$$= \sigma_1^2 + \sigma_2^2 - 2\rho\sigma_1\sigma_2.$$

From these considerations, a natural estimator for $\mu_1 - \mu_2$ is the average of the differences $\overline{D} = \overline{Y}_1 - \overline{Y}_2$, and

$$E(\overline{D}) = \mu_D = \mu_1 - \mu_2$$

$$\sigma_{\overline{D}}^2 = \mathrm{Var}(\overline{D}) = \frac{\sigma_D^2}{n} = \frac{1}{n}[\sigma_1^2 + \sigma_2^2 - 2\rho\sigma_1\sigma_2].$$

If the data had been obtained from an *independent* samples experiment and $n_1 = n_2 = n$,

$$E(\overline{Y}_1 - \overline{Y}_2) = \mu_1 - \mu_2$$

$$\sigma_{(\overline{Y}_1 - \overline{Y}_2)}^2 = \frac{1}{n}[\sigma_1^2 + \sigma_2^2].$$

If it is reasonable to believe that within the pairs (Y_{1i}, Y_{2i}), for $i = 1, 2, \ldots, n$, the values of Y_{1i} and Y_{2i} will tend to increase or decrease together ($\rho > 0$), then an examination of the preceding expressions for $\sigma_{\overline{D}}^2$ in the matched pairs experiment and $\sigma_{(\overline{Y}_1 - \overline{Y}_2)}^2$ in the independent samples experiment shows that the matched pairs experiment provides an estimator with smaller variance than does the independent samples experiment. In Exercise 12.11, you are asked to decide when the two experiments will yield estimators with the same variance and when the independent samples experiment will give the estimator with the smaller variance.

Because pairing samples makes the observations within each pair dependent, we cannot use the methods that were previously developed to compare populations based on independent samples from each. The analysis of a matched pairs experiment utilizes the n paired differences, D_i, for $i = 1, 2, \ldots, n$. Inferences regarding the differences in the means $\mu_1 - \mu_2$ are made by making inferences regarding the mean of the differences, μ_D. Define

$$\overline{D} = \frac{1}{n} \sum_{i=1}^{n} D_i \quad \text{and} \quad S_D^2 = \frac{1}{n-1} \sum_{i=1}^{n} (D_i - \overline{D})^2$$

and employ the appropriate one-sample procedure to complete the inference. If the number of pairs, and hence the number of differences, is large, say, $n > 30$, the large sample inferential methods developed in Chapters 8 and 10 can be utilized. If the number of differences, n, is small and it is reasonable to assume that the differences are approximately normally distributed, we can use inferential methods based on the t distribution. We illustrate with the following example.

EXAMPLE 12.2   We wish to compare two methods for determining the percentage of iron ore in ore samples. Because inherent differences in the ore samples would be likely to contribute unwanted variability in the measurements we observe, a matched pairs experiment was created by splitting each of 12 ore samples into two parts. One-half of each sample was randomly selected and subjected to method 1; the other half was subjected to method 2. The results are presented in Table 12.1. Do the data provide sufficient evidence that method 2 yields a higher average percentage than method 1? Test using $\alpha = .05$.

Solution   We have formed the differences in Table 12.1 by taking the method 1 measurement and subtracting the corresponding method 2 measurement. If the mean percentage for method 2 is larger then $\mu_D = \mu_1 - \mu_2 < 0$. Thus, we test

$$H_0 : \mu_D = 0 \quad \text{versus} \quad H_a : \mu_D < 0.$$

For these data

$$s_D^2 = \frac{\sum_{i=1}^{n} d_i^2 - \frac{1}{n} \left(\sum_{i=1}^{n} d_i \right)^2}{n-1} = \frac{.0112 - \frac{1}{12}(-.20)^2}{11} = .0007.$$

Table 12.1. Data for the matched pairs experiment in Example 12.2

Ore Sample	Method 1	Method 2	d_i
1	38.25	38.27	−.02
2	31.68	31.71	−.03
3	26.24	26.22	+.02
4	41.29	41.33	−.04
5	44.81	44.80	+.01
6	46.37	46.39	−.02
7	35.42	35.46	−.04
8	38.41	38.39	+.02
9	42.68	42.72	−.04
10	46.71	46.76	−.05
11	29.20	29.18	+.02
12	30.76	30.79	−.03
			$\bar{d} = -.0167$

If it is reasonable to assume that the differences are normally distributed, so it follows that

$$t = \frac{\bar{d} - 0}{s_D/\sqrt{n}} = \frac{-.0167}{\sqrt{.0007}/\sqrt{12}} = -2.1865$$

is the observed value of a statistic that, under the null hypothesis, has a t distribution with $n - 1 = 11$ degrees of freedom. Using Table 5, Appendix III, with $\alpha = .05$, we reject H_0 if $t < -1.796$. Hence we conclude that sufficient evidence exists to permit us to conclude that method 2 yields a higher average percentage than does method 1. Again, using Table 5, Appendix III, it follows that $.025 < p$-value $< .05$.

Although the results in Example 12.2 imply that the results of the experiment are statistically significant, we can assess the practical significance of the result by forming a confidence interval for μ_D. If it is reasonable to assume that the differences within each pair are approximately normally distributed, a $(1 - \alpha)100\%$ confidence interval for $\mu_D = \mu_1 - \mu_2$ is given by

$$\bar{D} \pm t_{\alpha/2}\left(\frac{s_D}{\sqrt{n}}\right)$$

where $t_{\alpha/2}$ is based on $n - 1$ degrees of freedom.

EXAMPLE **12.3** Use the data from Example 12.2 to form a 95% confidence interval for the difference in mean percentage readings using methods 1 and 2.

Solution   From Example 12.2, we observe that

$$\bar{d} = -.0167, \qquad s_D^2 = .0007, \qquad n - 1 = 11.$$

Because, with 11 degrees of freedom, $t_{0.025} = 2.201$, the desired interval is

$$-.0167 \pm (2.201)\frac{\sqrt{.0007}}{\sqrt{12}}, \quad \text{or} \quad (-.0335, +.0001).$$

The preceding methods based on the t distribution can be validly employed if it is reasonable to assume that the differences are normally distributed. When we compared two population means based on small independent samples, we required that the population variances be equal. The validity of the matched pair analysis does not require the assumption of equal population variances. The quantity S_D^2 provides an unbiased estimator for the variance of the differences, σ_D^2, regardless of the values of σ_1^2, σ_2^2, and ρ. The independent samples t test also required that both samples were taken from normally distributed populations. One way the differences within pairs can be normally distributed is if Y_{1i}, for $i = 1, 2, \ldots, n$, and Y_{2i}, for $i = 1, 2, \ldots, n$, are themselves normally distributed. However, it is possible that the pairwise differences will be normally distributed even if the Y_1's and Y_2's are not. Exercise 12.15 presents an example of such a situation. Thus the assumption that the differences be normally distributed is less restrictive than the assumption that both populations are normally distributed.

We have seen that the matched pairs experiment can be utilized to decrease the inherent variability present in the data. Further, in many situations, the assumptions required to validly employ a matched pairs analysis are less restrictive than the corresponding independent samples methods. Recall that the data from a matched pairs experiment are analyzed by focusing on the differences of the observations within each pair. Thus, some statisticians prefer to refer to the matched pairs experiment as a paired-difference experiment. In the next section, we present some common terminology associated with experimental designs and consider extensions of the independent samples experiment and the matched pairs experiment.

Exercises

12.9   Consider the data analyzed in Examples 12.2 and 12.3.

    a   Assuming that both the methods used to analyze the samples worked reasonably well, why do you think that the observations on the two halves of each ore sample will be positively correlated?

    b   Do you think that we should have taken independent observations using the two methods, or should we have conducted the paired analysis contained in the text? Why?

12.10   Two computers often are compared by running a collection of various "benchmark" programs and recording the difference in CPU time required to complete the same program. Six benchmark programs, run on two computers, produced the following table of CPU times (in minutes).

| | \multicolumn{6}{c}{Benchmark Program} |
Computer	1	2	3	4	5	6
1	1.12	1.73	1.04	1.86	1.47	2.10
2	1.15	1.72	1.10	1.87	1.46	2.15

a Do the data provide sufficient evidence to indicate a difference in mean CPU times required for the two computers to complete a job? Test using $\alpha = .05$.

b Give bounds for the associated p-value.

c Find a 95% confidence interval for the difference in mean CPU time required for the two computers to complete a job.

12.11 When Y_{1i}, for $i = 1, 2, \ldots, n$, and Y_{2i}, for $i = 1, 2, \ldots, n$, represent independent samples from two populations with means μ_1 and μ_2 and variances σ_1^2 and σ_2^2, respectively, we determined that $\sigma_{(\bar{Y}_1 - \bar{Y}_2)}^2 = \frac{1}{n}(\sigma_1^2 + \sigma_2^2)$. If the samples were paired and we computed the differences, D_i, for $i = 1, 2, \ldots, n$, we determined that $\sigma_{\bar{D}}^2 = \frac{1}{n}(\sigma_1^2 + \sigma_2^2 - 2\rho\sigma_1\sigma_2)$.

a When is $\sigma_{(\bar{Y}_1 - \bar{Y}_2)}^2$ equal to $\sigma_{\bar{D}}^2$?

b When is $\sigma_{(\bar{Y}_1 - \bar{Y}_2)}^2$ less than $\sigma_{\bar{D}}^2$?

c Based on the discussion in the text and your answers to (a) and (b), when would it be better to implement the paired experiment and when would it be better to implement the independent samples experiment?

12.12 Two procedures for sintering copper are to be compared by testing each procedure on six different types of powder. The measurement of interest is the porosity (volume percentage due to voids) of each test specimen. The results of the tests are as shown in the accompanying table.

Powder	Procedure I	Procedure II
1	21	23
2	27	26
3	18	21
4	22	24
5	26	25
6	19	16

Is there sufficient evidence to claim that Procedure II produces higher mean porosity values? Give bounds for the p-value. What would you conclude at the $\alpha = .05$ level?

12.13 A plant manager, in deciding whether to purchase a machine of design A or design B, checks the times for completing a certain task on each machine. Eight technicians were used in the experiment, with each technician using both machine A and machine B in a randomized order. The times (in seconds) required to complete the task are given in the accompanying table.

Technician	A	B
1	32	30
2	40	39
3	42	42
4	26	23
5	35	36
6	29	27
7	45	41
8	22	21

a Test to see if there is a significant difference between mean completion times, at the 5% significance level.

b Do you think pairing on technicians was worthwhile in this case? Explain.

c What assumptions are necessary for the test in (a)?

12.14 "Muck" is the rich, highly organic type of soil that serves as the primary growth medium for vegetation in the Florida Everglades. Because of the high concentration of organic material, muck can be destroyed by a variety of natural and human-made causes. Members of the Florida Game and Fresh Water Fish Commission, under the supervision of James Schortemeyer, staked out several plots in the Everglades in May 1972. The depth of muck at each location was measured when each plot was marked and again in October 1978. The following table identifies a portion of the data (given in inches) obtained.

Plot	1972 Reading	1978 Reading	Plot	1972 Reading	1978 Reading
1	34.5	31.5	9	44.0	35.2
2	44.0	37.9	10	40.5	37.2
3	37.5	35.5	11	27.0	24.7
4	27.0	23.0	12	29.5	25.8
5	37.0	34.5	13	31.5	29.0
6	40.0	31.1	14	35.0	36.8
7	47.2	46.0	15	44.0	36.5
8	35.2	31.0			

a Test to see if there is sufficient evidence to indicate a significant loss in average muck depth between 1972 and 1978. Give bounds on the associated p-value. What would you conclude if you desired to implement an $\alpha = .01$ level test? (Although you are free to take the necessary differences in any order you prefer, the answer provided at the back of the book assumes that the differences were formed by taking 1978 readings minus 1972 readings.)

b Give a 95% confidence interval for the difference in mean muck depths for 1978 and 1972. Interpret this interval. [See the remark following (a).]

c Give a 95% confidence interval for the 1972 mean muck depth in the portion of the Everglades in which the study was conducted.

d Repeat the instructions of (c) for the year 1978.

e What assumptions are necessary to apply the techniques you used in answering (a) and (b)? (c) and (d)?

12.15   Refer to the matched pairs experiment and assume that the ith measurement ($i = 1, 2$), in the jth pair, where $j = 1, 2, \ldots, n$, is

$$Y_{ij} = \mu_i + U_j + \varepsilon_{ij}$$

where

μ_i = expected response for population i, where $i = 1, 2$

U_j = a random variable that is uniformly distributed on the interval $(-1, +1)$

ε_{ij} = random error associated with the ith measurement in the jth pair.

Assume that ε_{ij} are independent normal random variables with $E(\varepsilon_{ij}) = 0$, and $V(\varepsilon_{ij}) = \sigma^2$ and that U_j and ε_{ij} are independent.

a   Find $E(Y_{ij})$.

b   Argue that Y_{1j}, for $j = 1, 2, \ldots, n$, are *not* normally distributed. (There is no need to actually find the distribution of the Y_1 values.)

c   Show that $\text{Cov}(Y_{1j}, Y_{2j}) = 1/3$, for $j = 1, 2, \ldots, n$.

d   Show that $D_j = Y_{1j} - Y_{2j}$ are independent, normally distributed random variables.

e   In (a) through (d), you verified that the differences within each pair can be normally distributed even though the individual measurements within the pairs are not. Can you come up with another example that illustrates this same phenomenon?

12.4 Some Elementary Experimental Designs

In Chapters 8 and 10, we considered methods to compare the means of two populations based on independent random samples obtained from each. Section 12.3 dealt with a comparison of two population means through the matched pairs experiment. In this section, we present general considerations associated with designing an experiment and specifically consider extensions of the independent samples and matched pairs methodologies when the objective is to compare the means of more than two populations.

Suppose that we wish to compare five teaching techniques, A, B, C, D, and E, and that we may utilize 125 students in the study. The objective is to compare the mean scores on a standardized test for students taught by each of the five methods. How would we proceed? Even though the 125 students are in some sense representative of the students that these teaching methods target, are the students all identical? The answer is obviously no. There are likely to be boys and girls in the group and the methods might not be equally effective for both genders. There are likely to be differences in the native abilities of the students in the group, resulting in some students performing better regardless of the teaching method used. Different students may come from families that place different emphases on education, and this could have an impact on the scores on the standardized test. In addition, there may be other differences among the 125 students that would have an unanticipated effect on the test scores. Based on these considerations, we decide that it might be wise to *randomly assign* 25 students to each of 5 groups. Each group will be taught using one of the techniques under study. The random division of the students into the five groups achieves two objectives. First, we eliminate the possible biasing effect of individual characteristics of the students on the measurements that we make. Second, it provides a probabilistic basis for the selection of the sample that permits the statistician to calculate probabilities associated with the observations in the sample and to use these probabilities in making inferences.

The preceding experiment illustrates the basic components of a designed experiment. The experimental units in this study are the individual students.

DEFINITION **12.1**   The objects upon which measurements are taken are called *experimental units*.

This experiment involves a single *factor*, namely, method of teaching. In this experiment the factor, method of teaching, has five *levels*; A, B, C, D, and E.

DEFINITION **12.2**   Variables completely controlled by the experimenter are called *factors*. The intensity level (distinct subcategory) of a factor is called its *level*.

In a single-factor experiment like the preceding one, each of the levels of the single factor represents a *treatment*. Thus, in our education example there are five treatments, one corresponding to each of the teaching methods. As another example, consider an experiment conducted to investigate the effect of various amounts of nitrogen and phosphate on the yield of a variety of corn. An *experimental unit* would be a specified acreage, say, 1 acre, of corn. A *treatment* would be a fixed number of pounds of nitrogen x_1 and of phosphate x_2 applied to a given acre of corn. For example, one treatment might be to use $x_1 = 100$ pounds of nitrogen per acre and $x_2 = 200$ pounds of phosphate. A second treatment might correspond to $x_1 = 150$ and $x_2 = 100$. Notice that the experimenter could use different amounts (x_1, x_2) of nitrogen and phosphate and that each *combination* would represent a different treatment.

DEFINITION **12.3**   A *treatment* is a specific combination of factor levels.

The preceding experiment for comparing teaching methods A, B, C, D, and E entailed *randomly* dividing the 125 students into five groups, each of size 25. Each group received exactly one of the *treatments*. This is an example of a *completely randomized design*.

DEFINITION **12.4**   A *completely randomized design* to compare k treatments is one in which a group of n relatively homogeneous experimental units are randomly divided into k subgroups of sizes $n_1, n_2, \ldots, n_k$ $(n_1 + n_2 + \cdots + n_k = n)$. All the experimental units in each subgroup receive the same treatment, so each treatment is applied to exactly one subgroup.

Associated with each treatment is a population (often conceptual) consisting of all observations that would have resulted if the treatment were repeatedly applied. In the teaching example, we could envision a population of all possible test scores if *all* students were taught using method A. Corresponding conceptual populations are associated with each of the other teaching methods. Thus, each treatment has a corresponding population of measurements. The observations obtained from a completely randomized design are typically viewed as being *independent random samples* taken from the populations corresponding to each of the treatments.

Suppose that we wish to compare five brands of aspirin, A, B, C, D, and E, regarding the mean amount of active ingredient per tablet for each of the brands. We decide to select 100 tablets randomly from the production of each manufacturer and use the results to implement the comparison. In this case, we physically sampled five distinct populations. Although we did not "apply" the different treatments to a homogeneous batch of blank tablets, it is common to refer to this experiment as involving a single factor (manufacturer) and five treatments (corresponding to the different manufacturers). Thus, in this example, for each population, we identify a corresponding treatment. Regardless of whether we have implemented a completely randomized design or taken independent samples from each of several existing populations, a one-to-one correspondence is established between the populations and the treatments. Both of these scenarios, in which independent samples are taken from each of k populations, are examples of a one-way layout.

DEFINITION **12.5**

A *one-way layout* to compare k populations is an arrangement in which independent random samples are obtained from each of the populations of interest.

Thus, a one-way layout, whether it corresponds to data obtained by using a completely randomized design or by taking independent samples from each of several existing populations, is the extension of the independent samples experiments that we considered in Chapters 8 and 10. Methods of analyzing data obtained from a one-way layout are presented in Sections 13.3 through 13.7.

In Section 12.3, we saw that a matched pairs design often yields a superior method for comparing the means of two populations or treatments. When we were interested in comparing the effectiveness of two drugs for controlling hypertension, we suggested forming matched pairs of individuals who were of the same sex and of similar age and weight. One randomly selected member of each pair received treatment 1, whereas the other received treatment 2. The objective was to control for extraneous sources of variability and thus to obtain a more precise analysis. Suppose that we wanted to compare three different medications instead of two. How would we proceed? Instead of forming several pairs of matched individuals, we could form several groups, each containing three members matched on sex, weight, and age. Within each group of three, we would randomly select one individual to receive treatment 1 and another to receive treatment 2, and then we would administer treatment 3 to the remaining member of each group. The objective of this design is identical to that of the matched pairs design—namely, to eliminate unwanted sources of variability that might creep into the observations in our experiment. This extension of the matched pairs design is called a *randomized block design*.

DEFINITION 12.6

A *randomized block design* containing b blocks and k treatments consists of b blocks of k experimental units each. The treatments are randomly assigned to the units in each block, with each treatment appearing exactly once in every block.

The difference between a randomized block design and the completely randomized design can be demonstrated by considering an experiment designed to compare subject reaction to a set of four stimuli (treatments) in a stimulus-response psychological experiment. We will denote the treatments as T_1, T_2, T_3, and T_4.

Suppose that eight subjects are to be randomly assigned to each of the four treatments. Random assignment of subjects to treatments (or vice versa) randomly distributes errors due to person-to-person variability in response to the four treatments and yields four samples that, for all practical purposes, are random and independent. This is a completely randomized experimental design.

The experimental error associated with a completely randomized design has a number of components. Some of these are due to the differences between subjects, to the failure of repeated measurements within a subject to be identical (due to the variations in physical and psychological conditions), to the failure of the experimenter to administer a given stimulus with exactly the same intensity in repeated measurements, and to errors of measurement. Reduction of any of these causes of error will increase the information in the experiment.

The subject-to-subject variation in the foregoing experiment can be eliminated by using subjects as blocks. Each subject would receive each of the four treatments assigned in a random sequence. The resulting randomized block design would appear as in Figure 12.3. Now only eight subjects are needed in order to obtain eight response measurements per treatment. Notice that each treatment occurs exactly once in each block.

The word *randomized* in the name of the design implies that the treatments are randomly assigned within a block. For our experiment, position in the block refers to the position in the sequence of stimuli assigned to a given subject over time. The purpose of the randomization (that is, position in the block) is to eliminate bias caused by fatigue or learning.

Blocks may represent time, location, or experimental material. If three treatments are to be compared and there is a suspected trend in the mean response over time, a substantial part of the time-trend variation may be removed by blocking. All three treatments would be randomly applied to experimental units in one small block of

FIGURE **12.3**
A randomized block design

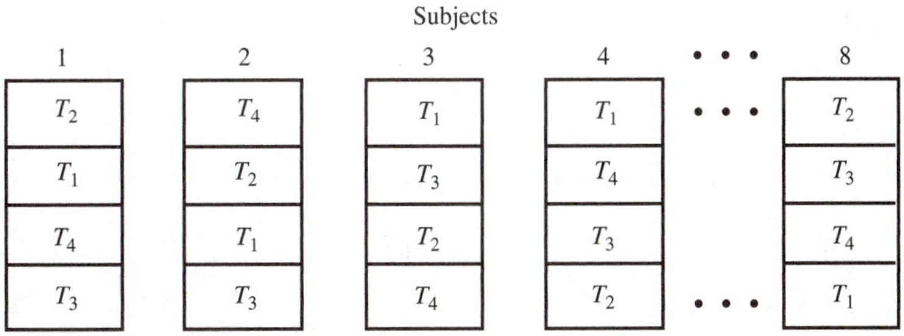

time. This procedure would be repeated in succeeding blocks of time until the required amount of data is collected. A comparison of the sale of competitive products in supermarkets should be made within supermarkets, thus using the supermarkets as blocks and removing store-to-store variability. Animal experiments in agriculture and medicine often utilize animal litters as blocks, applying all the treatments, one each, to animals within a litter. Because of heredity, animals within a litter are more homogeneous than those between litters. This type of blocking removes litter-to-litter variation. The analysis of data generated by a randomized block design is discussed in Sections 13.8 through 13.10.

The randomized block design is only one of many types of block designs. Blocking in two directions can be accomplished by using a Latin square design. Suppose that the subjects of the preceding example became fatigued as the stimuli were applied, so the last stimulus always produced a lower response than the first. If this trend (and consequent lack of homogeneity of the experimental units within a block) were true for all subjects, a Latin square design would be appropriate. The design would be constructed as shown in Figure 12.4. Each stimulus is applied once to each subject and occurs exactly once in each position of the order of presentation. All four stimuli occur in each row and in each column of the 4×4 configuration. The resulting design is a 4×4 Latin square. A Latin square design for three treatments requires a 3×3 configuration; in general, p treatments require a $p \times p$ array of experimental units. If more observations are desired per treatment, the experimenter should utilize several Latin square configurations in one experiment. In the preceding example, it would be necessary to run two Latin squares to obtain eight observations per treatment. The experiment would then contain the same number of observations per treatment as the randomized block design (Figure 12.3).

A comparison of means for any pair of stimuli would eliminate the effect of subject-to-subject variation, but it would also eliminate the effect of the fatigue trend within each stimulus, because each treatment was applied in each position of the stimuli-time administering sequence. Consequently, the effect of the trend would be canceled in comparing the means. A more extensive discussion of block designs and their analyses is contained in the texts listed in the references at the end of the chapter.

The objective of this section has been to present some of the basic considerations in designing experiments. We have discussed the role of randomization in all well-designed experiments and have focused on extensions of the independent samples

FIGURE **12.4**

A Latin square design

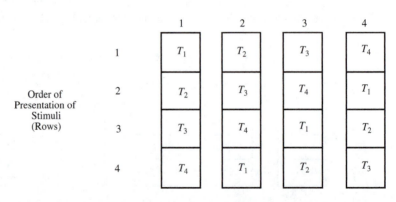

and matched pairs experiments to situations where we wish to compare more than two treatments. Particularly, we pointed out the existence of block designs, how they work, and how they can produce substantial increases in the quantity of information obtained from an experiment by reducing nuisance variation.

Exercises

12.16   Two drugs, A and B, are to be applied to five rats each. Suppose that the rats are numbered from 1 to 10. Use the random number table to assign the rats randomly to the two treatments.

12.17   Refer to Exercise 12.16. Suppose that the experiment involved three drugs, A, B, and C, with 5 rats assigned to each. Use the random number table to assign the 15 rats randomly to the three treatments.

12.18   A chemical engineer has two catalysts and three temperature settings that she wishes to use in a series of experiments.

    a   How many *treatments* (factor-level combinations) are there in this experiment? Carefully describe one of these treatments.

    b   Each experiment makes use of one catalyst–temperature combination. Show how you would use a random number table to randomize the order of the experiments.

12.19   Give two reasons for utilizing *randomization* in an experiment.

12.20   What is a *factor*?

12.21   What is a *treatment*?

12.22   Could a variable be a factor in one experiment and a nuisance variable (source of extraneous variation) in another?

12.23   If you were to design an experiment, what part of the design procedure would increase the accuracy of the experiment? What part of the design procedure would decrease the impact of extraneous sources of variability?

12.24   An experiment is to be conducted to compare the effect of digitalis on the contraction of the heart muscles of rats. The experiment is conducted by removing the heart from a live rat, slicing the heart into thin layers, and treating the layers with dosages of digitalis. The muscle contraction is then measured. If four dosages, A, B, C, and D, are to be employed, what advantage might be derived by applying A, B, C, and D to a slice of tissue from the heart of each rat? What principle of design is illustrated by this example?

12.25   Complete the assignment of treatments for the following 3×3 Latin square design.

	A	
C		

12.5 Summary

The objective of this chapter has been to identify the factors that affect the quantity of information in an experiment and to use this knowledge to design better experiments. The design of experiments is a very broad subject and certainly one not susceptible to condensation into a single chapter in an introductory text. But, the philosophy underlying design, some methods for varying information in an experiment, and some desirable strategies for design are easily explained.

We have seen that the amount of information pertinent to a parameter of interest depends on the selection of factor-level combinations (treatments) to be included in the experiment and on the allocation of the total number of experimental units to the treatments. Randomization is an important component of any designed experiment. The use of randomization helps to eliminate biases in experimental results and provides the theoretical basis for computing the probabilities that are key to the inference-making process. Blocking—comparing treatments within relatively homogeneous blocks of experimental material—can be used to eliminate block-to-block variation when comparing treatments. As such, it serves as a filter to reduce the effect of unwanted sources of variability.

The analysis of some elementary experimental designs is given in Chapter 13. A more extensive treatment of the design and analysis of experiments is a course in itself. If you are interested in exploring this subject, consult the texts listed in the references below.

References and Further Readings

Box, G. E. P., W. G. Hunter, and J. S. Hunter. 1978. *Statistics for Experimenters.* New York: John Wiley.

Cochran, W. G., and G. Cox. 1968. *Experimental Designs*, 2d ed. New York: John Wiley.

Graybill, F. 1976. *Theory and Application of the Linear Model.* North Scituate, Mass.: Duxbury Press.

Guenther, W. C. 1964. *Analysis of Variance.* Englewood Cliffs, N.J.: Prentice Hall.

Hicks, C. R., and K. V. Turner. 1999. *Fundamental Concepts in the Design of Experiments*, 5th ed. New York: Oxford University Press.

Scheaffer, R. L., W. Mendenhall, and L. Ott. 1996. *Elementary Survey Sampling*, 5th ed. Boston: PWS-Kent.

Scheffé, H. 1999. *The Analysis of Variance.* New York: Wiley Interscience.

Supplementary Exercises

12.26   How can one measure the information in a sample pertinent to a specific population parameter?

12.27   What is a *random sample*?

12.28   What factors affect the quantity of information in an experiment? What design procedures control these factors?

12.29   Refer to the matched pairs experiment of Section 12.3, and assume that the measurement receiving treatment i, where $i = 1, 2$, in the jth pair, where $j = 1, 2, \ldots, n$, is

$$Y_{ij} = \mu_i + P_j + \varepsilon_{ij}$$

where

μ_i = expected response for treatment i, for $i = 1, 2$

P_j = additive random effect (positive or negative) contribution by the jth pair of experimental units, for $j = 1, 2, \ldots, n$

ε_{ij} = random error associated with the experimental unit in the jth pair that receives treatment i.

Assume that ε_{ij} are independent normal random variables with $E(\varepsilon_{ij}) = 0$, $V(\varepsilon_{ij}) = \sigma^2$; and assume that P_j are independent normal random variables with $E(P_j) = 0$, $V(P_j) = \sigma_p^2$. Also, assume that P_j and ε_{ij} are independent.

a   Find $E(Y_{ij})$.

b   Find $E(\overline{Y}_i)$ and $V(\overline{Y}_i)$, where $\overline{Y}_i$ is the mean of the n observations receiving treatment i, where $i = 1, 2$.

c   Let $\overline{D} = \overline{Y}_1 - \overline{Y}_2$. Find $E(\overline{D})$, $V(\overline{D})$, and the probability distribution for $\overline{D}$.

12.30   Refer to Exercise 12.29. Prove that

$$\frac{\overline{D}\sqrt{n}}{S_D}$$

possesses a t distribution, under $H_0 : (\mu_1 - \mu_2) = 0$.

*12.31   Refer to Exercise 12.29. Suppose that a completely randomized design is employed for the comparison of the two treatment means. Then a response could be modeled by the expression

$$Y_{ij} = \mu_i + P_{ij} + \varepsilon_{ij},$$

but the "pair effect" P_{ij} (which will still affect an experimental unit) will be randomly selected and will likely differ from one of the $2n$ observations to another. Further, in contrast to the matched pairs experiment, the pair effects will not cancel when you calculate $(\overline{Y}_1 - \overline{Y}_2)$. Compare $V(\overline{Y}_1 - \overline{Y}_2) = V(\overline{D})$ for this design with the matched pairs design of Exercise 12.29. Why is the variance for the completely randomized design usually larger? [1]

12.32   Persons submitting computing jobs to a computer center usually are required to estimate the amount of computer time required to complete the job. This time is measured in CPUs, the amount of time that a job will occupy a portion of the computer's central processing unit's memory. A computer center decided to perform a comparison of the estimated versus actual CPU times for a particular customer. The corresponding times were available for 11 jobs. The sample data are given in the accompanying table.

---

[1] Exercises preceded by an asterisk are optional.

CPU Time	Job Number										
(minutes)	1	2	3	4	5	6	7	8	9	10	11
Estimated	.50	1.40	.95	.45	.75	1.20	1.60	2.6	1.30	.85	.60
Actual	.46	1.52	.99	.53	.71	1.31	1.49	2.9	1.41	.83	.74

a Why would you expect the observations within each of these pairs of data to be correlated?

b Do the data provide sufficient evidence to indicate that, *on the average*, the customer tends to underestimate the CPU time required for computing jobs? Test using $\alpha = .10$.

c Find the observed significance level for the test, and interpret its value.

d Find a 90% confidence interval for the difference in mean estimated CPU time versus mean actual CPU time.

12.33 The data that appear in the following table, gathered by Runzheimer and Co., Inc., give the annual operating cost comparisons for automatic and stick-shift versions of four different makes of automobiles (*USA Today*, January 10, 1985). If the data represent a random sampling of the comparative costs of operating automobiles, find a 90% confidence interval for the mean difference in operating costs between automatic and stick-shift versions of the same type of automobile. (Operating costs include oil, maintenance, tires, and fuel.)

Car Model	Annual Operating Cost	
	Automatic	Manual
Ford Tempo	$1520	$1440
Plymouth Reliant	1240	1150
AMC Encore S	1380	1150
Chevrolet Cavalier	1150	1100

12.34 An experiment was conducted to compare mean reaction time to two types of traffic signs: prohibitive (no left turn), and permissive (left turn only). Ten subjects were included in the experiment. Each subject was presented 40 traffic signs, 20 prohibitive and 20 permissive, in random order. The mean time to reaction and the number of correct actions were recorded for each subject. The mean reaction times to the 20 prohibitive and 20 permissive traffic signs for each of the ten subjects are reproduced in the following table.

Subject	Mean Reaction Times (ms) for 20 Traffic Signs	
	Prohibitive	Permissive
1	824	702
2	866	725
3	841	744
4	770	663
5	829	792
6	764	708
7	857	747
8	831	685
9	846	742
10	759	610

    a Explain why this is a matched pairs experiment, and give reasons why the pairing should be useful in increasing information on the difference between the mean reaction times to prohibitive and permissive traffic signs.

    b Do the data present sufficient evidence to indicate a difference in mean reaction times to prohibitive and permissive traffic signs? Test using $\alpha = .05$.

    c Find and interpret the approximate p-value for the test in (b).

    d Find a 95% confidence interval for the difference in mean reaction times to prohibitive and permissive traffic signs.

*12.35 Suppose that you wish to fit a model

$$Y = \beta_0 + \beta_1 x + \beta_2 x^2 + \varepsilon$$

to a set of n data points. If the n points are to be allocated at the design points $x = -1, 0, 1$, what fraction should be assigned to each value of x so as to minimize $V(\hat{\beta}_2)$? (Assume that n is large and that k_1, k_2, and k_3, $k_1 + k_2 + k_3 = 1$, are the fractions of the total number of observations to be assigned at $x = -1, 0$, and 1, respectively.)

CHAPTER **13**

The Analysis of Variance

13.1   Introduction

13.2   The Analysis of Variance Procedure

13.3   Comparison of More than Two Means: Analysis of Variance for a One-Way Layout

13.4   An Analysis of Variance Table for a One-Way Layout

13.5   A Statistical Model for the One-Way Layout

13.6   Proof of Additivity of the Sums of Squares and E(MST) for a One-Way Layout (Optional)

13.7   Estimation in the One-Way Layout

13.8   A Statistical Model for the Randomized Block Design

13.9   The Analysis of Variance for a Randomized Block Design

13.10 Estimation in the Randomized Block Design

13.11 Selecting the Sample Size

13.12 Simultaneous Confidence Intervals for More than One Parameter

13.13 Analysis of Variance Using Linear Models

13.14 Summary

References and Further Readings

13.1 Introduction

Most experiments involve a study of the effect of one or more independent variables on a response. Independent variables that can be controlled in an experiment are called *factors*, and the intensity level of a factor is called its *level*.

The analysis of data generated by a multivariable experiment requires identification of the independent variables in the experiment. These will not only be factors (controlled independent variables) but could also be directions of blocking. If one studies the wear for three types of tires, A, B, and C, on each of four automobiles, "tire types" is a factor representing a single *qualitative variable* (there is no quantitative or numerical value associated with the variable "tire type") at three levels. Automobiles are blocks and represent a single qualitative variable at four levels. The response for a Latin square design depends upon the factors that represent treatments, but it is also affected by two qualitative independent block variables, "rows" and "columns."

It is not possible to present a comprehensive treatment of the analysis of multivariable experiments in a single chapter of an introductory text. However, it is possible to introduce the reasoning underlying one method of analysis, the analysis of variance, and to show how the technique is applied to a few common experimental designs.

Methods for designing experiments to increase accuracy and to control for extraneous sources of variation were discussed in Chapter 12. In particular, the one-way layout and the randomized block design were shown to be generalizations of simple designs for the independent samples and matched pairs comparisons of means that were discussed in Chapters 8, 10, and 12. Treatments correspond to combinations of factor levels and identify the different populations of interest to the experimenter. This chapter presents an introduction to the analysis of variance and gives methods for the analysis of the one-way layout (including the completely randomized design) and randomized block designs. The analogous methods of analysis for the Latin square design are not presented in this chapter, but they can be found in the texts listed in the references at the end of the chapter.

13.2   The Analysis of Variance Procedure

The method of analysis for experiments involving several independent variables can be explained by intuitively developing the procedure or, more rigorously, through the linear models developed in Chapter 11. We begin by presenting an intuitive discussion of a procedure known as the *analysis of variance*. An outline of the linear model approach is presented in Section 13.13.

As the name implies, the analysis of variance procedure attempts to analyze the variation in a set of responses and assign portions of this variation to each of a set of independent variables. The reasoning is that response variables vary because of variation in a set of independent variables, some of which may be unknown. Because the experimenter rarely, if ever, includes all of the variables affecting the response in an experiment, random variation in the responses is observed even if all independent variables considered by the experimenter are held constant. The objective of the analysis of variance is to locate the important independent variables and determine how they affect the response.

The rationale underlying the analysis of variance can best be indicated with a symbolic discussion. The actual analysis—that is, how to do it—will be illustrated with an example.

The variability of a set of n measurements is proportional to the sum of squares of deviations $\sum_{i=1}^{n} (y_i - \bar{y})^2$, and this quantity is divided by $n - 1$ to calculate the

FIGURE **13.1**
Partitioning of the
total sum of squares
of deviations

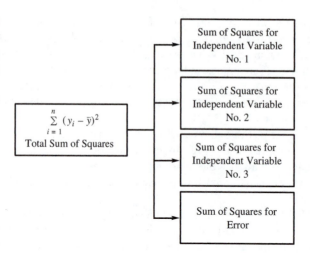

sample variance. The analysis of variance procedure partitions this sum of squares of deviations, called the *total sum of squares of deviations*, into parts, each of which is attributed to one of the independent variables in the experiment, plus a remainder that is associated with random error. Figure 13.1 illustrates such a partitioning for three independent variables. If a multivariable linear model were written for the response, as suggested in Chapter 11, the portion of the total sum of squares of deviations assigned to error would be labeled SSE.

For the cases that we consider, and under the hypothesis that the independent variables are unrelated to the response, each of the pieces of the total sum of squares of deviations, divided by an appropriate constant, provides an independent and un-biased estimator of σ^2, the variance of the experimental error. When a variable is highly related to the response, its portion of the total sum of squares (called the *sum of squares* for that variable) will be inflated. This condition can be detected by com-paring the sum of squares for that variable with the sum of squares for error, SSE. The test will be based on a statistic that possesses an F distribution and specifies that the hypothesis of no effect for the independent variable should be rejected if the value of F is large.

The mechanism involved in an analysis of variance can best be illustrated by considering a familiar example. Assume that we wish to compare the means of two normally distributed populations with means μ_1 and μ_2 and equal variances $\sigma_1^2 = \sigma_2^2 = \sigma^2$ on the basis of independent random samples of sizes $n_1 = n_2$. This experiment, formerly analyzed using the independent samples t test, will now be approached from another point of view. The total variation of the response measure-ments about their mean for the two samples is quantified by

$$\text{Total SS} = \sum_{i=1}^{2} \sum_{j=1}^{n_1} (Y_{ij} - \overline{Y})^2,$$

where Y_{ij} denotes the jth observation in the ith sample and where $\overline{Y}$ is the mean of all $2n_1 = n$ observations. This quantity can be partitioned into two parts, as follows:

$$\text{Total SS} = \sum_{i=1}^{2} \sum_{j=1}^{n_1} (Y_{ij} - \overline{Y})^2$$

$$= n_1 \underbrace{\sum_{i=1}^{2} (\overline{Y}_i - \overline{Y})^2}_{\text{SST}} + \underbrace{\sum_{i=1}^{2} \sum_{j=1}^{n_1} (Y_{ij} - \overline{Y}_i)^2}_{\text{SSE}}$$

(proof deferred to Section 13.6), where $\overline{Y}_i$ is the average of the observations in the ith sample, for $i = 1, 2$. Let us examine the quantity SSE more closely. Recall that we have assumed that the underlying population variances are equal and that $n_1 = n_2$.

$$\text{SSE} = \sum_{i=1}^{2} \sum_{j=1}^{n_1} (Y_{ij} - \overline{Y}_i)^2 = \sum_{i=1}^{2} (n_1 - 1) S_i^2$$

$$= (n_1 - 1)S_1^2 + (n_1 - 1)S_2^2$$

where

$$S_i^2 = \frac{1}{n_1 - 1} \sum_{j=1}^{n_1} (Y_{ij} - \overline{Y}_i)^2.$$

Recall that, in the case $n_1 = n_2$, the "pooled" estimator for the common variance σ^2 is given by

$$S_p^2 = \frac{(n_1 - 1)S_1^2 + (n_2 - 1)S_2^2}{n_1 + n_2 - 2} = \frac{(n_1 - 1)S_1^2 + (n_1 - 1)S_2^2}{n_1 + n_1 - 2} = \frac{\text{SSE}}{2n_1 - 2}.$$

We have partitioned the total sum of squares of deviations into two parts. One part, SSE, can be divided by $2n_1 - 2$ to obtain the pooled estimator of σ^2. Because there are only two treatments (or populations) and $n_1 = n_2$, the other part,

$$\text{SST} = n_1 \sum_{i=1}^{2} (\overline{Y}_i - \overline{Y})^2 = \frac{n_1}{2} (\overline{Y}_1 - \overline{Y}_2)^2,$$

which we will call the *sum of squares for treatments*, will be large if $|\overline{Y}_1 - \overline{Y}_2|$ is large. Hence, the larger SST is, the greater will be the weight of evidence to indicate a difference between μ_1 and μ_2. When will SST be large enough to indicate a real difference between μ_1 and μ_2?

Because we assume that Y_{ij} is normally distributed with $E(Y_{ij}) = \mu_i$, for $i = 1, 2$, and $V(Y_{ij}) = \sigma^2$, and because SSE/$(2n_1 - 2)$ is identical to the pooled estimator of σ^2 used in Chapters 8 and 10, it follows that

$$E\left(\frac{\text{SSE}}{2n_1 - 2}\right) = \sigma^2$$

and that

$$\frac{\text{SSE}}{\sigma^2} = \sum_{j=1}^{n_1} \frac{(Y_{1j} - \overline{Y}_1)^2}{\sigma^2} + \sum_{j=1}^{n_1} \frac{(Y_{2j} - \overline{Y}_2)^2}{\sigma^2}$$

has a χ^2 distribution with $2n_1 - 2$ degrees of freedom (see Section 8.8).

In Section 13.6, we will derive a result that implies that

$$E(\text{SST}) = \sigma^2 + \frac{n_1}{2}(\mu_1 - \mu_2)^2.$$

Notice that SST estimates σ^2 if $\mu_1 = \mu_2$ and a quantity larger than σ^2 if $\mu_1 \neq \mu_2$. Under the hypothesis that $\mu_1 = \mu_2$, it follows that

$$Z = \frac{\overline{Y}_1 - \overline{Y}_2}{\sqrt{2\sigma^2/n_1}}$$

has a standard normal distribution; hence,

$$Z^2 = \left(\frac{n_1}{2}\right)\left[\frac{(\overline{Y}_1 - \overline{Y}_2)^2}{\sigma^2}\right] = \frac{\text{SST}}{\sigma^2}$$

has a χ^2 distribution with 1 degree of freedom.

Notice that SST is a function of only the sample means $\overline{Y}_1$ and $\overline{Y}_2$, whereas SSE is a function of only the sample variances S_1^2 and S_2^2. Theorem 7.3 implies that, for $i = 1, 2$, the sample means, $\overline{Y}_i$, and sample variances, S_i^2, are independent. Because the samples are assumed to be independent, it follows that SST and SSE are independent random variables. Hence, from Definition 7.3, under the hypothesis that $\mu_1 = \mu_2$,

$$\frac{\dfrac{\text{SST}}{\sigma^2}\bigg/ 1}{\dfrac{\text{SSE}}{\sigma^2}\bigg/ (2n_1 - 2)} = \frac{\text{SST}/1}{\text{SSE}/(2n_1 - 2)}$$

has an F distribution with $\nu_1 = 1$ numerator degree of freedom and $\nu_2 = (2n_1 - 2)$ denominator degrees of freedom.

Sums of squares divided by their respective degrees of freedom are called *mean squares*. In this case, the mean square for error and the mean square for treatments are given by

$$\text{MSE} = \frac{\text{SSE}}{2n_1 - 2} \quad \text{and} \quad \text{MST} = \frac{\text{SST}}{1}.$$

Under $H_0 : \mu_1 = \mu_2$, both MST and MSE estimate σ^2. However, when H_0 is false and $\mu_1 \neq \mu_2$, MST estimates something larger than σ^2 and tends to be larger than

MSE. To test $H_0 : \mu_1 = \mu_2$ versus $H_a : \mu_1 \neq \mu_2$, we use

$$F = \frac{\text{MST}}{\text{MSE}}$$

as the test statistic.

Disagreement with the null hypothesis is indicated by a large value of F; hence, the rejection region for a test with significance level α is

$$F > F_\alpha.$$

Thus the analysis of variance test results in a one-tailed F test. The degrees of freedom for F are those associated with MST and MSE. In the present instance, as previously indicated, F is based upon $v_1 = 1$ and $v_2 = 2n_1 - 2$ numerator and denominator degrees of freedom, respectively.

For the two-sample problem under consideration, the F test just described is equivalent to the two-tailed t test of Chapter 10. As we will see in Section 13.3, the F test readily generalizes to allow comparison of any number of treatments.

EXAMPLE **13.1**   The coded values for the measure of elasticity in plastic, prepared by two different processes for samples of six drawn randomly from each of the two processes, are given in Table 13.1. Do the data present sufficient evidence to indicate a difference in mean elasticity for the two processes?

Solution   Although the two-sample t test of Section 10.8 could be used to analyze these data, we will use our analysis of variance F test discussed earlier in this section. The three desired sums of squares of deviations are

$$\text{Total SS} = \sum_{i=1}^{2} \sum_{j=1}^{6} (y_{ij} - \bar{y})^2 = \sum_{i=1}^{2} \sum_{j=1}^{6} y_{ij}^2 - \frac{1}{12} \left(\sum_{i=1}^{2} \sum_{j=1}^{6} y_{ij} \right)^2$$

$$= 711.35 - \frac{1}{12}(91.9)^2 = 7.5492$$

$$\text{SST} = n_1 \sum_{i=1}^{2} (\bar{y}_i - \bar{y})^2 = 6 \sum_{i=1}^{2} (\bar{y}_i - \bar{y})^2 = 1.6875$$

Table 13.1. Data for
Example 13.1

A	B
6.1	9.1
7.1	8.2
7.8	8.6
6.9	6.9
7.6	7.5
8.2	7.9

$$SSE = \sum_{i=1}^{2} \sum_{j=1}^{6} (y_{ij} - \bar{y}_i)^2 = 5.8617.$$

(You may verify that SSE is the pooled sum of squares of the deviations for the two samples. Also, Total SS = SST + SSE.) The mean squares for treatment and error, respectively, are

$$MST = \frac{SST}{1} = 1.6875$$

$$MSE = \frac{SSE}{2n_1 - 2} = \frac{5.8617}{10} = .58617.$$

To test the null hypothesis $\mu_1 = \mu_2$, we compute the value of the test statistic

$$F = \frac{MST}{MSE} = \frac{1.6875}{.58617} = 2.88$$

and reject H_0 if the calculated value of F exceeds F_α. The critical value of the F statistic with 1 numerator degree of freedom and 10 denominator degrees of freedom for $\alpha = .05$ is $F_{.05} = 4.96$. Although the mean square for treatments is almost three times larger than the mean square for error, it is not large enough to permit rejection of the null hypothesis. Consequently, at the $\alpha = .05$ level of significance, there is not sufficient evidence to indicate a difference between μ_1 and μ_2. The attained significance level is given by p-value $= P(F > 2.88)$. According to Table 7, Appendix III, p-value $> .10$.

The purpose of this example is to illustrate the computations involved in a simple analysis of variance. The F test for comparing two means is equivalent to a two-sample t test, because the square of a t-distributed random variable with v degrees of freedom has an F distribution with 1 numerator degree of freedom and v denominator degrees of freedom. You can easily verify that the square of $t_{.025} = 2.228$ (used for the two-tailed test with $\alpha = .05$ and $v = 10$ degrees of freedom) is equal to $F_{.05} = 4.96$. Had the t test been used for Example 13.1, we would have obtained $t = -1.6967$, which satisfies the relationship $t^2 = (-1.6967)^2 = 2.88 = F$.

Exercises

13.1   The reaction times for two different stimuli in a psychological word association experiment were compared by using each stimulus on independent random samples of size eight. Thus, a total of 16 people were used in the experiment. Do the following data present sufficient evidence to indicate that there is a difference in the mean reaction times for the two stimuli?

Stimulus 1	1	3	2	1	2	1	3	2
Stimulus 2	4	2	3	3	1	2	3	3

a   Use the analysis of variance approach to test the appropriate hypotheses. Test at the $\alpha = .05$ level of significance.

b   Test the appropriate hypotheses by using the two-sample t test for comparing population means, which we developed in Section 10.8. Compare the value of the t statistic to the value of the F statistic calculated in (a).

c   What assumptions are necessary for the test in (a)?

13.2   Refer to Exercises 8.76 and 10.65.

a   Use an F test to determine whether there is sufficient evidence to claim a difference in the mean verbal SAT scores for high school students who intend to major in engineering and language/literature. Give bounds for the associated p-value. What would you conclude at the $\alpha = .05$ level of significance?

b   How does the value of the F statistic obtained in (a) compare to the value of the t statistic that you obtained in Exercise 10.65?

c   What assumptions are necessary for the analyses performed in (a) and (b)?

13.3 Comparison of More than Two Means: Analysis of Variance for a One-Way Layout

An analysis of variance to detect a difference in a set of more than two population means is a simple generalization of the analysis of variance of Section 13.2. The random selection of independent samples from k populations is known as a *one-way layout*. As indicated in Section 12.4, the data in a one-way layout may correspond to data obtained from a completely randomized experimental design (see Definition 12.4) *or* from taking independent samples from each of several existing populations.

Assume that independent random samples have been drawn from k normal populations with means $\mu_1, \mu_2, \ldots, \mu_k$, respectively, and variance σ^2. All populations are assumed to possess equal variances. To be completely general, we will allow the sample sizes to be unequal and let n_i, for $i = 1, 2, \ldots, k$, be the number in the sample drawn from the ith population. The total number of observations in the experiment will be $n = n_1 + n_2 + \cdots + n_k$.

Let Y_{ij} denote the response on the jth experimental unit in the ith sample and let $Y_{i\bullet}$ and $\overline{Y}_{i\bullet}$ represent the total and mean, respectively, for the responses in the ith sample. The dot in the second position in the subscript of $Y_{i\bullet}$ is intended to remind you that this quantity is computed by summing over all possible values of the subscript that is replaced by the dot. Similarly the dot in the subscript of $\overline{Y}_{i\bullet}$ indicates that this mean is calculated by averaging over the values corresponding to the second subscript—j, in this case. Thus, for $i = 1, 2, \ldots, k$,

$$Y_{i\bullet} = \sum_{j=1}^{n_i} Y_{ij} \quad \text{and} \quad \overline{Y}_{i\bullet} = \left(\frac{1}{n_i}\right) \sum_{j=1}^{n_i} Y_{ij} = \left(\frac{1}{n_i}\right) Y_{i\bullet}.$$

This modification in the symbols for sample totals and averages will simplify the computing formulas for the sums of squares.

Then, as in the analysis of variance involving two means, we have

$$
\text{Total SS} = \text{SST} + \text{SSE}
$$

(proof deferred to Section 13.6), where

$$
\text{Total SS} = \sum_{i=1}^{k}\sum_{j=1}^{n_i}(Y_{ij}-\overline{Y})^2 = \sum_{i=1}^{k}\sum_{j=1}^{n_i}Y_{ij}^2 - \text{CM}
$$

$$
\text{CM} = \frac{(\text{total of all observations})^2}{n} = \frac{1}{n}\left(\sum_{i=1}^{k}\sum_{j=1}^{n_i}Y_{ij}\right)^2 = n\overline{Y}^2
$$

(the symbol CM denotes *correction for the mean*),

$$
\text{SST} = \sum_{i=1}^{k}n_i(\overline{Y}_{i\bullet}-\overline{Y})^2 = \sum_{i=1}^{k}\frac{Y_{i\bullet}^2}{n_i} - \text{CM}
$$

$$
\text{SSE} = \text{Total SS} - \text{SST}.
$$

Although the easy way to compute SSE is by subtraction, as shown earlier, it is interesting to observe that SSE is the pooled sum of squares for all k samples and is equal to

$$
\text{SSE} = \sum_{i=1}^{k}\sum_{j=1}^{n_i}(Y_{ij}-\overline{Y}_{i\bullet})^2
$$

$$
= \sum_{i=1}^{k}(n_i-1)S_i^2
$$

where

$$
S_i^2 = \frac{1}{n_i-1}\sum_{j=1}^{n_i}(Y_{ij}-\overline{Y}_{i\bullet})^2.
$$

Notice that SSE is a function of *only* the sample variances S_i^2, for $i=1,2,\ldots,k$. Because each of the S_i^2 values provides an unbiased estimator for $\sigma_i^2 = \sigma^2$ with n_i-1

degrees of freedom, an unbiased estimator of σ^2 based on $(n_1 + n_2 + \cdots + n_k - k) = n - k$ degrees of freedom is provided by

$$S^2 = \text{MSE} = \frac{\text{SSE}}{(n_1 - 1) + (n_2 - 1) + \cdots + (n_k - 1)} = \frac{\text{SSE}}{n - k}.$$

Because

$$\overline{Y} = \frac{1}{n} \sum_{i=1}^{k} \sum_{j=1}^{n_i} Y_{ij} = \frac{1}{n} \sum_{i=1}^{k} n_i \overline{Y}_{i\bullet},$$

it follows that SST is a function of *only* the sample means $\overline{Y}_{i\bullet}$, for $i = 1, 2, \ldots, k$. The mean square for treatments possesses $(k - 1)$ degrees of freedom, that is, one less than the number of means, and is

$$\text{MST} = \frac{\text{SST}}{k - 1}.$$

To test the null hypothesis,

$$H_0 : \mu_1 = \mu_2 = \cdots = \mu_k,$$

against the alternative that at least one of the equalities does not hold, we compare MST with MSE, using the F statistic based upon $\nu_1 = k - 1$ and $\nu_2 = n - k$ numerator and denominator degrees of freedom, respectively. The null hypothesis will be rejected if

$$F = \frac{\text{MST}}{\text{MSE}} > F_\alpha$$

where F_α is the critical value of F for a test of level α. In Exercise 13.6, you will prove that, under $H_0 : \mu_1 = \mu_2 = \cdots = \mu_k$, the statistic F possesses an F distribution with $k - 1$ and $n - k$ numerator and denominator degrees of freedom, respectively.

In keeping with our previous conventions, we will use the notation y_{ij} to denote the observed value of Y_{ij}. Similarly, we will use $y_{i\bullet}$ and $\overline{y}_{i\bullet}$ to denote the observed values of $Y_{i\bullet}$ and $\overline{Y}_{i\bullet}$, for $i = 1, 2, \ldots, k$, respectively. Intuitively, the greater the differences among the observed values of the treatment means, $\overline{y}_{1\bullet}, \overline{y}_{2\bullet}, \ldots, \overline{y}_{k\bullet}$, the greater is the evidence to indicate a difference among their corresponding population means. It can be seen from the preceding expression that the observed value of SST is 0 when all the treatment means are identical, because then $\overline{y}_{1\bullet} = \overline{y}_{2\bullet} = \cdots = \overline{y}_{k\bullet} = \overline{y}$ and the deviations appearing in SST, $(\overline{y}_{i\bullet} - \overline{y})$, for $i = 1, 2, \ldots, k$, all equal zero. As the treatment means get farther apart, the deviations $(\bar{y}_{i\bullet} - \bar{y})$ increase in absolute value and the observed value of SST increases in magnitude. Consequently, the larger the observed value of SST, the greater is the weight of evidence favoring rejection of the null hypothesis. This same line of reasoning applies to the F tests employed in the analyses of variance for all designed experiments.

The assumptions underlying the analysis of variance F tests deserve particular attention. The samples are assumed to have been randomly selected from the k populations independently. The populations are assumed to be normally distributed with equal variances σ^2 and means $\mu_1, \mu_2, \ldots, \mu_k$. Moderate departures from these assumptions will not seriously affect the properties of the test. This is particularly true of the normality assumption. The assumption of equal population variances is less critical if the sizes of the samples from the respective populations are all equal $(n_1 = n_2 = \cdots = n_k)$. A one-way layout with equal number of observations per treatment is said to be *balanced*.

EXAMPLE 13.2 Four groups of students were subjected to different teaching techniques and tested at the end of a specified period of time. As a result of dropouts from the experimental groups (due to sickness, transfer, and so on), the number of students varied from group to group. Do the data shown in Table 13.2 present sufficient evidence to indicate a difference in mean achievement for the four teaching techniques?

Solution The observed values of the quantities necessary to compute the value of the F statistic are

$$CM = \frac{1}{n}\left(\sum_{i=1}^{4}\sum_{j=1}^{n_i} y_{ij}\right)^2 = \frac{(1779)^2}{23} = 137,601.8$$

$$\text{Total SS} = \sum_{i=1}^{4}\sum_{j=1}^{n_i} y_{ij}^2 - CM = 139,511 - 137,601.8 = 1909.2$$

Table 13.2. Data for Example 13.2

	1	2	3	4
	65	75	59	94
	87	69	78	89
	73	83	67	80
	79	81	62	88
	81	72	83	
	69	79	76	
		90		
$y_{i\bullet}$	454	549	425	351
n_i	6	7	6	4
$\bar{y}_{i\bullet}$	75.67	78.43	70.83	87.75

$$SST = \sum_{i=1}^{4} \frac{y_{i\bullet}^2}{n_i} - CM = 138,314.4 - 137,601.8 = 712.6$$

$$SSE = \text{Total SS} - SST = 1196.6.$$

The observed values of the mean squares for treatment and error are

$$MST = \frac{SST}{k-1} = \frac{712.6}{3} = 237.5 \quad \text{and}$$

$$MSE = \frac{SSE}{n-k} = \frac{1196.6}{19} = 63.0.$$

Finally, the observed value of the test statistic for testing the null hypothesis H_0 : $\mu_1 = \mu_2 = \mu_3 = \mu_4$ is

$$F = \frac{MST}{MSE} = \frac{237.5}{63.0} = 3.77$$

where the appropriate numerator and denominator degrees of freedom are $v_1 = k - 1 = 3$ and $v_2 = n - k = (6 + 7 + 6 + 4) - 4 = 19$, respectively.

The attained significance level is given by p value $= P(F > 3.77)$. Using Table 7, Appendix III, with 3 numerator and 19 denominator degrees of freedom, we see that $.025 < p$-value $< .05$. Thus, if we choose $\alpha = .05$ (or any larger value), we reject the null hypothesis and conclude that there is sufficient evidence to indicate a difference in mean achievement among the four teaching procedures.

You may feel that this conclusion could have been made on the basis of visual observation of the treatment means. However, it is not difficult to construct a set of data that will lead the visual decision maker to erroneous results.

13.4   An Analysis of Variance Table for a One-Way Layout

The calculations for an analysis of variance are usually displayed in an analysis of variance (ANOVA or AOV) table. The table for the design in Section 13.3 for comparing k treatment means is shown in Table 13.3. The first column shows the source associated with each sum of squares of deviations; the second column gives the respective degrees of freedom; the third and fourth columns give the corresponding sums of squares and mean squares, respectively. A calculated value of F, comparing MST and MSE, is usually shown in the fifth column. Notice that the degrees of freedom and sums of squares sum to their respective totals.

The ANOVA table for Example 13.2, shown in Table 13.4, gives a compact presentation of the appropriate computed quantities for the analysis of variance.

Table 13.3. ANOVA table for a one-way layout

Source	d.f.	SS	MS	F
Treatments	$k-1$	SST	$MST = \dfrac{SST}{k-1}$	$\dfrac{MST}{MSE}$
Error	$n-k$	SSE	$MSE = \dfrac{SSE}{n-k}$	
Total	$n-1$	$\displaystyle\sum_{i=1}^{k}\sum_{j=1}^{n_i}(y_{ij}-\overline{y})^2$		

Table 13.4. ANOVA table for Example 13.2

Source	d.f.	SS	MS	PF
Treatments	3	712.6	237.5	3.77
Error	19	1196.6	63.0	
Total	22	1909.2		

Exercises

13.3 State the assumptions underlying the analysis of variance of a completely randomized design.

13.4 Refer to Example 13.2. Calculate the value of SSE by pooling the sums of squares of deviations within each of the four samples, and compare the answer with the value obtained by subtraction. This is an extension of the pooling procedure used in the two-sample case discussed in Section 13.2.

*13.5 In Exercise 6.51 we showed that if Y_1 and Y_2 are independent χ^2 distributed random variables with ν_1 and ν_2 degrees of freedom, respectively, then $Y_1 + Y_2$ has a χ^2 distribution with $\nu_1 + \nu_2$ degrees of freedom. Now suppose that $W = U + V$, where U and V are *independent* random variables, and that W and V have χ^2 distributions with r and s degrees of freedom, respectively, where $r > s$. Use moment-generating functions to prove that U must have a χ^2 distribution with $r - s$ degrees of freedom.[1]

13.6 Suppose that independent samples of sizes $n_1, n_2, \ldots, n_k$ are taken from each of k normally distributed populations with means $\mu_1, \mu_2, \ldots, \mu_k$ and common variances, all equal to σ^2. Let Y_{ij} denote the jth observation from population i, for $j = 1, 2, \ldots, n_i$ and $i = 1, 2, \ldots, k$, and let $n = n_1 + n_2 + \cdots + n_k$.

a Recall that

$$SSE = \sum_{i=1}^{k}(n_i - 1)S_i^2 \quad \text{where} \quad S_i^2 = \frac{1}{n_i - 1}\sum_{j=1}^{n_i}(Y_{ij} - \overline{Y}_{i\bullet})^2.$$

Argue that SSE/σ^2 has a χ^2 distribution with $n_1 + n_2 + \cdots + n_k - k = n - k$ degrees of freedom.

---

[1]Exercises preceded by an asterisk are optional.

b Argue that under the null hypothesis $H_0 : \mu_1 = \mu_2 = \cdots = \mu_k$, all the Y_{ij} are independent, normally distributed random variables with the *same mean and variance*. Use Theorem 7.3 to argue further that, under the null hypothesis,

$$\text{Total SS} = \sum_{i=1}^{k} \sum_{j=1}^{n_i} (Y_{ij} - \overline{Y})^2$$

is such that $(\text{Total SS})/\sigma^2$ has a χ^2 distribution with $n - 1$ degrees of freedom.

c In Section 13.3, we argued that SST is a function of only the sample means and that SSE is a function of only the sample variances. Hence, SST and SSE are independent. Recall that Total SS = SST + SSE. Use the results of Exercise 13.5 and (a) and (b) to show that, under the hypothesis $H_0 : \mu_1 = \mu_2 = \cdots = \mu_k$, SST/σ^2 has a χ^2 distribution with $k - 1$ degrees of freedom.

d Use the results of previous parts to argue that, under the hypothesis $H_0 : \mu_1 = \mu_2 = \cdots = \mu_k$, $F = \text{MST/MSE}$ has an F distribution with $k - 1$ and $n - k$ numerator and denominator degrees of freedom, respectively.

13.7 In a comparison of the strengths of concrete produced by four experimental mixes, three specimens were prepared from each type of mix. Each of the 12 specimens was subjected to increasingly compressive loads until breakdown. The accompanying table gives the compressive loads, in tons per square inch, attained at breakdown. Specimen numbers 1–12 are indicated in parentheses for identification purposes. Assuming that the requirements for a one-way layout are met, analyze the data. State whether there is statistical support at the $\alpha = .05$ level of significance for the conclusion that at least one of the concretes differs in average strength from the others.

Mix A	Mix B	Mix C	Mix D
(1) 2.30	(2) 2.20	(3) 2.15	(4) 2.25
(5) 2.20	(6) 2.10	(7) 2.15	(8) 2.15
(9) 2.25	(10) 2.20	(11) 2.20	(12) 2.25

13.8 A clinical psychologist wished to compare three methods for reducing hostility levels in university students. A certain psychological test (HLT) was used to measure the degree of hostility. High scores on this test indicate great hostility. Eleven students obtaining high and nearly equal scores were used in the experiment. Five were selected at random from among the eleven problem cases and treated by method A. Three were taken at random from the remaining six students and treated by method B. The other three students were treated by method C. All treatments continued throughout a semester. Each student was given the HLT test again at the end of the semester, with the results shown in the accompanying table. Do the data provide sufficient evidence to indicate that at least one of the methods of treatment produces a mean student response different from the other methods? Give bounds for the attained significance level. What would you conclude at the $\alpha = .05$ level of significance?

Method A	Method B	Method C
73	54	79
83	74	95
76	71	87
68		
80		

13.9 It is believed that women in the postmenopausal phase of life suffer from calcium deficiency. This phenomenon is associated with the relatively high proportion of bone fractures for women in that age group. Is this calcium deficiency associated with an estrogen deficiency, a condition that occurs after menopause? To investigate this theory, Richelson, Wahner, Melton, and Riggs compared the bone mineral density in three groups of women.

The first group of 14 women had undergone oophorectomy during young adult womanhood and had lived for a period of 15 to 25 years with an estrogen deficiency. A second group, identified as premenopausal, were approximately the same age (approximately 50 years) as the oophorectomy group except that the women had never suffered a period of estrogen deficiency. The third group of 14 women were postmenopausal and had suffered an estrogen deficiency for an average of 20 years. The mean and standard error of the mean for the three samples of lumbar spine bone-density measurements—14 measurements in each sample, one for each subject—are recorded in the following table.

Oophorectomized Group I		Premenopausal Group 2		Postmenopausal Group 3	
Mean	Standard Error	Mean	Standard Error	Mean	Standard Error
0.93	0.04	1.21	0.03	0.92	0.04

Source: L. S. Richelson, H. W. Wahner, L. J. Melton, III, and B. L. Riggs, "Relative Contributions of Aging and Estrogen Deficiency to Postmenopausal Bone Loss," *New England Journal of Medicine* 311(20) (1984), pp. 1273–75.

Is there sufficient evidence to permit us to conclude that the mean bone-density measurements differ for the three groups of women? What is the *p*-value associated with your test? What would you conclude at the $\alpha = .05$ level?

13.10 If vegetables intended for human consumption contain any pesticides at all, these pesticides should occur in minute quantities. Detection of pesticides in vegetables sent to market is accomplished by using solvents to extract the pesticides from the vegetables, and then performing tests on this extract to isolate and quantify the pesticides present. The extraction process is thought to be adequate because, if known amounts of pesticides are added to "clean" vegetables in a laboratory environment, essentially all of the pesticide can be recovered from the artificially contaminated extract.

The following data were obtained from a study by Wheeler et al., who sought to determine whether the extraction process is also effective when used in the more realistic situation where pesticides are applied to vegetable crops. Dieldrin (a commonly used pesticide) labeled with (radioactive) Carbon 14 was applied to growing radishes. Fourteen days later the extraction process was used, and these radishes were analyzed for pesticide content. A liquid scintillation counter was used to determine the amount of Carbon 14 present in the extract and also the amount left behind in the vegetable pulp. Because the vegetable pulp typically is discarded when analyzing for pesticides, if an appreciable proportion of pesticide remains in this pulp, a serious underassessment of the amount of pesticide could result. The pesticide was the only source of Carbon 14; thus the proportion of Carbon 14 in the pulp is likely to be indicative of the proportion of pesticide in the pulp. The following table shows a portion of the data that the researchers obtained when low, medium, and high concentrations of the solvent, acetonitrile, were used in the extraction process.

a Is there sufficient evidence that the mean percentage of Carbon 14 remaining in the vegetable pulp differs for the different concentrations of acetonitrile used in the extraction process? Give bounds for the attained significance level. What would you conclude at the $\alpha = .01$ level of significance?

b What assumptions are necessary to validly employ the analysis that you performed in (a)? Relate the necessary assumptions to the particular application represented in this exercise.

Percentage of Carbon 14 in Vegetable Pulp		
Concentration of Acetonitrile		
Low	Medium	High
23.37	20.39	18.87
25.13	20.87	19.69
23.78	20.78	19.29
27.74	20.19	18.10
25.30	20.01	18.42
25.21	20.23	19.33
22.12	20.73	17.26
20.96	19.53	18.09
23.11	18.87	18.69
22.57	18.17	18.82
24.59	23.34	18.72
23.70	22.45	18.75
Total 287.58	245.56	224.03

Source: W. B. Wheeler, N. P. Thompson, R. L. Edelstein, R. C. Littel, and R. T. Krause, "Influence of Various Solvent-Water Mixtures on the Extraction of Dieldrin and Methomyl Residues from Radishes," *Journal of the Association of Official Analytical Chemists* 65(5) (1982): 1112–17.

13.11 The Institute of Transportation Engineers National Student Paper Award for 1984 was given to Yean-Jye Lu for his report entitled "A Study of Left-Turn Maneuver Time for Signalized Intersections." One portion of the research described in this paper involved an evaluation of maneuver times for vehicles of various sizes that were involved in making a left turn at an intersection with a separate left-turn lane, but without a separate left-turn phase on the traffic light governing the intersection (an "unprotected" left-turn maneuver). The maneuver time was measured from the instant a vehicle entered the opposing lanes of traffic until it completely cleared the intersection. Four-cylinder automobiles were classified as "small cars," and six- or eight-cylinder automobiles as "large cars." Trucks and buses were combined to form a third category identified as "truck or bus." Other motorized vehicles (motorcycles, etc.) were ignored in the study. A portion of the data, giving maneuver times (in seconds) for vehicles that attempted the left-turn maneuver from a standing stop, appears in the accompanying table.

Vehicle Type	Sample Size	Mean	Standard Deviation
Small car	45	4.59	0.70
Large car	102	4.88	0.64
Truck or bus	18	6.24	0.90

Source: Reprinted from *ITE Journal* 54 (October 1984): 42–47 by permission of the Institute of Transportation Engineers, Washington, DC ©1984 I.T.E. All rights reserved.

Is there sufficient evidence to claim that the mean maneuver times differ for the three vehicle types? Give bounds for the attained significance level, and indicate the appropriate conclusion for an $\alpha = .05$ level test.

13.12 The Florida Game and Fish Commission desires to compare the amounts of residue from three chemicals found in the brain tissue of brown pelicans. Independent random samples of ten pelicans each yielded the accompanying results (measurements in parts per million). Is there evidence of sufficient differences among the mean residue amounts, at the 5% level of significance?

Statistic	Chemical		
	DDE	DDD	DDT
Mean	.032	.022	.041
Standard Deviation	.014	.008	.017

13.13 Water samples were taken at four different locations in a river to determine whether the quantity of dissolved oxygen, a measure of water pollution, varied from one location to another. Locations 1 and 2 were selected above an industrial plant, one near the shore and the other in midstream; location 3 was adjacent to the industrial water discharge for the plant; and location 4 was slightly downriver in midstream. Five water specimens were randomly selected at each location, but one specimen, corresponding to location 4, was lost in the laboratory. The data are shown in the accompanying table (the greater the pollution, the lower will be the dissolved oxygen readings). Do the data provide sufficient evidence to indicate a difference in mean dissolved oxygen content for the four locations? Give bounds for the attained significance level.

Location	Mean Dissolved Oxygen Content				
1	5.9	6.1	6.3	6.1	6.0
2	6.3	6.6	6.4	6.4	6.5
3	4.8	4.3	5.0	4.7	5.1
4	6.0	6.2	6.1	5.8	

13.14 An experiment was conducted to examine the effect of age on heart rate when a subject performs a specific amount of exercise. Ten male subjects were randomly selected from four age groups: 10–19, 20–39, 40–59, and 60–69. Each subject walked a treadmill at a fixed grade for a period of 12 minutes, and the increase in heart rate—the difference before and after exercise—was recorded (in beats per minute). These data are shown in the accompanying table. Do the data provide sufficient evidence to indicate a difference in mean increase in heart rate among the four age groups? Test by using $\alpha = .05$.

	Age			
	10–19	20–39	40–59	60–69
	29	24	37	28
	33	27	25	29
	26	33	22	34
	27	31	33	36
	39	21	28	21
	35	28	26	20
	33	24	30	25
	29	34	34	24
	36	21	27	33
	22	32	33	32
Total	309	275	295	282

13.5 A Statistical Model for the One-Way Layout

As earlier, we let Y_{ij} denote the random variables that generate the observed values y_{ij}, for $i = 1, 2, \ldots, k$ and $j = 1, 2, \ldots, n_i$. The Y_{ij} values correspond to independent random samples from normal populations with $E(Y_{ij}) = \mu_i$ and $V(Y_{ij}) = \sigma^2$, for $i = 1, 2, \ldots, k$ and $j = 1, 2, \ldots, n_i$. Let us consider the random sample drawn from population 1 and write

$$Y_{1j} = \mu_1 + \varepsilon_{1j}, \quad j = 1, 2, \ldots, n_1.$$

Equivalently,

$$\varepsilon_{1j} = Y_{1j} - \mu_1, \quad j = 1, 2, \ldots, n_1.$$

Because ε_{1j} simply is the difference between a normally distributed random variable and its mean, it follows that ε_{1j} is normally distributed with $E(\varepsilon_{1j}) = 0$ and $V(\varepsilon_{1j}) = V(Y_{1j}) = \sigma^2$. Further, the independence of Y_{1j}, for $j = 1, 2, \ldots, n_1$, implies that ε_{1j}, for $j = 1, 2, \ldots, n_1$, are mutually independent random variables. For each $i = 1, 2, \ldots, k$, we can proceed in an analogous manner to write

$$Y_{ij} = \mu_i + \varepsilon_{ij}, \quad j = 1, 2, \ldots, n_i$$

where the "error terms" ε_{ij} are independent, normally distributed random variables with $E(\varepsilon_{ij}) = 0$ and $V(\varepsilon_{ij}) = \sigma^2$, for $i = 1, 2, \ldots, k$ and $j = 1, 2, \ldots, n_i$. The error terms simply represent the difference between the observations in each sample and the corresponding population mean.

One more set of considerations will lead to the classical model for the one-way layout. Consider the means μ_i, for $i = 1, 2, \ldots, k$, and write

$$\mu_i = \mu + \tau_i \quad \text{where} \quad \tau_1 + \tau_2 + \cdots + \tau_k = 0.$$

Notice that $\sum_{i=1}^{k} \mu_i = k\mu + \sum_{i=1}^{k} \tau_i = k\mu$, and hence $\mu = k^{-1} \sum_{i=1}^{k} \mu_i$ is just the average of the k population means (the μ_i values). For this reason, μ is generally referred to as the *overall mean*. Since for $i = 1, 2, \ldots, k$,

$$\tau_i = \mu_i - \mu$$

quantifies the difference between the mean for population i and the overall mean, τ_i, is usually referred to as the *effect of population* (or treatment) i. Finally, we present the classical model for the one-way layout.

Statistical Model for a One-Way Layout

For $i = 1, 2, \ldots, k$ and $j = 1, 2, \ldots, n_i$,

$$Y_{ij} = \mu + \tau_i + \varepsilon_{ij}$$

where

Y_{ij} = the jth observation from population (treatment) i

μ = the overall mean

τ_i = the nonrandom effect of treatment i, where $\sum_{i=1}^{k} \tau_i = 0$

ε_{ij} = random error terms such that ε_{ij} are independent normally distributed random variables with $E(\varepsilon_{ij}) = 0$ and $V(\varepsilon_{ij}) = \sigma^2$.

The advantage of this model is that it very clearly summarizes all of the assumptions made in the analysis of the data obtained from a one-way layout. It also gives us a basis for presenting a precise statistical model for the randomized block design. (See Section 13.8.)

Notice that $H_0 : \mu_1 = \mu_2 = \cdots = \mu_k$ is true if and only if

$$H_0 : \tau_1 = \tau_2 = \cdots = \tau_k = 0$$

is true and that $H_a : \mu_i \neq \mu_{i'}$ for some $i \neq i'$ is true if and only if $H_a : \tau_i \neq 0$ for some i, where $i = 1, 2, \ldots, k$, is true. Thus, the F test for equality of means that we presented in Section 13.3 is the test of the null hypothesis

$$H_0 : \tau_1 = \tau_2 = \cdots = \tau_k = 0$$

versus the alternative that at least one $\tau_i \neq 0$.

Exercises

13.15 Let $\overline{Y}_{i\bullet}$ denote the average of all of the responses to treatment i. Use the model for the one-way layout to derive $E(\overline{Y}_{i\bullet})$ and $V(\overline{Y}_{i\bullet})$.

13.16 Refer to Exercise 13.15, and consider $\overline{Y}_{i\bullet} - \overline{Y}_{i'\bullet}$ for $i \neq i'$.

a Show that $E(\overline{Y}_{i\bullet} - \overline{Y}_{i'\bullet}) = \mu_i - \mu_{i'} = \tau_i - \tau_{i'}$. This result implies that $\overline{Y}_{i\bullet} - \overline{Y}_{i'\bullet}$ is an unbiased estimator of the difference in the effects of treatments i and i'.

b Derive $V(\overline{Y}_{i\bullet} - \overline{Y}_{i'\bullet})$.

13.17 Refer to the statistical model for the one-way layout.

a Show that $H_0 : \tau_1 = \tau_2 = \cdots = \tau_k = 0$ is true if and only if $H_0 : \mu_1 = \mu_2 = \cdots = \mu_k$ is true.

b Show that $H_a : \tau_i \neq 0$ for at least one i is true if and only if $H_a : \mu_i \neq \mu_{i'}$ for some $i \neq i'$ is true.

13.6 Proof of Additivity of the Sums of Squares and E(MST) for a One-Way Layout (Optional)

The proof that

$$\text{Total SS} = \text{SST} + \text{SSE}$$

for the one-way layout is presented in this section for the benefit of those who are interested. It may be omitted without loss of continuity.

The proof utilizes elementary results on summations that appear in the exercises for Chapter 1 and the device of adding and subtracting $\overline{Y}_{i\bullet}$ within the expression for the Total SS. Thus

$$\text{Total SS} = \sum_{i=1}^{k}\sum_{j=1}^{n_i}(Y_{ij}-\overline{Y})^2 = \sum_{i=1}^{k}\sum_{j=1}^{n_i}(Y_{ij}-\overline{Y}_{i\bullet}+\overline{Y}_{i\bullet}-\overline{Y})^2$$

$$= \sum_{i=1}^{k}\sum_{j=1}^{n_i}\left[(Y_{ij}-\overline{Y}_{i\bullet})+(\overline{Y}_{i\bullet}-\overline{Y})\right]^2$$

$$= \sum_{i=1}^{k}\sum_{j=1}^{n_i}\left[(Y_{ij}-\overline{Y}_{i\bullet})^2+2(Y_{ij}-\overline{Y}_{i\bullet})(\overline{Y}_{i\bullet}-\overline{Y})+(\overline{Y}_{i\bullet}-\overline{Y})^2\right].$$

Summing first over j, we obtain

$$\text{Total SS} = \sum_{i=1}^{k}\left[\sum_{j=1}^{n_i}(Y_{ij}-\overline{Y}_{i\bullet})^2+2(\overline{Y}_{i\bullet}-\overline{Y})\sum_{j=1}^{n_i}(Y_{ij}-\overline{Y}_{i\bullet})+n_i(\overline{Y}_{i\bullet}-\overline{Y})^2\right]$$

where

$$\sum_{j=1}^{n_i}(Y_{ij}-\overline{Y}_{i\bullet}) = Y_{i\bullet}-n_i\overline{Y}_{i\bullet} = Y_{i\bullet}-Y_{i\bullet} = 0.$$

Consequently, the middle term in the expression for the Total SS is equal to zero. Then, summing over i, we obtain

$$\text{Total SS} = \sum_{i=1}^{k}\sum_{j=1}^{n_i}(Y_{ij}-\overline{Y}_{i\bullet})^2+\sum_{i=1}^{k}n_i(\overline{Y}_{i\bullet}-\overline{Y})^2 = \text{SSE}+\text{SST}.$$

Proof of the additivity of the analysis of variance sums of squares for other experimental designs can be obtained in a similar manner, although the procedure is often tedious.

We now proceed with the derivation of the expected value of MST for a one-way layout (including a completely randomized design). Using the statistical model for the one-way layout presented in Section 13.5, it follows that

$$\overline{Y}_{i\bullet} = \frac{1}{n_i}\sum_{j=1}^{n_i} Y_{ij} = \frac{1}{n_i}\sum_{j=1}^{n_i}(\mu + \tau_i + \varepsilon_{ij}) = \mu + \tau_i + \overline{\varepsilon}_i \qquad \text{where} \qquad \overline{\varepsilon}_i = \frac{1}{n_i}\sum_{j=1}^{n_i}\varepsilon_{ij}.$$

Because the ε_{ij} are independent random variables with $E(\varepsilon_{ij}) = 0$ and $V(\varepsilon_{ij}) = \sigma^2$, Theorem 5.12 implies (see Example 5.27) that $E(\overline{\varepsilon}_i) = 0$ and $V(\overline{\varepsilon}_i) = \sigma^2/n_i$.

In a completely analogous manner, $\overline{Y}$ is given by

$$\overline{Y} = \frac{1}{n}\sum_{i=1}^{k}\sum_{j=1}^{n_i} Y_{ij} = \frac{1}{n}\sum_{i=1}^{k}\sum_{j=1}^{n_i}(\mu + \tau_i + \varepsilon_{ij}) = \mu + \overline{\tau} + \overline{\varepsilon}$$

where

$$\overline{\tau} = \frac{1}{n}\sum_{i=1}^{k} n_i \tau_i \quad \text{and} \quad \overline{\varepsilon} = \frac{1}{n}\sum_{i=1}^{k}\sum_{j=1}^{n_i}\varepsilon_{ij}.$$

Since the τ_i values are constants, $\overline{\tau}$ is simply a constant; again using Theorem 5.12, we obtain $E(\overline{\varepsilon}) = 0$ and $V(\overline{\varepsilon}) = \sigma^2/n$.

Therefore, with respect to the terms in the model for the one-way layout,

$$\text{MST} = \left(\frac{1}{k-1}\right)\sum_{i=1}^{k} n_i(\overline{Y}_{i\bullet} - \overline{Y})^2 = \left(\frac{1}{k-1}\right)\sum_{i=1}^{k} n_i(\tau_i + \overline{\varepsilon}_i - \overline{\tau} - \overline{\varepsilon})^2$$

$$= \left(\frac{1}{k-1}\right)\sum_{i=1}^{k} n_i(\tau_i - \overline{\tau})^2 + \left(\frac{1}{k-1}\right)\sum_{i=1}^{k} 2n_i(\tau_i - \overline{\tau})(\overline{\varepsilon}_i - \overline{\varepsilon})$$

$$+ \left(\frac{1}{k-1}\right)\sum_{i=1}^{k} n_i(\overline{\varepsilon}_i - \overline{\varepsilon})^2.$$

Because $\overline{\tau}$ and τ_i, for $i = 1, 2, \ldots, k$, are constants and $E(\varepsilon_{ij}) = E(\overline{\varepsilon}_i) = E(\overline{\varepsilon}) = 0$, it follows that

$$E(\text{MST}) = \left(\frac{1}{k-1}\right)\sum_{i=1}^{k} n_i(\tau_i - \overline{\tau})^2 + \left(\frac{1}{k-1}\right)E\left[\sum_{i=1}^{k} n_i(\overline{\varepsilon}_i - \overline{\varepsilon})^2\right].$$

Notice that

$$\sum_{i=1}^{k} n_i(\overline{\varepsilon}_i - \overline{\varepsilon})^2 = \sum_{i=1}^{k}(n_i\overline{\varepsilon}_i^2 - 2n_i\overline{\varepsilon}_i\overline{\varepsilon} + n_i\overline{\varepsilon}^2)$$

$$= \sum_{i=1}^{k} n_i\overline{\varepsilon}_i^2 - 2n\overline{\varepsilon}^2 + n\overline{\varepsilon}^2 = \sum_{i=1}^{k} n_i\overline{\varepsilon}_i^2 - n\overline{\varepsilon}^2.$$

Because $E(\bar{\varepsilon}_i) = 0$ and $V(\bar{\varepsilon}_i) = \sigma^2/n_i$, it follows that $E(\bar{\varepsilon}_i^2) = \sigma^2/n_i$, for $i = 1, 2, \ldots, k$. Similarly, $E(\bar{\varepsilon}^2) = \sigma^2/n$ and hence

$$E\left[\sum_{i=1}^{k} n_i (\bar{\varepsilon}_i - \bar{\varepsilon})^2\right] = \sum_{i=1}^{k} n_i E(\bar{\varepsilon}_i^2) - nE(\bar{\varepsilon}^2) = k\sigma^2 - \sigma^2 = (k-1)\sigma^2.$$

Summarizing, we obtain

$$E(\text{MST}) = \sigma^2 + \left(\frac{1}{k-1}\right)\sum_{i=1}^{k} n_i(\tau_i - \bar{\tau})^2 \qquad \text{where} \qquad \bar{\tau} = \frac{1}{n}\sum_{i=1}^{k} n_i \tau_i.$$

Under $H_0 : \tau_1 = \tau_2 = \cdots = \tau_k = 0$, it follows that $\bar{\tau} = 0$, and hence $E(\text{MST}) = \sigma^2$. Thus, when H_0 is true, MST/MSE is the ratio of two unbiased estimators for σ^2.

13.7 Estimation in the One-Way Layout

Confidence intervals for a single treatment mean and for the difference between a pair of treatment means based upon data obtained in a one-way layout (Section 13.3) are completely analogous to those given in Chapter 8. The only difference between the intervals in Chapter 8 and those given below is that the intervals appropriate for the one-way layout utilize $\sqrt{\text{MSE}}$ (with $n - k$ degrees of freedom) to estimate the population standard deviation(s) σ. The confidence interval for the mean of treatment i or the difference between treatments i and i' are, respectively, as follows:

$$\bar{Y}_{i\bullet} \pm t_{\alpha/2}\frac{S}{\sqrt{n_i}}$$

and

$$(\bar{Y}_{i\bullet} - \bar{Y}_{i'\bullet}) \pm t_{\alpha/2}S\sqrt{\frac{1}{n_i} + \frac{1}{n_{i'}}}$$

where

$$S = \sqrt{S^2} = \sqrt{\text{MSE}} = \sqrt{\frac{\text{SSE}}{n_1 + n_2 + \cdots + n_k - k}}$$

and $t_{\alpha/2}$ is based upon $(n - k)$ degrees of freedom.

The confidence intervals just stated are appropriate for a single treatment mean or a comparison of a pair of means selected prior to observation of the data. These intervals are likely to be shorter than the corresponding intervals from Chapter 8 because the value of $t_{\alpha/2}$ is based on a greater number of degrees of freedom ($n - k$ instead of $n_i - 1$ or $n_i + n_{i'} - 2$, respectively). The stated confidence coefficients are based on random sampling. If we were to look at the data and always compare the largest and smallest sample means, certainly we would expect the difference between the largest and smallest sample means to be larger than for a pair of means specified to be of interest before observing the data.

EXAMPLE **13.3**   Find a 95% confidence interval for the mean score for teaching technique 1, Example 13.2.

Solution   The 95% confidence interval for the mean score is

$$\overline{Y}_{1\bullet} \pm t_{.025}\frac{S}{\sqrt{n_1}}$$

where $t_{.025}$ is determined for $n - k = 19$ degrees of freedom, or

$$75.67 \pm (2.093)\frac{\sqrt{63}}{\sqrt{6}} \qquad \text{or} \qquad 75.67 \pm 6.78.$$

Notice that if we had analyzed only the data for teaching method 1, the value of $t_{.025}$ would have been based on only $n_1 - 1 = 5$ degrees of freedom, the number of degrees of freedom associated with s_1.

EXAMPLE **13.4**   Find a 95% confidence interval for the difference in mean score for teaching techniques 1 and 4, Example 13.2.

Solution   The 95% confidence interval is

$$(\overline{Y}_{1\bullet} - \overline{Y}_{4\bullet}) \pm (2.093)(7.94)\sqrt{1/6 + 1/4} \qquad \text{or} \qquad -12.08 \pm 10.73.$$

Hence the 95% confidence interval for $(\mu_1 - \mu_4)$ is -22.81 to -1.35. This suggests that $\mu_4 > \mu_1$.

Exercises

13.18   Refer to Examples 13.2 and 13.3.

a   Use the portion of the data in Table 13.2 that deals only with teaching technique 1 and the method of Section 8.8 to form a 95% confidence interval for the mean score of students taught using technique 1.

b   How does the length of the 95% confidence interval you found in (a) compare to the length of the 95% confidence interval obtained in Example 13.3?

c   What is the major reason that the interval that you found in (a) is longer than that found in Example 13.3?

13.19 Refer to Examples 13.2 and 13.4.

    a Use the portion of the data in Table 13.2 that deals only with teaching techniques 1 and 4 and the method of Section 8.8 to form a 95% confidence interval for the difference in mean score for students taught using techniques 1 and 4.

    b How does the length of the 95% confidence interval you found in (a) compare to the length of the 95% confidence interval obtained in Example 13.4?

    c What is the major reason that the interval that you found in (a) is longer than that found in Example 13.4?

13.20 **a** Based on your answers to Exercises 13.18 and 13.19 and the comments at the end of this section, how would you expect confidence intervals computed using the results of this section to compare with related intervals that make use of the data from only one or two of the samples obtained in a one-way layout? Why?

    b Refer to (a). Is it possible that a 95% confidence interval for the mean of a single population based only on the sample taken from that population will be shorter than the 95% confidence interval for the same population mean that would be obtained using the procedure of this section? How?

13.21 Refer to Exercise 13.9. As noted in the description of the experiment, the oophorectomized and the premenopausal groups of women were of approximately the same age, but those in the oophorectomized group suffered from an estrogen deficiency. Form a 95% confidence interval for the difference in mean bone densities for these two groups of women. Would you conclude that the mean bone densities for the oophorectomized and premenopausal women were significantly different? Why?

13.22 Refer to Exercise 13.7. Let μ_A and μ_B denote the mean strengths of concrete specimens prepared for mix A and mix B, respectively.

    a Find a 90% confidence interval for μ_A.

    b Find a 95% confidence interval for $(\mu_A - \mu_B)$.

13.23 Refer to Exercise 13.8. Let μ_A and μ_B, respectively, denote the mean scores at the end of the semester for the populations of extremely hostile students who were treated throughout that semester by method A and method B.

    a Find a 95% confidence interval for μ_A.

    b Find a 95% confidence interval for μ_B.

    c Find a 95% confidence interval for $(\mu_A - \mu_B)$.

13.24 Refer to Exercise 13.10.

    a Construct a 95% confidence interval for the mean percentage of Carbon 14 that remains in the vegetable pulp when the low level of acetonitrile is used.

    b Give a 90% confidence interval for the difference in mean percentages of Carbon 14 that remain in the vegetable pulp for low and medium levels of acetonitrile.

13.25 Refer to Exercise 13.11.

    a Give a 95% confidence interval for the mean left-turn maneuver time for buses and trucks.

    b Estimate the difference in mean maneuver times for small and large cars with a 95% confidence interval.

    c The study report by Lu involved vehicles that passed through the intersection of Guadalupe Avenue and 38th Street in Austin, Texas. Do you think that the results in (a) and (b) would be valid for a "nonprotected" intersection in your hometown? Why or why not?

13.26 It has been hypothesized that treatments (after casting) of a plastic used in optic lenses will improve wear. Four different treatments are to be tested. To determine whether any differences in mean wear exist among treatments, twenty-eight castings from a single formulation of the plastic were made and seven castings were randomly assigned to each of the treatments. Wear was determined by measuring the increase in "haze" after 200 cycles of abrasion (better wear being indicated by smaller increases). The data collected are reported in the accompanying table.

Treatment			
A	B	C	D
9.16	11.95	11.47	11.35
13.29	15.15	9.54	8.73
12.07	14.75	11.26	10.00
11.97	14.79	13.66	9.75
13.31	15.48	11.18	11.71
12.32	13.47	15.03	12.45
11.78	13.06	14.86	12.38

a Is there evidence of a difference in mean wear among the four treatments? Use $\alpha = .05$.

b Estimate the mean difference in haze increase between treatments B and C, using a 99% confidence interval.

c Find a 90% confidence interval for the mean wear for lenses receiving treatment A.

13.27 With the ongoing energy crisis, researchers for the major oil companies are attempting to find alternative sources of oil. It is known that some types of shale contain small amounts of oil that feasibly (if not economically) could be extracted. Four methods have been developed for extracting oil from shale, and the government has decided that some experimentation should be done to determine whether the methods differ significantly in the average amount of oil that each can extract from the shale. Method 4 is known to be the most expensive method to implement, and method 1 is the least expensive, so inferences about the differences in performance of these two methods are of particular interest. Sixteen bits of shale (of the same size) were randomly subjected to the four methods, with the results shown in the accompanying table (the units are in liters per cubic meter). All inferences are to be made with $\alpha = .05$.

Method 1	Method 2	Method 3	Method 4
3	2	5	5
2	2	2	2
1	4	5	4
2	4	1	5

a Assuming that the sixteen experimental units were as alike as possible, implement the appropriate analysis of variance to determine whether there is any significant difference among the mean amounts extracted by the four methods. Use $\alpha = .05$.

b Set up a 95% confidence interval for the difference in the mean amounts extracted by the two methods of particular interest. Interpret the result.

13.28 Refer to Exercise 13.12. Construct a 95% confidence interval for the mean amount of residue from DDT.

13.29   Refer to Exercise 13.13. Compare the mean dissolved oxygen content in midstream above the plant with the mean content adjacent to the plant (location 2 versus location 3). Use a 95% confidence interval.

13.30   Refer to Exercise 13.13. Compare the mean dissolved oxygen content for the two locations above the plant with the mean content slightly downriver from the plant, by finding a 95% confidence interval for $(1/2)(\mu_1 + \mu_2) - \mu_4$.

13.31   Refer to Exercise 13.14.

    a   Find a 90% confidence interval for the difference in mean increase in heart rate between the 10–19 age group and the 60–69 age group.

    b   Find a 90% confidence interval for the mean increase in heart rate for the 20–39 age group.

13.8   A Statistical Model for the Randomized Block Design

The method for constructing a randomized block design was presented in Section 12.4. As previously indicated in Definition 12.6, the randomized block design is a design for comparing k treatments utilizing b blocks. The treatments are randomly assigned to the experimental units in each block in such a way that each treatment appears exactly once in each of the b blocks. Thus, the total number of observations obtained in a randomized block design is $n = bk$. Implicit in the consideration of a randomized block design is the presence of two qualitative independent variables, "blocks" and "treatments." In this section we present a formal statistical model for the randomized block design.

Statistical Model for a Randomized Block Design

For $i = 1, 2, \ldots, k$ and $j = 1, 2, \ldots, b$,

$$Y_{ij} = \mu + \tau_i + \beta_j + \varepsilon_{ij}$$

where

    Y_{ij} = the observation on treatment i in block j

    μ = the overall mean

    τ_i = the nonrandom effect of treatment i, where $\displaystyle\sum_{i=1}^{k} \tau_i = 0$

    β_j = the nonrandom effect of block j, where $\displaystyle\sum_{j=1}^{b} \beta_j = 0$

    ε_{ij} = random error terms such that ε_{ij} are independent normally distributed random variables with $E(\varepsilon_{ij}) = 0$ and $V(\varepsilon_{ij}) = \sigma^2$.

Notice that $\mu, \tau_1, \tau_2, \ldots, \tau_k$, and $\beta_1, \beta_2, \ldots, \beta_b$ are all assumed to be unknown constants. This model differs from that for the completely randomized design (a specific type of one-way layout) only in containing parameters associated with the different blocks. Because the block effects are assumed to be fixed but unknown, this model usually is referred to as the *fixed block effects* model. Another model for the randomized block design, in which the β's are assumed to be random variables (called the *random block effects* model), is considered in the supplementary exercises. Our formal development in the body of this text is restricted to the fixed block effects model.

The statistical model just presented very clearly summarizes all of the assumptions made in the analysis of data in a randomized block design with fixed block effects. Let us consider the observation Y_{ij} made on treatment i in block j. Notice that the assumptions in the model imply that $E(Y_{ij}) = \mu + \tau_i + \beta_j$ and $V(Y_{ij}) = \sigma^2$ for $i = 1, 2, \ldots, k$ and $j = 1, 2, \ldots, b$. Let us consider the observations made on treatment i, and observe that two observations receiving treatment i have means that differ only by the difference of the block effects. For example,

$$E(Y_{i1}) - E(Y_{i2}) = \mu + \tau_i + \beta_1 - (\mu + \tau_i + \beta_2) = \beta_1 - \beta_2.$$

Similarly, two observations that are taken from the same block have means that differ only by the difference of the treatment effects. That is, if $i \neq i'$,

$$E(Y_{ij}) - E(Y_{i'j}) = \mu + \tau_i + \beta_j - (\mu + \tau_{i'} + \beta_j) = \tau_i - \tau_{i'}.$$

Observations that are taken on different treatments and in different blocks have means that differ by the difference in the treatment effects plus the difference in the block effects because if $i \neq i'$ and $j \neq j'$,

$$E(Y_{ij}) - E(Y_{i'j'}) = \mu + \tau_i + \beta_j - (\mu + \tau_{i'} + \beta_{j'}) = (\tau_i - \tau_{i'}) + (\beta_j - \beta_{j'}).$$

In the next section, we will proceed with an analysis of the data obtained from a randomized block design.

Exercises

13.32 State the assumptions underlying the analysis of variance for a randomized block design.

13.33 According to the model for the randomized block design given in this section, the expected response when treatment i is applied in block j is $E(Y_{ij}) = \mu + \tau_i + \beta_j$, for $i = 1, 2, \ldots, k$ and $j = 1, 2, \ldots, b$.

    a Use the model given in this section to calculate the average of the $n = bk$ expected responses associated with all of the blocks and treatments.

    b Give an interpretation for the parameter μ that appears in the model for the randomized block design.

13.34 Let $\overline{Y}_{i\bullet}$ denote the average of all of the responses to treatment i. Use the model for the randomized block design to derive $E(\overline{Y}_{i\bullet})$ and $V(\overline{Y}_{i\bullet})$. Is $\overline{Y}_{i\bullet}$ an unbiased estimator for the mean response to treatment i? Explain why or why not.

13.35   Refer to Exercise 13.34 and consider $\overline{Y}_{i\bullet} - \overline{Y}_{i'\bullet}$ for $i \neq i'$.

    a   Show that $E(\overline{Y}_{i\bullet} - \overline{Y}_{i'\bullet}) = \tau_i - \tau_{i'}$. This result implies that $\overline{Y}_{i\bullet} - \overline{Y}_{i'\bullet}$ is an unbiased estimator of the difference in the effects of treatment i and i'.

    b   Derive $V(\overline{Y}_{i\bullet} - \overline{Y}_{i'\bullet})$.

13.36   Refer to the model for the randomized block design, and let $\overline{Y}_{\bullet j}$ denote the average of all of the responses in block j.

    a   Derive $E(\overline{Y}_{\bullet j})$ and $V(\overline{Y}_{\bullet j})$.

    b   Show that $\overline{Y}_{\bullet j} - \overline{Y}_{\bullet j'}$ is an unbiased estimator for $\beta_j - \beta_{j'}$ the difference in the effects of blocks j and j'.

    c   Derive $V(\overline{Y}_{\bullet j} - \overline{Y}_{\bullet j'})$.

13.9 The Analysis of Variance for a Randomized Block Design

The analysis of variance for a randomized block design proceeds much like that for a completely randomized design (which is a special case of the one-way layout). In the randomized block design, the total sum of squares of deviations of the response measurements from their mean may be partitioned into three parts: the sum of squares for blocks, treatments, and error.

Denote the total and average of all observations in block j as $Y_{\bullet j}$ and $\overline{Y}_{\bullet j}$, respectively. Similarly, let $Y_{i\bullet}$ and $\overline{Y}_{i\bullet}$ represent the total and the average for all observations receiving treatment i. Again, the "dots" in the subscripts indicate which index is "summed over" to compute the totals and "averaged over" to compute the averages. Then for a randomized block design involving b blocks and k treatments, we have the following sums of squares:

$$\text{Total SS} = \text{SSB} + \text{SST} + \text{SSE}$$

$$= \sum_{i=1}^{k}\sum_{j=1}^{b}(Y_{ij} - \overline{Y})^2 = \sum_{i=1}^{k}\sum_{j=1}^{b} Y_{ij}^2 - \text{CM}$$

$$\text{SSB} = k\sum_{j=1}^{b}(\overline{Y}_{\bullet j} - \overline{Y})^2 = \sum_{j=1}^{b} \frac{Y_{\bullet j}^2}{k} - \text{CM}$$

$$\text{SST} = b\sum_{i=1}^{k}(\overline{Y}_{i\bullet} - \overline{Y})^2 = \sum_{i=1}^{k} \frac{Y_{i\bullet}^2}{b} - \text{CM}$$

$$\text{SSE} = \text{Total SS} - \text{SSB} - \text{SST}.$$

Table 13.5. ANOVA table for a randomized block design

Source	d.f.	SS	MS
Blocks	$b-1$	SSB	$\dfrac{\text{SSB}}{b-1}$
Treatments	$k-1$	SST	$\dfrac{\text{SST}}{k-1}$
Error	$n-b-k+1$	SSE	MSE
Total	$n-1$	Total SS	

In the formulas

$$\overline{Y} = (\text{average of all } n = bk \text{ observations}) = \frac{1}{bk}\sum_{j=1}^{b}\sum_{i=1}^{k}Y_{ij}$$

and

$$\text{CM} = \frac{(\text{total of all observations})^2}{n} = \frac{1}{bk}\left(\sum_{j=1}^{b}\sum_{i=1}^{k}Y_{ij}\right)^2.$$

The analysis of variance for the randomized block design is presented in Table 13.5. The degrees of freedom associated with each sum of squares are shown in the second column. Mean squares are calculated by dividing the sum of squares by their respective degrees of freedom.

To test the null hypothesis that there is no difference in treatment means, we use the F statistic,

$$F = \frac{\text{MST}}{\text{MSE}}$$

and reject the null hypothesis if $F > F_\alpha$, based on $v_1 = (k-1)$ and $v_2 = (n-b-k+1)$ numerator and denominator degrees of freedom, respectively.

As discussed in Section 12.4, blocking can be used to control for an extraneous source of variation (the variation between blocks). In addition, with blocking, we have the opportunity to see whether evidence exists to indicate a difference in the mean response for blocks. Under the null hypothesis that there is no difference in mean response for blocks (that is, $\beta_j = 0$, for $j = 1, 2, \ldots, b$), MSB provides an unbiased estimator for σ^2 based on $(b-1)$ degrees of freedom. Where real differences exist among block means, MSB will tend to be inflated in comparison with MSE, and

$$F = \frac{\text{MSB}}{\text{MSE}}$$

provides a test statistic. As in the test for treatments, the rejection region for the test will be

$$F > F_\alpha$$

based on $v_1 = b - 1$ and $v_2 = n - b - k + 1$ degrees of freedom.

EXAMPLE **13.5**   A stimulus-response experiment involving three treatments was laid out in a randomized block design using four subjects. The response was the length of time to reaction measured in seconds. The data, arranged in blocks, are shown in Figure 13.2. The treatment number is circled and shown above each observation. Do the data present sufficient evidence to indicate a difference in the mean responses for stimuli (treatments)? Subjects? Use $\alpha = .05$ for each test and give the associated p values.

Solution   The observed values of the sums of squares for the analysis of variance are shown jointly in Table 13.6, and individually as follows:

$$\text{CM} = \frac{(\text{total})^2}{n} = \frac{(21.2)^2}{12} = 37.45$$

$$\text{Total SS} = \sum_{j=1}^{4}\sum_{i=1}^{3}(y_{ij} - \overline{y})^2 = \sum_{j=1}^{4}\sum_{i=1}^{3} y_{ij}^2 - \text{CM} = 46.86 - 37.45 = 9.41$$

$$\text{SSB} = \sum_{j=1}^{4} \frac{Y_{\bullet j}^2}{3} - \text{CM} = 40.93 - 37.45 = 3.48$$

$$\text{SST} = \sum_{i=1}^{3} \frac{Y_{i\bullet}^2}{4} - \text{CM} = 42.93 - 37.45 = 5.48$$

$$\text{SSE} = \text{Total SS} - \text{SSB} - \text{SST} = 9.41 - 3.48 - 5.48 = .45.$$

FIGURE **13.2**
Randomized block
design for
Example 13.5

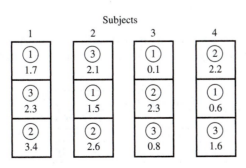

Table 13.6. ANOVA table for Example 13.5

Source	d.f.	SS	MS	F
Blocks	3	3.48	1.160	15.47
Treatments	2	5.48	2.740	36.53
Error	6	.45	.075	
Total	11	9.41		

We use the ratio of mean square for treatments to mean square for error to test a hypothesis of no difference in the mean response for treatments. Thus the calculated value of F is

$$F = \frac{MST}{MSE} = \frac{2.74}{.075} = 36.53.$$

The critical value of the F statistic ($\alpha = .05$) for $v_1 = 2$ and $v_2 = 6$ degrees of freedom is $F_{.05} = 5.14$. Because the computed value of F exceeds the critical value, there is sufficient evidence at the $\alpha = .05$ level to reject the null hypothesis and conclude that real differences do exist among the expected responses for the three stimuli. The corresponding p-value $= P(F > 36.53)$, which is such that p-value $< .005$.

A similar test may be conducted for the null hypothesis that no difference exists in the mean response for subjects. Rejection of this hypothesis would imply that there are significant differences among subjects and that blocking is desirable. The computed value of F based on $v_1 = 3$ and $v_2 = 6$ degrees of freedom is

$$F = \frac{MSB}{MSE} = \frac{1.16}{.075} = 15.47.$$

Because this value of F exceeds the corresponding tabulated critical value, $F_{.05} = 4.76$, we reject the null hypothesis and conclude that a real difference exists in the mean response among the group of subjects.

The associated p-value $= P(F > 15.47) < .005$.

Exercises

13.37   In Exercise 12.10, a matched pairs analysis was performed to compare the differences in mean CPU time to run benchmark programs on two computers. The data are reproduced in the following table.

Computer	Benchmark Program					
	1	2	3	4	5	6
1	1.12	1.73	1.04	1.86	1.47	2.10
2	1.15	1.72	1.10	1.87	1.46	2.15

a Treat the six programs as six blocks and test for a difference between the mean CPU times for the two computers by using a randomized block analysis. Use $\alpha = .05$. How does your decision compare to that reached in Exercise 12.10(a)?

b Give bounds for the associated p-value. How does your answer compare to your answer to Exercise 12.10(b)?

c How does the computed value of MSE compare to the value for s_D^2 that you used in your solution to Exercise 12.10?

13.38 The accompanying table presents data on yields relating to resistance to stain for three materials (M_1, M_2, and M_3) treated with four chemicals in a randomized block design. (A low value indicates good stain resistance.) Is there evidence of differences in mean resistance among the four chemicals? Give bounds for the p-value. What would you conclude at the $\alpha = .05$ level of significance?

		Material		
Chemical	M_1	M_2	M_3	Total
A	5	9	7	21
B	3	8	4	15
C	8	13	9	30
D	4	6	8	18
Total	20	36	28	84

$$\sum_i \sum_j y_{ij}^2 = 674, \qquad \tfrac{1}{12}\left(\sum_i \sum_j y_{ij}\right)^2 = 588.$$

13.39 An experiment was conducted to determine the effect of three methods of soil preparation on the first-year growth of slash pine seedlings. Four locations (state forest lands) were selected and each location was divided into three plots. Because soil fertility within a location was likely to be more homogeneous than between locations, a randomized block design was employed, using locations as blocks. The methods of soil preparation were A (no preparation), B (light fertilization), and C (burning). Each soil preparation was randomly applied to a plot within each location. On each plot the same number of seedlings was planted, and the observation recorded was the average first-year growth (in centimeters) of the seedlings on each plot. These observations are reproduced in the accompanying table.

Soil	Location			
Preparation	1	2	3	4
A	11	13	16	10
B	15	17	20	12
C	10	15	13	10

a Conduct an analysis of variance. Do the data provide sufficient evidence to indicate differences in the mean growth for the three soil preparations?

b Is there evidence to indicate differences in mean growth for the four locations?

13.40 Dudeck and Peacock reported on an experiment conducted to evaluate the performance of several cool season grasses for winter overseeding of golf greens in northern Florida. One of the variables of interest was the distance that a golf ball would roll on a green after being rolled down a ramp (used to induce a constant initial velocity to the ball). Because the distance the ball would roll was influenced by the slope of the green and the direction in which the grass was mowed, the experiment was set up in a randomized block design. The blocks were determined so that the slopes of the individual plots were constant within blocks (a transit was used to ensure accuracy), and all plots were mowed in the same direction and at the same height to eliminate mowing effects. The base grass was "Tiftgreen" bermuda grass in a semidormant state. The same method of seeding and rates of application were used for all of the ryegrasses that are represented in the following table of data. Measurements are average distances (in meters) from the base of the ramp to the stopping points for five balls rolled down the ramp and directly up the slope on each plot. Cultivars used in the study included A (Pennfine ryegrass), B (Dasher ryegrass), C (Regal ryegrass), D (Marvelgreen supreme), and E (Barry ryegrass). The grasses were planted within blocks, and yielded the measurements shown.

| Block | Variety | | | | | Total |
	A	B	C	D	E	
1	2.764	2.568	2.506	2.612	2.238	12.688
2	3.043	2.977	2.533	2.675	2.616	13.844
3	2.600	2.183	2.334	2.164	2.127	11.408
4	3.049	3.028	2.895	2.724	2.697	14.393
Total	11.456	10.756	10.268	10.175	9.678	52.333

Source: A. E. Dudeck and C. H. Peacock, "Effects of Several Overseeded Ryegrasses on Turf Quality, Traffic Tolerance and Ball Roll," *Proceedings of the Fourth International Turfgrass Research Conference,* R. W. Sheard, ed., pp. 75–81. Ontario Agricultural College, University of Guelph, Guelph, Ontario, and the International Turfgrass Society, 1981.

a Perform the appropriate analysis of variance to test for sufficient evidence to indicate that the mean distance of ball roll differs for the five cultivars. Give bounds for the attained significance level. What would you conclude at the $\alpha = .01$ level of significance?

b Is there evidence of a significant difference between the blocks used in the experiment? Test using $\alpha = .05$.

13.41 Refer to Exercise 13.27. Suppose that we now find out that the sixteen experimental units were obtained in the following manner. One sample was taken from each of four locations, each individual sample was split into four parts, and then each of the methods was applied to exactly one part from each location (with the proper randomization). The data are now presented more correctly in the form shown in the accompanying table. Does this new information suggest a more appropriate method of analysis than that used in Exercise 13.27? If so, perform the new analysis and answer the question in Exercise 13.27(a). Is this new information worthwhile?

Location	Method 1	Method 2	Method 3	Method 4
I	3	2	5	5
II	2	2	2	2
III	1	4	5	4
IV	2	4	1	5

13.42 Suppose that a randomized block design with b blocks and k treatments has each treatment measured *twice* in each block. Indicate how you would perform the computations for an analysis of variance.

13.43 An evaluation of diffusion bonding of zircaloy components is performed. The main objective is to determine which of three elements—nickel, iron, or copper—is the best bonding agent. A series of zircaloy components are bonded using each of the possible bonding agents. Due to significant variation in components machined from different ingots, a randomized block design is used, blocking on the ingots. Two components from each ingot are bonded together using each of the three agents, and the pressure (in units of 1000 pounds per square inch) required to separate the bonded components is measured. The data shown in the following table are obtained.

	Bonding Agent		
Ingot	Nickel	Iron	Copper
1	67.0	71.9	72.2
2	67.5	68.8	66.4
3	76.0	82.6	74.5
4	72.7	78.1	67.3
5	73.1	74.2	73.2
6	65.8	70.8	68.7
7	75.6	84.9	69.0

Is there evidence of a difference in mean pressures required to separate the components among the three bonding agents? Use $\alpha = .05$.

13.44 From time to time, one branch office of a company must make shipments to another branch office in another state. Three package delivery services operate between the two cities where the branch offices are located. Because the price structures for the three delivery services are quite similar, the company wants to compare the delivery times. The company plans to make several different types of shipments to its branch office. To compare the carriers, the company sends each shipment in triplicate, one with each carrier. The results listed in the accompanying table are the delivery times in hours.

	Carrier		
Shipment	I	II	III
1	15.2	16.9	17.1
2	14.3	16.4	16.1
3	14.7	15.9	15.7
4	15.1	16.7	17.0
5	14.0	15.6	15.5

Is there evidence of a difference in mean delivery times among the three carriers? Give bounds for the attained significance level.

*13.45 Refer to the model for the randomized block design presented in Section 13.8.

a Derive $E(MST)$.

b Derive $E(MSB)$.

c Derive $E(MSE)$.

Notice that these quantities appear in the F statistics used to test for differences in the mean response among the blocks and among the treatments.

13.10 Estimation in the Randomized Block Design

The confidence interval for the difference between a pair of treatment means in a randomized block design is completely analogous to that associated with the completely randomized design (a special case of the one-way layout) in Section 13.7. A $100(1 - \alpha)\%$ confidence interval for $\tau_i - \tau_{i'}$ is

$$(\overline{Y}_{i\bullet} - \overline{Y}_{i'\bullet}) \pm t_{\alpha/2} S \sqrt{\frac{2}{b}}$$

where $n_i = n_{i'} = b$, the number of observations contained in a treatment mean, and $S = \sqrt{\text{MSE}}$. The difference between the confidence intervals for the completely randomized and the randomized block designs is that the value $t_{\alpha/2}$ is based on $\nu = n - b - k + 1 = (b - 1)(k - 1)$ degrees of freedom and that S, appearing in the preceding expression, is obtained from the ANOVA table associated with the randomized block design.

EXAMPLE 13.6 Construct a 95% confidence interval for the difference between the mean responses for treatments 1 and 2, Example 13.5.

Solution The confidence interval for the difference in mean responses for a pair of treatments is

$$(\overline{Y}_{i\bullet} - \overline{Y}_{i'\bullet}) \pm t_{\alpha/2} S \sqrt{\frac{2}{b}}$$

where, for our example, $t_{.025}$ is based upon 6 degrees of freedom. For treatments 1 and 2 we have

$$(.98 - 2.63) \pm (2.447)(.27)\sqrt{\frac{2}{4}} \quad \text{or} \quad -1.65 \pm .47 = (-2.12, -1.18).$$

Thus, at the 95% confidence level we conclude that the mean reaction time to stimulus 1 is between 1.18 and 2.12 seconds shorter than the mean reaction time to stimulus 2.

Exercises

13.46 Refer to Exercises 13.37 and 12.10. Find a 95% confidence interval for the difference in mean CPU times required for the two computers to complete a job. How does your answer compare to that obtained in Exercise 12.10(c)?

13.47   Refer to Exercise 13.38. Construct a 95% confidence interval for the difference between mean resistances for chemicals A and B.

13.48   Refer to Exercise 13.39. Construct a 90% confidence interval for the differences in mean growth for methods A and B.

13.49   Refer to Exercise 13.40. Construct a 95% confidence interval for the difference in the mean distance of roll when Dasher ryegrass and Marvelgreen supreme are used for overseeding.

13.50   Refer to Exercise 13.41. Construct a 95% confidence interval for the difference between the mean amounts of oil extracted by methods 1 and 4. Compare the answer to that obtained in Exercise 13.27(b).

13.51   Refer to Exercise 13.43. Estimate the difference in mean pressures to separate components that are bonded with nickel and iron, using a 99% confidence interval.

13.11   Selecting the Sample Size

Selecting the sample size for the one-way layout (including the completely randomized) or the randomized block design is an extension of the procedures of Section 8.7. We confine our attention to the case of equal sample sizes, $n_1 = n_2 = \cdots = n_k$, for the treatments of the one-way layout. The number of observations per treatment is equal to the number of blocks b for the randomized block design. Thus the problem is to determine n_1 or b for these two designs so that the resulting experiment contains the desired amount of information.

The determination of sample sizes follows a similar procedure for both designs; we will outline a general method. First the experimenter must decide on the parameter (or parameters) of major interest. Usually, this involves comparing a pair of treatment means. Second, the experimenter must specify a bound on the error of estimation that can be tolerated. Once this has been determined, the next task is to select n_i (the size of the sample from population or treatment i) or, correspondingly, b (the number of blocks for a randomized block design) that will reduce the half-width of the confidence interval for the parameter so that it is less than or equal to the specified bound on the error of estimation. It should be emphasized that the sample size solution *always* will be an approximation, because σ is unknown and an estimate for σ is unknown until the sample is acquired. The best available value for σ will be used to produce an approximate solution. We will illustrate the procedure with an example.

EXAMPLE **13.7**   A completely randomized design is to be conducted to compare five teaching techniques in classes of equal size. Estimation of the differences in mean response on an achievement test is desired correct to within 30 test-score points, with probability equal to .95. It is expected that the test scores for a given teaching technique will possess a range approximately equal to 240. Find the approximate number of observations required for each sample in order to acquire the specified information.

Solution   The confidence interval for the difference between a pair of treatment means is

$$(\overline{Y}_{i\bullet} - \overline{Y}_{i'\bullet}) \pm t_{\alpha/2}S\sqrt{\frac{1}{n_i} + \frac{1}{n_{i'}}}.$$

Therefore, we wish to select n_i and $n_{i'}$ so that

$$t_{\alpha/2}S\sqrt{\frac{1}{n_i} + \frac{1}{n_{i'}}} \leq 30.$$

The value of σ is unknown, and S is a random variable. However, an approximate solution for $n_i = n_{i'}$ can be obtained by guessing that the observed value of s will be roughly equal to one-fourth of the range. Thus $s \approx 240/4 = 60$. The value of $t_{\alpha/2}$ will be based upon $(n_1 + n_2 + \cdots + n_5 - 5)$ degrees of freedom, and for even moderate values of n_i, $t_{.025}$ will approximately equal 2. Then

$$t_{.025}s\sqrt{\frac{1}{n_i} + \frac{1}{n_{i'}}} \approx (2)(60)\sqrt{\frac{2}{n_i}} = 30$$

or

$$n_i = 32, \quad i = 1, 2, \ldots, 5.$$

EXAMPLE 13.8   An experiment is to be conducted to compare the toxic effects of three chemicals on the skin of rats. The resistance to the chemicals was expected to vary substantially from rat to rat. Therefore, all three chemicals were to be tested on each rat, thereby blocking out rat-to-rat differences.

The standard deviation of the experimental error was unknown, but prior experimentation involving several applications of a given chemical on the same type of rat suggested a range of response measurements equal to five units.

Find a value for b such that the error of estimating the difference between a pair of treatment means is less than one unit, with probability equal to .95.

Solution   A very approximate value for s would be one-fourth of the range, or $s \approx 1.25$. Then we wish to select b so that

$$t_{.025}s\sqrt{\frac{1}{b} + \frac{1}{b}} = t_{.025}s\sqrt{\frac{2}{b}} \leq 1.$$

Because $t_{.025}$ will depend upon the degrees of freedom associated with s^2, which will be $(n - b - k + 1)$, we will use the approximation $t_{.025} \approx 2$. Then

$$(2)(1.25)\sqrt{\frac{2}{b}} = 1 \qquad \text{or} \qquad b \approx 13.$$

Approximately thirteen rats will be required to obtain the desired information. Since we will make 3 observations ($k = 3$) per rat, our experiment will require that a total of $n = bk = 13(3) = 39$ measurements be made.

The degrees of freedom associated with the resulting estimate s^2 will be $(n - b - k + 1) = 39 - 13 - 3 + 1 = 24$, based on this solution. Therefore, the guessed value of t would seem to be adequate for this approximate solution.

The sample size solutions for Examples 13.7 and 13.8 are very approximate and are intended to provide only a rough estimate of sample size and consequent costs of the experiment. The actual lengths of the resulting confidence intervals will depend on the data actually observed. These intervals may not have the exact lengths specified by the experimenter, but will have the required confidence coefficient. If the resulting intervals are still too long, the experimenter can obtain information on σ as the data are being collected and can recalculate a better approximation to the number of observations per treatment (n_i or b) as the experiment proceeds.

Exercises

13.52   Refer to Exercise 13.7. About how many specimens per concrete mix should be prepared to allow estimation of the difference in mean strengths for a preselected pair of specimen types to within .02 tons per square inch? Assume knowledge of the data given in Exercise 13.7. What is the total number of observations required in the entire experiment?

13.53   Refer to Exercises 13.8 and 13.23(a). Approximately how many observations would be necessary to estimate μ_A to within ten units? Use a 95% confidence coefficient.

13.54   Refer to Exercises 13.8 and 13.23(c). Assuming equal sample sizes for each treatment, approximately how many observations from method A and method B would be necessary to estimate $\mu_A - \mu_B$ to within 20 units? Use a 95% confidence coefficient.What is the total number of observations required in the entire experiment?

13.55   Refer to Exercise 13.39. How many locations would have to be used in order to estimate the difference between the mean growth for any two specified soil preparations to within one unit, with confidence coefficient .95? What is the total number of observations required in the entire experiment?

13.56   Refer to Exercise 13.41 and 13.49. How many locations should be used if it is desired to estimate $\mu_1 - \mu_4$ to within .5 unit, with confidence coefficient .95?

13.12   Simultaneous Confidence Intervals for More than One Parameter

The methods of Section 13.7 can be used to construct $(1 - \alpha)100\%$ confidence intervals for a single treatment mean or for the difference between a pair of treatment means in a one-way layout. Suppose that in the course of an analysis we wish to construct several of these confidence intervals. The method of Section 13.10 can be used to compare a pair of treatment means in a randomized block design. Although it is true that each interval will enclose the estimated parameter with probability $(1 - \alpha)$, what is the probability that *all* of the intervals will enclose their respective parameters? The objective of this section is to present a procedure for forming sets of confidence intervals where the simultaneous confidence coefficient is no smaller than $(1 - \alpha)$ for an arbitrary α.

Suppose that we want to find confidence intervals $I_1, I_2, \ldots, I_m$ for parameters $\theta_1, \theta_2, \ldots, \theta_m$ so that

$$P(\theta_j \in I_j \text{ for all } j = 1, 2, \ldots, m) \geq 1 - \alpha.$$

This goal can be achieved by using a simple probability inequality, known as the *Bonferroni inequality*. For any events $A_1, A_2, \ldots, A_m$, we have

$$\overline{A_1 \cap A_2 \cap \cdots \cap A_m} = \overline{A}_1 \cup \overline{A}_2 \cup \cdots \cup \overline{A}_m.$$

Therefore,

$$P(A_1 \cap A_2 \cap \cdots \cap A_m) = 1 - P(\overline{A}_1 \cup \overline{A}_2 \cup \cdots \cup \overline{A}_m).$$

Also, from the additive law of probability we know that

$$P(\overline{A}_1 \cup \overline{A}_2 \cup \cdots \cup \overline{A}_m) \leq \sum_{j=1}^{m} P(\overline{A}_j).$$

Hence, we obtain the *Bonferroni inequality*

$$P(A_1 \cap A_2 \cap \cdots \cap A_m) \geq 1 - \sum_{j=1}^{m} P(\overline{A}_j).$$

Suppose that $P(\theta_j \in I_j) = 1 - \alpha_j$, and let A_j denote the event $\{\theta_j \in I_j\}$. Then

$$P(\theta_1 \in I_1, \ldots, \theta_m \in I_m) \geq 1 - \sum_{j=1}^{m} P(\theta_j \notin I_j) = 1 - \sum_{j=1}^{m} \alpha_j.$$

If all α_j, for $j = 1, 2, \ldots, m$, are chosen equal to α, we can see that the simultaneous confidence coefficient of the intervals I_j, for $j = 1, 2, \ldots, m$, could be as small as $(1 - m\alpha)$, which is smaller than $(1 - \alpha)$ if $m > 1$. A simultaneous confidence coefficient of at least $(1 - \alpha)$ can be assured by choosing the confidence intervals I_j, for $j = 1, 2, \ldots, m$, so that $\sum_{j=1}^{m} \alpha_j = \alpha$. One way to achieve this objective is if each interval is constructed to have confidence coefficient $1 - (\alpha/m)$. We apply this technique in the following example.

EXAMPLE 13.9   For the four treatments given in Example 13.2, construct confidence intervals for all comparisons of the form $\mu_i - \mu_{i'}$, with simultaneous confidence coefficient no smaller than .95.

Solution   The appropriate $100(1 - \alpha)\%$ confidence interval for a single comparison, say, $\mu_1 - \mu_2$, would be of the form

$$(\overline{Y}_{1\bullet} - \overline{Y}_{2\bullet}) \pm t_{\alpha/2}S\sqrt{\frac{1}{n_1} + \frac{1}{n_2}}.$$

Because there are six such differences to consider, each interval should have confidence coefficient $1 - (\alpha/6)$. Thus the corresponding t value will be $t_{\alpha/2(6)} = t_{\alpha/12}$. Because we want $1 - \alpha = .95$, or $\alpha = .05$, the t value will be

$$t_{.05/12} = t_{.004}.$$

The closest table value is $t_{.005}$, so we will use this to approximate the desired result. The MSE for the data in Example 13.2 is based upon 19 degrees of freedom, so the table value is $t_{.005} = 2.861$.

Because $s = \sqrt{\text{MSE}} = \sqrt{63} = 7.937$, the intervals with a simultaneous confidence coefficient of approximately .95 have realization as follows:

$$\mu_1 - \mu_2: \quad (75.67 - 78.43) \pm 2.861(7.937)\sqrt{\frac{1}{6} + \frac{1}{7}}.$$

Similarly, we obtain

$$
\begin{aligned}
\mu_1 - \mu_2: & \quad -2.76 \pm 12.63 \\
\mu_1 - \mu_3: & \quad 4.84 \pm 13.11 \\
\mu_1 - \mu_4: & \quad -12.08 \pm 14.66 \\
\mu_2 - \mu_3: & \quad 7.60 \pm 12.63 \\
\mu_2 - \mu_4: & \quad -9.32 \pm 14.23 \\
\mu_3 - \mu_4: & \quad -16.92 \pm 14.66.
\end{aligned}
$$

We were unable to achieve our objective of obtaining a set of six confidence intervals with simultaneous confidence coefficient at least .95 because the t-tables in the text are too limited. Of course, more extensive tables of the t distributions are available. Because each of our six intervals has confidence coefficient .99, we can claim that the six intervals above have a simultaneous confidence coefficient of at least .94.

We emphasize that the technique presented in this section guarantees simultaneous coverage probabilities of at least $1 - \alpha$. The actual simultaneous coverage probability can be much larger than the nominal value $1 - \alpha$. Other methods for constructing simultaneous confidence intervals can be found in the books listed in the references at the end of the chapter.

Exercises

13.57 Refer to Exercise 13.11. Construct confidence intervals for all possible differences between mean maneuver times for the three vehicle classes so that the simultaneous confidence coefficient is at least .95. Interpret the results.

13.58 Refer to Exercise 13.10. After looking at the data, a reader of the report of Wheeler et al. elected to compare the mean proportions of Carbon 14 remaining in the vegetable pulp for high and low concentrations of acetonitrile, because that comparison provides the largest sample difference. How would you suggest this comparison be made?

13.59   Refer to Exercise 13.39. Construct confidence intervals on all possible differences among treatment (soil preparation) means so that the simultaneous confidence coefficient is at least .90.

13.60   Refer to Exercises 13.27 and 13.41. Because method 4 is the most expensive, it is desired to compare it to the other three. Construct confidence intervals for the differences $\mu_1 - \mu_4$, $\mu_2 - \mu_4$, and $\mu_3 - \mu_4$ so that the simultaneous confidence coefficient is at least .95.

13.13   Analysis of Variance Using Linear Models

The methods for analyzing linear models presented in Chapter 11 can be adapted for use in the analysis of variance. We will illustrate the method by formulating a linear model for data obtained through a completely randomized design involving $k = 2$ treatments.

Let Y_{ij} denote the random variable to be observed on the jth observation from treatment i, for $i = 1, 2$. Let us define a *dummy*, or *indicator*, *variable* x as follows:

$$x = \begin{cases} 1, & \text{if the observation is from population 1} \\ 0, & \text{otherwise.} \end{cases}$$

Although such dummy variables can be defined in many ways, this definition is consistent with the coding used in SAS and other statistical analysis computer programs. Notice that with this coding x is 1 if the observation is taken from population 1, and x is 0 if the observation is taken from population 2. If we use x as an independent variable in a linear model, we can model Y_{ij} as

$$Y_{ij} = \beta_0 + \beta_1 x + \varepsilon_{ij}$$

where ε_{ij} is a normally distributed random error with $E(\varepsilon_{ij}) = 0$ and $V(\varepsilon_{ij}) = \sigma^2$. In this model,

$$\mu_1 = E(Y_{1j}) = \beta_0 + \beta_1(1) = \beta_0 + \beta_1$$

and

$$\mu_2 = E(Y_{2j}) = \beta_0 + \beta_1(0) = \beta_0.$$

Thus it follows that $\beta_1 = \mu_1 - \mu_2$, and a test of the hypothesis $\mu_1 - \mu_2 = 0$ is equivalent to the test that $\beta_1 = 0$. Our intuition would suggest that $\hat{\beta}_0 = \overline{Y}_{2\bullet}$ and $\hat{\beta}_1 = \overline{Y}_{1\bullet} - \overline{Y}_{2\bullet}$ are good estimators of β_0 and β_1; indeed, it can be shown (proof omitted) that these are the least squares estimators obtained by fitting the preceding linear model. We illustrate the use of this technique through reanalyzing the data presented in Example 13.1.

EXAMPLE **13.10**   Fit an appropriate linear model to the data of Example 13.1, and test to see whether there is a significant difference between μ_1 and μ_2.

Solution   The model, as indicated earlier, is given by

$$Y_{ij} = \beta_0 + \beta_1 x + \varepsilon_{ij}$$

where

$$x = \begin{cases} 1, & \text{if the observation is from population 1} \\ 0, & \text{otherwise.} \end{cases}$$

The matrices used for the least squares estimators are then

$$\mathbf{Y} = \begin{bmatrix} 6.1 \\ 7.1 \\ 7.8 \\ 6.9 \\ 7.6 \\ 8.2 \\ 9.1 \\ 8.2 \\ 8.6 \\ 6.9 \\ 7.5 \\ 7.9 \end{bmatrix} \quad \text{and} \quad \mathbf{X} = \begin{bmatrix} 1 & 1 \\ 1 & 1 \\ 1 & 1 \\ 1 & 1 \\ 1 & 1 \\ 1 & 1 \\ 1 & 0 \\ 1 & 0 \\ 1 & 0 \\ 1 & 0 \\ 1 & 0 \\ 1 & 0 \end{bmatrix}$$

$$\mathbf{X'X} = \begin{bmatrix} 12 & 6 \\ 6 & 6 \end{bmatrix} \quad \text{and} \quad (\mathbf{X'X})^{-1} = \begin{bmatrix} 1/6 & -1/6 \\ -1/6 & 1/3 \end{bmatrix}.$$

The least squares estimates are given by

$$\hat{\beta} = (\mathbf{X'X})^{-1}\mathbf{X'Y} = \begin{bmatrix} 1/6 & -1/6 \\ -1/6 & 1/3 \end{bmatrix} \begin{bmatrix} 91.9 \\ 43.7 \end{bmatrix} = \begin{bmatrix} 8.033 \\ -.75 \end{bmatrix}.$$

Notice that $\hat{\beta}_0 = 8.033 = \overline{Y}_{2\bullet}$ and $\hat{\beta}_1 = -.75 = \overline{Y}_{1\bullet} - \overline{Y}_{2\bullet}$.
Further,

$$\text{SSE} = \mathbf{Y'Y} - \hat{\beta}'\mathbf{X'Y} = 5.8617,$$

is the same as the SSE calculated in Example 13.1. Therefore, $s^2 = \text{SSE}/(n-2) = .58617$, and $s = \sqrt{.58617} = .7656$.

To test $H_0 : \beta_1 = 0$, we construct the t statistic (see Section 11.12):

$$t = \frac{\hat{\beta}_1 - 0}{s\sqrt{c_{11}}} = \frac{-.75}{.7656\sqrt{1/3}} = -1.697.$$

Because we are interested in a two-tailed test, the associated p-value is $2P(t < -1.697) = 2P(t > 1.697)$, where t is based on 10 degrees of freedom. Thus, using

Table 5, Appendix III, we obtain $.05 < P(t > 1.697) < .10$ and $.10 <$ p-value $< .20$. Therefore, for any α value less than .1, we cannot reject H_0. That is, there is insufficient evidence to indicate that μ_1 and μ_2 differ.

This t test is equivalent to the F test of Example 13.1. In fact, the square of the observed t value is the observed F value of Example 13.1.

We will illustrate the linear model approach to a more complicated analysis of variance problem by considering a randomized block design.

EXAMPLE **13.11**   An experiment was conducted to compare the effects of four chemicals A, B, C, and D on water resistance in textiles. Three different bolts of material I, II, and III were used, with each chemical treatment being applied to one piece of material cut from each of the bolts. The data are given in Table 13.7. Write a linear model for this experiment and test the hypothesis that there are no differences among mean water resistances for the four chemicals. Use $\alpha = .05$.

Solution   In formulating the model, we define β_0 as the mean response for treatment D on material from bolt III, and then we introduce a distinct indicator variable for each treatment change and each bolt of material (block) change. The model is

$$Y = \beta_0 + \beta_1 x_1 + \beta_2 x_2 + \beta_3 x_3 + \beta_4 x_4 + \beta_5 x_5 + \varepsilon$$

where

$$x_1 = \begin{cases} 1, & \text{if material from bolt I is used} \\ 0, & \text{otherwise} \end{cases}$$

$$x_2 = \begin{cases} 1, & \text{if material from bolt II is used} \\ 0, & \text{otherwise} \end{cases}$$

$$x_3 = \begin{cases} 1, & \text{if treatment A is used} \\ 0, & \text{otherwise} \end{cases}$$

$$x_4 = \begin{cases} 1, & \text{if treatment B is used} \\ 0, & \text{otherwise} \end{cases} \quad \text{and}$$

$$x_5 = \begin{cases} 1, & \text{if treatment C is used} \\ 0, & \text{otherwise.} \end{cases}$$

Table 13.7. Data for Example 13.11

Bolt of Material	Treatments			
	A	B	C	D
I	10.1	11.4	9.9	12.1
II	12.2	12.9	12.3	13.4
III	11.9	12.7	11.4	12.9

We want to test the hypothesis that there are no differences among treatment means, which is equivalent to $H_0 : \beta_3 = \beta_4 = \beta_5 = 0$. Thus we must fit a complete and a reduced model. (See Section 11.14.)

For the complete model, we have

$$
Y = \begin{bmatrix} 10.1 \\ 12.2 \\ 11.9 \\ 11.4 \\ 12.9 \\ 12.7 \\ 9.9 \\ 12.3 \\ 11.4 \\ 12.1 \\ 13.4 \\ 12.9 \end{bmatrix} \quad \text{and} \quad X = \begin{bmatrix} 1 & 1 & 0 & 1 & 0 & 0 \\ 1 & 0 & 1 & 1 & 0 & 0 \\ 1 & 0 & 0 & 1 & 0 & 0 \\ 1 & 1 & 0 & 0 & 1 & 0 \\ 1 & 0 & 1 & 0 & 1 & 0 \\ 1 & 0 & 0 & 0 & 1 & 0 \\ 1 & 1 & 0 & 0 & 0 & 1 \\ 1 & 0 & 1 & 0 & 0 & 1 \\ 1 & 0 & 0 & 0 & 0 & 1 \\ 1 & 1 & 0 & 0 & 0 & 0 \\ 1 & 0 & 1 & 0 & 0 & 0 \\ 1 & 0 & 0 & 0 & 0 & 0 \end{bmatrix}.
$$

A little matrix algebra yields, for this complete model,

$$
\text{SSE}_C = Y'Y - \hat{\beta}'X'Y = 1721.760 - 1721.225 = .535.
$$

For the reduced model, we have

$$
Y = \beta_0 + \beta_1 x_1 + \beta_2 x_2 + \varepsilon
$$

and the corresponding X matrix has only the first three columns of the one given for the complete model. We then obtain

$$
\hat{\beta} = (X'X)^{-1}X'Y = \begin{bmatrix} 12.225 \\ -1.350 \\ .475 \end{bmatrix}
$$

and

$$
\text{SSE}_R = Y'Y - \hat{\beta}'X'Y = 1721.760 - 1716.025 = 5.735.
$$

It follows that the F ratio appropriate to compare these complete and reduced models is

$$
F = \frac{(\text{SSE}_R - \text{SSE}_C)/(k - g)}{\text{SSE}_C/(n - [k + 1])} = \frac{(5.735 - .535)/(5 - 2)}{(.535)/(12 - 6)} = \frac{1.733}{.0892} = 19.4.
$$

The tabulated F for $\alpha = .05$, $\nu_1 = 3$, and $\nu_2 = 6$ is 4.76. Hence, if we choose $\alpha = .05$, we reject the null hypothesis and conclude that the data present sufficient evidence to indicate that differences exist among the treatment means. The associated p-value is given by $P(F > 19.4) < .005$. The F test used in this example is equivalent to the one that would have been produced by the methods discussed in Section 13.9.

Although it provides a very useful technique, the linear model approach to analysis of variance calculation generally is used only when the computations are being done on a computer. The calculation formulas given earlier in the chapter are more convenient for hand calculation. Notice that if there are k treatments involved in a study, the "dummy variables" approach requires that we define $k - 1$ dummy variables if we wish to use the linear model approach to analyze the data.

Exercises

13.61 Refer to Example 13.11. In Exercise 13.33, you interpreted the parameters in the model for a randomized block design in terms of the mean response for each treatment in each block. In terms of the model with dummy variables given in Example 13.11, β_0 is the mean response to treatment D for bolt of material (block) III.

a In terms of the β values, what is the mean response to treatment A in block III?

b Based on your answers to (a) and (b), what is an interpretation of the parameter β_3?

13.62 Refer to Exercise 13.8.

a Answer the question posed in Exercise 13.8 by fitting complete and reduced linear models. Test using $\alpha = .05$.

b Use the calculations for the complete model from (a) to test the hypothesis that there is no difference between the means for methods A and C. Test using $\alpha = .05$.

c Give the attained significance levels for the tests implemented in (a) and (b).

13.63 Refer to Exercise 13.38. Answer the question regarding possible differences among the treatment means by fitting complete and reduced models.

13.64 Refer to Exercise 13.39. Answer (b) by constructing an F test, using complete and reduced linear models.

13.14 Summary

The one-way layout (including the completely randomized design) and the randomized block designs are illustrations of experiments involving one and two qualitative independent variables, respectively. The analysis of variance partitions the total sum of squares of deviations of the response measurements about their mean into portions associated with each independent variable and the experimental error. The former may be compared with the sum of squares for error, using mean squares and the F statistics, to see whether the mean squares for the independent variables are unusually large and thereby indicative of an effect on the response.

In this chapter we have presented a very brief introduction to the analysis of variance and its associated subject, the design of experiments. Experiments can be designed to investigate the effect of many quantitative and qualitative variables on a response. These may be variables of primary interest to the experimenter, as well as nuisance variables, such as blocks, which may contribute unwanted variation that we attempt to separate from the experimental error. These experiments are subject to an analysis of variance when properly designed. A more extensive coverage of the basic concepts of experimental design and the analysis of experiments will be found in the references.

References and Further Readings

Box, G. E. P., W. G. Hunter, and J. S. Hunter. 1978. *Statistics for Experimenters.* New York: John Wiley.

Cochran, W. G., and G. Cox, 1968. *Experimental Designs*, 2d ed. New York: John Wiley.

Graybill, F., 1976. *Theory and Application of the Linear Model.* North Scituate, MA: Duxbury Press.

Guenther, W. C. 1964. *Analysis of Variance.* Englewood Cliffs, NJ: Prentice-Hall.

Hicks, C. R., and K. V. Turner. 1999. *Fundamental Concepts in the Design of Experiments*, 5th ed. New York: Oxford University Press.

Scheffé, H. 1999. *The Analysis of Variance.* New York: Wiley Interscience.

Supplementary Exercises

13.65 Three cleansing agents for the skin were used on three persons. For each person three patches of skin were exposed to a contaminant and afterward cleansed by using one of the three cleansing agents. After 8 hours the residual contaminant was measured, with the following results:

$$SST = 1.18 \qquad SSB = .78 \qquad SSE = 2.24.$$

a What are the experimental units and what are the blocks in this experiment?

b Test the hypothesis that there are no differences among the treatment means, using $\alpha = .05$.

13.66 Refer to Exercise 13.7. Suppose that the sand used in the mixes for samples 1, 2, 3, and 4 came from pit A, that the sand used for samples 5, 6, 7, and 8 came from pit B, and that the sand for the other samples came from pit C. Analyze the data, assuming that the requirements for a randomized block are met with three blocks consisting, respectively, of samples 1, 2, 3, and 4; samples 5, 6, 7, and 8; and samples 9, 10, 11, and 12.

a At the 5% significance level, is there evidence of differences in concrete strength due to the sand used?

b Is there evidence, at the 5% significance level, of differences in average strength among the four types of concrete used?

c Does the conclusion of (b) contradict the conclusion that was obtained in Exercise 13.7?

13.67 Refer to Exercise 13.66. Let μ_A and μ_B, respectively, denote the mean strengths of concrete specimens prepared from mix A and mix B.

a Find a 95% confidence interval for $(\mu_A - \mu_B)$.

b Is the interval found in (a) the same interval found in Exercise 13.22(b)? Why or why not?

13.68 A study was initiated to investigate the effect of two drugs, administered simultaneously, on reducing human blood pressure. It was decided to use three levels of each drug and to include all nine combinations in the experiment. Nine high-blood-pressure patients were selected for the experiment, and one was randomly assigned to each of the nine drug combinations. The response observed was a drop in blood pressure over a fixed interval of time.

a Is this a randomized block design?

b Suppose that two patients were assigned to each of the nine drug combinations. What type of experimental design is this?

13.69 Refer to Exercise 13.68. Suppose that prior experimentation suggests that $\sigma = 20$.

a How many replications would be required to estimate any treatment (drug combination) mean correct to within ± 10 with probability .95?

b How many degrees of freedom will be available for estimating σ^2 when using the number of replications determined in (a)?

c Give the approximate half width of a confidence interval for the difference in mean response for two treatments when using the number of replications determined in (a).

13.70 A dealer has in stock three cars (model A, model B, and model C) of the same make but different models. Wishing to compare mileage obtained for these different models, a customer arranged to test each car with each of three brands of gasoline (brand X, brand Y, and brand Z). In each trial a gallon of gasoline was added to an empty tank and the car was driven without stopping until it ran out of gasoline. The accompanying table shows the number of miles covered in each of the nine trials.

Brand of Gasoline	Distance (miles)		
	Model A	Model B	Model C
X	22.4	17.0	19.2
Y	20.8	19.4	20.2
Z	21.5	18.7	21.2

a Should the customer conclude that the different car models differ in mean gas mileage? Test at the $\alpha = .05$ level.

b Do the data indicate that the brand of gasoline affects gas mileage?

13.71 Refer to Exercise 13.70. Suppose that the gas mileage is unrelated to brand of gasoline. Carry out an analysis of the data appropriate for a completely randomized design with three treatments.

a Should the customer conclude that the three cars differ in gas mileage? Test at the $\alpha = .05$ level.

b Comparing your answer for Exercise 13.70(a) with your answer for (a), can you suggest a reason why blocking may be unwise in certain cases?

c Why might it be *WRONG* to analyze the data in the manner suggested in (a)?

13.72 In the hope of attracting more riders, a city transit company plans to have express bus service from a suburban terminal to the downtown business district. These buses should save travel time. The city decides to perform a study of the effect of four different plans (such as a special bus lane and traffic signal progression) on the travel time for the buses. Travel times (in minutes) are measured for several weekdays during a morning rush-hour trip while each plan is in effect. The results are recorded in the following table.

Plan			
1	2	3	4
27	25	34	30
25	28	29	33
29	30	32	31
26	27	31	
	24	36	

a   What type of experimental design was employed?

b   Is there evidence of a difference in the mean travel times for the four plans? Use $\alpha = 0.01$.

c   Form a 95% confidence interval for the difference between plan 1 (express lane) and plan 3 (a control: no special travel arrangements).

13.73   A study was conducted to compare the effect of three levels of digitalis on the level of calcium in the heart muscle of dogs. A description of the actual experimental procedure is omitted, but it is sufficient to note that the general level of calcium uptake varies from one animal to another so that comparison of digitalis levels (treatments) had to be blocked on heart muscles. That is, the tissue for a heart muscle was regarded as a block and comparisons of the three treatments were made within a given muscle. The calcium uptakes for the three levels of digitalis, A, B, and C, were compared based on the heart muscles of four dogs. The results are shown in the accompanying table.

Dogs			
1	2	3	4
A	C	B	A
1342	1698	1296	1150
B	B	A	C
1608	1387	1029	1579
C	A	C	B
1881	1140	1549	1319

a   Calculate the sums of squares for this experiment and construct an ANOVA table.

b   How many degrees of freedom are associated with SSE?

c   Do the data present sufficient evidence to indicate a difference in the mean uptake of calcium for the three levels of digitalis?

d   Do the data indicate a difference in the mean uptake in calcium for the four heart muscles?

e   Give the standard deviation of the difference between the mean calcium uptakes for two levels of digitalis.

f   Find a 95% confidence interval for the difference in mean responses between treatments A and B.

13.74   Refer to Exercise 13.73. Approximately how many replications are required for each level of digitalis (how many blocks), so that the error of estimating the difference in mean response for a pair of digitalis levels is less than 20, with probability .95? Assume that additional observations would be made within a randomized block design.

13.75   A completely randomized design was conducted to compare the effects of five stimuli on reaction time. Twenty-seven people were employed in the experiment, which was conducted using a completely randomized design. Regardless of the results of the analysis of variance, it is desired to compare stimuli A and D. The reaction times (in seconds) were as shown in the accompanying table.

| | \multicolumn{5}{c}{Stimulus} |
	A	B	C	D	E
	.8	.7	1.2	1.0	.6
	.6	.8	1.0	.9	.4
	.6	.5	.9	.9	.4
	.5	.5	1.2	1.1	.7
		.6	1.3	.7	.3
		.9	.8		
		.7			
Total	2.5	4.7	6.4	4.6	2.4
Mean	.625	.671	1.067	.920	.480

a Conduct an analysis of variance and test for a difference in mean reaction times due to the five stimuli. Give bounds for the *p*-value.

b Compare stimuli A and D to see if there is a difference in mean reaction times. What can be said about the attained significance level?

13.76 Because we would expect mean reaction time to vary from one person to another, the experiment in Exercise 13.75 might have been conducted more effectively by using a randomized block design with people as blocks. Hence, four people were used in a new experiment, and each person was subjected to each of the five stimuli in a random order. The reaction times (in seconds) were as shown in the accompanying table. Conduct an analysis of variance and test for differences in mean reaction times for the four stimuli.

| | \multicolumn{5}{c}{Stimulus} |
Subject	A	B	C	D	E
1	.7	.8	1.0	1.0	.5
2	.6	.6	1.1	1.0	.6
3	.9	1.0	1.2	1.1	.6
4	.6	.8	.9	1.0	.4

13.77 Refer to Exercise 13.40. Construct confidence intervals to compare each of the ryegrass cultivars with Marvelgreen supreme in such a way that the simultaneous confidence coefficient is at least .95. Interpret the results.

13.78 Show that

$$\text{Total SS} = \text{SST} + \text{SSB} + \text{SSE}$$

for a randomized block design, where

$$\text{SSE} = \sum_{j=1}^{b} \sum_{i=1}^{k} (Y_{ij} - \overline{Y}_{\bullet j} - \overline{Y}_{i \bullet} + \overline{Y})^2.$$

*13.79   Consider the following model for the responses measured in a randomized block design containing b blocks and k treatments:

$$Y_{ij} = \mu + \tau_i + \beta_j + \varepsilon_{ij}$$

where

$Y_{ij} =$ response to treatment i in block j

$\mu =$ overall mean

$\tau_i =$ nonrandom effect of treatment i, where $\displaystyle\sum_{i=1}^{k} \tau_i = 0$

$\beta_j =$ random effect of block j, where β_j are independent, normally distributed random variables with $E(\beta_j) = 0$ and $V(\beta_j) = \sigma_\beta^2$, for $j = 1, 2, \ldots, b$.

$\varepsilon_{ij} =$ random error terms where ε_{ij} are independent, normally distributed random variables with $E(\varepsilon_{ij}) = 0$ and $V(\varepsilon_{ij}) = \sigma_\varepsilon^2$, for $i = 1, 2, \ldots, k$ and $j = 1, 2, \ldots, b$.

Further, assume that the β_j and ε_{ij} also are independent. This model differs from that presented in Section 13.8 in that the block effects are assumed to be *random variables*, instead of fixed but unknown constants.

a   If the model just described is appropriate, show that observations taken from different blocks are independent of one another. That is, show that Y_{ij} and $Y_{ij'}$ are independent if $j \neq j'$, as are Y_{ij} and $Y_{i'j'}$ if $i \neq i'$ and $j \neq j'$.

b   Under the model just described, derive the covariance of two observations from the same block. That is find $\text{Cov}(Y_{ij}, Y_{i'j})$ if $i \neq i'$.

c   Two random variables that have a joint normal distribution are independent if and only if their covariance is 0. Use the result from (b) to determine conditions under which two observations from the same block are independent of one another.

*13.80   Refer to the model for the randomized block design with random block effect given in Exercise 13.79.

a   Give the expected value and variance of Y_{ij}.

b   Let $\overline{Y}_{i\bullet}$ denote the average of all of the responses to treatment i. Use the model for the randomized block design to derive $E(\overline{Y}_{i\bullet})$ and $V(\overline{Y}_{i\bullet})$. Is $\overline{Y}_{i\bullet}$ an unbiased estimator for the mean response to treatment i? Explain why or why not. Notice that $V(\overline{Y}_{i\bullet})$ depends on b and both σ_β^2 and σ_ε^2.

c   Consider $\overline{Y}_{i\bullet} - \overline{Y}_{i'\bullet}$ for $i \neq i'$. Show that $E(\overline{Y}_{i\bullet} - \overline{Y}_{i'\bullet}) = \tau_i - \tau_{i'}$. This result implies that $\overline{Y}_{i\bullet} - \overline{Y}_{i'\bullet}$ is an unbiased estimator of the difference in the effects of treatments i and i'.

d   Derive $V(\overline{Y}_{i\bullet} - \overline{Y}_{i'\bullet})$. Notice that $V(\overline{Y}_{i\bullet} - \overline{Y}_{i'\bullet})$ depends only on b and σ_ε^2.

*13.81   Refer to the model for the randomized block design with random block effect given in Exercise 13.79, and let $\overline{Y}_{\bullet j}$ denote the average of all the responses in block j.

a   Derive $E(\overline{Y}_{\bullet j})$ and $V(\overline{Y}_{\bullet j})$.

b   Derive $E(\text{MST})$.

    c Derive $E(\text{MSB})$.

    d Derive $E(\text{MSE})$.

*13.82 Refer to the model for the randomized block design with random block effect given in Exercise 13.79 and the results obtained in Exercise 13.81 (c) and (d).

    a Give an unbiased estimator for σ_ε^2.

    b Give an unbiased estimator for σ_β^2.

*13.83 Suppose that $Y_1, Y_2, \ldots, Y_n$ is a random sample from a normal distribution with mean μ and variance σ^2. The independence of $\sum_{i=1}^{n}(Y_i - \overline{Y})^2$ and $\overline{Y}$ can be shown as follows. Define an $n \times n$ matrix $\mathbf{A}$ by

$$
\mathbf{A} = \begin{bmatrix}
\dfrac{1}{\sqrt{n}} & \dfrac{1}{\sqrt{n}} & \dfrac{1}{\sqrt{n}} & \dfrac{1}{\sqrt{n}} & \cdots & \dfrac{1}{\sqrt{n}} & \dfrac{1}{\sqrt{n}} \\[2mm]
\dfrac{1}{\sqrt{2}} & \dfrac{-1}{\sqrt{2}} & 0 & 0 & \cdots & 0 & 0 \\[2mm]
\dfrac{1}{\sqrt{2\cdot 3}} & \dfrac{1}{\sqrt{2\cdot 3}} & \dfrac{-2}{\sqrt{2\cdot 3}} & 0 & \cdots & 0 & 0 \\[2mm]
\vdots & \vdots & \vdots & \vdots & \vdots & \vdots & \vdots \\[2mm]
\dfrac{1}{\sqrt{(n-1)n}} & \dfrac{1}{\sqrt{(n-1)n}} & & & \cdots & \dfrac{1}{\sqrt{(n-1)n}} & \dfrac{-(n-1)}{\sqrt{(n-1)n}}
\end{bmatrix}
$$

and notice that $\mathbf{A}'\mathbf{A} = \mathbf{I}$, the identity matrix. Then

$$
\sum_{i=1}^{n} Y_i^2 = \mathbf{Y}'\mathbf{Y} = \mathbf{Y}'\mathbf{A}'\mathbf{A}\mathbf{Y}
$$

where $\mathbf{Y}$ is the vector of Y_i values.

    a Show that

$$
\mathbf{A}\mathbf{Y} = \begin{bmatrix}
\overline{Y}\sqrt{n} \\
U_1 \\
U_2 \\
\vdots \\
U_{n-1}
\end{bmatrix}
$$

where $U_1, U_2, \ldots, U_{n-1}$ are linear functions of $Y_1, Y_2, \ldots, Y_n$. Thus

$$
\sum_{i=1}^{n} Y_i^2 = n\overline{Y}^2 + \sum_{i=1}^{n-1} U_i^2.
$$

    b Show that the linear functions $\overline{Y}\sqrt{n}, U_1, U_2, \ldots, U_{n-1}$ are pairwise orthogonal and hence independent under the normality assumption. (See Exercise 5.110.)

    c Show that

$$
\sum_{i=1}^{n} (Y_i - \overline{Y})^2 = \sum_{i=1}^{n-1} U_i^2
$$

and conclude that this quantity is independent of $\overline{Y}$.

d   Using the results of (c), show that

$$\frac{\sum_{i=1}^{n}(Y_i - \overline{Y})^2}{\sigma^2} = \frac{(n-1)S^2}{\sigma^2}$$

has a χ^2 distribution with $(n-1)$ degrees of freedom.

13.84   Consider a one-way layout with k treatments. Assume that Y_{ij} is the jth response for treatment (population) i and that Y_{ij} is normally distributed with mean μ_i and variance σ^2, for $i = 1, 2, \ldots, k$ and $j = 1, 2, \ldots, n_i$.

a   Use Exercise 13.83 to justify that $\overline{Y}_1, \overline{Y}_2, \ldots, \overline{Y}_k$ are independent of SSE.

b   Show that MST/MSE has an F distribution with $v_1 = k-1$ and $v_2 = n_1 + n_2 + \cdots + n_k - k$ degrees of freedom under $H_0 : \mu_1 = \mu_2 = \cdots = \mu_k$. (You may assume, for simplicity, that $n_1 = n_2 = \cdots = n_k$.)

CHAPTER 14

Analysis of Categorical Data

14.1 A Description of the Experiment

14.2 The Chi-Square Test

14.3 A Test of a Hypothesis Concerning Specified Cell Probabilities: A Goodness-of-Fit Test

14.4 Contingency Tables

14.5 $r \times c$ Tables with Fixed Row or Column Totals

14.6 Other Applications

14.7 Summary and Concluding Remarks

References and Further Readings

14.1 A Description of the Experiment

Many experiments, particularly in the social sciences, result in *categorical* (or count) *data*. For instance, the classification of people into five income brackets would result in the numbers of individuals that fall into each of the five income classes (categories). Or we might be interested in studying the reactions of mice to a particular stimulus in a psychological experiment. If each mouse will react in one of three ways when the stimulus is applied and if a large number of mice were subjected to the stimulus, the experiment would yield three counts, indicating the number of mice falling in each of the reaction categories. Similarly, a traffic study might require a count and classification of the type of motor vehicles using a section of highway. An industrial process manufactures items that fall into one of three quality categories: acceptable, seconds, and rejects. A student of the arts might classify paintings in one of k categories according to style and period in order to study trends in style over time. We might wish to classify ideas in a philosophical study or styles in the field of literature. The results of an advertising campaign would yield count data indicating a classification of consumer reaction. Indeed, many observations in the physical sciences as well are not amenable to measurement on a continuous scale and hence result in categorical data.

The illustrations in the preceding paragraph exhibit, to a reasonable degree of approximation, the following characteristics, which define a *multinomial experiment* (see Section 5.9):

1. The experiment consists of n identical trials.
2. The outcome of each trial falls into one of k categories or cells.
3. The probability that the outcome of a single trial will fall in a particular cell, say, cell i, is p_i, where $i = 1, 2, \ldots, k$, and remains the same from trial to trial. Notice that

$$p_1 + p_2 + p_3 + \cdots + p_k = 1.$$

4. The trials are independent.
5. We are interested in $n_1, n_2, n_3, \ldots, n_k$, where n_i for $i = 1, 2, \ldots, k$ is equal to the number of trials in which the outcome falls into cell i. Notice that $n_1 + n_2 + n_3 + \cdots + n_k = n$.

This experiment is analogous to tossing n balls at k boxes, where each ball must fall into exactly one of the boxes. The boxes are arranged such that the probability that a ball will fall into a box varies from box to box but remains the same for each box in repeated tosses. Finally, the balls are tossed in such a way that the trials are independent. At the conclusion of the experiment, we observe n_1 balls in the first box, n_2 in the second, ..., and n_k in the kth. The total number of balls is equal to

$$\sum_{i=1}^{k} n_i = n.$$

Notice the similarity between the binomial and the multinomial experiments and, in particular, that the binomial experiment represents the special case for the multinomial experiment when $k = 2$. The two cell probabilities, p and $q = 1 - p$, of the binomial experiment are replaced by the k cell probabilities, $p_1, p_2, \ldots, p_k$, of the multinomial experiment. The objective of this chapter is to make inferences about the cell probabilities $p_1, p_2, \ldots, p_k$. The inferences will be expressed in terms of a statistical test of hypotheses concerning their specific numerical values or their relationship one to another.

Because the calculation of multinomial probabilities is somewhat cumbersome, it would be difficult to calculate the exact significance levels (probabilities of type I errors) for hypotheses regarding the values of $p_1, p_2, \ldots, p_k$. Fortunately, we have been relieved of this chore by the British statistician Karl Pearson, who proposed a very useful test statistic for testing hypotheses concerning $p_1, p_2, \ldots, p_k$ and gave the approximate sampling distribution of this statistic. We will outline the construction of Pearson's test statistic in the following section.

14.2 The Chi-Square Test

Suppose that $n = 100$ balls were tossed at the cells (boxes) and that we knew that p_1 was equal to .1. How many balls would be expected to fall in the first cell? Referring to Section 5.9 and applying our knowledge of the multinomial experiment, we would calculate

$$E(n_1) = np_1 = (100)(.1) = 10.$$

In like manner, the expected number falling in the remaining cells may be calculated by using the formula

$$E(n_i) = np_i, \qquad i = 1, 2, \ldots, k.$$

Now suppose that we hypothesize values for $p_1, p_2, \ldots, p_k$ and calculate the expected value for each cell. Certainly if our hypothesis is true, the cell counts n_i should not deviate greatly from their expected values np_i for $i = 1, 2, \ldots, k$. Hence it would seem intuitively reasonable to use a test statistic involving the k deviations,

$$n_i - np_i, \qquad \text{for } i = 1, 2, \ldots, k.$$

In 1900 Karl Pearson proposed the following test statistic, which is a function of the squares of the deviations of the observed counts from their expected values, weighted by the reciprocals of their expected values:

$$X^2 = \sum_{i=1}^{k} \frac{[n_i - E(n_i)]^2}{E(n_i)} = \sum_{i=1}^{k} \frac{[n_i - np_i]^2}{np_i}.$$

Although the mathematical proof is beyond the scope of this text, it can be shown that when n is large, X^2 will possess approximately a chi-square probability distribution in repeated sampling. We can easily show this result for the case $k = 2$, as follows. If $k = 2$, then $n_2 = n - n_1$ and $p_1 + p_2 = 1$. Thus,

$$X^2 = \sum_{i=1}^{2} \frac{[n_i - E(n_i)]^2}{E(n_i)} = \frac{(n_1 - np_1)^2}{np_1} + \frac{(n_2 - np_2)^2}{np_2}$$

$$= \frac{(n_1 - np_1)^2}{np_1} + \frac{[(n - n_1) - n(1 - p_1)]^2}{n(1 - p_1)}$$

$$= \frac{(n_1 - np_1)^2}{np_1} + \frac{(-n_1 + np_1)^2}{n(1 - p_1)}$$

$$= (n_1 - np_1)^2 \left(\frac{1}{np_1} + \frac{1}{n(1 - p_1)} \right) = \frac{(n_1 - np_1)^2}{np_1(1 - p_1)}.$$

We have seen (Section 7.5) that

$$\frac{n_1 - np_1}{\sqrt{np_1(1 - p_1)}}$$

has approximately a standard normal distribution for large n. Thus, for large n, X^2 as just given is approximately a χ^2 random variable with 1 degree of freedom. Recall that the square of a standard normal random variable has a χ^2 distribution (see Example 6.11).

Experience has shown that the cell counts n_i should not be too small in order that the chi-square distribution provide an adequate approximation to the distribution of X^2. As a rule of thumb we will require that all expected cell counts equal or exceed 5, although Cochran (1952) has noted that this value can be as low as 1 for some situations.

You will recall the use of the χ^2 probability distribution for testing a hypothesis concerning a population variance σ^2 in Section 10.9. In particular, we have seen that the shape of the χ^2 distribution varies depending upon the number of degrees of freedom, and we discussed the use of Table 6, Appendix III, which presents the critical values of χ^2 corresponding to various right-hand tail areas of the distribution. Therefore, we must know which χ^2 distribution to use—that is, the number of degrees of freedom—in approximating the distribution of X^2, and we must know whether to use a one-tailed or two-tailed test in locating the rejection region for the test. The latter problem may be solved directly. Because large deviations of the observed cell counts from those expected would tend to contradict the null hypothesis concerning the cell probabilities $p_1, p_2, \ldots, p_k$, we would reject the null hypothesis when X^2 is large and employ a one-tailed statistical test, using the upper-tail values of χ^2 to locate the rejection region.

The determination of the appropriate number of degrees of freedom to be employed for the test can be a little tricky and, therefore, will be specified for the physical applications described in the following sections. In addition, we will state the principle involved (which is fundamental to the mathematical proof of the approximation) so that you will understand why the number of degrees of freedom changes with various applications. This principle states that *the appropriate number of degrees of freedom will equal the number of cells k less 1 degree of freedom for each independent linear restriction placed upon the observed cell counts.* For example, one linear restriction is present because the sum of the cell counts must equal n; that is,

$$n_1 + n_2 + n_3 + \cdots + n_k = n.$$

Other restrictions will be introduced for some applications because of the necessity for estimating unknown parameters required in the calculation of the expected cell frequencies or because of the method by which the sample is collected. When unknown parameters must be estimated in order to compute X^2, a maximum-likelihood estimator should be employed. The degrees of freedom for the approximating chi-square distribution will be reduced by 1 for each parameter estimated. These cases will arise as we consider various practical examples.

14.3 A Test of a Hypothesis Concerning Specified Cell Probabilities: A Goodness-of-Fit Test

The simplest hypothesis concerning the cell probabilities would be one that specifies numerical values for each. In this case we are testing $H_0 : p_1 = p_{1,0}, p_2 = p_{2,0}, \ldots, p_k = p_{k,0}$, where $p_{i,0}$ denotes a specified value for p_i. The alternative is the general one that states that at least one of the equalities does not hold. Because the only restriction on the observations is that $\sum_{i=1}^{k} n_i = n$, the X^2 test statistic will have approximately a χ^2 distribution with $k - 1$ degrees of freedom.

EXAMPLE 14.1   A group of rats, one by one, proceed down a ramp to one of three doors. We wish to test the hypothesis that the rats have no preference concerning the choice of a door and, therefore, that

$$H_0: p_1 = p_2 = p_3 = 1/3,$$

where p_i is the probability that a rat will choose door i, for $i = 1, 2,$ or 3.

Suppose that the rats were sent down the ramp $n = 90$ times and that the three observed cell frequencies were $n_1 = 23, n_2 = 36,$ and $n_3 = 31$. The expected cell frequency would be the same for each cell: $E(n_i) = np_i = (90)(1/3) = 30$. The observed and expected cell frequencies are presented in Table 14.1. Notice the discrepancy between the observed and expected cell frequencies. Do the data present sufficient evidence to warrant rejection of the hypothesis of no preference?

Solution   The chi-square test statistic for our example will possess $(k - 1) = 2$ degrees of freedom since the only linear restriction on the cell frequencies is that

$$n_1 + n_2 + n_3 = 90.$$

Therefore, if we choose $\alpha = .05$ we would reject the null hypothesis when $X^2 > 5.991$ (see Table 6, Appendix III).

Substituting into the formula for X^2, we obtain

$$X^2 = \sum_{i=1}^{k} \frac{[n_i - E(n_i)]^2}{E(n_i)} = \sum_{i=1}^{k} \frac{(n_i - np_i)^2}{np_i}$$

$$= \frac{(23 - 30)^2}{30} + \frac{(36 - 30)^2}{30} + \frac{(31 - 30)^2}{30} = 2.87.$$

Table 14.1. Observed and expected cell counts

Value	Door		
	1	2	3
Observed cell frequency	$n_1 = 23$	$n_2 = 36$	$n_3 = 31$
Expected cell frequency	(30)	(30)	(30)

Because X^2 is less than the tabulated critical value of χ^2, the null hypothesis is not rejected, and we conclude that the data do not present sufficient evidence to indicate that the rats have a preference for any of the doors. In this case, the p-value is given by p-value $= P(\chi^2 > 2.87)$ where χ^2 possesses a chi-square distribution with $k - 1 = 2$ degrees of freedom. Using Table 6, Appendix III, it follows that p-value > 0.10.

The χ^2 statistic also can be used to test whether sample data indicate that a particular model for a population distribution does not fit the data. An example of such a test, called the *goodness-of-fit test*, is given in the following example.

EXAMPLE 14.2 The number of accidents Y per week at an intersection was checked for $n = 50$ weeks, with the results as shown in Table 14.2. Test the hypothesis that the random variable Y has a Poisson distribution, assuming the observations to be independent. Use $\alpha = .05$.

Solution The null hypothesis H_0 states that Y has the Poisson distribution, given by

$$p(y \mid \lambda) = \frac{\lambda^y e^{-\lambda}}{y!}, \qquad y = 0, 1, 2, \ldots.$$

Because λ is unknown we must find its maximum-likelihood estimator, which turns out to be $\hat{\lambda} = \overline{Y}$ (see Exercise 9.72). For the given data $\hat{\lambda}$ has the value $\overline{y} = 24/50 = .48$.

We have, for the given data, three cells with five or more observations—the cells defined by $Y = 0$, $Y = 1$, and $Y \geq 2$. Under H_0, the probabilities for these cells are

$$p_1 = P(Y = 0) = e^{-\lambda}, \qquad p_2 = P(Y = 1) = \lambda e^{-\lambda},$$

$$\text{and} \quad p_3 = P(Y \geq 2) = 1 - e^{-\lambda} - \lambda e^{-\lambda}.$$

These probabilities are estimated by replacing λ with $\hat{\lambda}$, which gives

$$\hat{p}_1 = e^{-.48} = .619, \qquad \hat{p}_2 = .48e^{-.48} = .297, \quad \text{and} \quad \hat{p}_3 = 1 - \hat{p}_1 - \hat{p}_2 = .084.$$

If the observations are independent, the cell frequencies n_1, n_2, and n_3 have a multinomial distribution with parameters p_1, p_2, and p_3. Thus, $E(n_i) = np_i$, and

Table 14.2. Data for Example 14.2

y	Frequency
0	32
1	12
2	6
3 or more	0

the estimated expected cell frequencies are given by

$$\widehat{E(n_1)} = n\hat{p}_1 = 30.95, \qquad \widehat{E(n_2)} = n\hat{p}_2 = 14.85, \quad \text{and} \quad \widehat{E(n_3)} = n\hat{p}_3 = 4.20.$$

Thus, the test statistic is given by

$$X^2 = \sum_{i=1}^{3} \frac{[n_i - \widehat{E(n_i)}]^2}{\widehat{E(n_i)}},$$

which has approximately a χ^2 distribution with $(k - 2) = 1$ degree of freedom. (One degree of freedom is lost because λ had to be estimated, the other, because $\sum_{i=1}^{3} n_i = n$.)

On computing X^2 we find

$$X^2 = \frac{(32 - 30.95)^2}{30.95} + \frac{(12 - 14.85)^2}{14.85} + \frac{(6 - 4.20)^2}{4.20} = 1.354.$$

Because $\chi^2_{.05} = 3.841$, with 1 degree of freedom, we do not reject H_0. The data do not present sufficient evidence to contradict our hypothesis that Y possesses a Poisson distribution.

Exercises

14.1 A city expressway with four lanes in each direction was studied to see whether drivers preferred to drive on the inside lanes. A total of 1000 automobiles were observed during the heavy early morning traffic and their respective lanes recorded. The results were as shown in the accompanying table. Do the data present sufficient evidence to indicate that some lanes are preferred over others? (Test the hypothesis that $p_1 = p_2 = p_3 = p_4 = 1/4$, using $\alpha = .05$.) Give bounds for the associated p value.

Lane	Observed Count
1	294
2	276
3	238
4	192

14.2 Do you hate Mondays? Researchers in Germany have provided another reason for you: they concluded that the risk of heart attack on a Monday for a working person may be as much as 50% greater than on any other day (*Riverside Press-Enterprise*, November 17, 1992). The researchers kept track of heart attacks and coronary arrests over a period of 5 years among 330,000 people who lived near Augsberg, Germany. In an attempt to verify the researcher's claim, 200 working people who had recently had heart attacks were surveyed. The day on which their heart attacks occurred appear in the following table.

Sunday	Monday	Tuesday	Wednesday	Thursday	Friday	Saturday
24	36	27	26	32	26	29

Do these data present sufficient evidence to indicate that there is a difference in the percentages of heart attacks that occur on different days of the week? Test using $\alpha = .05$.

14.3 After inspecting the data in Exercise 14.2, you might wish to test the hypothesis that the probability that a heart attack victim suffered a heart attack on Monday is 1/7 against the alternative that this probability is greater than 1/7.

   a Carry out the test above, using $\alpha = .05$.

   b What tenet of good statistical practice is violated in the test in (a)?

   c Prior to looking at the current data, is there a reason that you might legitimately consider the hypotheses from (a)?

14.4 The Mendelian theory states that the number of a certain type of peas falling into the classifications round and yellow, wrinkled and yellow, round and green, and wrinkled and green should be in the ratio 9:3:3:1. Suppose that 100 such peas revealed 56, 19, 17, and 8 in the respective categories. Are these data consistent with the model? Use $\alpha = .05$. (The expression 9:3:3:1 means that 9/16 of the peas should be round and yellow, 3/16 should be wrinkled and yellow, etc.)

14.5 Portable personal computers, sometimes called laptops, represent a fast-growing segment of the PC market. According to the Market Intelligence Research Company, the use of portable laptop computers can be classified into the following user segments ("Laptop's Three Musts for Success in Sales," *Sales and Marketing Management,* February 1988, p. 91).

User Segment	1988 Percentages	Current Survey Frequency
Business–professional	69%	102
Government	21%	32
Education	7%	12
Home	3%	4
Total	100%	150

Do the data provide sufficient evidence to indicate that the figures obtained in the 1988 study are not accurate today? Give the approximate p-value associated with the test and use it to make your decision.

14.6 Two types of defects, A and B, are frequently seen in the output of a certain manufacturing process. Each item can be classified into one of the four classes $A \cap B$, $A \cap \overline{B}$, $\overline{A} \cap B$, $\overline{A} \cap \overline{B}$, where $\overline{A}$ denotes the absence of the type A defect. For 100 inspected items the following frequencies were observed:

$$A \cap B : 48 \qquad A \cap \overline{B} : 18 \qquad \overline{A} \cap B : 21 \qquad \overline{A} \cap \overline{B} : 13.$$

Is there sufficient evidence to indicate that the four categories, in the order listed, do not occur in the ratio 5:2:2:1? (Use $\alpha = .05$.)

14.7 The data in the following table are the frequency counts for 400 observations on the number of bacterial colonies within the field of a microscope, using samples of milk film. Is there sufficient evidence to claim that the data do not fit the Poisson distribution? (Use $\alpha = .05$.)

Number of Colonies per Field	Frequency of Observation
0	56
1	104
2	80
3	62
4	42
5	27
6	9
7	9
8	5
9	3
10	2
11	0
19	1
	400

Source: C. A. Bliss and R. A. Fisher, "Fitting the Negative Binomial Distribution to Biological Data," *Biometrics* 9 (1953): 176–200. Biometrics Society. Reprinted by permission. All rights reserved.

14.8 The numbers of accidents experienced by machinists in a certain industry were observed for a fixed period of time, with the results as shown in the accompanying table. Test, at the 5% level of significance, the hypothesis that the data come from a Poisson distribution.

Accidents per Machinist	Frequency of Observation (Number of Machinists)
0	296
1	74
2	26
3	8
4	4
5	4
6	1
7	0
8	1

Source: C. A. Bliss and R. A. Fisher, "Fitting the Negative Binomial Distribution to Biological Data," *Biometrics* 9 (1953): 176–200. Biometrics Society. Reprinted by permission. All rights reserved.

14.4 Contingency Tables

A problem frequently encountered in the analysis of count data concerns the independence of two methods of classification of observed events. For example, we might classify a sample of people by gender and by opinion on a political issue in order to test the hypothesis that opinions on this issue are independent of gender. Analogously, we might classify patients suffering from a certain disease according to the type of medication and the rate of recovery in order to see if recovery rate depends upon the type of medication. In each of these examples we wish to investigate a *dependency* (or *contingency*) between two classification criteria.

Suppose that we wish to classify defects found on furniture produced in a manufacturing plant according to (1) the type of defect and (2) the production shift. A total of $n = 309$ furniture defects was recorded and the defects were classified as one of four types, A, B, C, or D. At the same time each piece of furniture was identified according to the production shift in which it was manufactured. These counts are presented in Table 14.3, which is known as a *contingency table*. (Numbers in parentheses are the estimated expected cell frequencies.) Our objective is to test the null hypothesis that type of defect is independent of shift against the alternative that the two categorization schemes are dependent. That is, we wish to test H_0: column classification is independent of row classification.

Let p_A equal the unconditional probability that a defect will be of type A. Similarly, define p_B, p_C, and p_D as the probabilities of observing the three other types of defects. Then these probabilities, which we will call the *column probabilities* of Table 14.3, will satisfy the requirement

$$p_A + p_B + p_C + p_D = 1.$$

In like manner, let p_i for $i = 1, 2,$ or 3 equal the *row probabilities* that a defect will have occurred on shift i, where

$$p_1 + p_2 + p_3 = 1.$$

If the two classifications are independent of each other, a cell probability will equal the product of its respective row and column probabilities. For example, the probability that a particular defect will occur on shift 1 and be of type A is $p_1 \times p_A$. We observe that the numerical values of the cell probabilities are unspecified in the problem under consideration. The null hypothesis specifies only that each cell probability will equal the product of its respective row and column probabilities and therefore will imply independence of the two classifications.

The analysis of the data obtained from a contingency table differs from the analysis in Example 14.1 because we must *estimate* the row and column probabilities in order to estimate the expected cell frequencies.

Table 14.3. A contingency table

Shift	A	B	C	D	Total
1	15(22.51)	21(20.99)	45(38.94)	13(11.56)	94
2	26(22.99)	31(21.44)	34(39.77)	5(11.81)	96
3	33(28.50)	17(26.57)	49(49.29)	20(14.63)	119
Total	74	69	128	38	309

Type of Defect spans columns A–D.

The estimated expected cell frequencies may be substituted for the $E(n_i)$ in X^2, and X^2 will continue to possess a distribution in repeated sampling that is approximated by the chi-square probability distribution.

The maximum-likelihood estimator for any row or column probability is found as follows. Let n_{ij} denote the observed frequency in row i and column j of the contingency table, and let p_{ij} denote the probability of an observation falling into this cell. If observations are independently selected, then the cell frequencies have a multinomial distribution, and the maximum-likelihood estimator of p_{ij} is simply the observed relative frequency for that cell. That is,

$$\hat{p}_{ij} = \frac{n_{ij}}{n}, \qquad i = 1, 2, \ldots, r; \quad j = 1, 2, \ldots, c.$$

(see Exercise 9.79).

Likewise, viewing row i as a single cell, the probability for row i is given by p_i, and hence

$$\hat{p}_i = \frac{r_i}{n}$$

(where r_i denotes the number of observations in row i) is the maximum-likelihood estimator of p_i.

By analogous arguments, the maximum-likelihood estimator of the jth column probability is c_j/n, where c_j denotes the number of observations in column j.

Under the null hypothesis, the maximum-likelihood estimate of the expected value of n_{11} is

$$E(\widehat{n_{11}}) = n(\hat{p}_1 \times \hat{p}_A) = n\left(\frac{r_1}{n}\right)\left(\frac{c_1}{n}\right) = \frac{r_1 \cdot c_1}{n}.$$

In other words, we observe that the estimated expected value of the cell frequency, n_{ij} for a contingency table is equal to the product of its respective row and column totals divided by the total sample size. That is,

$$E(\widehat{n_{ij}}) = \frac{r_i c_j}{n}.$$

The estimated expected cell frequencies for our example are shown in parentheses in Table 14.3. For example,

$$E(\widehat{n_{11}}) = \frac{r_1 c_1}{n} = \frac{94(74)}{309} = 22.51.$$

We may now use the expected and observed cell frequencies shown in Table 14.3 to calculate the value of the test statistic:

$$X^2 = \sum_{j=1}^{4} \sum_{i=1}^{3} \frac{[n_{ij} - E(\widehat{n_{ij}})]^2}{E(\widehat{n_{ij}})}$$

$$= \frac{(15 - 22.51)^2}{22.51} + \frac{(26 - 22.99)^2}{22.99} + \cdots + \frac{(20 - 14.63)^2}{14.63}$$
$$= 19.17.$$

The only remaining obstacle involves the determination of the appropriate number of degrees of freedom associated with the test statistic. We will give this as a rule, which we will subsequently justify. *The degrees of freedom associated with a contingency table possessing r rows and c columns will always equal $(r-1)(c-1)$.* For our example we will compare X^2 with the critical value of χ^2 with $(r-1)(c-1) = (3-1)(4-1) = 6$ degrees of freedom.

You will recall that the number of degrees of freedom associated with the χ^2 statistic will equal the number of cells (in this case, $k = rc$) less 1 degree of freedom for each independent linear restriction placed upon the observed cell frequencies. The total number of cells for the data of Table 14.3 is $k = 12$. From this we subtract 1 degree of freedom because the sum of the observed cell frequencies must equal n; that is,

$$n_{11} + n_{12} + \cdots + n_{34} = 309.$$

In addition, we used the cell frequencies to estimate three of the four column probabilities. Notice that estimate of the fourth column probability will be determined once we have estimated p_A, p_B, and p_C, because

$$p_A + p_B + p_C + p_D = 1.$$

Thus we lose $c - 1 = 3$ degrees of freedom for estimating the column probabilities.

Finally, we used the cell frequencies to estimate $(r-1) = 2$ row probabilities, and therefore we lose $(r-1) = 2$ additional degrees of freedom. The total number of degrees of freedom remaining is

$$\text{d.f.} = 12 - 1 - 3 - 2 = 6.$$

In general, we see that the total number of degrees of freedom associated with an $r \times c$ contingency table is

$$\text{d.f.} = rc - 1 - (c-1) - (r-1) = (r-1)(c-1).$$

Therefore, in our example relating shift to type of furniture defect, if we use $\alpha = .05$, we will reject the null hypothesis that the two classifications are independent if $X^2 > 12.592$. Because the value of the test statistic, $X^2 = 19.17$, exceeds the critical value of χ^2, we will reject the null hypothesis at the $\alpha = .05$ level of significance. The associated p-value is given by $p \text{ value} = P(\chi^2 > 19.17)$. Bounds on this probability can be obtained using Table 6, Appendix III, from which it follows that p-value $< .005$. Thus, for any value of α greater than or equal to .005, the data present sufficient evidence to indicate dependence between defect type and manufacturing shift. A study of the production operations for the three shifts would probably reveal the cause.

EXAMPLE **14.3**   A survey was conducted to evaluate the effectiveness of a new flu vaccine that had been administered in a small community. The vaccine was provided free of charge in a two-shot sequence over a period of two weeks to those wishing to avail themselves of it. Some people received the two-shot sequence, some appeared only for the first shot, and the others received neither.

A survey of 1000 local inhabitants in the following spring provided the information shown in Table 14.4. Do the data present sufficient evidence to indicate a dependence between the two classifications—vaccine category and occurrence or nonoccurrence of flu?

Table 14.4. Data tabulation for Example 14.3

Status	No Vaccine	One Shot	Two Shots	Total
Flu	24 (14.4)	9 (5.0)	13 (26.6)	46
No flu	289 (298.6)	100 (104.0)	565 (551.4)	954
Total	313	109	578	1000

Solution   The question asks whether the data provide sufficient evidence to indicate a dependence between vaccine category and occurrence or nonoccurrence of flu. We therefore analyze the data as a contingency table.

The estimated expected cell frequencies may be calculated by using the appropriate row and column totals,

$$E(\widehat{n_{ij}}) = \frac{r_i c_j}{n}.$$

Thus, for example,

$$E(\widehat{n_{11}}) = \frac{r_1 c_1}{n} = \frac{(46)(313)}{1000} = 14.4$$

$$E(\widehat{n_{12}}) = \frac{r_1 c_2}{n} = \frac{(46)(109)}{1000} = 5.0.$$

These and the remaining estimated expected cell frequencies are shown in parentheses in Table 14.4.

The value of the test statistic X^2 will now be computed and compared with the critical value of χ^2 possessing $(r-1)(c-1) = (1)(2) = 2$ degrees of freedom. Then for $\alpha = .05$, we will reject the null hypothesis when $X^2 > 5.991$. Substituting into the formula for X^2, we obtain

$$X^2 = \frac{(24-14.4)^2}{14.4} + \frac{(289-298.6)^2}{298.6} + \cdots + \frac{(565-551.4)^2}{551.4}$$

$$= 17.35.$$

Observing that X^2 falls in the rejection region, we reject the null hypothesis of independence of the two classifications. If we choose to use the attained significance level approach to making our inference, it would follow that p-value $< .005$. As is always the case, we find agreement between our fixed α level approach to testing, and the proper interpretation of the p-value.

A comparison of the percentage incidence of flu for each of the three categories would suggest that those receiving the two-shot sequence were less susceptible to the disease. Further analysis of the data could be obtained by deleting one of three categories, the second column, for example, to compare the effect of the vaccine with that of no vaccine. This could be done either by using a 2×2 contingency table or by treating the two categories as two binomial populations and using the methods of Section 10.3. Or we might wish to analyze the data by comparing the results of the two-shot vaccine sequence with those of the combined no-vaccine, one-shot group. That is, we would combine the first two columns of the 2×3 table into one.

We have considered only the simplest hypothesis connected with a contingency table, that of independence between rows and columns. Many other hypotheses are possible, and numerous techniques have been devised to test these hypotheses. For further information on this topic, consult Agresti (1996) and Fienberg (1980).

Exercises

14.9   According to one theory, people with type A behavior—hard-driving, impatient, competitive people—are much more susceptible to heart problems than those classified as low-key, type B individuals. In a study conducted by Robert Case and colleagues (1985),[1] 516 heart attack patients were administered the Jenkins Activity Survey (JAS), a questionnaire that purports to provide an approximate measure of individual type A behavior. The scores range from a low of -21 (type B behavior) to a high of 23 (type A behavior). The higher the score, the greater the level of type A behavior exhibited by the individual. Each of the 516 patients were categorized according to whether their JAS scores were less than -5 (type B), between -5 and 5 (neutral) or greater than 5 (type A). The number of patients in each of the three classes and information about whether they died during a 3-year follow-up period are given in the following table.

	JAS Score		
3-year Follow-up	Less than -5	-5 to 5	Greater than 5
Died	21	17	11
Alive	159	154	154

a   Do the data provide sufficient evidence to indicate a dependence between the results of the 3-year follow-up of heart attack patients and their JAS scores? Test using $\alpha = .05$.

b   Give bounds for the associated p-value and interpret the result.

[1]*Source:* Robert B. Case, et al., "Type A Behavior and Survival After Acute Myocardial Infarction," *New England Journal of Medicine* 312(12) (1985).

14.10   A study was conducted by Joseph Jacobson and Diane Wille to determine the effect of early child care on infant-mother attachment patterns.[2] In the study, 93 infants were classified as either "secure" or "anxious" using the Ainsworth strange-situation paradigm. In addition, the infants were classified according to the average number of hours per week that they spent in child care. The data appear in the accompanying table.

Attachment Pattern	Hours in Child Care		
	Low (0–3 hours)	Moderate (4–19 hours)	High (20–54 hours)
Secure	24	35	5
Anxious	11	10	8

   a   Do the data indicate a dependence between attachment patterns and the number of hours spent in child care? Test using $\alpha = .05$.

   b   Give bounds for the attained significance level.

14.11   Suppose that the entries in a contingency table that appear in row i and column j are denoted n_{ij}, for $i = 1, 2, \ldots, r$ and $j = 1, 2, \ldots, c$, that the row and column totals are denoted r_i, for $i = 1, 2, \ldots, r$, and c_j, for $j = 1, 2, \ldots, c$, and that the total sample size is n.

   a   Show that

$$X^2 = \sum_{j=1}^{c} \sum_{i=1}^{r} \frac{[n_{ij} - E(\widehat{n_{ij}})]^2}{E(\widehat{n_{ij}})} = n \left\{ \sum_{j=1}^{c} \sum_{i=1}^{r} \frac{n_{ij}^2}{r_i c_j} - 1 \right\}.$$

   Notice that this formula provides a computationally more efficient way to compute the value of X^2.

   b   Using the preceding formula, what happens to the value of X^2 if every entry in the contingency table is multiplied by the same integer constant $k > 0$?

14.12   According to a Sports Participation Survey by the National Sporting Goods Association reported in *American Demographics* (May 1989, p. 41), the fitness activities in which people participate change as they get older. The following table gives the results of a survey of 1103 men and women in which individuals who were frequent participants in fitness activities were classified by gender and type of exercise. Do these data provide sufficient evidence to indicate that gender and type of activity are dependent? Give bounds for the p-value. What would you conclude at the $\alpha = .05$ level of significance?

Gender	Type of Activity					
	Walking	Cycling	Aerobics	Running	Calisthenics	Swimming
Male	60	85	28	113	79	179
Female	106	81	138	55	89	90

14.13   In the academic world, students and their faculty often collaborate on research papers, producing works in which publication credit can take several forms. Many feel that the first authorship

[2] *Source:* Linda Schmittroth (ed.), *Statistical Record of Women Worldwide* (Detroit and London: Gale Research, 1991), pp. 8, 9, 335.

of a student's paper should be given to the student unless the input from the faculty advisor was substantial. In an attempt to see whether this is, in fact, the case, authorship credit was studied for several different levels of faculty input and two objectives (dissertations versus nondegree research). The frequency of authorship assignment decisions for published dissertations is given in the accompanying table as assigned by 60 faculty members and 161 students:[3]

Faculty Respondents Authorship Assignment	High Input	Medium Input	Low Input
Faculty first author, student mandatory second author	4	0	0
Student first author, faculty mandatory second author	15	12	3
Student first author, faculty courtesy second author	2	7	7
Student sole author	2	3	5
Student Respondents Authorship Assignment	High Input	Medium Input	Low Input
Faculty first author, student mandatory second author	19	6	2
Student first author, faculty mandatory second author	19	41	27
Student first author, faculty courtesy second author	3	7	31
Student sole author	0	3	3

a Is there sufficient evidence to indicate a dependence between the authorship assignment and the input of the faculty advisor as judged by faculty members? Test using $\alpha = .01$.

b Is there sufficient evidence to indicate a dependence between the authorship assignment and the input of the faculty advisor as judged by students? Test using $\alpha = .01$.

c Have any of the assumptions necessary for a valid analysis in (a) and (b) been violated? What effect might this have on the validity of your conclusions?

14.14 A study of the amount of violence viewed on television as it relates to the age of the viewer yielded the results shown in the accompanying table for 81 people. (Each person in the study was classified, according to the person's TV viewing habits, as a low-violence or high-violence viewer.) Do the data indicate that viewing of violence is not independent of age of viewer, at the 5% significance level?

---

[3] *Source:* M. Martin Costa and M. Gatz, "Determining of Authorship Credit in Published Dissertations," *Psychological Science* 3(6) (1992).

	Age		
Viewing	16–34	35–54	55 and Over
Low violence	8	12	21
High violence	18	15	7

14.15 The results of a study[4] suggest that the initial electrocardiogram (ECG) of a suspected heart attack victim can be used to predict in-hospital complications of an acute nature. The study included 469 patients with suspected myocardial infarction (heart attack). Each patient was categorized according to whether their initial ECG was positive or negative and whether the person suffered life-threatening complications subsequently in the hospital. The results are summarized in the following table.

	Subsequent In-Hospital Life-Threatening Complications		
ECG	No	Yes	Total
Negative	166	1	167
Positive	260	42	302
Total	426	43	469

a Is there sufficient evidence to indicate that whether or not a heart attack patient suffers complications depends on the outcome of the initial ECG? Test using $\alpha = .05$.

b Give bounds for the observed significance level.

14.16 Refer to Exercise 14.6. Test the hypothesis, at the 5% significance level, that the type A defects occur independently of the type B defects.

14.17 An interesting and practical use of the χ^2 test comes about in testing for segregation of species of plants or animals. Suppose that two species of plants, say A and B, are growing on a test plot. To assess whether the species tend to segregate, a researcher randomly samples n plants from the plot; the species of each sampled plant *and* the species of its nearest neighbor are recorded. The data are then arranged in a table, as shown here.

	Nearest Neighbor	
Sampled Plant	A	B
A	a	b
B	c	d
		n

If a and d are large relative to b and c, we would be inclined to say that the species tend to segregate. (Most of A's neighbors are of type A, and most of B's neighbors are of type B.) If b and c are large compared to a and d, we would say that the species tend to be overly mixed. In either of these cases (segregation or overmixing), a χ^2 test should yield a large value and the hypothesis of random mixing would be rejected. For each of the following cases, test the hypothesis of random mixing (or, equivalently, the hypothesis that the species of a sample plant is independent of the species of its nearest neighbor). Use $\alpha = .05$ in each case.

[4] *Source:* J. E. Brush et al., "Use of the Initial Electrocardiogram to Predict In-Hospital Complications of Acute Myocardial Infarction," *New England Journal of Medicine* (May 1985).

a $a = 20, b = 4, c = 8, d = 18$
b $a = 4, b = 20, c = 18, d = 8$
c $a = 20, b = 4, c = 18, d = 8.$

14.5 $r \times c$ Tables with Fixed Row or Column Totals

In the previous section we described the analysis of an $r \times c$ contingency table by using examples that, for all practical purposes, fit the multinomial experiment described in Section 14.1. Although the methods of collecting data in many surveys may meet the requirements of a multinomial experiment, other methods do not. For example, we might not wish to randomly sample the population described in Example 14.3 because we might find that, due to chance, one category is completely missing. People who have received no flu shots might fail to appear in the sample. We might decide beforehand to interview a specified number of people in each column category, thereby fixing the column totals in advance. We would then have three separate and independent binomial experiments, corresponding to "no vaccine," "one shot," and "two shots," with respective probabilities p_1, p_2, and p_3 that a person contracts the flu. In this case we are interested in testing the null hypothesis

$$H_0: p_1 = p_2 = p_3.$$

(We actually are testing the equivalence of three binomial distributions.) Under this hypothesis the maximum-likelihood estimators of the expected cell frequencies are the same as in Section 14.4, namely,

$$E\widehat{(n_{ij})} = \frac{r_i c_j}{n}.$$

It can be shown that the resulting X^2 will possess a probability distribution in repeated sampling that is approximated by a chi-square distribution with $(r-1)(c-1)$ degrees of freedom.

To illustrate, suppose that we wish to test a hypothesis concerning the equivalence of four binomial populations, as indicated in the following example.

EXAMPLE **14.4**   A survey of voter sentiment was conducted in four midcity political wards to compare the fraction of voters favoring candidate A. Random samples of 200 voters were polled in each of the four wards, with results as shown in Table 14.5. Do the data present sufficient evidence to indicate that the fractions of voters favoring candidate A differ in the four wards?

Solution   You will observe that the method for testing hypotheses concerning the equivalence of the parameters of the four binomial populations that correspond to the four wards is identical to that for testing the hypothesis of independence of the row and column

Table 14.5. Data tabulation for Example 14.4

Opinion	Ward 1	Ward 2	Ward 3	Ward 4	Total
Favor A	76 (59)	53 (59)	59 (59)	48 (59)	236
Do not favor A	124 (141)	147 (141)	141 (141)	152 (141)	564
Total	200	200	200	200	800

classifications. If we denote the fraction of voters favoring A as p and hypothesize that p is the same for all four wards, we imply that the first-row probabilities are all equal to p and and that the second-row probabilities are all equal to $1 - p$. The maximum-likelihood estimate (combining the results from all four samples) for the common value of p is $\hat{p} = 236/800 = r_1/n$. The expected number of individuals who favor candidate A in Ward 1 is $E(n_{11}) = 200p$, which is estimated by the value

$$\widehat{E(n_{11})} = 200\hat{p} = 200(236/800) = (c_1 r_1)/n.$$

Notice that even though we are considering a very different experiment than that considered in Section 14.4, the estimated mean cell frequencies are computed the same way as they were in Section 14.4. The other estimated expected cell frequencies, calculated by using the row and column totals, appear in parentheses in Table 14.5. We see that

$$X^2 = \sum_{j=1}^{4} \sum_{i=1}^{2} \frac{[n_{ij} - \widehat{E(n_{ij})}]^2}{\widehat{E(n_{ij})}}$$

$$= \frac{(76 - 59)^2}{59} + \frac{(124 - 141)^2}{141} + \cdots + \frac{(152 - 141)^2}{141} = 10.72.$$

The critical value of χ^2 for $\alpha = .05$ and $(r - 1)(c - 1) = (1)(3) = 3$ degrees of freedom is 7.815. Because X^2 exceeds this critical value, we reject the null hypothesis and conclude that the fraction of voters favoring candidate A is not the same for all four wards. The associated p-value is given by $P(\chi^2 > 10.72)$ when χ^2 has 3 degrees of freedom. Thus, $.01 \leq p\text{-value} \leq .025$.

This example was worked out in Exercise 10.94 by the likelihood ratio method. Notice that the conclusions are the same.

Exercises

14.18 A study to determine the effectiveness of a drug (serum) for the treatment of arthritis resulted in the comparison of two groups each consisting of 200 arthritic patients. One group was inoculated with the serum, whereas the other received a placebo (an inoculation that appears to contain serum but actually is not active). After a period of time each person in the study was asked whether his or her arthritic condition had improved. The results in the accompanying table were observed. Do these data present sufficient evidence to indicate that the proportions of arthritic individuals who said their condition had improved differed depending on whether they received the serum?

Condition	Treated	Untreated
Improved	117	74
Not improved	83	126

a Test by using the X^2 statistic. Use $\alpha = .05$.

a Test by using the Z test of Section 10.3 and $\alpha = .05$. Compare your result with that in (a).

a Give bounds for the attained significance level associated with the test in (a).

14.19 The chi-square test used in Exercise 14.18 is equivalent to the two-tailed Z test of Section 10.3 provided α is the same for the two tests. Show algebraically that the chi-square test statistic X^2 is the square of the test statistic Z for the equivalent test.

14.20 If you find yourself forgetting names and appointments, 2 tablespoons daily of the common health-food staple lecithin may improve your memory, according to results presented by Florence Safford and Barry Baumel to the Gerontological Society of America (*Gainesville Sun*, February 9, 1989). Their experiment involved 61 people age 50 to 80 years and excluded Alzheimer's patients. Of the 41 subjects receiving lecithin, 37 reported a significant decrease in memory lapses. Among the 20 who received a placebo, 8 reported decreases in memory lapses. The data summary is given in the following table.

Group	Memory Lapses Decrease	No Decrease	Total
Lecithin	37	4	41
Placebo	8	12	20

a Use the χ^2 test to determine whether the proportions of individuals in the memory lapse categories differ for the two groups. Use $\alpha = .05$.

b Use the Z test of Section 10.3 to implement the test. How does the value of z compare to the value of X^2 that you obtained in (a)?

14.21 Are baby-boomers more likely to increase their investing now that they are reaching middle age? A poll was conducted by Hal Riney & Partners (*Los Angeles Times*, June 11, 1990)[5] in which 200 investors in each of two age groups were asked about their likely investment pattern over the next 5 years as compared to the last 5 years. Do the data that follow provide sufficient information to conclude that anticipated investment patterns of the baby-boomer age group differ from those of the older age group? Test using $\alpha = .01$.

Age Group	More	Less	Same	Total
35–54	90	18	92	200
55 +	40	60	100	200

[5] *Source:* Adapted from C. J. Miko and E. Weilant (eds.), *Opinions '90 Cumulation* (Detroit and London: Gale Research, 1991).

14.22 A manufacturer of buttons wished to determine whether the fraction of defective buttons produced by three machines varied from machine to machine. Samples of 400 buttons were selected from each of the three machines and the number of defectives counted for each sample. The results were as shown in the accompanying table. Do these data present sufficient evidence to indicate that the fraction of defective buttons varied from machine to machine?

Machine Number	Number of Defectives
1	16
2	24
3	9

a Test, using $\alpha = .05$, with a χ^2 test.
b Test, using $\alpha = .05$, with a likelihood ratio test.

14.23 H. W. Menard[6] conducted research involving manganese nodules, a mineral-rich agglomeration found abundantly on the deep-sea floor. In one portion of his report, Menard provided data relating the magnetic age of the earth's crust to "the probability of finding manganese nodules." The data shown in the accompanying table give the number of samples of the earth's core and the percentage of those that contain manganese nodules for each of a set of magnetic-crust ages. Do the data provide sufficient evidence to indicate that the probability of finding manganese nodules in the deep-sea earth's crust differs depending upon the magnetic-age classification? Test with $\alpha = .05$.

Age	Number of Samples	Percentage with Nodules
Miocene–recent	389	5.9
Oligocene	140	17.9
Eocene	214	16.4
Paleocene	84	21.4
Late Cretaceous	247	21.1
Early and middle Cretaceous	1,120	14.2
Jurassic	99	11.0

14.24 In this age of VCRs and videotaped movies, a new type of advertising has evolved whereby the advertisement is included on a commercially produced videocassette. Advertisers have found that, although many consumers fast-forward past the ads, the best chance of getting a person to watch the ads lies in the 18–34 age group. Overall, however, most people find the ads annoying. The data in the accompanying table show the responses of two groups of consumers, 50 in the 18-to-34-year-old category and 50 in the 35-to-50-year-old group:[7]

Age Group	Annoyance Level		
	High	Moderate	None
18–34	18	15	17
35–54	21	19	10

[6] *Source:* H. W. Menard, "Time, Chance and the Origin of Manganese Nodules," *American Scientist*, September–October 1976. ©1976 Scientific Research Company of North America. All rights reserved.
[7] *Source:* Adapted from C. J. Miko and E. Weilant (eds.), *Opinions '90 Cumulation* (Detroit and London: Gale Research, 1991), p. 19.

Is there a significant difference in the two age groups regarding the proportions who exhibit the different levels of annoyance? What can be said about the attained significance level?

14.25 A survey was conducted to study the relationship between lung disease and air pollution. Four areas were chosen for the survey, two cities frequently plagued with smog and-two nonurban areas in states that possessed low air-pollution counts. Only adult permanent residents of the area were included in the study. Random samples of 400 adult permanent residents from each area gave the results listed in the accompanying table.

Area	Number with Lung Disease
City A	34
City B	42
Nonurban area 1	21
Nonurban area 2	18

a Do the data provide sufficient evidence to indicate a difference in the proportions with lung disease for the four locations?

b Should cigarette smokers have been excluded from the samples? How would this affect inferences drawn from the data?

14.26 Refer to Exercise 14.25. Estimate the difference in the fractions of adult permanent residents with lung disease between cities A and B. Use a 95% confidence interval.

14.27 A survey was conducted to investigate interest of middle-aged adults in physical-fitness programs in Rhode Island, Colorado, California, and Florida. The objective of the investigation was to determine whether adult participation in physical-fitness programs varies from one region of the United States to another. Random samples of people were interviewed in each state and the data reproduced in the accompanying table were recorded. Do the data indicate differences among the rates of adult participation in physical fitness programs from one state to another? What would you conclude with $\alpha = .01$?

Participation	Rhode Island	Colorado	California	Florida
Yes	46	63	108	121
No	149	178	192	179

14.6 Other Applications

The applications of the chi-square test in analyzing categorical data described in Sections 14.3, 14.4, and 14.5 represent only a few of the interesting classification problems that may be approximated by the multinomial experiment and for which our method of analysis is appropriate. Generally, these applications are complicated to a greater or lesser degree because the numerical values of the cell probabilities are unspecified and hence require the estimation of one or more population parameters. Then, as in Sections 14.4 and 14.5, we can estimate the cell probabilities. Although we omit the mechanics of the statistical tests, several additional applications of the chi-square test are worth mention as a matter of interest.

For example, suppose that we wish to test a hypothesis stating that a population possesses a normal probability distribution. The cells of a sample frequency histogram would correspond to the k cells of the multinomial experiment, and the observed cell frequencies would be the number of measurements falling into each cell of the histogram. Given the hypothesized normal probability distribution for the population, we could use the areas under the normal curve to calculate the theoretical cell probabilities and, hence, the expected cell frequencies. Maximum-likelihood estimators must be employed when μ and σ are unspecified for the normal population, and these parameters must be estimated to obtain the estimated cell probabilities.

The construction of a two-way table to investigate dependency between two classifications can be extended to three or more classifications. For example, if we wish to test the mutual independence of three classifications, we would employ a three-dimensional "table." The reasoning and methodology associated with the analysis of both the two- and three-way tables are identical, although the analysis of the three-way table is a bit more complex.

A third and interesting application of our methodology would be its use in the investigation of the rate of change of a multinomial (or binomial) population as a function of time. For example, we might study the problem-solving ability of a human (or any animal) subjected to an educational program and tested over time. If, for instance, the human is tested at prescribed intervals of time and the test is of the yes or no type, yielding a number of correct answers, y, that would follow a binomial probability distribution, we would be interested in the behavior of the probability of a correct response, p, as a function of time. If the number of correct responses was recorded for c time periods, the data would fall in a $2 \times c$ table similar to that in Example 14.4 (Section 14.5). We would then be interested in testing the hypothesis that p is equal to a constant—that is, that no learning has occurred—and we would then proceed to more interesting hypotheses to determine whether the data present sufficient evidence to indicate a gradual (say, linear) change over time as opposed to an abrupt change at some point in time. The procedures we have described could be extended to decisions involving more than two alternatives.

You will observe that our change over time example is common to business, to industry, and to many other fields, including the social sciences. For example, we might wish to study the rate of consumer acceptance of a new product for various types of advertising campaigns as a function of the length of time that the campaign has been in effect. Or we might wish to study the trend in the lot fraction defective in a manufacturing process as a function of time. Both these examples, as well as many others, require a study of the behavior of a binomial (or multinomial) process as a function of time.

The examples just described are intended to suggest the relatively broad application of the chi-square analysis of categorical data, a fact that should be borne in mind by the experimenter concerned with this type of data. The statistical test employing X^2 as a test statistic is often called a *goodness-of-fit test*. Its application for some of these examples requires care in the determination of the appropriate estimates and the number of degrees of freedom for X^2, which, for some of these problems, may be rather complex.

14.7 Summary and Concluding Remarks

The material in this chapter has been concerned with a test of a hypothesis regarding the cell probabilities associated with a multinomial experiment (Sections 14.2 and 14.3) or several independent multinomial experiments (Section 14.5). When the number of observations n is large, the test statistic X^2 can be shown to possess, approximately, a chi-square probability distribution in repeated sampling, the number of degrees of freedom being dependent upon the particular application. In general we assume that n is large and that the minimum expected cell frequency is equal to or greater than 5.

Several words of caution concerning the use of the X^2 statistic as a method of analyzing categorical data are appropriate. The determination of the correct number of degrees of random associated with the X^2 statistic is critical in locating the rejection region. If the number is specified incorrectly, erroneous conclusions might result. Notice, too, that nonrejection of the null hypothesis does not imply that it should be accepted. We would have difficulty in stating a meaningful alternative hypothesis for many practical applications, and therefore we would lack knowledge of the probability of making a type II error. For example, we hypothesize that the two classifications of a contingency table are independent. A specific alternative must specify a measure of dependence that may or may not possess practical significance to the experimenter. Finally, if parameters are missing and the expected cell frequencies must be estimated, missing parameters should be estimated by the method of maximum likelihood in order that the test be valid. In other words, the application of the chi-square test for other than the simple applications outlined in Sections 14.3, 14.4, and 14.5 will require experience beyond the scope of this introductory presentation of the subject.

References and Further Readings

Agresti, Alan. 1996. *Categorical Data Analysis*. New York: John Wiley.

Cochran, W. G. 1952. "The χ^2 Test of Goodness of Fit," *Annals of Mathematical Statistics* 23:315–45.

Conover, W. J. 1999. *Practical Nonparametric Statistics*, 3d ed. New York: John Wiley.

Daniel, W. W. 1990. *Applied Nonparametric Statistics*, 2d ed. Boston: PWS-Kent.

Dixon, W. J., and F. J. Massey, Jr. 1983. *Introduction to Statistical Analysis*, 4th ed. New York: McGraw-Hill.

Fienberg, Stephen E. 1980. *The Analysis of Cross-Classified Categorical Data*, 2d ed. Cambridge, Mass.: MIT Press.

Kendall, M. G., and A. Stuart. 1979. *The Advanced Theory of Statistics*, 4th ed., vol. 2. New York: Hafner Press, chapter 30.

Supplementary Exercises

14.28   List the characteristics of a multinomial experiment.

14.29   A survey was conducted to determine student, faculty, and administration attitudes on a new university parking policy. The distribution of those favoring or opposing the policy was as shown in the accompanying table. Do the data provide sufficient evidence to indicate that attitudes regarding the parking policy are independent of student, faculty, or administration status?

Opinion	Student	Faculty	Administration
Favor	252	107	43
Oppose	139	81	40

14.30   How would you rate yourself as a driver? According to a recent survey conducted by the Field Institute (*Riverside Press-Enterprise*, May 15, 1991), most Californians think that they are good drivers but have little respect for the driving ability of others. The data in the following tables show the distribution of opinions, according to gender, for two different questions. Data in the first table give the results obtained when drivers rated themselves, the second table gives the results obtained when drivers rated others. Although not stated in the source, we assume that there were 100 men and 100 women in each of the surveyed groups.

Rating Self as Driver

Gender	Excellent	Good	Fair
Male	43	48	9
Female	44	53	3

Rating Others as Drivers

Gender	Excellent	Good	Fair	Poor
Male	4	42	41	13
Female	3	48	35	14

a   Is there sufficient evidence to indicate that there is a difference in the proportions associated with the ratings categories for male and female drivers when drivers rate themselves? Give bounds for the p-values associated with the test.

b   Is there sufficient evidence to indicate that there is a difference in the proportions associated with the ratings categories for male and female drivers when drivers rate others? Give bounds for the p-values associated with the test.

c   Have you violated any assumptions in your analyses in (a) and (b)? What effect might these violations have on the validity of your conclusions?

14.31   The following table shows the categorization of 204 men who were awaiting bypass heart surgery according to the relative degree of coronary artery obstruction and according to the perceived level of discomfort due to angina pectoris.[8] Do the data present sufficient evidence to indicate that the level of angina is dependent on the level of coronary artery obstruction? The authors report the p-value for a chi-square test to be $p = .01$.

---

[8] *Source:* C. David Jenkins et al., "Correlates of Angina Pectoris Among Men Awaiting Coronary Bypass Surgery," *Psychomatic Medicine* 45(2) (1983).

Level of Angina	Arteries Obstructed 75% or More			
	0 or 1	2	3 to 6	Total
None	3	21	20	44
Mild	2	12	9	23
Moderate	26	20	31	77
Moderate/severe	13	10	18	41
Severe	7	5	7	19
Total	51	68	85	204

a Compute the value of X^2 for the data.

b Give bounds for the p-value associated with the value for X^2 obtained in (a) and compare with the p-value reported by Jenkins et al.

c What conclusion would you reach based on your analysis?

14.32 Has the rise over the last 20 years in the number of children involved in organized sports such as Little League baseball, organized football, and soccer teams produced a decrease in the numbers of children who participate in self-directed physical games? To study this problem, researchers[9] interviewed 78 male and 77 female fourth- and fifth-grade students from public grammar schools in the Champaign–Urbana area of Illinois. Each student was questioned to determine whether he or she actively participated in one or more organized sports and the frequency of involvement in self-directed games. The following table was obtained by classifying the girls into one of three levels of participation in self-directed games and according to whether they participated in organized sports.

Frequency of Participation in Self-Directed Physical Games	Organized Sports	
	Participant	Nonparticipant
Less than once a week	4	16
More than once a week but less than daily	15	20
Every day	7	15
Total	26	51

a Do the data provide sufficient evidence to indicate dependence between level of involvement in self-directed games and participation in organized sports? Test using $\alpha = .05$.

b Give bounds for the p-value associated with the value for X^2 obtained in (a). In the paper of Kleiber and Roberts, they gave $X^2 = 3.02$, d.f. $= 2$, p-value $= .22$. Do you agree?

14.33 It is often not clear whether all properties of a binomial experiment are actually met in a given application. A goodness-of-fit test is desirable for such cases. Suppose that an experiment consisting of four trials was repeated 100 times. The number of repetitions on which a given number of successes was obtained is recorded in the accompanying table. Estimate p (assuming that the experiment was binomial), obtain estimates of the expected cell frequencies, and test for goodness of fit. To determine the appropriate number of degrees of freedom for X^2, notice that p had to be estimated.

[9]*Source:* D. A. Kleiber and G. C. Roberts, "The Relationship Between Game and Sport Involvement in Later Childhood: A Preliminary Investigation," *Research Quarterly for Exercise and Sport* 54(2) (1983).

Possible Results (no. of successes)	No. of Times Obtained
0	11
1	17
2	42
3	21
4	9

14.34 Counts on the number of items per cluster (or colony or group) must necessarily be greater than or equal to 1. Thus, the Poisson distribution generally does not fit these kinds of counts. For modeling counts on phenomena such as number of bacteria per colony, number of people per household, and number of animals per litter, the *logarithmic series* distribution often proves useful. This discrete distribution has probability function given by

$$p(y \mid \theta) = -\frac{1}{\ln(1-\theta)} \frac{\theta^y}{y}, \qquad y = 1, 2, 3, \ldots, 0 < \theta < 1$$

where θ is an unknown parameter.

a Show that the maximum-likelihood estimator, $\hat{\theta}$, of θ satisfies the equation

$$\overline{Y} = \frac{\hat{\theta}}{-(1-\hat{\theta})\ln(1-\hat{\theta})}, \qquad \text{where} \quad \overline{Y} = \frac{1}{n}\sum_{i=1}^{n} Y_i.$$

b The data in the following table give frequencies of observation for counts on the number of bacteria per colony, for a certain type of soil bacteria.[10]

Bacteria per colony	1	2	3	4	5	6	7+
Number of colonies observed	359	146	57	41	26	17	29

Test the hypothesis that these data fit a logarithmic series distribution. Use $\alpha = .05$. (Notice that the value $\overline{y}$ must be approximated because we do not have exact information on counts greater than six.)

14.35 Refer to the $r \times c$ contingency table of Section 14.4. Show that the maximum-likelihood estimator of the probability p_i for row i is $\hat{p}_i = r_i/n$, for $i = 1, 2, \ldots, r$.

*14.36 A genetic model states that the proportions of offspring in three classes should be p^2, $2p(1-p)$, and $(1-p)^2$ for a parameter p, $0 \le p \le 1$. An experiment yielded frequencies of 30, 40, and 30 for the respective classes.[11]

a Does the model fit the data? (Use maximum likelihood to estimate p.)

b Suppose that the hypothesis states that the model holds with $p = .5$. Do the data contradict this hypothesis?

*14.37 According to the genetic model for the relationship between sex and color blindness, the four categories, male and normal, female and normal, male and color blind, female and color blind, should have probabilities given by $p/2$, $(p^2/2) + pq$, $q/2$, and $q^2/2$, respectively, where

---

[10] *Source:* C. A. Bliss and R. A. Fisher, "Fitting the Negative Binomial Distribution to Biological Data," *Biometrics*, 9, 1953, pp. 176–200. Reprinted by permission. Biometrics Society. All rights reserved.
[11] Exercises preceded by an asterisk are optional.

$q = 1 - p$. A sample of 2000 people revealed 880, 1032, 80, and 8 in the respective categories. Do these data agree with the model? Use $\alpha = .05$. (Use maximum likelihood to estimate p.)

*14.38 Suppose that $(Y_1, Y_2, \ldots, Y_k)$ has a multinomial distribution with parameters $n, p_1, p_2, \ldots, p_k$, and $(X_1, X_2, \ldots, X_k)$ has a multinomial distribution with parameters $m, p_1^*, p_2^*, \ldots, p_k^*$. Construct a test of the null hypothesis that the two multinomial distributions are identical; that is, test $H_0: p_1 = p_1^*, p_2 = p_2^*, \ldots, p_k = p_k^*$.

*14.39 In an experiment to evaluate an insecticide, the probability of insect survival was expected to be linearly related to the dosage D over the region of experimentation. That is, $p = 1 + \beta D$. An experiment was conducted using four levels of dosage, 1, 2, 3, and 4, and 1000 insects in each group. The resulting data were as shown in the following table. Do these data contradict the hypothesis that $p = 1 + \beta D$? (*Hint*: Write the cell probabilities in terms of β, and find the maximum-likelihood estimator of β.)

Dosage	No. of Survivors
1	820
2	650
3	310
4	50

CHAPTER 15

Nonparametric Statistics

15.1 Introduction

15.2 A General Two-Sample Shift Model

15.3 The Sign Test for a Matched Pairs Experiment

15.4 The Wilcoxon Signed-Rank Test for a Matched Pairs Experiment

15.5 The Use of Ranks for Comparing Two Population Distributions: Independent Random Samples

15.6 The Mann–Whitney U Test: Independent Random Samples

15.7 The Kruskal–Wallis Test for the One-Way Layout

15.8 The Friedman Test for Randomized Block Designs

15.9 The Runs Test: A Test for Randomness

15.10 Rank Correlation Coefficient

15.11 Some General Comments on Nonparametric Statistical Tests

References and Further Readings

15.1 Introduction

Some experiments yield response measurements that defy quantification. That is, they generate response measurements that can be ordered (ranked), but the location of the response on a scale of measurement is arbitrary. Although experiments of this type occur in almost all fields of study, they are particularly evident in social science research and in studies of consumer preference. For example, suppose that a judge is employed to evaluate and rank the instructional abilities of four teachers or the edibility and taste characteristics of five brands of cornflakes. Because it clearly is impossible to give an exact measure of teacher competence or food taste, the response measurements are of a completely different character than those presented in preceding chapters. Nonparametric statistical methods are useful for analyzing this type of data.

Nonparametric statistical procedures apply not only to observations that are difficult to quantify, they are particularly useful in making inferences in situations where serious doubt exists about the assumptions that underlie standard methodology. For example, the t test for comparing a pair of means based on independent samples, Section 10.8, is based on the assumption that both populations are normally distributed with equal variances. The experimenter will never know whether these assumptions hold in a practical situation but often will be reasonably certain that departures from the assumptions will be small enough that the properties of the statistical procedure will be undisturbed. That is, α and β will be approximately what the experimenter thinks they are. On the other hand, it is not uncommon for the experimenter seriously to question the assumptions and wonder whether he or she is using a valid statistical procedure. This difficulty may be circumvented by using a nonparametric statistical test and thereby avoiding reliance on a very uncertain set of assumptions.

The term *nonparametric statistics* has no standard definition that is agreed upon by all statisticians. However, most would agree that nonparametric statistical methods work well under fairly general assumptions about the nature of any probability distributions or parameters that are involved in an inferential problem. As a working definition we will define *parametric methods* as those that apply to problems where the distribution(s) from which the sample(s) is (are) taken is (are) specified except for the values of a finite number of parameters. Nonparametric methods apply in all other instances. For example, the one-sample t test developed in Chapter 10 applies to problems in which the population is normally distributed with unknown mean and variance. Because the distribution from which the sample is taken is specified except for the values of two parameters, μ and σ^2, this t test is a parametric procedure. Alternatively, suppose that independent samples are taken from two populations and we wish to test the hypothesis that the two population distributions are identical but of unspecified form. In this case the distribution is unspecified and the hypothesis must be tested by using nonparametric methods.

Research has shown that nonparametric statistical tests are almost as capable of detecting differences among populations as the parametric methods of preceding chapters when normality and other assumptions are satisfied. They may be, and often are, more powerful in detecting population differences when the assumptions are not satisfied. For this reason many statisticians advocate the use of nonparametric statistical procedures in preference to their parametric counterparts.

15.2   A General Two-Sample Shift Model

Many times an experimenter takes observations from two populations with the objective of testing whether the populations have the same distribution. For example, if independent random samples $X_1, X_2, \ldots, X_{n_1}$ and $Y_1, Y_2, \ldots, Y_{n_2}$ are taken from normal populations with equal variances and respective means μ_X and μ_Y, the experimenter may wish to test $H_0 : \mu_X - \mu_Y = 0$ versus $H_a : \mu_X - \mu_Y < 0$. In this case, if H_0 is true, both populations are normally distributed with the same mean and

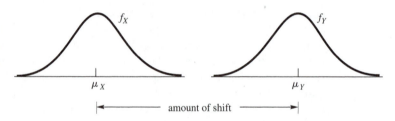

the same variance; that is, the population distributions are identical. If H_a is true, then $\mu_Y > \mu_X$ and the distributions of X_1 and Y_1 are the same, except that the location parameter (μ_Y) for Y_1 is larger than the location parameter (μ_X) for X_1. Hence, the distribution of Y_1 is shifted to the right of the distribution of X_1 (see Figure 15.1).

This is an example of a two-sample parametric *shift* (or *location*) *model*. The model is parametric because the distributions are specified (normal) except for the values of the parameters μ_X, μ_Y, and σ^2. The amount that the distribution of Y_1 is shifted to the right of the distribution of X_1 is $\mu_Y - \mu_X$ (see Figure 15.1). In the remainder of this section we define a shift model that applies for any distribution, normal or otherwise.

Let $X_1, X_2, \ldots, X_{n_1}$ be a random sample from a population with distribution function $F(x)$, and let $Y_1, Y_2, \ldots, Y_{n_2}$ be a random sample from a population with distribution function $G(y)$. If we wish to test whether the two populations have the same distribution—that is, $H_0: F(z) = G(z)$ versus $H_a: F(z) \neq G(z)$, with the actual form of $F(z)$ and $G(z)$ unspecified—a nonparametric method is required. Notice that H_a is a very broad hypothesis. Many times an experimenter may wish to consider the more specific alternative hypothesis that Y_1 has the same distribution as X_1 shifted by an (unknown) amount θ (see Figure 15.2); that is, that the distributions *differ in location*. Then $G(y) = P(Y_1 \leq y) = P(X_1 \leq y - \theta) = F(y - \theta)$ for some unknown parameter value θ. Notice that the particular form of $F(x)$ remains unspecified.

Throughout this chapter if we refer to the two-sample shift (location) model, we assume that $X_1, X_2, \ldots, X_{n_1}$ constitutes a random sample from distribution function $F(x)$ and that $Y_1, Y_2, \ldots, Y_{n_2}$ constitutes a random sample from distribution function $G(y) = F(y - \theta)$ for some unknown value θ. For the two-sample shift model, $H_0 : F(z) = G(z)$ is equivalent to $H_0 : \theta = 0$. If θ is greater (less) than 0, then the distribution of the Y values is located to the right (left) of the distribution of the X values.

F I G U R E   **15.2**
Two density
functions, with the
density for Y shifted
θ units to the right
of that for X

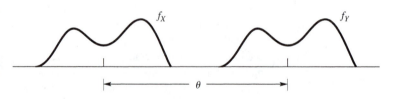

15.3 The Sign Test for a Matched Pairs Experiment

Suppose that we have n pairs of observations of the form (X_i, Y_i) and that we wish to test the hypothesis that the distribution of the Xs is the same as that of the Ys versus the alternative that the distributions differ in location (see Section 15.2). Much as we did in Section 12.3, we let $D_i = X_i - Y_i$. One of the simplest nonparametric tests is based on the signs of these differences and, reasonably enough, is called the *sign test*. Under the null hypothesis that X_i and Y_i come from the same continuous probability distributions, the probability that D_i is positive is equal to $1/2$ (as is the probability that D_i is negative). Let M denote the total number of positive (or negative) differences. Then if the variables X_i and Y_i values have the same distribution, M will have a binomial distribution with $p = 1/2$, and the rejection region for a test based on M can be obtained by using the binomial probability distribution of Chapter 3. The sign test is summarized as follows.

The Sign Test for a Matched Pairs Experiment

Let $p = P(X > Y)$.

Null hypothesis:	$H_0 : p = 1/2$
Alternative hypothesis:	$H_a : p > 1/2$ or $(p < 1/2$ or $p \neq 1/2)$
Test statistic:	$M =$ number of positive differences where $D_i = X_i - Y_i$
Rejection region:	For $H_a : p > 1/2$, reject H_0 for the largest values of M; for $H_a : p < 1/2$, reject H_0 for the smallest values of M; for $H_a : p \neq 1/2$, reject H_0 for very large or very small values of M.
Assumptions:	The pairs (X_i, Y_i) are randomly and independently selected.

The following example illustrates the use of the sign test.

EXAMPLE 15.1   The number of defective electrical fuses proceeding from each of two production lines, A and B, was recorded daily for a period of 10 days, with the results shown in Table 15.1. Assume that both production lines produced the same daily output. Compare the number of defectives produced by A and B each day, and let M equal the number of days when A exceeded B. Do the data present sufficient evidence to indicate that either production line produces more defectives than the other? State the null hypothesis to be tested, and use M as a test statistic.

Table 15.1. Data for Example 15.1

Day	A	B
1	172	201
2	165	179
3	206	159
4	184	192
5	174	177
6	142	170
7	190	182
8	169	179
9	161	169
10	200	210

Solution   Pair the observations as they appear in the data tabulation and let M be the number of times that the observation for production line A exceeds that for B in a given day. Under the null hypothesis that the two distributions of defectives are identical, the probability p that A exceeds B for a given pair is $p = .5$, given that there are no ties. Consequently, the null hypothesis is equivalent to the hypothesis that the binomial parameter $p = .5$.

Very large or very small values of M are most contradictory to the null hypothesis. Therefore, the rejection region for the test will be located by including the most extreme values of M that at the same time provide a value of α that is suitable for the test.

Suppose that we would like the value of α to be on the order of .05 or .10. We commence the selection of the rejection region by including $M = 0$ and $M = 10$ and calculate the α associated with this region, using $p(y)$, the probability distribution for the binomial random variable (see Chapter 3). With $n = 10$, $p = .5$, we have

$$\alpha = p(0) + p(10) = \binom{10}{0}(.5)^{10} + \binom{10}{10}(.5)^{10} = .002.$$

Because this value of α is too small, the region will be expanded by including the next pair of M values most contradictory to the null hypothesis, $M = 1$ and $M = 9$. The value of α for this region ($M = 0, 1, 9, 10$) can be obtained from Table 1, Appendix III:

$$\alpha = p(0) + p(1) + p(9) + p(10) = .022.$$

This also is too small, so we again expand the region to include $M = 0, 1, 2, 8, 9, 10$. You can verify that the corresponding value of α is .11. Suppose that this value of α is acceptable to the experimenter; then we employ $M = 0, 1, 2, 8, 9, 10$ as the rejection region for the test.

From the data we observe that $m = 2$, so we reject the null hypothesis. We conclude that sufficient evidence exists to indicate that the population distributions for numbers of defective fuses are not identical. The probability of our rejecting the null hypothesis when it is true is only $\alpha = .11$, and, therefore, we are reasonably confident of our conclusion.

The experimenter in this example is using the test procedure as a rough tool for detecting faulty production lines. The rather large value of α is not likely to disturb him because he can easily collect additional data if he is concerned about making a type I error in reaching his conclusion.

Attained significance levels (p-values) for the sign test are calculated as outlined in Section 10.6. Specifically, if $n = 15$ and we wish to test $H_0 : p = 1/2$ versus $H_a : p < 1/2$ based on the observed value of $M = 3$, Table 1 of Appendix III can be used to determine that (because $n = 15$, $p = 1/2$)

$$p\text{-value} = P(M \leq 3) = .018.$$

For the two-tailed test ($H_a : p \neq 1/2$), p-value $= 2(.018) = .036$.

One problem that may arise in connection with a sign test is that the observations associated with one or more pairs may be equal and therefore may result in ties. When this situation occurs, delete the tied pairs and reduce n, the total number of pairs.

You will also encounter situations where n, the number of pairs, is large. Then the values of α associated with the sign test can be approximated by using the normal approximation to the binomial probability distribution discussed in Section 7.5. You can verify (by comparing exact probabilities with their approximations) that these approximations will be quite adequate for n as small as 10 or 15. This result is due to the symmetry of the binomial probability distribution for $p = .5$. For $n \geq 25$, the Z test of Chapter 10 will suffice, where

$$Z = \frac{M - np}{\sqrt{npq}} = \frac{M - n/2}{(1/2)\sqrt{n}}.$$

This statistic would be used for testing the null hypothesis $p = .5$ against the alternative $p \neq .5$ for a two-tailed test or against the alternative $p > .5$ (or $p < .5$) for a one-tailed test. The tests would use the familiar rejection regions of Chapter 10.

The data of Example 15.1 are the result of a matched pairs experiment. Suppose that the paired differences are normally distributed with a common variance σ^2. Will the sign test detect a shift in location of the two populations as effectively as the Student's t test? Intuitively, we would suspect that the answer is no, and this is correct because the Student's t test uses comparatively more information. In addition to giving the sign of the difference, the t test uses the magnitudes of the observations to obtain a more accurate value of the sample means and variances. Thus, we might say that the sign test is not as "efficient" as the Student's t test; but this statement is meaningful only if the populations conform to the assumption just stated: that the differences in paired observations are normally distributed with a common variance σ_D^2. The sign test might be more efficient when these assumptions are not satisfied.

Sign Test for Large Samples $n > 25$

Null hypothesis:	$H_0 : p = .5$ (neither treatment is preferred to the other).
Alternative hypothesis:	$H_a : p \neq .5$ for a two-tailed test (*Note:* We use the two-tailed test for an example. Many analyses require a one-tailed test.)
Test statistic:	$Z = \dfrac{M - n/2}{(1/2)\sqrt{n}}$
Rejection region:	Reject H_0 if $z \geq z_{\alpha/2}$ or if $z \leq -z_{\alpha/2}$, where $z_{\alpha/2}$ is obtained from Table 3, Appendix III.

The sign test actually tests the null hypothesis that the *median* of the variables D_i is zero versus the alternative that it is different from zero. [The median of the variables D_i being zero does imply that $P(D_i < 0) = P(D_i > 0)$.] If the variables X_i and Y_i values have the same distribution, the median of the variables D_i will be zero, as previously discussed. However, for models other than the shift model, there are other situations in which the median of the variables D_i is zero. In these instances the null hypothesis for the sign test is slightly more general than the statement that X_i and Y_i have the same distribution.

Summarizing, the sign test is an easily applied nonparametric procedure for comparing two populations. No assumptions are made concerning the underlying population distributions. The value of the test statistic can be obtained quickly by a visual count, and the rejection region can be located easily by using a table of binomial probabilities. Furthermore, we need not know the exact values of pairs of responses, just whether $X_i > Y_i$ for each pair (X_i, Y_i). Exercise 15.3 provides an example of the use of the sign test for data of this sort.

Exercises

15.1 What significance levels between $\alpha = .01$ and $\alpha = .15$ are available for a two-tailed sign test with 25 paired observations? (Make use of tabulated values in Table 1, Appendix III, $n = 25$.) What are the corresponding rejection regions?

15.2 For a comparison of the academic effectiveness of two junior high schools A and B, an experiment was designed using ten sets of identical twins, each twin having just completed the sixth grade. In each case the twins in the same set had obtained their previous schooling in the same classrooms at each grade level. One child was selected at random from each set and assigned to school A. The other was sent to school B. Near the end of the ninth grade, an achievement test was given to each child in the experiment. The results are shown in the accompanying table.

Twin Pair	A	B	Twin Pair	A	B
1	67	39	6	50	52
2	80	75	7	63	56
3	65	69	8	81	72
4	70	55	9	86	89
5	86	74	10	60	47

a Using the sign test, test the hypothesis that the two schools are the same in academic effectiveness, as measured by scores on the achievement test, against the alternative that the schools are not equally effective. Give the attained significance level. What would you conclude with $\alpha = .05$?

b Suppose it is suspected that junior high school A has a superior faculty and better learning facilities. Test the hypothesis of equal academic effectiveness against the alternative that school A is superior. What is the p-value associated with this test?

15.3 New food products are frequently subjected to taste tests by a panel of judges. The judges are usually asked to state a preference for one food over another, so that no quantitative scale need be employed. Suppose that two new mixtures, A and B, of an orange-flavored drink are presented to ten judges. The preferences of the judges are given in the accompanying table. Does this evidence indicate a significant difference between the tastes of A and B, at the 5% significance level?

Judge	Preference	Judge	Preference
1	A	6	A
2	A	7	B
3	A	8	A
4	A	9	B
5	A	10	A

15.4 On clear, cold nights in the central Florida citrus region, the precise location of below-freezing temperatures is important, because the methods of protecting trees from freezing conditions are very expensive. One method of locating likely cold spots is by relating temperature to elevation. It is conjectured that, on calm nights, the cold spots will be at low elevations. The highest and lowest spots in a particular grove yielded the minimum temperatures listed in the accompanying table for ten cold nights in a recent winter.

Night	High Elevation	Low Elevation
1	32.9	31.8
2	33.2	31.9
3	32.0	29.2
4	33.1	33.2
5	33.5	33.0
6	34.6	33.9
7	32.1	31.0
8	33.1	32.5
9	30.2	28.9
10	29.1	28.0

a Is there sufficient evidence to support the conjecture that low elevations tend to be colder? (Use the sign test. Give the associated p-value.)

b Would it be reasonable to use a t test on the data? Why or why not?

15.5 A psychological experiment was conducted to compare the lengths of response time (in seconds) for two different stimuli. To remove natural person-to-person variability in the responses, both stimuli were applied to each of nine subjects, thus permitting an analysis of the difference between response times within each person. The results are given in the following table.

Subject	Stimulus 1	Stimulus 2
1	9.4	10.3
2	7.8	8.9
3	5.6	4.1
4	12.1	14.7
5	6.9	8.7
6	4.2	7.1
7	8.8	11.3
8	7.7	5.2
9	6.4	7.8

a Use the sign test to determine whether sufficient evidence exists to indicate a difference in mean response for the two stimuli. Use a rejection region for which $\alpha \leq .05$.

b Test the hypothesis of no difference in mean response, using Student's t test.

15.6 Refer to Exercise 12.13. Using the sign test, do you find sufficient evidence to support concluding that completion times differ for the two populations? Use $\alpha = .10$.

15.7 The data set in the accompanying table represents the number of industrial accidents in 12 manufacturing plants for 1-week periods before and after an intensive promotion on safety.

a Do the data support the claim that the campaign was successful? What is the attained significance level? What would you conclude with $\alpha = .01$?

b Discuss the problems associated with a parametric analysis designed to answer the question in (a).

Plant	Before	After	Plant	Before	After
1	3	2	7	5	3
2	4	1	8	3	3
3	6	3	9	2	0
4	3	5	10	4	3
5	4	4	11	4	1
6	5	2	12	5	2

15.4 The Wilcoxon Signed-Rank Test for a Matched Pairs Experiment

As in Section 15.3, assume that we have n paired observations of the form (X_i, Y_i) and that $D_i = X_i - Y_i$. Again we assume that we are interested in testing the hypothesis that the Xs and the Ys have the same distribution versus the alternative that the distributions differ in location. Under the null hypothesis of no difference in the distributions of the Xs and Ys, you would expect (on the average) half of the differences in pairs to be negative and half to be positive. That is, the expected number of negative differences between pairs is $n/2$ (where n is the number of pairs). Further, it would follow that positive and negative differences of equal absolute magnitude should occur with equal probability. If we were to order the differences according to their absolute values and rank them from smallest to largest, the expected rank sums for the negative and positive differences would be equal. Sizable differences in the sums of the ranks assigned to the positive and negative differences would provide evidence to indicate a shift in location between the distributions.

To carry out the Wilcoxon test, we calculate the differences (D_i) for each of the n pairs. Differences equal to zero are eliminated, and the number of pairs, n, is reduced accordingly. Then we rank the *absolute values* of the differences, assigning a 1 to the smallest, a 2 to the second smallest, and so on. If two or more absolute differences are tied for the same rank, then the average of the ranks that would have been assigned to these differences is assigned to each member of the tied group. For example, if two absolute differences are tied for ranks 3 and 4, then each would receive rank 3.5 and the next highest absolute difference would be assigned rank 5. Then we calculate the rank sum for the negative differences and also calculate the rank sum for the positive differences. For a two-tailed test we use T, the *smaller* of these two quantities, as a test statistic to test the null hypothesis that the two population relative frequency histograms are identical. The smaller the value of T is, the greater will be the weight of evidence favoring rejection of the null hypothesis. Hence we will reject the null hypothesis if T is less than or equal to some value, say, T_0.

To detect the one-sided alternative, that the distribution of the Xs is shifted to the right of that of the Ys, we use the rank sum T^- of the negative differences, and we reject the null hypothesis for small values of T^-, say, $T^- \leq T_0$. If we wish to detect a shift of the distribution of the Ys to the right of the Xs, we use the rank sum T^+ of the positive differences as a test statistic, and we reject small values of T^+, say, $T^+ \leq T_0$.

The probability that T is less than or equal to some value T_0 has been calculated for a combination of sample sizes and values of T_0. These probabilities, given in Table 9, Appendix III, can be used to find the rejection region for the test based on T.

For example, suppose that you have $n = 7$ pairs and wish to conduct a two-tailed test of the null hypothesis that the two population relative frequency distributions are identical. Then, with $\alpha = .05$, you would reject the null hypothesis for all values of T less than or equal to 2. The rejection region for the Wilcoxon rank-sum test for a paired experiment is always of this form: reject the null hypothesis if $T \leq T_0$ where T_0 is the critical value for T. Bounds for the attained significance level (p value) are determined as follows. For a two-tailed test, if $T = 3$ is observed when $n = 7$, Table 9, Appendix III, indicates that H_0 would be rejected if $\alpha = .1$, but not if $\alpha = .05$. Thus, $.05 < p$ value $< .1$. For the one-sided alternative that the Xs are shifted to the right of the Ys with $n = 7$ and $\alpha = .05$, H_0 is rejected if $T = T^- \leq 4$. In this case if $T = T^- = 1$, then $.01 < p$ value $< .025$. The test based on T, called the *Wilcoxon signed-rank test*, is summarized as follows.

Wilcoxon Signed-Rank Test for a Matched Pairs Experiment

H_0: the population distributions for the Xs and Ys are identical.

H_a: the two population distributions differ in location (two-tailed); or

H_a: the population relative frequency distribution for the Xs is shifted to the right of that for the Ys (one-tailed).

Test statistic:

1. For a two-tailed test, use $T = \min(T^+, T^-)$, where $T^+ = $ sum of the ranks of the positive differences and $T^- = $ sum of the ranks of the negative differences.

2. For a one-tailed test (to detect the one-tailed alternative just given), use the rank sum T^- of the negative differences.[1]

Rejection region:

1. For a two-tailed test, reject H_0 if $T \le T_0$, where T_0 is the critical value for the two-sided test given in Table 9, Appendix III.

2. For a one-tailed test (as described earlier), reject H_0 if $T^- \le T_0$, where T_0 is the critical value for the one-sided test.

EXAMPLE **15.2**   Test the hypothesis that there is no difference in population distributions of cake density for a matched pairs experiment involving six baked cakes, one using mix A and one using mix B in each pair. What can be said about the attained significance level?

Solution   The original data and differences in density (in ounces per cubic inch) for the six pairs of cakes are shown in Table 15.2.

Table 15.2. Paired data and their differences for Example 15.2

A	B	Difference, A − B	Absolute Difference	Rank of Absolute Difference
.135	.129	.006	.006	3
.102	.120	−.018	.018	5
.108	.112	−.004	.004	1.5
.141	.152	−.011	.011	4
.131	.135	−.004	.004	1.5
.144	.163	−.019	.019	6

---

[1]To detect a shift of the distribution of the Ys to the right of the distribution of the Xs, use the rank sum T^+, the sum of the ranks of the positive differences, and reject H_0 if $T^+ \le T_0$.

As with our other nonparametric tests, the null hypothesis to be tested is that the two population frequency distributions of cake densities are identical. The alternative hypothesis, which implies a two-tailed test, is that the distributions differ in location.

Because the amount of data is small, we will conduct our test by using $\alpha = .10$. From Table 9, Appendix III, the critical value of T for a two-tailed test, $\alpha = .10$, is $T_0 = 2$. Hence, we will reject H_0 if $T \leq 2$.

Because there is only one positive difference that has rank 3, $T^+ = 3$ and $T^- = 18$, and therefore $T = 3$. The observed value of T exceeds the critical value of T, and so there is not sufficient evidence to indicate a difference in the two population frequency distributions of cake densities. Because we cannot reject H_0 for $\alpha = .10$, we can only say that p-value $> .10$.

Although Table 9, Appendix III, is applicable for values of n (the number of data pairs) as large as $n = 50$, it is worth noting that T^+ (or T^-) will be approximately normally distributed when the null hypothesis is true and n is large (say 25 or more). This enables us to construct a large-sample Z test, where if $T = T^+$,

$$E(T^+) = \frac{n(n + 1)}{4} \quad \text{and} \quad V(T^+) = \frac{n(n + 1)(2n + 1)}{24}.$$

Then the Z statistic,

$$Z = \frac{T^+ - E(T^+)}{\sqrt{V(T^+)}} = \frac{T^+ - [n(n + 1)/4]}{\sqrt{n(n + 1)(2n + 1)/24}}$$

can be used as a test statistic. Thus for a two-tailed test and $\alpha = .05$, we would reject the hypothesis of identical population distributions when $|z| \geq 1.96$. For a one-tailed test that the distribution of the Xs is shifted to the right (left) of the distribution of the Ys, reject H_0 when $z > z_\alpha$ ($z < -z_\alpha$).

A Large-Sample Wilcoxon Signed-Rank Test for a Matched Pairs Experiment: $n > 25$

Null hypothesis:	H_0: The population relative frequency distributions for the Xs and Ys are identical.
Alternative hypothesis:	H_a: The two population relative frequency distributions differ in location (a two-tailed test); or H_a: The population relative frequency distribution for the Xs is shifted to the right (or left) of the relative frequency distribution of the Ys (one-tailed tests)

Test statistic: $Z = \dfrac{T^+ - [n(n+1)/4]}{\sqrt{n(n+1)(2n+1)/24}}$

Rejection region: Reject H_0 if $z \geq z_{\alpha/2}$ or $z \leq -z_{\alpha/2}$ for a two-tailed test. To detect a shift in the distributions of the Xs to the right of the Ys, reject H_0 when $z \geq z_\alpha$. To detect a shift in the opposite direction, reject H_0 if $z \leq -z_\alpha$.

Exercises

15.8 The accompanying table gives the scores of a group of fifteen students in mathematics and art.

Student	Math	Art	Student	Math	Art
1	22	53	9	62	55
2	37	68	10	65	74
3	36	42	11	66	68
4	38	49	12	56	64
5	42	51	13	66	67
6	58	65	14	67	73
7	58	51	15	62	65
8	60	71			

a Use Wilcoxon's signed-rank test to determine if the locations of the distributions of scores for these students differ significantly for the two subjects. Give bounds for the p-value and indicate the appropriate conclusion with $\alpha = .05$.

b State specific null and alternative hypotheses for the test you conducted in (a).

15.9 Refer to Exercise 15.2. What answers are obtained if Wilcoxon's signed-rank test is used in analyzing the data? Compare these answers with the answers obtained in Exercise 15.2.

15.10 Refer to Exercise 15.4(a). Answer the question by using the Wilcoxon signed-rank test.

15.11 Eight subjects were asked to perform a simple puzzle assembly task under normal conditions and under conditions of stress. During the stress condition the subjects were told that a mild shock would be delivered 3 minutes after the start of the experiment and every 30 seconds thereafter until the task was completed. Blood pressure readings were taken under both conditions. Data in the accompanying table represent the highest reading during the experiment. Do the data present sufficient evidence to indicate higher blood pressure readings during conditions of stress? Analyze the data by using the Wilcoxon signed-rank test for a matched pairs experiment. Give the appropriate p-value.

Subject	Normal	Stress
1	126	130
2	117	118
3	115	125
4	118	120
5	118	121
6	128	125
7	125	130
8	120	120

15.12 Two methods, A and B, for controlling traffic were employed at each of $n = 12$ intersections for a period of 1 week. The numbers of accidents occurring during this time period are recorded in the following table. The order of use (which method was employed for the first week) was randomly chosen for each intersection.

Intersection	Method A	Method B	Intersection	Method A	Method B
1	5	4	7	2	3
2	6	4	8	4	1
3	8	9	9	7	9
4	3	2	10	5	2
5	6	3	11	6	5
6	1	0	12	1	1

a Analyze these data using the sign test.

b Analyze these data using the Wilcoxon signed-rank test for a matched pairs experiment

15.13 Dental researchers have developed a new material for preventing cavities, a plastic sealant that is applied to the chewing surfaces of teeth. To determine whether the sealant is effective, it was applied in half of the teeth of each of 12 school-age children. After 5 years, the number of cavities in the sealant-coated teeth and in the untreated teeth were counted. The results are given in the accompanying table. Is there sufficient evidence to indicate that sealant-coated teeth are less prone to cavities than are untreated teeth? Test using $\alpha = 0.05$.

Child	Sealant-Coated	Untreated
1	3	3
2	1	3
3	0	2
4	4	5
5	1	0
6	0	1
7	1	5
8	2	0
9	1	6
10	0	0
11	0	3
12	4	3

15.14 Refer to Exercise 12.14. With $\alpha = .01$, use the Wilcoxon signed-rank test to see if there was a significant loss in muck depth between 1972 and 1978.

15.15 Suppose that $Y_1, Y_2, \ldots, Y_n$ is a random sample from a continuous distribution function $F(y)$. It is desired to test a hypothesis concerning the median ξ of $F(y)$. Construct a test of $H_0 : \xi = \xi_0$ against $H_a : \xi \neq \xi_0$, where ξ_0 is a specified constant.

a Use the sign test.

b Use the Wilcoxon signed-rank test.

15.16 The spokesperson for an organization supporting property tax reductions in a certain section of a city stated that the median annual income for household heads in that section was $15,000. A random sample of ten household heads from that section revealed the following annual incomes:

$$14,800 \quad 16,900 \quad 18,000 \quad 19,100 \quad 13,200$$
$$18,500 \quad 20,000 \quad 19,200 \quad 15,100 \quad 16,500.$$

With $\alpha = .10$, test the hypothesis that the median income for the population from that section is $15,000 against the alternative that it is greater than $15,000.

a Use the sign test.

b Use the Wilcoxon signed-rank test.

15.5 The Use of Ranks for Comparing Two Population Distributions: Independent Random Samples

A statistical test for comparing two populations based on independent random samples, the *rank-sum* test, was proposed by F. Wilcoxon in 1945. Again, we assume that we are interested in testing whether the two populations have the same distribution versus the shift (or location) alternative (see Section 15.2). Suppose that you were to select independent random samples of n_1 and n_2 observations, each from two populations; call them I and II. Wilcoxon's idea was to combine the $n_1 + n_2 = n$ observations and rank them, in order of magnitude, from 1 (the smallest) to n (the largest). Ties are treated as in Section 15.4. That is, if two or more observations are tied for the same rank, the average of the ranks that would have been assigned to these observations is assigned to each member of the tied group. If the observations were selected from identical populations, the *rank sums* for the samples should be more or less proportional to the sample sizes n_1 and n_2. For example, if n_1 and n_2 were equal, you would expect the rank sums to be nearly equal. In contrast, if the observations in one population, say, population I, tended to be larger than those in population II, the observations in sample I would tend to receive the highest ranks and would have a larger than expected rank sum. Thus (sample sizes being equal), if one rank sum is very large (and, correspondingly, the other is very small), it may indicate a statistically significant difference between the two populations.

Mann and Whitney proposed an equivalent statistical test in 1947 that also used the rank sums of two samples. Because the Mann–Whitney U test and tables of critical values of U occur so often in the literature, we will explain its use in Section 15.6 and will give several examples of its applications. In this section we illustrate the logic of the rank-sum test and demonstrate how to determine the rejection region for the test and the value of α.

EXAMPLE **15.3**   The bacteria counts per unit volume are shown in Table 15.3 for two types of cultures, A and B. Four observations were made for each culture. Let n_1 and n_2 represent the number of observations in samples A and B, respectively.

For the data given in Table 15.3, the corresponding ranks are as shown in Table 15.4. Do these data present sufficient evidence to indicate a difference in the population distributions for cultures A and B?

Solution   Let W equal the rank sum for sample A (for this sample, $W = 12$). Certainly, very small or very large values of W provide evidence to indicate a difference between the two population distributions; and hence W, the *rank sum*, can be employed as a test statistic.

The rejection region for a given test is obtained in the same manner as for the sign test. We start by selecting the most contradictory values of W as the rejection region and add to these until α is of acceptable size.

The minimum rank sum would include the ranks 1, 2, 3, 4, or $W = 10$. Similarly, the maximum would include 5, 6, 7, 8, with $W = 26$. Therefore, we include these two values of W in the rejection region. What is the corresponding value of α?

Finding the value of α is a probability problem that can be solved by using the methods of Chapter 2. If the populations are identical, every permutation of the eight ranks represents a sample point and is equally likely. Then α represents the sum of the probabilities of the sample points (arrangements) that imply $W = 10$ or $W = 26$ in sample A. The total number of permutations of the eight ranks is 8! The number of different arrangements of the ranks 1, 2, 3, 4 in sample A with the 5, 6, 7, 8 of sample B is $4! \times 4!$. Similarly, the number of arrangements that place the maximum

Table 15.3. Data for Example 15.3

A	B
27	32
31	29
26	35
25	28

Table 15.4. Ranks

	A	B
	3	7
	6	5
	2	8
	1	4
Rank Sum	12	24

value of W in sample A (ranks 5, 6, 7, 8) is $4! \times 4!$. Then the probability that $W = 10$ or $W = 26$ is

$$p(10) + p(26) = \frac{(2)(4!)(4!)}{8!} = \frac{2}{\binom{8}{4}} = \frac{1}{35} = .029.$$

If this value of α is too small, the rejection region can be enlarged to include the next smallest and next largest rank sums, $W = 11$ and $W = 25$. The rank sum $W = 11$ includes the ranks 1, 2, 3, 5, and

$$p(11) = \frac{4!\,4!}{8!} = \frac{1}{70}.$$

Similarly,

$$p(25) = 1/70.$$

Then

$$\alpha = p(10) + p(11) + p(25) + p(26) = 2/35 = .057.$$

Expansion of the rejection region to include 12 and 24 substantially increases the value of α. The set of sample points giving a rank of 12 includes all sample points associated with rankings of (1, 2, 3, 6) and (1, 2, 4, 5). Thus,

$$p(12) = \frac{(2)(4!)(4!)}{8!} = \frac{1}{35}.$$

and

$$\alpha = p(10) + p(11) + p(12) + p(24) + p(25) + p(26)$$
$$= \frac{1}{70} + \frac{1}{70} + \frac{1}{35} + \frac{1}{35} + \frac{1}{70} + \frac{1}{70} = \frac{4}{35} = .114.$$

This value of α might be considered too large for practical purposes. Hence we are better satisfied with the rejection region $W = 10, 11, 25,$ and 26.

The rank sum for the sample, $W = 12$, does not fall in this preferred rejection region, so we do not have sufficient evidence to reject the hypothesis that the population distributions of bacteria counts for the two cultures are identical.

15.6 The Mann–Whitney U Test: Independent Random Samples

The Mann–Whitney statistic U is obtained by ordering all $(n_1 + n_2)$ observations according to their magnitude and counting the number of observations in sample A that precede each observation in sample B. The statistic U is the sum of these counts.

For example, the eight observations of Example 15.3 are

25	26	27	28	29	31	32	35
A	A	A	B	B	A	B	B.

The smallest B observation is 28, and $u_1 = 3$ observations from sample A precede it. Similarly, $u_2 = 3$ A observations precede the second B observation and $u_3 = 4$ and $u_4 = 4$ A observations precede the third and fourth B observations, respectively (32 and 35). Then

$$U = u_1 + u_2 + u_3 + u_4 = 14.$$

Or you could count the number of B observations that precede each A observation and use the sum U_B of these counts as the U statistic. In either case, very large or small values of U imply a separation of the ordered A and B observations and thus provide evidence to indicate a difference (a shift of location) between the population distributions of A and B.

As noted in Section 15.5, the Mann–Whitney U statistic is related to Wilcoxon's rank sum. In fact, it can be shown (proof omitted) that

Formulas for the Mann–Whitney U Statistic

$$U_A = n_1 n_2 + \frac{n_1(n_1 + 1)}{2} - W_A$$

$$U_B = n_1 n_2 + \frac{n_2(n_2 + 1)}{2} - W_B$$

where

$$n_1 = \text{number of observations in sample } A$$

$$n_2 = \text{number of observations in sample } B$$

$$U_A + U_B = n_1 n_2$$

$$W_A = \text{rank sum for sample } A$$

$$W_B = \text{rank sum for sample } B.$$

As you can see from the formulas for U_A and U_B, U_A is small when W_A is large, a situation likely to occur when the population distribution of the A measurements is shifted to the right of the population distribution for the B measurements. Consequently, to conduct a one-tailed test to detect a shift in the A distribution to the right of the B distribution, you will reject the null hypothesis of no difference in the population distributions if U_A is less than some specified value U_0. That is, you will reject H_0 for small values of U_A. Similarly, to conduct a one-tailed test to detect a shift of the B distribution to the right of the A distribution, you would reject H_0 if U_B is less than some specified value U_0.

Table 8, Appendix III, gives the probability that an observed value of U will be less than some specified value U_0. This is the value of α for a one-tailed test. To conduct a two-tailed test—that is, to detect a shift in the population distributions for the A and B measurements in either direction—we agree always to use U, the smaller of U_A or U_B, as the test statistic and to reject H_0 for $U < U_0$. The value of α for the two-tailed test is double the tabulated value given in Table 8, Appendix III.

To see how to locate the rejection region for the Mann–Whitney U test, suppose that $n_1 = 4$ and $n_2 = 5$. Then you would consult the third section of Table 8, Appendix III (the one corresponding to $n_2 = 5$). Notice that the table is constructed on the assumption that $n_1 \le n_2$. That is, you must always identify the smaller sample as sample 1. From the table we see, for example, $P(U \le 2) = .0317$ and $P(U \le 3) = .0556$. So if you want to conduct a one-tailed Mann–Whitney U test with $n_1 = 4$ and $n_2 = 5$ for α near .05, you should reject the null hypothesis of equality of population relative frequency distributions when $U \le 3$. The probability of a type I error for the test is $\alpha = .0556$. If you use this same rejection for a two-tailed test, that is, $U \le 3$, α is double the tabulated value, or $\alpha = 2(.0556) = .1112$.

When applying the test to a set of data, you may find that some of the observations are of equal value. Ties in the observations can be handled by averaging the ranks that would have been assigned to the tied observations and assigning this average to each. Thus, if three observations are tied and are due to receive ranks 3, 4, and 5, we assign rank 4 to all three. The next observation in the sequence receives rank 6, and ranks 3 and 5 do not appear. Similarly, if two observations are tied for ranks 3 and 4, each receives rank 3.5, and ranks 3 and 4 do not appear.

Table 8, Appendix III, can also be used to find the observed significance level for a test. For example, if $n_1 = 5$, $n_2 = 5$, and $U = 4$, the p-value for a one-tailed test is

$$P\{U \le 4\} = .0476.$$

If the test is two-tailed, the p value becomes

$$2(.0476), \quad \text{or} \quad .0952.$$

The Mann–Whitney U Test

Null hypothesis: H_0: The population relative frequency distributions for A and B are identical.

Alternative hypothesis: H_a: The two populations' relative frequency distributions are shifted in respect to their relative locations (a two-tailed test); or

H_a: The population relative frequency distribution for A is shifted to the right of the relative

Test statistic:

frequency distribution for population B (a one-tailed test).[2]

For a two-tailed test, use U, the smaller of

$$U_A = n_1 n_2 + \frac{n_1(n_1 + 1)}{2} - W_A$$

and

$$U_B = n_1 n_2 + \frac{n_2(n_2 + 1)}{2} - W_B,$$

where W_A and W_B are the rank sums for samples A and B, respectively. For a one-tailed test, use U_A.

Rejection region:

1. For the two-tailed test and a given value of α, reject H_0 if $U \leq U_0$, where $P(U \leq U_0) = \alpha/2$. [*Note:* Observe that U_0 is the value such that $P(U \leq U_0)$ is equal to half of α.]

2. For a one-tailed test and a given value of α, reject H_0 if $U_A \leq U_0$, where $P(U_A \leq U_0) = \alpha$.

Assumptions:

Samples have been randomly and independently selected from their respective populations. Ties in the observations can be handled by averaging the ranks that would have been assigned to the tied observations and assigning this average to each. Thus, if three observations are tied and are due to receive ranks 3, 4, and 5, we assign rank 4 to all three.

EXAMPLE **15.4**   Test the hypothesis that there is no difference in the population distributions for the bacteria count data of Example 15.3.

Solution   We have already noted that the Mann–Whitney U test and the Wilcoxon rank-sum test are equivalent, so we should reach the same conclusions here as we did in Example 15.3. Recall that the alternative hypothesis was that the distributions of bacteria counts for cultures A and B differed and that this implied a two-tailed test. Thus, because Table 8, Appendix III, gives values of $P(U \leq U_0)$ for specified sample sizes and values of U_0, we must double the tabulated value to find α. Suppose, as in Example 15.3, that we desire a value of α near .05. Checking Table 8 for

[2]For the sake of convenience, we describe the one-tailed test as being designed to detect a shift in the distribution of the A measurements to the right of the distribution of the B measurements. To detect a shift in the B distribution to the right of the A distribution, just interchange the letters A and B in the discussion.

$n_1 = n_2 = 4$, we find $P(U \leq 1) = .0286$. When $U \leq 1$ is the rejection region, α will equal $2(.0286) = .0572$ or, rounding to three decimal places, $\alpha = .057$ (the same value of α obtained for Example 15.3).

For the bacteria data the rank sums are $W_A = 12$ and $W_B = 24$. Then

$$U_A = n_1 n_2 + \frac{n_1(n_1 + 1)}{2} - W_A = (4)(4) + \frac{4(4+1)}{2} - 12 = 14$$

and

$$U_B = n_1 n_2 + \frac{n_2(n_2 + 1)}{2} - W_B = (4)(4) + \frac{4(4+1)}{2} - 24 = 2.$$

Because the smaller observed value of U is 2 (just calculated), U does not fall in the rejection region. Hence, there is not sufficient evidence to show a difference in the population distributions of bacteria counts for cultures A and B. The p-value is given by $2P(U \leq 2) = 2(.0571) = .1142$.

EXAMPLE 15.5 An experiment was conducted to compare the strengths of two types of kraft papers, one a standard kraft paper of a specified weight and the other the same standard kraft paper treated with a chemical substance. Ten pieces of each type of paper, randomly selected from production, produced the strength measurements shown in Table 15.5. Test the hypothesis of no difference in the distributions of strengths for the two types of paper against the alternative hypothesis that the treated paper tends to be stronger.

Solution In Table 15.5, the ranks are shown in parentheses alongside the $n_1 + n_2 = 10 + 10 = 20$ strength measurements, and the rank sums W_A and W_B are shown below the columns. Because we wish to detect a shift in the distribution of the B measurements to the right of the distribution of the A measurements, we will reject the null hypothesis of no difference in population strength distributions when W_B is excessively large. Because this situation occurs when U_B is small, we will conduct a one-tailed statistical test and reject the null hypothesis when $U_B \leq U_0$.

Table 15.5. Data for Example 15.5

	Standard, A	Treated, B
	1.21 (2)	1.49 (15)
	1.43 (12)	1.37 (7.5)
	1.35 (6)	1.67 (20)
	1.51 (17)	1.50 (16)
	1.39 (9)	1.31 (5)
	1.17 (1)	1.29 (3.5)
	1.48 (14)	1.52 (18)
	1.42 (11)	1.37 (7.5)
	1.29 (3.5)	1.44 (13)
	1.40 (10)	1.53 (19)
Rank Sums	$W_A = 85.5$	$W_B = 124.5$

Suppose that we choose a value of α near .05. Then we can find U_0 by consulting the portion of Table 8, Appendix III, corresponding to $n_2 = 10$. The probability $P(U \le U_0)$ nearest .05 is .0526 and corresponds to $U_0 = 28$. Hence, we will reject if $U_B \le 28$.

Calculating U_B, we have

$$U_B = n_1 n_2 + \frac{n_2(n_2 + 1)}{2} - W_B = (10)(10) + \frac{(10)(11)}{2} - 124.5 = 30.5.$$

As you can see, U_B is not less than $U_0 = 28$. Therefore, we cannot reject the null hypothesis. At the $\alpha = .05$ level of significance, there is not sufficient evidence to indicate that the treated kraft paper is stronger than the standard. The p-value is given by $P(U_B \le 30.5) = .0716$.

A simplified large-sample test ($n_1 > 10$ and $n_2 > 10$) can be obtained by using the familiar Z statistic of Chapter 10. When the population distributions are identical, it can be shown that the U statistic has the following expected value and variance when $U = U_A$ (or $U = U_B$):

$$E(U_A) = \frac{n_1 n_2}{2} \quad \text{and} \quad V(U_A) = \frac{n_1 n_2(n_1 + n_2 + 1)}{12}.$$

Also, when n_1 and n_2 are large,

$$Z = \frac{U_A - E(U_A)}{\sigma_{U_A}}$$

has approximately a standard normal distribution. This approximation is adequate when n_1 and n_2 both are greater than or equal to 10. Thus, for a two-tailed test with $\alpha = .05$, we will reject the null hypothesis if $|z| \ge 1.96$.

The Z statistic yields the same conclusion as the exact U test for Example 15.5:

$$z = \frac{30.5 - [(10)(10)/2]}{\sqrt{[(10)(10)(10 + 10 + 1)]/12}} = \frac{30.5 - 50}{\sqrt{2100/12}} = -\frac{19.5}{\sqrt{175}}$$

$$= -\frac{19.5}{13.23} = -1.47.$$

For a one-tailed test with $\alpha = .05$ located in the lower tail of the z distribution, we will reject the null hypothesis if $z < -1.645$. You can see that $z = -1.47$ does not fall in the rejection region and that this test reaches the same conclusion as the exact U test of Example 15.5.

The Mann–Whitney U Test for Large Samples, $n_1 > 10$ and $n_2 > 10$

Null hypothesis: H_0: The population relative frequency distributions for A and B are identical.

Alternative hypothesis: H_a: The two populations' relative frequency distributions are not identical (a two-tailed test); or

H_a: The population relative frequency distribution for A is shifted to the right (or left) of the relative frequency distribution for population B (a one-tailed test).

Test statistic:

$$Z = \frac{U_A - (n_1 n_2 / 2)}{\sqrt{n_1 n_2 (n_1 + n_2 + 1)/12}}$$

Rejection region: Reject H_0 if $z > z_{\alpha/2}$ or $z < -z_{\alpha/2}$ for a two-tailed test. For a one-tailed test, place all α in one tail of the z distribution. To detect a shift in the distribution of the A observations to the right of the distribution of the B observations, reject H_0 when $z < -z_\alpha$. To detect a shift in the opposite direction, reject H_0 when $z > z_\alpha$. Tabulated values of z are given in Table 4, Appendix III.

It may seem to you that the Mann–Whitney U test and the equivalent Wilcoxon rank-sum test are not very efficient, because they do not appear to use all the information in the sample. Actually, theoretical studies have shown that this is not the case. Suppose, for example, that all of the assumptions for a two-sample t test are met in testing $H_0 : \mu_1 - \mu_2 = 0$ versus $H_a : \mu_1 - \mu_2 = d_0$, for some specific d_0. Because the two-sample t test simply tests for a difference in location (see Section 15.2), we can use the Mann–Whitney U statistic to test these same hypotheses. For a given α and β, the total sample size required for the t test is approximately .95 times the total sample size required for the Mann–Whitney U. Thus, the nonparametric procedure is almost as good as the t test for the situation in which the t test is optimal. For many nonnormal distributions the nonparametric procedure requires fewer observations than a corresponding parametric procedure would require to produce the same values of α and β.

Exercises

15.17   Two plastics, each produced by a different process, were tested for ultimate strength. The measurements in the accompanying table represent breaking loads in units of 1000 pounds per square inch. Do the data present evidence of a difference between the locations of the distributions of ultimate strengths for the two plastics? Solve by using the Mann–Whitney U test with a level of significance as near as possible to $\alpha = .10$.

Plastic 1	Plastic 2
15.3	21.2
18.7	22.4
22.3	18.3
17.6	19.3
19.1	17.1
14.8	27.7

15.18 The coded values for a measure of brightness in paper (light reflectivity), prepared by two different processes, are as shown in the accompanying table for samples of size 9 drawn randomly from each of the two processes. Do the data present sufficient evidence to indicate a difference in locations of brightness measurements for the two processes? Give bounds for the attained significance level.

A	B
6.1	9.1
9.2	8.2
8.7	8.6
8.9	6.9
7.6	7.5
7.1	7.9
9.5	8.3
8.3	7.8
9.0	8.9

a Use the Mann–Whitney U test.

b Use Student's t test.

c Give specific null and alternative hypotheses, along with any assumptions, for the tests used in (a) and (b).

15.19 Fifteen experimental batteries were selected at random from a lot at pilot plant A, and fifteen standard batteries were selected at random from production at plant B. All thirty batteries were simultaneously placed under an electrical load of the same magnitude. The first battery to fail was an A, the second a B, the third a B, and so on. The following sequence shows the order of failure for the 30 batteries:

A B B B A B A A B B B B A B A
B B B B A A B A A A B A A A A

Using the large-sample theory for the U test, determine whether there is sufficient evidence to permit the experimenter to conclude that the lengths of life for the experimental batteries tend to be greater than the lengths of life for the standard batteries. Use $\alpha = .05$.

15.20 Refer to Exercises 8.74 and 8.75. Is there sufficient evidence to indicate a difference in the populations of LC50 measurements for DDT and Diazinon? Give bounds for the attained significance level associated with the U statistic. What do you conclude when $\alpha = .10$?

15.21 Given below are wing stroke frequencies[3] for samples of two species of Euglossine bees. Four bees of the species *Euglossa mandibularis* Friese and six of the species *Euglossa imperialis* Cockerell are shown in the accompanying table.

[3] *Source:* T. M. Casey, M. L. May, and K. R. Morgan, "Flight Energetics of Euglossine Bees in Relation to Morphology and Wing Stroke Frequency," *Journal of Experimental Biology* 116 (1985).

Wing Stroke Frequencies	
Euglossa mandibularis Friese	*Euglossa imperialis* Cockerell
235	180
225	169
190	180
188	185
	178
	183

Do the data present sufficient evidence to indicate that the distributions of wing stroke frequencies differ for the two species? Use the test based on the Mann–Whitney U statistic with α as close to, but not exceeding, .10. Give the approximate p-value associated with the test.

15.22 Cancer treatment using chemotherapy employs chemicals that kill both cancer cells and normal cells. In some instances the toxicity of the cancer drug, that is, its effect on normal cells, can be reduced by the simultaneous injection of a second drug. A study was conducted to determine whether a particular drug injection was beneficial in reducing the harmful effects of a chemotherapy treatment on the survival time for rats. Two randomly selected groups of rats, 12 rats in each group, were used for the experiment. Both groups, call them A and B, received the toxic drug in a dosage large enough to cause death, but group B also received the antitoxin that was intended to reduce the toxic effect of the chemotherapy on normal cells. The test was terminated at the end of 20 days, or 480 hours. The lengths of survival time for the two groups of rats, to the nearest 4 hours, are shown in the following table. Do the data provide sufficient evidence to indicate that rats receiving the antitoxin tended to survive longer after chemotherapy than those not receiving the toxin? Use the Mann–Whitney U test with a value of α near .05.

Only Chemotherapy (A)	Chemotherapy plus Drug (B)
84	140
128	184
168	368
92	96
184	480
92	188
76	480
104	244
72	440
180	380
144	480
120	196

15.7 The Kruskal–Wallis Test for the One-Way Layout

In Section 13.3, we presented an analysis of variance procedure to compare the means of k populations. The resultant F test was based on the assumption that we obtained independent samples from normal populations with equal variances. That is, as discussed in Section 15.2, we were interested in testing whether all the populations had the same distribution versus the alternative that the populations differed in location. A key element in the development of the procedure was the quantity identified as the sum of squares for treatments, SST. As we pointed out in the discussion in Section 13.3, the larger the values of SST, the greater the weight of evidence favoring rejection of the null hypothesis that the means are all equal. In this section we present a nonparametric technique to test whether the populations differ in location. Like the other nonparametric techniques discussed in this chapter, the Kruskal–Wallis procedure requires no assumptions about the actual form of the probability distributions.

As in Section 13.3, we assume that independent random samples have been drawn from k populations that differ only in location. However, we need not assume that these populations possess normal distributions. For complete generality, we permit the sample sizes to be unequal, and we let n_i, for $i = 1, 2, \ldots, k$, represent the size of the sample drawn from the ith population. Analogously to the procedure of Section 15.5, combine all the $n_1 + n_2 + \cdots + n_k = n$ observations and rank them from 1 (the smallest) to n (the largest). Ties are treated as in previous sections. That is, if two or more observations are tied for the same rank, then the average of the ranks that would have been assigned to these observations is assigned to each member of the tied group. Let R_i denote the sum of the ranks of the observations from population i, and let $\overline{R}_i = R_i / n_i$ denote the corresponding average of the ranks. If $\overline{R}$ equals the overall average of all of the ranks, consider the rank analogue of SST, which is computed by using the ranks rather than the actual values of the measurements:

$$V = \sum_{i=1}^{k} n_i (\overline{R}_i - \overline{R})^2.$$

If the null hypothesis is true and the populations do not differ in location, we would expect the $\overline{R}_i$ values to be approximately equal and the resulting value of V to be relatively small. If the alternative hypothesis is true, we would expect this to be exhibited in differences among the values of the $\overline{R}_i$ values, leading to a large value for V. Notice that $\overline{R} = (\text{sum of the first } n \text{ integers})/n = [n(n + 1)/2]/n = (n + 1)/2$ and thus that

$$V = \sum_{i=1}^{k} n_i \left(\overline{R}_i - \frac{n + 1}{2} \right)^2.$$

Instead of focusing on V, Kruskal and Wallis (1952) considered the statistic $H = 12V/[n(n + 1)]$, which may be rewritten (see Exercise 15.27) as

$$H = \frac{12}{n(n + 1)} \sum_{i=1}^{k} \frac{R_i^2}{n_i} - 3(n + 1).$$

As previously noted, the null hypothesis of equal locations is rejected in favor of the alternative that the populations differ in location if the value of H is large. Thus, the corresponding α-level test calls for rejection of the null hypothesis in favor of the alternative if $H > h(\alpha)$, where $h(\alpha)$ is such that, when H_0 is true, $P[H > h(\alpha)] = \alpha$.

If the underlying distributions are continuous and if there are no ties among the n observations, the null distribution of H can (tediously) be found by using the methods of Chapter 2. We can find the distribution of H for any values of k and $n_1, n_2, \ldots, n_k$ by calculating the value of H for each of the $n!$ equally likely permutations of the ranks of the n observations (see Exercise 15.28). These calculations have been performed and tables developed for some relatively small values of k and for $n_1, n_2, \ldots, n_k$ (see, for example, Table A.12 of Hollander and Wolfe (1999).)

Kruskal and Wallis showed that, if the n_i values are "large," the null distribution of H can be approximated by a chi-square distribution with $k-1$ degrees of freedom. This approximation is generally accepted to be adequate if each of the n_i values is greater than or equal to 5. Our examples and exercises are all such that this large sample approximation is adequate. If you wish to use the Kruskal–Wallis analysis for smaller data sets, where this large-sample approximation is not adequate, refer to Hollander and Wolfe (1999) to obtain the appropriate critical values.

We summarize the large sample Kruskal–Wallis procedure as follows.

Kruskal–Wallis Test Based on H for Comparing k Population Distributions

Null hypothesis: H_0: The k population distributions are identical.

Alternative hypothesis: H_a: At least two of the population distributions differ in location.

Test statistic:

$$H = \frac{12}{n(n+1)} \sum_{i=1}^{k} \frac{R_i^2}{n_i} - 3(n+1)$$

where

n_i = number of measurements in sample from population i

R_i = rank sum for sample i, where the rank of each measurement is computed according to its relative size in the overall set of $n = n_1 + n_2 + \cdots + n_k$ observations formed by combining the data from all k samples.

Rejection region: Reject H_0 if $H > \chi_\alpha^2$ with $(k-1)$ degrees of freedom.

Assumptions: The k samples are randomly and independently drawn. There are five or more measurements in each sample.

EXAMPLE **15.6**   A quality control engineer has selected independent samples from the output of three assembly lines in an electronics plant. For each line, the output of ten randomly selected hours of production was examined for defects. Do the data in Table 15.6 provide evidence that the probability distributions of the number of defects per hour of output differ in location for at least two of the lines? Use $\alpha = .05$. Also give the p-value associated with the test.

Table 15.6. Data for Example 15.6

Line 1		Line 2		Line 3	
Defects	Rank	Defects	Rank	Defects	Rank
6	5	34	25	13	9.5
38	27	28	19	35	26
3	2	42	30	19	15
17	13	13	9.5	4	3
11	8	40	29	29	20
30	21	31	22	0	1
15	11	9	7	7	6
16	12	32	23	33	24
25	17	39	28	18	14
5	4	27	18	24	16
	$R_1 = 120$		$R_2 = 210.5$		$R_3 = 134.5$

Solution   In this case, $n_1 = 10 = n_2 = n_3$ and $n = 30$. Thus

$$H = \frac{12}{30(31)}\left[\frac{(120)^2}{10} + \frac{(210.5)^2}{10} + \frac{(134.5)^2}{10}\right] - 3(31) = 6.097.$$

Because all of the n_i values are greater than or equal to 5, we may use the approximation for the null distribution of H and reject the null hypothesis of equal locations if $H > \chi_\alpha^2$ based on $k - 1 = 2$ degrees of freedom. We consult Table 6, Appendix III, to determine that $\chi_{.05}^2 = 5.99147$. Thus, we reject the null hypothesis at the $\alpha = .05$ level and conclude that at least one of the three lines tends to produce a greater number of defects than the others.

According to Table 6, Appendix III, the value of $H = 6.097$ leads to rejection of the null hypothesis if $\alpha = .05$ but not if $\alpha = .025$. Thus, $.025 < p$-value $< .05$.

It can be shown (proof omitted) that, if we wish to compare only $k = 2$ populations, the Kruskal–Wallis test is equivalent to the two-sample Wilcoxon two-sided test presented in Section 15.5. If data are obtained from a one-way layout involving $k > 2$ populations, but we wish to compare a particular pair of populations, the Wilcoxon two-sample test (or the equivalent Mann–Whitney U test of Section 15.6) can be used for this purpose. Notice that the analysis based on the Kruskal–Wallis H

statistic does not require knowledge of the actual values of the observations. We need only know the ranks of the observations to complete the analysis. Exercise 15.24 illustrates the use of the Kruskal–Wallis analysis for such a case.

Exercises

15.23   Three different brands of magnetron tubes (the key components in microwave ovens) were subjected to stressful testing, and the number of hours each operated without repair was recorded (see the accompanying table). Although these times do not represent typical life lengths, they do indicate how well the tubes can withstand extreme stress.

Brand		
A	B	C
36	49	71
48	33	31
5	60	140
67	2	59
53	55	42

a   Use the F test for a one-way layout (Chapter 13) to test the hypothesis that the mean length of life under stress is the same for the three brands. Use $\alpha = .05$. What assumptions are necessary for the validity of this procedure? Is there any reason to doubt these assumptions?

b   Use the Kruskal–Wallis test to determine whether evidence exists to conclude that the brands of magnetron tubes tend to differ in length of life under stress. Test using $\alpha = .05$.

15.24   An experiment was conducted to compare the length of time it takes a person to recover from each of the three types of influenza—Victoria A, Texas, and Russian. Twenty-one human subjects were selected at random from a group of volunteers and divided into three groups of seven each. Each group was randomly assigned a strain of the virus and the influenza was induced in the subjects. All of the subjects were then cared for under identical conditions, and the recovery time (in days) was recorded. The ranks of the results appear in the following table.

Victoria A	Texas	Russian
20	14.5	9
6.5	16.5	1
21	4.5	9
16.5	2.5	4.5
12	14.5	6.5
18.5	12	2.5
9	18.5	12

a   Do the data provide sufficient evidence to indicate that the recovery times for one (or more) type(s) of influenza tend(s) to be longer than for the other types? Give the associated p-value.

b Do the data provide sufficient evidence to indicate a difference in locations of the distributions of recovery times for the Victoria A and Russian types? Give the associated p-value.

15.25 The EPA wants to determine whether temperature changes in the ocean's water caused by a nuclear power plant will have a significant effect on the animal life in the region. Recently hatched specimens of a certain species of fish are randomly divided into four groups. The groups are placed in separate simulated ocean environments that are identical in every way except for water temperature. Six months later, the specimens are weighed. The results (in ounces) are given in the accompanying table. Do the data provide sufficient evidence to indicate that one (or more) of the temperatures tend(s) to produce larger weight increases than the other temperatures? Test using $\alpha = .10$.

Weights of Specimens			
38°F	42°F	46°F	50°F
22	15	14	17
24	21	28	18
16	26	21	13
18	16	19	20
19	25	24	21
	17	23	

15.26 Weevils cause millions of dollars worth of damage each year to cotton crops. Three chemicals designed to control weevil populations are applied, one to each of three cotton fields. After 3 months, ten plots of equal size are randomly selected within each field and the percentage of cotton plants with weevil damage is recorded for each. Do the data in the accompanying table provide sufficient evidence to indicate a difference in location among the distributions of damage rates corresponding to the three treatments? Give bounds for the associated p-value.

A	B	C
10.8	22.3	9.8
15.6	19.5	12.3
19.2	18.6	16.2
17.9	24.3	14.1
18.3	19.9	15.3
9.8	20.4	10.8
16.7	23.6	12.2
19.0	21.2	17.3
20.3	19.8	15.1
19.4	22.6	11.3

15.27 The Kruskal–Wallis statistic is $H = \frac{12}{n(n+1)} \sum_{i=1}^{k} n_i \left(\overline{R}_i - \frac{n+1}{2} \right)^2$. Perform the indicated squaring of each term in the sum and add the resulting values to show that $H = \frac{12}{n(n+1)} \sum_{i=1}^{k} \frac{R_i^2}{n_i} - 3(n+1)$. [*Hint:* Recall that $\overline{R}_i = R_i/n_i$ and that $\sum_{i=1}^{k} R_i = $ sum of the first n integers $= n(n+1)/2$.]

15.28 Assuming no ties, obtain the exact null distribution of the Kruskal–Wallis H statistic for the case $k = 3$, $n_1 = n_2 = n_3 = 2$. (Because the sample sizes are all equal, if ranks 1 and 2 are assigned to treatment 1, ranks 3 and 4 are assigned to treatment 2, and ranks 5 and 6 are assigned to treatment 3, the value of H is exactly the same as if ranks 3 and 4 are assigned to treatment 1, ranks 5 and 6 are assigned to treatment 2, and ranks 1 and 2 are assigned to treatment 3. That is, for any particular set of ranks, we may interchange the roles of the k populations and obtain the same values of the H statistic. Thus, the number of cases that we must consider can be reduced by a factor of $1/k!$. Consequently, H must be evaluated only for $(6!/[2! \cdot 2! \cdot 2!])/3! = 15$ distinct arrangements of ranks.)

15.8 The Friedman Test for Randomized Block Designs

In Section 12.4, we discussed the merits of a randomized block design for an experiment to compare the performance of several treatments. We assume that b blocks are used in the experiment, which is designed to compare the locations of the distributions of the responses corresponding to each of k treatments. The analysis of variance, discussed in Section 13.9, was based on the assumptions that the observations in each block-treatment combination were normally distributed with equal variances. As in the case of the one-way layout, SST was the key quantity in the analysis.

The Friedman test, developed by Nobel Prize–winning economist Milton Friedman (1937), is designed to test the null hypothesis that the probability distributions of the k treatments are identical versus the alternative that at least two of the distributions differ in location. The test is based on a statistic that is a rank analogue of SST for the randomized block design (see Section 13.9) and is computed in the following manner. After the data from a randomized block design are obtained, the observed values of the responses to each of the k treatments *for each block* are ranked from 1 (the smallest in the block) to k (the largest in the block). If two or more observations in the same block are tied for the same rank, then the average of the ranks that would have been assigned to these observations is assigned to each member of the tied group. However, ties need to be dealt with in this manner only if they occur within the same block.

Let R_i denote the sum of the ranks of the observations corresponding to treatment i, and let $\overline{R}_i = R_i/b$ denote the corresponding average of the ranks (recall that in a randomized block design, each treatment is applied exactly once in each block, resulting in a total of b observations per treatment and, hence, in a total of bk total observations). Because ranks of 1 to k are assigned within each block, the sum of the ranks assigned in each block is $1 + 2 + \cdots + k = k(k+1)/2$. Thus the sum of all the ranks assigned in the analysis is $bk(k+1)/2$. If $\overline{R}$ denotes the overall average of the ranks of all the bk observations, it follows that $\overline{R} = (k+1)/2$. Consider the rank analog of SST for a randomized block design given by

$$W = b \sum_{i=1}^{k} (\overline{R}_i - \overline{R})^2.$$

If the null hypothesis is true and the probability distributions of the treatment responses do not differ in location, we expect the $\overline{R}_i$ values to be approximately equal and the resulting value for W to be small. If the alternative hypothesis were true, we would expect this to lead to differences among the $\overline{R}_i$ values and corresponding large values of W. Instead of W, Friedman considered the statistic $F_r = 12W/[k(k+1)]$, which may be rewritten (see Exercise 15.34) as

$$F_r = \frac{12}{bk(k+1)} \sum_{i=1}^{k} R_i^2 - 3b(k+1).$$

As previously noted, the null hypothesis of equal locations is rejected in favor of the alternative that the treatment distributions differ in location for larger values of F_r. That is, the corresponding α-level test rejects the null hypothesis in favor of the alternative if $F_r > f(\alpha)$ where $f(\alpha)$ is such that, when H_0 is true, $P[F_r > f(\alpha)] = \alpha$.

If there are no ties among the observations within the blocks, the null distribution of F_r can (tediously) be found by using the methods of Chapter 2. For any values of b and k, the distribution of F_r is found as follows. If the null hypothesis is true, then each of the $k!$ permutations of the ranks $1, 2, \ldots, k$ within each block is equally likely. Further, because we assume that the observations in different blocks are mutually independent, it follows that each of the $(k!)^b$ possible combinations of the b sets of permutations for the within-block ranks are equally likely when H_0 is true. Consequently, we can evaluate the value of F_r for each possible case and thereby give the null distribution of F_r (see Exercise 15.35). Selected values for $f(\alpha)$ for various choices of k and b are given in Table A.22 of Hollander and Wolfe (1999). Like the other nonparametric procedures discussed in this chapter, the real advantage of this procedure is that it can be used regardless of the form of the actual distributions of the populations corresponding to the treatments.

As with the Kruskal–Wallis statistic, the null distribution of the Friedman F_r statistic can be approximated by a chi-square distribution with $k-1$ degrees of freedom as long as b is "large." Empirical evidence indicates that the approximation is adequate if either b (the number of blocks) or k (the number of treatments) exceeds 5. Again, our examples and exercises deal with situations where this large-sample approximation is adequate. If you need to implement a Friedman analysis for small samples, refer to Hollander and Wolfe (1999) to obtain appropriate critical values.

Friedman Test Based on F_r for a Randomized Block Design

Null hypothesis: H_0: The probability distributions for the k treatments are identical.

Alternative hypothesis: H_a: At least two of the distributions differ in location.

Test statistic: $F_r = \dfrac{12}{bk(k+1)} \displaystyle\sum_{i=1}^{k} R_i^2 - 3b(k+1)$

where

$$b = \text{number of blocks}$$

$$k = \text{number of treatments}$$

$$R_i = \text{sum of the ranks for the } i\text{th treatment,}$$
where the rank of each measurement is computed relative to its size within its own block.

Rejection region: $F_r > \chi_\alpha^2$ with $(k-1)$ degrees of freedom.

Assumptions: The treatments are randomly assigned to experimental units within blocks. Either the number of blocks (b) or the number of treatments (k) exceeds 5.

EXAMPLE **15.7**   An experiment to compare completion times for three technical tasks was performed in the following manner. Because completion times may vary considerably from person to person, each of the six technicians was asked to perform all three tasks. The tasks were presented to each technician in a random order with suitable time lags between the tasks. Do the data in Table 15.7 present sufficient evidence to indicate that the distributions of completion times for the three tasks differ in location? Use $\alpha = .05$. Give bounds for the associated p-value.

Solution   The experiment was run according to a randomized block design with technicians playing the role of blocks. In this case $k = 3$ treatments are compared using $b = 6$ blocks. Because the number of blocks exceeds 5, we may use the Friedman analysis and compare the value of F_r to χ_α^2, based on $k - 1 = 2$ degrees of freedom. Consulting Table 6, Appendix III, we find $\chi_{.05}^2 = 5.99147$. For the data given in Table 15.7,

Table 15.7. Completion times for three tasks

Technician	Task A	Rank	Task B	Rank	Task C	Rank
1	1.21	1	1.56	3	1.48	2
2	1.63	1.5	2.01	3	1.63	1.5
3	1.42	1	1.70	2	2.06	3
4	1.16	1	1.27	2.5	1.27	2.5
5	2.43	2	2.64	3	1.98	1
6	1.94	1	2.81	3	2.44	2
		$R_1 = 7.5$		$R_2 = 16.5$		$R_3 = 12$

$$F_r = \frac{12}{6(3)(4)}[(7.5)^2 + (16.5)^2 + (12)^2] - 3(6)(4) = 6.75.$$

Because $F_r = 6.75$, which exceeds 5.99147, we conclude at the $\alpha = .05$ level that the completion times of at least two of the three tasks possess probability distributions that differ in location.

Because $F_r = 6.75$ is the observed value of a statistic that has approximately a χ^2 distribution with 2 degrees of freedom, it follows that (approximately) $.025 < p$-value $< .05$.

It can be seen (see Exercise 15.33) that, if we wish to compare only $k = 2$ treatments using a randomized block design (so that the blocks would be of size 2), the Friedman statistic is the square of the standardized sign statistic (that is, the square of the Z statistic given in Section 15.3). Thus, for $k = 2$, the Friedman analysis is equivalent to a two-tailed sign test.

Exercises

15.29 Corrosion of metals is a problem in many mechanical devices. Three sealant used to help retard the corrosion of metals were tested to see whether there were any differences among them. Samples from ten different ingots of the same metal composition were treated with each of the three sealant and the amount of corrosion was measured after exposure to the same environmental conditions for 1 month. The data are given in the accompanying table. Is there any evidence of a difference in the abilities of the sealant to prevent corrosion? Test using $\alpha = .05$.

	Sealant		
Ingot	I	II	III
1	4.6	4.2	4.9
2	7.2	6.4	7.0
3	3.4	3.5	3.4
4	6.2	5.3	5.9
5	8.4	6.8	7.8
6	5.6	4.8	5.7
7	3.7	3.7	4.1
8	6.1	6.2	6.4
9	4.9	4.1	4.2
10	5.2	5.0	5.1

15.30 A serious drought-related problem for farmers is the spread of aflatoxin, a highly toxic substance caused by mold, which contaminates field corn. In higher levels of contamination, aflatoxin is hazardous to animal and possibly human health. (Officials of the FDA have set a maximum limit of 20 parts per billion aflatoxin as safe for interstate marketing.) Three sprays, A, B, and C, have been developed to control aflatoxin in field corn. To determine whether differences exist among the sprays, ten ears of corn are randomly chosen from a contaminated corn field, and each is divided into three pieces of equal size. The sprays are then randomly assigned to the pieces for each ear of corn, thus setting up a randomized block design. The accompanying table gives the amount (in parts per billion) of aflatoxin present in the corn samples after spraying. Use the Friedman test based on F_r to determine whether there are differences among the sprays for control of aflatoxin. Give approximate bounds for the p-value.

	Spray				Spray		
Ear	A	B	C	Ear	A	B	C
1	21	23	15	6	5	12	6
2	29	30	21	7	18	18	12
3	16	19	18	8	26	32	21
4	20	19	18	9	17	20	9
5	13	10	14	10	4	10	2

15.31 A study was performed to compare the preferences of eight "expert listeners" regarding 15 models (with approximately equal list prices) of a particular component in a stereo system. Every effort was made to ensure that differences perceived by the listeners were due to the particular component of interest and no other cause (all of the other components in the system were identical, the same type of music was used, the music was played in the same room, etc.). Thus the results of the listening test reflect the audio preferences of the judges and not judgments regarding quality, reliability, or other variables. Further, the results pertain only to the models of the components used in the study and not to any other models that may be offered by the various manufacturers. The data in the accompanying table give the results of the listening tests. The models are depicted simply as models $A, B, \ldots, O$. Under each column heading are the numbers of judges who ranked each brand of component from 1 (lowest rank) to 15 (highest rank).

							Rank								
Model	1	2	3	4	5	6	7	8	9	10	11	12	13	14	15
A	0	0	0	0	0	0	0	0	0	0	0	0	0	0	8
B	0	0	0	1	0	2	1	1	1	0	0	0	0	2	0
C	0	1	1	1	4	0	0	1	0	0	0	0	0	0	0
D	1	0	1	1	0	1	0	0	1	0	1	0	1	1	0
E	0	2	1	3	0	2	0	0	0	0	0	0	0	0	0
F	0	0	0	0	0	0	0	0	1	2	2	3	0	0	0
G	0	0	0	0	0	0	0	0	0	1	0	2	4	1	0
H	1	2	1	1	0	0	2	1	0	0	0	0	0	0	0
I	3	2	1	0	0	0	0	0	0	1	0	1	0	0	0
J	0	0	1	0	2	0	2	0	0	0	2	0	1	0	0
K	0	0	0	0	0	0	1	1	0	2	1	1	1	1	0
L	0	0	0	0	0	0	1	1	4	0	1	0	1	0	0
M	1	1	2	1	1	2	0	0	0	0	0	0	0	0	0
N	2	0	0	0	0	0	1	1	0	0	0	1	0	3	0
O	0	0	0	0	1	1	0	2	1	2	1	0	0	0	0

a Use the Friedman procedure to test whether the distributions of the preference scores differ in location for the 15 component models. Give bounds for the attained significance level. What would you conclude at the $\alpha = .01$ level of significance? (*Hint:* The sum of the ranks associated with the component of "model O" is $5 + 6 + 8 + 8 + 9 + 10 + 10 + 11 = 67$; other rank sums can be computed in an analogous manner.)

b If, prior to running the experiment, we desired to compare components of models H and G, this comparison could be made by using the sign test presented in Section 15.3. Using the information just given, we can determine that model G was preferred to model H by all eight judges. Explain why. Give the attained significance level if the sign test is used to compare these two brands of speakers.

c Explain why there is not enough information given to use the sign test in a comparison of only models H and M.

15.32 An experiment is conducted to investigate the toxic effect of three chemicals, A, B, and C, on the skin of rats. Three adjacent 1-inch squares are marked on the backs of eight rats and each of the three chemicals is applied to each rat. The squares of skin are then scored from 0 to 10, depending on the degree of irritation. The resulting data are given in the accompanying table. Is there sufficient evidence to support the research hypothesis that the probability distributions of skin irritation scores corresponding to the three chemicals differ in location? Use $\alpha = .01$.

	Chemical		
Rat	A	B	C
1	6	5	3
2	9	8	4
3	6	9	3
4	5	8	6
5	7	8	9
6	5	7	6
7	6	7	5
8	6	5	7

15.33 Consider the Friedman statistic F_r when $k = 2$ and $b = $ (number of blocks) $= n$. Then $F_r = \frac{2}{n}(R_1^2 + R_2^2) - 9n$. Let M be the number of blocks (pairs) in which treatment one has rank 1. If there are no ties, then treatment 1 has rank 2 in the remaining $n - M$ pairs. Thus, $R_1 = M + 2(n - M) = 2n - M$. Analogously, $R_2 = n + M$. Substitute these values into the preceding expression for F_r, and show that the resulting value is $4(M - .5n)^2/n$. Compare this result with the square of the Z statistic in Section 15.3. This procedure demonstrates that $F_r = Z^2$.

15.34 Consider the Friedman statistic $F_r = \frac{12b}{k(k+1)} \sum_{i=1}^{k} (\overline{R}_i - \overline{R})^2$. Square each term in the sum, and show that an alternative form of F_r is $F_r = \frac{12}{bk(k+1)} \sum_{i=1}^{k} R_i^2 - 3b(k + 1)$. [*Hint:* Recall that $\overline{R}_i = R_i/b$, $\overline{R} = (k + 1)/2$ and note that $\sum_{i=1}^{k} R_i = $ sum of all of the ranks $= bk(k + 1)/2$].

15.35 If there are no ties and $b = 2, k = 3$, derive the exact null distribution of F_r.

15.9 The Runs Test: A Test for Randomness

Consider a production process in which manufactured items emerge in sequence and each is classified as either defective (D) or nondefective (N). We have studied how we might compare the fraction defective over two equal time intervals by using the normal deviate test (Chapter 10) and extended this to a test of a hypothesis of constant p over two or more time intervals, using the chi-square test of Chapter 14. The purpose of these tests was to detect a change or trend in the fraction defective, p. Evidence to indicate an increasing fraction defective might indicate the need for a process study to locate the source of difficulty. A decreasing value might suggest that a process quality control program was having a beneficial effect in reducing the fraction defective.

Trends in the fraction of defective items (or other quality measures) are not the only indication of lack of process control. A process might be causing periodic runs of defective items even though the average fraction of defective items remains constant, for all practical purposes, over long periods of time. For example, spotlight bulbs are manufactured on a rotating machine with a fixed number of positions for bulbs. A bulb is placed on the machine at a given position, the air is removed, gases are pumped into the bulb, and the glass base is flame-sealed. If a machine contains 20 positions and several adjacent positions are faulty (perhaps due to too much heat used in the sealing process), surges of defective bulbs will emerge from the process in a periodic manner. Tests that compare the process fraction of defective items produced during equal intervals of time will not detect this periodic difficulty in the process. This periodicity, indicated by runs of defectives, indicates nonrandomness in the occurrence of defective items over time and can be detected by a *test for randomness*. The statistical test we present, known as the *runs test*, is discussed in detail by Wald and Wolfowitz (1940). Other practical applications of the runs test will follow.

As the name implies, the runs test is used to study a sequence of events where each element in the sequence can assume one of two outcomes, success (S) or failure (F). If we think of the sequence of items emerging from a manufacturing process as defective (F) or nondefective (S), the observation of twenty items might yield

$$S \quad S \quad S \quad S \quad S \quad F \quad F \quad S \quad S \quad S$$
$$F \quad F \quad F \quad S \quad S \quad S \quad S \quad S \quad S \quad S.$$

We notice the groupings of defectives and nondefectives and wonder whether this grouping implies nonrandomness and, consequently, lack of process control.

DEFINITION 15.1   A *run* is a maximal subsequence of like elements.

For example, the first five successes constitute a maximal subsequence of five like elements (that is, it includes the maximum number of like elements before encountering an F. (The first four elements form a subsequence of like elements, but it is not maximal because the fifth element also could be included.) Consequently, the 20 elements are arranged in five runs, the first containing five Ss, the second containing two Fs, and so on.

A very small or very large number of runs in a sequence indicates nonrandomness. Therefore, let R (the number of runs in a sequence) be the test statistic, and let the rejection region be $R \le k_1$ and $R \ge k_2$, as indicated in Figure 15.3. We must then

FIGURE **15.3**
The rejection region
for the runs test

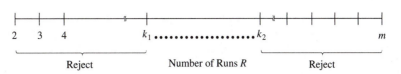

FIGURE **15.4**
The distribution of
n_1 S elements in y_1
cells (none empty)

$$|S|SSSS|SS\ldots|SS|SSS|S|$$

find the probability distribution for R, $P(R = r)$, in order to calculate α and to locate a suitable rejection region for the test.

Suppose that the complete sequence contains n_1 S elements and n_2 F elements, resulting in Y_1 runs of Ss and Y_2 runs of Fs, where $(Y_1 + Y_2) = R$. Then, for a given Y_1, Y_2 can equal Y_1, $(Y_1 - 1)$, or $(Y_1 + 1)$. Let m denote the maximum possible number of runs. Notice that $m = 2n_1$ if $n_1 = n_2$, and that $m = (2n_1 + 1)$ if $n_1 < n_2$. We will suppose that every distinguishable arrangement of the $(n_1 + n_2)$ elements in the sequence constitutes a simple event for the experiment and that the sample points are equiprobable. It then remains for us to count the number of sample points that imply R runs.

The total number of distinguishable arrangements of n_1 S elements and n_2 F elements is

$$\binom{n_1 + n_2}{n_1}$$

and therefore the probability per sample point is

$$\frac{1}{\binom{n_1 + n_2}{n_1}}.$$

The number of ways of achieving y_1 S runs is equal to the number of distinguishable arrangements of n_1 indistinguishable elements in y_1 cells, none of which is empty, as represented in Figure 15.4. This is equal to the number of ways of distributing the $(y_1 - 1)$ identical inner bars in the $(n_1 - 1)$ spaces between the S elements (the outer two bars remain fixed). Consequently, it is equal to the number of ways of selecting $(y_1 - 1)$ spaces (for the bars) out of the $(n_1 - 1)$ spaces available, or

$$\binom{n_1 - 1}{y_1 - 1}.$$

The number of ways of observing y_1 S runs and y_2 F runs, obtained by applying the mn rule, is

$$\binom{n_1 - 1}{y_1 - 1}\binom{n_2 - 1}{y_2 - 1}.$$

This gives the number of sample points in the event "y_1 runs of Ss and y_2 runs of Fs." Then multiplying this number by the probability per sample point, we obtain the probability of exactly y_1 runs of Ss and y_2 runs of Fs:

$$p(y_1, y_2) = \frac{\binom{n_1 - 1}{y_1 - 1}\binom{n_2 - 1}{y_2 - 1}}{\binom{n_1 + n_2}{n_1}}.$$

Then $P(R = r)$ equals the sum of $p(y_1, y_2)$ over all values of y_1 and y_2 such that $(y_1 + y_2) = r$.

To illustrate the use of the formula, the event $R = 4$ could occur when $y_1 = 2$ and $y_2 = 2$ with either the S or F elements commencing the sequences. Consequently,

$$P(R = 4) = 2P(Y_1 = 2, Y_2 = 2).$$

On the other hand, $R = 5$ could occur when $y_1 = 2$ and $y_2 = 3$, or when $y_1 = 3$ and $y_2 = 2$, and these occurrences are mutually exclusive. Then

$$P(R = 5) = P(Y_1 = 3, Y_2 = 2) + P(Y_1 = 2, Y_2 = 3).$$

EXAMPLE 15.8   Suppose that a sequence consists of $n_1 = 5$ S elements and $n_2 = 3$ F elements. Calculate the probability of observing $R = 3$ runs. Also, calculate $P(R \leq 3)$.

Solution   Three runs could occur when $y_1 = 2$ and $y_2 = 1$, or when $y_1 = 1$ and $y_2 = 2$. Then

$$P(R = 3) = P(Y_1 = 2, Y_2 = 1) + P(Y_1 = 1, Y_2 = 2)$$

$$= \frac{\binom{4}{1}\binom{2}{0}}{\binom{8}{5}} + \frac{\binom{4}{0}\binom{2}{1}}{\binom{8}{5}} = \frac{4}{56} + \frac{2}{56} = .107.$$

Next we require that $P(R \leq 3) = P(R = 2) + P(R = 3)$. Accordingly,

$$P(R = 2) = 2P(Y_1 = 1, Y_2 = 1) = (2)\frac{\binom{4}{0}\binom{2}{0}}{\binom{8}{5}} = \frac{2}{56} = .036.$$

Thus the probability of 3 or fewer runs is $.107 + .036 = .143$.

The values of $P(R \leq a)$ are given in Table 10, Appendix III, for all combinations of n_1 and n_2 where n_1 and n_2 are less than or equal to 10. These can be used to locate the rejection regions of one- or two-tailed tests. We will illustrate with an example.

EXAMPLE **15.9**   A true–false examination was constructed with the answers running in the following sequence:

$$T \ F \ F \ T \ F \ T \ F \ T \ T \ F \ T \ F \ F \ T \ F \ T \ F \ T \ T \ F.$$

Does this sequence indicate a departure from randomness in the arrangement of T and F answers?

Solution   The sequence contains $n_1 = 10$ T and $n_2 = 10$ F answers, with $y = 16$ runs. Nonrandomness can be indicated by either an unusually small or an unusually large number of runs; consequently, we will be using a two-tailed test.

Suppose that we wish to use α approximately equal to .05 with .025 or less in each tail of the rejection region. Then from Table 10, Appendix III, with $n_1 = n_2 = 10$, we see that $P(R \leq 6) = .019$ and $P(R \leq 15) = .981$. Then $P(R \geq 16) = 1 - P(R \leq 15) = .019$, and we would reject the hypothesis of randomness at the $\alpha = .038$ significance level if $R \leq 6$ or $R \geq 16$. Because $R = 16$ for the observed data, we conclude that evidence exists to indicate nonrandomness in the professor's arrangement of answers. The attempt to mix the answers was overdone.

A second application of the runs test is in detecting nonrandomness of a sequence of quantitative measurements over time. These sequences, known as *time series*, occur in many fields. For example, the measurement of a quality characteristic of an industrial product, blood pressure of a human, and the price of a stock on the stock market all vary over time. Departures from randomness in a series, caused either by trends or periodicities, can be detected by examining the deviations of the time series measurements from their average. Negative and positive deviations could be denoted by S and F, respectively, and we could then test this time sequence of deviations for nonrandomness. We will illustrate with an example.

EXAMPLE **15.10**   Paper is produced in a continuous process. Suppose that a brightness measurement Y is made on the paper once every hour and that the results appear as shown in Figure 15.5.

FIGURE **15.5**
Paper brightness
versus time

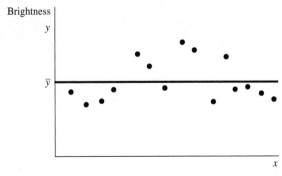

The average $\bar{y}$ for the 15 sample measurements appears as shown. Notice the deviations about $\bar{y}$. Do these data indicate a lack of randomness and thereby suggest periodicity and lack of control in the process?

Solution   The sequence of negative (S) and positive (F) deviations as indicated in Figure 15.5 is

$$S\ S\ S\ S\ F\ F\ S\ F\ F\ S\ F\ S\ S\ S\ S.$$

Then $n_1 = 10$, $n_2 = 5$, and $R = 7$. Consulting Table 10 in Appendix III, we find $P(R \leq 7) = .455$. This value of R is not improbable, assuming the hypothesis of randomness to be true. Consequently, there is not sufficient evidence to indicate nonrandomness in the sequence of brightness measurements.

The runs test can also be used to compare two population frequency distributions for a two-sample unpaired experiment. Thus it provides an alternative to the Mann–Whitney U test (Section 15.6). If the measurements for the two samples are arranged in order of magnitude, they form a sequence. The measurements for samples 1 and 2 can be denoted as S and F, respectively, and once again we are concerned with a test for randomness. If all measurements for sample 1 are smaller than those for sample 2, the sequence will result in $SSSS \ldots SFFF \ldots F$, or $R = 2$ runs. A small value of R provides evidence of a difference in population frequency distributions, and the rejection region chosen is $R \leq a$. This rejection region implies a one-tailed statistical test. An illustration of the application of the runs test to compare two population frequency distributions is left as an exercise.

As in the case of the other nonparametric test statistics studied in earlier sections of this chapter, the probability distribution for R tends toward normality as n_1 and n_2 become large. The approximation is good when n_1 and n_2 are both greater than 10. Consequently, we may use the Z statistic as a large-sample test statistic, where

$$Z = \frac{R - E(R)}{\sqrt{V(R)}}$$

and

$$E(R) = \frac{2n_1 n_2}{n_1 + n_2} + 1$$

$$V(R) = \frac{2n_1 n_2 (2n_1 n_2 - n_1 - n_2)}{(n_1 + n_2)^2 (n_1 + n_2 - 1)}$$

are the expected value and variance of R, respectively. The rejection region for a two-tailed test, with $\alpha = .05$, is $|z| \geq 1.96$. If α is the desired probability of a type I error, for an upper-tail test, we reject the null hypothesis if $z > z_\alpha$ (for a lower-tail test, we reject H_0 if $z < -z_\alpha$).

Exercises

15.36 Consider a runs test based on $n_1 = n_2 = 5$ elements. Assuming H_0 to be true, use Table 10, Appendix III, to find the following:

   a $P(R = 2)$.

   b $P(R \leq 3)$.

   c $P(R \leq 4)$.

15.37 A union supervisor claims that applicants for jobs are selected without regard to race. The hiring records of the local—one that contains all male members—gave the following sequence of white and black hirings:

W W W W B W W W B B W B B.

Do these data suggest a nonrandom racial selection in the hiring of the union's members?

15.38 The conditions (D for diseased, S for sound) of the individual trees in a row of ten poplars were found to be, from left to right:

S S D D S D D D S S.

Is there sufficient evidence to indicate nonrandomness in the sequence and, therefore, the possibility of contagion?

15.39 Items emerging from a continuous production process were classified as defective or nondefective. A sequence of items, observed over time, was as follows:

D N N N N N D D N N N N N N D D
D N N N N N D N N N D D N N N D D.

   a   Compute the probability that $R \leq 11$, where $n_1 = 11$ and $n_2 = 23$.

   b   Do these data suggest lack of randomness in the occurrence of defectives (D) and nondefectives (N)? Use the large-sample approximation for the runs test.

15.40 A quality control chart has been maintained for a measurable characteristic of items taken from a conveyor belt at a fixed point in a production line. The measurements obtained today, in order of time, are as follows:

68.2	71.6	69.3	71.6	70.4	65.0	63.6	64.7
65.3	64.2	67.6	68.6	66.8	68.9	66.8	70.1.

   a   Classify the measurements in this time series as above or below the sample mean, and determine (using the runs test) whether consecutive observations suggest lack of stability in the production process.

   b   Divide the time period into two equal parts and compare the means, using Student's t test. Do the data provide evidence of a shift in the mean level of the quality characteristics?

15.41 Refer to Exercise 15.18. Use the runs test to analyze the data. Compare your answer here with your answer to Exercise 15.18.

15.42 Refer to Exercise 15.19. If, indeed, the experimental batteries have a greater mean life, what would be the effect of this on the expected number of runs? Using the large-sample theory for the runs test, test (using $\alpha = .05$) whether there is a difference in the distributions of battery life for the two populations. Give the approximate p-value.

15.10 Rank Correlation Coefficient

In the preceding sections we used ranks to indicate the relative magnitude of observations in nonparametric tests for comparison of treatments. We now employ the same technique in testing for a correlation between two ranked variables. Two common rank correlation coefficients are Spearman's statistic, r_S, and Kendall's τ. We will present the Spearman r_S because its computation is identical to that for the sample correlation coefficient r of Section 11.8. Kendall's rank correlation coefficient is discussed in detail in Kendall and Stuart (1979).

Suppose that eight elementary-science teachers have been ranked by a judge according to their teaching ability, and all have taken a national teachers examination. The data are given in Table 15.8. Do the data suggest agreement between the judge's ranking and the examination score? Alternatively, we might express this question by asking whether a correlation exists between the judge's ranking and the examination scores.

The two variables of interest are rank and test score. The former is already in rank form, and the test scores may be ranked similarly, as shown in parentheses in Table 15.8. The ranks for tied observations are obtained by averaging the ranks that the tied observations would occupy, as is done for the Mann–Whitney U statistic.

The *Spearman rank correlation coefficient, r_S*, is calculated by using the ranks as the paired measurements on the two variables X and Y in the formula for r in Chapter 11. Thus,

$$r_S = \frac{S_{xy}}{\sqrt{S_{xx}S_{yy}}} = \frac{\sum_{i=1}^{n} x_i y_i - \frac{1}{n}\left(\sum_{i=1}^{n} x_i\right)\left(\sum_{i=1}^{n} y_i\right)}{\sqrt{\left[\sum_{i=1}^{n} x_i^2 - \frac{1}{n}\left(\sum_{i=1}^{n} x_i\right)^2\right]\left[\sum_{i=1}^{n} y_i^2 - \frac{1}{n}\left(\sum_{i=1}^{n} y_i\right)^2\right]}}.$$

When there are no ties in either the x observations or the y observations, this expression for r_S algebraically reduces to a simpler expression:

Table 15.8. Data for science teachers

Teacher	Judge's Rank	Examination Score
1	7	44 (1)
2	4	72 (5)
3	2	69 (3)
4	6	70 (4)
5	1	93 (8)
6	3	82 (7)
7	8	67 (2)
8	5	80 (6)

$$r_S = 1 - \frac{6 \sum\limits_{i=1}^{n} d_i^2}{n(n^2 - 1)} \quad \text{where} \quad d_i = x_i - y_i.$$

If the number of ties is small in comparison with the number of data pairs, little error will result from using this shortcut formula. We will leave proof of this simplification as an exercise and will illustrate the use of the formula by an example.

EXAMPLE **15.11**   Calculate r_S for the judge's ranking and examination score data from Table 15.8.

Solution   The differences and squares of differences between the two rankings are shown in Table 15.9.

Substituting into the formula for r_S, we obtain

$$r_S = 1 - \frac{6 \sum\limits_{i=1}^{n} d_i^2}{n(n^2 - 1)} = 1 - \frac{6(144)}{8(64 - 1)} = -.714.$$

The Spearman rank correlation coefficient may be employed as a test statistic to test a hypothesis of no association between two populations. We assume that the n pairs of observations (x_i, y_i) have been randomly selected and, therefore, the absence of any association between the populations implies a random assignment of the n ranks within each sample. Each random assignment (for the two samples) represents a sample point associated with the experiment, and a value of r_S can be calculated for each. It is possible to calculate the probability that r_S assumes a large absolute value due solely to chance and thereby suggests an association between populations when none exists.

Table 15.9. Data and calculations for Example 15.11

Teacher	x_i	y_i	d_i	d_i^2
1	7	1	6	36
2	4	5	−1	1
3	2	3	−1	1
4	6	4	2	4
5	1	8	−7	49
6	3	7	−4	16
7	8	2	6	36
8	5	6	−1	1
Total				144

The rejection region for a two-tailed test includes values of r_S near $+1$ and near -1. If the alternative is that the correlation between X and Y is negative, we reject H_0 for values of r_S near -1. Similarly, if the alternative is that the correlation between X and Y is positive, we reject H_0 for large positive values of r_S.

The critical values of r_S are given in Table 11, Appendix III. Recorded across the top of the table are values of α that you might wish to use for a one-tailed test of the null hypothesis of no association between X and Y. The number of rank pairs n appears at the left side of the table. The table entries give the critical value r_0 for a one-tailed test. Thus, $P(r_S \geq r_0) = \alpha$. For example, suppose that you have $n = 8$ rank pairs and the research hypothesis is that the correlation between the ranks is positive. Then you want to reject the null hypothesis of no association only for large positive values of r_S, and you will use a one-tailed test. Referring to Table 11 and using the row corresponding to $n = 8$ and the column for $\alpha = .05$, you read $r_0 = .643$. Therefore, you reject H_0 for all values of r_S greater than or equal to .643.

If you wish to give the p-value associated with an observed value of $r = .82$, Table 11 gives that H_0 would be rejected with $\alpha = .025$ but not with $\alpha = .01$. Thus, $.01 < p$ value $< .025$.

The test is conducted in exactly the same manner if you wish to test the alternative hypothesis that the ranks are negatively correlated. The only difference is that you reject the null hypothesis if $r_S \leq -.643$. That is, you just place a minus sign in front of the tabulated value of r_0 to get the lower-tail critical value. Similarly, if $r = -.82$, then $.01 < p$ value $< .025$.

To conduct a two-tailed test, you reject the null hypothesis if $r_S \geq r_0$ or $r_S \leq -r_0$. The value of α for the test is double the value shown at the top of the table. For example, if $n = 8$ and you choose the .025 column, you reject H_0 if $r_S \geq .738$ or $r_S \leq -.738$. The α value for the test is $2(.025) = .05$.

The p-value associated with a two-tailed test based on an observed value of $r = .82$ is twice (because of the two tails) the one-tailed p-value; that is, $.02 < p$-value $< .05$.

EXAMPLE 15.12 Test the hypothesis of no association between populations for Example 15.11. Give bounds for the associated p-value.

Solution The critical value of r_S for a one-tailed test with $\alpha = .05$ and $n = 8$ is .643. Let us assume that a correlation between judge's rank and the teachers' examination scores could not possibly be positive. (Low rank means good teaching and should be associated with a high test score if the judge and the test both measure teaching ability.) The alternative hypothesis is that the population rank correlation coefficient, ρ_S, is less than zero, so we are concerned with a one-tailed statistical test. Thus α for the test is the tabulated value .05, and we reject the null hypothesis if $r_S \leq -.643$.

The calculated value of the test statistic, $r_S = -.714$, is less than the critical value for $\alpha = .05$. Because H_0 is rejected for $\alpha = .05$ but not for $\alpha = .025$, the p-value associated with the test lies in the interval $.025 < p$-value $< .05$. Hence, the null hypothesis is rejected at the $\alpha = .05$ level of significance. It appears that some agreement does exist between the judge's rankings and the test scores. However, this agreement could exist when *neither* provides an adequate yardstick for measuring teaching ability. For example, the association could exist if both the judge and those who constructed the teachers examination possessed a completely erroneous, but similar, concept of the characteristics of good teaching.

Spearman's Rank Correlation Test

Null hypothesis: H_0: There is no association between the rank pairs.

Alternative hypothesis: H_a: There is an association between the rank pairs (a two-tailed test); or

                    H_a: The correlation between the rank pairs is positive (or negative) (a one-tailed test).

Test statistic:

$$r_S = \frac{n \sum_{i=1}^{n} x_i y_i - \left(\sum_{i=1}^{n} x_i \right) \left(\sum_{i=1}^{n} y_i \right)}{\sqrt{\left[n \sum_{i=1}^{n} x_i^2 - \left(\sum_{i=1}^{n} x_i \right)^2 \right] \left[n \sum_{i=1}^{n} y_i^2 - \left(\sum_{i=1}^{n} y_i \right)^2 \right]}}$$

where x_i and y_i represent the ranks of the ith pair of observations.

Rejection region: For a two-tailed test, reject H_0 if $r_S \geq r_0$ or $r_S \leq -r_0$, where r_0 is given in Table 11, Appendix III. Double the tabulated probability to obtain the value of α for the two-tailed test. For a one-tailed test, reject H_0 if $r_S \geq r_0$ (for an upper-tailed test) or $r_S \leq -r_0$ (for a lower-tailed test). The α value for a one-tailed test is the value shown in Table 11, Appendix III.

Exercises

15.43   An experiment was conducted to study the relationship between the ratings of tobacco leaf graders and the moisture content of the corresponding tobacco leaves. Twelve leaves were rated by the grader on a scale of 1 to 10, and corresponding measurements on moisture content were made on the same leaves. The data are shown in the following table.

Calculate r_S. Do the data provide sufficient evidence to indicate an association between the grader's rating and the moisture content of the leaves?

Leaf	Grader's Rating	Moisture Content
1	9	.22
2	6	.16
3	7	.17
4	7	.14
5	5	.12
6	8	.19
7	2	.10
8	6	.12
9	1	.05
10	10	.20
11	9	.16
12	3	.09

15.44   Manufacturers of perishable foods often use preservatives to retard spoilage. One concern is that too much preservative will change the flavor of the food. An experiment is conducted using portions of food products with varying amounts of preservative added. The length of time until the food begins to spoil and a taste rating are recorded for each portion of food. The taste rating is the average rating for three tasters, each of whom rated each food portion on a scale from 1 (bad) to 5 (good). Twelve measurements are shown in the following table. Use a nonparametric test to determine whether spoilage times and taste ratings are correlated. Give the associated p value, and indicate the appropriate conclusion for an $\alpha = .05$–level test.

Food Portion	Days Until Spoilage	Taste Rating
1	30	4.3
2	47	3.6
3	26	4.5
4	94	2.8
5	67	3.3
6	83	2.7
7	36	4.2
8	77	3.9
9	43	3.6
10	109	2.2
11	56	3.1
12	70	2.9

15.45   A large corporation selects graduates for employment by using both interviews and a psychological achievement test. Interviews conducted at the home office of the company were far more expensive than the test, which could be conducted on campus. Consequently, the personnel office was interested in determining whether the test scores were correlated with interview ratings and whether the tests could be substituted for interviews. The idea was not to eliminate interviews but to reduce their number. Ten prospects were ranked during interviews and then tested. The paired scores were as shown in the accompanying table.

a   Calculate the Spearman rank correlation coefficient, r_S. Rank 1 is assigned to the candidate judged to be the best.

Subject	Interview Rank	Test Score
1	8	74
2	5	81
3	10	66
4	3	83
5	6	66
6	1	94
7	4	96
8	7	70
9	9	61
10	2	86

b Do the data present sufficient evidence to indicate that the correlation between interview rankings and test scores is less than zero? If such evidence does exist, can we say that tests could be used to reduce the number of interviews?

15.46 A political scientist wished to examine the relationship of the voter image of a conservative political candidate and the distance in miles between the residence of the voter and the residence of the candidate. Each of 12 voters rated the candidate on a scale of 1 to 20. The resulting data are shown in the following table.

Voter	Rating	Distance
1	12	75
2	7	165
3	5	300
4	19	15
5	17	180
6	12	240
7	9	120
8	18	60
9	3	230
10	8	200
11	15	130
12	4	130

a Calculate the Spearman rank correlation coefficient, r_S.

b Do these data provide sufficient evidence to indicate a negative correlation between rating and distance?

15.47 Refer to Exercise 15.8. Compute Spearman's rank correlation coefficient for these data, and test $H_0 : \rho_S = 0$ at the 10% level of significance.

15.48 The data shown in the accompanying table give measures of bending and twisting stiffness as measured by engineering tests for 12 tennis racquets.

a Calculate the value of the rank correlation coefficient, r_S, between bending stiffness and twisting stiffness.

b Use the test based on the rank correlation coefficient to determine whether there is a significant positive relationship between bending stiffness and twisting stiffness. Use $\alpha = .05$.

Racquet	Bending Stiffness	Twisting Stiffness
1	419	227
2	407	231
3	363	200
4	360	211
5	257	182
6	622	304
7	424	384
8	359	194
9	346	158
10	556	225
11	474	305
12	441	235

15.49 Refer to Exercise 11.2. Regard both book and audited values as random variables, and test for positive correlation between the two by using Spearman's rank correlation coefficient. Give bounds for the p-value associated with the test.

15.50 Refer to Exercise 11.4. Treating both flow-through and static values as random variables, test for the presence of a correlation between the two by using Spearman's rank correlation coefficient, with $\alpha = .10$.

15.11 Some General Comments on Nonparametric Statistical Tests

The nonparametric statistical tests presented in the preceding pages represent only a few of the many nonparametric statistical methods of inference available. A much larger collection of nonparametric procedures, along with worked examples, is given in the texts listed in the references [for instance, see Conover (1999), Daniel (1990), Hollander and Wolfe (1999), and Siegel (1988)]. Nonparametric estimation procedures are available for, among other things, estimating location parameters and handling the inferential problems associated with the linear model.

We have indicated that nonparametric statistical procedures are particularly useful when experimental observations are susceptible to ordering but cannot be measured on a quantitative scale. Parametric statistical procedures can rarely be applied to this type of data. Hence any inferential procedures must be based on nonparametric methods.

A second application of nonparametric statistical methods is in testing hypotheses associated with populations of quantitative data when uncertainty exists concerning the satisfaction of assumptions about the form of the population distributions. Just how useful are nonparametric methods for this situation? Nonparametric statistical methods are rapid and often lead to an immediate decision in testing hypotheses. When experimental conditions depart substantially from the basic assumptions underlying parametric tests, the response measurements often can be transformed to alleviate the condition, but an unfortunate consequence frequently develops: the transformed response is no longer meaningful from a practical point of view, and analysis of the transformed data no longer answers the objectives of the experimenter. The use of nonparametric methods often circumvent this difficulty. Finally, notice that many nonparametric methods are nearly as efficient as their parametric counterparts when the assumptions underlying the parametric procedures are true; and as noted earlier, they could be more efficient when the assumptions are not satisfied. These reasons suggest that nonparametric techniques play a very useful role in statistical methodology.

References and Further Readings

Conover, W. J. 1999. *Practical Nonparametric Statistics*, 3d ed. New York: John Wiley.

Daniel, W. W. 1990. *Applied Nonparametric Statistics*, 2d ed. Boston: PWS–Kent.

Friedman, M. 1937. "The Use of Ranks to Avoid the Assumption of Normality Implicit in the Analysis of Variance," *Journal of the American Statistical Association* 32:675–701.

Hajek, J. 1969. *A Course in Nonparametric Statistics*. San Francisco: Holden-Day.

Hollander, M., and D. A. Wolfe. 1999. *Nonparametric Statistical Methods*, 2d ed. New York: John Wiley.

Kendall, M. G., and A. Stuart. 1979. *The Advanced Theory of Statistics*, 4th ed., vol. 2. New York: Hafner Press.

Kruskal, W. H., and W. A. Wallis. 1952. "Use of Ranks in One-Criterion Variance Analysis," *Journal of the American Statistical Association* 47:583–621.

Savage, I. R. 1953. "Bibliography of Nonparametric Statistics and Related Topics," *Journal of the American Statistical Association* 48:844–906.

Siegel, S. 1988. *Nonparametric Statistics for the Behavioral Sciences*. New York: McGraw-Hill.

Wald, A., and J. Wolfowitz. 1940. "On a Test Whether Two Samples Are from the Same Population," *Annals of Mathematical Statistics* 2:147–62.

Supplementary Exercises

15.51 Clinical data on the effectiveness of two drugs in treating a particular disease were collected from ten hospitals. The number of patients treated with the drugs varied from one hospital to another as well as between drugs within a given hospital. The data, expressed in terms of percentage recovery, are shown in the accompanying table. Do the data present sufficient evidence to indicate a higher recovery rate with one of the two drugs?

a Test by using the sign test. Choose your rejection region so that α is near .10.

b Why would it be inappropriate to use a Student's t test to analyze the data?

Hospital	Drug A			Drug B		
	Number in Group	Number Recovered	Percentage Recovered	Number in Group	Number Recovered	Percentage Recovered
1	84	63	75.0	96	82	85.4
2	63	44	69.8	83	69	83.1
3	56	48	85.7	91	73	80.2
4	77	57	74.0	47	35	74.5
5	29	20	69.0	60	42	70.0
6	48	40	83.3	27	22	81.5
7	61	42	68.9	69	52	75.4
8	45	35	77.8	72	57	79.2
9	79	57	72.2	89	76	85.4
10	62	48	77.4	46	37	80.4

15.52   Two gourmets rated 20 meals on a scale of 1 to 10. The data are shown in the accompanying table. Do the data provide sufficient evidence to indicate that one of the gourmets tends to give higher ratings than the other? Test by using the sign test with a value of α near .05.

Meal	A	B	Meal	A	B
1	6	8	11	6	9
2	4	5	12	8	5
3	7	4	13	4	2
4	8	7	14	3	3
5	2	3	15	6	8
6	7	4	16	9	10
7	9	9	17	9	8
8	7	8	18	4	6
9	2	5	19	4	3
10	4	3	20	5	5

15.53   Refer to the comparison of gourmet meal ratings in Exercise 15.52, and use the Wilcoxon signed-rank test to determine whether the data provide sufficient evidence to indicate a difference in the ratings of the two gourmets. Test by using a value of α near .05. Compare the results of this test with the results of the sign test in Exercise 15.52. Are the test conclusions consistent?

15.54   In an investigation of the visual scanning behavior of deaf children, measurements of eye-movement rates were taken on nine deaf and nine hearing children. From the data given in the

	Deaf Children A	Hearing Children B
	2.75 (15)	.89 (1)
	2.14 (11)	1.43 (7)
	3.23 (18)	1.06 (4)
	2.07 (10)	1.01 (3)
	2.49 (14)	.94 (2)
	2.18 (12)	1.79 (8)
	3.16 (17)	1.12 (5.5)
	2.93 (16)	2.01 (9)
	2.20 (13)	1.12 (5.5)
Rank Sum	126	45

table, is there sufficient evidence to justify claiming that the distributions of eye-movement rates differ for deaf children, A and hearing children, B?

15.55 A comparison of reaction (in seconds) to two different stimuli in a psychological word association experiment produced the results in the accompanying table when applied to a random sample of 16 people. Do the data present sufficient evidence to indicate a difference in location for the distributions of reaction times for the two stimuli? Use the Mann–Whitney U statistic and test with $\alpha = .05$. [*Note:* This test was conducted by using Student's t in Exercise 13.1. Compare your results.]

Stimulus 1	Stimulus 2
1	4
3	2
2	3
1	3
2	1
1	2
3	3
2	3

15.56 If (as in the case of measurements produced by two well-calibrated instruments) the means of two populations are equal, the Mann–Whitney U statistic can be used to test hypotheses concerning the population variances (or more general measures of variability) as follows. Rank the combined sample. Number the ranked observations from the outside in; that is, number the smallest observation 1; the largest 2; the next to smallest 3; the next to largest 4; and so on. This final sequence of numbers induces an ordering on the symbols A (population A items) and B (population B items). If $\sigma_A^2 > \sigma_B^2$, one would expect to find a preponderance of As with low ranks and thus a relatively small sum of ranks for the A observations.

a Given the measurements in the accompanying table, produced by well-calibrated precision instruments, A and B, test at near the $\alpha = .05$ level to determine whether the more expensive instrument, B is more precise than A. (Notice that this implies a one-tailed test.) Use the Mann–Whitney U test.

A	B
1060.21	1060.24
1060.34	1060.28
1060.27	1060.32
1060.36	1060.30
1060.40	

b Test by using the F statistic of Section 10.9.

15.57 Calculate the probability that $U \leq 2$ for $n_1 = n_2 = 5$. Assume that no ties occur and that H_0 is true.

15.58 Calculate the probability that the Wilcoxon T (Section 15.4) is less than or equal to 2 for $n = 3$ pairs. Assume that no ties occur and that H_0 is true.

15.59 To investigate possible differences among production rates for three production lines turning out similar items, examiners took independent random samples of total production figures for 7 days for each line. The resulting data appear in the following table.

Line 1	Line 2	Line 3
48	41	18
43	36	42
39	29	28
57	40	38
21	35	15
47	45	33
58	32	31

Do the data provide sufficient evidence to indicate any differences in location for the three sets of production figures, at the 5% significance level?

15.60 **a** Suppose that a company wants to study how personality relates to leadership. Four supervisors with different types of personalities are selected. Several employees are then selected from the group supervised by each, and these employees are asked to rate the leader of their group on a scale from 1 to 20 (20 signifies highly favorable). The accompanying table shows the resulting data. Is there sufficient evidence to indicate that one or more of the supervisors tend to receive higher ratings than the others? Use $\alpha = 0.05$.

	Supervisor		
I	II	III	IV
20	17	16	8
19	11	15	12
20	13	13	10
18	15	18	14
17	14	11	9
	16		10

b Suppose that the company is particularly interested in comparing the ratings of the personality types represented by supervisors I and III. Make this comparison, using $\alpha = .05$.

15.61 The leaders of a labor union want to determine its members' preferences before negotiating with management. Ten union members are randomly selected, and each member completed an extensive questionnaire. The responses to the various aspects of the questionnaire will enable the union to rank in order of importance the items to be negotiated. The sample rankings are shown in the accompanying table. Is there sufficient evidence to indicate that one or more of the items are preferred to the others? Test using $\alpha = .05$.

Person	More Pay	Job Stability	Fringe Benefits	Shorter Hours
1	2	1	3	4
2	1	2	3	4
3	4	3	2	1
4	1	4	2	3
5	1	2	3	4
6	1	3	4	2
7	2.5	1	2.5	4
8	3	1	4	2
9	1.5	1.5	3	4
10	2	3	1	4

15.62 Six groups of three children matched for IQ and age were formed. Each child was taught the concept of time by using one of three methods: lecture, demonstration, or teaching machine. The scores shown in the following table indicate the students' performance when they were tested on how well they had grasped the concept.

Group	Lecture	Demonstration	Teaching Machine
1	20	22	24
2	25	25	27
3	30	40	39
4	37	26	41
5	24	20	21
6	16	18	25

Is there sufficient evidence to indicate that the teaching methods differ in effectiveness? Give bounds for the p-value.

15.63 Calculate $P(R \leq 6)$ for the runs test, where $n_1 = n_2 = 8$ and H_0 is true.

15.64 Consider a Wilcoxon rank-sum test for the comparison of two probability distributions based on independent random samples of $n_1 = n_2 = 5$. Find $P(W \leq 17)$, assuming that H_0 is true.

*15.65 For the sample from population A, let U denote the Mann–Whitney statistic, and let W_A denote the Wilcoxon rank-sum statistic.[4] Show that

$$U = n_1 n_2 + (1/2)n_1(n_1 + 1) - W_A.$$

*15.66 Refer to Exercise 15.65.
  a Show that $E(U) = (1/2)n_1 n_2$ when H_0 is true.
  b Show that $V(U) = (1/12)[n_1 n_2(n_1 + n_2 + 1)]$ when H_0 is true, where H_0 states that the two populations are identical.

*15.67 Let T denote the Wilcoxon signed-rank test for n pairs of observations. Show that $E(T) = (1/4)n(n + 1)$ and $V(T) = (1/24)[n(n + 1)(2n + 1)]$ when the two populations are identical. Observe that these properties do not depend on whether T is constructed from negative or positive differences.

*15.68 Refer to the Spearman rank correlation coefficient of Section 15.10. Show that, when there are no ties in either the x observations or the y observations, then

$$r_S = \frac{n \sum_{i=1}^{n} x_i y_i - \left(\sum_{i=1}^{n} x_i \right)\left(\sum_{i=1}^{n} y_i \right)}{\sqrt{\left[n \sum_{i=1}^{n} x_i^2 - \left(\sum_{i=1}^{n} x_i \right)^2 \right]\left[n \sum_{i=1}^{n} y_i^2 - \left(\sum_{i=1}^{n} y_i \right)^2 \right]}}$$

$$= 1 - \frac{6 \sum_{i=1}^{n} d_i^2}{n(n^2 - 1)},$$

where $d_i = x_i - y_i$.

[4]Exercises preceded by an asterisk are optional.

APPENDIX **ONE**

Matrices and Other Useful Mathematical Results

A1.1   Matrices and Matrix Algebra

A1.2   Addition of Matrices

A1.3   Multiplication of a Matrix by a Real Number

A1.4   Matrix Multiplication

A1.5   Identity Elements

A1.6   The Inverse of a Matrix

A1.7   The Transpose of a Matrix

A1.8   A Matrix Expression for a System of Simultaneous Linear Equations

A1.9   Inverting a Matrix

A1.10 Solving a System of Simultaneous Linear Equations

A1.11 Other Useful Mathematical Results

A1.1 Matrices and Matrix Algebra

The following presentation represents a very elementary and condensed discussion of matrices and matrix operations. If you seek a more comprehensive introduction to the subject, consult the books listed in the references indicated at the end of Chapter 11.

We will define a *matrix* as a rectangular array (arrangement) of real numbers and will indicate specific matrices symbolically with bold capital letters. The number

in the matrix, *elements*, appear in specific row-column positions, all of which are
filled. The number of rows and columns may vary from one matrix to another, so
we conveniently describe the size of a matrix by giving its *dimensions*—that is, the
number of its rows and columns. Thus matrix $\mathbf{A}$

$$\underset{2\times3}{\mathbf{A}} = \begin{bmatrix} 6 & 0 & -1 \\ 4 & 2 & 7 \end{bmatrix}$$

possesses dimensions 2×3 because it contains two rows and three columns. Similarly, for

$$\underset{4\times1}{\mathbf{B}} = \begin{bmatrix} 1 \\ -3 \\ 0 \\ 7 \end{bmatrix} \quad \text{and} \quad \underset{2\times2}{\mathbf{C}} = \begin{bmatrix} 2 & 0 \\ -1 & 4 \end{bmatrix}$$

the dimensions of $\mathbf{B}$ and $\mathbf{C}$ are 4×1 and 2×2, respectively. Note that the row
dimension always appears first and that the dimensions may be written below the
identifying symbol of the matrix as indicated for matrices $\mathbf{A}, \mathbf{B}$, and $\mathbf{C}$.

As in ordinary algebra, an element of a matrix may be indicated by a symbol, *a, b,*
..., and its row-column position identified by means of a double subscript. Thus a_{21}
would be the element in the second row, first column. Rows are numbered in order
from top to bottom and columns from left to right. In matrix $\mathbf{A}$, $a_{21} = 4$, $a_{13} = -1$,
and so on.

Elements in a particular row are identified by their column subscript and hence
are numbered from left to right. The first element in a row is on the left. Likewise,
elements in a particular column are identified by their row subscript and therefore are
identified from the top element in the column to the bottom. For example, the first
element in column 2 of matrix $\mathbf{A}$ is 0, the second is 2. The first, second, and third
elements of row 1 are 6, 0, and -1, respectively.

The term *matrix algebra* involves, as the name implies, an algebra dealing with
matrices, much as the ordinary algebra deals with real numbers or symbols representing real numbers. Hence, we will wish to state rules for the addition and multiplication of matrices as well as to define other elements of an algebra. In so doing
we will point out the similarities as well as the dissimilarities between matrix and
ordinary algebra. Finally, we will use our matrix operations to state and solve a very
simple *matrix equation*. This, as you may suspect, will be the solution that we desire
for the least squares equations.

A1.2 Addition of Matrices

Two matrices, say $\mathbf{A}$ and $\mathbf{B}$, can be added *only* if they are of the same dimensions. The
sum of the two matrices will be a matrix obtained by adding *corresponding* elements
of matrices $\mathbf{A}$ and $\mathbf{B}$—that is, elements in corresponding positions. This being the
case, the resulting sum will be a matrix of the same dimensions as $\mathbf{A}$ and $\mathbf{B}$.

EXAMPLE **A1.1**   Find the indicated sum of matrices **A** and **B**:

$$\underset{2\times3}{\mathbf{A}} = \begin{bmatrix} 2 & 1 & 4 \\ -1 & 6 & 0 \end{bmatrix} \qquad \underset{2\times3}{\mathbf{B}} = \begin{bmatrix} 0 & -1 & 1 \\ 6 & -3 & 2 \end{bmatrix}$$

Solution

$$\mathbf{A} + \mathbf{B} = \begin{bmatrix} 2 & 1 & 4 \\ -1 & 6 & 0 \end{bmatrix} + \begin{bmatrix} 0 & -1 & 1 \\ 6 & -3 & 2 \end{bmatrix}$$

$$= \begin{bmatrix} (2+0) & (1-1) & (4+1) \\ (-1+6) & (6-3) & (0+2) \end{bmatrix} = \begin{bmatrix} 2 & 0 & 5 \\ 5 & 3 & 2 \end{bmatrix}$$

EXAMPLE **A1.2**   Find the sum of the matrices

$$\underset{3\times3}{\mathbf{A}} = \begin{bmatrix} 1 & 0 & 3 \\ 1 & -1 & 4 \\ 2 & -1 & 0 \end{bmatrix} \qquad \text{and} \qquad \underset{3\times3}{\mathbf{B}} = \begin{bmatrix} 4 & 2 & -1 \\ 1 & 0 & 6 \\ 3 & 1 & 4 \end{bmatrix}$$

Solution

$$\mathbf{A} + \mathbf{B} = \begin{bmatrix} 5 & 2 & 2 \\ 2 & -1 & 10 \\ 5 & 0 & 4 \end{bmatrix}$$

Note that $(\mathbf{A} + \mathbf{B}) = (\mathbf{B} + \mathbf{A})$, as in ordinary algebra, and remember that we never add matrices of unlike dimensions.

A1.3 Multiplication of a Matrix by a Real Number

We desire a rule for multiplying a matrix by a real number, for example, 3**A**, where

$$\mathbf{A} = \begin{bmatrix} 2 & 1 \\ 4 & 6 \\ -1 & 0 \end{bmatrix}.$$

Certainly we would want 3**A** to equal $(\mathbf{A} + \mathbf{A} + \mathbf{A})$, to conform with the addition rule. Hence, 3**A** would mean that each element in the **A** matrix must be multiplied by the multiplier 3, and

$$3\mathbf{A} = \begin{bmatrix} 3(2) & 3(1) \\ 3(4) & 3(6) \\ 3(-1) & 3(0) \end{bmatrix} = \begin{bmatrix} 6 & 3 \\ 12 & 18 \\ -3 & 0 \end{bmatrix}.$$

In general, given a real number c and a matrix **A** with elements a_{ij}, the product $c\mathbf{A}$ will be a matrix whose elements are equal to ca_{ij}.

A1.4 Matrix Multiplication

The rule for matrix multiplication requires "row-column multiplication," which we will define subsequently. The procedure may seem a bit complicated to the novice but should not prove too difficult after practice. We will illustrate with an example. Let **A** and **B** be

$$\mathbf{A} = \begin{bmatrix} 2 & 0 \\ 1 & 4 \end{bmatrix} \qquad \mathbf{B} = \begin{bmatrix} 5 & 2 \\ -1 & 3 \end{bmatrix}.$$

An element in the *ith row* and *jth column* of the product **AB** is obtained by multiplying the *ith row of* **A** by the *jth column of* **B**. Thus the element in the first row, first column of **AB** is obtained by multiplying the first row of **A** by the first column of **B**. Likewise, the element in the first row, second column would be the product of the first row of **A** and the second column of **B**. Notice that we always use the rows of **A** and the columns of **B**, where **A** is the matrix to the left of **B** in the product **AB**.

Row-column multiplication is relatively easy. Obtain the products, first-row element by first-column element, second-row element by second-column element, third by third, and so on, and then sum. Remember that row and column elements are marked from left to right and top to bottom, respectively.

Applying these rules to our example, we obtain

$$\underset{2\times2}{\mathbf{A}}\;\underset{2\times2}{\mathbf{B}} = \begin{bmatrix} 2 & 0 \\ 1 & 4 \end{bmatrix}\begin{bmatrix} 5 & 2 \\ -1 & 3 \end{bmatrix} = \begin{bmatrix} \textcircled{10} & 4 \\ 1 & 14 \end{bmatrix}$$

The first-row-first-column product would be $(2)(5)+(0)(-1) = 10$, which is located (and circled) in the first row, first column of **AB**. Likewise, the element in the first row, second column is equal to the product of the first row of **A** and the second column of **B**, or $(2)(2)+(0)(3) = 4$. The second-row-first-column product is $(1)(5)+(4)(-1) = 1$ and is located in the second row, first column of **AB**. Finally, the second-row-second-column product is $(1)(2) + (4)(3) = 14$.

EXAMPLE A1.3   Find the products **AB** and **BA**, where

$$\mathbf{A} = \begin{bmatrix} 2 & 1 \\ 1 & -1 \\ 0 & 4 \end{bmatrix} \qquad \text{and} \qquad \mathbf{B} = \begin{bmatrix} 4 & -1 & -1 \\ 2 & 0 & 2 \end{bmatrix}$$

Solution

$$\underset{3\times2}{\mathbf{A}}\;\underset{2\times3}{\mathbf{B}} = \begin{bmatrix} 2 & 1 \\ 1 & -1 \\ 0 & 4 \end{bmatrix}\begin{bmatrix} 4 & -1 & -1 \\ 2 & 0 & 2 \end{bmatrix} = \begin{bmatrix} 10 & -2 & 0 \\ 2 & -1 & -3 \\ 8 & 0 & 8 \end{bmatrix}$$

and

$$\underset{2\times3}{\mathbf{B}}\;\underset{3\times2}{\mathbf{A}} = \begin{bmatrix} 4 & -1 & -1 \\ 2 & 0 & 2 \end{bmatrix}\begin{bmatrix} 2 & 1 \\ 1 & -1 \\ 0 & 4 \end{bmatrix} = \begin{bmatrix} 7 & 1 \\ 4 & 10 \end{bmatrix}$$

Note that in matrix algebra, unlike ordinary algebra, **AB** does not equal **BA**. Because **A** contains three rows and **B** contains three columns, we can form $(3)(3) = 9$ row-column combinations and hence nine elements for **AB**. In contrast, **B** contains only two rows, **A** two columns, and hence the product **BA** will possess only $(2)(2) = 4$ elements, corresponding to the four different row-column combinations.

Furthermore, we observe that row-column multiplication is predicated on the assumption that the rows of the matrix on the left contain the same number of elements as the columns of the matrix on the right, so that corresponding elements will exist for the row-column multiplication. What do we do when this condition is not satisfied? We agree never to multiply two matrices, say **AB**, where the rows of **A** and the columns of **B** contain an unequal number of elements.

An examination of the dimensions of the matrices will tell whether they can be multiplied as well as give the dimensions of the product. Writing the dimensions underneath the two matrices,

$$\underset{m \times p}{\mathbf{A}} \ \underset{p \times q}{\mathbf{B}} = \underset{m \times q}{\mathbf{AB}}$$

we observe that the inner two numbers, giving the number of elements in a row of **A** and column of **B**, respectively, must be equal. The outer two numbers, indicating the number of rows of **A** and columns of **B**, give the dimensions of the product matrix. You may verify the operation of this rule for Example A1.3.

EXAMPLE **A1.4**   Obtain the product **AB**:

$$\underset{1 \times 3}{\mathbf{A}} \ \underset{3 \times 2}{\mathbf{B}} = \begin{bmatrix} 2 & 1 & 0 \end{bmatrix} \begin{bmatrix} 2 & 0 \\ 0 & 3 \\ -1 & 0 \end{bmatrix} = \begin{bmatrix} 4 & 3 \end{bmatrix}$$

Note that product **AB** is (1×2) and that **BA** is undefined because of the respective dimensions of **A** and **B**.

EXAMPLE **A1.5**   Find the product **AB**, where

$$\mathbf{A} = \begin{bmatrix} 1 & 2 & 3 & 4 \end{bmatrix} \qquad \text{and} \qquad \mathbf{B} = \begin{bmatrix} 1 \\ 2 \\ 3 \\ 4 \end{bmatrix}$$

Solution

$$\underset{1 \times 4}{\mathbf{A}} \ \underset{4 \times 1}{\mathbf{B}} = \begin{bmatrix} 1 & 2 & 3 & 4 \end{bmatrix} \begin{bmatrix} 1 \\ 2 \\ 3 \\ 4 \end{bmatrix} = \begin{bmatrix} 30 \end{bmatrix}$$

Note that this example produces a different method for writing a sum of squares.

A1.5 Identity Elements

The identity elements for addition and multiplication in ordinary algebra are 0 and 1, respectively. In addition, 0 plus any other element, say a, is identically equal to a; that is,

$$0 + 2 = 2, \qquad 0 + (-9) = -9.$$

Similarly, the multiplication of the identity element 1 by any other element, say a, is equal to a; that is,

$$(1)(5) = 5, \qquad (1)(-4) = -4.$$

In matrix algebra two matrices are said to be equal when all corresponding elements are equal. With this in mind we will define the identity matrices in a manner similar to that employed in ordinary algebra. Hence, if A is any matrix, a matrix B will be an identity matrix for addition if

$$A + B = A \qquad \text{and} \qquad B + A = A.$$

It easily can be seen that the identity matrix for addition is one in which every element is equal to zero. This matrix is of interest but of no practical importance in our work.

Similarly, if A is any matrix, the identity matrix for multiplication is a matrix I that satisfies the relation

$$AI = A \qquad \text{and} \qquad IA = A$$

This matrix, called the *identity matrix*, is the *square matrix*

$$\mathop{\mathbf{I}}_{n \times n} = \begin{bmatrix} 1 & 0 & 0 & 0 & \cdots & 0 \\ 0 & 1 & 0 & 0 & \cdots & 0 \\ 0 & 0 & 1 & 0 & \cdots & 0 \\ 0 & 0 & 0 & 1 & \cdots & 0 \\ \vdots & \vdots & \vdots & \vdots & & \vdots \\ 0 & 0 & 0 & 0 & \cdots & 1 \end{bmatrix}$$

That is, all elements in the *main diagonal* of the matrix, running from top left to bottom right, are equal to 1; all other elements equal zero. Note that the identity matrix is always indicated by the symbol $\mathbf{I}$.

Unlike ordinary algebra, which contains only one identity element for multiplication, matrix algebra must contain an infinitely large number of identity matrices. Thus we must have matrices with dimensions $1 \times 1, 2 \times 2, 3 \times 3, 4 \times 4$, and so on, so as to provide an identity of the correct dimensions to permit multiplication. All will be of this pattern.

That the $\mathbf{I}$ matrix satisfies the relation

$$IA = AI = A$$

can be shown by an example.

EXAMPLE **A1.6**   Let

$$\mathbf{A} = \begin{bmatrix} 2 & 1 & 0 \\ -1 & 6 & 3 \end{bmatrix}$$

Show that $\mathbf{IA} = \mathbf{A}$ and $\mathbf{AI} = \mathbf{A}$.

Solution $$\underset{2\times 2}{\mathbf{I}}\ \underset{2\times 3}{\mathbf{A}} = \begin{bmatrix} 1 & 0 \\ 0 & 1 \end{bmatrix}\begin{bmatrix} 2 & 1 & 0 \\ -1 & 6 & 3 \end{bmatrix} = \begin{bmatrix} 2 & 1 & 0 \\ -1 & 6 & 3 \end{bmatrix} = \mathbf{A}$$

and

$$\underset{2\times 3}{\mathbf{A}}\ \underset{3\times 3}{\mathbf{I}} = \begin{bmatrix} 2 & 1 & 0 \\ -1 & 6 & 3 \end{bmatrix}\begin{bmatrix} 1 & 0 & 0 \\ 0 & 1 & 0 \\ 0 & 0 & 1 \end{bmatrix} = \begin{bmatrix} 2 & 1 & 0 \\ -1 & 6 & 3 \end{bmatrix} = \mathbf{A}$$

A1.6   The Inverse of a Matrix

For matrix algebra to be useful, we must be able to construct and solve matrix equations for a matrix of unknowns in a manner similar to that employed in ordinary algebra. This, in turn, requires a method of performing division.

For example, we would solve the simple equation in ordinary algebra,

$$2x = 6$$

by dividing both sides of the equation by 2 and obtaining $x = 3$. Another way to view this operation is to define the reciprocal of each element in an algebraic system and to think of division as multiplication by the reciprocal of an element. We could solve the equation $2x = 6$ by multiplying both sides of the equation by the reciprocal of 2. Because every element in the real number system possesses a reciprocal, with the exception of 0, the multiplication operation eliminates the need for division.

The reciprocal of a number c in ordinary algebra is a number b that satisfies the relation

$$cb = 1$$

that is, the product of a number by its reciprocal must equal the identity element for multiplication. For example, the reciprocal of 2 is $1/2$ and $(2)(1/2) = 1$.

A reciprocal in matrix algebra is called the *inverse* of a matrix and is defined as follows:

DEFINITION **A1.1**   Let $\mathbf{A}_{n\times n}$ be a square matrix. If a matrix $\mathbf{A}^{-1}$ can be found such that

$$\mathbf{AA}^{-1} = \mathbf{I} \quad \text{and} \quad \mathbf{A}^{-1}\mathbf{A} = \mathbf{I}$$

then $\mathbf{A}^{-1}$ is called the *inverse* of $\mathbf{A}$.

Note that the requirement for an inverse in matrix algebra is the same as in ordinary algebra—that is, the product of **A** by its inverse must equal the identity matrix for multiplication. Furthermore, the inverse is undefined for nonsquare matrices, and hence many matrices in matrix algebra do not have inverses (recall that 0 was the only element in the real number system without an inverse). Finally, we state without proof that many square matrices do not possess inverses. Those that do will be identified in Section A1.9, and a method will be given for finding the inverse of a matrix.

A1.7 The Transpose of a Matrix

We have just discussed a relationship between a matrix and its inverse. A second useful matrix relationship defines the *transpose* of a matrix.

DEFINITION A1.2   Let $A_{p \times q}$ be a matrix of dimensions $p \times q$. Then $\mathbf{A}'$, called the *transpose* of **A**, is defined to be a matrix obtained by interchanging corresponding rows and columns of **A**; that is, first with first, second with second, and so on.

For example, let

$$\mathbf{A}_{3 \times 2} = \begin{bmatrix} 2 & 0 \\ 1 & 1 \\ 4 & 3 \end{bmatrix}$$

Then

$$\mathbf{A}'_{2 \times 3} = \begin{bmatrix} 2 & 1 & 4 \\ 0 & 1 & 3 \end{bmatrix}$$

Note that the first and second rows of $\mathbf{A}'$ are identical with the first and second columns, respectively, of **A**.

As a second example, let

$$\mathbf{Y} = \begin{bmatrix} y_1 \\ y_2 \\ y_3 \end{bmatrix}$$

Then $\mathbf{Y}' = [y_1 \ y_2 \ y_3]$. As a point of interest, we observe that $\mathbf{Y}'\mathbf{Y} = \sum_{i=1}^{3} y_i^2$. Finally, if

$$\mathbf{A} = \begin{bmatrix} 2 & 1 & 4 \\ 0 & 2 & 3 \\ 1 & 6 & 9 \end{bmatrix}$$

then

$$\mathbf{A}' = \begin{bmatrix} 2 & 0 & 1 \\ 1 & 2 & 6 \\ 4 & 3 & 9 \end{bmatrix}$$

A1.8 A Matrix Expression for a System of Simultaneous Linear Equations

We will now introduce you to one of the very simple and important applications of matrix algebra. Let

$$2v_1 + v_2 = 5$$

$$v_1 - v_2 = 1$$

be a pair of simultaneous linear equations in the two variables, v_1 and v_2. We will then define three matrices:

$$\underset{2\times 2}{\mathbf{A}} = \begin{bmatrix} 2 & 1 \\ 1 & -2 \end{bmatrix} \qquad \underset{2\times 1}{\mathbf{V}} = \begin{bmatrix} v_1 \\ v_2 \end{bmatrix} \qquad \underset{2\times 1}{\mathbf{G}} = \begin{bmatrix} 5 \\ 1 \end{bmatrix}$$

Note that $\mathbf{A}$ is the matrix of coefficients of the unknowns when the equations are each written with the variables appearing in the same order, reading left to right, and with the constants on the right-hand side of the equality sign. The $\mathbf{V}$ matrix gives the unknowns in a column and in the same order as they appear in the equations. Finally, the $\mathbf{G}$ matrix contains the constants in a column exactly as they occur in the set of equations.

The simultaneous system of two linear equations may now be written in matrix notation as

$$\mathbf{AV} = \mathbf{G}$$

a statement that can easily be verified by multiplying $\mathbf{A}$ and $\mathbf{V}$ and then comparing the answer with $\mathbf{G}$.

$$\mathbf{AV} = \begin{bmatrix} 2 & 1 \\ 1 & -1 \end{bmatrix} \begin{bmatrix} v_1 \\ v_2 \end{bmatrix} = \begin{bmatrix} 2v_1 + v_2 \\ v_1 - v_2 \end{bmatrix} = \begin{bmatrix} 5 \\ 1 \end{bmatrix} = \mathbf{G}$$

Observe that corresponding elements in $\mathbf{AV}$ and $\mathbf{G}$ are equal—that is, $2v_1 + v_2 = 5$ and $v_1 - v_2 = 1$. Therefore, $\mathbf{AV} = \mathbf{G}$.

The method for writing a pair of linear equations in two unknowns as a matrix equation can easily be extended to a system of r equations in r unknowns. For example, if the equations are

$$a_{11}v_1 + a_{12}v_2 + a_{13}v_3 + \cdots + a_{1r}v_r = g_1$$

$$a_{21}v_1 + a_{22}v_2 + a_{23}v_3 + \cdots + a_{2r}v_r = g_2$$

$$a_{31}v_1 + a_{32}v_2 + a_{33}v_3 + \cdots + a_{3r}v_r = g_3$$

$$\vdots \qquad \vdots \qquad \vdots \qquad \qquad \vdots = \vdots$$

$$a_{r1}v_1 + a_{r2}v_2 + a_{r3}v_3 + \cdots + a_{rr}v_r = g_r$$

define

$$\mathbf{A} = \begin{bmatrix} a_{11} & a_{12} & a_{13} & \cdots & a_{1r} \\ a_{21} & a_{22} & a_{23} & \cdots & a_{2r} \\ a_{31} & a_{32} & a_{33} & \cdots & a_{3r} \\ \vdots & \vdots & \vdots & & \vdots \\ a_{r1} & a_{r2} & a_{r3} & \cdots & a_{rr} \end{bmatrix} \qquad \mathbf{V} = \begin{bmatrix} v_1 \\ v_2 \\ v_3 \\ \vdots \\ v_r \end{bmatrix} \qquad \mathbf{G} = \begin{bmatrix} g_1 \\ g_2 \\ g_3 \\ \vdots \\ g_r \end{bmatrix}$$

Observe that, once again, $\mathbf{A}$ is a square matrix of variable coefficients, whereas $\mathbf{V}$ and $\mathbf{G}$ are column matrices containing the variables and constants, respectively. Then $\mathbf{AV} = \mathbf{G}$.

Regardless of how large the system of equations, if we possess n linear equations in n unknowns, the system may be written as the simple matrix equation $\mathbf{AV} = \mathbf{G}$.

You will observe that the matrix $\mathbf{V}$ contains all the unknowns, whereas $\mathbf{A}$ and $\mathbf{G}$ are constant matrices.

Our objective, of course, is to solve for the matrix of unknowns, $\mathbf{V}$, where the equation $\mathbf{AV} = \mathbf{G}$ is similar to the equation

$$2v = 6$$

in ordinary algebra. This being true, we would not be too surprised to find that the methods of solution are the same. In ordinary algebra both sides of the equation are multiplied by the reciprocal of 2; in matrix algebra both sides of the equation are multiplied by $\mathbf{A}^{-1}$. Then

$$\mathbf{A}^{-1}(\mathbf{AV}) = \mathbf{A}^{-1}\mathbf{G}$$

or

$$\mathbf{A}^{-1}\mathbf{AV} = \mathbf{A}^{-1}\mathbf{G}$$

But $\mathbf{A}^{-1}\mathbf{A} = \mathbf{I}$ and $\mathbf{IV} = \mathbf{V}$. Therefore, $\mathbf{V} = \mathbf{A}^{-1}\mathbf{G}$. In other words, the solution to the system of simultaneous linear equations can be obtained by finding $\mathbf{A}^{-1}$ and then obtaining the product $\mathbf{A}^{-1}\mathbf{G}$. The solution values of $v_1, v_2, v_3, \ldots, v_r$ will appear in sequence in the column matrix $\mathbf{V} = \mathbf{A}^{-1}\mathbf{G}$.

A1.9   Inverting a Matrix

We have indicated in Section A1.8 that the key to the solution of a system of simultaneous linear equations by the method of matrix algebra rests on the acquisition of the inverse of the $\mathbf{A}$ matrix. Many methods exist for inverting matrices. The method that we present is not the best from a computational point of view, but it works very well for the matrices associated with most experimental designs and it is one of the easiest to present to the novice. It depends upon a theorem in matrix algebra and the use of *row operations*.

Before defining *row operations* on matrices, we must state what is meant by the addition of two rows of a matrix and the multiplication of a row by a constant. We will illustrate with the **A** matrix for the system of two simultaneous linear equations,

$$\mathbf{A} = \begin{bmatrix} 2 & 1 \\ 1 & -1 \end{bmatrix}$$

Two rows of a matrix may be added by adding corresponding elements. Thus if the two rows of the **A** matrix are added, one obtains a new row with elements $[(2+1)$ $(1-1)] = [3\ 0]$. Multiplication of a row by a constant means that each element in the row is multiplied by the constant. Twice the first row of the **A** matrix would generate the row $[4\ 2]$. With these ideas in mind, we will define three ways to operate on a row in a matrix:

1. A row may be multiplied by a constant.
2. A row may be multiplied by a constant and added to or subtracted from another row (which is identified as the one upon which the operation is performed).
3. Two rows may be interchanged.

Given matrix **A**, it is quite easy to see that we might perform a series of row operations that would yield some new matrix **B**. In this connection we state without proof a surprising and interesting theorem from matrix algebra; namely, there exists some matrix **C** such that

$$\mathbf{CA} = \mathbf{B}.$$

In other words, a series of row operations on a matrix **A** is equivalent to multiplying **A** by a matrix **C**. We will use this principle to invert a matrix.

Place the matrix **A**, which is to be inverted, alongside an identity matrix of the same dimensions:

$$\mathbf{A} = \begin{bmatrix} 2 & 1 \\ 1 & -1 \end{bmatrix} \quad \mathbf{I} = \begin{bmatrix} 1 & 0 \\ 0 & 1 \end{bmatrix}.$$

Then perform the same row operations on **A** *and* **I** *in such a way that* **A** *changes to an identity matrix.* In doing so, we must have multiplied **A** by a matrix **C** so that **CA** = **I**. Therefore, **C** must be the inverse of **A**! The problem, of course, is to find the unknown matrix **C** and, fortunately, this proves to be of little difficulty. Because we performed the same row operations on **A** and **I**, the identity matrix must have changed to $\mathbf{CI} = \mathbf{C} = \mathbf{A}^{-1}$.

$$\mathbf{A} = \begin{bmatrix} 2 & 1 \\ 1 & -1 \end{bmatrix} \qquad \mathbf{I} = \begin{bmatrix} 1 & 0 \\ 0 & 1 \end{bmatrix}$$

$$\downarrow \text{ (same row operations) } \downarrow$$

$$\mathbf{CA} = \mathbf{I} \qquad\qquad \mathbf{CI} = \mathbf{C} = \mathbf{A}^{-1}$$

We will illustrate with the following example.

EXAMPLE **A1.7**   Invert the matrix

$$A = \begin{bmatrix} 2 & 1 \\ 1 & -1 \end{bmatrix}$$

Solution

$$A = \begin{bmatrix} 2 & 1 \\ 1 & -1 \end{bmatrix} \qquad I = \begin{bmatrix} 1 & 0 \\ 0 & 1 \end{bmatrix}$$

Step 1. Operate on row 1 by multiplying row 1 by 1/2. (Note: It is helpful to the beginner to identify the row upon which he or she is operating because all other rows will remain unchanged, even though they may be used in the operation. We will star the row upon which the operation is being performed.)

$$*\begin{bmatrix} 1 & 1/2 \\ 1 & -1 \end{bmatrix} \qquad \begin{bmatrix} 1/2 & 0 \\ 0 & 1 \end{bmatrix}$$

Step 2. Operate on row 2 by subtracting row 1 from row 2.

$$\begin{bmatrix} 1 & 1/2 \\ 0 & -3/2 \end{bmatrix}* \qquad \begin{bmatrix} 1/2 & 0 \\ -1/2 & 1 \end{bmatrix}$$

(Note that row 2 is simply used to operate on row 1 and hence remains unchanged.)
Step 3. Multiply row 2 by $(-2/3)$.

$$\begin{bmatrix} 1 & 1/2 \\ 0 & 1 \end{bmatrix}* \qquad \begin{bmatrix} 1/2 & 0 \\ 1/3 & -2/3 \end{bmatrix}$$

Step 4. Operate on row 1 by multiplying row 2 by 1/2 and subtracting from row 1.

$$*\begin{bmatrix} 1 & 0 \\ 0 & 1 \end{bmatrix} \qquad \begin{bmatrix} 1/3 & 1/3 \\ 1/3 & -2/3 \end{bmatrix}$$

(Note that row 2 is simply used to operate on row 1 and hence remains unchanged.)
Hence the inverse of **A** must be

$$A^{-1} = \begin{bmatrix} 1/3 & 1/3 \\ 1/3 & -2/3 \end{bmatrix}$$

A ready check on the calculations for the inversion procedure is available because $A^{-1}A$ must equal the identity matrix **I**. Thus

$$A^{-1}A = \begin{bmatrix} 1/3 & 1/3 \\ 1/3 & -2/3 \end{bmatrix}\begin{bmatrix} 2 & 1 \\ 1 & -1 \end{bmatrix} = \begin{bmatrix} 1 & 0 \\ 0 & 1 \end{bmatrix}$$

EXAMPLE **A1.8** Invert the matrix

$$A = \begin{bmatrix} 2 & 0 & 1 \\ 1 & -1 & 2 \\ 1 & 0 & 0 \end{bmatrix}$$

and check the results.

Solution

$$A = \begin{bmatrix} 2 & 0 & 1 \\ 1 & -1 & 2 \\ 1 & 0 & 0 \end{bmatrix} \quad I = \begin{bmatrix} 1 & 0 & 0 \\ 0 & 1 & 0 \\ 0 & 0 & 1 \end{bmatrix}$$

Step 1. Multiply row 1 by 1/2.

$$\begin{matrix} * \\ \\ \end{matrix} \begin{bmatrix} 1 & 0 & 1/2 \\ 1 & -1 & 2 \\ 1 & 0 & 0 \end{bmatrix} \begin{bmatrix} 1/2 & 0 & 0 \\ 0 & 1 & 0 \\ 0 & 0 & 1 \end{bmatrix}$$

Step 2. Operate on row 2 by subtracting row 1 from row 2.

$$\begin{matrix} \\ * \\ \\ \end{matrix} \begin{bmatrix} 1 & 0 & 1/2 \\ 0 & -1 & 3/2 \\ 1 & 0 & 0 \end{bmatrix} \begin{bmatrix} 1/2 & 0 & 0 \\ -1/2 & 1 & 0 \\ 0 & 0 & 1 \end{bmatrix}$$

Step 3. Operate on row 3 by subtracting row 1 from row 3.

$$\begin{matrix} \\ \\ * \end{matrix} \begin{bmatrix} 1 & 0 & 1/2 \\ 0 & -1 & 3/2 \\ 0 & 0 & -1/2 \end{bmatrix} \begin{bmatrix} 1/2 & 0 & 0 \\ -1/2 & 1 & 0 \\ -1/2 & 0 & 1 \end{bmatrix}$$

Step 4. Operate on row 2 by multiplying row 3 by 3 and adding to row 2.

$$\begin{matrix} \\ * \\ \\ \end{matrix} \begin{bmatrix} 1 & 0 & 1/2 \\ 0 & -1 & 0 \\ 0 & 0 & -1/2 \end{bmatrix} \begin{bmatrix} 1/2 & 0 & 0 \\ -2 & 1 & 3 \\ -1/2 & 0 & 1 \end{bmatrix}$$

Step 5. Multiply row 2 by (−1).

$$\begin{matrix} \\ * \\ \\ \end{matrix} \begin{bmatrix} 1 & 0 & 1/2 \\ 0 & 1 & 0 \\ 0 & 0 & -1/2 \end{bmatrix} \begin{bmatrix} 1/2 & 0 & 0 \\ 2 & -1 & -3 \\ -1/2 & 0 & 1 \end{bmatrix}$$

Step 6. Operate on row 1 by adding row 3 to row 1.

$$\begin{matrix} * \\ \\ \\ \end{matrix} \begin{bmatrix} 1 & 0 & 0 \\ 0 & 1 & 0 \\ 0 & 0 & -1/2 \end{bmatrix} \begin{bmatrix} 0 & 0 & 1 \\ 2 & -1 & -3 \\ -1/2 & 0 & 1 \end{bmatrix}$$

Step 7. Multiply row 3 by (-2).

$$\begin{bmatrix} 1 & 0 & 0 \\ 0 & 1 & 0 \\ {}_{*}0 & 0 & 1 \end{bmatrix} \begin{bmatrix} 0 & 0 & 1 \\ 2 & -1 & -3 \\ 1 & 0 & -2 \end{bmatrix} = \mathbf{A}^{-1}$$

The seven row operations have changed the $\mathbf{A}$ matrix to the identity matrix and, barring errors of calculation, have changed the identity to $\mathbf{A}^{-1}$.

Checking, we have

$$\mathbf{A}^{-1}\mathbf{A} = \begin{bmatrix} 0 & 0 & 1 \\ 2 & -1 & -3 \\ 1 & 0 & -2 \end{bmatrix} \begin{bmatrix} 2 & 0 & 1 \\ 1 & -1 & 2 \\ 1 & 0 & 0 \end{bmatrix} = \begin{bmatrix} 1 & 0 & 0 \\ 0 & 1 & 0 \\ 0 & 0 & 1 \end{bmatrix}$$

We see that $\mathbf{A}^{-1}\mathbf{A} = \mathbf{I}$ and hence that the calculations are correct.

Note that the sequence of row operations required to convert $\mathbf{A}$ to $\mathbf{I}$ is not unique. One person might achieve the inverse by using five row operations whereas another might require ten, but the end result will be the same. However, in the interests of efficiency it is desirable to employ a system.

Observe that the inversion process utilizes row operations to change off-diagonal elements in the $\mathbf{A}$ matrix to 0s and the main diagonal elements to 1s. One systematic procedure is as follows. Change the top left element into a 1 and then perform row operations to change all other elements in the *first* column to 0. Then move to the diagonal element in the second row, second column, change it into a 1, and change all elements in the *second* column *below* the main diagonal to 0. This process is repeated, moving down the main diagonal from top left to bottom right, until all elements below the main diagonal have been changed to 0s. To eliminate nonzero elements above the main diagonal, operate on all elements in the last column, changing each to 0; then move to the next to last column and repeat the process. Continue this procedure until you arrive at the first element in the first column, which was the starting point. This procedure is indicated diagrammatically in Figure A1.1.

Matrix inversion is a tedious process, at best, and requires every bit as much labor as the solution of a system of simultaneous equations by elimination or substitution. You will be pleased to learn that we do not expect you to develop a facility for matrix inversion. Fortunately, most matrices associated with designed experiments follow patterns and are easily inverted.

It will be beneficial to you to invert a few 2×2 and 3×3 matrices. Matrices lacking pattern, particularly large matrices, are inverted most efficiently and economically by using a computer. (Programs for matrix inversion have been developed for most computers.)

We emphasize that obtaining the solution for the least squares equations (Chapter 11) by matrix inversion has distinct advantages that may or may not be apparent. Not the least of these is the fact that the inversion procedure is systematic and hence is particularly suitable for electronic computation. However, the major advantage is that the inversion procedure will automatically produce the variances of the estimators of all parameters in the linear model.

FIGURE **A1.1**
Procedure for
matrix inversion

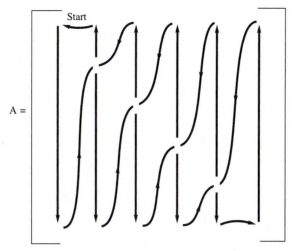

Before leaving the topic of matrix inversion, we ask how one may identify a matrix that has an inverse. Reference to a discussion of linear equations in ordinary algebra should reveal the answer.

Clearly, a unique solution for a system of simultaneous linear equations cannot be obtained unless the equations are independent. Thus if one of the equations is a linear combination of the others, the equations are dependent. Coefficient matrices associated with dependent systems of linear equations do not possess an inverse.

A1.10 Solving a System of Simultaneous Linear Equations

We have finally obtained all the ingredients necessary for solving a system of simultaneous linear equations,

$$2v_1 + v_2 = 5$$
$$v_1 - v_2 = 1$$

Recalling that the matrix solution to the system of equations $\mathbf{AV} = \mathbf{G}$ is $\mathbf{V} = \mathbf{A}^{-1}\mathbf{G}$, we obtain

$$\mathbf{V} = \mathbf{A}^{-1}\mathbf{G} = \begin{bmatrix} 1/3 & 1/3 \\ 1/3 & -2/3 \end{bmatrix} \begin{bmatrix} 5 \\ 1 \end{bmatrix} = \begin{bmatrix} 2 \\ 1 \end{bmatrix}$$

Hence the solution is

$$\mathbf{V} = \begin{bmatrix} v_1 \\ v_2 \end{bmatrix} = \begin{bmatrix} 2 \\ 1 \end{bmatrix}$$

that is, $v_1 = 2$ and $v_2 = 1$, a fact that may be verified by substitution of these values in the original linear equations.

EXAMPLE **A1.9** Solve the system of simultaneous linear equations

$$2v_1 + v_3 = 4$$

$$v_1 - v_2 + 2v_3 = 2$$

$$v_1 = 1$$

Solution The coefficient matrix for these equations,

$$\mathbf{A} = \begin{bmatrix} 2 & 0 & 1 \\ 1 & -1 & 2 \\ 1 & 0 & 0 \end{bmatrix}$$

appeared in Example A1.8. In that example we found that

$$\mathbf{A}^{-1} = \begin{bmatrix} 0 & 0 & 1 \\ 2 & -1 & -3 \\ 1 & 0 & -2 \end{bmatrix}$$

Solving, we obtain

$$\mathbf{V} = \mathbf{A}^{-1}\mathbf{G} = \begin{bmatrix} 0 & 0 & 1 \\ 2 & -1 & -3 \\ 1 & 0 & -2 \end{bmatrix} \begin{bmatrix} 4 \\ 2 \\ 1 \end{bmatrix} = \begin{bmatrix} 1 \\ 3 \\ 2 \end{bmatrix}.$$

Thus $v_1 = 1$, $v_2 = 3$ and $v_3 = 2$ give the solution to the set of three simultaneous linear equations.

A1.11 Other Useful Mathematical Results

The purpose of this section is to provide the reader with a convenient reference to some of the key mathematical results that are used frequently in the body of the text.

The Binomial Expansion of $(x + y)^n$ Let x and y be any real numbers, then

$$(x + y)^n = \binom{n}{0} x^n y^0 + \binom{n}{1} x^{n-1} y^1 + \binom{n}{2} x^{n-2} y^2 + \cdots + \binom{n}{n} x^0 y^n$$

$$= \sum_{i=0}^{n} \binom{n}{i} x^{n-i} y^i$$

The Sum of a Geometric Series   Let r be a real number such that $|r| < 1$, and m be any integer $m \geq 1$

$$\sum_{i=0}^{\infty} r^i = \frac{1}{1-r}, \quad \sum_{i=1}^{\infty} r^i = \frac{r}{1-r}, \quad \sum_{i=0}^{m} r^i = \frac{1 - r^{m+1}}{1 - r}$$

The (Taylor) Series Expansion of e^x   Let x be any real number, then

$$e^x = \sum_{i=0}^{\infty} \frac{x^i}{i!}$$

Some useful formulas for particular summations follow. The proofs (omitted) are most easily established by using mathematical induction.

$$\sum_{i=1}^{n} i = \frac{n(n+1)}{2}$$

$$\sum_{i=1}^{n} i^2 = \frac{n(n+1)(2n+1)}{6}$$

$$\sum_{i=1}^{n} i^3 = \left(\frac{n(n+1)}{2}\right)^2$$

Gamma Function   Let $t > 0$, then $\Gamma(t)$ is defined by the following integral:

$$\Gamma(t) = \int_0^{\infty} y^{t-1} e^{-y} dy.$$

Using the technique of integration by parts, it follows that for any $t > 0$

$$\Gamma(t+1) = t\Gamma(t)$$

and if $t = n$, where n is an integer,

$$\Gamma(n) = (n-1)!.$$

Further,

$$\Gamma(1/2) = \sqrt{\pi}.$$

If $\alpha, \beta > 0$, the *Beta function*, $B(\alpha, \beta)$, is defined by the following integral,

$$B(\alpha, \beta) = \int_0^1 y^{\alpha-1}(1-y)^{\beta-1} dy$$

and is related to the gamma function as follows:

$$B(\alpha, \beta) = \frac{\Gamma(\alpha)\Gamma(\beta)}{\Gamma(\alpha+\beta)}.$$

Common Probability Distributions, Means, Variances, and Moment-Generating Functions

Table A2.1. Discrete distributions

Distribution	Probability Function	Mean	Variance	Moment-Generating Function
Binomial	$p(y) = \binom{n}{y} p^y (1-p)^{n-y}$; $y = 0, 1, \ldots, n$	np	$np(1-p)$	$[pe^t + (1-p)]^n$
Geometric	$p(y) = p(1-p)^{y-1}$; $y = 1, 2, \ldots$	$\dfrac{1}{p}$	$\dfrac{1-p}{p^2}$	$\dfrac{pe^t}{1 - (1-p)e^t}$
Hypergeometric	$p(y) = \dfrac{\binom{r}{y}\binom{N-r}{n-y}}{\binom{N}{n}}$; $y = 0, 1, \ldots, n$ if $n \leq r$, $y = 0, 1, \ldots, r$ if $n > r$	$\dfrac{nr}{N}$	$n\left(\dfrac{r}{N}\right)\left(\dfrac{N-r}{N}\right)\left(\dfrac{N-n}{N-1}\right)$	does not exist in closed form
Poisson	$p(y) = \dfrac{\lambda^y e^{-\lambda}}{y!}$; $y = 0, 1, 2, \ldots$	λ	λ	$\exp[\lambda(e^t - 1)]$
Negative binomial	$p(y) = \binom{y-1}{r-1} p^r (1-p)^{y-r}$; $y = r, r+1, \ldots$	$\dfrac{r}{p}$	$\dfrac{r(1-p)}{p^2}$	$\left[\dfrac{pe^t}{1 - (1-p)e^t}\right]^r$

Table A2.2. Continuous distributions

Distribution	Probability Function	Mean	Variance	Moment-Generating Function
Uniform	$f(y) = \dfrac{1}{\theta_2 - \theta_1}; \theta_1 \le y \le \theta_2$	$\dfrac{\theta_1 + \theta_2}{2}$	$\dfrac{(\theta_2 - \theta_1)^2}{12}$	$\dfrac{e^{t\theta_2} - e^{t\theta_1}}{t(\theta_2 - \theta_1)}$
Normal	$f(y) = \dfrac{1}{\sigma\sqrt{2\pi}}\exp\left[-\left(\dfrac{1}{2\sigma^2}\right)(y-\mu)^2\right]$ $-\infty < y < +\infty$	μ	σ^2	$\exp\left(\mu t + \dfrac{t^2\sigma^2}{2}\right)$
Exponential	$f(y) = \dfrac{1}{\beta}e^{-y/\beta}; \quad \beta > 0$ $0 < y < \infty$	β	β^2	$(1-\beta t)^{-1}$
Gamma	$f(y) = \left[\dfrac{1}{\Gamma(\alpha)\beta^\alpha}\right]y^{\alpha-1}e^{-y/\beta};$ $0 < y < \infty$	$\alpha\beta$	$\alpha\beta^2$	$(1-\beta t)^{-\alpha}$
Chi-square	$f(y) = \dfrac{(y)^{(v/2)-1}e^{-y/2}}{2^{v/2}\Gamma(v/2)};$ $y > 0$	v	$2v$	$(1-2t)^{-v/2}$
Beta	$f(y) = \left[\dfrac{\Gamma(\alpha+\beta)}{\Gamma(\alpha)\Gamma(\beta)}\right]y^{\alpha-1}(1-y)^{\beta-1};$ $0 < y < 1$	$\dfrac{\alpha}{\alpha+\beta}$	$\dfrac{\alpha\beta}{(\alpha+\beta)^2(\alpha+\beta+1)}$	does not exist in closed form

APPENDIX **THREE**

Tables

Table 1. Binomial Probabilities

Tabulated values are $P(Y \leq a) = \sum\limits_{y=0}^{a} p(y)$. (Computations are rounded at third decimal place.)

(a) $n = 5$

							p							
a	0.01	0.05	0.10	0.20	0.30	0.40	0.50	0.60	0.70	0.80	0.90	0.95	0.99	a
0	.951	.774	.590	.328	.168	.078	.031	.010	.002	.000	.000	.000	.000	0
1	.999	.977	.919	.737	.528	.337	.188	.087	.031	.007	.000	.000	.000	1
2	1.000	.999	.991	.942	.837	.683	.500	.317	.163	.058	.009	.001	.000	2
3	1.000	1.000	1.000	.993	.969	.913	.812	.663	.472	.263	.081	.023	.001	3
4	1.000	1.000	1.000	1.000	.998	.990	.969	.922	.832	.672	.410	.226	.049	4

(b) $n = 10$

							p							
a	0.01	0.05	0.10	0.20	0.30	0.40	0.50	0.60	0.70	0.80	0.90	0.95	0.99	a
0	.904	.599	.349	.107	.028	.006	.001	.000	.000	.000	.000	.000	.000	0
1	.996	.914	.736	.376	.149	.046	.011	.002	.000	.000	.000	.000	.000	1
2	1.000	.988	.930	.678	.383	.167	.055	.012	.002	.000	.000	.000	.000	2
3	1.000	.999	.987	.879	.650	.382	.172	.055	.011	.001	.000	.000	.000	3
4	1.000	1.000	.998	.967	.850	.633	.377	.166	.047	.006	.000	.000	.000	4
5	1.000	1.000	1.000	.994	.953	.834	.623	.367	.150	.033	.002	.000	.000	5
6	1.000	1.000	1.000	.999	.989	.945	.828	.618	.350	.121	.013	.001	.000	6
7	1.000	1.000	1.000	1.000	.998	.988	.945	.833	.617	.322	.070	.012	.000	7
8	1.000	1.000	1.000	1.000	1.000	.998	.989	.954	.851	.624	.264	.086	.004	8
9	1.000	1.000	1.000	1.000	1.000	1.000	.999	.994	.972	.893	.651	.401	.096	9

Table 1. (*Continued*)

(c) $n = 15$

							p							
a	0.01	0.05	0.10	0.20	0.30	0.40	0.50	0.60	0.70	0.80	0.90	0.95	0.99	a
0	.860	.463	.206	.035	.005	.000	.000	.000	.000	.000	.000	.000	.000	0
1	.990	.829	.549	.167	.035	.005	.000	.000	.000	.000	.000	.000	.000	1
2	1.000	.964	.816	.398	.127	.027	.004	.000	.000	.000	.000	.000	.000	2
3	1.000	.995	.944	.648	.297	.091	.018	.002	.000	.000	.000	.000	.000	3
4	1.000	.999	.987	.836	.515	.217	.059	.009	.001	.000	.000	.000	.000	4
5	1.000	1.000	.998	.939	.722	.403	.151	.034	.004	.000	.000	.000	.000	5
6	1.000	1.000	1.000	.982	.869	.610	.304	.095	.015	.001	.000	.000	.000	6
7	1.000	1.000	1.000	.996	.950	.787	.500	.213	.050	.004	.000	.000	.000	7
8	1.000	1.000	1.000	.999	.985	.905	.696	.390	.131	.018	.000	.000	.000	8
9	1.000	1.000	1.000	1.000	.996	.966	.849	.597	.278	.061	.002	.000	.000	9
10	1.000	1.000	1.000	1.000	.999	.991	.941	.783	.485	.164	.013	.001	.000	10
11	1.000	1.000	1.000	1.000	1.000	.998	.982	.909	.703	.352	.056	.005	.000	11
12	1.000	1.000	1.000	1.000	1.000	1.000	.996	.973	.873	.602	.184	.036	.000	12
13	1.000	1.000	1.000	1.000	1.000	1.000	1.000	.995	.965	.833	.451	.171	.010	13
14	1.000	1.000	1.000	1.000	1.000	1.000	1.000	1.000	.995	.965	.794	.537	.140	14

(d) $n = 20$

							p							
a	0.01	0.05	0.10	0.20	0.30	0.40	0.50	0.60	0.70	0.80	0.90	0.95	0.99	a
0	.818	.358	.122	.012	.001	.000	.000	.000	.000	.000	.000	.000	.000	0
1	.983	.736	.392	.069	.008	.001	.000	.000	.000	.000	.000	.000	.000	1
2	.999	.925	.677	.206	.035	.004	.000	.000	.000	.000	.000	.000	.000	2
3	1.000	.984	.867	.411	.107	.016	.001	.000	.000	.000	.000	.000	.000	3
4	1.000	.997	.957	.630	.238	.051	.006	.000	.000	.000	.000	.000	.000	4
5	1.000	1.000	.989	.804	.416	.126	.021	.002	.000	.000	.000	.000	.000	5
6	1.000	1.000	.998	.913	.608	.250	.058	.006	.000	.000	.000	.000	.000	6
7	1.000	1.000	1.000	.968	.772	.416	.132	.021	.001	.000	.000	.000	.000	7
8	1.000	1.000	1.000	.990	.887	.596	.252	.057	.005	.000	.000	.000	.000	8
9	1.000	1.000	1.000	.997	.952	.755	.412	.128	.017	.001	.000	.000	.000	9
10	1.000	1.000	1.000	.999	.983	.872	.588	.245	.048	.003	.000	.000	.000	10
11	1.000	1.000	1.000	1.000	.995	.943	.748	.404	.113	.010	.000	.000	.000	11
12	1.000	1.000	1.000	1.000	.999	.979	.868	.584	.228	.032	.000	.000	.000	12
13	1.000	1.000	1.000	1.000	1.000	.994	.942	.750	.392	.087	.002	.000	.000	13
14	1.000	1.000	1.000	1.000	1.000	.998	.979	.874	.584	.196	.011	.000	.000	14
15	1.000	1.000	1.000	1.000	1.000	1.000	.994	.949	.762	.370	.043	.003	.000	15
16	1.000	1.000	1.000	1.000	1.000	1.000	.999	.984	.893	.589	.133	.016	.000	16
17	1.000	1.000	1.000	1.000	1.000	1.000	1.000	.996	.965	.794	.323	.075	.001	17
18	1.000	1.000	1.000	1.000	1.000	1.000	1.000	.999	.992	.931	.608	.264	.017	18
19	1.000	1.000	1.000	1.000	1.000	1.000	1.000	1.000	.999	.988	.878	.642	.182	19

Table 1. (*Continued*)

(e) $n = 25$

a	0.01	0.05	0.10	0.20	0.30	0.40	0.50	0.60	0.70	0.80	0.90	0.95	0.99	a
							p							
0	.778	.277	.072	.004	.000	.000	.000	.000	.000	.000	.000	.000	.000	0
1	.974	.642	.271	.027	.002	.000	.000	.000	.000	.000	.000	.000	.000	1
2	.998	.873	.537	.098	.009	.000	.000	.000	.000	.000	.000	.000	.000	2
3	1.000	.966	.764	.234	.033	.002	.000	.000	.000	.000	.000	.000	.000	3
4	1.000	.993	.902	.421	.090	.009	.000	.000	.000	.000	.000	.000	.000	4
5	1.000	.999	.967	.617	.193	.029	.002	.000	.000	.000	.000	.000	.000	5
6	1.000	1.000	.991	.780	.341	.074	.007	.000	.000	.000	.000	.000	.000	6
7	1.000	1.000	.998	.891	.512	.154	.022	.001	.000	.000	.000	.000	.000	7
8	1.000	1.000	1.000	.953	.677	.274	.054	.004	.000	.000	.000	.000	.000	8
9	1.000	1.000	1.000	.983	.811	.425	.115	.013	.000	.000	.000	.000	.000	9
10	1.000	1.000	1.000	.994	.902	.586	.212	.034	.002	.000	.000	.000	.000	10
11	1.000	1.000	1.000	.998	.956	.732	.345	.078	.006	.000	.000	.000	.000	11
12	1.000	1.000	1.000	1.000	.983	.846	.500	.154	.017	.000	.000	.000	.000	12
13	1.000	1.000	1.000	1.000	.994	.922	.655	.268	.044	.002	.000	.000	.000	13
14	1.000	1.000	1.000	1.000	.998	.966	.788	.414	.098	.006	.000	.000	.000	14
15	1.000	1.000	1.000	1.000	1.000	.987	.885	.575	.189	.017	.000	.000	.000	15
16	1.000	1.000	1.000	1.000	1.000	.996	.946	.726	.323	.047	.000	.000	.000	16
17	1.000	1.000	1.000	1.000	1.000	.999	.978	.846	.488	.109	.002	.000	.000	17
18	1.000	1.000	1.000	1.000	1.000	1.000	.993	.926	.659	.220	.009	.000	.000	18
19	1.000	1.000	1.000	1.000	1.000	1.000	.998	.971	.807	.383	.033	.001	.000	19
20	1.000	1.000	1.000	1.000	1.000	1.000	1.000	.991	.910	.579	.098	.007	.000	20
21	1.000	1.000	1.000	1.000	1.000	1.000	1.000	.998	.967	.766	.236	.034	.000	21
22	1.000	1.000	1.000	1.000	1.000	1.000	1.000	1.000	.991	.902	.463	.127	.002	22
23	1.000	1.000	1.000	1.000	1.000	1.000	1.000	1.000	.998	.973	.729	.358	.026	23
24	1.000	1.000	1.000	1.000	1.000	1.000	1.000	1.000	1.000	.996	.928	.723	.222	24

Table 2. Table of e^{-x}

x	e^{-x}	x	e^{-x}	x	e^{-x}	x	e^{-x}
0.00	1.000000	2.60	.074274	5.10	.006097	7.60	.000501
0.10	.904837	2.70	.067206	5.20	.005517	7.70	.000453
0.20	.818731	2.80	.060810	5.30	.004992	7.80	.000410
0.30	.740818	2.90	.055023	5.40	.004517	7.90	.000371
0.40	.670320	3.00	.049787	5.50	.004087	8.00	.000336
0.50	.606531	3.10	.045049	5.60	.003698	8.10	.000304
0.60	.548812	3.20	.040762	5.70	.003346	8.20	.000275
0.70	.496585	3.30	.036883	5.80	.003028	8.30	.000249
0.80	.449329	3.40	.033373	5.90	.002739	8.40	.000225
0.90	.406570	3.50	.030197	6.00	.002479	8.50	.000204
1.00	.367879	3.60	.027324	6.10	.002243	8.60	.000184
1.10	.332871	3.70	.024724	6.20	.002029	8.70	.000167
1.20	.301194	3.80	.022371	6.30	.001836	8.80	.000151
1.30	.272532	3.90	.020242	6.40	.001661	8.90	.000136
1.40	.246597	4.00	.018316	6.50	.001503	9.00	.000123
1.50	.223130	4.10	.016573	6.60	.001360	9.10	.000112
1.60	.201897	4.20	.014996	6.70	.001231	9.20	.000101
1.70	.182684	4.30	.013569	6.80	.001114	9.30	.000091
1.80	.165299	4.40	.012277	6.90	.001008	9.40	.000083
1.90	.149569	4.50	.011109	7.00	.000912	9.50	.000075
2.00	.135335	4.60	.010052	7.10	.000825	9.60	.000068
2.10	.122456	4.70	.009095	7.20	.000747	9.70	.000061
2.20	.110803	4.80	.008230	7.30	.000676	9.80	.000056
2.30	.100259	4.90	.007447	7.40	.000611	9.90	.000050
2.40	.090718	5.00	.006738	7.50	.000553	10.00	.000045
2.50	.082085						

Table 3. Poisson Probabilities

$$P(Y \leq a) = \sum_{y=0}^{a} e^{-\lambda} \frac{\lambda^y}{y!}$$

λ \ a	0	1	2	3	4	5	6	7	8	9
0.02	0.980	1.000								
0.04	0.961	0.999	1.000							
0.06	0.942	0.998	1.000							
0.08	0.923	0.997	1.000							
0.10	0.905	0.995	1.000							
0.15	0.861	0.990	0.999	1.000						
0.20	0.819	0.982	0.999	1.000						
0.25	0.779	0.974	0.998	1.000						
0.30	0.741	0.963	0.996	1.000						
0.35	0.705	0.951	0.994	1.000						
0.40	0.670	0.938	0.992	0.999	1.000					
0.45	0.638	0.925	0.989	0.999	1.000					
0.50	0.607	0.910	0.986	0.998	1.000					
0.55	0.577	0.894	0.982	0.988	1.000					
0.60	0.549	0.878	0.977	0.997	1.000					
0.65	0.522	0.861	0.972	0.996	0.999	1.000				
0.70	0.497	0.844	0.966	0.994	0.999	1.000				
0.75	0.472	0.827	0.959	0.993	0.999	1.000				
0.80	0.449	0.809	0.953	0.991	0.999	1.000				
0.85	0.427	0.791	0.945	0.989	0.998	1.000				
0.90	0.407	0.772	0.937	0.987	0.998	1.000				
0.95	0.387	0.754	0.929	0.981	0.997	1.000				
1.00	0.368	0.736	0.920	0.981	0.996	0.999	1.000			
1.1	0.333	0.699	0.900	0.974	0.995	0.999	1.000			
1.2	0.301	0.663	0.879	0.966	0.992	0.998	1.000			
1.3	0.273	0.627	0.857	0.957	0.989	0.998	1.000			
1.4	0.247	0.592	0.833	0.946	0.986	0.997	0.999	1.000		
1.5	0.223	0.558	0.809	0.934	0.981	0.996	0.999	1.000		
1.6	0.202	0.525	0.783	0.921	0.976	0.994	0.999	1.000		
1.7	0.183	0.493	0.757	0.907	0.970	0.992	0.998	1.000		
1.8	0.165	0.463	0.731	0.891	0.964	0.990	0.997	0.999	1.000	
1.9	0.150	0.434	0.704	0.875	0.956	0.987	0.997	0.999	1.000	
2.0	0.135	0.406	0.677	0.857	0.947	0.983	0.995	0.999	1.000	

Table 3. (*Continued*)

λ \ a	0	1	2	3	4	5	6	7	8	9
2.2	0.111	0.355	0.623	0.819	0.928	0.975	0.993	0.998	1.000	
2.4	0.091	0.308	0.570	0.779	0.904	0.964	0.988	0.997	0.999	1.000
2.6	0.074	0.267	0.518	0.736	0.877	0.951	0.983	0.995	0.999	1.000
2.8	0.061	0.231	0.469	0.692	0.848	0.935	0.976	0.992	0.998	0.999
3.0	0.050	0.199	0.423	0.647	0.815	0.916	0.966	0.988	0.996	0.999
3.2	0.041	0.171	0.380	0.603	0.781	0.895	0.955	0.983	0.994	0.998
3.4	0.033	0.147	0.340	0.558	0.744	0.871	0.942	0.977	0.992	0.997
3.6	0.027	0.126	0.303	0.515	0.706	0.844	0.927	0.969	0.988	0.996
3.8	0.022	0.107	0.269	0.473	0.668	0.816	0.909	0.960	0.984	0.994
4.0	0.018	0.092	0.238	0.433	0.629	0.785	0.889	0.949	0.979	0.992
4.2	0.015	0.078	0.210	0.395	0.590	0.753	0.867	0.936	0.972	0.989
4.4	0.012	0.066	0.185	0.359	0.551	0.720	0.844	0.921	0.964	0.985
4.6	0.010	0.056	0.163	0.326	0.513	0.686	0.818	0.905	0.955	0.980
4.8	0.008	0.048	0.143	0.294	0.476	0.651	0.791	0.887	0.944	0.975
5.0	0.007	0.040	0.125	0.265	0.440	0.616	0.762	0.867	0.932	0.968
5.2	0.006	0.034	0.109	0.238	0.406	0.581	0.732	0.845	0.918	0.960
5.4	0.005	0.029	0.095	0.213	0.373	0.546	0.702	0.822	0.903	0.951
5.6	0.004	0.024	0.082	0.191	0.342	0.512	0.670	0.797	0.886	0.941
5.8	0.003	0.021	0.072	0.170	0.313	0.478	0.638	0.771	0.867	0.929
6.0	0.002	0.017	0.062	0.151	0.285	0.446	0.606	0.744	0.847	0.916

λ \ a	10	11	12	13	14	15	16
2.8	1.000						
3.0	1.000						
3.2	1.000						
3.4	0.999	1.000					
3.6	0.999	1.000					
3.8	0.998	0.999	1.000				
4.0	0.997	0.999	1.000				
4.2	0.996	0.999	1.000				
4.4	0.994	0.998	0.999	1.000			
4.6	0.992	0.997	0.999	1.000			
4.8	0.990	0.996	0.999	1.000			
5.0	0.986	0.995	0.998	0.999	1.000		
5.2	0.982	0.993	0.997	0.999	1.000		
5.4	0.977	0.990	0.996	0.999	1.000		
5.6	0.972	0.988	0.995	0.998	0.999	1.000	
5.8	0.965	0.984	0.993	0.997	0.999	1.000	
6.0	0.957	0.980	0.991	0.996	0.999	0.999	1.000

Table 3. (*Continued*)

λ \ a	0	1	2	3	4	5	6	7	8	9
6.2	0.002	0.015	0.054	0.134	0.259	0.414	0.574	0.716	0.826	0.902
6.4	0.002	0.012	0.046	0.119	0.235	0.384	0.542	0.687	0.803	0.886
6.6	0.001	0.010	0.040	0.105	0.213	0.355	0.511	0.658	0.780	0.869
6.8	0.001	0.009	0.034	0.093	0.192	0.327	0.480	0.628	0.755	0.850
7.0	0.001	0.007	0.030	0.082	0.173	0.301	0.450	0.599	0.729	0.830
7.2	0.001	0.006	0.025	0.072	0.156	0.276	0.420	0.569	0.703	0.810
7.4	0.001	0.005	0.022	0.063	0.140	0.253	0.392	0.539	0.676	0.788
7.6	0.001	0.004	0.019	0.055	0.125	0.231	0.365	0.510	0.648	0.765
7.8	0.000	0.004	0.016	0.048	0.112	0.210	0.338	0.481	0.620	0.741
8.0	0.000	0.003	0.014	0.042	0.100	0.191	0.313	0.453	0.593	0.717
8.5	0.000	0.002	0.009	0.030	0.074	0.150	0.256	0.386	0.523	0.653
9.0	0.000	0.001	0.006	0.021	0.055	0.116	0.207	0.324	0.456	0.587
9.5	0.000	0.001	0.004	0.015	0.040	0.089	0.165	0.269	0.392	0.522
10.0	0.000	0.000	0.003	0.010	0.029	0.067	0.130	0.220	0.333	0.458

λ	10	11	12	13	14	15	16	17	18	19
6.2	0.949	0.975	0.989	0.995	0.998	0.999	1.000			
6.4	0.939	0.969	0.986	0.994	0.997	0.999	1.000			
6.6	0.927	0.963	0.982	0.992	0.997	0.999	0.999	1.000		
6.8	0.915	0.955	0.978	0.990	0.996	0.998	0.999	1.000		
7.0	0.901	0.947	0.973	0.987	0.994	0.998	0.999	1.000		
7.2	0.887	0.937	0.967	0.984	0.993	0.997	0.999	0.999	1.000	
7.4	0.871	0.926	0.961	0.980	0.991	0.996	0.998	0.999	1.000	
7.6	0.854	0.915	0.954	0.976	0.989	0.995	0.998	0.999	1.000	
7.8	0.835	0.902	0.945	0.971	0.986	0.993	0.997	0.999	1.000	
8.0	0.816	0.888	0.936	0.966	0.983	0.992	0.996	0.998	0.999	1.000
8.5	0.763	0.849	0.909	0.949	0.973	0.986	0.993	0.997	0.999	0.999
9.0	0.706	0.803	0.876	0.926	0.959	0.978	0.989	0.995	0.998	0.999
9.5	0.645	0.752	0.836	0.898	0.940	0.967	0.982	0.991	0.996	0.998
10.0	0.583	0.697	0.792	0.864	0.917	0.951	0.973	0.986	0.993	0.997

λ	20	21	22
8.5	1.000		
9.0	1.000		
9.5	0.999	1.000	
10.0	0.998	0.999	1.000

Table 3. (*Continued*)

λ \ a	0	1	2	3	4	5	6	7	8	9
10.5	0.000	0.000	0.002	0.007	0.021	0.050	0.102	0.179	0.279	0.397
11.0	0.000	0.000	0.001	0.005	0.015	0.038	0.079	0.143	0.232	0.341
11.5	0.000	0.000	0.001	0.003	0.011	0.028	0.060	0.114	0.191	0.289
12.0	0.000	0.000	0.001	0.002	0.008	0.020	0.046	0.090	0.155	0.242
12.5	0.000	0.000	0.000	0.002	0.005	0.015	0.035	0.070	0.125	0.201
13.0	0.000	0.000	0.000	0.001	0.004	0.011	0.026	0.054	0.100	0.166
13.5	0.000	0.000	0.000	0.001	0.003	0.008	0.019	0.041	0.079	0.135
14.0	0.000	0.000	0.000	0.000	0.002	0.006	0.014	0.032	0.062	0.109
14.5	0.000	0.000	0.000	0.000	0.001	0.004	0.010	0.024	0.048	0.088
15.0	0.000	0.000	0.000	0.000	0.001	0.003	0.008	0.018	0.037	0.070

λ \ a	10	11	12	13	14	15	16	17	18	19
10.5	0.521	0.639	0.742	0.825	0.888	0.932	0.960	0.978	0.988	0.994
11.0	0.460	0.579	0.689	0.781	0.854	0.907	0.944	0.968	0.982	0.991
11.5	0.402	0.520	0.633	0.733	0.815	0.878	0.924	0.954	0.974	0.986
12.0	0.347	0.462	0.576	0.682	0.772	0.844	0.899	0.937	0.963	0.979
12.5	0.297	0.406	0.519	0.628	0.725	0.806	0.869	0.916	0.948	0.969
13.0	0.252	0.353	0.463	0.573	0.675	0.764	0.835	0.890	0.930	0.957
13.5	0.211	0.304	0.409	0.518	0.623	0.718	0.798	0.861	0.908	0.942
14.0	0.176	0.260	0.358	0.464	0.570	0.669	0.756	0.827	0.883	0.923
14.5	0.145	0.220	0.311	0.413	0.518	0.619	0.711	0.790	0.853	0.901
15.0	0.118	0.185	0.268	0.363	0.466	0.568	0.664	0.749	0.819	0.875

λ \ a	20	21	22	23	24	25	26	27	28	29
10.5	0.997	0.999	0.999	1.000						
11.0	0.995	0.998	0.999	1.000						
11.5	0.992	0.996	0.998	0.999	1.000					
12.0	0.988	0.994	0.997	0.999	0.999	1.000				
12.5	0.983	0.991	0.995	0.998	0.999	0.999	1.000			
13.0	0.975	0.986	0.992	0.996	0.998	0.999	1.000			
13.5	0.965	0.980	0.989	0.994	0.997	0.998	0.999	1.000		
14.0	0.952	0.971	0.983	0.991	0.995	0.997	0.999	0.999	1.000	
14.5	0.936	0.960	0.976	0.986	0.992	0.996	0.998	0.999	0.999	1.000
15.0	0.917	0.947	0.967	0.981	0.989	0.994	0.997	0.998	0.999	1.000

Table 3. (*Continued*)

λ \ a	4	5	6	7	8	9	10	11	12	13
16	0.000	0.001	0.004	0.010	0.022	0.043	0.077	0.127	0.193	0.275
17	0.000	0.001	0.002	0.005	0.013	0.026	0.049	0.085	0.135	0.201
18	0.000	0.000	0.001	0.003	0.007	0.015	0.030	0.055	0.092	0.143
19	0.000	0.000	0.001	0.002	0.004	0.009	0.018	0.035	0.061	0.098
20	0.000	0.000	0.000	0.001	0.002	0.005	0.011	0.021	0.039	0.066
21	0.000	0.000	0.000	0.000	0.001	0.003	0.006	0.013	0.025	0.043
22	0.000	0.000	0.000	0.000	0.001	0.002	0.004	0.008	0.015	0.028
23	0.000	0.000	0.000	0.000	0.000	0.001	0.002	0.004	0.009	0.017
24	0.000	0.000	0.000	0.000	0.000	0.000	0.001	0.003	0.005	0.011
25	0.000	0.000	0.000	0.000	0.000	0.000	0.001	0.001	0.003	0.006

	14	15	16	17	18	19	20	21	22	23
16	0.368	0.467	0.566	0.659	0.742	0.812	0.868	0.911	0.942	0.963
17	0.281	0.371	0.468	0.564	0.655	0.736	0.805	0.861	0.905	0.937
18	0.208	0.287	0.375	0.469	0.562	0.651	0.731	0.799	0.855	0.899
19	0.150	0.215	0.292	0.378	0.469	0.561	0.647	0.725	0.793	0.849
20	0.105	0.157	0.221	0.297	0.381	0.470	0.559	0.644	0.721	0.787
21	0.072	0.111	0.163	0.227	0.302	0.384	0.471	0.558	0.640	0.716
22	0.048	0.077	0.117	0.169	0.232	0.306	0.387	0.472	0.556	0.637
23	0.031	0.052	0.082	0.123	0.175	0.238	0.310	0.389	0.472	0.555
24	0.020	0.034	0.056	0.087	0.128	0.180	0.243	0.314	0.392	0.473
25	0.012	0.022	0.038	0.060	0.092	0.134	0.185	0.247	0.318	0.394

	24	25	26	27	28	29	30	31	32	33
16	0.978	0.987	0.993	0.996	0.998	0.999	0.999	1.000		
17	0.959	0.975	0.985	0.991	0.995	0.997	0.999	0.999	1.000	
18	0.932	0.955	0.972	0.983	0.990	0.994	0.997	0.998	0.999	1.000
19	0.893	0.927	0.951	0.969	0.980	0.988	0.993	0.996	0.998	0.999
20	0.843	0.888	0.922	0.948	0.966	0.978	0.987	0.992	0.995	0.997
21	0.782	0.838	0.883	0.917	0.944	0.963	0.976	0.985	0.991	0.994
22	0.712	0.777	0.832	0.877	0.913	0.940	0.959	0.973	0.983	0.989
23	0.635	0.708	0.772	0.827	0.873	0.908	0.936	0.956	0.971	0.981
24	0.554	0.632	0.704	0.768	0.823	0.868	0.904	0.932	0.953	0.969
25	0.473	0.553	0.629	0.700	0.763	0.818	0.863	0.900	0.929	0.950

	34	35	36	37	38	39	40	41	42	43
19	0.999	1.000								
20	0.999	0.999	1.000							
21	0.997	0.998	0.999	0.999	1.000					
22	0.994	0.996	0.998	0.999	0.999	1.000				
23	0.988	0.993	0.996	0.997	0.999	0.999	1.000			
24	0.979	0.987	0.992	0.995	0.997	0.998	0.999	0.999	1.000	
25	0.966	0.978	0.985	0.991	0.991	0.997	0.998	0.999	0.999	1.000

Table 4. Normal curve areas
Standard normal probability in right-hand tail
(for negative values of z areas are found by symmetry)

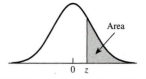

Area

0   z

				Second decimal place of z					
z	.00	.01	.02	.03	.04	.05	.06	.07	.08	.09
0.0	.5000	.4960	.4920	.4880	.4840	.4801	.4761	.4721	.4681	.4641
0.1	.4602	.4562	.4522	.4483	.4443	.4404	.4364	.4325	.4286	.4247
0.2	.4207	.4168	.4129	.4090	.4052	.4013	.3974	.3936	.3897	.3859
0.3	.3821	.3783	.3745	.3707	.3669	.3632	.3594	.3557	.3520	.3483
0.4	.3446	.3409	.3372	.3336	.3300	.3264	.3228	.3192	.3156	.3121
0.5	.3085	.3050	.3015	.2981	.2946	.2912	.2877	.2843	.2810	.2776
0.6	.2743	.2709	.2676	.2643	.2611	.2578	.2546	.2514	.2483	.2451
0.7	.2420	.2389	.2358	.2327	.2296	.2266	.2236	.2206	.2177	.2148
0.8	.2119	.2090	.2061	.2033	.2005	.1977	.1949	.1922	.1894	.1867
0.9	.1841	.1814	.1788	.1762	.1736	.1711	.1685	.1660	.1635	.1611
1.0	.1587	.1562	.1539	.1515	.1492	.1469	.1446	.1423	.1401	.1379
1.1	.1357	.1335	.1314	.1292	.1271	.1251	.1230	.1210	.1190	.1170
1.2	.1151	.1131	.1112	.1093	.1075	.1056	.1038	.1020	.1003	.0985
1.3	.0968	.0951	.0934	.0918	.0901	.0885	.0869	.0853	.0838	.0823
1.4	.0808	.0793	.0778	.0764	.0749	.0735	.0722	.0708	.0694	.0681
1.5	.0668	.0655	.0643	.0630	.0618	.0606	.0594	.0582	.0571	.0559
1.6	.0548	.0537	.0526	.0516	.0505	.0495	.0485	.0475	.0465	.0455
1.7	.0446	.0436	.0427	.0418	.0409	.0401	.0392	.0384	.0375	.0367
1.8	.0359	.0352	.0344	.0336	.0329	.0322	.0314	.0307	.0301	.0294
1.9	.0287	.0281	.0274	.0268	.0262	.0256	.0250	.0244	.0239	.0233
2.0	.0228	.0222	.0217	.0212	.0207	.0202	.0197	.0192	.0188	.0183
2.1	.0179	.0174	.0170	.0166	.0162	.0158	.0154	.0150	.0146	.0143
2.2	.0139	.0136	.0132	.0129	.0125	.0122	.0119	.0116	.0113	.0110
2.3	.0107	.0104	.0102	.0099	.0096	.0094	.0091	.0089	.0087	.0084
2.4	.0082	.0080	.0078	.0075	.0073	.0071	.0069	.0068	.0066	.0064
2.5	.0062	.0060	.0059	.0057	.0055	.0054	.0052	.0051	.0049	.0048
2.6	.0047	.0045	.0044	.0043	.0041	.0040	.0039	.0038	.0037	.0036
2.7	.0035	.0034	.0033	.0032	.0031	.0030	.0029	.0028	.0027	.0026
2.8	.0026	.0025	.0024	.0023	.0023	.0022	.0021	.0021	.0020	.0019
2.9	.0019	.0018	.0017	.0017	.0016	.0016	.0015	.0015	.0014	.0014
3.0	.00135									
3.5	.000 233									
4.0	.000 031 7									
4.5	.000 003 40									
5.0	.000 000 287									

From R. E. Walpole, *Introduction to Statistics* (New York: Macmillan, 1968).

Table 5. Percentage points of the _t_ distributions

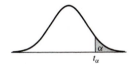

$t_{.100}$	$t_{.050}$	$t_{.025}$	$t_{.010}$	$t_{.005}$	d.f.
3.078	6.314	12.706	31.821	63.657	1
1.886	2.920	4.303	6.965	9.925	2
1.638	2.353	3.182	4.541	5.841	3
1.533	2.132	2.776	3.747	4.604	4
1.476	2.015	2.571	3.365	4.032	5
1.440	1.943	2.447	3.143	3.707	6
1.415	1.895	2.365	2.998	3.499	7
1.397	1.860	2.306	2.896	3.355	8
1.383	1.833	2.262	2.821	3.250	9
1.372	1.812	2.228	2.764	3.169	10
1.363	1.796	2.201	2.718	3.106	11
1.356	1.782	2.179	2.681	3.055	12
1.350	1.771	2.160	2.650	3.012	13
1.345	1.761	2.145	2.624	2.977	14
1.341	1.753	2.131	2.602	2.947	15
1.337	1.746	2.120	2.583	2.921	16
1.333	1.740	2.110	2.567	2.898	17
1.330	1.734	2.101	2.552	2.878	18
1.328	1.729	2.093	2.539	2.861	19
1.325	1.725	2.086	2.528	2.845	20
1.323	1.721	2.080	2.518	2.831	21
1.321	1.717	2.074	2.508	2.819	22
1.319	1.714	2.069	2.500	2.807	23
1.318	1.711	2.064	2.492	2.797	24
1.316	1.708	2.060	2.485	2.787	25
1.315	1.706	2.056	2.479	2.779	26
1.314	1.703	2.052	2.473	2.771	27
1.313	1.701	2.048	2.467	2.763	28
1.311	1.699	2.045	2.462	2.756	29
1.282	1.645	1.960	2.326	2.576	inf.

From "Table of Percentage Points of the _t_-Distribution." Computed by Maxine Merrington, _Biometrika_, Vol. 32 (1941), p. 300. Reproduced by permission of Professor E. S. Pearson.

Table 6. Percentage points of the χ^2 distributions

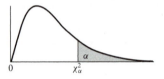

d.f.	$\chi^2_{0.995}$	$\chi^2_{0.990}$	$\chi^2_{0.975}$	$\chi^2_{0.950}$	$\chi^2_{0.900}$
1	0.0000393	0.0001571	0.0009821	0.0039321	0.0157908
2	0.0100251	0.0201007	0.0506356	0.102587	0.210720
3	0.0717212	0.114832	0.215795	0.351846	0.584375
4	0.206990	0.297110	0.484419	0.710721	1.063623
5	0.411740	0.554300	0.831211	1.145476	1.61031
6	0.675727	0.872085	1.237347	1.63539	2.20413
7	0.989265	1.239043	1.68987	2.16735	2.83311
8	1.344419	1.646482	2.17973	2.73264	3.48954
9	1.734926	2.087912	2.70039	3.32511	4.16816
10	2.15585	2.55821	3.24697	3.94030	4.86518
11	2.60321	3.05347	3.81575	4.57481	5.57779
12	3.07382	3.57056	4.40379	5.22603	6.30380
13	3.56503	4.10691	5.00874	5.89186	7.04150
14	4.07468	4.66043	5.62872	6.57063	7.78953
15	4.60094	5.22935	6.26214	7.26094	8.54675
16	5.14224	5.81221	6.90766	7.96164	9.31223
17	5.69724	6.40776	7.56418	8.67176	10.0852
18	6.26481	7.01491	8.23075	9.39046	10.8649
19	6.84398	7.63273	8.90655	10.1170	11.6509
20	7.43386	8.26040	9.59083	10.8508	12.4426
21	8.03366	8.89720	10.28293	11.5913	13.2396
22	8.64272	9.54249	10.9823	12.3380	14.0415
23	9.26042	10.19567	11.6885	13.0905	14.8479
24	9.88623	10.8564	12.4011	13.8484	15.6587
25	10.5197	11.5240	13.1197	14.6114	16.4734
26	11.1603	12.1981	13.8439	15.3791	17.2919
27	11.8076	12.8786	14.5733	16.1513	18.1138
28	12.4613	13.5648	15.3079	16.9279	18.9392
29	13.1211	14.2565	16.0471	17.7083	19.7677
30	13.7867	14.9535	16.7908	18.4926	20.5992
40	20.7065	22.1643	24.4331	26.5093	29.0505
50	27.9907	29.7067	32.3574	34.7642	37.6886
60	35.5346	37.4848	40.4817	43.1879	46.4589
70	43.2752	45.4418	48.7576	51.7393	55.3290
80	51.1720	53.5400	57.1532	60.3915	64.2778
90	59.1963	61.7541	65.6466	69.1260	73.2912
100	67.3276	70.0648	74.2219	77.9295	82.3581

Table 6. (*Continued*)

$\chi^2_{0.100}$	$\chi^2_{0.050}$	$\chi^2_{0.025}$	$\chi^2_{0.010}$	$\chi^2_{0.005}$	d.f.
2.70554	3.84146	5.02389	6.63490	7.87944	1
4.60517	5.99147	7.37776	9.21034	10.5966	2
6.25139	7.81473	9.34840	11.3449	12.8381	3
7.77944	9.48773	11.1433	13.2767	14.8602	4
9.23635	11.0705	12.8325	15.0863	16.7496	5
10.6446	12.5916	14.4494	16.8119	18.5476	6
12.0170	14.0671	16.0128	18.4753	20.2777	7
13.3616	15.5073	17.5346	20.0902	21.9550	8
14.6837	16.9190	19.0228	21.6660	23.5893	9
15.9871	18.3070	20.4831	23.2093	25.1882	10
17.2750	19.6751	21.9200	24.7250	26.7569	11
18.5494	21.0261	23.3367	26.2170	28.2995	12
19.8119	22.3621	24.7356	27.6883	29.8194	13
21.0642	23.6848	26.1190	29.1413	31.3193	14
22.3072	24.9958	27.4884	30.5779	32.8013	15
23.5418	26.2962	28.8454	31.9999	34.2672	16
24.7690	27.5871	30.1910	33.4087	35.7185	17
25.9894	28.8693	31.5264	34.8053	37.1564	18
27.2036	30.1435	32.8523	36.1908	38.5822	19
28.4120	31.4104	34.1696	37.5662	39.9968	20
29.6151	32.6705	35.4789	38.9321	41.4010	21
30.8133	33.9244	36.7807	40.2894	42.7956	22
32.0069	35.1725	38.0757	41.6384	44.1813	23
33.1963	36.4151	39.3641	42.9798	45.5585	24
34.3816	37.6525	40.6465	44.3141	46.9278	25
35.5631	38.8852	41.9232	45.6417	48.2899	26
36.7412	40.1133	43.1944	46.9630	49.6449	27
37.9159	41.3372	44.4607	48.2782	50.9933	28
39.0875	42.5569	45.7222	49.5879	52.3356	29
40.2560	43.7729	46.9792	50.8922	53.6720	30
51.8050	55.7585	59.3417	63.6907	66.7659	40
63.1671	67.5048	71.4202	76.1539	79.4900	50
74.3970	79.0819	83.2976	88.3794	91.9517	60
85.5271	90.5312	95.0231	100.425	104.215	70
96.5782	101.879	106.629	112.329	116.321	80
107.565	113.145	118.136	124.116	128.299	90
118.498	124.342	129.561	135.807	140.169	100

From "Tables of the Percentage Points of the χ^2-Distribution." *Biometrika*, Vol. 32 (1941), pp. 188–189, by Catherine M. Thompson. Reproduced by permission of Professor E. S. Pearson.

Table 7. Percentage points of the F distributions

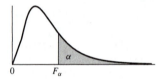

Denominator d.f.	α	Numerator d.f.								
		1	2	3	4	5	6	7	8	9
1	.100	39.86	49.50	53.59	55.83	57.24	58.20	58.91	59.44	59.86
	.050	161.4	199.5	215.7	224.6	230.2	234.0	236.8	238.9	240.5
	.025	647.8	799.5	864.2	899.6	921.8	937.1	948.2	956.7	963.3
	.010	4052	4999.5	5403	5625	5764	5859	5928	5982	6022
	.005	16211	20000	21615	22500	23056	23437	23715	23925	24091
2	.100	8.53	9.00	9.16	9.24	9.29	9.33	9.35	9.37	9.38
	.050	18.51	19.00	19.16	19.25	19.30	19.33	19.35	19.37	19.38
	.025	38.51	39.00	39.17	39.25	39.30	39.33	39.36	39.37	39.39
	.010	98.50	99.00	99.17	99.25	99.30	99.33	99.36	99.37	99.39
	.005	198.5	199.0	199.2	199.2	199.3	199.3	199.4	199.4	199.4
3	.100	5.54	5.46	5.39	5.34	5.31	5.28	5.27	5.25	5.24
	.050	10.13	9.55	9.28	9.12	9.01	8.94	8.89	8.85	8.81
	.025	17.44	16.04	15.44	15.10	14.88	14.73	14.62	14.54	14.47
	.010	34.12	30.82	29.46	28.71	28.24	27.91	27.67	27.49	27.35
	.005	55.55	49.80	47.47	46.19	45.39	44.84	44.43	44.13	43.88
4	.100	4.54	4.32	4.19	4.11	4.05	4.01	3.98	3.95	3.94
	.050	7.71	6.94	6.59	6.39	6.26	6.16	6.09	6.04	6.00
	.025	12.22	10.65	9.98	9.60	9.36	9.20	9.07	8.98	8.90
	.010	21.20	18.00	16.69	15.98	15.52	15.21	14.98	14.80	14.66
	.005	31.33	26.28	24.26	23.15	22.46	21.97	21.62	21.35	21.14
5	.100	4.06	3.78	3.62	3.52	3.45	3.40	3.37	3.34	3.32
	.050	6.61	5.79	5.41	5.19	5.05	4.95	4.88	4.82	4.77
	.025	10.01	8.43	7.76	7.39	7.15	6.98	6.85	6.76	6.68
	.010	16.26	13.27	12.06	11.39	10.97	10.67	10.46	10.29	10.16
	.005	22.78	18.31	16.53	15.56	14.94	14.51	14.20	13.96	13.77
6	.100	3.78	3.46	3.29	3.18	3.11	3.05	3.01	2.98	2.96
	.050	5.99	5.14	4.76	4.53	4.39	4.28	4.21	4.15	4.10
	.025	8.81	7.26	6.60	6.23	5.99	5.82	5.70	5.60	5.52
	.010	13.75	10.92	9.78	9.15	8.75	8.47	8.26	8.10	7.98
	.005	18.63	14.54	12.92	12.03	11.46	11.07	10.79	10.57	10.39
7	.100	3.59	3.26	3.07	2.96	2.88	2.83	2.78	2.75	2.72
	.050	5.59	4.74	4.35	4.12	3.97	3.87	3.79	3.73	3.68
	.025	8.07	6.54	5.89	5.52	5.29	5.12	4.99	4.90	4.82
	.010	12.25	9.55	8.45	7.85	7.46	7.19	6.99	6.84	6.72
	.005	16.24	12.40	10.88	10.05	9.52	9.16	8.89	8.68	8.51

Table 7. (*Continued*)

$$F_\alpha$$

Numerator d.f.											
10	12	15	20	24	30	40	60	120	∞	α	d.f.
60.19	60.71	61.22	61.74	62.00	62.26	62.53	62.79	63.06	63.33	.100	1
241.9	243.9	245.9	248.0	249.1	250.1	251.1	252.2	253.3	254.3	.050	
968.6	976.7	984.9	993.1	997.2	1001	1006	1010	1014	1018	.025	
6056	6106	6157	6209	6235	6261	6287	6313	6339	6366	.010	
24224	24426	24630	24836	24940	25044	25148	25253	25359	25465	.005	
9.39	9.41	9.42	9.44	9.45	9.46	9.47	9.47	9.48	9.49	.100	2
19.40	19.41	19.43	19.45	19.45	19.45	19.47	19.48	19.49	19.50	.050	
39.40	39.41	39.43	39.45	39.46	39.46	39.47	39.48	39.49	39.50	.025	
99.40	99.42	99.43	99.45	99.46	99.47	99.47	99.48	99.49	99.50	.010	
199.4	199.4	199.4	199.4	199.5	199.5	199.5	199.5	199.5	199.5	.005	
5.23	5.22	5.20	5.18	5.18	5.17	5.16	5.15	5.14	5.13	.100	3
8.79	8.74	8.70	8.66	8.64	8.62	8.59	8.57	8.55	8.53	.050	
14.42	14.34	14.25	14.17	14.12	14.08	14.04	13.99	13.95	13.90	.025	
27.23	27.05	26.87	26.69	26.60	26.50	26.41	26.32	26.22	26.13	.010	
43.69	43.39	43.08	42.78	42.62	42.47	42.31	42.15	41.99	41.83	.005	
3.92	3.90	3.87	3.84	3.83	3.82	3.80	3.79	3.78	3.76	.100	4
5.96	5.91	5.86	5.80	5.77	5.75	5.72	5.69	5.66	5.63	.050	
8.84	8.75	8.66	8.56	8.51	8.46	8.41	8.36	8.31	8.26	.025	
14.55	14.37	14.20	14.02	13.93	13.84	13.75	13.65	13.56	13.46	.010	
20.97	20.70	20.44	20.17	20.03	19.89	19.75	19.61	19.47	19.32	.005	
3.30	3.27	3.24	3.21	3.19	3.17	3.16	3.14	3.12	3.10	.100	5
4.74	4.68	4.62	4.56	4.53	4.50	4.46	4.43	4.40	4.36	.050	
6.62	6.52	6.43	6.33	6.28	6.23	6.18	6.12	6.07	6.02	.025	
10.05	9.89	9.72	9.55	9.47	9.38	9.29	9.20	9.11	9.02	.010	
13.62	13.38	13.15	12.90	12.78	12.66	12.53	12.40	12.27	12.14	.005	
2.94	2.90	2.87	2.84	2.82	2.80	2.78	2.76	2.74	2.72	.100	6
4.06	4.00	3.94	3.87	3.84	3.81	3.77	3.74	3.70	3.67	.050	
5.46	5.37	5.27	5.17	5.12	5.07	5.01	4.96	4.90	4.85	.025	
7.87	7.72	7.56	7.40	7.31	7.23	7.14	7.06	6.97	6.88	.010	
10.25	10.03	9.81	9.59	9.47	9.36	9.24	9.12	9.00	8.88	.005	
2.70	2.67	2.63	2.59	2.58	2.56	2.54	2.51	2.49	2.47	.100	7
3.64	3.57	3.51	3.44	3.41	3.38	3.34	3.30	3.27	3.23	.050	
4.76	4.67	4.57	4.47	4.42	4.36	4.31	4.25	4.20	4.14	.025	
6.62	6.47	6.31	6.16	6.07	5.99	5.91	5.82	5.74	5.65	.010	
8.38	8.18	7.97	7.75	7.65	7.53	7.42	7.31	7.19	7.08	.005	

Table 7. (*Continued*)

$$F_\alpha$$

Denominator d.f.		Numerator d.f.								
	α	1	2	3	4	5	6	7	8	9
8	.100	3.46	3.11	2.92	2.81	2.73	2.67	2.62	2.59	2.56
	.050	5.32	4.46	4.07	3.84	3.69	3.58	3.50	3.44	3.39
	.025	7.57	6.06	5.42	5.05	4.82	4.65	4.53	4.43	4.36
	.010	11.26	8.65	7.59	7.01	6.63	6.37	6.18	6.03	5.91
	.005	14.69	11.04	9.60	8.81	8.30	7.95	7.69	7.50	7.34
9	.100	3.36	3.01	2.81	2.69	2.61	2.55	2.51	2.47	2.44
	.050	5.12	4.26	3.86	3.63	3.48	3.37	3.29	3.23	3.18
	.025	7.21	5.71	5.08	4.72	4.48	4.32	4.20	4.10	4.03
	.010	10.56	8.02	6.99	6.42	6.06	5.80	5.61	5.47	5.35
	.005	13.61	10.11	8.72	7.96	7.47	7.13	6.88	6.69	6.54
10	.100	3.29	2.92	2.73	2.61	2.52	2.46	2.41	2.38	2.35
	.050	4.96	4.10	3.71	3.48	3.33	3.22	3.14	3.07	3.02
	.025	6.94	5.46	4.83	4.47	4.24	4.07	3.95	3.85	3.78
	.010	10.04	7.56	6.55	5.99	5.64	5.39	5.20	5.06	4.94
	.005	12.83	9.43	8.08	7.34	6.87	6.54	6.30	6.12	5.97
11	.100	3.23	2.86	2.66	2.54	2.45	2.39	2.34	2.30	2.27
	.050	4.84	3.98	3.59	3.36	3.20	3.09	3.01	2.95	2.90
	.025	6.72	5.26	4.63	4.28	4.04	3.88	3.76	3.66	3.59
	.010	9.65	7.21	6.22	5.67	5.32	5.07	4.89	4.74	4.63
	.005	12.23	8.91	7.60	6.88	6.42	6.10	5.86	5.68	5.54
12	.100	3.18	2.81	2.61	2.48	2.39	2.33	2.28	2.24	2.21
	.050	4.75	3.89	3.49	3.26	3.11	3.00	2.91	2.85	2.80
	.025	6.55	5.10	4.47	4.12	3.89	3.73	3.61	3.51	3.44
	.010	9.33	6.93	5.95	5.41	5.06	4.82	4.64	4.50	4.39
	.005	11.75	8.51	7.23	6.52	6.07	5.76	5.52	5.35	5.20
13	.100	3.14	2.76	2.56	2.43	2.35	2.28	2.23	2.20	2.16
	.050	4.67	3.81	3.41	3.18	3.03	2.92	2.83	2.77	2.71
	.025	6.41	4.97	4.35	4.00	3.77	3.60	3.48	3.39	3.31
	.010	9.07	6.70	5.74	5.21	4.86	4.62	4.44	4.30	4.19
	.005	11.37	8.19	6.93	6.23	5.79	5.48	5.25	5.08	4.94
14	.100	3.10	2.73	2.52	2.39	2.31	2.24	2.19	2.15	2.12
	.050	4.60	3.74	3.34	3.11	2.96	2.85	2.76	2.70	2.65
	.025	6.30	4.86	4.24	3.89	3.66	3.50	3.38	3.29	3.21
	.010	8.86	6.51	5.56	5.04	4.69	4.46	4.28	4.14	4.03
	.005	11.06	7.92	6.68	6.00	5.56	5.26	5.03	4.86	4.72

Table 7. (*Continued*)

$$F_\alpha$$

				Numerator d.f.								
10	12	15	20	24	30	40	60	120	∞	α	d.f.
2.54	2.50	2.46	2.42	2.40	2.38	2.36	2.34	2.32	2.29	.100	8
3.35	3.28	3.22	3.15	3.12	3.08	3.04	3.01	2.97	2.93	.050	
4.30	4.20	4.10	4.00	3.95	3.89	3.84	3.78	3.73	3.67	.025	
5.81	5.67	5.52	5.36	5.28	5.20	5.12	5.03	4.95	4.86	.010	
7.21	7.01	6.81	6.61	6.50	6.40	6.29	6.18	6.06	5.95	.005	
2.42	2.38	2.34	2.30	2.28	2.25	2.23	2.21	2.18	2.16	.100	9
3.14	3.07	3.01	2.94	2.90	2.86	2.83	2.79	2.75	2.71	.050	
3.96	3.87	3.77	3.67	3.61	3.56	3.51	3.45	3.39	3.33	.025	
5.26	5.11	4.96	4.81	4.73	4.65	4.57	4.48	4.40	4.31	.010	
6.42	6.23	6.03	5.83	5.73	5.62	5.52	5.41	5.30	5.19	.005	
2.32	2.28	2.24	2.20	2.18	2.16	2.13	2.11	2.08	2.06	.100	10
2.98	2.91	2.85	2.77	2.74	2.70	2.66	2.62	2.58	2.54	.050	
3.72	3.62	3.52	3.42	3.37	3.31	3.26	3.20	3.14	3.08	.025	
4.85	4.71	4.56	4.41	4.33	4.25	4.17	4.08	4.00	3.91	.010	
5.85	5.66	5.47	5.27	5.17	5.07	4.97	4.86	4.75	4.64	.005	
2.25	2.21	2.17	2.12	2.10	2.08	2.05	2.03	2.00	1.97	.100	11
2.85	2.79	2.72	2.65	2.61	2.57	2.53	2.49	2.45	2.40	.050	
3.53	3.43	3.33	3.23	3.17	3.12	3.06	3.00	2.94	2.88	.025	
4.54	4.40	4.25	4.10	4.02	3.94	3.86	3.78	3.69	3.60	.010	
5.42	5.24	5.05	4.86	4.76	4.65	4.55	4.44	4.34	4.23	.005	
2.19	2.15	2.10	2.06	2.04	2.01	1.99	1.96	1.93	1.90	.100	12
2.75	2.69	2.62	2.54	2.51	2.47	2.43	2.38	2.34	2.30	.050	
3.37	3.28	3.18	3.07	3.02	2.96	2.91	2.85	2.79	2.72	.025	
4.30	4.16	4.01	3.86	3.78	3.70	3.62	3.54	3.45	3.36	.010	
5.09	4.91	4.72	4.53	4.43	4.33	4.23	4.12	4.01	3.90	.005	
2.14	2.10	2.05	2.01	1.98	1.96	1.93	1.90	1.88	1.85	.100	13
2.67	2.60	2.53	2.46	2.42	2.38	2.34	2.30	2.25	2.21	.050	
3.25	3.15	3.05	2.95	2.89	2.84	2.78	2.72	2.66	2.60	.025	
4.10	3.96	3.82	3.66	3.59	3.51	3.43	3.34	3.25	3.17	.010	
4.82	4.64	4.46	4.27	4.17	4.07	3.97	3.87	3.76	3.65	.005	
2.10	2.05	2.01	1.96	1.94	1.91	1.89	1.86	1.83	1.80	.100	14
2.60	2.53	2.46	2.39	2.35	2.31	2.27	2.22	2.18	2.13	.050	
3.15	3.05	2.95	2.84	2.79	2.73	2.67	2.61	2.55	2.49	.025	
3.94	3.80	3.66	3.51	3.43	3.35	3.27	3.18	3.09	3.00	.010	
4.60	4.43	4.25	4.06	3.96	3.86	3.76	3.66	3.55	3.44	.005	

Table 7. (*Continued*)

$$F_\alpha$$

Denominator d.f.		Numerator d.f.								
	α	1	2	3	4	5	6	7	8	9
15	.100	3.07	2.70	2.49	2.36	2.27	2.21	2.16	2.12	2.09
	.050	4.54	3.68	3.29	3.06	2.90	2.79	2.71	2.64	2.59
	.025	6.20	4.77	4.15	3.80	3.58	3.41	3.29	3.20	3.12
	.010	8.68	6.36	5.42	4.89	4.56	4.32	4.14	4.00	3.89
	.005	10.80	7.70	6.48	5.80	5.37	5.07	4.85	4.67	4.54
16	.100	3.05	2.67	2.46	2.33	2.24	2.18	2.13	2.09	2.06
	.050	4.49	3.63	3.24	3.01	2.85	2.74	2.66	2.59	2.54
	.025	6.12	4.69	4.08	3.73	3.50	3.34	3.22	3.12	3.05
	.010	8.53	6.23	5.29	4.77	4.44	4.20	4.03	3.89	3.78
	.005	10.58	7.51	6.30	5.64	5.21	4.91	4.69	4.52	4.38
17	.100	3.03	2.64	2.44	2.31	2.22	2.15	2.10	2.06	2.03
	.050	4.45	3.59	3.20	2.96	2.81	2.70	2.61	2.55	2.49
	.025	6.04	4.62	4.01	3.66	3.44	3.28	3.16	3.06	2.98
	.010	8.40	6.11	5.18	4.67	4.34	4.10	3.93	3.79	3.68
	.005	10.38	7.35	6.16	5.50	5.07	4.78	4.56	4.39	4.25
18	.100	3.01	2.62	2.42	2.29	2.20	2.13	2.08	2.04	2.00
	.050	4.41	3.55	3.16	2.93	2.77	2.66	2.58	2.51	2.46
	.025	5.98	4.56	3.95	3.61	3.38	3.22	3.10	3.01	2.93
	.010	8.29	6.01	5.09	4.58	4.25	4.01	3.84	3.71	3.60
	.005	10.22	7.21	6.03	5.37	4.96	4.66	4.44	4.28	4.14
19	.100	2.99	2.61	2.40	2.27	2.18	2.11	2.06	2.02	1.98
	.050	4.38	3.52	3.13	2.90	2.74	2.63	2.54	2.48	2.42
	.025	5.92	4.51	3.90	3.56	3.33	3.17	3.05	2.96	2.88
	.010	8.18	5.93	5.01	4.50	4.17	3.94	3.77	3.63	3.52
	.005	10.07	7.09	5.92	5.27	4.85	4.56	4.34	4.18	4.04
20	.100	2.97	2.59	2.38	2.25	2.16	2.09	2.04	2.00	1.96
	.050	4.35	3.49	3.10	2.87	2.71	2.60	2.51	2.45	2.39
	.025	5.87	4.46	3.86	3.51	3.29	3.13	3.01	2.91	2.84
	.010	8.10	5.85	4.94	4.43	4.10	3.87	3.70	3.56	3.46
	.005	9.94	6.99	5.82	5.17	4.76	4.47	4.26	4.09	3.96
21	.100	2.96	2.57	2.36	2.23	2.14	2.08	2.02	1.98	1.95
	.050	4.32	3.47	3.07	2.84	2.68	2.57	2.49	2.42	2.37
	.025	5.83	4.42	3.82	3.48	3.25	3.09	2.97	2.87	2.80
	.010	8.02	5.78	4.87	4.37	4.04	3.81	3.64	3.51	3.40
	.005	9.83	6.89	5.73	5.09	4.68	4.39	4.18	4.01	3.88

Table 7. (*Continued*)

$$F_\alpha$$

				Numerator d.f.							
10	12	15	20	24	30	40	60	120	∞	α	d.f.
2.06	2.02	1.97	1.92	1.90	1.87	1.85	1.82	1.79	1.76	.100	15
2.54	2.48	2.40	2.33	2.29	2.25	2.20	2.16	2.11	2.07	.050	
3.06	2.96	2.86	2.76	2.70	2.64	2.59	2.52	2.46	2.40	.025	
3.80	3.67	3.52	3.37	3.29	3.21	3.13	3.05	2.96	2.87	.010	
4.42	4.25	4.07	3.88	3.79	3.69	3.58	3.48	3.37	3.26	.005	
2.03	1.99	1.94	1.89	1.87	1.84	1.81	1.78	1.75	1.72	.100	16
2.49	2.42	2.35	2.28	2.24	2.19	2.15	2.11	2.06	2.01	.050	
2.99	2.89	2.79	2.68	2.63	2.57	2.51	2.45	2.38	2.32	.025	
3.69	3.55	3.41	3.26	3.18	3.10	3.02	2.93	2.84	2.75	.010	
4.27	4.10	3.92	3.73	3.64	3.54	3.44	3.33	3.22	3.11	.005	
2.00	1.96	1.91	1.86	1.84	1.81	1.78	1.75	1.72	1.69	.100	17
2.45	2.38	2.31	2.23	2.19	2.15	2.10	2.06	2.01	1.96	.050	
2.92	2.82	2.72	2.62	2.56	2.50	2.44	2.38	2.32	2.25	.025	
3.59	3.46	3.31	3.16	3.08	3.00	2.92	2.83	2.75	2.65	.010	
4.14	3.97	3.79	3.61	3.51	3.41	3.31	3.21	3.10	2.98	.005	
1.98	1.93	1.89	1.84	1.81	1.78	1.75	1.72	1.69	1.66	.100	18
2.41	2.34	2.27	2.19	2.15	2.11	2.06	2.02	1.97	1.92	.050	
2.87	2.77	2.67	2.56	2.50	2.44	2.38	2.32	2.26	2.19	.025	
3.51	3.37	3.23	3.08	3.00	2.92	2.84	2.75	2.66	2.57	.010	
4.03	3.86	3.68	3.50	3.40	3.30	3.20	3.10	2.99	2.87	.005	
1.96	1.91	1.86	1.81	1.79	1.76	1.73	1.70	1.67	1.63	.100	19
2.38	2.31	2.23	2.16	2.11	2.07	2.03	1.98	1.93	1.88	.050	
2.82	2.72	2.62	2.51	2.45	2.39	2.33	2.27	2.20	2.13	.025	
3.43	3.30	3.15	3.00	2.92	2.84	2.76	2.67	2.58	2.49	.010	
3.93	3.76	3.59	3.40	3.31	3.21	3.11	3.00	2.89	2.78	.005	
1.94	1.89	1.84	1.79	1.77	1.74	1.71	1.68	1.64	1.61	.100	20
2.35	2.28	2.20	2.12	2.08	2.04	1.99	1.95	1.90	1.84	.050	
2.77	2.68	2.57	2.46	2.41	2.35	2.29	2.22	2.16	2.09	.025	
3.37	3.23	3.09	2.94	2.86	2.78	2.69	2.61	2.52	2.42	.010	
3.85	3.68	3.50	3.32	3.22	3.12	3.02	2.92	2.81	2.69	.005	
1.92	1.87	1.83	1.78	1.75	1.72	1.69	1.66	1.62	1.59	.100	21
2.32	2.25	2.18	2.10	2.05	2.01	1.96	1.92	1.87	1.81	.050	
2.73	2.64	2.53	2.42	2.37	2.31	2.25	2.18	2.11	2.04	.025	
3.31	3.17	3.03	2.88	2.80	2.72	2.64	2.55	2.46	2.36	.010	
3.77	3.60	3.43	3.24	3.15	3.05	2.95	2.84	2.73	2.61	.005	

Table 7. (*Continued*)

$$F_\alpha$$

Denominator d.f.		Numerator d.f.								
	α	1	2	3	4	5	6	7	8	9
22	.100	2.95	2.56	2.35	2.22	2.13	2.06	2.01	1.97	1.93
	.050	4.30	3.44	3.05	2.82	2.66	2.55	2.46	2.40	2.34
	.025	5.79	4.38	3.78	3.44	3.22	3.05	2.93	2.84	2.76
	.010	7.95	5.72	4.82	4.31	3.99	3.76	3.59	3.45	3.35
	.005	9.73	6.81	5.65	5.02	4.61	4.32	4.11	3.94	3.81
23	.100	2.94	2.55	2.34	2.21	2.11	2.05	1.99	1.95	1.92
	.050	4.28	3.42	3.03	2.80	2.64	2.53	2.44	2.37	2.32
	.025	5.75	4.35	3.75	3.41	3.18	3.02	2.90	2.81	2.73
	.010	7.88	5.66	4.76	4.26	3.94	3.71	3.54	3.41	3.30
	.005	9.63	6.73	5.58	4.95	4.54	4.26	4.05	3.88	3.75
24	.100	2.93	2.54	2.33	2.19	2.10	2.04	1.98	1.94	1.91
	.050	4.26	3.40	3.01	2.78	2.62	2.51	2.42	2.36	2.30
	.025	5.72	4.32	3.72	3.38	3.15	2.99	2.87	2.78	2.70
	.010	7.82	5.61	4.72	4.22	3.90	3.67	3.50	3.36	3.26
	.005	9.55	6.66	5.52	4.89	4.49	4.20	3.99	3.83	3.69
25	.100	2.92	2.53	2.32	2.18	2.09	2.02	1.97	1.93	1.89
	.050	4.24	3.39	2.99	2.76	2.60	2.49	2.40	2.34	2.28
	.025	5.69	4.29	3.69	3.35	3.13	2.97	2.85	2.75	2.68
	.010	7.77	5.57	4.68	4.18	3.85	3.63	3.46	3.32	3.22
	.005	9.48	6.60	5.46	4.84	4.43	4.15	3.94	3.78	3.64
26	.100	2.91	2.52	2.31	2.17	2.08	2.01	1.96	1.92	1.88
	.050	4.23	3.37	2.98	2.74	2.59	2.47	2.39	2.32	2.27
	.025	5.66	4.27	3.67	3.33	3.10	2.94	2.82	2.73	2.65
	.010	7.72	5.53	4.64	4.14	3.82	3.59	3.42	3.29	3.18
	.005	9.41	6.54	5.41	4.79	4.38	4.10	3.89	3.73	3.60
27	.100	2.90	2.51	2.30	2.17	2.07	2.00	1.95	1.91	1.87
	.050	4.21	3.35	2.96	2.73	2.57	2.46	2.37	2.31	2.25
	.025	5.63	4.24	3.65	3.31	3.08	2.92	2.80	2.71	2.63
	.010	7.68	5.49	4.60	4.11	3.78	3.56	3.39	3.26	3.15
	.005	9.34	6.49	5.36	4.74	4.34	4.06	3.85	3.69	3.56
28	.100	2.89	2.50	2.29	2.16	2.06	2.00	1.94	1.90	1.87
	.050	4.20	3.34	2.95	2.71	2.56	2.45	2.36	2.29	2.24
	.025	5.61	4.22	3.63	3.29	3.06	2.90	2.78	2.69	2.61
	.010	7.64	5.45	4.57	4.07	3.75	3.53	3.36	3.23	3.12
	.005	9.28	6.44	5.32	4.70	4.30	4.02	3.81	3.65	3.52

Table 7. (*Continued*)

$$F_\alpha$$

10	12	15	20	24	30	40	60	120	∞	α	d.f.
					Numerator d.f.						
1.90	1.86	1.81	1.76	1.73	1.70	1.67	1.64	1.60	1.57	.100	22
2.30	2.23	2.15	2.07	2.03	1.98	1.94	1.89	1.84	1.78	.050	
2.70	2.60	2.50	2.39	2.33	2.27	2.21	2.14	2.08	2.00	.025	
3.26	3.12	2.98	2.83	2.75	2.67	2.58	2.50	2.40	2.31	.010	
3.70	3.54	3.36	3.18	3.08	2.98	2.88	2.77	2.66	2.55	.005	
1.89	1.84	1.80	1.74	1.72	1.69	1.66	1.62	1.59	1.55	.100	23
2.27	2.20	2.13	2.05	2.01	1.96	1.91	1.86	1.81	1.76	.050	
2.67	2.57	2.47	2.36	2.30	2.24	2.18	2.11	2.04	1.97	.025	
3.21	3.07	2.93	2.78	2.70	2.62	2.54	2.45	2.35	2.26	.010	
3.64	3.47	3.30	3.12	3.02	2.92	2.82	2.71	2.60	2.48	.005	
1.88	1.83	1.78	1.73	1.70	1.67	1.64	1.61	1.57	1.53	.100	24
2.25	2.18	2.11	2.03	1.98	1.94	1.89	1.84	1.79	1.73	.050	
2.64	2.54	2.44	2.33	2.27	2.21	2.15	2.08	2.01	1.94	.025	
3.17	3.03	2.89	2.74	2.66	2.58	2.49	2.40	2.31	2.21	.010	
3.59	3.42	3.25	3.06	2.97	2.87	2.77	2.66	2.55	2.43	.005	
1.87	1.82	1.77	1.72	1.69	1.66	1.63	1.59	1.56	1.52	.100	25
2.24	2.16	2.09	2.01	1.96	1.92	1.87	1.82	1.77	1.71	.050	
2.61	2.51	2.41	2.30	2.24	2.18	2.12	2.05	1.98	1.91	.025	
3.13	2.99	2.85	2.70	2.62	2.54	2.45	2.36	2.27	2.17	.010	
3.54	3.37	3.20	3.01	2.92	2.82	2.72	2.61	2.50	2.38	.005	
1.86	1.81	1.76	1.71	1.68	1.65	1.61	1.58	1.54	1.50	.100	26
2.22	2.15	2.07	1.99	1.95	1.90	1.85	1.80	1.75	1.69	.050	
2.59	2.49	2.39	2.28	2.22	2.16	2.09	2.03	1.95	1.88	.025	
3.09	2.96	2.81	2.66	2.58	2.50	2.42	2.33	2.23	2.13	.010	
3.49	3.33	3.15	2.97	2.87	2.77	2.67	2.56	2.45	2.33	.005	
1.85	1.80	1.75	1.70	1.67	1.64	1.60	1.57	1.53	1.49	.100	27
2.20	2.13	2.06	1.97	1.93	1.88	1.84	1.79	1.73	1.67	.050	
2.57	2.47	2.36	2.25	2.19	2.13	2.07	2.00	1.93	1.85	.025	
3.06	2.93	2.78	2.63	2.55	2.47	2.38	2.29	2.20	2.10	.010	
3.45	3.28	3.11	2.93	2.83	2.73	2.63	2.52	2.41	2.29	.005	
1.84	1.79	1.74	1.69	1.66	1.63	1.59	1.56	1.52	1.48	.100	28
2.19	2.12	2.04	1.96	1.91	1.87	1.82	1.77	1.71	1.65	.050	
2.55	2.45	2.34	2.23	2.17	2.11	2.05	1.98	1.91	1.83	.025	
3.03	2.90	2.75	2.60	2.52	2.44	2.35	2.26	2.17	2.06	.010	
3.41	3.25	3.07	2.89	2.79	2.69	2.59	2.48	2.37	2.25	.005	

Table 7. (*Continued*)

$$F_\alpha$$

Denominator d.f.		Numerator d.f.								
	α	1	2	3	4	5	6	7	8	9
29	.100	2.89	2.50	2.28	2.15	2.06	1.99	1.93	1.89	1.86
	.050	4.18	3.33	2.93	2.70	2.55	2.43	2.35	2.28	2.22
	.025	5.59	4.20	3.61	3.27	3.04	2.88	2.76	2.67	2.59
	.010	7.60	5.42	4.54	4.04	3.73	3.50	3.33	3.20	3.09
	.005	9.23	6.40	5.28	4.66	4.26	3.98	3.77	3.61	3.48
30	.100	2.88	2.49	2.28	2.14	2.05	1.98	1.93	1.88	1.85
	.050	4.17	3.32	2.92	2.69	2.53	2.42	2.33	2.27	2.21
	.025	5.57	4.18	3.59	3.25	3.03	2.87	2.75	2.65	2.57
	.010	7.56	5.39	4.51	4.02	3.70	3.47	3.30	3.17	3.07
	.005	9.18	6.35	5.24	4.62	4.23	3.95	3.74	3.58	3.45
40	.100	2.84	2.44	2.23	2.09	2.00	1.93	1.87	1.83	1.79
	.050	4.08	3.23	2.84	2.61	2.45	2.34	2.25	2.18	2.12
	.025	5.42	4.05	3.46	3.13	2.90	2.74	2.62	2.53	2.45
	.010	7.31	5.18	4.31	3.83	3.51	3.29	3.12	2.99	2.89
	.005	8.83	6.07	4.98	4.37	3.99	3.71	3.51	3.35	3.22
60	.100	2.79	2.39	2.18	2.04	1.95	1.87	1.82	1.77	1.74
	.050	4.00	3.15	2.76	2.53	2.37	2.25	2.17	2.10	2.04
	.025	5.29	3.93	3.34	3.01	2.79	2.63	2.51	2.41	2.33
	.010	7.08	4.98	4.13	3.65	3.34	3.12	2.95	2.82	2.72
	.005	8.49	5.79	4.73	4.14	3.76	3.49	3.29	3.13	3.01
120	.100	2.75	2.35	2.13	1.99	1.90	1.82	1.77	1.72	1.68
	.050	3.92	3.07	2.68	2.45	2.29	2.17	2.09	2.02	1.96
	.025	5.15	3.80	3.23	2.89	2.67	2.52	2.39	2.30	2.22
	.010	6.85	4.79	3.95	3.48	3.17	2.96	2.79	2.66	2.56
	.005	8.18	5.54	4.50	3.92	3.55	3.28	3.09	2.93	2.81
∞	.100	2.71	2.30	2.08	1.94	1.85	1.77	1.72	1.67	1.63
	.050	3.84	3.00	2.60	2.37	2.21	2.10	2.01	1.94	1.88
	.025	5.02	3.69	3.12	2.79	2.57	2.41	2.29	2.19	2.11
	.010	6.63	4.61	3.78	3.32	3.02	2.80	2.64	2.51	2.41
	.005	7.88	5.30	4.28	3.72	3.35	3.09	2.90	2.74	2.62

Table 7. (*Continued*)

$$F_\alpha$$

10	12	15	20	24	30	40	60	120	∞	α	d.f.
					Numerator d.f.						
1.83	1.78	1.73	1.68	1.65	1.62	1.58	1.55	1.51	1.47	.100	29
2.18	2.10	2.03	1.94	1.90	1.85	1.81	1.75	1.70	1.64	.050	
2.53	2.43	2.32	2.21	2.15	2.09	2.03	1.96	1.89	1.81	.025	
3.00	2.87	2.73	2.57	2.49	2.41	2.33	2.23	2.14	2.03	.010	
3.38	3.21	3.04	2.86	2.76	2.66	2.56	2.45	2.33	2.21	.005	
1.82	1.77	1.72	1.67	1.64	1.61	1.57	1.54	1.50	1.46	.100	30
2.16	2.09	2.01	1.93	1.89	1.84	1.79	1.74	1.68	1.62	.050	
2.51	2.41	2.31	2.20	2.14	2.07	2.01	1.94	1.87	1.79	.025	
2.98	2.84	2.70	2.55	2.47	2.39	2.30	2.21	2.11	2.01	.010	
3.34	3.18	3.01	2.82	2.73	2.63	2.52	2.42	2.30	2.18	.005	
1.76	1.71	1.66	1.61	1.57	1.54	1.51	1.47	1.42	1.38	.100	40
2.08	2.00	1.92	1.84	1.79	1.74	1.69	1.64	1.58	1.51	.050	
2.39	2.29	2.18	2.07	2.01	1.94	1.88	1.80	1.72	1.64	.025	
2.80	2.66	2.52	2.37	2.29	2.20	2.11	2.02	1.92	1.80	.010	
3.12	2.95	2.78	2.60	2.50	2.40	2.30	2.18	2.06	1.93	.005	
1.71	1.66	1.60	1.54	1.51	1.48	1.44	1.40	1.35	1.29	.100	60
1.99	1.92	1.84	1.75	1.70	1.65	1.59	1.53	1.47	1.39	.050	
2.27	2.17	2.06	1.94	1.88	1.82	1.74	1.67	1.58	1.48	.025	
2.63	2.50	2.35	2.20	2.12	2.03	1.94	1.84	1.73	1.60	.010	
2.90	2.74	2.57	2.39	2.29	2.19	2.08	1.96	1.83	1.69	.005	
1.65	1.60	1.55	1.48	1.45	1.41	1.37	1.32	1.26	1.19	.100	120
1.91	1.83	1.75	1.66	1.61	1.55	1.50	1.43	1.35	1.25	.050	
2.16	2.05	1.94	1.82	1.76	1.69	1.61	1.53	1.43	1.31	.025	
2.47	2.34	2.19	2.03	1.95	1.86	1.76	1.66	1.53	1.38	.010	
2.71	2.54	2.37	2.19	2.09	1.98	1.87	1.75	1.61	1.43	.005	
1.60	1.55	1.49	1.42	1.38	1.34	1.30	1.24	1.17	1.00	.100	∞
1.83	1.75	1.67	1.57	1.52	1.46	1.39	1.32	1.22	1.00	.050	
2.05	1.94	1.83	1.71	1.64	1.57	1.48	1.39	1.27	1.00	.025	
2.32	2.18	2.04	1.88	1.79	1.70	1.59	1.47	1.32	1.00	.010	
2.52	2.36	2.19	2.00	1.90	1.79	1.67	1.53	1.36	1.00	.005	

Table 8. Distribution function of U

$P(U \leq U_0)$; U_0 is the
argument; $n_1 \leq n_2$;
$3 \leq n_2 \leq 10$.

$n_2 = 3$

	n_1		
U_0	1	2	3
0	.25	.10	.05
1	.50	.20	.10
2		.40	.20
3		.60	.35
4			.50

$n_2 = 4$

	n_1			
U_0	1	2	3	4
0	.2000	.0667	.0286	.0143
1	.4000	.1333	.0571	.0286
2	.6000	.2667	.1143	.0571
3		.4000	.2000	.1000
4		.6000	.3143	.1714
5			.4286	.2429
6			.5714	.3429
7				.4429
8				.5571

Table 8. (*Continued*)

$n_2 = 5$

U_0	1	2	3	4	5
			n_1		
0	.1667	.0476	.0179	.0079	.0040
1	.3333	.0952	.0357	.0159	.0079
2	.5000	.1905	.0714	.0317	.0159
3		.2857	.1250	.0556	.0278
4		.4286	.1964	.0952	.0476
5		.5714	.2857	.1429	.0754
6			.3929	.2063	.1111
7			.5000	.2778	.1548
8				.3651	.2103
9				.4524	.2738
10				.5476	.3452
11					.4206
12					.5000

$n_2 = 6$

U_0	1	2	3	4	5	6
				n_1		
0	.1429	.0357	.0119	.0048	.0022	.0011
1	.2857	.0714	.0238	.0095	.0043	.0022
2	.4286	.1429	.0476	.0190	.0087	.0043
3	.5714	.2143	.0833	.0333	.0152	.0076
4		.3214	.1310	.0571	.0260	.0130
5		.4286	.1905	.0857	.0411	.0206
6		.5714	.2738	.1286	.0628	.0325
7			.3571	.1762	.0887	.0465
8			.4524	.2381	.1234	.0660
9			.5476	.3048	.1645	.0898
10				.3810	.2143	.1201
11				.4571	.2684	.1548
12				.5429	.3312	.1970
13					.3961	.2424
14					.4654	.2944
15					.5346	.3496
16						.4091
17						.4686
18						.5314

Table 8. (*Continued*)

$n_2 = 7$

U_0	\multicolumn{7}{c}{n_1}
	1	2	3	4	5	6	7
0	.1250	.0278	.0083	.0030	.0013	.0006	.0003
1	.2500	.0556	.0167	.0061	.0025	.0012	.0006
2	.3750	.1111	.0333	.0121	.0051	.0023	.0012
3	.5000	.1667	.0583	.0212	.0088	.0041	.0020
4		.2500	.0917	.0364	.0152	.0070	.0035
5		.3333	.1333	.0545	.0240	.0111	.0055
6		.4444	.1917	.0818	.0366	.0175	.0087
7		.5556	.2583	.1152	.0530	.0256	.0131
8			.3333	.1576	.0745	.0367	.0189
9			.4167	.2061	.1010	.0507	.0265
10			.5000	.2636	.1338	.0688	.0364
11				.3242	.1717	.0903	.0487
12				.3939	.2159	.1171	.0641
13				.4636	.2652	.1474	.0825
14				.5364	.3194	.1830	.1043
15					.3775	.2226	.1297
16					.4381	.2669	.1588
17					.5000	.3141	.1914
18						.3654	.2279
19						.4178	.2675
20						.4726	.3100
21						.5274	.3552
22							.4024
23							.4508
24							.5000

Table 8. (*Continued*)

$n_2 = 8$

U_0	n_1							
	1	2	3	4	5	6	7	8
0	.1111	.0222	.0061	.0020	.0008	.0003	.0002	.0001
1	.2222	.0444	.0121	.0040	.0016	.0007	.0003	.0002
2	.3333	.0889	.0242	.0081	.0031	.0013	.0006	.0003
3	.4444	.1333	.0424	.0141	.0054	.0023	.0011	.0005
4	.5556	.2000	.0667	.0242	.0093	.0040	.0019	.0009
5		.2667	.0970	.0364	.0148	.0063	.0030	.0015
6		.3556	.1394	.0545	.0225	.0100	.0047	.0023
7		.4444	.1879	.0768	.0326	.0147	.0070	.0035
8		.5556	.2485	.1071	.0466	.0213	.0103	.0052
9			.3152	.1414	.0637	.0296	.0145	.0074
10			.3879	.1838	.0855	.0406	.0200	.0103
11			.4606	.2303	.1111	.0539	.0270	.0141
12			.5394	.2848	.1422	.0709	.0361	.0190
13				.3414	.1772	.0906	.0469	.0249
14				.4040	.2176	.1142	.0603	.0325
15				.4667	.2618	.1412	.0760	.0415
16				.5333	.3108	.1725	.0946	.0524
17					.3621	.2068	.1159	.0652
18					.4165	.2454	.1405	.0803
19					.4716	.2864	.1678	.0974
20					.5284	.3310	.1984	.1172
21						.3773	.2317	.1393
22						.4259	.2679	.1641
23						.4749	.3063	.1911
24						.5251	.3472	.2209
25							.3894	.2527
26							.4333	.2869
27							.4775	.3227
28							.5225	.3605
29								.3992
30								.4392
31								.4796
32								.5204

Table 8. (*Continued*)

$n_2 = 9$

U_0	1	2	3	4	5	6	7	8	9
					n_1				
0	.1000	.0182	.0045	.0014	.0005	.0002	.0001	.0000	.0000
1	.2000	.0364	.0091	.0028	.0010	.0004	.0002	.0001	.0000
2	.3000	.0727	.0182	.0056	.0020	.0008	.0003	.0002	.0001
3	.4000	.1091	.0318	.0098	.0035	.0014	.0006	.0003	.0001
4	.5000	.1636	.0500	.0168	.0060	.0024	.0010	.0005	.0002
5		.2182	.0727	.0252	.0095	.0038	.0017	.0008	.0004
6		.2909	.1045	.0378	.0145	.0060	.0026	.0012	.0006
7		.3636	.1409	.0531	.0210	.0088	.0039	.0019	.0009
8		.4545	.1864	.0741	.0300	.0128	.0058	.0028	.0014
9		.5455	.2409	.0993	.0415	.0180	.0082	.0039	.0020
10			.3000	.1301	.0559	.0248	.0115	.0056	.0028
11			.3636	.1650	.0734	.0332	.0156	.0076	.0039
12			.4318	.2070	.0949	.0440	.0209	.0103	.0053
13			.5000	.2517	.1199	.0567	.0274	.0137	.0071
14				.3021	.1489	.0723	.0356	.0180	.0094
15				.3552	.1818	.0905	.0454	.0232	.0122
16				.4126	.2188	.1119	.0571	.0296	.0157
17				.4699	.2592	.1361	.0708	.0372	.0200
18				.5301	.3032	.1638	.0869	.0464	.0252
19					.3497	.1942	.1052	.0570	.0313
20					.3986	.2280	.1261	.0694	.0385
21					.4491	.2643	.1496	.0836	.0470
22					.5000	.3035	.1755	.0998	.0567
23						.3445	.2039	.1179	.0680
24						.3878	.2349	.1383	.0807
25						.4320	.2680	.1606	.0951
26						.4773	.3032	.1852	.1112
27						.5227	.3403	.2117	.1290
28							.3788	.2404	.1487
29							.4185	.2707	.1701
30							.4591	.3029	.1933
31							.5000	.3365	.2181
32								.3715	.2447
33								.4074	.2729
34								.4442	.3024
35								.4813	.3332
36								.5187	.3652
37									.3981
38									.4317
39									.4657
40									.5000

Table 8. (*Continued*)

$n_2 = 10$

U_0	n_1									
	1	2	3	4	5	6	7	8	9	10
0	.0909	.0152	.0035	.0010	.0003	.0001	.0001	.0000	.0000	.0000
1	.1818	.0303	.0070	.0020	.0007	.0002	.0001	.0000	.0000	.0000
2	.2727	.0606	.0140	.0040	.0013	.0005	.0002	.0001	.0000	.0000
3	.3636	.0909	.0245	.0070	.0023	.0009	.0004	.0002	.0001	.0000
4	.4545	.1364	.0385	.0120	.0040	.0015	.0006	.0003	.0001	.0001
5	.5455	.1818	.0559	.0180	.0063	.0024	.0010	.0004	.0002	.0001
6		.2424	.0804	.0270	.0097	.0037	.0015	.0007	.0003	.0002
7		.3030	.1084	.0380	.0140	.0055	.0023	.0010	.0005	.0002
8		.3788	.1434	.0529	.0200	.0080	.0034	.0015	.0007	.0004
9		.4545	.1853	.0709	.0276	.0112	.0048	.0022	.0011	.0005
10		.5455	.2343	.0939	.0376	.0156	.0068	.0031	.0015	.0008
11			.2867	.1199	.0496	.0210	.0093	.0043	.0021	.0010
12			.3462	.1518	.0646	.0280	.0125	.0058	.0028	.0014
13			.4056	.1868	.0823	.0363	.0165	.0078	.0038	.0019
14			.4685	.2268	.1032	.0467	.0215	.0103	.0051	.0026
15			.5315	.2697	.1272	.0589	.0277	.0133	.0066	.0034
16				.3177	.1548	.0736	.0351	.0171	.0086	.0045
17				.3666	.1855	.0903	.0439	.0217	.0110	.0057
18				.4196	.2198	.1099	.0544	.0273	.0140	.0073
19				.4725	.2567	.1317	.0665	.0338	.0175	.0093
20				.5275	.2970	.1566	.0806	.0416	.0217	.0116
21					.3393	.1838	.0966	.0506	.0267	.0144
22					.3839	.2139	.1148	.0610	.0326	.0177
23					.4296	.2461	.1349	.0729	.0394	.0216
24					.4765	.2811	.1574	.0864	.0474	.0262
25					.5235	.3177	.1819	.1015	.0564	.0315
26						.3564	.2087	.1185	.0667	.0376
27						.3962	.2374	.1371	.0782	.0446
28						.4374	.2681	.1577	.0912	.0526
29						.4789	.3004	.1800	.1055	.0615
30						.5211	.3345	.2041	.1214	.0716
31							.3698	.2299	.1388	.0827
32							.4063	.2574	.1577	.0952
33							.4434	.2863	.1781	.1088
34							.4811	.3167	.2001	.1237
35							.5189	.3482	.2235	.1399
36								.3809	.2483	.1575
37								.4143	.2745	.1763
38								.4484	.3019	.1965
39								.4827	.3304	.2179

Table 8. (*Continued*)

$n_2 = 10$

U_0	1	2	3	4	5	6	7	8	9	10
40								.5173	.3598	.2406
41									.3901	.2644
42									.4211	.2894
43									.4524	.3153
44									.4841	.3421
45									.5159	.3697
46										.3980
47										.4267
48										.4559
49										.4853
50										.5147

Computed by M. Pagano, Department of Statistics, University of Florida.

Table 9.   Critical values of T in the Wilcoxon matched-pairs, signed-ranks test; $n = 5(1)50$

One-sided	Two-sided	$n = 5$	$n = 6$	$n = 7$	$n = 8$	$n = 9$	$n = 10$
$P = .05$	$P = .10$	1	2	4	6	8	11
$P = .025$	$P = .05$		1	2	4	6	8
$P = .01$	$P = .02$			0	2	3	5
$P = .005$	$P = .01$				0	2	3

One-sided	Two-sided	$n = 11$	$n = 12$	$n = 13$	$n = 14$	$n = 15$	$n = 16$
$P = .05$	$P = .10$	14	17	21	26	30	36
$P = .025$	$P = .05$	11	14	17	21	25	30
$P = .01$	$P = .02$	7	10	13	16	20	24
$P = .005$	$P = .01$	5	7	10	13	16	19

One-sided	Two-sided	$n = 17$	$n = 18$	$n = 19$	$n = 20$	$n = 21$	$n = 22$
$P = .05$	$P = .10$	41	47	54	60	68	75
$P = .025$	$P = .05$	35	40	46	52	59	66
$P = .01$	$P = .02$	28	33	38	43	49	56
$P = .005$	$P = .01$	23	28	32	37	43	49

One-sided	Two-sided	$n = 23$	$n = 24$	$n = 25$	$n = 26$	$n = 27$	$n = 28$
$P = .05$	$P = .10$	83	92	101	110	120	130
$P = .025$	$P = .05$	73	81	90	98	107	117
$P = .01$	$P = .02$	62	69	77	85	93	102
$P = .005$	$P = .01$	55	68	68	76	84	92

Table 9. (*Continued*)

One-sided	Two-sided	$n = 29$	$n = 30$	$n = 31$	$n = 32$	$n = 33$	$n = 34$
$P = .05$	$P = .10$	141	152	163	175	188	201
$P = .025$	$P = .05$	127	137	148	159	171	183
$P = .01$	$P = .02$	111	120	130	141	151	162
$P = .005$	$P = .01$	100	109	118	128	138	149

One-sided	Two-sided	$n = 35$	$n = 36$	$n = 37$	$n = 38$	$n = 39$
$P = .05$	$P = .10$	214	228	242	256	271
$P = .025$	$P = .05$	195	208	222	235	250
$P = .01$	$P = .02$	174	186	198	211	224
$P = .005$	$P = .01$	160	171	183	195	208

One-sided	Two-sided	$n = 40$	$n = 41$	$n = 42$	$n = 43$	$n = 44$	$n = 45$
$P = .05$	$P = .10$	287	303	319	336	353	371
$P = .025$	$P = .05$	264	279	295	311	327	344
$P = .01$	$P = .02$	238	252	267	281	297	313
$P = .005$	$P = .01$	221	234	248	262	277	292

One-sided	Two-sided	$n = 46$	$n = 47$	$n = 48$	$n = 49$	$n = 50$
$P = .05$	$P = .10$	389	408	427	446	466
$P = .025$	$P = .05$	361	379	397	415	434
$P = .01$	$P = .02$	329	345	362	380	398
$P = .005$	$P = .01$	307	323	339	356	373

From "Some Rapid Approximate Statistical Procedures" (1964), 28, F. Wilcoxon and R. A. Wilcox. Reproduced with the kind permission of R. A. Wilcox and the Lederle Laboratories.

Table 10. Distribution of the total number of runs R in samples of size (n_1, n_2); $P(R \le a)$

					a				
(n_1, n_2)	2	3	4	5	6	7	8	9	10
(2, 3)	.200	.500	.900	1.000					
(2, 4)	.133	.400	.800	1.000					
(2, 5)	.095	.333	.714	1.000					
(2, 6)	.071	.286	.643	1.000					
(2, 7)	.056	.250	.583	1.000					
(2, 8)	.044	.222	.533	1.000					
(2, 9)	.036	.200	.491	1.000					
(2, 10)	.030	.182	.455	1.000					
(3, 3)	.100	.300	.700	.900	1.000				
(3, 4)	.057	.200	.543	.800	.971	1.000			
(3, 5)	.036	.143	.429	.714	.929	1.000			
(3, 6)	.024	.107	.345	.643	.881	1.000			
(3, 7)	.017	.083	.283	.583	.833	1.000			
(3, 8)	.012	.067	.236	.533	.788	1.000			
(3, 9)	.009	.055	.200	.491	.745	1.000			
(3, 10)	.007	.045	.171	.455	.706	1.000			
(4, 4)	.029	.114	.371	.629	.886	.971	1.000		
(4, 5)	.016	.071	.262	.500	.786	.929	.992	1.000	
(4, 6)	.010	.048	.190	.405	.690	.881	.976	1.000	
(4, 7)	.006	.033	.142	.333	.606	.833	.954	1.000	
(4, 8)	.004	.024	.109	.279	.533	.788	.929	1.000	
(4, 9)	.003	.018	.085	.236	.471	.745	.902	1.000	
(4, 10)	.002	.014	.068	.203	.419	.706	.874	1.000	
(5, 5)	.008	.040	.167	.357	.643	.833	.960	.992	1.000
(5, 6)	.004	.024	.110	.262	.522	.738	.911	.976	.998
(5, 7)	.003	.015	.076	.197	.424	.652	.854	.955	.992
(5, 8)	.002	.010	.054	.152	.347	.576	.793	.929	.984
(5, 9)	.001	.007	.039	.119	.287	.510	.734	.902	.972
(5, 10)	.001	.005	.029	.095	.239	.455	.678	.874	.958
(6, 6)	.002	.013	.067	.175	.392	.608	.825	.933	.987
(6, 7)	.001	.008	.043	.121	.296	.500	.733	.879	.966
(6, 8)	.001	.005	.028	.086	.226	.413	.646	.821	.937
(6, 9)	.000	.003	.019	.063	.175	.343	.566	.762	.902
(6, 10)	.000	.002	.013	.047	.137	.288	.497	.706	.864
(7, 7)	.001	.004	.025	.078	.209	.383	.617	.791	.922
(7, 8)	.000	.002	.015	.051	.149	.296	.514	.704	.867
(7, 9)	.000	.001	.010	.035	.108	.231	.427	.622	.806
(7, 10)	.000	.001	.006	.024	.080	.182	.355	.549	.743
(8, 8)	.000	.001	.009	.032	.100	.214	.405	.595	.786
(8, 9)	.000	.001	.005	.020	.069	.157	.319	.500	.702
(8, 10)	.000	.000	.003	.013	.048	.117	.251	.419	.621
(9, 9)	.000	.000	.003	.012	.044	.109	.238	.399	.601
(9, 10)	.000	.000	.002	.008	.029	.077	.179	.319	.510
(10, 10)	.000	.000	.001	.004	.019	.051	.128	.242	.414

Table 10. (*Continued*)

(n_1, n_2)	11	12	13	14	15	16	17	18	19	20
					a					
(2, 3)										
(2, 4)										
(2, 5)										
(2, 6)										
(2, 7)										
(2, 8)										
(2, 9)										
(2, 10)										
(3, 3)										
(3, 4)										
(3, 5)										
(3, 6)										
(3, 7)										
(3, 8)										
(3, 9)										
(3, 10)										
(4, 4)										
(4, 5)										
(4, 6)										
(4, 7)										
(4, 8)										
(4, 9)										
(4, 10)										
(5, 5)										
(5, 6)	1.000									
(5, 7)	1.000									
(5, 8)	1.000									
(5, 9)	1.000									
(5, 10)	1.000									
(6, 6)	.998	1.000								
(6, 7)	.992	.999	1.000							
(6, 8)	.984	.998	1.000							
(6, 9)	.972	.994	1.000							
(6, 10)	.958	.990	1.000							
(7, 7)	.975	.996	.999	1.000						
(7, 8)	.949	.988	.998	1.000	1.000					
(7, 9)	.916	.975	.994	.999	1.000					
(7, 10)	.879	.957	.990	.998	1.000					
(8, 8)	.900	.968	.991	.999	1.000	1.000				
(8, 9)	.843	.939	.980	.996	.999	1.000	1.000			
(8, 10)	.782	.903	.964	.990	.998	1.000	1.000			
(9, 9)	.762	.891	.956	.988	.997	1.000	1.000	1.000		
(9, 10)	.681	.834	.923	.974	.992	.999	1.000	1.000	1.000	
(10, 10)	.586	.758	.872	.949	.981	.996	.999	1.000	1.000	1.000

From "Tables for Testing Randomness of Grouping in a Sequence of Alternatives," C. Eisenhart and F. Swed, *Annals of Mathematical Statistics*, Volume 14 (1943). Reproduced with the kind permission of the Editor, *Annals of Mathematical Statistics*.

Table 11. Critical values of Spearman's rank correlation coefficient

n	$\alpha = .05$	$\alpha = .025$	$\alpha = .01$	$\alpha = .005$
5	0.900	—	—	—
6	0.829	0.886	0.943	—
7	0.714	0.786	0.893	—
8	0.643	0.738	0.833	0.881
9	0.600	0.683	0.783	0.833
10	0.564	0.648	0.745	0.794
11	0.523	0.623	0.736	0.818
12	0.497	0.591	0.703	0.780
13	0.475	0.566	0.673	0.745
14	0.457	0.545	0.646	0.716
15	0.441	0.525	0.623	0.689
16	0.425	0.507	0.601	0.666
17	0.412	0.490	0.582	0.645
18	0.399	0.476	0.564	0.625
19	0.388	0.462	0.549	0.608
20	0.377	0.450	0.534	0.591
21	0.368	0.438	0.521	0.576
22	0.359	0.428	0.508	0.562
23	0.351	0.418	0.496	0.549
24	0.343	0.409	0.485	0.537
25	0.336	0.400	0.475	0.526
26	0.329	0.392	0.465	0.515
27	0.323	0.385	0.456	0.505
28	0.317	0.377	0.448	0.496
29	0.311	0.370	0.440	0.487
30	0.305	0.364	0.432	0.478

From "Distribution of Sums of Squares of Rank Differences for Small Samples," E. G. Olds, *Annals of Mathematical Statistics*, Volume 9 (1938). Reproduced with the kind permission of the Editor, *Annals of Mathematical Statistics*.

Table 12. Random numbers

Line/Col.	(1)	(2)	(3)	(4)	(5)	(6)	(7)	(8)	(9)	(10)	(11)	(12)	(13)	(14)
1	10480	15011	01536	02011	81647	91646	69179	14194	62590	36207	20969	99570	91291	90700
2	22368	46573	25595	85393	30995	89198	27982	53402	93965	34095	52666	19174	39615	99505
3	24130	48360	22527	97265	76393	64809	15179	24830	49340	32081	30680	19655	63348	58629
4	42167	93003	06243	61680	07856	16376	39440	53537	71341	57004	00849	74917	97758	16379
5	37570	39975	81837	16656	06121	91782	60468	81305	49684	60672	14110	06927	01263	54613
6	77921	06907	11008	42751	27756	53498	18602	70659	90655	15053	21916	81825	44394	42880
7	99562	72095	56420	69994	98872	31016	71194	18738	44013	48840	63213	21069	10634	12952
8	96301	91977	05463	07972	18876	20922	94595	56869	69014	60045	18425	84903	42508	32307
9	89579	14342	63661	10281	17453	18103	57740	84378	25331	12566	58678	44947	05585	56941
10	85475	36857	53342	53988	53060	59533	38867	62300	08158	17983	16439	11458	18593	64952
11	28918	69578	88231	33276	70997	79936	56865	05859	90106	31595	01547	85590	91610	78188
12	63553	40961	48235	03427	49626	69445	18663	72695	52180	20847	12234	90511	33703	90322
13	09429	93969	52636	92737	88974	33488	36320	17617	30015	08272	84115	27156	30613	74952
14	10365	61129	87529	85689	48237	52267	67689	93394	01511	26358	85104	20285	29975	89868
15	07119	97336	71048	08178	77233	13916	47564	81056	97735	85977	29372	74461	28551	90707
16	51085	12765	51821	51259	77452	16308	60756	92144	49442	53900	70960	63990	75601	40719
17	02368	21382	52404	60268	89368	19885	55322	44819	01188	65255	64835	44919	05944	55157
18	01011	54092	33362	94904	31273	04146	18594	29852	71585	85030	51132	01915	92747	64951
19	52162	53916	46369	58586	23216	14513	83149	98736	23495	64350	94738	17752	35156	35749
20	07056	97628	33787	09998	42698	06691	76988	13602	51851	46104	88916	19509	25625	58104
21	48663	91245	85828	14346	09172	30168	90229	04734	59193	22178	30421	61666	99904	32812
22	54164	58492	22421	74103	47070	25306	76468	26384	58151	06646	21524	15227	96909	44592
23	32639	32363	05597	24200	13363	38005	94342	28728	35806	06912	17012	64161	18296	22851
24	29334	27001	87637	87308	58731	00256	45834	15398	46557	41135	10367	07684	36188	18510
25	02488	33062	28834	07351	19731	92420	60952	61280	50001	67658	32586	86679	50720	94953

Table 12. (*Continued*)

Line/Col.	(1)	(2)	(3)	(4)	(5)	(6)	(7)	(8)	(9)	(10)	(11)	(12)	(13)	(14)
26	81525	72295	04839	96423	24878	82651	66566	14778	76797	14780	13300	87074	79666	95725
27	29676	20591	68086	26432	46901	20849	89768	81536	86645	12659	92259	57102	80428	25280
28	00742	57392	39064	66432	84673	40027	32832	61362	98947	96067	64760	64584	96096	98253
29	05366	04213	25669	26422	44407	44048	37937	63904	45766	66134	75470	66520	34693	90449
30	91921	26418	64117	94305	26766	25940	39972	22209	71500	64568	91402	42416	07844	69618
31	00582	04711	87917	77341	42206	35126	74087	99547	81817	42607	43808	76655	62028	76630
32	00725	69884	62797	56170	86324	88072	76222	36086	84637	93161	76038	65855	77919	88006
33	69011	65795	95876	55293	18988	27354	26575	08625	40801	59920	29841	80150	12777	48501
34	25976	57948	29888	88604	67917	48708	18912	82271	65424	69774	33611	54262	85963	03547
35	09763	83473	73577	12908	30883	18317	28290	35797	05998	41688	34952	37888	38917	88050
36	91567	42595	27958	30134	04024	86385	29880	99730	55536	84855	29080	09250	79656	73211
37	17955	56349	90999	49127	20044	59931	06115	20542	18059	02008	73708	83517	36103	42791
38	46503	18584	18845	49618	02304	51038	20655	58727	28168	15475	56942	53389	20562	87338
39	92157	89634	94824	78171	84610	82834	09922	25417	44137	48413	25555	21246	35509	20468
40	14577	62765	35605	81263	39667	47358	56873	56307	61607	49518	89656	20103	77490	18062
41	98427	07523	33362	64270	01638	92477	66969	98420	04880	45585	46565	04102	46880	45709
42	34914	63976	88720	82765	34476	17032	87589	40836	32427	70002	70663	88863	77775	69348
43	70060	28277	39475	46473	23219	53416	94970	25832	69975	94884	19661	72828	00102	66794
44	53976	54914	06990	67245	68350	82948	11398	42878	80287	88267	47363	46634	06541	97809
45	76072	29515	40980	07391	58745	25774	22987	80059	39911	96189	41151	14222	60697	59583
46	90725	52210	83974	29992	65831	38857	50490	83765	55657	14361	31720	57375	56228	41546
47	64364	67412	33339	31926	14883	24413	59744	92351	97473	89286	35931	04110	23726	51900
48	08962	00358	31662	25388	61642	34072	81249	35648	56891	69352	48373	45578	78547	81788
49	95012	68379	93526	70765	10592	04542	76463	54328	02349	17247	28865	14777	62730	92277
50	15664	10493	20492	38391	91132	21999	59516	81652	27195	48223	46751	22923	32261	85653

Table 12. *(Continued)*

Line/Col.	(1)	(2)	(3)	(4)	(5)	(6)	(7)	(8)	(9)	(10)	(11)	(12)	(13)	(14)
51	16408	81899	04153	53381	79401	21438	83035	92350	36693	31238	59649	91754	72772	02338
52	18629	81953	05520	91962	04739	13092	97662	24822	94730	06496	35090	04822	86774	98289
53	73115	35101	47498	87637	99016	71060	88824	71013	18735	20286	23153	72924	35165	43040
54	57491	16703	23167	49323	45021	33132	12544	41035	80780	45393	44812	12515	98931	91202
55	30405	83946	23792	14422	15059	45799	22716	19792	09983	74353	68668	30429	70735	25499
56	16631	35006	85900	98275	32388	52390	16815	69298	82732	38480	73817	32523	41961	44437
57	96773	20206	42559	78985	05300	22164	24369	54224	35083	19687	11052	91491	60383	19746
58	38935	64202	14349	82674	66523	44133	00697	35552	35970	19124	63318	29686	03387	59846
59	31624	76384	17403	53363	44167	64486	64758	75366	76554	31601	12614	33072	60332	92325
60	78919	19474	23632	27889	47914	02584	37680	20801	72152	39339	34806	08930	85001	87820
61	03931	33309	57047	74211	63445	17361	62825	39908	05607	91284	68833	25570	38818	46920
62	74426	33278	43972	10119	89917	15665	52872	73823	73144	88662	88970	74492	51805	99378
63	09066	00903	20795	95452	92648	45454	09552	88815	16553	51125	79375	97596	16296	66092
64	42238	12426	87025	14267	20979	04508	64535	31355	86064	29472	47689	05974	52468	16834
65	16153	08002	26504	41744	81959	65642	74240	56302	00033	67107	77510	70625	28725	34191
66	21457	40742	29820	96783	29400	21840	15035	34537	33310	06116	95240	15957	16572	06004
67	21581	57802	02050	89728	17937	37621	47075	42080	97403	48626	68995	43805	33386	21597
68	55612	78095	83197	33732	05810	24813	86902	60397	16489	03264	88525	42786	05269	92532
69	44657	66999	99324	51281	84463	60563	79312	93454	68876	25471	93911	25650	12682	73572
70	91340	84979	46949	81973	37949	61023	43997	15263	80644	43942	89203	71795	99533	50501
71	91227	21199	31935	27022	84067	05462	35216	14486	29891	68607	41867	14951	91696	85065
72	50001	38140	66321	19924	72163	09538	12151	06878	91903	18749	34405	56087	82790	70925
73	65390	05224	72958	28609	81406	39147	25549	48542	42627	45233	57202	94617	23772	07896
74	27504	96131	83944	41575	10573	08619	64482	73923	36152	05184	94142	25299	84387	34925
75	37169	94851	39117	89632	00959	16487	65536	49071	39782	17095	02330	74301	00275	48280

Table 12. (*Continued*)

Line/Col.	(1)	(2)	(3)	(4)	(5)	(6)	(7)	(8)	(9)	(10)	(11)	(12)	(13)	(14)
76	11508	70225	51111	38351	19444	66499	71945	05422	13442	78675	84081	66938	93654	59894
77	37449	30362	06694	54690	04052	53115	62757	95348	78662	11163	81651	50245	34971	52924
78	46515	70331	85922	38329	57015	15765	97161	17869	45349	61796	66345	81073	49106	79860
79	30986	81223	42416	58353	21532	30502	32305	86482	05174	07901	54339	58861	74818	46942
80	63798	64995	46583	09785	44160	78128	83991	42885	92520	83531	80377	35909	81250	54238
81	82486	84846	99254	67632	43218	50076	21361	64816	51202	88124	41870	52689	51275	83556
82	21885	32906	92431	09060	64297	51674	64126	62570	26123	05155	59194	52799	28225	85762
83	60336	98782	07408	53458	13564	59089	26445	29789	85205	41001	12535	12133	14645	23541
84	43937	46891	24010	25560	86355	33941	25786	54990	71899	15475	95434	98227	21824	19585
85	97656	63175	89303	16275	07100	92063	21942	18611	47348	20203	18534	03862	78095	50136
86	03299	01221	05418	38982	55758	92237	26759	86367	21216	98442	08303	56613	91511	75928
87	79626	06486	03574	17668	07785	76020	79924	25651	83325	88428	85076	72811	22717	50585
88	85636	68335	47539	03129	65651	11977	02510	26113	99447	68645	34327	15152	55230	93448
89	18039	14367	61337	06177	12143	46609	32989	74014	64708	00533	35398	58408	13261	47908
90	08362	15656	60627	36478	65648	16764	53412	09013	07832	41574	17639	82163	60859	75567
91	79556	29068	04142	16268	15387	12856	66227	38358	22478	73373	88732	09443	82558	05250
92	92608	82674	27072	32534	17075	27698	98204	63863	11951	34648	88022	56148	34925	57031
93	23982	25835	40055	67006	12293	02753	14827	23235	35071	99704	37543	11601	35503	85171
94	09915	96306	05908	97901	28395	14186	00821	80703	70426	75647	76310	88717	37890	40129
95	59037	33300	26695	62247	69927	76123	50842	43834	86654	70959	79725	93872	28117	19233
96	42488	78077	69882	61657	34136	79180	97526	43092	04098	73571	80799	76536	71255	64239
97	46764	86273	63003	93017	31204	36692	40202	35275	57306	55543	53203	18098	47625	88684
98	03237	45430	55417	63282	90816	17349	88298	90183	36600	78406	06216	95787	42579	90730
99	86591	81482	52667	61582	14972	90053	89534	76036	49199	43716	97548	04379	46370	28672
100	38534	01715	94964	87288	65680	43772	39560	12918	86537	62738	19636	51132	25739	56947

Abridged from *Handbook of Tables for Probability and Statistics*, 2nd edition, edited by William H. Beyer (Cleveland: The Chemical Rubber Company, 1968). Reproduced by permission of the publishers, The Chemical Rubber Company.

ANSWERS

Chapter 1

1.5
  a   2.45–2.65, 2.65–2.85
  b   7/30
  c   16/30

1.7
  a   approx. .68
  b   approx. .95
  c   approx. .815
  d   approx. 0

1.11
  a $\bar{y} = 1252.44$; $s = 393.75$
  b $k = 1 : (858.69, 1646.19)$
        $k = 2 : (464.94, 2039.94)$
        $k = 3 : (71.19, 2433.69)$

1.13
  a $\bar{y} = 5.676$; $s = 7.48$
  b $k = 1 : (-1.80, 13.15)$
        $k = 2 : (-9.23, 20.63)$
        $k = 3 : (-16.75, 28.11)$

1.15
   Ex. 1.2: approx. 650.75
   Ex. 1.3: approx. 3.04
   Ex. 1.4: approx. 9.125

1.17 $\bar{y} - s = -19 < 0$

1.19   .84

1.21
  a   approx. 16%
  b   approx. 81.5%
  c   No, $225 > \mu + 3\sigma$

1.23
  a   177
  b $\bar{y} = 210.80$; $s = 162.17$

1.25   231, 323

1.27   .05

1.29   .025

1.31   (.5, 10.5)

Chapter 2

2.7 $S = \{A^+, A^-, B^+, B^-, AB^+,$
     $AB^-, O^+, O^-\}$

2.9
  a $P(E_4) = .2$; $P(E_5) = .1$
  b $p = .2$

2.11
  a $E_1 =$ very likely, $E_2 =$ somewhat
     likely, $E_3 =$ unlikely, $E_4 =$ other
  b   no, $P(E_1) = .24$, $P(E_2) = .24$,
     $P(E_3) = .40$, $P(E_4) = .12$
  c   .48

2.13
  a   .09
  b   .19

2.15
  a   .08
  b   .16
  c   .14
  d   .84

2.17
  a $P(A) = 1/3$; $P(B) = 5/9$;
     $P(A \cup B) = 7/9$; $P(A \cap B) = 1/9$

2.19 a 9
b 5/9
c 5/9

2.21 b 1/15

2.23 a 3/5; 1/15
b 14/15; 2/5

2.25 b 11/16; 3/8; 1/4

2.27 42

2.29 a 6! = 720
b 15 × 24/720 = 1/2

2.31 a 36
b 1/6

2.33 $9(10)^6$

2.35 504

2.37 408, 408

2.39 a 8385
b 18,252
c 8515

2.41 a 4/19,600
b 276/19,600
c 4140/19,600
d 15180/19,600

2.43 a 60
b .6

2.45 a $\binom{90}{10} = 5.7206 \times 10^{12}$
b $\binom{20}{4}\binom{70}{6} / \binom{90}{10} = .111$

2.47 $48 / \binom{52}{2}$

2.49 1/56

2.51 5/162

2.53 a .0605
b .00134
c .00029

2.57 a 1/3
b 1/5
c 5/7
d 1
e 1/7

2.59 a 3/4
b 3/4
c 2/3

2.61 a .40
b .37
c .10
d .67
e .60
f .33
g .90
h .27
i .25

2.63 no

2.69 .364

2.71 a .1
b .9
c .6
d 2/3

2.73 a .999
b .9009

2.77 .05

2.79 a 1/1000
b 1/8000

2.81 ≥ .90

2.87 .149

2.89 $(.98)^3(.02)$

2.91 $(.75)^4$

2.93 a 1/4
b 1/16

2.95 a 1/4
b 1/3

2.97 a $1/n$
b $1/n$; $1/n$
c 3/7

2.99 1/12

2.101 .3130

2.103 .4

2.105 .02143

2.109   .25

2.111   .9412

2.113   **a**   .57
     b   .18
     c   .3158
     d   .9

2.115   **a**   2/5
     b   3/20

2.117 $P(Y = 0) = (.02)^3$,
     $P(Y = 1) = 3(.98)(.02)^2$;
     $P(Y = 2) = 3(.98)^2(.02)$;
     $P(Y = 3) = (.98)^3$

2.119 $P(Y = 2) = 1/15$;
     $P(Y = 3) = 2/15$;
     $P(Y = 4) = 3/15$;
     $P(Y = 5) = 4/15$;
     $P(Y = 6) = 5/15$

2.123   18!

2.125 $(13 \times 12) \binom{4}{3} \binom{4}{2} \Big/ \binom{52}{5}$

2.127   **a**   .4
     b   .6
     c   .25

2.129 $4(p^4q + q^4p)$

2.131   **a**   .5
     b   .15
     c   .1
     d   .875

2.133   **a**   .87
     b   .31
     c   .29
     d   .94
     e   .12
     f   .21

2.135   2/11

2.137   design A: P(current flows) = .9801
     design B: P(current flows) = .9639

2.139 $P(Y = 1) = 35/70$
     $P(Y = 2) = 20/70$
     $P(Y = 3) = 10/70$
     $P(Y = 4) = 4/70$
     $P(Y = 5) = 1/70$

2.141   **a** $(4!)^3 = 13824$
     b   1/4

2.145   no

2.147   **a** $\binom{6}{3}(1/2)^6$
     b $27(1/2)^{10}$

Chapter 3

3.1 $p(0) = .2$, $p(1) = .7$, $p(2) = .1$

3.3 $p(2) = 1/6$, $p(3) = 1/3$,
     $p(4) = 1/2$

3.5 $p(0) = 1/3$, $p(1) = 1/2$,
     $p(3) = 1/6$;

3.7 $p(0) = .857375$
     $p(1) = .135375$
     $p(2) = .007125$
     $p(3) = .000125$

3.9 $P(X = 0) = 8/27$;
     $P(X = 1) = 12/27$;
     $P(X = 2) = 6/27$;
     $P(X = 3) = 1/27$;
     $P(Y = 0) = 2744/3375$;
     $P(Y = 1) = 588/3375$;
     $P(Y = 2) = 42/3375$;
     $P(Y = 3) = 1/3375$;
     $P(X + Y = 0) = 729/3375$;
     $P(X + Y = 1) = 1458/3375$;
     $P(X + Y = 2) = 972/3375$;
     $P(X + Y = 3) = 216/3375$;

3.11 $E(Y) = 1/4$, $V(Y) = 27/16$,
     cost $= .25$

3.13 $85

3.15   13,800.388

3.17 $.31

3.19   firm I: E(profit) = $60,000
     E(total profit) = $120,000

3.21 $510

3.25   .4,   .3999

3.27   **a** .1536
         b .9728

3.29   .000

3.31   **a** .1681
         b .5282

3.33   .992

3.37   **c** .151
         d .302

3.41 $185,000

3.43 $840

3.45   **a** .672
         b .672
         c 8

3.49   **a** p
         b pq/n,   goes to 0

3.51   .072

3.53   geometric, $p = .27$

3.57   **a** .009
         b .01

3.59   **a** .081
         b .81

3.61 $p(y) = (.94)^{y-1}(.06)$,
         $y = 1, 2, 3, \ldots$

3.63   2

3.65 $(n-1)^5/n^6$

3.69 $-(\ln(p))p/q$

3.73   150,   4500,   no

3.75   **a** .04374
         b .99144

3.77   .1

3.79   **a** .128
         b .049
         d $\mu = 15; \sigma^2 = 60$

3.81 $p(y) = \binom{y}{r-1} p^r q^{y+1-r}$,
         $y = r-1, r, r+1, \ldots$

3.83   **a** 5/11
         b r/y_0

3.85   1/42

3.87 $p(y) = \binom{4}{y}\binom{2}{2-y} \big/ \binom{6}{2}$, $zy = 0, 1, 2$.

3.89   **a** $p(0) = 14/30$, $p(1) = 14/30$,
               $p(2) = 2/30$
         b $p(0) = 5/30$, $p(1) = 15/30$,
               $p(2) = 9/30$, $p(3) = 1/30$

3.91 $P(Y \leq 1) = .187$

3.93 $p(0) = .2$, $p(1) = .6$, $p(2) = .2$

3.95   **a** .553
         b $\mu = 9.5$, $\sigma = 5.362$

3.97   **a** .090
         b .143
         c .857
         d .241

3.99 $E(S) = 70$, $V(S) = 700$, no

3.101 .6288

3.103 $2.5e^{-1.5}$

3.105 $1 - [(8/3)e^{-1}]^{10}$

3.107 $1 - (.97)^{100}$

3.109 $61e^{-6}$

3.111 40

3.113 $1300

3.119 binomial, $n = 3$, $p = .6$

3.121 **a** binomial, $n = 5$, $p = 1/3$
         b geometric, $p = 1/2$
         c Poisson, $\lambda = 2$

3.123 **a** 7/3
         b 5/9
         c $p(1) = 1/6$, $p(2) = 2/6$,
               $p(3) = 3/6$

3.129 $e^{\lambda(t-1)}$

3.131 **a** .64
         b $C = 10$

3.133 **a** $p(-1) = \dfrac{1}{2k^2}$, $p(0) = 1 - \dfrac{1}{k^2}$,
               $p(1) = \dfrac{1}{2k^2}$

3.135 (85, 115)

3.137 **a** $p(0) = 1/8$, $p(1) = 3/8$,
$p(2) = 3/8$, $p(3) = 1/8$
c $E(Y) = 3/2$, $V(Y) = 3/4$

3.139 **a** 24
b 4.898
c no; 40 is more than 3 standard
deviations below the mean

3.141 (61.03, 98.97)

3.143 no, using Tchebysheff's theorem
$P(Y \geq 350) \leq .1126$

3.145 **a** 1.0
b .5905
c .1681
d .0321
e 0

3.147 **a** $n = 25$, $a = 5$
b $n = 25$, $a = 5$

3.149 **a** .023
b 1.2
c \$1.25

3.151 $1 - (.99999)^{10,000}$

3.153 $V(Y) = .4$

3.155 .476

3.157 **a** .982
b $1 - (.018)^3$

3.159 **a** .9997
b $n = 2$

3.161 (18.35, 181.65)

3.163 **a** $E[Y(t)] = ke^{\lambda t}$,
$V[Y(t)] = K(e^{2\lambda t} - e^{\lambda t})$
b 3.2974, 2.139

3.165 .00722

3.167 **a** $12.5e^{-5}$, $18.5e^{-5}$
b $P(Y > 10) = .014$

3.169 .0837

3.171 3

3.173 **a** .119
b .117

3.175 **a** $n[1 + k(1 - .95^k)]$
b 5
c approximately $.57N$

Chapter 4

4.3 **a** .289, .591, not equal
b .316, .618, not equal
c no contradiction, binomial
distribution is discrete

4.5 **a** $c = 1/2$
b $y^2/4$, $0 \leq y \leq 2$
d 3/4
e 3/4

4.7 **a** $F(y) = \begin{cases} 0, & y < 0 \\ y^2/2, & 0 \leq y \leq 1 \\ y - 1/2, & 1 < y \leq 1.5 \\ 1, & y > 1.5 \end{cases}$
b .125
c .575

4.9 **b** $F(y) = \begin{cases} 0, & y < b \\ 1 - b/y, & y \geq b \end{cases}$
c $b/(b + c)$
d $(b + c)/(b + d)$

4.11 **a** $c = 3/2$
c $(y^3 + y^2)/2$, $0 \leq y \leq 1$
d 0; 0; 1
e 3/16
f 104/123

4.13 **a** $f(y) = \begin{cases} 1/8, & 0 < y < 2 \\ y/8, & 2 \leq y < 4 \end{cases}$
b 7/16
c 13/16
d 7/9

4.15 $E(Y) = .708$; $V(Y) = .049$

4.19 $E(Y) = 2.583$, $V(Y) = 1.160$

4.21 \$4.65; .012

4.23 $E(Y) = 60$, $V(Y) = .333$

4.25 4

4.27 **a** $E(Y) = 5.5$, $V(Y) = .1475$

b using Tchebysheff's theorem; (5, 6.268)

c yes, $P(Y < 5.5) = .5775$

4.29 1/2, 1/4

4.33 **a** 2/5
b 1/5

4.35 **a** 3/4
b $c_0 + (10.333)c_1$

4.37 3/4

4.39 1/3

4.41 **a** 1/8
b 1/8
c 1/4

4.43 **a** 2/7
b $\mu = .015, \ \sigma^2 = .00041$

4.45 $E(\text{volume}) = (6.5 \times 10^{-6})\pi,$
$V(\text{volume}) = (3.524 \times 10^{-11})\pi^2$

4.47 **a** 0
b 1.1
c 1.645
d 2.575

4.49 .1587

4.51 \$426

4.53 3.000

4.55 .2660

4.57 **a** .9544
b .83

4.59 **a** .4060
b 960.5

4.61 7.301

4.63 **a** .7580
b 22.2

4.69 **a** .2865
b .1481

4.71 **a** .1353
b 460.52 cfs

4.73 **a** .5057
b 1936

4.77 e^{-1}

4.79 $2e^{-1}$

4.81 $E(\text{area}) = 200\pi, \ V(\text{area}) = 200000\pi^2$

4.83 $\mu = 3.2, \ \sigma^2 = 6.4$

4.85 (0, 9.6569)

4.87 $E(L) = 276, \ V(L) = 47,664$

4.89 **d** $\sqrt{\beta}\Gamma(\alpha + \frac{1}{2})/\Gamma(\alpha)$ if $\alpha > 0$
e $\dfrac{1}{\beta(\alpha - 1)}$ if $\alpha > 1$, $\dfrac{\Gamma(\alpha - \frac{1}{2})}{\sqrt{\beta}\Gamma(\alpha)}$
if $\alpha > \frac{1}{2}$, $\dfrac{1}{\beta^2(\alpha - 1)(\alpha - 2)}$ if $\alpha > 2$

4.91 $k = 60$

4.93 $E(Y) = 3/5, \ V(Y) = 1/25$

4.97 52/3, 29.96

4.99 **a** .75
b 1/18

4.101 **a** $c = 105$
b 3/8

4.107 $m_X(t) = \exp\{t(4 - 3\mu) + (1/2)(9\sigma^2 t^2)\}$
normal, $E(X) = 4 - 3\mu, \ V(X) = 9\sigma^2$,
uniqueness of moment-generating functions

4.109 $m(t) = \dfrac{e^{t\theta_2} - e^{t\theta_1}}{t(\theta_2 - \theta_1)}$

4.111 $\alpha\beta, \ \alpha\beta^2$

4.113 **a** 2/5
b $1/(t + 1)$
c 1

4.115 .5

4.117 1.0

4.119 could be!, 2000 is only .53 standard deviations above mean

4.121 (6.39, 28.28)

4.123 \$113.33

4.125 **a** $F(x) =$
$$\begin{cases} 0, & x < 0 \\ (1/100)e^{-x/100}, & 0 \le x < 200 \\ 1, & x \ge 200 \end{cases}$$
b 86.47

4.127 **a** $F_1(y) = \begin{cases} 0, & y < 0 \\ .4, & 0 \le y < .5 \\ 1, & y \ge .5 \end{cases}$

$F_2(y) = \begin{cases} 0, & y < 0 \\ 4y^2/3, & 0 \le y < .5 \\ (4y-1)/3, & .5 \le y < 1 \\ 1, & y \ge 1 \end{cases}$

b $F(y) = (.25)F_1(y) + (.75)F_2(y)$

c $E(Y) = .5333, \quad V(Y) = .076$

4.129 85.36

4.131 $1 - (.927)^5$

4.133 **a** $c = 4$
b 1; 1/2
c $(1 - t/2)^{-2}$

4.135 $\Gamma(\alpha + \beta)\Gamma(\alpha + k) / [\Gamma(\alpha)\Gamma(\alpha + \beta + k)]$

4.137 $e^{-2.5}$

4.139 **a** mean $= 1/2$, variance $= 1/4$
b $1 - e^{-6}$

4.141 $f(r) = (2\lambda \pi r) \exp(-\lambda \pi r^2)$, $r > 0$

4.143 $\sqrt{2}$

4.145 $k = (.4)^{1/3}$

4.147 $m(t) = \exp(t^2/2); 0; 1$

4.149 **a** $E(Y) = 598.74g$,
$V(Y) = 3.1856 \times 10^{12}g^2$
b $(0, 3.570 \times 10^6)$
c .8023

4.153 **a** $e^{-2.5}$
b .019

4.155 $E(Y) = 0, \quad V(Y) = 1/(n-1)$

4.157 **c** $1 - e^{-4}$

4.159 150

4.161 **a** 12
b 120

Chapter 5

5.1

y_2	y_1 0	1	2
0	1/9	2/9	1/9
1	2/9	2/9	0
2	1/9	0	0

5.3 $\dfrac{\binom{4}{y_1}\binom{3}{y_2}\binom{2}{3-y_1-y_2}}{\binom{9}{3}}$

5.5 **a** .1065
b 1/2

5.7 **a** $k = 6$
b 31/64

5.9 **a** 29/32
b 1/4

5.11 **a** .5625
b .8125
c .6563

5.13 **a** $e^{-1} - 2e^{-2}$
b 1/2
c e^{-1}

5.15 1/2

5.17 **a** $p(0) = 4/9, \quad p(1) = 4/9, \quad p(2) = 1/9$
b no

5.19 **a** $p(0) = 5/42, \quad p(1) = 20/42,$
$p(2) = 15/42, \quad p(3) = 2/42$
b 2/3
c 8/15

5.21 **a** $f_2(y_2) = 3(1 - y_2^2)/2, \quad 0 \le y_2 \le 1$
b $0 \le y_2 < 1$
c 1/3

5.23 **a** $f_1(y_1) = 3(1 - y_1)^2, \quad 0 \le y_1 \le 1;$
$f_2(y_2) = 6y_2(1 - y_2), \quad 0 \le y_2 \le 1$
b 32/63
c $1/y_2, \quad 0 \le y_1 \le y_2$
d $2(1 - y_2)/(1 - y_1)^2, \quad y_1 \le y_2 \le 1$
e 1/4

5.25 **a** $f_1(y_1) = \begin{cases} 1 + y_1, & -1 \le y_1 < 0 \\ 1 - y_1, & 0 \le y_1 \le 1 \end{cases}$
$f_2(y_2) = 2(1 - y_2), \quad 0 \le y_2 \le 1$
b 1/3

5.27 **b** $f_2(y_2) =$
$$\begin{cases} 15(1 + y_2)^2 y_2^2, & -1 \le y_2 < 0 \\ 15(1 - y_2)^2 y_2^2, & 0 \le y_2 \le 1 \end{cases}$$
c if $0 < y_1 < 1$,
$$f(y_2 \mid y_1) = \frac{3y_2^2}{2(1 - y_1)^3},$$
$$y_1 - 1 \le y_2 \le 1 - y_1$$
d $1/2$

5.29 $1/2$

5.31 e^{-1}

5.35 $1/4$

5.39 no

5.41 no

5.45 no

5.47 no

5.49 no

5.51 no

5.53 yes

5.55 $1/4$

5.57 exponential, mean 1

5.59 **a** $f(y_1, y_2) = (1/9)e^{-(y_1+y_2)/3}$, $y_1 >$
$0, y_2 > 0$
b $1 - (4/3)e^{-1/3}$

5.61 **a** $1/4$
b $23/144$

5.63 $4/3$

5.65 **a** $1/4; 1/2$
b $3/80; 1/20$
c $-5/4$

5.67 0

5.69 1

5.71 1

5.73 **a** $1, 1$
b $1, 1$
c 0
d $1 - (\alpha/4)$
e $\left(-2\sqrt{\dfrac{4+\alpha}{2}}, \ 2\sqrt{\dfrac{4+\alpha}{2}} \right)$

5.75 $-2/9$, no, Y_1 larger implies Y_2 smaller

5.77 0

5.79 $0, 0$

5.81 0

5.83 **a** 2
b no
c 4, perfect increasing linear relationship
d -4, perfect decreasing linear relationship

5.85 **a** $-\alpha/4$

5.87 -22, 480

5.89 $1/9$

5.91 $2/3$, $1/18$

5.93 $(11.48, 52.68)$

5.95 42, 25, no

5.97 $p_1 - p_2$, $\left\{ \dfrac{N - n}{n(N - 1)} \right\} \{p_1 + p_2 -$
$(p_1 - p_2)^2\}$

5.99 **a** $.0823$
b $n/3$, $2n/9$
c $-n/9$
d 0, $2n/3$

5.101 **a** $.0972$
b $.2, .072$

5.103 $.08953$

5.105 **a** $.046$
b $.2262$

5.107 **a** $.2759$
b $.8031$

5.113 **a** $y_2/2$
b $1/4$

5.115 **a** $3/2$
b 1.25

5.117 $3/8$

5.119 $m_U(t) = (1 - t^2)^{-1/2}$, $E(U) = 0$,
$V(U) = 1$

5.121 11/36

5.123 **a** $f(y_1) = 3y_1^2, \ 0 \le y_1 \le 1$
$f(y_2) = (3/2)(1 - y_2^2),$
$0 \le y_2 \le 1$

b 23/44

c if $0 < y_2 \le 1$, $f(y_1|y_2) =$
$2y_1/(1 - y_2^2), \ y_2 \le y_1 \le 1$

d 2/3

5.125 **a** $\dfrac{1}{\beta^2} e^{-(y_1+y_2)/\beta},$
$y_1, y_2 > 0$

b $1 - \left(1 + \dfrac{a}{\beta}\right) e^{-a/\beta}$

5.127 **a** λp

b λp

5.129 $p(y) = \dbinom{y+\alpha-1}{y}\left(\dfrac{\beta}{\beta+1}\right)^y \left(\dfrac{1}{\beta+1}\right)^\alpha,$
$y = 0, 1, 2, \ldots$

5.133 $\mu_1 - \mu_2, \ (\sigma_1^2/n) + (\sigma_2^2/m)$

5.135 **a** $m(t_1, t_2, t_3) = (p_1 e^{t_1} + p_2 e^{t_2} + p_3 e^{t_3})^n$

c $-np_1 p_2$

Chapter 6

6.1 **a** $(1 - u)/2, \ -1 \le u \le 1$

b $(1 + u)/2, \ -1 \le u \le 1$

c $(1 - \sqrt{u})/\sqrt{u}, \ 0 < u \le 1$

d $E(U_1) = -1/3, \ E(U_2) = 1/3,$
$E(U_3) = 1/6$

6.3 **a** $f(u) =$
$\begin{cases} (u + 4)/100, & -4 \le u \le 6 \\ 1/10, & 6 < u \le 11 \end{cases}$

b 5.5833

6.5 $1/8\sqrt{2(u - 3)}, \ 5 \le u \le 53$

6.7 **a** $2u, \ 0 \le u \le 1$

b 2/3

6.9 **a** $4ue^{-2u}, \ u \ge 0$

b 1, 1/2

6.11 $[-\ln(1 - U)]^{1/2}$

6.13 **a** $\alpha y^{\alpha-1}/\theta^\alpha \ \ 0 \le y \le \theta$

b $\theta(U)^{1/\alpha}$

c $y_i = 4\sqrt{u_i}$; 2.0785, 3.3229,
1.5036, 1.5610, 2.4030

6.23 **b** $\Gamma[1 + (k/2)]\beta^{k/2}$

6.25 **a** $\left[\dfrac{1}{\Gamma(3/2)(kT)^{3/2}}\right] w^{1/2} e^{-w/kT},$
$w > 0$

b $(3/2)kT$

6.27 $2/(1 + u)^3, \ u > 0$

6.29 $4(80 - 31u + 3u^2),$
$4.5 \le u \le 5$

6.31 $-\ln u, \ 0 \le u \le 1$

6.33 $4ue^{-2u}, \ u > 0$

6.35 normal, $E(U) = \mu \sum\limits_{i=1}^{n} a_i, \ V(U) =$
$\sigma^2 \sum\limits_{i=1}^{n} a_i^2$

6.37 **a** normal, $E(\overline{Y}) = \mu, \ V(\overline{Y}) = \sigma^2/n$

b .7888

c .8664, .9544, .9756; increases as n
increases

6.39 \$190.27

6.41 .025

6.43 binomial $(n_1 + n_2, p)$

6.45 **a** binomial (n^2, p)

b binomial $(n_1 + n_2 + \cdots + n_n, p)$

c hypergeometric $(r = n_1, N = n_1 + n_2$
$+ \cdots + n_n)$

6.47 .077

6.55 **a** $\left(y_1^2 y_2^2\right)^{-1}, \ 1 < y_1, \ 1 < y_2$

e no

6.59 **a** $2u, \ 0 \le u \le 1$

b $E(U_2) = 2/3, \ V(U_2) = 1/18$

6.61 $(2/3)^5$

6.63 **a** $\dfrac{\dfrac{n!}{(j-1)!(k-1-j)!(n-k)!} y_j^{j-1} [y_k - y_j]^{k-1-j}[\theta - y_k]^{n-k}}{\theta^n},$
$0 \le y_j < y_k \le \theta$

b $\dfrac{(n-k+1)j}{(n+1)^2(n+2)}\theta^2$

c $\dfrac{(n-k+j+1)(k-j)}{(n+1)^2(n+2)}\theta^2$

6.65 **b** $1-e^{-9}$

6.67 **a** $e^{-(u-4)},\ u \geq 4$

     b 5

6.69 $m(n-1)r^{n-2}(1-r),\ 0 < r < 1$

6.73 $(2/3)(w^{-1/2}-w),\ 0 \leq w \leq 1$

6.75 **a** $1/(2\sqrt{u}),\ 0 \leq u \leq 1$

     b $f(u) = \begin{cases} 1/2, & 0 \leq u \leq 1 \\ 1/(2u^2), & u > 1 \end{cases}$

c $ue^{-u},\ u > 0$

d $-\ln u,\ 0 \leq u \leq 1$

6.77 $f(d) = 1,\ 0 \leq d \leq 1$

6.79 $f(u) = 1/[\pi(1+u^2)],\ -\infty < u < \infty$

6.81 $f(u) = (1/B(\alpha,\beta))\,(1-u)^{\alpha-1}u^{\beta-1},$
$0 < u < 1$

6.83 $f(u) = \begin{cases} 1/(4\sqrt{u}), & 0 \leq u \leq 1 \\ 1/(8\sqrt{u}), & 1 < u \leq 9 \end{cases}$

6.85 $p(u) = \begin{cases} .4156, & u = C_1 - C_3 \\ .5844, & u = C_2 - C_3 \end{cases}$

6.87 $1-(1/2)^4$

6.89 **b** $\exp[\mu + (\sigma^2/2)]$

Chapter 7

7.1 **a** .7698
     b .8664, .9284, .9642, .9836
     c increases as n increases
     d yes

7.3 .8664

7.5 .9876

7.7 **a** $\mu_1 - \mu_2$
     b $(\sigma_1^2/m) + (\sigma_2^2/n)$
     c $n = 18$

7.9 $P(S^2 > .065) =$
$P(\chi^2 > 14.625) \approx .1$

7.11 **a** $b = 2.421$
     b $a = .657$
     c .95

7.17 **a** $E(F) = 1.029$
     b $V(F) = .076$
     c $3 = \mu + 7.15\sigma$, not likely

7.19 **a** χ^2 with 5 d.f.
     b χ^2 with 4 d.f.
     c χ^2 with 5 d.f.

7.21 **a** normal, $E(\hat{\theta}) = \theta = c_1\mu_1$
$+ c_2\mu_2 + \cdots + c_k\mu_k$
$V(\hat{\theta}) = \left(\dfrac{c_1^2}{n_1} + \dfrac{c_2^2}{n_2} + \cdots + \dfrac{c_k^2}{n_k}\right)\sigma^2$

     b χ^2 with $n_1 + n_2 + \cdots + n_k - k$ d.f.

     c t with $n_1 + n_2 + \cdots + n_k - k$ d.f.

7.23 .9544

7.25 .0548

7.27 $n = 153$

7.29 .0217

7.31 $n = 664$

7.33 **a** no
     b approx. .0132

7.35 **a** random sample, approx. 1
     b approx. .1271

7.37 .0062

7.39 .0062

7.41 $n = 51$

7.43 $n = 56$

7.45 **a** $n > 9$
     b $n > 14, n > 14, n > 36, n > 36,$
$n > 891, n > 8991$

7.47 .8980

7.49 .7698

7.51 $n = 61$

7.53 **a** .7486
     b .729

7.55 **a** approx .5948
     b approx .0559, approx .0017

7.57 **a** approx .0630
         b approx .0630

7.59 .8414

7.61 .0041

7.63 $\mu = 10.15$

7.65 normal, mean $= .4\mu_1 + .2\mu_2 + .4\mu_3$,
         variance $= (.16\sigma_1^2/n_1) + (.04\sigma_2^2/n_2)$
         $+ (.16\sigma_3^2/n_3)$.

7.69 **a** F with num. d.f. $= 1$, denom. d.f. $= 9$
         b F with num. d.f. $= 9$, denom. d.f. $= 1$
         c $c = 49.04$

7.71 **b** .1587

7.75 .8413

7.77 .1587

7.79 .264

Chapter 8

8.3 $a = [\sigma_2^2 - c]/[\sigma_1^2 + \sigma_2^2 - 2c]$

8.5 $\hat{\theta} = \overline{Y} - 1$

8.7 $\hat{\theta}_3 - 9\hat{\theta}_2 + 54$

8.9 **c** $[n^2/(n-1)](Y/n)[1 - (Y/n)]$

8.11 **a** $\beta/[3n-1]$
         b $2\beta^2/[(3n-1)(3n-2)]$

8.13 **a** $(1-2p)/(n+2)$
         b $[np(1-p) + (1-2p)^2]/$
             $[(n+2)^2]$
         c p near $1/2$

8.15 MSE $= \beta^2$

8.17 11.5 ± 0.99

8.19 **a** 11.3 ± 1.54
         b 1.3 ± 1.7
         c $.17 \pm .08$

8.21 -0.7, $b = .404$

8.23 **a** .0017
         b .0016
         c no

8.25 **a** .06, .04
         b yes

8.27 .70, .26, no

8.29 $.7 \pm .205$

8.31 **a** 20 ± 1.265
         b -3 ± 1.855, yes

8.33 1020 ± 645.10

8.35 **b** $\left(\dfrac{2Y}{9.48773}, \dfrac{2Y}{.710721} \right)$

8.37 **a** $(Y^2/5.02389, Y^2/.0009821)$
         b $Y^2/.0039321$
         c $Y^2/3.84146$

8.39 **b** $Y_{(n)}/[(.95)^{1/n}]$

8.41 **a** $Y/.0513$
         b 80% confidence interval

8.43 $.805 \pm .021$

8.45 $.667 \pm .052$

8.47 26.2 ± 10.74

8.49 **a** $.73 \pm .022$
         b random sample

8.51 $-.51 \pm .028$, yes

8.53 $.06 \pm .12$

8.55 **a** $7.2 \pm .751$
         b $2.5 \pm .738$

8.57 $(-.12, .56)$

8.59 $n = 100$

8.61 3687

8.63 136

8.65 497

8.67 **a** 64
         b 647
         c 647

8.69 60.8 ± 5.701

8.71 **a** 3.4 ± 3.7
         b $.7 \pm 3.32$

8.73 -1 ± 4.72

8.75 **a** $(0, 9.76)$
b $(-1.52, 12.38)$

8.77 $(-84.39, -28.93)$

8.79 **a** $2\overline{X} + \overline{Y} \pm 1.96\sigma\sqrt{\frac{4}{n} + \frac{3}{m}}$

b $2\overline{X} + \overline{Y} \pm t_{\alpha/2}S\sqrt{\frac{4}{n} + \frac{3}{m}}$,
where $S^2 =$
$$\frac{\sum(Y_i - \overline{Y})^2 + 1/3\sum(X_i - \overline{X})^2}{n + m - 2}$$

8.81 $(.227, 2.194)$

8.83 **a** $(n-1)S^2/\chi_\alpha^2$, d.f. $= n-1$
b $(n-1)S^2/\chi_{1-\alpha}^2$, d.f. $= n - 1$

8.85 **a** $\sqrt{(n-1)S^2/\chi_\alpha^2}$
b $\sqrt{(n-1)S^2/\chi_{1-\alpha}^2}$

8.87 $s^2 = .02855$; $(.0129, .1246)$

8.89 $(1.407, 31.264)$; no

8.91 765

8.93 **a** $.0625 \pm .0237$
b 563

8.95 38,416

8.97 $n = 768$

8.99 $(29.30, 391.15)$

8.101 11.13 ± 1.44

8.103 3 ± 3.63

8.105 $-.75 \pm .77$

8.107 $.832 \pm .0153$

8.109 **a** $\dfrac{S_1^2}{\sigma_1^2} \Big/ \dfrac{S_2^2}{\sigma_2^2}$

b $\left(\dfrac{S_2^2}{S_1^2 F_{v_2, v_1, \alpha/2}}, \dfrac{S_2^2}{S_1^2} F_{v_1, v_2, \alpha/2}\right)$
$v_i = n_i - 1, i = 1, 2$

8.111 $2(n-1)\sigma^4/n^2$

8.113 **b** $V(S_p^2) = 2\sigma^4/(n_1 + n_2 - 2)$

Chapter 9

9.1 $1/3; 2/3; 3/5$

9.3 **b** $12n^2/[(n+1)^2(n+2)]$

9.5 $n - 1$

9.7 $1/n$

9.17 **c** need $\text{Var}(X_{2i} - X_{2i-1}) < \infty$

9.19 **b** $.6826$
c no

9.25 $\alpha\beta$

9.27 $\overline{X}/(\overline{X} + \overline{Y})$

9.39 $\sum\limits_{i=1}^{n} \ln(Y_i)$; no

9.49 yes

9.51 $3\overline{Y}(\overline{Y} + 1 - \frac{1}{n})$

9.53 $(n+1)Y_{(n)}/n$

9.55 **b** $(3n+1)Y_{(n)}/(3n)$

9.61 $\hat{\theta} = \dfrac{2\overline{Y} - 1}{1 - \overline{Y}}$, no, not MVUE

9.63 $\sigma^2 = \dfrac{1}{n}\sum\limits_{i=1}^{n} Y_i^2$

9.67 $\hat{\theta} = \dfrac{1 - 2m_2'}{4m_2' - 1}$, where $m_2' = \dfrac{1}{n}\sum\limits_{i=1}^{n} Y_i^2$

9.69 $\hat{\theta} = 2\overline{Y}/3$

9.73 $(\overline{Y})^2$

9.75 **a** $\hat{\theta} = (Y_{(n)} - 1)/2$
b $(Y_{(n)})^2/12$

9.77 **a** $\hat{\theta} = \overline{Y}/\alpha$
b $E(\hat{\theta}) = \theta$; $V(\hat{\theta}) = \theta^2/(\alpha n)$
d $\sum\limits_{i=1}^{n} Y_i$
e $\left(\dfrac{2\sum\limits_{i=1}^{n} Y_i}{31.41}, \dfrac{2\sum\limits_{i=1}^{n} Y_i}{10.85}\right)$

9.79 $\hat{p}_A = .30, \hat{p}_B = .38$
$\hat{p}_C = .32; -.08 \pm .1641$

9.81 $y = 0, \quad \hat{p} = 1/4$
$y = 1, \quad \hat{p} = 1/4$ or $3/4$
$y = 2, \quad \hat{p} = 3/4$

9.83 $Y_{(n)}/2$

9.85 **a** $Y_{(1)}$
b $Y_{(1)}/\theta$
c $\left((\alpha/2)^{1/2n} Y_{(1)}, \right.$
$\left. (1 - (\alpha/2))^{1/2n} Y_{(1)} \right)$

9.87 $\hat{R} = (1 - \hat{p})/\hat{p}$ where
$\hat{p} = \dfrac{\text{no. of defectives}}{n}$

9.89 $\hat{p} \pm z_{\alpha/2} \sqrt{\dfrac{\hat{p}(1 - \hat{p})}{n}}$

9.91 $e^{-\overline{Y}} \pm z_{\alpha/2} \sqrt{\dfrac{\overline{Y}e^{-2\overline{Y}}}{n}}$

9.93 $\hat{\sigma}^2 = \dfrac{1}{n} \sum (Y_i - \mu)^2$

9.95 $e^{-t/\overline{Y}}$

9.97 **a** $\hat{N}_1 = 2\overline{Y} - 1$
b $E(\hat{N}_1) = N;$
$V(\hat{N}_1) = \left(\dfrac{1}{3n} \right) (N^2 - 1)$

9.99 252 ± 85.193

Chapter 10

10.3 **a** $c = 11$
b .596
c .057

10.5 $c = 1.684$

10.7 **a** $H_0 : \mu = 900, \quad H_a : \mu \neq 900$
b $|z| > 2.58$
c $z = -1.77$

10.9 $z = -4.22$

10.11 $z = 3.65$

10.13 **a** $H_0 : \mu_1 - \mu_2 = 0,$
$H_a : \mu_1 - \mu_2 \neq 0$
b two-tailed
c $z = -.954,$ do not reject H_0

10.15 $H_0 : p = .5, \quad H_a : p > .5$
$z = 7.56,$ yes

10.17 $H_0 : p_1 - p_2 = 0,$
$H_a : p_1 - p_2 \neq 0, \quad z = 10.65,$ yes

10.19 $z = 4.47$

10.23 $z = 1.50,$ no

10.25 $z = -1.48$ (1 = homeless), no

10.27 approx. 0

10.29 .67

10.31 .025

10.33 **a** $z = .49$
b .1056

10.35 $885 \pm 21.80,$ no

10.37 .57, yes

10.39 129.146, yes

10.41 $z = 1.58,$ p-value $= .1142,$ do not reject

10.43 **a** $z = -.996,$ p-value $= .1587$
b no
c $z = -1.826,$ p-value $= .0336$
d yes

10.45 $z = -1.538,$ p-value $= .0618,$
do not reject

10.47 $z = -1.732,$ p-value $= .0836$

10.51 $t = -1.34,$ do not reject, p-value $> .10$

10.53 **a** $t = -3.24,$ p-value $< .005,$ yes
b 39.556 ± 3.55

10.55 **a** $t = -3.4,$ p-value $< .005,$ reject
b 365 ± 21.53
c $\mu \neq 400$

10.57 $t = -1.57,$ $.10 < p$-value $< .20,$
do not reject

10.59 $t = .989,$ p-value $> .10$

10.61 **a** $t = 1.92$, do not reject, $.05 < p$-value $< .10$

b $t = .365$, do not reject, p-value $> .20$

10.63 $t = -.647$, do not reject

10.65 **a** $t = -5.54$, p-value $< .01$

b yes

c $t = 1.56$, $.10 < p$-value $< .20$

d yes

10.67 $\chi^2 = 12.6$, do not reject, $.05 < p$-value $< .10$

10.71 **a** $\sigma_1^2 \neq \sigma_2^2$

b $\sigma_1^2 < \sigma_2^2$

c $\sigma_1^2 > \sigma_2^2$

10.73 $\chi^2 = 22.45$, p-value $< .005$

10.75 $F = 8.03$, yes

10.77 **a** .15

b .45

c .75

d 1

10.79 **a** reject if $\overline{Y} > 7.82$

b .2611, .6406, .9131, .9909

10.81 $n = 16$

10.83 **a** $U = \dfrac{2}{\beta_0} \sum\limits_{i=1}^{4} Y_i$ has $\chi_{(24)}^2$ distribution under H_0: reject H_0 if $U > \chi_\alpha^2$

b yes

10.85 **c** yes, is UMP

10.87 **a** $\sum\limits_{i=1}^{n} Y_i \geq k$

b use Poisson table to find k such that $P(\sum Y_i \geq k) = \alpha$

c yes

d $\sum\limits_{i=1}^{n} Y_i \leq k$

10.89 **a** $\sum Y_i <$ const.

b yes

10.91 **a** reject H_0 if $Y_{(n)} \leq \theta_0 \sqrt[n]{\alpha}$

b yes

10.95 $\chi^2 = \dfrac{(n-1)S_1^2 + (m-1)S_2^2}{\sigma_0^2}$ has $\chi_{(n+m-2)}^2$ distribution under H_0; reject if $\chi^2 > \chi_\alpha^2$

10.97 **a** $\lambda = \dfrac{(\overline{X})^m (\overline{Y})^m}{\left(\dfrac{m\overline{X} + n\overline{Y}}{m+n}\right)^{m+n}}$

b $\overline{X}/\overline{Y}$ distributed as F with $2m$ and $2n$ degrees of freedom

10.103 **a** $t = -22.17$, p-value $< .01$

b $-.0105 \pm .001$

c yes

d no

10.105 **a** $H_0: p = .2$, $H_a: p > .2$

b $\alpha = .075$

10.107 $z = 5.24$, p-value approx. 0

10.109 **a** $F = 2.904$, no

b (.050, .254)

10.111 **a** $t = -2.657$, $.02 < p$-value $< .05$

b -4.542 ± 3.046

10.113 $T = \dfrac{(\overline{X}+\overline{Y}-\overline{W})-(\mu_1-\mu_2-\mu_3)}{\left\{\left(\frac{1+a+b}{n(3n-3)}\right)\left[\sum(X_i-\overline{X})^2 + \frac{1}{a}\sum(Y_i-\overline{Y})^2 + \frac{1}{b}\sum(W_i-\overline{W})^2\right]\right\}}$ with $(3n-3)$ degrees of freedom

10.115 $\lambda = \left(\dfrac{\sum(y_i - y_{(1)})}{n\theta_{1,0}}\right)^n \exp\{-\left(\sum(y_i - y_{(1)})/\theta_{1,0}\right) + n\}$

Chapter 11

11.1 $\hat{y} = 1.5 - .6x$

11.3 $\hat{y} = 21.6 + 4.8x$

11.5 **b** $\hat{y} = 34.0092 - .0837x$

d 23.5467

11.7 $\hat{\beta}_1 = 2.514$

11.9 **a** $\hat{y} = 452.11935 - 29.40184x$

11.13 **a** SSE $= 18.29$; $s^2 = 3.05$

b $\hat{y} = 43.36 + 2.42x^*$, SSE $= 18.29$

11.15 **a** $\hat{y} = 3 + .475x$
 c 5.025

11.17 $\text{cov}(\hat{\beta}_0, \hat{\beta}_1) = \dfrac{-\bar{x}\sigma^2}{\sum(x_i - \bar{x})^2}$

11.19 **a** $t = 5.20$; reject
 b $(-.97, -.23)$

11.21 **a** $t = 3.791,\ \ p < .01$
 b reject
 c no
 d $.475 \pm .289$

11.25 $t = \dfrac{\hat{\beta}_1 - \hat{\gamma}_1}{\sqrt{S^2\left[\frac{1}{\sum(x_i-\bar{x})^2} + \frac{1}{\sum(c_i-\bar{c})^2}\right]}}$,
 where $S^2 = \dfrac{\text{SSE}_1 + \text{SSE}_2}{n + m - 4}$

11.27 $t = -10.72,\ \ $yes

11.29 $t = 9.62,\ \ $yes

11.31 $x^* = \bar{x}$

11.33 $(4.67, 9.63)$

11.35 25.395 ± 2.875

11.37 **b** $(72.39, 75.77)$

11.39 $(59.73, 70.57)$

11.41 $(-.86, 15.16)$

11.43 $(.27, .51)$

11.45 $t = 9.608,\ \ p$-value $< .01$

11.49 $.979 \pm .104$

11.51 **a** $\hat{\alpha}_0 = 36.70, \hat{\alpha}_1 = .0095$
 b $(36.17, 37.23)$

11.55 $\hat{y} = 2.1 - .6x$

11.57 **a** $\hat{y} = 29.99 + 1.39x$
 b $\hat{y} = 36.54 + 1.39x - .20x^2$

11.61 $t = 1.31$

11.63 21.94 ± 3.01

11.65 21.94 ± 6.17

11.67 $t^2 = 114.92$

11.69 $F = 10.21$

11.71 90.38 ± 8.42

11.73 **a** $\hat{y} = -13.54 - .053x$
 b $t = -7.01,\ \ $reject
 c $.93 \pm .33$

11.75 **a** $\hat{y} = 1.4285 + .5000x_1 + .1190x_2 - .5000x_3$
 b $\hat{y} = 2.0715$
 c $t = -13.7,\ \ $reject
 d $(1.88, 2.26)$
 e $(1.73, 2.41)$

11.77 $n/2$ points at $x = -9$ and at $x = 9$

11.79 **a** $\hat{y} = 9.34 + 2.46\,x_1 + .60\,x_2 + .41\,x_1x_2$
 c $9.34,\ \ 11.80$
 d $t = 2.63,\ \ $reject
 e $(12.44, 13.18)$
 f $(12.03, 13.59)$

11.85 $r = .89,\ \ t = 4.81,\ \ p$-value $< .01$, reject

Chapter 12

12.1 $n_1 = 34,\ \ n_2 = 56$

12.3 $n = 246,\ \ n_1 = 92,\ \ n_2 = 154$

12.5 three observations each at $x = 2$ and $x = 5$

12.11 **a** $\rho = 0$
 b $\rho > 0$
 c paired better when $\rho > 0$, independent better when $p < 0$.

12.13 **a** $t = 2.65,\ \ $reject

12.15 **a** μ_i

12.29 **a** μ_i
 b $\mu_i,\ \ (\sigma_p^2 + \sigma^2)/n$
 c $\mu_1 - \mu_2,\ \ 2\sigma^2/n$, normal

12.33 112.5 ± 94.304

12.35 $k_1 = 1/4,\ \ k_2 = 1/2,\ \ k_3 = 1/4$

Chapter 13

13.1 **a** $F = 2.93$, do not reject
   b $|t| = 1.71$, do not reject
      $F = t^2$

13.7 SSE = .020; $F = 2.0$, do not reject

13.9 SST = .7588, SSE = .7462, $F = 19.83$, $p < .005$ (approx.)

13.11 SST = 36.2855, SSE = 76.6996, $F = 38.3164$, p-value $< .005$, reject

13.13 F = 63.66, yes, p-value $< .005$

13.19 **a** -12.08 ± 10.96
    b longer
    c fewer degrees of freedom

13.21 $-.28 \pm .102$

13.23 $67.86 \le \mu_A \le 84.14$, $55.82 \le \mu_B \le 76.48$, $-3.63 \le \mu_A - \mu_B \le 22.96$

13.25 **a** $6.24 \pm .318$
    b $-.29 \pm .241$

13.27 **a** $F = 1.32$, no
    b $(-.21, 4.21)$

13.29 $(1.39, 1.93)$

13.31 **a** 2.7 ± 3.75
    b 27.5 ± 2.65

13.33 **a** μ
    b overall mean

13.35 **b** $2\sigma^2/b$

13.37 **a** $F = 3.04$, do not reject
    c MSE $= 2S_D^2$ (except for round-off)

13.39 **a** $F = 10.05$, yes
    b $F = 10.88$, yes

13.41 $F = 1.40$, no

13.43 $F = 6.36$, yes

13.47 2 ± 2.83

13.49 $.145 \pm .179$

13.51 -4.8 ± 5.259

13.53 $n_A = 3$

13.55 $b = 16$; $n = 48$

13.57 $m = 3$, $\alpha/m = .0167$, $t_{.0084} \approx z_{.0084} = 2.39$
    $\mu_S - \mu_L : -.29 \pm .294$,
    $\mu_S - \mu_T : -1.65 \pm .459$,
    $\mu_L - \mu_T : -1.36 \pm .420$

13.59 $-6.56 \le \mu_A - \mu_B \le -.44$, $-2.56 \le \mu_A - \mu_C \le 3.56$, $0.94 \le \mu_B - \mu_C \le 7.06$

13.61 **a** $\beta_0 + \beta_3$

13.63 $F = 7$, reject

13.65 **b** $F = 1.05$, do not reject

13.67 $.02 \le \mu_A - \mu_B \le .14$, no

13.69 16, 135, 14.1

13.71 $F = 7.33$; yes; blocking induces loss in degrees of freedom for estimating σ^2; could result in sight loss of information if block to block variation is small

13.73 **a**

Source	d.f.	S.S.	M.S.	F
Treatments	2	524,177.17	262,088.58	258.24
Blocks	3	173,415	57,805	56.95
Error	6	6,090.5	1,015.08	
Total	11	703,681.67		

   b 6
   c yes; $F = 258.19$, p-value $< .005$
   d yes; $F = 56.95$, p-value $< .005$
   e 22.529
   f -237.25 ± 55.13

13.75 **a** SSE = .571, SST = 1.212, $F = 11.68$, p-value $< .005$
    b $t = -2.73$, 22 degrees of freedom, $.01 < p$-value $< .02$

13.77 $.05/4 \approx .01$,
    $\mu_A - \mu_D : .320 \pm .251$
    $\mu_B - \mu_D : .145 \pm .251$
    $\mu_C - \mu_D : .023 \pm .251$
    $\mu_E - \mu_D : -.124 \pm .251$

13.79 **b** σ_β^2
    c $\sigma_\beta^2 = 0$

13.81   **a** $\mu; \sigma_\beta^2 + (\sigma_\epsilon^2/k)$

        b $\sigma_\beta^2 + \left(\dfrac{b}{k-1}\right)\sum\limits_{i=1}^{k}\tau_i^2$

        c $\sigma_\epsilon^2 + k\sigma_\beta^2$

        d σ_ϵ^2

Chapter 14

14.1 $X^2 = 24.48$, p-value $< .005$

14.3   **a** $z = 1.50$,   do not reject
      b   hypothesis suggested by observed data

14.5 $X^2 = .2995$,   no, p-value $> .10$

14.7 $X^2 = 69.42$ (using 8 cells); reject

14.9   **a** $X^2 = 2.56$; do not reject
      b p-value $> .10$

14.11   **a** X^2 multiplied by k also

14.13   **a** $X^2 = 19.045$,   reject
       b $X^2 = 60.139$,   reject
       c   some expected counts < 5

14.15   **a** $X^2 = 22.94$,   reject
       b p-value $< .005$

14.17   **a** $X^2 = 13.99$,   reject

       b $X^2 = 13.99$,   reject
       c $X^2 = 1.36$,   do not reject

14.21 $X^2 = 42.179$,   reject

14.23 $X^2 = 38.43$,   yes

14.25   **a** $X^2 = 14.19$,   reject

14.27 $X^2 = 21.51$,   reject

14.29 $X^2 = 6.18$,   reject, $.025 < p$-value $< .05$

14.31   **a** $X^2 = 20.513$
       b p-value $< .01$
       c   reject

14.33 $X^2 = 8.56$,   d.f. $= 3$; reject

14.37 $X^2 = 3.26$,   do not reject

14.39 $X^2 = 74.85$,   reject

Chapter 15

15.1 $.014$, $m \le 6$ and $m \ge 19$
      $.044$, $m \le 7$ and $m \ge 18$
      $.108$, $m \le 8$ and $m \ge 17$

15.3 $P(M \le 2$ or $M \ge 8) = .11$; no

15.5   **a** $P(M \le 2$ or $M \ge 7) = .18$; do not reject
      b $t = 1.65$; do not reject

15.7   **a** p-value $= .011$,   do not reject

15.9   **a** $T = 6$, $.02 < p$-value $< .05$
      b $T = 6$, $.01 < p$-value $< .025$

15.11 $T = 3.51$, $.025 < p$-value $< .05$

15.13 $T^- = 11$,   reject

15.17 $U = 9$,   do not reject

15.19 $z = 1.80$,   yes

15.21 $U = 0$, p-value $= .0096$

15.23   **a**   SST $= 2586.133$, SSE $= 11702.8$, $F = 1.33$,   do not reject
       b $H = 1.22$,   do not reject

15.25 $H = 2.03$,   do not reject

15.29 $F_r = 6.35$,   reject

15.31   **a** $F_r = 65.675$, p-value $< .005$,   reject
       b $m = 0$, $P(M = 0) = (1/2)^8 = 1/256$, p-value $= 1/128$

15.35 $P(F_r = 4) = P(F_r = 0) = 1/6$
      $P(F_r = 3) = P(F_r = 1) = 1/3$

15.37 $R = 6$,   no

15.39   **a** $.0256$
      b   an usually small number of runs (judged at $\alpha = .05$) would imply a clustering of defective items in time; $z = -1.95$; do not reject

15.41 $R = 13$,   do not reject

15.43 $r_s = .912$, yes

15.45 **a** $r_s = -.845$
b reject

15.47 use two-tailed test; $r_s = .6768$, reject

15.49 $r_s = 1.000$, p-value $< .005$

15.51 **a** $m = 2$, yes
b variance not constant

15.53 $T = 73.5$, no, consistent with Exercise 15.52

15.55 $U = 17.5$, do not reject

15.57 .0159

15.59 $H = 7.154$, reject

15.61 $F_r = 6.21$, do not reject

15.63 .10

INDEX

Acceptance region, 481
Addition of matrices, 764–65
Additive law of probability, 55–56,
    58, 61
  for conditional probabilities, 58
  effect of mutually exclusive events
    on, 61
Allometric equation, 575
α
  probability of a type I error, 463
  shape parameter, 176
Alternative hypothesis, 461
  choice of, 472
  lower-tail, 471
  simple, 509
  two-tailed test, 472
  upper-tail, 469
Analysis of categorical data,
    680–707
Analysis of variance, 628–79
  assumptions for, 638
  F test for, 637–38, 639
  for one-way layout, 635–39
  procedure for, 629–35
  for randomized block design,
    655–61
  selecting the sample size for,
    663–65
  sums of squares, 647
  uses for, 629
  using linear models for, 668–72
Analysis of variance table
  for one-way layout, 639–44
  for randomized block design, 658

ANOVA table. *See* analysis of
    variance table
AOV table. *See* Analysis of variance
    table
Arithmetic mean. *See* Mean
Association between populations,
    test for, 751–52
Asymptotically normal distribution,
    348
Attained significance level, 483

Bayes' Rule, 68
Bell-shaped distribution. *See* Normal
    distribution
Bernoulli variable, 437
β
  scale parameter, 176
  type II error. *See* Type II error
Beta density function, 182
  uses for, 182–83
Beta distribution, 182–87
  mean and variance, 184
  moment-generating function for,
    781
  of the second kind, 327
  related to binomial distribution,
    184
Beta function
  incomplete, 183
  related to gamma function, 779
Beta probability function, 779
Bias of a point estimator, 366
Binomial coefficients, 45
Binomial distribution, 97–109

cumulative form for, 183
formula for, 99
hypergeometric distribution and,
    123
moment-generating function for,
    781
normal approximation to, 354–60
Poisson distribution and, 127–29
tables, 102, 128
uses for, 101–3
Binomial distribution function, 184
Binomial expansion, formula for, 45,
    778
Binomial experiment, 97, 101, 265
Binomial probability function
  relationship to incomplete beta
    function, 183
  tables, 184
Binomial random variable, mean and
    variance for, 103
Bivariate density function, 214–15,
    268
Bivariate distribution, 210–22
  definition of, 211
Bivariate distribution function,
    212–14
Bivariate normal distribution, 268–70
  testing for independence in,
    567–68
Bivariate probability function, 211
Bivariate transformation, 297–300
Bivariate transformation method,
    310–11
Bivariate uniform distribution, 215

Block, 621
Block designs, 621–22
  randomized, 621–22, 653–63
Block effects, 654, 655
Blocks, 653
Bonferroni inequality, 59, 666

Categorical data
  analysis of, 680–707
  definition of, 680
  experiments with, 680–81
  methods of analyzing, 701–2
Cell frequencies, estimated expected,
  690
Cell probabilities, testing hypotheses
  concerning, 684–88, 702
Central limit theorem, 346–63
  binomial distribution and, 354–57
  moment-generating function and,
    352–54
  formal statement of, 348
  proof of, 352–54
  uses for, 346, 354
Central moment, 188
Central tendency, measures of, 8,
  131, 170
Characteristic curve, operating, 144
χ^2. *See* Chi-square
Chi-square density function, 408
Chi-square distribution, 178
  degrees of freedom associated
    with, 307
  mean and variance, 178
  moment-generating function for,
    781
Chi-square random variable, 178
Chi-square tables, 178, 334–35, 337
Chi-square test
  concerning variance, 498–501
  for goodness of fit, 684–88
  test statistic, 500
  test statistic (Pearson's), 682
  uses for, 411, 685
  using for categorical data, 701–2
CM (Correction for the mean), 636
Combination, 44
Combinatorial analysis, 37, 39–50
  results from, 39, 43
Complement, 23
  of the rejection region, 481
  probability of, 56–57

Complementary events, 63
Completely randomized design, 619
  difference from randomized block
    design, 621, 654
  experimental error of, 621
Complete model, 597–99
Completeness, 436
Composite hypothesis, 509
Compound event, 26
  in a discrete sample space
    experiment, 28
Conditional density function, 228
Conditional discrete probability
  function, 226
Conditional distribution, 273–74
Conditional distribution function,
  227
Conditional expectation, 270–75
Conditional mean, 271
Conditional probability, 50–55,
  98–99
Conditional probability distribution,
  225–29
  continuous, 227–28
  discrete, 225–26
Conditional variance, 272
Confidence bands, 565–66
Confidence coefficient, 380
  simultaneous, 666
Confidence interval, 380–84
  for β_1, 554
  compared with prediction interval,
    564–65
  for difference between means, 385,
    401–4, 649–53, 662–63
  for $E(Y)$, 560, 564–65, 566
  hypothesis testing and, 480
  large-sample, 385–95
  matched pairs experiment, 614–15
  for mean of treatment, 649–53
  multiple linear regression, 587
  one-sided, 381, 386, 400
  for one-way layout, 649
  for p of binomial population,
    385
  for $(p_1 - p_2)$, 385
  for parameter β_i, 554–55
  pivotal method for, 381–83
  for population mean, 385,
    399–401, 404
  for population variance, 407–10

randomized block design, 662
relationship with
  hypothesis-testing, 480
relationship with t test, 494
for sample variance, 337
simple linear regression, 559
simultaneous, 665–67
small-sample, 399–407
sufficient statistic and, 439
two-sided, 380, 386, 401, 480–81
unbiased, 415
width of, 607
Confidence level, 396
Confidence limits, 380, 386–87, 400,
  481
Consistency, 420–29
Consistent estimator, 421
Contingency tables, 688–97
  degrees of freedom associated
    with, 691
  with fixed row or column totals,
    697–701
  in tests for independence, 688
Continuous random variable(s)
  beta, 182–85
  definition of, 151, 154, 234
  density function for, 154
  distributions, 150–208
  expected value for, 162–66
  gamma, 175–78, 190
  jointly continuous, 212–13, 214
  kth moment about the origin, 188
  marginal density function of, 223
  moment-generating function of,
    189
  moments for, 188–90
  normal, 170–75
  Tchebysheff's Theorem for,
    194–97
  uniform, 166–67, 189
Correction for the mean (CM), 636
Correlation, 567–71
Correlation coefficient, 250
  notation for, 568
  Spearman rank, 750–56
Covariance, 249–55
  definition of, 250
  formula for, 251
  zero, 252–53
Cov (notation for covariance), 250
Cramer-Rao inequality, 420

Cumulative distribution function. *See* Distribution function
Curvature, detecting, 610
Curve, area under distribution, 6

Decomposition of an event, 67
Degrees of freedom
  appropriate number of, 683
  associated with $r \times c$ contingency table, 691
  for a chi-square distribution, 178, 307
  for F distribution, 340–41, 595
  for sum of squares, 656
  for t distribution, 338, 553
DeMorgan's laws, 24
Density function
  beta, 182–83
  bivariate, 214–15, 216, 268
  chi-square, 408
  definition of, 154
  exponential, 178–79, 346
  F, 340
  for function of random variables (U), 281, 282
  gamma, 176
  identification of, 155–58
  joint, 214–15, 216, 218
  of the kth-order statistic, 320–22
  log-normal, 204–5
  marginal, 223
  minimum/maximum, 318
  notation for, 431
  properties of, 155
  Rayleigh, 302
  for the second-order statistic, 322
  t, 338
  as a theoretical model, 187
  uniform, 166–67
  uses for, 155
  Weibull, 301, 437
Dependence, measures of, 249
Dependency between two classification criteria, 689
Dependent event, 52
Dependent random variable, 4, 234
Dependent variables, 4, 234, 532
Design of an experiment. *See* Experimental design
  accuracy-increasing, 608–11
  block, 621–23

completely randomized, 619, 621
Latin square, 622
matched pairs, 612–18
optimal, 610
randomized block design, 621–22, 653–63
sample size and, 395–96, 663–65
Designs, 608
Deterministic model, 532–33, 534
Deviation
  standard, 9
  total sum of squares of, 630–31, 647, 655
Difference between means, 423, 608–12
  confidence interval for, 385, 401–3, 404
  matched pairs experiment for, 613–14
  in one-way layout, 649–50
  in randomized block design, 662
  tests for, 491–93
Discontinuous function, 197–98
Discrete distribution, 84–88
Discrete random variable(s), 84
  definition of, 234
  distribution function for, 152
  distributions, 83–149
  expected value of, 88–97
  independence of, 234
  integer-valued, 136
  jointly discrete, 212–13
  joint probability function of, 211
  marginal probability function of, 223
  mean and variance, 93–94
  moment-generating functions, 131
  probability-generating function for, 137
  Tchebysheff's theorem for, 139–42
  variance of, 91, 93
Discrete sample space, 27
Disjoint sets, 24
Dispersion, measures of, 8, 131
Distribution, 4
  bell-shaped or normal, 5
  beta, 182–87, 327
  binomial, 97–109
  bivariate, 210–22
  bivariate normal, 268–70

chi-square, 178, 411
conditional, 225–33, 273–74
continuous, 151–62
discrete, 84–88
exponential, 178–79
F, 340
of function of random variables, 280–81
gamma, 175–82
geometric, 110–16
hypergeometric, 99, 119–24
joint, 210–22
log-normal, 328
marginal, 222–25, 229–33
mixed, 198
multinomial, 263–68, 264
multivariate, 209–78
negative binomial, 116–19
normal, 170–75
Pareto, 293
Poisson, 124–31
relative frequency, 4
sampling, 330–63
skewed, 175
t, 338
Tchebysheff's theorem for any, 139–42
uniform, 166–70
unique characterization of, 131
Weibull, 206, 439
Distribution-free tests. *See* Nonparametric statistics
Distribution function
  binomial, 184
  bivariate, 212–14
  conditional, 227
  of continuous random variables, 153
  cumulative, 151, 183, 214
  of discrete random variables, 151–53
  of a gamma-distributed random variable, 176
  identification of, 155–58
  joint, 212–14
  method of, 281–94
  multivariate, 218
  properties of, 153
  of a random variable, 151
  of t, 425
  of U, 290

Distributive laws, 24
Dummy variable, 668

e^{-x}, table of, 786
$E(Y)$, 88
Effect of population (or treatment) i, 645
Efficiency, 417–20
Empirical rule, 10
  and Tchebysheff's Theorem, 140
Empty set, 22
ε, 373, 533, 536
Equal matrices, 768
Error, 537
  in prediction, 562–63, 564
  random, 536, 549–50
  in repeated sampling, 562
  standard, 371, 397
  sum of squares for, 538, 582, 593, 630, 636
Error of estimation, 280, 373
  good approximate bound on, 374–75
  two-standard error bound on, 375
Estimated expected cell frequencies, 690
Estimation, 364–415
  for an unknown value of p, 113
  least squares method of, 531–606
  methods of, 444–59, 531–606
  minimum-variance unbiased, 435–55
  in the one-way layout, 649–53
  of population mean, 8
  in randomized block design, 662–63
Estimator(s)
  biased, 366, 372
  for comparing two population means, 423
  confidence interval, 380–84
  consistent, 421
  consistent sequence of, 426
  criterion for comparing, 419
  efficient, 420
  interval, 380
  large-sample confidence interval, 385–95
  least squares, 539, 540, 544–50, 600
  maximum likelihood, 449, 450

mean square error of, 367
method of moments, 441, 444–48
minimum-variance unbiased, 436
point, 366–73, 416–35
of population variance, 335
relative efficiency of unbiased, 417
sampling distribution of, 416
unbiased, 366, 367, 370–73
Estimator $\hat{\theta}$
  hypothesis-testing procedures for, 467–77
  standard error of, 480
Event-composition method, 34, 59–67
  erroneous application of, 61
Event(s), 26
  complementary, 63
  compound, 26, 28
  decomposition of, 67
  dependent, 52
  in discrete sample space, 28
  independent, 50–55
  intersection of two or more, 209–22
  intersection of n, 218
  mutually exclusive, 56–57
  simple, 26, 27, 28
  stochastic, 20
  symmetric difference between, 71
Expectation
  conditional, 270–75
  of continuous functions and mixed distributions, 197–201
  theorems, 92–93
Expected value
  conditional, 270–75
  of constant, 244
  of constant times function, 92
  for continuous random variables, 162–66, 188–94
  definition of, 88
  of discontinuous functions, 197–98
  for discrete random variables, 88–97
  of functions of random variables, 241–44
  for hypergeometric random variable, 260
  for least squares estimators, 544–47
  of linear functions, 255–63

of mean square for treatments for a one-way layout, 647–49
for mixed distributions, 198–200
of nonrandom quantity, 92
of point estimators, 371
for Poisson random variable, 128
for a runs test, 748
standard deviation as, 91
of sum of functions, 92, 163, 244
theorems for multivariate random variables, 244–49
theorems for univariate random variables, 92–93
of U test statistic, 729
variance as, 91, 163
Experiments, 25–33
  binomial, 265
  categorical data, 680–81
  factors and levels, 619, 628
  independent samples, 612
  matched pairs, 711–16
  multinomial, 263–65, 681
  paired-difference, 615
  probabilistic model for, 25–33
  single-factor, 619
Experimental design, 75, 607–27. See also Hypothesis tests
  accuracy in, 608–12
  completely randomized, 619, 654
  definition of, 395
  elementary designs, 618–23
  Latin square, 622, 629
  matched pairs experiment, 612–18
  randomized block, 620–21, 653–61, 663, 670
  selecting sample size for, 395–99
Experimental unit, 619
Exponential density function, 178
  graph of, 346
  mean and variance, 179
  uses for, 178
Exponential distribution with mean β, 290–91
Exponential models, 574
Exponential probability distribution, 178, 187–88, 290, 320, 346, 347, 374
  mean and variance of, 781
  moment-generating function for, 781
  probability function for, 781

F_r (test statistic), 739
F (test statistic)
  analysis of variance, 637
  F test, 503
  hypothesis-testing concerning
    variances, 501–7
  one-tailed, from analysis of
    variance test, 633
  test statistic, 503
$F(y)$ and $f(y)$, 154
Factor, 619, 628
Factorial moment, 137
Factorization criterion, 432
  uses for, 437, 439
F density function, 340
F distribution, 340
  table, 340–41
Fit, lack of, 601
Fitted models, 598, 599
Fixed block effects model, 654
Frequency distribution, 10
Friedman test, 738–43
  relationship with the sign test,
    741
  summary, 739–40
F test, 503–5
  analysis of variance and, 633, 634,
    656–57
Function
  of continuous random variable,
    expected value of, 162–63
  decreasing, 296
  density. *See* Density function
  discontinuous, 197–98
  distribution. *See* Distribution
    function
  expected value of, 81
  gamma, 176
  increasing, 294
  likelihood, 510, 521–22
  linear. *See* Linear function
  methods for finding probability
    distribution of, 280–309
  moment-generating, 132–36
  of normally distributed random
    variables, 305–7
  probability. *See* Probability
    function
  probability-generating, 136–39
  of random variable, expected value
    of, 90–97, 191

  of random variable, finding the
    moments of, 191
  of random variables, 279–329
  of random variables, expected
    value of, 241–244
  reliability, 327
  step, 152

Gamma density function, 175–79,
  445–46
  beta function related to, 779
Gamma distribution, 175–82
  mean and variance, 177
  moment-generating function for,
    190
  relationship with Poisson
    distribution, 176
  tables, 178
Gamma function, 176, 779
Gamma random variable, 176–79
  chi-square, 178
Geometric distribution, 110–16
  mean and variance for, 112
  uses for, 111
Geometric probability function, 110
Geometric probability distribution,
  110–116
  mean and variance of, 781
  moment-generating function for,
    781
  probability function for, 781
Geometric random variable
  probability-generating function
    for, 138
  variance, 143
Geometric representation of a joint
    density function, 216
Geometric series, formula for sum of,
  64, 778
Goodness
  of an estimate, 280
  of an estimation procedure, 525
  of a point estimator, 366, 373–80
  of a statistical test, 525
Goodness of a test, 463, 507
Goodness-of-fit test, 684–88
  care in using, 702
Graphical descriptive methods, 3–8

H (test statistic), 733, 734
H_0, 462

Hierarchical model, 273–74
High-influence points, 601
Histogram
  area under, 6
  binomial probability, 100
  bivariate distribution function, 152
  construction of, 4–5
  geometric probability distribution,
    110–11
  guidelines for constructing, 5–6
  probabilistic interpretation of,
    5–6
  probability, 86, 89, 92
  relative frequency, 5, 9, 188, 347
  three-dimensional, 211
Histogram examples
  binomial distribution, 100
  discrete distribution, 86, 92
  relative frequency, 5, 7
Hooke's law, 555
Hypergeometric distribution, 99,
  119–24
  mean and variance, 121–22
  moment-generating function for,
    781
Hypergeometric probability function,
  121
Hypothesis
  alternative (or research), 461, 462,
    469, 471, 509
  composite, 509
  null, 461, 462, 481
  simple, 509
Hypothesis-testing, 460–530
  calculating type II error
    probabilities, 477–80
  concerning population variances,
    498–507
  confidence intervals and, 480–81
  elements of a statistical test,
    461–67
  importance of probability to, 21
  power of tests, 507–17
  reporting results of a test, 482–87
  theory of, 487–89
  use of probability in, 349
Hypothesis tests
  attained significance levels in,
    482–87
  for β_i, 533
  for categorical data, 680–707

Hypothesis tests (*cont.*)
chi-square, 175, 682–83, 685, 701–2
choice of appropriate, 472
concerning cell probabilities, 684–88, 702
elements of, 461–67
errors in, 463–66, 477–80, 488–89
Friedman, 738–43
F test, 498–501, 633
goodness-of-fit, 684–88
Kruskal-Wallis, 732–38
large-sample, 467–77
level of, 463
likelihood ratio, 517–24
Mann-Whitney U, 724–32, 748
most powerful, 509–10
Neyman-Pearson lemma for, 507–17
nonparametric, 708–61
one-tailed test, 471, 479, 487
power of, 507–17
rank-sum, 722–24, 730, 734
robust, 494, 505
sign, 711–16
small sample, 489–98
Spearman's rank correlation, 750–56
t, 494, 533, 634
uniformly most powerful, 511, 514
upper-tail, 470
Wilcoxon signed-rank, 716–22
Z, 477–80, 481, 569

Identity elements, 768–69
Identity matrix, 768
Incomplete beta function, 183
Increasing function, 294
Independence, 52, 233–34
testing for, 567
Independent event, 51–52, 99
and the multiplicative law of probability, 61
Independent random sample, 620
Mann-Whitney U test, 724–32
rank-sum test, 722–24
Independent random variable, 4, 233–49
advantages of using, 245

definition of, 233–34, 237
moment-generating functions of, 304
testing for, 568
Independent samples experiment, 620
Independent variable, 532
controlled, 628
regression with, 534
rescaling, 597
sum of squares for, 630, 733
Indicator variable, 668
Inequality
Bonferroni, 59, 666
Cramer-Rao, 420
Inference, 2, 331
Inference-making, 12–13
concerning the parameters β_i, 552–57
degree of goodness, 2
errors in, 188
multiple linear regression, 585–91
role of probability in, 20–22
simple linear regression, 557–62
Integer-valued random variable, 136–37
Integration
limits of, 236
region of, 218
Intersection(s)
of events, 55, 209–10
probability of, 55
of sets, 23
Intersection (set theory), 22
Interval. *See* Confidence interval; Prediction interval
Interval estimate, 365
Interval estimator, 380
that works well, 399–400
Invariance property, 452
Inverse of distribution function, 290–91
Inverse of a matrix, 769–70
Inversion of a matrix, 772–77

Jacobian, 310–17
Joint density function, 214–15, 268
geometric representation of, 216
minimum/maximum, 318–19
Joint distribution, 210–22
definition of, 211

Joint distribution function, 212–14
for continuous random variables, 213–14
for discrete random variables, 213
Jointly continuous random variables, 212–13, 214
Jointly discrete random variables, 212–13
Joint probability function, 211
Joint relative frequency histogram, 214

Kendall's rank correlation coefficient, 750
Kruskal-Wallis test, 732–38
relationship with rank-sum test, 734
summary, 734
tables for, 734
kth factorial moment, 137
kth moment of a random variable, 132
kth-order statistic, 320

Lack of fit, 601
λ, 126, 518
Large samples
Friedman test for, 739–40
Kruskall-Wallis test for, 734
Mann-Whitney U test for, 729–30
maximum-likelihood estimators and, 455–56
sign test for, 714
Wilcoxon signed-rank test for, 719–20
Large-sample confidence interval, 385–95
Large-sample hypothesis tests, 467–77
summary, 472
Latin square design, 622
Law of large numbers, 423
Law of total probability, 67
Laws of probability
additive law, 55–56, 58, 61
law of total probability, 67
multiplicative law, 55–56, 61
Laws of sets
DeMorgan's, 24
distributive, 24
Layout, one-way, 620

Least squares equations, 538
  and solutions for a general linear model, 579
  solving using matrix inversion, 776
Least squares estimators
  covariance for, 545–47
  means of, 549
  for the multiple linear regression model, 584–85
  notation for, 544, 550
  properties of, 544–52, 584–85
  for the simple linear regression model, 539, 544–52, 581–82
  unbiased, 544
  variances of, 547–50
Least squares method, 537–44
Level of a factor, 619, 628
Level of the test, 463
Likelihood. *See also* Maximum-likelihood estimator
Likelihood function
  notation for, 517
  supremum, 518
Likelihood ratio test, 517–24
  large-sample distribution for, 521
  rejection region in, 518
  summary, 518
Limits. *See* Confidence limits
Linear correlation, simple coefficient of, 250
Linear dependence, 250
Linear equations
  matrix expression for a system of simultaneous, 771–72
  solving a system of simultaneous, 777–78
Linear functions
  correlation and, 567–71
  covariance of, 255–58
  expected value, 255–63
  least squares estimators as, 550
  of model parameters, 557–62, 585–91
  of random variables, 255–63
  variance, 255–63
Linear models, 531–606, 609
  definition of, 536
  examples, 571–78
  fitting, by using matrices, 578–85

least squares method for estimating parameters, 537–44
slope of the line in, 609
solutions for a general linear model, 579
test for a null hypothesis equal to zero, 593–600
uses for, 567, 672
using for analysis of variance, 668–72
Linear regression models
  multiple, 535, 584–93
  simple, 534–35, 539, 544–52, 557–67
Linear statistical models, 534–37
  analysis of variance and, 668–72
  applications of, 571–77
  correlation and, 567–71
  expected values and, 536
ln, 299
Location model, 710
Logarithmic series distribution, 706
Log-normal density function, 204–5
Log-normal distribution, 328
Lower confidence bound, 481
Lower confidence limit, 380
Lower-tail alternative, 471
Lower-tail rejection region, 471

M (test statistic), 711
$m(t)$, 132
Mann-Whitney U test, 724–32
  efficiency of, 730
  formula for, 725
  relationship with Runs test, 748
  summary, 726–27
  summary for large samples, 729
  usefulness of, 730
Marginal density function, 223
Marginal distribution, 222–25, 229–33
  continuous, 223
  discrete, 223
Marginal probability function, 223
Matched pairs experiment, 612–18
  sign test for, 711–16
  usefulness of, 615
  Wilcoxon signed-rank test for, 716–22
Mathematical models. *See* Model(s)

Matrix algebra, 764
  non-commutative nature of, 767
Matrix (matrices)
  addition, 764–65
  definition of, 763
  dimensions, 764
  elements, 764
  equality, 768
  expression for system of simultaneous linear equations, 771–72
  fitting the linear model by using, 578–85
  identity elements, 768–69
  identity matrix, 768
  inverse, 769–70
  inverse, determining whether it exists, 777
  inversion of, 772–77
  main diagonal of, 768
  multiplication, 766–67
  multiplication, by a real number, 765
  notation for, 763–64
  square, 768
  for system of simultaneous linear equations, 771–72
  transpose of, 770
max, 317
Maximum-likelihood estimate, 106
Maximum-likelihood estimator (MLE), 449, 450
  for any row or column probability, 690
  invariance property of, 452
  large-sample properties of, 455–57
Maxwell distribution, 207
Mean
  beta distribution, 184
  binomial distribution, 103
  chi-square distribution, 178
  conditional, 271
  confidence intervals for, 649–53
  correction for, 636
  difference in, 385, 401–4, 404, 423, 467, 491–93, 608–9, 613–4
  of discrete random variable, 93–94
  estimating, 279–80
  of exponential density function, 179

Mean (*cont.*)
  F distribution, 340
  formula for, 8
  gamma distribution, 177
  geometric distribution, 112
  hypergeometric distribution,
    121–22
  of least squares estimators, 549
  mixed distribution, 200
  negative binomial distribution, 117
  normal distribution, 170, 173
  Poisson distribution, 126, 128–29,
    134–35
  population. *See* Population mean
  sampling distribution, 332
  small-sample test for, 489
  uniform distribution, 168
Mean square error of point estimator,
    367
Mean square for blocks (MSB),
    656–57
Mean square for error (MSE), 632
Mean square for treatments (MST),
    632
Measures of dependence, 249
Measures of dispersion (or variation),
    8
Median, 417
Memoryless property, 179
  of exponential distribution, 179
  of geometric distribution, 114
Method of distribution functions,
    281–94
  summary, 288
  transformation and, 289–90
Method of least squares, 536,
    537–44, 600–1, 609–10
  fitting a straight line by, 610
Method of maximum likelihood, 106,
    441, 448–55
  definition of, 449
  example of, 113
Method of moment-generating
    functions, 281, 302–9
  summary, 307
  uses for, 302, 304, 305
Method of moments, 444–48
Method of sampling, 74
Method of transformations, 281,
    294–302
  summary, 300

min, 317
Minimal sufficiency, 429–35
Minimal sufficient statistics, 436,
    437, 443
Minimum of the random variables,
    317
Minimum-variance unbiased
    estimators (MVUE), 435–55
  and the method of maximum
    likelihood, 448
  unique, 444
Mixed distribution, 198–200
  finding expected values with,
    199–200
  mean and variance, 199–200
MLE. *See* maximum-likelihood
    estimator (MLE)
mn rule, 39
Model parameters, 557–62, 585–91
Model(s)
  allometric equation, 574
  complete, 593, 597–99
  deterministic, 532–33, 534
  fixed block effects, 654
  hierarchical, 273–74
  linear. *See* Linear models
  linearized, 574
  mathematical, 14
  multiple linear regression, 535,
    537, 544, 578, 584–85, 591
  of nature, 534, 571–78
  no-intercept, 542
  nonlinear, 574
  for one-way layout, 645–47
  planar, 599
  probabilistic, 532, 533–534
  probabilistic, for experiments,
    25–33
  quadratic, 583
  random block effects, 654
  for randomized block design,
    653–55
  reduced, 593, 597–99
  regression. *See* Regression
    model
  second-order, 598, 599
  selection of, 187–88
  simple linear regression, 534
  two-sample shift, 709–11
  validation of, 167
  ways to select, 187–88

Moment-generating function,
    131–36
  for beta distribution, 781
  for binomial distribution, 781
  for chi-square distribution, 781
  of continuous random variable,
    188–90
  definition of, 132
  for exponential distribution, 781
  extracting moments from,
    190–91
  for function of a random variable,
    192
  for gamma distribution, 781
  for geometric distribution, 781
  for hypergeometric distribution,
    781
  *k*th derivative of, 133
  method of, 281, 302–9
  for negative binomail distribution,
    781
  for normal distribution, 781
  for Poisson distribution, 781
  probability distributions and, 135
  probability-generating function
    and, 137
  for random variable, 132–36, 188,
    302–7, 354
  summary, 307
  of uniform distribution, 781
Moments, 131–36
  central, 188
  for continuous random variables,
    188
  factorial, 137
  method of, 444–48
  population, 444–45
  of random variable(s), 131–32,
    444
  sample, 444–45
  taken about mean, 132, 188
  taken about origin, 131–32,
    188–89
Most powerful test, 509–10
  uniformly, 511, 514
MSB. *See* Mean square for blocks
MSE. *See* Mean square for error
    (MSE)
MST (mean square for treatments),
    632
μ. *See* Population mean

Multicollinearity, 601–2
Multinomial coefficients, 43
Multinomial distribution, 264
Multinomial experiment, 263–64,
    681
  binomial experiment and, 681
  characteristics of, 681
Multinomial term, 43
Multiple linear regression model,
    535, 585–91
  least squares estimators for,
    584–85
  predicting a particular value of Y
    using, 591–93
Multiplication
  of matrices, 765–67
  of matrix by a real number, 765
Multiplicative law of probability, 55
  for an independent event, 61
Multivariable experiment, 629
Multivariable transformation method,
    310–17
  uses for, 312
Multivariate density function, 218
Multivariate distribution, 209–78
  bivariate, 210–22
  normal, 268
Multivariate distribution function,
    218
Multivariate probability function,
    218
Multivariate transformation,
    297–300
Mutually exclusive events, 27, 29,
    56–57
  and the additive law of probability,
    61
  pairwise, 29
Mutually exclusive sets, 24
MVUE (minimum-variance unbiased
    estimators), 435–44
  and the method of maximum
    likelihood, 448
  unique, 444

Negative binomial distribution,
    116–19
  mean and variance, 117
  moment-generating function of,
    781
  probability function, 117

Neyman-Pearson lemma, 507–17
  theorem, 510
  when to use, 511, 514
No-intercept model, 542
Nonparametric statistics, 708–761
  definition of, 709
  Friedman test for randomized
    block designs, 738–43
  Kruskal-Wallis test for the
    one-way layout, 732–38
  Mann-Whitney U test, 724–32
  rank-sum test, 722–24
  runs test, 743–49
  sign test for a matched pairs
    experiment, 711–16
  sources for further information
    about, 756
  Spearman rank correlation
    coefficient, 750–56
  two-sample shift model, 709–11
  uses for, 708–9, 756–57
  when to use, 709
  Wilcoxon rank-sum test, 722–24
  Wilcoxon signed-rank test for a
    matched pairs experiment,
    716–22
Nonrandomness, test for, 747
Normal approximation to the
    binomial distribution, 354–60
  when to use, 355–56, 358
Normal (bell-shaped) probability
    distribution, 5, 9, 170–75,
    331–32, 425, 467–68
  asymptotically, 348
  bivariate, 268–69
  mean and variance of, 331–32
  moment-generating function of,
    781
  sampling distributions related to,
    331–45
Normal curve, 10
  areas under, 10, 170–71, 792
  empirical rule and, 10
Normal density function, 170
  areas under, 171
  bivariate, 268
  graph, 171
Normal distribution, 5, 10, 170–75
  mean, 170, 173
  relationship with sampling
    distribution, 331–45

table, 171
  variance, 171
Normal random variables, 170–73
  mean and variance of, 170–71
  standard, 334
  standardized, 173
Nuisance parameter, 517
Null hypothesis, 461
  choice of, 489
  composite, 513
  power of test and, 508
  p-value and, 483
  relationship with confidence
    interval, 481
  simple, 509
  test for, 593–600
Null set, 22
Numerical descriptive measures,
    8–12
Numerical event, 72

Observed life, total, 325
One-sided confidence interval, 381
One-tailed test, 471
  sample size for, 479
  when to use, 487
One-way layout, 620
  additivity of sums of squares for,
    647–49
  analysis of variance for, 635–39
  analysis of variance table for,
    639–44
  balanced, 638
  estimation in, 649–53
  expected value of mean square for
    treatments for, 647–49
  Kruskal-Wallis test for, 732–38
  sample size selection for, 663
  statistical model for, 645–47
Operating characteristic curve, 144
Order statistics, 317–25
Or (set theory), 22
Origin, moment taken about,
    131–32
Outcomes. *See* Observed value(s)
Outlier, 601
Overall mean, 645

p, 113
$p(y)$, 85, 86
Paired data, 612–18

Paired-difference experiment. *See*
    Matched pairs experiment
Parameter(s)
    α and β, 176
    β_i, 552–57
    of bivariate normal density
        function, 268
    definition of, 88, 90
    of density function, 167
    estimated, 415
    model, 557–62, 585–91
    of normal density function, 170
    nuisance, 517
    scale, 176
    shape, 176
Parametric methods, definition of,
    709
Parametric statistical methods, 709
Pareto distributions, 293
Partition of S, 67
Partitioning
    of objects into groups, 43
    of sample space, 72–73
    of total sum of squares of
        deviation, 630–31, 655
Pearson's chi-square tests, 682–83
Pearson test statistic, 682
Permutation, 41
Pivotal method, 381
Pivotal quantity, 384, 401, 402, 414,
    456
Planar model, fitted, 598
Point estimate, 365
Point estimation, 366
    method of maximum likelihood,
        448–55
    minimum-variance unbiased,
        435–44
Point estimator(s)
    bias and mean square error of,
        366–70
    consistency of, 420–29
    error of estimation and, 373
    expected values of, 371
    goodness of, 373–80
    method of moments, 441, 444–48
    minimal sufficient, 437
    properties of, 416–44, 457–59
    relative efficiency of, 417–20
    sample sizes for, 371
    standard error of, 371

sufficiency of, 429–35
unbiased, 366, 370–73
variance of, 371
Poisson distribution, 124–31
    mean, 126, 128–29
    moment-generating function of,
        781
    partial sums for, 127–28
    relationship with gamma
        distribution, 176
    table of partial sums for, 127–28
    uses for, 126
    variance, 128–29
Poisson probability function, 126
Pooled estimator, 402
Pooled sum of squares, 636
Population
    definition of, 2, 74
    nonparametric procedure for
        comparisons, 714
Population distributions
    testing for identical, 709–11
    that differ in location, 710
    use of ranks for comparing two,
        722–24
Population frequency distribution,
    188
Population mean
    large-sample confidence interval
        for, 385
    maximum-likelihood estimator of,
        450–51
    minimum-variance unbiased
        estimator of, 438–39
    notation for, 8
    overall, 645
    relationship to expected value, 89
    small-sample confidence interval
        for, 399–401
    small-sample hypothesis-testing
        for, 489–91, 494–98
    small-sample tests for comparing,
        492
Population mean comparisons,
    399–407
    analysis of variance, 630
    estimating differences between
        means, 608–9
    more than two means, 635–39
    summary of small-sample
        hypothesis tests for, 492

Population standard deviation, 91
Population variance, 91
    confidence intervals for, 407–10
    consistent estimator of, 423
    maximum-likelihood estimator of,
        450–51
    minimum-variance unbiased
        estimator (MVUE) for,
        438–39
    notation for, 9
    pooled estimator for, 402, 492
    tests of hypotheses concerning,
        498–507
Population variance comparisons
    the hypothesized larger variance,
        503
    of two normal distributions, 501–2
Power curve, 508
Power distributions, 293, 428, 433
Power series, 190
Power of the test, 507–17
    definition of, 508
    most powerful test, 509–10
    relationship with type II error, 508
    uniformly most powerful test, 511,
        514
Practical significance, 488
Prediction, quality of, 14
Prediction bands, 565–66
Prediction error, placing a bound on,
    564
Prediction interval
    multiple linear regression, 592
    simple linear regression, 564–65
Prediction problems, 562
Probabilistic model, 532, 533–534
Probability, 19–82
    additive law of, 55–56, 58, 61
    axioms, 29
    Bayes' rule, 68
    conditional, 50–55
    conditional discrete, 226
    counting rules for, 39–50
    definition of, 19–20
    of error, 463–66, 477–80, 482,
        485
    event-composition method for
        calculating, 56–67
    of event(s), 34, 55
    experiments, 25–33, 74–76
    histogram, 86, 89, 92, 100

inference and, 13, 20–22
of intersection of events, 55
joint, 211, 212
laws of, 55–57, 67
law of total, 67–72
marginal, 222
multiplicative law of, 55, 61
notation for, 29
Poisson, 124–131
and random variables, 72–74
relative frequency concept of,
    20
role in inference-making, 13,
    20–22
sample-point method for
    calculating, 34–39
type I error, 463
type II error, 463, 477–80
unconditional, 50
of union of events, 55–56
uses for, 84
Probability density function. *See*
    Density function
Probability distribution. *See*
    Distribution
Probability function, 85
    beta, 183, 779
    binomial, 183, 184
    bivariate, 211
    chi-square, 408
    conditional discrete, 226
    of exponential distribution,
        781
    gamma, 176, 779
    geometric, 110
    hypergeometric, 99
    joint, 211
    logarithmic series, 706
    marginal, 223
    negative binomial, 117
    normal, 171
    Poisson, 126
    uniform, 167
    uniquely-determined, 131–36
Probability-generating function,
    136–39
    definition of, 137
    for geometric random variable,
        138
    moment-generating function and,
        137

Products of two matrices, 766–77
p-value(s), 483
    tables for computing, 485
    uses for, 483, 484

Quadratic model, 583
Qualitative variable, 629
Quantity, pivotal, 384, 401, 402, 414,
    456
Queuing theory, 137–38

R_S (test statistic), 753
r (test statistic), 568
Random assignment, 44, 618–19
Random block effects model, 654
Random error
    assumption of normality of the
        distribution of, 549–50
    notation for, 536
Random event, 20
Randomization, importance of, 624
Randomized block design, 620–21
    analysis of variance for, 655–61
    estimation in, 662–63
    Friedman test for, 738–43
    linear model approach to, 670
    model for, 653–55
    sample size selection for, 663
Randomized design, completely, 619,
    620
Randomness, a test for, 743–49
Random number table, 75–76
Random sample, 75
    independent, 620, 722–32
    simple, 75
    as sufficient statistic, 432
Random sampling, 74–76
    from a relatively large population,
        395
Random variable(s), 73
    Bernoulli, 437
    beta, 183
    binomial, 100
    chi-square, 178
    conditional density of, 228
    conditional discrete probability
        functions of, 226
    continuous. *See* Continuous
        random variable
    covariance of two, 249–55
    dependent, 4, 234

discrete. *See* discrete random
    variable
distribution function of, 151
expected value for multivariate,
    241–49
expected value for univariate,
    88–97
exponential, 178
factorial moment for, 137
functions of, 279–329
gamma, 176
geometric, 110
hypergeometric, 121
independent, 4, 233–49
integer-valued, 136–37
joint distribution function of, 212
jointly continuous, 212–13, 214
jointly discrete, 212–13
kth factorial moment, 137
kth moment of, taken about the
    mean, 132
kth moment of, taken about the
    origin, 131
linear functions of, 255–63
marginal probability functions of,
    223
maximum of, 317–18
means for, 143
measures of dependence, 249
minimum of, 317–18
mixed, 199
moment of, 131
moment-generating functions of,
    132–36
negative binomial, 117
normal, 170
ordered, 317
Poisson, 126
possessing a specified probability
    distribution with mean zero,
    533
probability density function for,
    154–158, 163–64, 218
probability-generating functions
    for, 136–39
predicting values of, using
    multiple regression, 591–93
predicting values of, using simple
    linear regression, 562–67
special theorems for, 244–49
standard deviation of, 91

Random variables (*cont.*)
  standard normal, 173
  t-distributed, compared with
      normal, 338
  testing for independence of, 567
  uniform, 166–67
  univariate, 212
  variance, 91
  vector, 567
  Weibull, 206
Range of a set of measurements, 12
Rank correlation coefficient, 750–56
Ranks, use for comparing two
      population distributions,
      722–24
Rank sums, 722
Rank-sum test, 722–24
  relationship with Kruskal-Wallis
      test, 734
  usefulness of, 730
Rao-Blackwell theorem, 435–36
Rayleigh density function, 302
$r \times c$ contingency tables, 688–97
  with fixed row or column totals,
      697–701
Reduced model, 593
  compared with complete model,
      597–99
Region where the experiment was
      conducted, 573
Regression
  multiple linear, 585–93
  simple linear, 557–67
Regression model, 534
  lack of fit, 601
  multiple linear, 534, 584–85
  simple linear, 534–35, 539, 544–54
Rejection region (RR), 462
  complement of, 481
  F test, 501
  form of, 511
  graphs, 499
  likelihood ratio test and, 518
  lower-tail, 471
  for runs test, 744–45
  two-tailed, 471, 472, 553
  upper-tail, 469
Relative efficiency, 417–20
Relative frequency, 20
  and conditional probability, 50–51
Relative frequency distribution, 4

Relative frequency histogram, 4
  joint, 214
Reliability function, 327
Rescaling independent variables,
      597
Research hypothesis, 461, 462
Residuals, 601
Response variable, 534
Robust statistical test, 494, 505
Row operations, 773
Row probability, 689
Row totals, fixed, 697–98
Rows (qualitative independent block
      variables), 629
RR. *See* Rejection region (RR)
Run, 744
Runs test, 743–49
  expected value and variance for,
      748
  rejection region for, 745
  relationship with Mann-Whitney
      U test, 748

S. See Sample space
S. See Success
S^2. *See* sample variance
Sample
  definition of, 2
  elements affecting the information
      in, 607–8
  independent random, 620
  likelihood of the, 431
  paired, 612–18
  random, 75
Sample correlation coefficient, 568
  nonparametric analogue to, 750
Sample mean, formula and notation
      for, 8
Sample median, 417
Sample point, 26
  as an ordered pair of numbers,
      40
  in a binomial experiment
      counting, 39–50
  equiprobable, 120
  representations of, 41
Sample-point method, 34–39
  and combinatorial analysis,
      39–50
Sample size
  for one-way layout, 663–65

for randomized block design,
      663–65
  runs distribution and, 814–15
  selection of, 395–99
  selection of, for analysis of
      variance, 663–65
  for *Z* test, 477–80
Sample space, 27
  discrete, 27, 28
  partition of, 67
  that is not discrete, 31
Sample standard deviation, 388
Sample variance, 335
  as an estimator, 372
  formula for, 9
Sampling
  error in repeated, 562
  random, 74–76
Sampling distribution, 330–45,
      360–63
  central limit theorem and, 346
  definition of, 331
  mean of, 332
  mean and variance, 332
  normal distributions and,
      331–45
  relationship with normal
      distribution, 331–45
  of sum of squares, 334
  of unbiased point estimator,
      374
  variance of, 332
Sampling method, 74
  matched pairs, 612
  random, 74–76
  replacement in, 75, 459
  simple random, 75
Sampling procedure. *See*
      Experimental design
Sampling without replacement, 74,
      260
Sampling with replacement, 74
Scale parameter, 176
Second-order model, 598
  fitted, 598
Series
  Taylor series expansion, 779
Set notation, 22–25
Set of confidence intervals, 666
Set of elements, 22
  countably infinite, 84

Set of measurements, 3–8
  graphical methods for
    characterizing. *See* Graphical
    descriptive methods
  numerical methods for
    characterizing, 8–12
  range, 12
  variability of, 629
Set(s), 22–25
  complement of, 23
  DeMorgan's Laws of, 24
  disjoint, 24
  distributive laws of, 24
  empty, 22
  intersection of, 23
  mutually exclusive, 24
  notation for, 22–25
  null, 22
  subset of, 22
  union of, 22–23, 517–18
  universal, 22
  Venn diagrams and, 22–24
Shape parameter, 176
Shift model, 710
σ^2, 91
Signed-rank test. *See* Wilcoxon
    signed-rank test
Significance, statistical versus
    practical, 488
Significance level, 482
  attained, 482–87, 490–91, 493,
    500–1, 503, 504–5, 713
Sign test for a matched pairs
    experiment, 711–16
  compared with the t test, 713
  relationship with the Friedman
    test, 741
  summary for a large sample,
    714
  uses for, 714
Signed rank test for matched pairs
    experiment, 716–22
Simple event, 26
Simple hypothesis, 509
Simple linear regression model,
    534–35, 539, 544–52,
    557–67, 581–82
  least squares estimators for, 539,
    544–52, 581–82
  predicting a particular value of Y
    using, 562–67

summary, 559
Simple null hypothesis, 509
Simple random sampling, 75
Simulations, computer, 290
Simultaneous confidence coefficient,
    666
Simultaneous linear equations
  matrix expression for system of,
    771–72
  solving system of, 777–78
Single-factor experiment, 619
Skewed data distribution, 175
Slope, estimating, 610
Slutsky's theorem, 424
Small-sample confidence intervals,
    399–407
  summary, 404
Small-sample hypothesis-testing,
    489–98
  for comparing two population
    means, 492
  for a population mean, 490
Spearman's rank correlation
    coefficient, 750–56
  summary of the test, 753
  table, 752
SSB (sum of squares for blocks),
    655
SSE. *See* Sum of squares for error
SST. *See* Sum of squares for
    treatments (SST)
Standard deviation
  of β_1, 610
  definition of, 9
  population, 9, 395–96
  of random variable, 91
  sample, 9
  units of, 173
Standard error, 371
  of the estimator $\hat{\theta}$, 480
  minimizing, 612
Standard normal density function,
    338
Standard normal distribution
    function, 348
Standard normal random variable,
    173
  expected value and variance of,
    338
Statistic, 331
  sufficient. *See* Sufficient statistic

Statistical experiment. *See*
    Experiment
Statistical inference. *See* Inference
Statistical test, 461–67
  elements of, 462
  reporting results of, 482–87
  robust, 494, 505
Statistics
  definition of, 1–2
  F, 503–5
  and hypothesis testing, 461
  kth-order, 320, 321
  minimal sufficient, 436, 437, 443
  nonparametric, 709
  objective of, 3
  order, 317–25
  parametric, 709
  sufficient, 417, 429, 430, 432,
    436
Step function, 152
Stochastic event, 20
Student's t distribution. *See* t
    distribution
Subset, 22
  unordered, 44
Sufficiency, 429–35
  definition of, 430
  and likelihood, 431–432
Sufficient statistic, 429, 431–32
  confidence interval and, 439
  function of, 436
  minimal, 436, 443
  uses for, 439
Sum of functions, expected value of,
    93, 162–63, 244
Summations, formulas for, 779
Sum of a geometric series, formula
    for, 64, 778
Sum of squares for blocks (SSB),
    655
Sum of squares for error (SSE), 538
  formula for, 655
  pooled, 636
  as portion of total sum of squares
    of deviations, 630
  proof of additivity of, 647
  for reduced and complete models,
    593–94
Sum of squares for independent
    variables, 630, 733
  sampling distribution of, 334

Sum of squares for treatments (SST), 631
  formula for, 655
  proof of additivity of, 647
  rank analogue of, 733
  rank analogue of, for the randomized block design, 738
Sum of squares of deviations, 610
  additivity of, 647–48
  adjusted total, 594
  partition of total, 630–31, 655
  total, 630, 655
Support of the density, 296

T (test statistic)
  multiple linear regression, 587
  simple linear regression, 559
  small sample test for comparing two population means, 492
  small-sample test for the population mean, 490
  test for β_i, 533
  t test, 338
  Wilcoxon signed-rank test, 718
Tables
  analysis of variance (ANOVA), 639–44, 658
  binomial, 102, 128
  chi-square, 178, 334–35, 337
  for computing p-values, 485
  F, 340–41
  gamma, 178
  Kruskal-Wallis test, 734
  of means and variances for common random variables, 143
  normal distribution, 717
  of partial sums for Poisson distribution, 127–28
  random number, 75–76
  Spearman rank correlation coefficient, 752
  t, 338
  three-way, 702
Tables of the Binomial Probability Distribution, 184
Target parameter, 365
  minimum-variance unbiased estimators for, 436–41
Taylor series expansion, formula for, 779

of e^x, 779
Tchebysheff's theorem, 17
  bounds for probabilities, 375
  for continuous random variables, 194–97
  for discrete random variables, 139–42
  error of estimation and, 373–74
  formal statement of, 140, 194
  point estimators and, 421–22
  uses for, 195
t density function, 338
t-distributed random variable, 338
t distribution, 338
  table, 338
t distribution function relationship to standard normal distribution function, 425
Test statistic, 462
  chi-square, 500
  chi-square (Pearson's), 682
  F, 503
  F_r, 739
  H, 734
  M, 711
  r, 568
  r_s, 753
  T. See T (test statistic)
  $\hat{\theta}$, 468
  U. See U (test statistic)
  W, 723
  Z. See Z (test statistic)
Theory, 461
  and reality, 13–14
θ
  estimator of, 325
  maximum-likelihood estimator of, 451–52
  in probability function, 430
$\hat{\theta}$ (point estimator), 366
Three-way ANOVA table, 702
Three-way table, 702
Ties
  in paired experiments, 713, 716–17
  in rank correlation, 750–1
Time series, 747
Total sum of squares of deviations (Total SS), 630
  formula for, 647
Transformation method, 281, 294–302

multivariable, 310–17
  summary, 300
Transpose of a matrix, 770
Treatment, 619
Treatment, experimental, 618–19, 653
  effects of, 654
  in randomized block design, 621–22
  sum of squares for, 630–1
t test
  from the analysis of variance test, 634
  and least squares estimators, 533
  two-sample, 494
  usefulness of, 494
Trials, experimental, 97
t statistic, 568, 569, 573
t tests, Student's, 489–90, 495
Two-sample shift model, 709–11
  assumptions for, 710
Two-sample t test, 494
Two-sided confidence interval, 380
Two-tailed alternative hypothesis, 472
Two-tailed rejection region, 472
Two-tailed test, 471
  p-value for, 484–85
  when to use, 472, 487
Two-way ANOVA table, 702
Type I error, 463
  probability of, 463, 465–66
  relationship with type II error, 465
Type II error, 463
  probabilities, 477–80
  relationship with power of the test, 508
  relationship with type I error, 465

U (test statistic), 724, 727, 730
  expected value and variance of, 729
Unbiased confidence interval, 415
Unbiased estimator
  consistent, 421
  minimum-variance, 436–41, 448, 451–52
  Rao-Blackwell theorem for, 435
  relative efficiency of, 417–20
  sample variance as, 372
  unique minimum-variance, 444

Unbiased point estimator, 366, 370–73
  examples of, 370–73
  of p, 470
Unconditional probability, 50
Uniform density function, 166–67
Uniform distribution, 166–70
  bivariate analogue of, 215
  mean and variance, 168
  uses for, 167
Uniform probability distribution, 166–70, 290, 374
  moment-generating function of, 781
Uniform probability function, 167
Uniformly most powerful test, 511, 514
Union(s)
  of events, 55–56
  probability of, 55–56
  of sets, 22–23, 517–18
Union (set theory), 22
Unique minimum-variance unbiased Estimator (UMVUE), 444
Uniqueness theorem, 302
Univariate random variables, 212
Universal set, 22
Unrestricted maximum-likelihood estimator, 519
Upper confidence bound, 481
Upper confidence limit, 380
Upper-tail alternative, 469
Upper-tail rejection region, 469, 472
Upper-tail test, 470

Variable(s)
  Bernoulli, 437
  dependent, 532
  dummy, 668
  independent, 532, 534
  indicator, 668
  of integration, 154

  nonrandom, 532
  qualitative, 629
  quantitative, 629
  random. *See* Random variable(s)
  rescaled, 597
  response, 534
Variance
  analysis of, 628–79
  beta distribution, 184
  binomial distribution, 103
  chi-square distribution, 178
  comparison of, 501–5
  conditional, 272
  confidence interval for, 407–10
  of continuous random variable, 163
  definition of, 9
  of discrete random value, 93–94
  of exponential density function, 179
  gamma distribution, 177
  geometric distribution, 112
  hypergeometric distribution, 121–22
  importance of, 607
  of least squares estimators, 547–50
  of linear functions, 255–63
  maximum-likelihood estimator for, 452
  minimizing, 609
  minimum, 436
  mixed distribution, 200
  negative binomial distribution, 117
  normal distribution, 170, 335
  of point estimators, 367
  Poisson distribution, 128–29, 134–35
  pooled estimator for, 402, 492
  Population. *See* Population variance
  of random variable, 91
  relative efficiency of, 417

  for runs test, 748
  sample, 9, 372
  sampling distribution, 332
  sampling distribution of the point estimator, 371
  standard normal distribution, 338
  t distribution, 338
  test statistic U, 729
  tests of hypotheses concerning, 498–507
  unbiased estimator for, 544
  uniform distribution, 168
Variation
  coefficient of, 361
  measures of, 8, 131
Vector random variable, 567
Venn diagram, 22
Venn diagram examples, 22, 23, 24, 56, 60
  for sample spaces, 27, 28, 31

W (test statistic), 723
Weibull density function, 301, 437
Weibull distribution, 206, 439
Wilcoxon rank-sum test. *See* Rank-sum test
Wilcoxon signed-rank test, 716–22

Y value, predicting, 562–66, 591–93

Zero probability, 154
Z (test statistic)
  large-sample α-level hypothesis test, 472
  runs test, 748
  sign test for large samples, 714
  Wilcoxon signed-rank test, 720
  Z test, 481
  for a correlation, 569
  finding the sample size for, 477–80
  test statistic for, 481

Continuous distributions

Distribution	Probability Function	Mean	Variance	Moment-Generating Function
Uniform	$f(y) = \dfrac{1}{\theta_2 - \theta_1}; \theta_1 \leq y \leq \theta_2$	$\dfrac{\theta_1 + \theta_2}{2}$	$\dfrac{(\theta_2 - \theta_1)^2}{12}$	$\dfrac{e^{t\theta_2} - e^{t\theta_1}}{t(\theta_2 - \theta_1)}$
Normal	$f(y) = \dfrac{1}{\sigma\sqrt{2\pi}} \exp\left[-\left(\dfrac{1}{2\sigma^2}\right)(y-\mu)^2\right]$ $-\infty < y < +\infty$	μ	σ^2	$\exp\left(\mu t + \dfrac{t^2\sigma^2}{2}\right)$
Exponential	$f(y) = \dfrac{1}{\beta}e^{-y/\beta}; \quad \beta > 0$ $0 < y < \infty$	β	β^2	$(1 - \beta t)^{-1}$
Gamma	$f(y) = \left[\dfrac{1}{\Gamma(\alpha)\beta^\alpha}\right]y^{\alpha-1}e^{-y/\beta};$ $0 < y < \infty$	$\alpha\beta$	$\alpha\beta^2$	$(1 - \beta t)^{-\alpha}$
Chi-square	$f(y) = \dfrac{(y)^{(v/2)-1}e^{-y/2}}{2^{v/2}\Gamma(v/2)};$ $y^2 > 0$	v	$2v$	$(1 - 2t)^{-v/2}$
Beta	$f(y) = \left[\dfrac{\Gamma(\alpha+\beta)}{\Gamma(\alpha)\Gamma(\beta)}\right]y^{\alpha-1}(1-y)^{\beta-1};$ $0 < y < 1$	$\dfrac{\alpha}{\alpha+\beta}$	$\dfrac{\alpha\beta}{(\alpha+\beta)^2(\alpha+\beta+1)}$	does not exist in closed form